# Urban Operations

# Plus

# Mine Countermine Operations

## Department of Defense

Pentagon
Publishing

Field Manual
No. 3-06

Headquarters
Department of the Army
Washington, DC, 26 October 2006

# Urban Operations

## Contents



**Distribution Restriction:** Approved for public release; distribution is unlimited.

***This publication supersedes FM 3-06, 1 June 2003.**

# Figures

## Historical Vignettes

# Preface

Doctrine provides a military organization with a common philosophy, a language, a purpose, and unity of effort. Rather than establishing a set of hard and fast rules, the objective of doctrine is to foster initiative and creative thinking. To this end, FM 3-06 discusses major Army operations in an urban environment. This environment, consisting of complex terrain, a concentrated population, and an infrastructure of systems, is an operational environment in which Army forces will operate. In the future, it may be the predominant operational environment. Each urban operation is unique and will differ because of the multitude of combinations presented by the threat, the urban area itself, the major operation of which it may be part (or the focus), and the fluidity of societal and geopolitical considerations. Therefore, there will always exist an innate tension between Army doctrine, the actual context of the urban operation, and future realities. Commanders must strike the proper balance between maintaining the capability to respond to current threats and preparing for future challenges.

## PURPOSE

This manual provides the analytical tools for evaluating an urban operation to determine if the operation is necessary for overall mission success. It also provides the means to understanding and determining the impacts of the urban environment on military operations and provides information on managing, taking advantage of, and mitigating the effects of those impacts as appropriate. As such, this manual demonstrates how to apply the doctrinal principles in FM 3-0 to this unique environment.

## SCOPE

Chapter 1 introduces theoretical and historical perspectives of urban operations that serve as the underlying basis for the rest of the manual. While this manual has incorporated lessons learned from recent and ongoing operations, it has deliberately taken a broad and varied historical perspective in order to remain relevant to future threats and circumstance differing from those that the Army currently faces. Chapter 2 discusses the characteristics of urban centers and populations as well as their impact on operations. It is unlikely that Army forces will ever operate in a benign urban environment; therefore, Chapter 3 discusses the varied nature of potential urban threats. Chapter 4 describes the effects of the urban environment on warfighting functions and tactics. An understanding of the complexities of the urban environment, the nature of the enemy as an adaptive, learning opponent, and the effects of the environment on warfighting capabilities and skills is essential to sound decision making. Chapters 5 and 6 discuss the potential costs of urban operations and risk reduction measures that the commander and his staff must consider early in their planning. These chapters also outline an urban operational framework and specific urban considerations that create the foundations necessary for successfully applying operational doctrine to an urban environment.

The second part of the manual (Chapters 7 – 10) discusses how urban operations are conducted and resourced. Urban operations include major offensive and defensive operations in urban environments as well as stability or civil support operations ranging from peace operations and counterterrorism to disaster relief and humanitarian assistance. For the different types of operations—offense, defense, and stability or civil support—the purpose, characteristics, organization, and considerations are discussed. However, commanders consider that most urban operations will normally involve the simultaneous execution of offense and defense with stability or civil support (although proportional emphasis will shift over time) and plan accordingly.

## APPLICABILITY

This manual is intended for Army commanders and their staffs at the brigade level and above. It addresses full spectrum operations that Army units will execute in urban settings. However, users should also consult JP 3-06 for specific joint information. Additionally, users should be familiar with FM 3-06.1, FM 3-06.11, TC 90-1, and urban operations chapters, appendices, or sections found in other infantry, armor, combined arms, and proponent field manuals for the tactics, techniques, and procedures (TTP) and appropriate proponent information necessary to conduct tactical urban operations at the brigade level and below. This publication applies to the Active Army, the Army National Guard (ARNG)/Army National Guard of the United States (ARNGUS), and United States Army Reserve unless otherwise stated.

## ADMINISTRATIVE INSTRUCTIONS

When this FM 3-06 was approved, doctrine was incomplete for incorporating stability operations and civil support operations in place of stability operations and support operations. This manual incorporates stability and reconstruction and civil support operations doctrine found in FM 1 and FMI 5-0.1; it will be revised to reflect the future updates of FM 3-0 and FM 3-07 and, when published, FM 3-28. (Note: Stability and reconstruction operations has been redesignated stability operations to comply with Department of Defense Directive 3000.05.)

Chapter 2 defines "city" according to a population size. However, in historical vignettes and accounts, the term "city" may be applied in its common usage without specific regard to size to maintain conformity with most other historical reports.

In this manual, the term "threat" is applied broadly to include an enemy force (conventional or unconventional), an armed belligerent in a peace operation, antagonistic or unfriendly elements of the civilian population, or some other hazardous condition in the urban environment that negatively influences mission accomplishment.

The term military operations on urban terrain (MOUT) is replaced by urban operations (UO). MOUT is an acronym from FM 90-10, which is superseded by this manual.

Otherwise, the glossary lists most terms used in FM 3-06 that have joint or Army definitions. Where Army and joint definitions are different, (Army) follows the term. Definitions for which FM 3-06 is the proponent manual (the authority) are marked with an asterisk (*). The proponent or amplifying manual for other terms is listed in parentheses after the definition.

The manual attempts to incorporate a broad range of historical vignettes into each chapter where the account supports the doctrinal line of reasoning. Two historical vignettes, however, were included as appendices (A and C) because of their longer lengths.

Unless this publication states otherwise, masculine nouns or pronouns do not refer exclusively to men.

This publication contains copyrighted material.

The proponent for this publication is the United States Army Training and Doctrine Command. Send written comments and recommended changes on DA Form 2028 (Recommended Changes to Publications and Blank Forms) directly to Commander, U.S. Army Combined Arms Center and Fort Leavenworth, Combined Arms Doctrine Directorate, ATTN: ATZL-CD, (FM 3-06), 201 Reynolds Avenue, Fort Leavenworth, KS 66027-2337. Send comments and recommendations by e-mail to webb-cadd@leavenworth.army.mil. Follow the DA Form 2028 format or submit an electronic DA Form 2028.

This page intentionally left blank.

This page intentionally left blank.

## Chapter 1

# The Urban Outlook

*Today's security environment demands more from Army leaders than ever before. Army leaders must not only be able to lead Soldiers but also influence other people. They must be able to work with members of other Services and governmental agencies. They must win the willing cooperation of multinational partners, both military and civilian. But ultimately, the Army demands self-aware and adaptive leaders who can compel enemies to surrender in war and master the circumstances facing them in peace. Victory and success depend on the effectiveness of these leaders' organizations. Developing effective organizations requires hard, realistic, and relevant training.*

FM 1

Given the prevalence of large cities throughout the world, Army forces will likely be required to conduct operations in, around, and over large urban areas. These operations will normally be in support of a joint force commander (JFC) conducting military operations pursuant to U.S. national security policy. This manual is designed to facilitate the planning and conduct of the full range and spectrum of land operations in a complex urban environment. Each urban environment and urban operation is unique; prescribing a specific doctrinal "solution" for each situation is impossible. Instead, this manual provides a framework to commanders and their staffs for understanding the urban environment, for analyzing and deciding whether or not to initiate urban operations (UO), and for applying operational doctrine to this complex environment. It also provides a broad base of historical vignettes and examples to help develop a refined analytical perspective and stimulate thought. The manual also includes some planning points and tactics and techniques to assist in preparing for and conducting UO. This information provides a foundation for approaching major UO. Combined with other joint and Army doctrine, this information also will help commanders and their staffs learn to adapt and succeed in this challenging environment.

## THE PROSPECT OF URBAN OPERATIONS

1-1. The world is undergoing massive urbanization. Although exceptions exist, an overall trend of migration from rural to urban areas is occurring throughout the globe. (Australia, one of the world's most urbanized countries, is actually becoming less urbanized.) This trend is especially evident in developing nations. Combined with the exponential growth of the global population in the last quarter century, this migration has created massive urban areas that hold the centers of population, government, and economics in their respective regions. In many cases, rapid urbanization has overburdened already weak infrastructures, scarce resources, and a fragile economic base. As urbanization has changed the demographic landscape, potential enemies recognize the inherent danger and complexity of this environment to the attacker, and may view it as their best chance to negate the technological and firepower advantages of modernized opponents. Given the global population trends and the likely strategies and tactics of future threats, Army forces will likely conduct operations in, around, and over urban areas—not as a matter of fate, but as a deliberate choice linked to national security objectives and strategy, and at a time, place, and method of the commander's choosing.

### Army Urban Operations

Army forces conduct UO either as one component of a larger operation or as a single operation focused totally on a specific urban environment. Major Army UO are often part of a joint and multinational effort requiring rigorous interagency and civil-military coordination that typically includes the full spectrum of military operations. Commanders of Army major operations must determine if UO are essential to mission accomplishment. If so, commanders must carefully integrate the operations into campaign planning to support the operational objectives of the JFC.
Army leaders conducting UO must—

- Understand the urban environment to determine decisive points.
- Shape the operation to set the conditions for success.
- Precisely mass the effects of combat power to thoroughly engage the decisive points that lead to centers of gravity.
- Continually consolidate gains essential to the retention of the initiative.
- Transition the urban area to the control of another force or agency or, ultimately, back to legitimate and functioning civilian control.

# AN URBAN PERSPECTIVE

1-2. As a subset of all Army operations, UO are operations focused on an urban environment. UO include full spectrum operations—offensive, defensive, and stability or civil support—that may be executed, either sequentially or (more likely) simultaneously, during the conduct of a single urban operation (see figure 1-1). UO may be the sole mission of the commander or one of several tasks nested in a larger operation. Regardless of the types of operations conducted or whether the urban area is the single focus of the operation or only one component of a larger operation or campaign, the complex urban environment significantly affects the overall conduct of the mission.

1-3. When conceptualizing urban operations, commanders must understand two important terms: urban area and urban environment. The first is a subset of the second. An ***urban area*** **is a topographical complex where man-made construction or high population density is the dominant feature**. Focusing on urban areas means concentrating on the physical aspects of the area and their effects on weapons, equipment, line-of-sight, and tactics, techniques, and procedures. The ***urban environment*** **includes the physical aspects of the urban area as well as the complex and dynamic interaction and relationships between its key components—the terrain (natural and man-made), the society, and the supporting infrastructure—as an overlapping and interdependent system of systems.**

1-4. Importantly, commanders must also understand and consider that critical elements of the infrastructure may lie far beyond the area's physical confines. For example, the generating source providing power to the urban energy system is part of that system but may be located well outside of the urban area. Similarly, effects of the interaction between components of the infrastructure, located both inside and outside the urban area, extend well into smaller, neighboring urban areas and surrounding rural areas and often form their political, economic, and cultural focus. Understanding the total urban environment is essential to planning and conducting full spectrum urban operations.

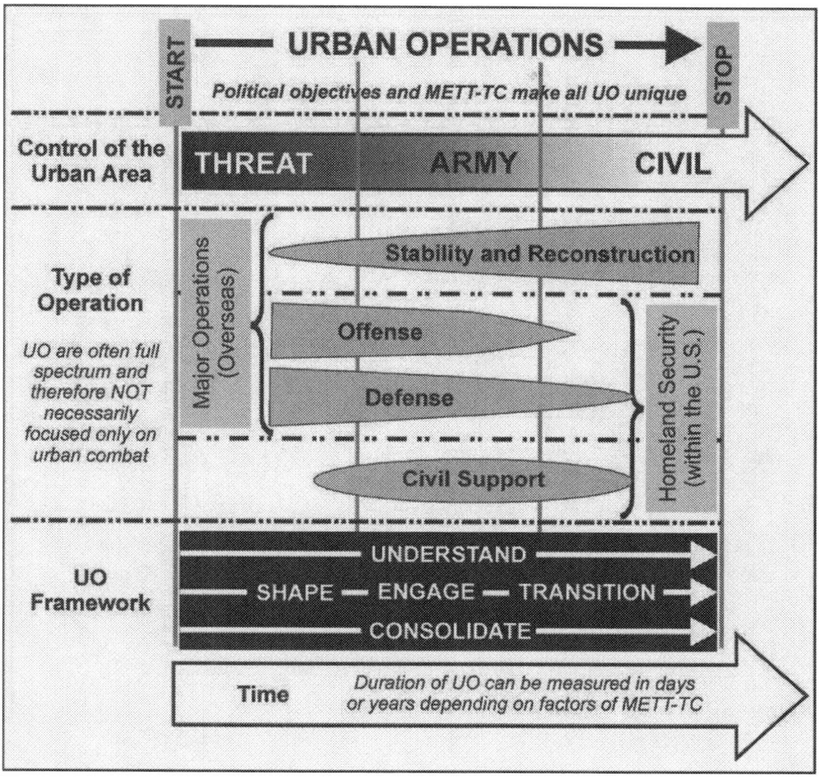


**Figure 1-1. Full spectrum urban operations**

# HISTORICAL SIGNIFICANCE OF URBAN AREAS IN WARFARE

1-5. Urban areas always have been central to, or have significantly influenced, military operations. One of the first urban-centered battles was the siege of Troy at the beginning of Greek history. Moreover, much of the history of early Greece revolved around wars between its city-states or with Persia and centered on the conquest, siege, or blockade of cities. Five hundred years later, the Roman Empire replaced Greece as the dominant world power; although, urban areas remained central to the Roman method of warfare. Even Rome's history can be viewed as a microcosm of urban warfare over the past two thousand years. Though military operations within the physical confines of many of these historic urban areas were not the norm, the focus of these operations was their conquest or control.

1-6. Although Rome last saw combat in 1944, urban areas have been no less prominent in warfare since that time. Seoul in Korea, Beirut in Lebanon, Panama City in Panama, Grozny in Chechnya, Sarajevo in Bosnia-Herzegovina, Kabul in Afghanistan, and Baghdad in Iraq have been centers of conflict in the last 50 years. Urban areas, now more pervasive than ever before, will continue to be essential to successful operational and strategic warfighting. Today, armies cannot expect to execute major military operations without the influence of the urban environments within their area of operations.

> **Rome: A Microcosm of Urban Warfare**
>
> During two millennia, Rome has been the center of at least 12 battles. The Gauls lay siege to Rome first in 387 BC. That first siege lasted six months and ended after the barbarians burnt much of the city. The surviving patrician families paid a ransom for the withdrawal of Brennus' army. From 408 to 410 AD, the Goth leader, Alaric, successfully besieged Rome no less than three times. The Byzantine General Belisarius captured Rome twice from the Goths and withstood siege inside the city once between 536 and 549. Five hundred years later in 1084, Norman adventurer Robert Guiscard captured medieval Rome and sacked the city during a dispute between the Pope and the Holy Roman Empire. Forces of the Holy Roman Empire again stormed and captured the city to punish the Pope in 1527. During the Italian Revolution in 1849, a French army supporting the Pope captured the city from the Italian revolutionary army under Garibaldi. In 1944, the last military action took place in and around Rome when the U.S. Fifth Army captured the city from the retreating German army. Rome's turbulent history—fought over ethnic and religious differences, prestige, and military necessity—demonstrates the importance of urban areas in warfare and the various causes and combatants within this complex environment.

## STRATEGIC IMPORTANCE OF URBAN AREAS

1-7. Several reasons have attracted (and continue to attract) armies to combat in urban areas:

- A military force chooses to position itself in an urban area to capitalize on the perceived defensive advantages offered by the environment. In contrast, an opposing force, by analyzing the factors of the situation, determines that it must enter the urban area to attack and destroy its enemy (or devote essential combat power to their isolation).
- The urban environment's people (their allegiance and support), infrastructure, capabilities, or other resources have or can be of significant operational or strategic value.
- The urban area has significant symbolic importance.
- The urban area's geographical location dominates a region or avenue of approach.

1-8. Russia's 1994 experience in Chechnya illustrates an increasingly important motivation for conducting urban operations. The Chechen rebels, after failing to engage Russian forces outside the city, chose to turn Grozny into the main battlefield. Leaders of the defeated Chechen conventional forces recognized that fighting in the urban area provided them their best chance for success. The complexities of urban combat and the perceived advantages of defending an urban area mitigated their numerical and technological inferiority. The urban area provided the Chechens protection from fires, resources, interior lines, and covered and concealed positions and movement. Given such advantages offered by the environment, smaller or less-sophisticated military forces have similarly chosen to fight in urban areas either as a deliberate strategy or to escape certain destruction in open terrain.

1-9. Such advantages of operating in an urban environment also prompt forces to conduct an urban operation to facilitate a larger campaign plan and decisive battle in another location. The urban operation can focus the enemy on the urban area and allow other forces to conduct operations elsewhere. From a defensive perspective, an urban defense may gain time and space to reorganize forces in new defensive positions, to divert enemy forces from other critical tasks, or to prepare to conduct offensive operations. To some extent, these reasons motivated Soviet forces defending Leningrad and Stalingrad from the Germans in World War II. The stubborn defense permitted the Soviets to reorganize for later offensive operations. From an offensive perspective, an attack on an urban area may be a shaping operation used to divert enemy resources from the decisive operation that will follow.

1-10. Armies also fight in an urban area to obtain some critical feature or resource in the area, such as a port facility. The desire to control an important seaport and access to the Persian Gulf largely motivated the Iranian and Iraqi struggle for Basra in the 1980s. Earlier, in 1944, British forces fought German units in Arnhem for control of the Rhine River Bridge. Other key infrastructure of the urban environment may have operational or strategic significance and can compel military forces to attack or defend the area. As urban areas account for an increasing share of a country's national income, often generating over 50 percent of gross national product, the strategic implications for their control or influence become even greater.

1-11. Urban areas are often located on terrain that dominates a region or an avenue of approach. In these cases, offensive armies capture these areas to proceed with security to another objective. Conversely, defensive forces commonly defend the area to deny the area of operations. To illustrate, Cassino, Italy during World War II stood astride the critical highway approach up the Liri valley to Rome. The allies had to attack and capture the monastery to facilitate the allied offensive north. Cassino's location made bypassing virtually impossible. Likewise, in the early 1980's, Israeli army urban operations in Beirut were a result of its strategic location near the Israeli security zone; various Arab insurgent and terrorist groups used Beirut as a base for attacks against Israel. Beirut evolved as the major base of the Palestine Liberation Organization, a major opponent of Israel. Beirut's location made it a security threat to Israel and thus compelled several major Israeli operations in the urban area (see Appendix A).

1-12. Another reason for engaging in urban operations is the symbolic—historical, cultural, political, and even economic—importance of many urban areas. Often, capital cities—such as Rome, Paris, Seoul, Berlin, and Baghdad—are identified as the strategic centers of gravity of their respective nations. Possessing or threatening these urban areas may impact directly on the outcome of a conflict. The objective of Germany's wars with France in 1870 and 1914 was ultimately Paris. Napoleon's 1812 campaign had as its objective Moscow, as did Hitler's 1941 offensive into Russia. The objective of the Soviet 1945 offensive was Berlin, and the North Vietnamese 1975 offensive had as its objective the South's capital of Saigon. Still, history also reminds us that commanders must assess the sustainability and decisiveness of operations directed toward these "prestige" objectives. For example, in 1812, Napoleon captured Moscow but had to evacuate it within 30 days. He lacked supplies and shelter, failed to destroy the Russian Army, and failed to defeat the political will of the Czar and the people. Similarly, the North Korean occupation of Seoul during the Korean War was equally indecisive.

## U.S. Army's Experience in Urban Operations

1-13. The U.S. Army has a varied history of conducting urban operations. The American Revolution saw the Army conduct several urban operations. These operations included the unsuccessful defense of New York, the successful attack on Trenton, and the decisive siege and attack on British forces at Yorktown. The Mexican War also had a successful assault on the fortified city of Monterey and the decisive siege of Mexico City. During the American Civil War, the armies, in the tradition of Napoleonic maneuver warfare, avoided urban areas and fought in the open. However, the opposing armies frequently made urban areas their objective because of their importance as railheads. Success in the siege of several key urban areas—Vicksburg, Atlanta, and Petersburg—contributed to the Northern victory.

1-14. Following the Civil War, the U.S. Army faced no large-scale urban combat for several generations. The Indian Wars, the Spanish-American War, the Philippine Insurrection, and even World War I did not require the Army to fight in large urban areas. Between the Civil War and World War II, the U.S. Army fought in several urban areas worldwide supporting U.S. commitments. These limited urban combat operations were small but essential parts of what are currently called urban stability operations. From 1900 to 1901, the Army provided public security for a sector of Peking, China of around 50,000 inhabitants. The Army conducted UO and, in the course of the operation, the 9th U.S. Infantry suffered 20-percent casualties while fighting in Tientsin. Punitive expeditions to places such as Siberia, Cuba, Philippines, Central America, and Mexico put the Army in various urban situations that required using military power, notably, the occupation and security of Vera Cruz, Mexico in 1914. In the context of these smaller-scale contingencies, UO became a staple of U.S. Army employment.

1-15. World War II forced the Army to grapple with the issues of large-scale urban combat almost immediately. In his 1941 defense of the Philippines, General MacArthur examined how to defend Manila.

Manila represented a large, modern, friendly urban area, which was the capital city of a close U.S. ally. Defending the urban area posed numerous challenges. Ultimately General MacArthur determined that he could best conduct its defense outside the city by defeating the enemy forces in combat on the invasion beaches or shortly after they landed. When Japanese forces defeated MacArthur's Philippine Army in a series of engagements, MacArthur had to decide how best to protect the friendly populace of Manila. He had two choices: abandoning the city or waging a costly defense that would likely result in the city's destruction, thousands of noncombatant casualties, and no operational advantage. He had little choice but to declare Manila an open city and move his forces to Bataan to wage an operational defense in the vain hope that a counteroffensive could relieve his isolated force. On 2 January 1942, Japanese forces entered Manila unopposed.

1-16. Had General MacArthur decided to defend Manila, his forces would have found scant doctrine in the Army regarding how to fight in an urban area. Doctrine for urban operations did not appear until early 1944, when faced with the possibility of fighting through the larger urban areas of Western Europe. At this time the U.S. Army published FM 31-50. This manual had the first formal discussion of how the Army viewed urban combat. It was based on the Army's limited experiences in the Mediterranean theater and the study of German and Soviet experiences on the Eastern front.

1-17. FM 31-50 emphasized a deliberate pace, individual and small unit initiative, the liberal use of direct and indirect firepower, and decentralized command and execution. It focused on the urban area (as opposed to the environment); however, it did include policies towards the noncombatants. The manual was also focused at the regimental combat team level. Complementing the doctrine of FM 31-50 was the 1944 operations manual, FM 100-5. This latter manual rightly emphasized the importance of combined arms actions and the need for extensive reconnaissance of prepared and defended cities. The Army successfully implemented this doctrine in several major instances of urban combat, most notably the capture of the first German city, Aachen, and hundreds of small-scale urban assaults on cities, towns, and villages across France, the Benelux, and Germany. Army forces also successfully employed this urban combat doctrine during the liberation of Manila in 1945.

1-18. The legacy of this era of Army operations was an effective tactical solution to urban offensive combat: isolate the urban area, seize a foothold, and expand the foothold block by block until occupying the entire urban area and destroying the enemy. The doctrine's emphasis on firepower kept friendly casualties to a minimum. Unfortunately, when enemy forces stoutly defended the urban area, the emphasis on firepower resulted in the area's virtual destruction and high casualties among noncombatants.

1-19. The doctrinal approach honed in World War II remained the accepted Army approach to urban combat to the century's end. The last successful implementation occurred during the Korean War and the final liberation of Seoul. The Vietnam conflict did not offer the Army opportunities or the requirement to practice urban combat or test and refine doctrine on a large scale. The largest urban battle, Hue, was a chaotic tactical battle that validated most of the historical lessons of urban combat without generating any new doctrinal insights for large-scale urban warfare.

1-20. From the mid-1950s through the 1990s, the Army conducted UO in the U.S. in support of civil authorities during civil unrest and anti-Vietnam protests. Some operations involved numerous active and Reserve Component forces engaged in restoring public order. The Detroit riots of 1967 and the Los Angeles riots of 1992 required the commitments of active and National Guard units. In 1968, the Army deployed over 35,000 troops to Washington D.C., Chicago, and Baltimore following the death of Dr. Martin Luther King, Jr.

1-21. In the 1970s and 1980s, Army doctrine predominantly focused on urban areas and successfully fighting a conventional ground war against Soviet and Warsaw Pact forces in Central Europe. FM 90-10 (1979) described how to conduct urban operations against Soviet forces in Germany. Its concepts were never tested other than in simulation, and its approach to urban combat was not substantially different from that practiced by the Army since World War II. Despite previous doctrine's admonition to avoid cities, the Army has had to fight in them in diverse circumstances.

1-22. In the 1990s and early 21st century, commanders and planners began to more carefully consider the impact of urban environments on the overall conduct of campaigns and major operations. Maneuver warfare characterized Operation Desert Storm in 1991; however, urban environments became prominent during subsequent operations in Somalia, the Balkans, and again in Iraq. (During Operation Iraqi Freedom in 2003, for example, Baghdad was considered as a strategic center of gravity for planning.) This was a result of adversaries recognizing the asymmetric advantages to be gained by fighting superior U.S. forces in an urban environment as well as a general worldwide trend of increasing urbanization. The evolving nature of Army operations in the urban environment was recognized in the Army doctrine of full spectrum operations. No longer could combat operations be considered exclusively as a separate phase of the major operation. Instead, full spectrum operations—combat and stability or civil support—would be conducted simultaneously. It also became clear to Army commanders that while one type of operation can often dominate for a period, simultaneous, full-spectrum urban operations were now the norm and an in-depth understanding of complex urban societies essential to overall success.

# MODERN ARMY URBAN OPERATIONS

1-23. Modern urban operations will be full spectrum operations executed jointly and will often contain a multinational and interagency component. They will span the entire range of possible applications of military power. At higher echelons, these separations are often viewed as levels or scales of intensity. For the tactical units conducting urban operations, these divisions appear indistinct, as the intensity is often high despite where the operation falls within the range of military operations.

## JOINT, INTERAGENCY, AND MULTINATIONAL

1-24. The urban environment is too multifaceted for a single-service, single-agency, or single-dimensional solution. Generating desired effects and avoiding unintended negative consequences in this complex environment requires careful integration of joint (and often multinational) forces and interagency capabilities throughout all phases of the operation. Effective interagency collaboration will help plan effects, supporting actions, and measures of effectiveness to ensure that military actions complement diplomatic, economic, and informational activities.

1-25. Joint urban operations (JUOs) in which Army forces are a major component will be primarily land operations. These operations may take place within the context of a joint campaign conducted by a joint force land component commander or a joint task force (JTF) commander. Or they may be an Army operation under an ARFOR commander who himself operates for a JFC, depending on the organization of the theater's joint command structure. In the later case, the JFC will manage joint issues in the urban area.

1-26. The JFC conducting JUOs will focus on effectively organizing his forces for UO and tasking them in accordance with their service capabilities. His guide for the conduct of the JUO will be the joint operational tasks described in JP 3-0. JP 3 06 will provide the JTF commander specific guidance regarding the conduct of joint operational tasks in the urban environment. Army commanders will execute tasks assigned by the JFC and advise him on using Army forces and capabilities. Army commanders will also ensure that Army UO are nested within the JFC's concept of operations. Also, the ARFOR commander will request support through the JFC from other service and functional commanders who have urban capabilities critical to the success of Army UO. See Appendix D for more information on joint capabilities in an urban environment.

1-27. The military conflict will not be an end to itself. It is inevitably a means of transition from a perceived unsatisfactory state of affairs to an improved end. Therefore, urban operations will require the careful orchestration of military and civilian capabilities to achieve success. An early identification of potential requirements will allow for the proper allocation of responsibilities between military and civilian agencies and a clear understanding of what is possible for each to achieve. Such understanding will preclude the establishment of false perceptions from all concerned, but particularly from the urban population and the public at large.

1-28. Interagency cooperation will include the Army as part of the Department of Defense, elements of the Department of State, and other various governmental and nongovernmental organizations. Urban operations demand the expertise and abilities of various organizations and a synergistic unity of effort if strategic goals aimed at bringing about a more satisfactory political, social, diplomatic, economic, and military situation are to be achieved. However, there will often be a gap in time before bringing many governmental capabilities to bear that the Army and other military forces—to include coalition partners—will have to fill. Routine interagency training, planning, and coordination will help to decrease that gap in time to its minimum. As it is likely that civilian organization leaders will be in charge during some phases of the overall operation, they should lead planning at those times. A careful survey of urban requirements and rapid communication to participating civilian organizations will also help speed civilian capabilities to where they are needed most.

**Winning the Peace in Iraq:**
**The Requirement for Full Spectrum Operations**

*We found that if we concentrated solely on establishing a large security force and targeted counterinsurgent combat operations—and only after that was accomplished, worked toward establishing a sustainable infrastructure supported by a strong government developing a free-market system—we would have waited too long. The outcome of a sequential plan allowed insurgent leaders to gain a competitive advantage through solidifying the psychological and structural support of the populace.*

*Further, those who viewed the attainment of security solely as a function of military action alone were mistaken. A gun on every street corner, although visually appealing, provides only a short-term solution and does not equate to long-term security grounded in democratic process. Our observation was born not from idealism, but because it creates the essence of true security, protecting not only our soldiers, but Iraq, the region, and, consequently, our homeland.*

Major General Peter Chiarelli
Commander, 1st Cavalry Division

## FULL SPECTRUM OPERATIONS

1-29. Army forces will conduct full spectrum operations within urban areas. Army commanders conduct full spectrum operations abroad by executing offensive, defensive, and stability urban operations as part of an the joint, interagency, and multinational effort described above. Army forces within the United States and its territories conduct full spectrum operations by combining offensive, defensive, and civil support operations to support homeland security. The situation will mandate that one type of operation—offense, defense, stability, or civil support—dominates the urban operation. Depending on whether the operation is overseas or within the United States, however, commanders will often find themselves executing offensive and defensive operations and stability or civil support operations simultaneously. In fact, waiting until all combat operations are concluded before beginning stability or civil support operations often results in lost, sometimes irretrievable, opportunities. Unsurprisingly, the mission determines the dominant type of operation, with the other types of Army operations conducted to shape the AO for mission success. The dominant type of operation will vary between different urban areas even in the same campaign.

## Offense

1-30. Against a large conventional enemy in a major urban area with a large civil population present, offensive operations will likely require a greater commitment of Army resources than in other environments. Urban offensive operations will also incur the greatest risks to Army forces and noncombatants. Within defensive and stability or civil support operations, forces may conduct tactical offensive UO, such as counterattacks to maintain the initiative or raids to eliminate elements disrupting the stability or civil support operation.

## Defense

1-31. Defensive UO are generally conducted as a shaping operation within a larger major operation. These temporary operations often set conditions for successful offensive operations, stability operations, or civil support operations. Commanders often conduct defensive UO within other types of operations to protect essential facilities in the urban area, protect flanks against counterattack, prevent the breakout of isolated enemies, or protect valuable supply bases or vulnerable convoy routes. Army forces conducting defensive UO must creatively use the environment to enhance their combat power.

1-32. In UO, essential facilities will likely include urban ports and airfields required by sister services to support ground operations. Therefore, Army commanders will need to carefully plan integrated air and base defense operations with air and naval component commanders. Protection requirements increase dramatically as air and naval assets are most vulnerable when aircraft are on the ground and ships are docked at port. Without a carefully synchronized and integrated base defense plan, Army commanders risk sporadic or total loss of support by air and naval forces—forces essential to successful UO.

### *Stability*

1-33. Stability operations in an urban environment require offensive and defensive operations, combined with other tasks unique to each stability operation. Army forces conduct urban stability operations for various reasons, including noncombatant evacuation operations, peace operations, or support to insurgencies (see Chapter 9). Urban stability operations will require an offensive capability to destroy any military capability that overtly threatens its objectives before that military threat can adversely affect the operation. Army forces employ defensive capabilities to safeguard themselves as well as secure critical places, populations, or infrastructure in the urban area. Commanders may also employ defensive capabilities to separate and protect one faction from another. Various stability operations will also require the distribution of food or aid and the protection or assistance of agencies conducting economic or humanitarian activities.

### *Civil Support*

1-34. Army civil support operations in an urban environment aid other U.S. agencies and organizations in mitigating the consequences of natural and man-made disasters. In response to the disaster, civil support operations require the equipment, personnel, or organizational abilities of Army forces rather than the Army's combat capabilities. In a civil support mission, these capabilities often involve Army transportation, medical, quartermaster, or engineer forces. Unless conducted in conjunction with homeland defense, urban civil support operations will seldom require combat; however, commanders must determine if force protection threats exist that could hamper Army civil support operations. During homeland security operations, defensive and offensive capabilities will be required to defeat hostile armed forces.

## PREPARING FOR FUTURE URBAN OPERATIONS

1-35. To operate successfully in a complex urban environment requires a thorough understanding of the urban environment and rigorous, realistic UO training. Training should be conducted by the complete combined arms team and cover full spectrum operations to include appropriate tactics, techniques, and procedures (see FM 3-06.11 and TC 90-1). It should also replicate—

- The psychological impact of intense, close combat against a well-trained, relentless, and adaptive enemy.
- The effects of noncombatants—including governmental and nongovernmental organizations and agencies—in close proximity to Army forces. This necessitates—
  - An in-depth understanding of culture and its effects on perceptions.
  - An understanding of civil administration and governance.
  - The ability to mediate and negotiate with civilians including the ability to effectively communicate through an interpreter.
  - The development and use of flexible, effective, and understandable rules of engagement.
- A complex intelligence environment requiring lower-echelon units to collect and forward essential information to higher echelons for rapid synthesis into timely and useable intelligence for all levels of command. Understanding the multifaceted urban environment necessitates a bottom-fed approach to developing intelligence (instead of a top-fed approach more common and efficient for open terrain and conventional threats). It also emphasizes the need for intelligence reach and a truly collaborative approach to the development and sharing of intelligence.
- The communications challenges imposed by the environment as well as the need to transmit large volumes of information and data.
- The medical and logistic problems associated with operations in an urban area including constant threat interdiction against lines of communications and sustainment bases.

### Developing Joint, Interagency, and Multinational Operability

1-36. UO training extends from the individual Soldier to the joint level including the integration of conventional and special operations forces. Preparedness also includes enhancing interoperability in regards to urban multinational and interagency operations. Training, as well as campaign and contingency plan creation, should include significant and sustained participation by civilian agencies. Fruitful multinational and interagency relationships must be cultivated before the onset of operations, that is, before Soldiers and their coalition counterparts are making decisions—many with strategic implications and often when they are under fire. Joint, interagency, and multinational collaboration will help design effects, supporting actions, and measures of effectiveness necessary to ensure that military actions in urban environments complement the diplomatic, informational, and economic activities necessary to achieve strategic objectives.

### Conducting Live, Virtual, and Constructive Training

1-37. Force preparedness mandates integrating the actual use of urban terrain, exercises at urban training sites, simulations, or any combination into tactical- and operational-level intra- and interservice training. This type of multi-faceted training will help commanders develop a better understanding of the complexity of the urban environment and enable them to execute missions across full spectrum operations. Careful use of these facilities will also allow Army forces to fully integrate urban operations within self-development, institutional, and operational training.

### Ensuring Every Soldier is an Urban Warrior

1-38. In a complex urban environment, every Soldier—regardless of branch or military occupational specialty—must be committed and prepared to close with and kill or capture threat forces in an urban environment. Every Soldier must also be prepared to effectively interact with the urban area's noncombatant population and assist in his unit's intelligence collection efforts. Each urban operation will be unique and commanders—given an opportunity to surge and conduct mission-specific post-mobilization and in-theater training—will need to conduct an analysis of the tasks requiring emphasis or modification to fit the conditions of the operational environment. In UO, every soldier will likely be required to—

- Perform advanced rifle marksmanship to include advanced firing positions, short-range marksmanship, and night firing techniques (unassisted and with the use of optics).
- Operate their unit's crew-served weapons.
- Conduct urban reconnaissance and combat patrolling.
- Enter and clear buildings and rooms as part of an urban attack or cordon and search operation.
- Defend an urban area.
- Act as a member of a combat convoy (including specific drivers training).
- Control civil disturbances.
- Navigate in an urban area.

1-39. While not all-inclusive and necessarily urban-specific, other critical individual and collective UO tasks (often modified for the urban environment) might include—

- Conduct troop-leading procedures.
- React to contact, ambush, snipers, indirect fire, and improvised explosive devices.
- Establish an observation point, personnel or vehicle checkpoint, or roadblock.
- Secure a disabled vehicle or downed aircraft.
- Call for indirect fire and close air support.
- Create and employ explosive charges.
- Handle detainees and enemy prisoners of war.
- Treat and evacuate casualties.
- Accurately report information.
- Understand the society and culture specific to the area of operations.
- Use basic commands and phrases in the region's dominant language.
- Conduct tactical questioning.
- Interact with the media.
- Conduct thorough after-action reviews.

1-40. Commanders must understand that all Soldiers will require urban-specific equipment to conduct many urban-specific tasks. Critically, commanders cannot expect supporting forces to fight alongside and support ground maneuver units in an urban environment without the appropriate equipment and training.

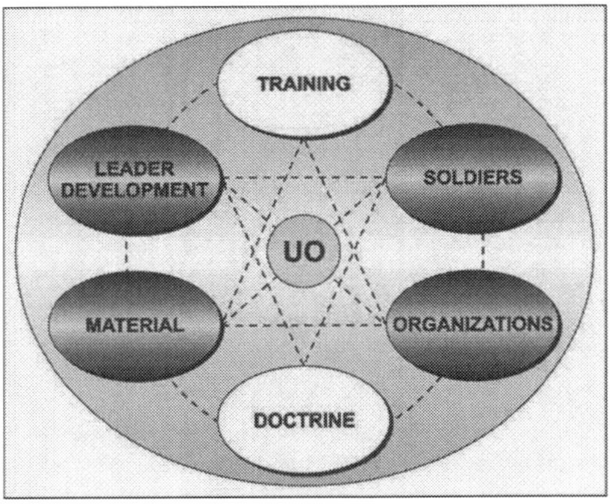


**Figure 1-2. UO and DOTMLPF**

## Continual Adaptation and Innovation

1-41. Realistic UO training (as well as the conduct of real world operations) has the added benefit of identifying operational requirements and resultant changes necessary in our doctrine, organizations, training, materiel, leadership and education, personnel, and facilities (DOTMLPF) (see figure 1-2 and FM 100-11). While technology (materiel) and organizational changes are important, confident Soldiers (personnel) remain the decisive means for success. Technology and organizational changes will be a critical enabler to attain better understanding of the urban environment, enhance command and control on the noncontiguous battlefield, and achieve the agile, simultaneous, and precise effects required in UO. In the future, technology may lead to a radically new operational concept and approach to UO. Still, competent leaders and well-trained, disciplined, and adaptive Soldiers will remain the decisive means for the Army to succeed in this complex and multidimensional urban environment.

## Chapter 2

# Understanding the Urban Environment

*From a planning perspective, commanders view cities not just as a topographic feature but as dynamic entities that include hostile forces, local population, and infrastructure. Planning for urban operations requires careful IPB, with particular emphasis on the three-dimensional nature of the topography and the intricate social structure of the population.*

FM 3-0

Of all the environments in which to conduct operations, the urban environment confronts Army commanders with a combination of difficulties rarely found elsewhere. Its distinct characteristics result from an intricate topography and high population density. The topography's complexity stems from the man-made features and supporting infrastructure superimposed on the natural terrain. Hundreds, thousands, or millions of civilians may be near or intermingled with soldiers—friendly and enemy. This second factor, and the human dimension it represents, is potentially the most important and perplexing for commanders and their staffs to understand and evaluate. To this end, this chapter provides information essential to understanding the urban environment and conducting an effective intelligence preparation of the battlefield (see FM 2-01.3, FMI 2-91.4, and Appendix B).

Although urban areas possess general similarities, each environment is distinct and will react to and affect the presence and operations of Army forces differently. A tactical technique effective in one area may not be effective in another area due to physical differences, such as street patterns or the type of building construction. An Army policy popular with one urban group may cause resentment and hostility in another due to diverse cultural differences. All difficulties potentially exist, and they increase the complexity for Army forces operating in urban areas. These difficulties range from conventional military forces to disease and starvation (see Chapter 3) to a pervasive media—often acutely present in intricate combinations. Thus, commanders at all levels must make extraordinary efforts to assess and understand their particular urban environment to plan, prepare for, and execute effective urban operations (UO).

## A COMPLEX ENVIRONMENT

2-1. Urban areas vary depending on their history, the cultures of their inhabitants, their economic development, the local climate, available building materials, and many other factors. This variety exists not only among urban areas but also within any particular area. The ever-changing mix of natural and man-made features in urban areas present commanders with some of the most difficult terrain in which to conduct military operations.

2-2. Although urban areas possess similar characteristics, no two are identical. The sprawl of Los Angeles, for example, bears little physical resemblance to New Delhi. Societal characteristics most significantly affect each area's uniqueness and complexity. While complex, information about the terrain, its potential effects on operations, and how it changes over time may be determined with some degree of certainty. However, the human dimension is much more difficult to understand and assess, particularly its effects on military operations. Like any environment, the side that can best understand and exploit the positive and mitigate the negative effects of the urban environment has the best chance of success.

2-3. Whether a large metropolis or a small village, each urban environment has identifiable components that constantly change and interact. This "system of systems" consists of the terrain, the society, and the infrastructure that links the two (see figure 2-1).

2-4. These systems are not separate and distinct categories, but rather overlapping and interdependent systems, acting dynamically with each other. Thoroughly analyzing these elements, along with the other factors of mission, enemy, weather, troops and support available, and time—

- Contributes to the accuracy of the commanders' situational understanding.
- Potentially lessens the number and cost of close combat engagements.
- Allows commanders to develop courses of action that apply appropriate resources against decisive points.

2-5. In stability operations and civil support operations, this understanding allows commanders to engage and dominate the decisive points critical to maintaining peace or restoring normalcy to the urban environment. Although each system is categorized into subordinate components or subsystems, commanders must often "step back" and visualize each system, the complex urban environment, and their area of operations (AO) as a unified whole. This "systems thinking" aids commanders in uncovering key relationships and intersections that can help reveal centers of gravity (COGs) and decisive points.

2-6. To comprehend the urban environment and its components to the fullest extent possible, commanders must carefully integrate and employ tactical reconnaissance forces, special operations forces (SOF)—to include psychological operations (PSYOP) and civil affairs units—and a myriad of other human intelligence (HUMINT) assets and regional, language, and cultural experts. The societal aspects and integrating infrastructure will challenge commanders' assessment and understanding. These aspects will also require greater dependence on nonmilitary and nongovernmental organizations (NGOs) and host-nation agencies for their information, knowledge, and expertise. This last consideration requires commanders to develop effective techniques and procedures for coordinating, interacting, and, to the greatest extent possible, synchronizing activities with these agencies.

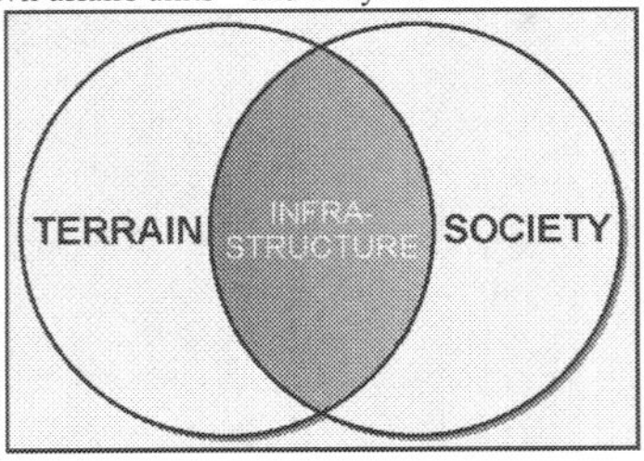


**Figure 2-1. Keys to understanding the urban environment**

# URBAN TERRAIN

2-7. Although complex and difficult to penetrate with many intelligence, surveillance, and reconnaissance (ISR) assets, the terrain is the most recognizable aspect of an urban area. Truly understanding it, however, requires comprehending its multidimensional nature. The terrain consists of natural and man-made features, with man-made features dominating; an analysis considers both. Buildings, streets, and other infrastructure have varied patterns, forms, and sizes. The infinite ways in which these factors can intertwine make it difficult to describe a "typical" urban area. However, these various factors provide a framework for understanding the complex terrain in an urban area. Furthermore, man-made features significantly affect military systems and Soldiers, and thus tactics and operations. General effects on urban operations are discussed in this chapter. Specific effects on warfighting functions (see Chapters 4 and 10) and the spectrum of operations (see Chapters 7, 8, and 9) are interwoven throughout the manual.

## MULTIDIMENSIONAL BATTLEFIELD

2-8. Urban areas present an extraordinary blend of horizontal, vertical, interior, exterior, and subterranean forms superimposed on the natural relief, drainage, and vegetation. An urban area may appear dwarfed on a map by the surrounding countryside. In fact, the size and extent of the urban area of operations is many times that of a similarly sized portion of undeveloped natural terrain. A multi-storied building may take up the same surface area as a small field, but each story or floor contains approximately an equal area as the ground upon which it sits. In effect, a ten-story building can have eleven times more defensible area than "bare" ground—ten floors and the roof. It is the sheer volume and density created by this urban geometry that makes UO resource intensive in time, manpower, and materiel.

2-9. Like natural disasters, UO can radically alter the physical characteristics of the urban terrain in ways not experienced in other environments. They may cause (either intentionally or not) uncontrollable fires or the loss of electricity. A power outage can cause flooding (especially in subsurface areas) by shutting down pumping stations. Entire buildings may be destroyed, eliminating reference points, leaving large piles of rubble, altering fields of fire, and making movement and transportation extremely difficult. Additionally, buildings and other urban structures, damaged but not destroyed, can become (or remain) effective obstacles and possible booby traps. Even without enemy exploitation, their weakened construction and unstable structure increase the risk of injury to Soldiers and civilians moving within them. (Engineer expertise will often be needed to determine whether the buildings can support occupation by Army forces or civilians.) Yet, even the total collapse of a building may not eliminate its defenders. Of additional concern, the likely presence of toxic industrial material (TIM) can create additional obstacles and health hazards.

2-10. Commanders in other environments normally address the depth, breadth, and height of their AO in terms of two areas: airspace and surface. In an urban environment, they broaden their scope to include supersurface and subsurface areas (see figure 2-2) that voluminously extend the commanders area of operations. Although spatially separated, each area may be used as an avenue of approach or mobility corridor, line of communications (LOC), or engagement area.

2-11. Supersurface and subsurface areas magnify the complexity of the urban physical environment. Commanders must consider activities that occur outside buildings and subterranean areas (the external space) as well as the activities happening unseen in buildings and subterranean systems (the internal space). This internal volume further challenges command, control, and intelligence collection activities and increases the combat power required to conduct UO. Commanders must develop methods and techniques to help themselves, their staffs, and their subordinate commanders and staffs to represent, visualize, and reference these multiple dimensions. Increasing the difficulty, such dimensions can change rapidly simply due to continued urban growth or, as described earlier, the effects of nature and UO themselves.

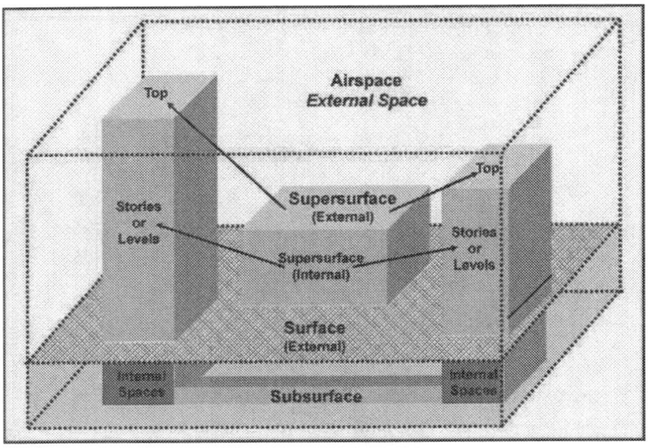


**Figure 2-2. The multidimensional urban battlefield**

## Airspace

2-12. As in all other environments, aircraft and aerial munitions use the airspace as rapid avenues of approach in urbanized areas. Forces can use aviation assets for observation and reconnaissance, aerial attack, or high-speed insertion and extraction of Soldiers, supplies, and equipment. Some surface obstacles in an urban area, such as rubble, do not affect flight (though they may prevent the take-off and landing of aircraft). Buildings of varying height and the increased density of towers, signs, power lines, and other urban constructions, however, create obstacles to flight and the trajectory of many munitions (masking). Similarly, these obstacles can restrict a pilot's line of sight as well as physically limit low-altitude maneuverability in the urban airspace. Excellent cover and concealment afforded enemy gunners in an urban area increases aviation vulnerability to small arms and man-portable air defense systems (MANPADS), particularly when supporting ground forces. The potential for a high volume of air traffic (military and civilian) over and within urban airspace (including fixed-wing, rotary-wing, and unmanned aircraft systems) may become another significant hazard and necessitate increased airspace command and control measures.

## Surface

2-13. Surface areas apply to exterior ground-level areas, such as parking lots, airfields, highways, streets, sidewalks, fields, and parks. They often provide primary avenues of approach and the means for rapid advance. However, buildings and other structures often canalize forces moving along them. As such, obstacles on urban surface areas usually have more effect than those in open terrain since bypass often requires entering and transiting buildings or radical changes to selected routes. Where urban areas border the ocean or sea, large lakes, and major rivers, the surface of these bodies of water may provide key friendly and threat avenues of approach or essential LOCs—a significant consideration for Army commanders. As such, amphibious, river-crossing, and river operations may be integral parts of the overall urban operation (see FM 3-05.212 and FM 3-90.12).

2-14. Larger open areas—such as stadiums, sports fields, school playgrounds, and parking lots—are often critical areas during urban operations. They can provide locations for displaced civilians, interrogation centers, and prisoner of war holding facilities. These areas also can afford suitable aircraft landing and pickup zones and artillery firing locations. They can provide logistic support areas and aerial resupply possibilities because they are often centrally located. Finally, large open areas (and immense or unusually shaped structures) within urban areas are often easier to see—especially from the air—and can serve as excellent target reference points from which to shift or control fires.

### Supersurface

2-15. These areas include the internal floors or levels (intrasurface areas) and external roofs or tops of buildings, stadiums, towers, or other vertical structures. They can provide cover and concealment; limit or enhance observation and fields of fire; and restrict, canalize, or block movement. However, forces can move within and between supersurface areas creating additional, though normally secondary, avenues of approach. Rooftops may offer ideal locations for landing helicopters for small-scale air assaults and aerial resupply. First, however, engineers must analyze buildings for their structural integrity and obstacles. Such obstacles include electrical wires, antennas, and enemy-emplaced mines (although personnel may be inserted by jumping, rappelling, or fast roping from a hovering helicopter and extracted by hoist mechanisms). Some rooftops are designed as helipads. Roofs and other supersurface areas may also provide excellent locations for snipers; lightweight, handheld antitank weapons; MANPADS; and communications retransmission sites. They enable top-down attacks against the weakest points of armored vehicles and unsuspecting aircraft. Overall, elevated firing positions reduce the value of any cover in surrounding open areas and permit engagement at close range with less risk of immediate close assault. This area (and the subsurface area) requires commanders to think, plan, and execute ground operations vertically as well as horizontally. In this latter regard, UO share strong similarities with mountain operations (see FM 3-97.6).

### Subsurface

2-16. Subsurface areas are below the surface level. They may serve as secondary and, in fewer instances, primary avenues of approach at lower tactical levels. When thoroughly reconnoitered and controlled, they offer excellent covered and concealed LOCs for moving supplies and evacuating casualties. They may also provide sites for caching and stockpiling supplies. Subsurface areas include subterranean areas such as subways, mines, tunnels, sewers, drainage systems, cellars, civil defense shelters, and other various underground utility systems. In older cities, they may include ancient hand-dug tunnels and catacombs. Both attacker and defender can use subsurface areas to gain surprise and maneuver against the rear and flanks of a threat and to conduct ambushes. However, these areas are often the most restrictive and easiest to defend or block. Their effectiveness depends on superior knowledge of their existence and overall design. Army commanders may also need to consider potential avenues of approach afforded by the subsurface areas of rivers and major bodies of water that border urban areas. This particularly applies when operating as part of a joint task force (JTF) task organized with SOF or when opposing a threat with similar capabilities.

## Major Urban Patterns

2-17. Four major urban patterns (satellite, network, linear, and segment) can influence UO (see figure 2-3). Central to two of the patterns (satellite and network) is the hub or dominant urban area or pattern around which outlying urban areas or patterns radiate. (A segmented urban area, because it tends to be a larger urban area, can often be a hub.) In offensive and defensive operations, the hub serves as a pivot or strong point; as such, it can become a major obstacle to an attacker. If the attacker chooses to bypass the urban area (hub) located along his axis of advance without first isolating the area, he may expose his flank or LOC to attack from the hub as well as dependent urban areas or subordinate satellite patterns. Because the focus of stability and civil support operations is normally on people, commanders should understand the value and influence of the hub to the economic, political, or cultural well being of the surrounding area. Generally the larger the hub, the greater influence it has on satellite urban areas and surrounding rural areas. Commanders must remember that urban areas are not islands; all are connected to the surrounding rural (and other urban) areas through fluid and permeable boundaries and LOCs.

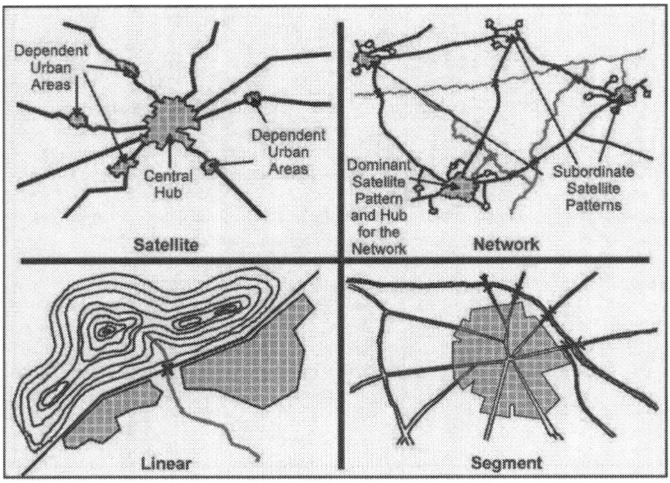


**Figure 2-3. Major urban patterns**

## Satellite Pattern

2-18. This common pattern consists of a central hub surrounded by smaller, dependent urban areas. LOCs tend to converge on the hub. Outlying areas often support the principal urban area at the hub with means of reinforcement, resupply, and evacuation. In some instances, they may serve as mutually supporting battle positions. Commanders should consider the effects of the outlying urban areas on operations within the hub, and, conversely, the effects of operations within the hub on outlying urban areas. Information operations (IO), for example, targeted primarily at key leaders and other civilians located within the hub of a satellite pattern may subsequently influence civilians in outlying urban areas and achieve necessary effects without having to commit specific resources to these outlying areas.

## Network Pattern

2-19. The network pattern represents the interlocking of the primary hubs of subordinate satellite patterns. Its elements are more self-sufficient and less supportive of each other, although a dominant hub may exist. Major LOCs in a network extend more than in a satellite pattern and take more of a rectangular rather than a convergent form. Its natural terrain may vary more than in a single satellite array. Operations in one area may or may not easily influence, or be influenced by, other urban areas in the pattern.

## Linear Pattern

2-20. Potentially a subelement of the previous two patterns, the linear pattern may form one ray of the satellite pattern or be found along connecting links between the hubs of a network. Most frequently, this pattern results from the stringing of minor urban areas along a confined natural terrain corridor, such as an elongated valley, a body of water, or a man-made communications route. In offensive and defensive operations, this latter form of the linear pattern facilitates developing a series of strong defensive positions in depth, effectively blocking or delaying an attacking force moving along the canalized terrain.

### Segment Pattern

2-21. When dominant natural terrain, such as a river or man-made features (canals, major highways, or railways), divides an urban area, it creates a segmented pattern. This pattern often makes it easier for commanders to assign areas of operations to subordinate commanders. However, this pattern may fragment operations and increase risk to an operation requiring mutual support between subordinate units. Still, the segmented urban areas may allow commanders to isolate threats more easily in these areas and focus operations within segments that contain their decisive points. Although an integral part of the whole (the urban area), each segment may develop distinct social, economic, cultural, and political characteristics. This social segmenting may benefit commanders faced with limited assets to influence or control the urban populace. After thoroughly analyzing the society, they may be able to focus IO and populace and resources control measures against only specific segments that affect decisive operations. Commanders may need only to isolate other segments or may need to just monitor for any significant changes in the attitudes, beliefs, or actions of the civilians located there.

## LESSER STREET PATTERNS

2-22. Lesser patterns in the urban area result from the layout of the streets, roads, highways, and other thoroughfares. They evolve from influences of natural terrain, the original designer's personal prejudices, and the changing needs of the inhabitants. Street patterns (and widths) influence all warfighting functions; however, they greatly affect movement and maneuver, command and control, and sustainment. (In some portions of older Middle Eastern urban areas, the labyrinths of streets were designed only to allow two loaded donkeys to pass each other; tanks are too wide.) Urban areas can display any of three basic patterns and their combinations: radial, grid, and irregular (see figure 2-4).

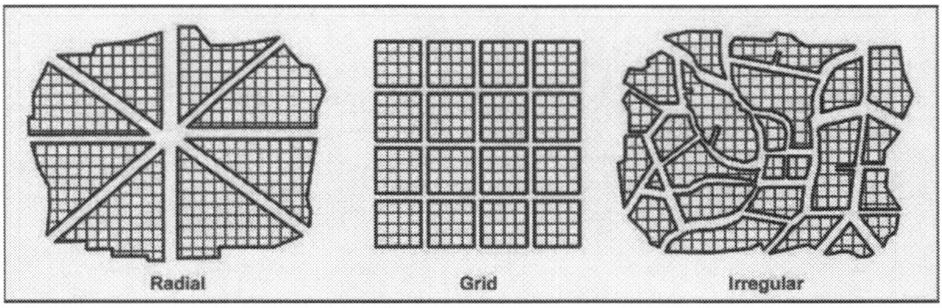


**Figure 2-4. Basic internal street patterns**

### Radial

2-23. Societies of highly concentrated religious or secular power often construct urban areas with a radial design: all primary thoroughfares radiating out from the center of power. Urban areas with this design may signal an important historical aspect in the overall analysis of the urban society. Terrain permitting, these streets may extend outward in a complete circle or may form a semicircle or arc when a focal point abuts a natural barrier, such as a coastline or mountain. To increase mobility and traffic flow, societies often add concentric loops or rings to larger radial patterns. Unless commanders carefully plan boundaries, routes, and axes of advance, their subordinate units' movement or maneuver may be inadvertently funneled toward the center of urban areas with this pattern resulting in congestion, loss of momentum, and an increased potential for ambush or fratricide.

### Grid

2-24. The most adaptable and universal form for urban areas is the grid or rectangular pattern: lines of streets at right angles to one another forming blocks similar to the pattern of a chessboard. A grid pattern can fill in and eventually take over an original radial pattern. Grid patterns often appear to ease the assignment of boundaries for subordinates units. However, commanders also consider how the natural terrain influences operations and the establishment of graphic control measures. They also consider the influence of the buildings and other structures lining these streets, such as their height and construction, before assigning boundaries and developing other control measures. Commanders should also consider the following when developing urban graphic control measures:

- Describing boundaries, phase lines, checkpoints, and other graphic control measures by easily recognizable features is as important in urban areas as elsewhere. While easily identifiable urban structures such as unusually tall or oddly-shaped buildings, cemeteries, stadiums, or prominent rail or highway interchanges can be useful references, available natural features are a better descriptor than man-made features that may be altered or unrecognizable. As an aid to air-to-ground coordination, commanders should select features that can be identified by both ground and air forces. Those that help in controlling ground forces may not be easily visible from the air and vice versa.
- Commanders should also consider whether a boundary along an easily-recognizable terrain feature, such as a river, will also be easy to identify by the threat who may seek to "find the seam" and exploit the likely control and coordination difficulties associated with boundaries, especially between higher-level units. This often requires commanders to carefully position a control measure away from the key feature to provide a designated subordinate force with the terrain and space necessary to control the feature. On the other hand, commanders working closely with local authorities during stability and civil support operations may not need to thoroughly understand the physical effect of street patterns on the assignment of boundaries as they might for combat UO. Instead, commanders may choose to assign boundaries overlaid on existing geopolitical boundaries used by local agencies to increase interoperability and aid in unity of effort.

### Irregular

2-25. In most urban areas, regardless of the original intent, plan, or vision, existing street patterns emerge from successive plans overlaid one on another. Some are well planned to fit with previous plans while others a haphazard response to explosive urban growth. The result may mix patterns. Urban engineers and planners may specifically design irregular patterns for aesthetic reasons (as in many suburban housing developments) or to conform to marked terrain relief. Irregular street patterns may alert commanders and analysts that the underlying natural terrain may exert greater influence over operations than in other portions of the urban area. Finally, irregular street patterns make the movement and maneuver of forces less predictable. However, a labyrinth of irregular or "twisting" street patterns may increase the possibility of fratricide particularly for units that are trained or accustomed only to grid patterns.

## An Urban Model

2-26. Throughout the world, urban areas have similar form and function. In form, urban areas contain like characteristics, readily divisible into distinct sections or areas. Functionally, they tend to be the centers of population, finance, politics, transportation, industry, and culture. While urban areas may be modeled by several different means, figure 2-5 illustrates the general forms and internal functions. Some forms and functions may overlap. For example, high-rise buildings are located in core areas as well as in outlying areas and may be used for residential purposes. With the rapid urbanization associated with developing nations, the areas displayed in this urban model often manifest themselves less clearly there than in developed nations.

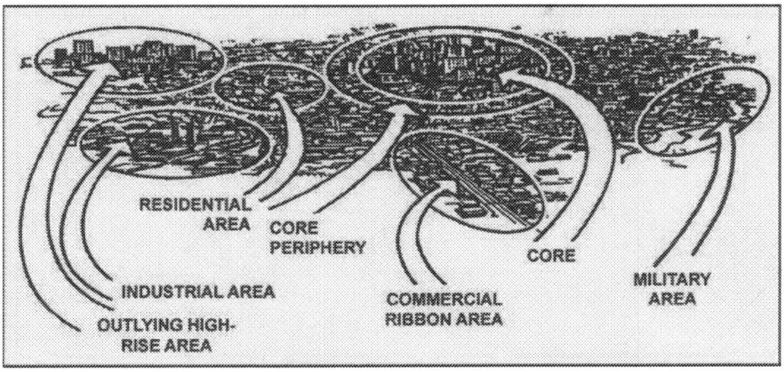


**Figure 2-5. Urban functional zones**

2-27. This analysis helps to determine, in general terms, potential advantages and disadvantages each portion of the urban area may have toward accomplishing the urban operation. However, construction materials and methods can vary drastically. Commanders and their staff will often need to identify specific building types and construction and understand weapons effects on them. If a commander desires precise effects, the chosen munitions or weapons system must be sufficiently accurate, capable of penetrating the target structure (without exiting the other side), and achieve effects within. Often noncombatants, critical infrastructure, or protected targets are in the vicinity. Commanders may need to determine if the surrounding walls or structures will sufficiently absorb or negate the blast or thermal effects of the weapon. Regardless, understanding the structure and composition of buildings and other structures in the urban AO may be necessary to allow commanders to determine the best means to accomplish the mission.

## Core

2-28. The core is the heart of the urban area, the downtown or central business district. Relatively small and compact, it contains a large percentage of the urban area's shops, offices, and public institutions. Often, it houses the headquarters for commercial and financial activities and contains important cultural, historical, and governmental buildings. These activities prefer the core because of its accessibility. As the focal point of the transportation network, residents find the core the easiest part of the urban area to reach. It normally has the densest concentration of multistory buildings and subterranean features (underground parking garages, underground shopping centers, and basements).

2-29. High-rise buildings, varying greatly in height (possibly 50 stories above ground and four stories below ground), make up the cores of many of today's urban areas. Buildings routinely abut one another, with little or no setback from the sidewalks. Building height and density (except in outlying high-rise areas) often decreases from the core to the edge of the residential areas, while the amount of open areas frequently increases. Modern urban planning allows for more open spaces between buildings than found in the cores of older urban areas. Most core areas have undergone constant redevelopment resulting in various types of construction. Commonly, brick buildings abound in the oldest part of the core; framed, heavy-clad structures in the next oldest part; and a concentration of framed, light-clad buildings in the newest part. The outer edge of the core, the core periphery, has ordinarily undergone less change than the core resulting in buildings of uniform height (commonly two to three stories in towns and five to ten stories in larger urban areas).

2-30. Generally, offensive operations focused in core areas (even when effectively isolated) will require greater resources—particularly manpower, time, and information—than in many other parts of the urban area. Mounted maneuver often proves more difficult in core areas because of fewer open areas, buildings closer to the streets, and an increased density of civilian vehicles. Razed buildings in central core areas (especially high-rise buildings) become greater obstacles to mobility as they can collapse on and easily block thoroughfares. Rubble piles provide excellent covered and concealed positions for dismounted threat forces. Consequently, commanders often use more dismounted forces as part of their combined arms operations. Conversely, the core may be critical to urban defensive operations, particularly older areas of heavier construction that afford greater protection. Despite potential difficulties, the core area may be key to accomplishing many stability or civil support missions since it houses much of the human activity that occurs in the urban area.

## Industrial Area

2-31. Industrial areas often develop on the outskirts of the urban areas where commercial transportation is easiest (along airfields and major sea, river, rail, and highway routes). The road networks in and around industrial areas are generally more developed and suitable for transportation assets. These areas will likely displace farther from the core and residential areas as urban planners recognize the potential threat of TIM. The dispersed pattern of the buildings provides sufficient space for large cargoes, trucks, and materiel handling equipment. These areas may provide ideal sites for sustainment bases and maintenance sites. While older heavier-clad structures may be found, new construction generally consists of low, large, flat-roofed factory and warehouse buildings with large parking areas and work yards. These structures generally have steel frame and lightweight exterior walls. Multistory structures usually have reinforced concrete floors and ceilings.

| Toxic Industrial Chemical | Industrial/Commercial Uses |
|---|---|
| Ammonia | Commercial Refrigerant, Fertilizer and Food Production, Petroleum, Explosives, Other Chemicals |
| Arsine | Semiconductor Industry |
| Boron Trichloride | Organic Catalyst, Soldering Magnesium |
| Boron Trifluoride | Chemical Catalyst, Aluminum Refining |
| Carbon Disulfide | Industrial Solvent, Dry Cleaning, Agriculture, Petroleum, Electroplating |
| Chlorine | Potable Water, Disinfectants, Metal Treatment, Plastics & Rubber |
| Diborane | Plastics and Rubber |
| Ethylene Oxide | Industrial Alcohols, Fumigant, Industrial Sterilant |
| Fluorine | Uranium Processing, Rocket Fuel |
| Formaldehyde | Plastics, Fertilizers, Preservative/Corrosion Inhibitor, Fungicide and Germicide, Pesticide, Pharmaceuticals |
| Fuming Nitric Acid | Fertilizers, Explosives, Metal Processing, Pesticides, Rocket Fuel |
| Hydrogen Bromide | Chemical Industry, Pharmaceuticals |
| Hydrogen Chloride | Fabrics, Semiconductors |
| Hydrogen Cyanide | Pesticides, Other Chemicals, Pharmaceuticals, Electroplating |
| Hydrogen Fluoride | Glass Production, Chemical Catalyst |
| Hydrogen Sulfide | Metallurgy, Agricultural Disinfectant |
| Phosgene | Dyes, Pharmaceuticals, Herbicides & Insecticides |
| Phosphorus Trichloride | Metallurgy, Pesticides and Germicides, Gasoline Additive |
| Sulfur Dioxide | Paper, Food Processing, Ice Production, Disinfectant, Leather Processing |
| Sulfuric Acid | Fertilizers, Petroleum, Iron and Steel Production, Battery Electrolyte |
| Tungsten | Electronics, Other Chemicals |

**Figure 2-6. Some toxic industrial chemicals and their industrial or commercial uses**

2-32. Toxic industrial chemicals and other TIM may be transported through an urban area (by rail, barge, truck, or pipeline) or found stored throughout. However, larger concentrations will exist in industrial areas, and their presence should concern Army forces operating near them. Some TIM are heavier than air and tend to settle in low-lying and subsurface areas.

2-33. Each year, over 70,000 different chemicals are produced, processed, or consumed globally. An estimated 25,000 commercial facilities around the world produce, process, or store chemicals that have a legitimate industrial use yet are also classified as chemical warfare agents. Many other chemicals (not classified as weapons) may still be sufficiently hazardous to pose a considerable threat to Army forces and civilians in urban areas as choking agents or asphyxiates, flammables or incendiaries, water contaminants, low-grade blister or nerve agents, or debilitating irritants. These chemicals can be released either accidentally or deliberately. For example, on 2 December 1984, nearly 40 tons of methylisocyanate used to produce pesticides leaked from a storage tank at Bhopal, India, killing thousands and injuring hundreds of thousands. Figure 2-6 contains a small sampling of other toxic industrial chemicals along with their industrial or commercial usage that commanders may encounter in an urban area. The most common chemicals that may pose an immediate risk to Army forces are highly toxic irritant gases such as ammonia, chlorine, hydrogen chloride, and sulfur dioxide.

2-34. Standard chemical defense equipment may not protect against (and chemical detection devices may fail to detect) many toxic industrial chemicals. Therefore, the risk to Soldiers operating near the chemicals may increase. Commanders must vigilantly identify these potential hazards, carefully consider them as part of their overall vulnerability analysis, factor the analysis into their risk assessment, and execute necessary contamination avoidance measures. (Local urban fire fighters may be a critical source of information for determining the likely locations of TIM.) Any assessment includes the chance that toxic industrial chemicals may be deliberately released by a threat to gain advantage or accidentally released by friendly actions (see FM 3-11.9, FM 3-11.14, FM 3-11.21, and the U.S. Department of Transportation's current version of the Emergency Response Guidebook).

## Outlying High-Rise Area

2-35. High-rise areas consist of multistoried apartments, commercial offices, and businesses separated by large open areas, such as parking lots, parks, and individual one-story buildings. High-rise buildings are generally of framed, light-clad construction with thin walls of brick, lightweight concrete, or glass. The automobile, mass transit systems, and improved road networks encourage these areas to grow and function further from the urban core.

2-36. Similar to the urban core, units given the mission to clear these areas, or even portions therein, will need more resources—most notably personnel and time—to accomplish their mission. Commanders should consider courses of action that isolate these entire areas, multiple sections within these areas, or even individual buildings before assigning tasks. Without careful consideration and analysis, some tasks in these areas could unintentionally—but rapidly—drain a unit's resources or unhinge other portions of the major operation. When defending, commanders who can integrate these areas in the defense will present the attacker with similar resource problems and may be appropriate in a defense to delay. However, defending commanders must ensure that the defense is arranged so that this portion cannot be easily isolated and bypassed. Defensive positions in structures may require extensive reinforcement due to light-clad construction.

## Residential Area

2-37. Residential areas can be found dispersed throughout the urban area; however, large suburban areas (or sprawl) normally form on the outskirts. Residential areas often consist of row houses or single-family dwellings set in a grid or ringed pattern in a planned development project. Yards, gardens, trees, and fences may separate the buildings in a residential area. In some areas of the world, residential areas may be located in high walled compounds with houses built right up to the edge of the street. Modern residential construction is more often of light-clad, framed wood construction, or brick, however; residential homes formed by poured or pre-cast concrete can be found throughout many parts of the world. The combined population of surrounding suburban areas often far outnumbers that of the urban area proper. Specific suburbs tend toward homogeneity based on ethnicity, religion, economics, or some other social aspect. Commanders must locate and analyze these areas to determine their impact on operations—often the most critical importance is the people located there (see the subsequent discussion in this chapter on the urban society).

2-38. In offensive and defensive operations, commanders should determine whether operations pose an unacceptable physical risk to civilians. If so, they may have to relocate civilians to a safer area, perhaps another residential area. If not, commanders may implement a "stay-put" policy for that area and attempt to isolate the effects of the operation from them. During civil support and stability operations, residential locations may be the initial focal point for operations since most of the permanent population is located there.

2-39. This area also contains an urban phenomenon known as shantytowns. These areas are commonly on unoccupied, low-value land in and around many urban areas in underdeveloped countries. Shantytowns may contain over 50-percent of the total urban population. They usually lack streets and public utilities. The lean-to structures tend to be irregularly laid out, connected by walking paths, and made of any scrap material available: lumber, brick, sheet metal, cardboard, cloth, or vegetation. The random arrangement of structures, the absence of formal street naming and numbering, and often the lack of easily identifiable buildings and terrain create challenges. These challenges include navigation, coordination, and the transmission of accurate information and intelligence. Depending on the operation, the temporary nature of the structures can also mean that mobility can be either more or less restricted than other sections of the urban area. A military force may easily knock down and traverse structures without affecting mobility at all. However, their destruction may cause unacceptable civilian casualties, in which case mobility becomes more restrictive as the narrow paths often do not accommodate vehicular traffic. Similarly, the makeshift materials inhibit weapons effects less than many other parts of the urban area built more solidly. A tank round, for example, may go much farther and injure many more noncombatants than in an area where the primary building material is stone. Regardless, commanders should consider the effects of their operations in this area, to include vehicles and weapons, as the weak structures increase the risk of fratricide, civilian casualties, and large, rapidly spreading fires.

## Commercial Ribbon Area

2-40. Commercial ribbon areas are rows of stores, shops, and restaurants built along both sides of major streets that run through and between urban areas. These same types of areas often develop along the roads that connect one urban area to another (strip areas). The buildings uniformly stand two to three stories tall (about one story taller than the dwellings on the streets behind them).

## Military Area

2-41. Fortifications and military installations may be found in or near urban areas throughout the world. Historically, they may have been the "seed" responsible for initiating the growth of the present-day urban area. Many countries possess long coastlines and borders with potentially hostile neighbors. To meet their defensive needs, they developed coastal and border defense works that include extensive subsurface facilities, many contiguous to urban areas. North Korea, for example, has built numerous hardened artillery, missile, and command and control facilities along both its coasts and the demilitarized zone which separates them from South Korea. While some fortifications may be inactive (particularly ancient fortifications), they can be rapidly activated and modified by threats (or Army forces) to accomplish their original role. Even if not activated, they may still serve as choke points and major obstacles to movement and maneuver.

2-42. Permanent-type fortifications can be made of earth, wood, rock, brick, concrete, steel-reinforced concrete, or any combination of the above. Some contemporary constructions have been built subsurface and employ heavy armor, major caliber weapons, internal communications, service facilities, and chemical, biological, radiological, and nuclear overpressure systems. Because they have been built specifically for military purposes, commanders and planners should carefully consider the effects of these military constructions on the conduct of UO.

# URBAN SOCIETY

2-43. Although intricate, understanding the urban terrain is relatively straightforward in comparison to comprehending the multifaceted nature of urban society. UO often require Army forces to operate in close proximity to a high density of civilians. Even evacuated areas can have a stay-behind population in the tens of thousands. This population's presence, attitudes, actions, communications with the media, and needs may affect the conduct of operations. Homogeneity decreases drastically as the size of the urban area increases. Commanders must take into account the characteristics of a population whose beliefs and interests vary based on factors (see figure 2-7). Analysis and understanding of these societal factors is critical to a successful IO campaign and, thus, the entire operation.

2-44. Civilian populations continually influence, to varying degrees, operations conducted in an urban area. Thoroughly understanding these societal aspects and avoiding "mirror-imaging"—overlaying one's own values and thought processes on top of the person or group one is trying to understand—will help to accurately anticipate civilian actions and response. For example, while bribery and nepotism are not accepted norms in U.S. society, they may be customary practices in the foreign urban environment in which Army forces are operating. If commanders attempt to understand incidences of bribery and nepotism using their own societal norms and values as a basis, they are likely to misinterpret the significance of the events and implement an inappropriate course of action. Critically, however, the development of an in-depth understanding of the urban society does not mean that commanders and their Soldiers should adopt the culture of the foreign urban society. While understanding and respect of the urban area's culture is crucial to successful UO, Soldiers function best in dealing with the urban society when they act in accordance with American culture and values.

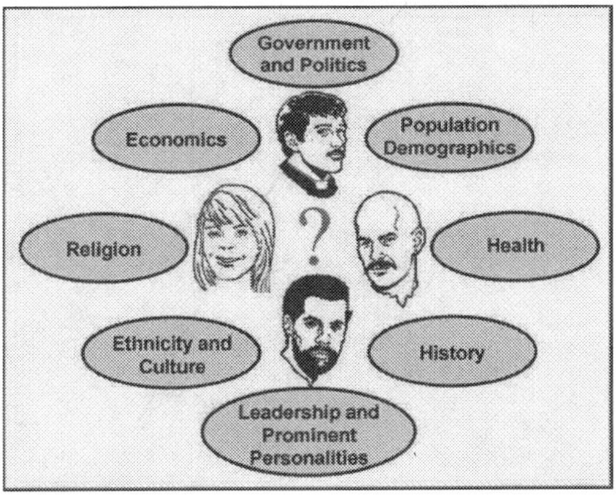


Figure 2-7. Key aspects of the urban society

## POTENTIAL CENTER OF GRAVITY

2-45. A COG during an urban operation, particularly in stability and civil support operations, may be the civilian inhabitants themselves—specifically their behavior. However, supportive behavior is generally an advantage in any type of operation. Correspondingly, neutral behavior toward friendly forces is an advantage over hostile behavior. (However, a neutral population normally will not readily provide information to Army forces—often critical to understanding the urban environment and threat. In this light, neutrality can be seen as a disadvantage.) To influence or control their behavior, commanders must first understand the society's complex nature and character. Second, they must understand and accept that every military action (or inaction) may influence the relationship between the urban population and Army forces, and, by extension, mission success or failure. Lastly, they must understand that Army forces may play only a supporting (but essential) role as part of an integrated and synchronized multiagency effort focusing all aspects of national power—diplomatic, informational, military, and economic. With this awareness, commanders can take one or more actions:

- Coordinate and plan operations.
- Implement effective civil-military programs.

- Take the immediate action necessary to maintain support of a friendly populace, neutralize or gain the support of hostile or neutral factions, or do any combination of these activities to achieve precise effects and accomplish the mission.

Without this understanding, commanders increase the risk that their actions, particularly concerning the urban population, may not have the intended and necessary effects.

2-46. Although the factor of civil considerations takes on added significance in UO, it is just one that commanders evaluate. Sometimes, as in most urban counterinsurgency operations, it may be the most important factor to consider as a COG. At other times it may be the least important as to be almost negligible. Its importance is not constant; it changes over time (like all factors). At the beginning of the operation, civil considerations may not be essential to mission accomplishment, but as the operation progresses this factor's importance to success may increase. In other circumstances, the opposite may be true. Overall, commanders should consider three objectives regarding the civilians of the urban area:

- Minimize their interference with urban operations. In offensive and defensive operations this means moving them away from combat operations or establishing measures to shield them from its effects. In all operations, it often requires centralizing them in one or more locations—which may mean keeping them where they are.
- Maximize their support of Army, joint, and multinational forces and government agencies.
- Observe the necessary legal, moral, and humanitarian obligations.

## General Population Size

2-47. Urban areas are commonly classified according to the general size of their population instead of landmass. Figure 2-8 lists one method of defining and categorizing urban areas by various population sizes.

2-48. These categories are useful to establish commonality and standardize terms that shape ideas, discussion, and concepts. Smaller populations usually suggest homogeneity among the inhabitants. Homogeneity can make consensus or compromise easier to achieve because fewer opposing viewpoints exist. Given this homogeneity, effects of change are more certain and often easier to determine. However, homogenous does not mean identical. If major social divisions do exist (either physical or ideological), commanders can more easily determine those divisions and their fundamental causes with smaller populations. Treating an urban population as a completely homogenous entity can lead to false assumptions, cultural misunderstandings, and poor situational understanding.

| Category | Population |
|---|---|
| Village | 3,000 or less. |
| Town | Over 3,000 to 100,000. |
| City | Over 100,000 to 1 million. |
| Metropolis | Over 1 million to 10 million. |
| Megalopolis | Over 10 million. |

**Figure 2-8. Urban areas by population size**

2-49. As urban areas expand, the urban patterns begin to blur and the social complexity increases. For example, as satellite patterns continue to grow, the LOCs between a central hub and outlying urban areas may develop and begin to assume a linear urban pattern. Simultaneously, a hub and outlying urban areas may continue to expand until they merge into a single, large metropolis. On a larger scale, a network pattern can grow and unite as a single megalopolis. This growth physically unites smaller urban areas but cannot force conformity of needs and beliefs. It also increases the physical and social complexity of an urban area.

## Group Size, Location, and Composition

2-50. Understanding how specific elements of the urban society affect operations (and vice versa), begins with analyzing their size, location, and composition (see figure 2-9). Because commanders must minimize civilian casualties, size and location (without regard to composition) are important initial demographic considerations. After determining the presence and numbers of civilians relative to decisive points, commanders can then decide whether civilian proximity and density represent a significant risk to their mission—refugees clogging LOCs, for example. If civilians are the primary focus of the operation, as in many stability and civil support operations, this same analysis may help to determine decisive points. In this analysis, commanders should consider that urban areas, on many levels, are in constant motion. The densities of circulating people and other traffic often vary according to cultural events or the time of day, such as religious holidays or sporting events and rush hours or market times. Therefore in planning urban operations, commanders must consider the timing, rhythms, or patterns of the population and their vehicular movements in the urban area. Identifying and understanding trends and patterns of activity (and disruptions to them) may provide critical information to commanders.

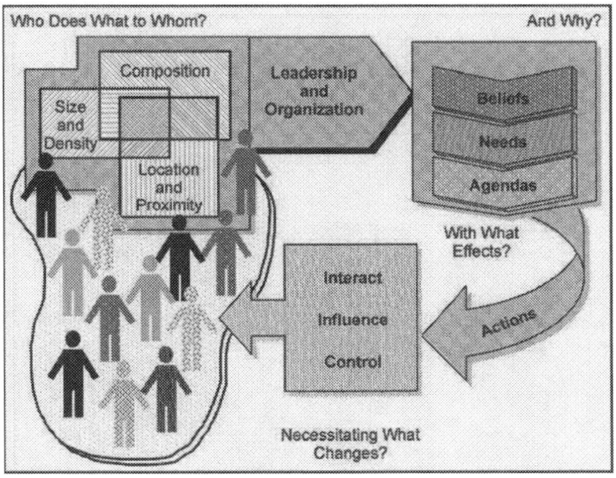


**Figure 2-9. Simplified analysis of urban society**

2-51. Commanders normally determine the composition of, or the identifiable groups or organizations within, the civilian urban population. Groups may be categorized by race, religion, national origin, tribe, clan, economic or social class, party affiliation, education level, union memberships, age, gender, occupation, or any other significant social demographic. Physical and ideological overlaps (and divisions) often exist between groups. Overlaps may provide early focus for analysis and suggest ways to affect more than one group simultaneously. In some cases, groups may have radically different ideologies but are (or can be) united by a single characteristic. Commanders must understand the intricacies of "who does what to whom." Such understanding furthers identifying the urban society's sources of power, influence (both formal and informal), and decisive points that hold the keys to controlling or protecting this potential COG. (See also the discussion of competing power structures in Chapter 3.) Commanders should have expert, detailed, and current knowledge and information to avoid developing simple formulas of social interaction that may actively mislead and add to a flawed course of action.

## Leadership and Organization

2-52. Commanders must also understand how authority and responsibility is held or shared within and between each of the identified groups; they must understand leadership and the social hierarchy. For groups to exert meaningful influence, leadership provides vision, direction, and organized coherence. This leadership can be a function of personality as well as organization. Some groups depend on a charismatic leader to provide cohesion; although in some cultures, the spokesman is not the leader. Others deemphasize individual leadership and provide redundancy and replacement in decision making. Others combine elements of both these types of leadership and organization. Based solely on personality, a leader may centralize power or, while still being in ultimate control, decentralize decision making and execution to subordinates. In contrast, a single person may head a group while a ruling council actually makes and executes policy. Groups centered on one leader (which may or may not be the officially designated leader) can often produce decisions and initiate actions rapidly but are vulnerable to disruptions if key personalities are removed or co-opted. Groups with shared or redundant leadership take longer to make decisions yet are more resistant to change and outside influence.

2-53. In UO, particularly stability operations, Army commanders and leaders at all levels will devote considerable effort to identifying and cultivating relationships with civilian leaders in their AO. This civilian leadership will include political, religious, tribal or clan, ethnic, and economic leaders. Commanders should consider that their attention toward and discussion with identified leaders may increase (or, some instance, decrease) the targeted leaders' prestige and power. While this may be intentional, commanders must often ensure that the leaders that they chose to deal with are the legitimate and accepted in the eyes of the urban population. Otherwise, they may further imbalance an already weak power structure and exacerbate the situation. In unique circumstances, commanders may need to identify and interact with the leadership of criminal organizations. These civilian leaders will be the means to affect change and allow the urban populace to understand and accept the purpose behind Army operations. As importantly, they will be the conduit for understanding the urban society's sentiments, perceptions, and reactions to Army operations. As such, reliable and trustworthy linguists that can not only interpret the language but can serve as cultural advisors to Army forces will be in great demand. Keeping the local population objectively informed not only of current Army operations but also of the intent and desired end state will often be a key task for commanders. A poor communications effort and a lack of understanding of intent may alienate a commander and cause significant problems in future negotiations. It is imperative that communications be clear and effective and that all Army leaders are aware of its implications. See FM 4-05.401 and FM 3-07 for the effective use of interpreters and the conduct of productive meetings and negotiations.

## INTERESTS AND ACTIONS

*Me and Somalia against the world, me and my clan against Somalia, me and my family against the clan, me and my brother against my family, me against my brother.*

Somali Proverb

2-54. Identifying and analyzing groups also helps commanders focus on specific segments of the urban society to determine their beliefs, needs, and agendas. It also helps commanders determine how those interests motivate groups to future action (or inaction)—previous patterns of activity are critical in this regard. This analysis seeks to determine why groups (and their leaders) act as they do. Commanders should consider political, economic, cultural, and religious factors in this analysis. These factors affect all groups to some extent and often provide the basis for their beliefs, needs (actual or perceived), and subsequent behavior. Size and location considerations also apply to each group to help determine to what extent its beliefs or ideologies, needs, and actions may impact the urban operation. However, size and proximity may not accurately indicate actual or potential capabilities. Individuals, small groups, and groups located some distance from the actual conduct of the urban operation may be able to influence large portions of the population. These individuals or groups may have a capability disproportionate to their size and proximity—especially against objectives that are not terrain oriented (as in the case of many stability operations).

## INTERACTION, INFLUENCE, OR CONTROL

2-55. Commanders must cultivate an understanding of a group's—

- Size, location (and proximity to operations), and composition (to include leadership and organization).
- Interests.
- Capabilities.
- Potential actions (intent) and their effects—if any—on operations.

2-56. Then they can develop or modify courses of action as appropriate. Certain courses of action may be needed to improve the interaction between Army forces and civilians (and between other agencies) to accomplish common goals. Others may be needed to influence favorable support, stabilize neutral groups, or neutralize hostile groups. Still others may require more forceful means to control and protect—never to punish—civilians. Courses of action may include—

- Establishing buffer zones and restricted areas.
- Setting up checkpoints and roadblocks with other travel restrictions on people and goods.
- Screening civilians.
- Conducting negotiations (directly or as a mediator).
- Providing or protecting rations, water, and other critical resources.
- Restoring or improving specific, key infrastructure.
- Enforcing curfews.
- Inspecting facilities.
- Directing amnesty programs.
- Conducting internment and resettlement operations.
- Maintaining a "stay-put" policy.

2-57. Commanders must remember that many measures will require significant resources that may initially be beyond the capabilities of the Army force to impose and enforce. (Where possible, commanders should attempt to use host-nation security forces and local urban law enforcement to control activities.) Other elements of the environment, terrain and infrastructure, may fragment efforts and make it difficult to impose controls throughout the area. A careful assessment and understanding of the urban society's interests (beliefs, needs, and agendas) is essential before implementing any populace and resources control measures. Otherwise inappropriate controls, particularly if they are improperly perceived as punishment, may only aggravate the situation. Finally, an appropriate course of action may require no specific action towards the urban society. In most cases, training and discipline, grounded in cultural understanding and sensitivity, will help mitigate many potential adverse effects resulting from military-civilian interaction. (Commanders should review FM 3-19.15 and FM 3-19.40 for additional civilian control measures and considerations.)

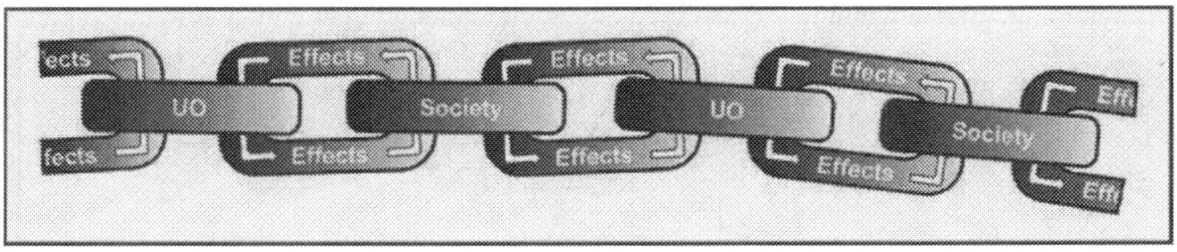


**Figure 2-10. UO society chain of effects**

## A Chain of Effects

2-58. Since urban society is so dynamic and the relationship between various elements so complex, commanders must continually assess how their operations will affect the society and vice versa (see figure 2-10). Specifically, they assess how effectively their measures improve interaction with, influence of, and control over civilians. There is always a difference between intended and actual effects of a specific course of action. Nowhere is this more prominent than dealing with the urban society. This chain of effects frustrates understanding during UO. Therefore, commanders must continuously monitor these effects to make decisions and modifications while planning, preparing, executing, sustaining, and transitioning UO. (Commanders may conduct informal and formal polling activities to assess and understand changing civilian perceptions.) Initially certain aspects of the society, such as religion, may not affect the operation. However, if the threat successfully shapes the perceptions of the urban populace that Army forces are biased against their religion or religious beliefs (or at least critical segments are affected by propaganda), this element may become extremely important. In this instance, the urban commander may need to adjust his IO (to include PSYOP), public affairs (PA) activities, and CMO to counter this propaganda while diverting other combat power to control the populace. Overall, commanders must understand and account for second, third, and higher order effects of their actions and decisions.

**Understanding the Effects of Unit and Soldier Actions in Iraq**

During OPERATION IRAQI FREEDOM, an armor task force commander described his methodology as that of plotting and measuring everything. After a positive or negative event, he would have his staff evaluate all actions they had conducted before, during, and after the event. This would allow him to correlate activities with outcomes and develop tactics, techniques, and procedures for future success. While not all actions easily correlated to events, this methodological approach helped establish a base line of comparative success for his task force.

Some units assessed the Iraqi culture and developed practical guidelines or rules of interaction for their soldiers to follow as they conducted their missions. These guidelines encompassed a list of actions that their Soldiers should and should not do when dealing with the Iraqi populace. The guiding principle behind these actions was not only to "win the hearts and minds" of the Iraqi people but, more accurately, to establish legitimacy and credibility through appropriate Soldier actions that reflected a combined understanding of the mission and the Iraqi culture.

DO:

- Separate the men from the women and children during cordon and searches.
- Talk to the oldest or most senior male when civilian cooperation is needed.
- Bring something (for example, water, food, or replacement locks) to give residents to compensate for damages when the search ends up being a "dry hole."
- Learn Arabic phrases; say hello and greet people on the street.
- Use female Soldiers to search and talk to Iraqi females.
- Cordon and knock whenever possible instead of kicking in doors needlessly.
- Understand that in Iraq, women do not have the same social status as men.

DO NOT:

- Force a detainee's head to the ground or use your feet to hold someone to the ground.
- Stare at or touch Iraqi women.
- Show obvious disrespect to suspects in front of family members.
- Manhandle the Koran or other religious objects.
- Commit to anything; instead of agreeing to (or denying) an Iraqi's request, it is good practice to say "In'sha Allah" (If God wills it).

# URBAN INFRASTRUCTURE

2-59. Urban infrastructures are those systems that support urban inhabitants and their economy. They form the essential link between the physical terrain and the urban society. During urban stability and civil support operations, restoration or repair of urban infrastructure will often be decisive to mission accomplishment. During urban combat operations, destroying, controlling, or protecting vital parts of the infrastructure may be a necessary shaping operation that can isolate a threat from potential sources of support. A threat force operating in an urban area may rely on the area's water, electricity, and sources of bulk fuel to support his forces. This is true particularly when his bases or facilities are physically located in or near the area. Isolating this threat from these sources may require him to generate his own electricity and transport his own water and fuel from outside the urban area. To transport supplies, the threat may rely on roads, airfields, sea or river lanes, and rail lines. Controlling these critical transportation nodes may prevent the threat from resupplying his forces. The control of key radio, television, and newspaper facilities may isolate him from the urban populace (another potential source of support).

## INTERDEPENDENCE

2-60. Commanders must understand that destroying or disrupting any portion of the urban infrastructure can have a cascading effect (either intentional or unintentional) on the other elements of the infrastructure. Yet, they may be able to gain an operational advantage while minimizing unwanted and unintended effects. Commanders can control, seize, or secure an essential facility or structure by using precision munitions, electronic disruption of communications, or SOF and conventional ground forces. To gain this advantage, commanders will rely more on the expertise of Army engineer and civil affairs units; local urban engineers, planners, and public works employees; and others with infrastructure-specific expertise. After understanding the technical aspects of the area's systems and subsystems, commanders can then develop the best course of action.

## A SYSTEM OF SYSTEMS

2-61. Hundreds of systems may exist. Each system has a critical role in the smooth functioning of the urban area. Simple or complex, all systems fit into six broad categories (see figure 2-11). Commanders should analyze key facilities in each category and determine their role and importance throughout all phases of the urban operation. As there is much overlap between infrastructure systems, this analysis considers each system individually and in relation to others to determine an appropriate course of action toward it.

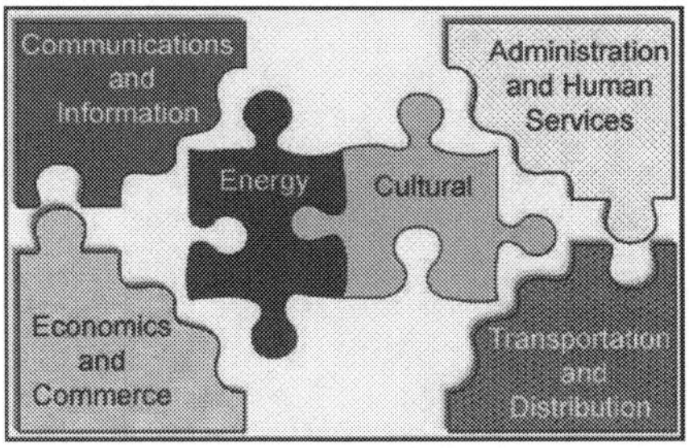


**Figure 2-11. Urban infrastructure**

## A COMBINATION OF STRUCTURES AND PEOPLE

2-62. As depicted earlier in figure 2-1, each element of the infrastructure consists of both a physical (terrain) and human component. For example, the physical component of the electrical segment of the energy infrastructure consists of power stations, substations, a distribution network of lines and wires, and necessary vehicles and repair supplies and equipment. The human component of this same segment consists of the supervisors, engineers, linemen, electricians, and others who operate the system. Commanders must understand and recognize the physical and human components in their assessments.

## POTENTIAL IMPACT ON FUTURE OPERATIONS

2-63. Destroying or incapacitating of any of these elements may impact future operations and inhabitants of the urban area. Destroying urban infrastructure during initial phases of an operation may require commanders to assume responsibility for repair, maintenance and clean up, and operation of those same facilities later. Although exceptions will exist, commanders cannot destroy or significantly damage the infrastructure of a foreign urban center during operations and expect the population to remain friendly to U.S. or allied forces. On the other hand, early repair or restoration of critical or essential infrastructure may improve civil-military relations, speed transition back to competent civilian authorities, and, overall, aid in successful mission accomplishment. Still, support from the urban society (albeit of increased importance in UO) is only one factor that commanders weigh while developing appropriate courses of action.

## Resource Intensive

2-64. Requirements to protect, restore, or maintain critical infrastructure may divert substantial amounts of resources and manpower needed elsewhere and place additional constraints on subordinate commanders. Civilian infrastructure is often more difficult to secure and defend than military infrastructure. The potentially large and sprawling nature of many systems (such as water, power, transportation, communications, and government), make their protection a challenge. Yet, the infrastructure of an urban area may provide commanders with essential logistics and support. Therefore, the initial expenditure of time and other resources may be necessary to support concurrent or future operations. Legal considerations, however, may affect using the infrastructure and acquiring the urban area's goods and services. Commanders, their staffs, and subordinates (often down to the individual soldier) must know their limits concerning Army authority to commandeer civilian supplies or equipment to facilitate mission accomplishment (see the legal support discussion in Chapter 9). In stability and civil support operations, the safeguard or restoration of critical urban infrastructure for military or civilian use may be a decisive point in the overall operation.

2-65. Keys to understanding the magnitude of the resources and manpower required to restore the infrastructure are an initial infrastructure assessment and, as soon as practical afterward, a detailed infrastructure survey. An initial assessment provides the commander immediate feedback concerning the status of basic services needed to meet the urgent needs of the urban population. The systems assessed are based on mission, enemy, terrain and weather, troops and support available, time available, civil considerations, and the commander's vision of the overall end state. The infrastructure assessment, while typically performed by engineers, may be accomplished by, or in conjunction with, others with sufficient expertise to provide the type and quality of information required. These others may include civil affairs, medical, and chemical personnel. Those tasked with this assessment should routinely consult other Army and coalition forces and governmental and nongovernmental agencies currently operating in the urban area as well as the urban civilian leadership for their informed input.

2-66. While an infrastructure assessment functions to support the resolution of immediate challenges to urban reconstruction and restoration, it also provides the initial basis for determining the conditions for successful transition. However, commanders and planners must continually expand and refine their understanding. As a necessary follow-on, commanders initiate a detailed infrastructure survey. This survey is normally conducted by U.S. Army Corps of Engineers personnel assigned to forward engineer support teams. As with the assessment, the commander should incorporate other technical specialty personnel in the survey team to enhance the quality and accuracy of the product (see FM 3-34.250).

## Communications and Information

2-67. This system is comprised of the facilities and the formal and informal means to transmit information and data from place to place. Understanding communication and information infrastructure of an urban area is important because it ultimately controls the flow of information to the population and the enemy. It includes—

- Telecommunications, such as telephone (to include wireless), telegraph, radio, television, and computer systems.
- Police, fire, and rescue communications systems.
- Public address, loudspeaker, and emergency alert systems.
- The postal system.
- Newspapers, magazines, billboards and posters, banners, graffiti, and other forms of print media.
- The informal human interaction that conveys information such as messengers, open-air speeches and protests, and everyday conversations.
- Other inventive informal means such as burning tires and honking horns.

2-68. Perhaps more than any other element of the infrastructure, communications and information link all the other elements in an interdependent "system of systems." It is a critical enabler that helps coordinate, organize, and manage urban activities and influence and control the urban society. Army commanders are acutely aware of the impact that a loss or degradation in communications has on their own operations. The urban environment experiences similar impacts to communication failures; however, urban governments and administrations are generally less prepared to deal with a collapsed communications and information infrastructure than are trained Army forces.

2-69. Militarily, a functioning urban communications and information system can serve as an alternate for both friendly and threat forces and can be easily secured with civilian, off-the-shelf technologies. Threats may make use of commercial systems intertwined with legitimate civilian users, making it unpalatable to prevent use of these assets. Forces can also use these systems to influence public opinion, gain intelligence information, support deception efforts, or otherwise support IO.

## Increasing Impact of Computers

2-70. In many urban areas, computers link other elements of the urban infrastructure. They link functions and systems in the urban area and connect the area to other parts of the world. This latter aspect creates important implications for commanders of a major operation. Operations involving this cybernetic function may produce undesirable effects on a greater scale than initially intended. For example, commanders may be able to close or obstruct an urban area's banking system; however, this system may impact the international monetary exchange with unwanted or even unknown effects. The authority to conduct these types of IO will often be retained at the strategic level.

## Pervasive Media

2-71. The media is central to the communications and information infrastructure and a critical operational concern. Compared to other operational environments (jungles, deserts, mountains, and cold weather areas), the media has more access to urban operations. This is due largely to airports, sea and river ports, and major road networks; ready access to power sources and telecommunications facilities; as well as access to existing local media structures. Hence, media presence may be pervasive and IO even more critical to success in UO than operations in many other environments.

### *A Complex Relationship*

2-72. A complex relationship exists among information, the public, and policy formulation. Although the degree and manner in which public opinion shapes government policy are difficult to accurately determine, negative visual images of military operations presented by the media can change political objectives and, subsequently, military objectives. As important, media reporting can influence civilian activity in an urban AO to either the advantage or disadvantage of the commander.

### *Induce Cooperation Through Credibility*

2-73. Commanders do not control the media; however, they monitor the flow of information that the news media receives and subsequently reports. Consequently, commanders should plan and execute PA operations that will induce cooperation between the media and Army forces. Successful relations between urban Army forces and the news media evolve from regular interaction based on credibility and trust. More information is usually better than less, except when the release of such information may jeopardize security and the success of the operations and threaten the safety of Soldiers. However, commanders cannot simply withhold information to protect the command from embarrassment. They consider media interests as part of the normal planning process and work to ensure that information presented to the news media is accurate, timely, and consistent with operations security. Since the media will likely arrive in the urban area before the conduct of operations, early deployment of PA assets may be critical. Commanders should synchronize PA activities with CMO and PSYOP. Such action eliminates duplicated effort and ensures a unity of purpose consistent with the IO concept of support (see Chapter 4).

2-74. Failure to provide sufficient information can hamper a commander's ability to conduct the mission. Commanders cannot refuse to deal with particular news media because they consistently report a negative image of Army forces and operations. Poor relationships with any media can result in inaccurate and even biased reporting. Such reporting can cause a public reaction that influences the ability to achieve operational objectives. During the Russian 1994-95 battle against Chechen separatists in Grozny, for example, the Russian military refused to communicate with reporters. The media reported primarily from the perspective of the Chechen rebels. This encouraged both local and international support for the rebels. It also allowed the Chechens, who lacked sophisticated information systems, to use the media to broadcast operational guidance to their forces. (During their second Chechnya campaign of 1999-2000, Russia learned this lesson well and the Russian view of the war dominated domestic public opinion.) On the other hand, successfully engaging the media can serve as a force multiplier. The Army's open and responsive interaction with the media during peacekeeping operations in Bosnian urban areas helped to explain the challenges and successes of Army forces in the Balkans to the public. This helped maintain domestic, international, and local political support for NATO operations and, with a successful command information program, helped maintain Soldiers' morale.

### Transportation and Distribution

2-75. This element of the infrastructure consists of—

- Networked highways and railways to include bridges, subways and tunnels, underpasses and overpasses, ferries, and fords.
- Ports, harbors, and inland waterways.
- Airports, seaplane stations, and heliports.
- Mass transit.
- Cableways and tramways.
- Transport companies and delivery services that facilitate the movement of supplies, equipment, and people.

Similar to communications and information, this facet provides the physical link to all other elements of the infrastructure.

2-76. Army forces deploying into a theater of operations depend on ports and airfields; seizure and protection of these critical transportation nodes may impact the projection of combat power. Once in theater, transportation and distribution systems in the urban area can contribute greatly to the movement of forces, maneuver, and logistic operations throughout the entire AO. Control of decisive points in this infrastructure may be important to the military operation and to the normal functioning of the urban area (and surrounding rural areas). Supplies traveling through the transportation and distribution system may be military-specific supplies (such as ammunition and repair parts) and supplies for both the military and urban population (such as food, medicine, oil, and gas). The system may also support the movement of military forces and the urban area's population (for which it was designed). Therefore, commanders of a major operation may have to develop innovative methods that limit the transit of threat supplies and reinforcements while facilitating the movement of their own resources and those of civilians. This last consideration attempts to minimize hardship and promote normalcy in the urban area and will increase in significance as the need for legitimacy increases.

2-77. Most urban areas (particularly in developing countries) have two forms of transportation and distribution systems that exist simultaneously: a formal system and an informal or paratransit system. Large organizations, bureaucracy, imported technology, scheduled services, and fixed fares or rates characterize formal systems. Low barriers to entry; family and individual entrepreneur organizations; adapted technology; flexible routes, destinations, and times of service; and negotiated prices characterize the informal system. The informal system is more decentralized and covers a much greater portion of the urban area than the formal system. The informal transportation and distribution system often includes a waterborne element, is more likely to function through turbulence and conflict, and can extend hundreds of kilometers beyond the urban area. Accordingly, commanders should understand both systems to establish effective movement control.

## ENERGY

2-78. The energy system provides the power to run the urban area. It consists of the industries and facilities that produce, store, and distribute electricity, coal, oil, wood, and natural gas. This area also encompasses alternate energy sources, such as nuclear, solar, hydroelectric, and geothermal power. Energy is needed for industrial production and is therefore vital to economics and commerce. Among many other things, this system also provides the fuels to heat, cool, and light homes and hospitals, cook and preserve food, power communications, and run the transportation necessary to move people and their supplies throughout the urban area. Loss of an important energy source such as electricity or gasoline, especially for those accustomed to having it, will become an immense area of discontent that the commander of a major urban operation will need to quickly address. Therefore some threats, particularly terrorists and insurgents, may actively target this element of the urban area's infrastructure to erode support for civilian authorities and Army forces.

2-79. Sources of energy may be tens or hundreds of miles away from the urban area itself. Therefore, commanders may exert control without applying combat power directly to the urban area itself by controlling or destroying the source (power generation or refinement plant) or the method of distribution (pipelines or power lines). With electrical energy that cannot be stored in any sizable amount, the latter may be the best means as most major urban areas receive this energy from more than one source in a network of power grids. However, control may be as simple as securing a power station or plant and turning off switches or removing a vital component that could later be restored. On the other hand, lengthy pipelines and power lines may compound security and protection of this element of the infrastructure.

2-80. The number of nations that have invested in nuclear power and nuclear research is increasing. With this increase, the potential for Army forces to operate in urban areas that include (or are near) these facilities also increases. Damage to one of these facilities and potential radiation hazards will present special challenges to commanders of a major operation. To safeguard friendly forces and civilians, commanders will need to employ a blend of peacetime and tactical nuclear contamination avoidance principles (see FM 3-11.14).

## ECONOMICS AND COMMERCE

2-81. This system encompasses—

- Business and financial centers to include stores, shops, restaurants, hotels, marketplaces, banks, trading centers, and business offices.
- Recreational facilities such as amusement parks, golf courses, and stadiums.
- Outlying industrial, mineral, and agricultural features to include strip malls, farms, food processing and storage centers, manufacturing plants, mines, and mills.

2-82. An essential aspect of this area during operations may be the political sensitivity of U.S. or allied industries investing and operating in a foreign country, particularly during stability operations. An enemy or a disgruntled civilian population may attack or disrupt commercial activities as a political statement against the United States or our allies. Food production facilities also may assist commanders in Army food services and may be an essential concern during relief operations. During long-term stability operations, visible, material, and tangible economic progress consisting of the creation or restoration (and protection) of businesses, agriculture, and overall jobs will often be critical to—

- Generating or maintaining the urban population's support to Army forces and operations.
- Reducing support to threat forces and operations to include eliminating civilians as a potential manpower pool for insurgent or terrorist organizations and activities.
- Lowering other hostile civilian activities such as protests and riots.
- Transitioning the urban area back to legitimate civilian responsibility and control.

2-83. This element of the infrastructure also consists of the production and storage of toxic industrial chemicals used in agriculture (insecticides, herbicides, and fertilizers), manufacturing, cleaning, and research (to include biological agents). Fertilizer plants may be of specific concern as they contribute to providing a key material in terrorist and insurgent bomb-making activities. A thorough analysis of this element of the infrastructure may also be essential to understanding how urban insurgencies are funded and supported. This helps commanders to understand the true organization of the insurgency as well as to suggest methods to isolate insurgents from their economic or financial support. In their overall assessment of this area of the infrastructure, commanders should also consider the activities and influence of criminal organizations or elements.

### ADMINISTRATION AND HUMAN SERVICES

2-84. This wide-ranging system covers urban administrative organizations and service functions concerned with an urban area's public governance, health, safety, and welfare. Together, it encompasses—

- Governmental services that include embassies and diplomatic organizations.
- Activities that manage vital records, such as birth certificates and deeds.
- The judicial system.
- Hospitals and other medical services and facilities.
- Public housing and shelter.
- Water supply systems.
- Waste and hazardous material storage and processing facilities.
- Emergency and first-responder services such as police, fire, and rescue.
- Prisons.
- Welfare and social service systems.

### CULTURAL

2-85. This system encompasses many organizations and structures that provide the urban populace with its social identity and reflect its culture. (This infrastructure system overlaps with many recreational facilities included under the economics and commerce infrastructure. For example, an urban society may radically follow soccer matches and teams. Hence, soccer stadiums relate to the society's cultural infrastructure.) Some of these facilities, particularly religious structures, will be protected targets and others may require security and law enforcement protection from looting and pilferage. However, commanders will need to quickly educate, inform, and continually remind the urban populace (and the media) that cultural infrastructure may lose its protected status when used by threats for military purposes. Cultural infrastructure may include—

- Religious organizations, places of worship, and shrines.
- Schools and universities.
- Museums and archeological sites.
- Historic monuments.
- Libraries.
- Theaters.

## RESTORING AND PROTECTING ESSENTIAL SERVICES

2-86. Losing the support of essential elements of the infrastructure will have an immediate, destabilizing, and life-threatening impact on the inhabitants of the urban area. In stability and civil support operations, numerous parts of the administrative and human services and energy infrastructure often rise to critical importance before all other elements. Again, however, complete restoration of these essential services is often a lengthy, resource-intensive civil-military operation. (Following the end of major combat operations of OPERATION IRAQI FREEDOM, most units developed one or more of their logical lines of operations oriented along the restoration or improvement of urban infrastructure and essential services. An acronym used by many units to focus and track critical activities within this line of operation was sewer, water, electricity, and trash (SWET). Later, other units modified this acronym to SWEAT and then SWEAT-MS to include concerns for restoring academics [or schools], revitalizing medical facilities, and establishing security [police and host-nation security forces].) Of critical importance will be a simultaneous IO campaign that includes efforts to help ensure that the urban population develops realistic expectations about Army abilities to restore their essential urban infrastructure.

### Understanding the Urban Environment: Paris – 1944

The summer of 1944 confronted German General Dietrich von Choltitz with a dilemma. As military commander of greater Paris, he was to eliminate French Resistance internal to the city while defending against approaching Allied units, missions for which he had insufficient forces. (Notably, General Eisenhower, the commander of the Allied forces, wanted to bypass Paris to sustain the offensive. Seizing the French capital would task his forces with the support of tens of thousands of civilians. Eisenhower was nevertheless ordered to capture the city. A political decision resulted in civilian assistance taking precedence over combat operations.)

Choltitz's situation was further complicated by Hitler's demand that he destroy the city, an action the general saw as needlessly destructive (and infeasible given his scant resources). Choltitz's seniors directed the preparation, and later the destruction, of Paris's 45 Seine River bridges. They were the only remaining crossing points over that waterway given Allied bombing of others outside the French capital. Premature destruction would trap German forces defending to their north, a second-order effect that Choltitz used to justify his disobedience of orders demanding the bridges' demolition.

The German general also recognized that some mission-critical elements were part of Paris's social rather than physical infrastructure: the leadership of the various resistance groups and the relationships between them. Choltitz understood that he lacked resources to defeat the many separate factions; he therefore chose the unorthodox (asymmetric) approach of accepting an intermediary's offer of a truce with these groups. Such an agreement provided some measure of the stability needed while Choltitz awaited promised reinforcements. Further, he realized that the resistance factions were by no means united in their goals. Communist elements sought a much different end than those looking toward a de Gaulle-led postwar government. A truce thus set the French Communists (who sought an uprising so as to legitimize their claims to power) against others trying to buy time until Allied forces arrived, forces that included Free French units supportive of de Gaulle.

Although his defense of the capital failed, Choltitz succeeded in harboring his available resources, reducing the effectiveness of the resistance organizations fighting his soldiers, and maintaining withdrawal routes for units north of the Seine. The German commander's analysis in support of these efforts was effective in part because of his insightful (1) identification of critical points that included elements of terrain, citizenry, and infrastructure; (2) understanding of the relationships between these parts; and (3) use of an asymmetric approach to address his lack of sufficient force to otherwise handle the urban densities that challenged him.

This page intentionally left blank.

# Chapter 3

# Understanding the Urban Threat

*... [T]he United States could be forced to intervene in unexpected crises against opponents with a wide range of capabilities. Moreover, these interventions may take place in distant regions where urban environments, other complex terrain, and varied climatic conditions present major operational challenges.*

*Quadrennial Defense Review Report*, 30 September 2001

As the strategic environment has become less stable, more uncertain, and more dangerous, Army forces must be trained and ready to address persistent and evolving urban threats. These threats range from regional conventional military forces, paramilitary forces, guerrillas, and insurgents to terrorists, criminal groups, and angry crowds. These threats can hide in plain sight and become indistinguishable from the noncombatant urban population that may help shield, protect, and sustain them. Although uncertain about events, Army forces can be clear about trends. Increasingly, the Army will face threats that severely differ in doctrine, organization, and equipment, are skilled at developing and adapting techniques to counter Army tactics, techniques, and procedures (TTP), and can fully interact with the three other components of the urban battlefield—terrain, society, and infrastructure. In urban operations, commanders must broaden their concept of the threat to include natural disasters, hunger and starvation, and rampant disease. Further, commanders must plan to contend with many passive urban threats, such as psychological illnesses and toxic industrial materials (TIM). These threats may be found in isolation, but most likely commanders will encounter them in various combinations. Moreover, each new threat will pose a different combination and likely have new capabilities that previous opponents lacked.

## ASYMMETRICAL AND ADAPTIVE

3-1. An emphasis on asymmetry to offset U.S. military capabilities has emerged as a significant trend among potential threats and become an integral part of threat objectives and tactics (discussed below). Though, this is not to imply that all future urban threats will fight and operate asymmetrically. Some may be asymmetrical; others may not. Future threats may be a symmetric conventional military; others may be paramilitary, insurgent, or even a nonmilitary threat such as a criminal organization. Asymmetry results when one opponent has dissimilar capabilities—values, organization, training, or equipment—that the other cannot immediately counter. It is not a new concept. It naturally evolves from a sound mission, enemy, terrain and weather, troops and support available, time available, civil considerations (METT TC) analysis by an intelligent, freethinking, and adaptive threat. These asymmetric approaches will include the most advanced, commercially-available technology innovatively applied and mixed with crude, simple, and unsophisticated weapons and TTP. The success of these approaches often depends upon the predictability of the forces on which they are employed. Thus, Army commanders must consider whether their operations and activities lend themselves to pattern analysis by threat forces and, if so, they may need to vary their battle rhythm and take other steps necessary to become less predictable.

# WEAPONS OF MASS DESTRUCTION

3-2. A chief asymmetric means of engaging the national power of the United States is to employ weapons of mass destruction (WMD) against the United States or its allies. These weapons can be used against military forces and include high-yield explosives as well as nuclear, biological, and chemical weapons. Operations in urban areas may require concentrating forces and may create a lucrative target for a threat that possesses fewer numbers and less equipment.

3-3. A threat's WMD use will adversely affect the Army's abilities to conduct urban operations (UO) to various degrees. For example, the intervening structures and the effects of urban microclimates complicate the ability to detect and identify radiological, chemical, or biological attacks from a standoff distance. Also, the individual Soldier's ability to recognize his leaders, understand oral and visual commands, and operate increasingly sophisticated equipment is difficult when wearing protective clothing and equipment—particularly if his training proficiency is low. Despite the increased challenges and complexity, Army forces have the best (or better) training and equipment necessary to respond to such an attack compared with most armies around the world or with any civilian organization.

3-4. Although initial casualties could be high, the public can accept military casualties before those of civilians. Therefore, threats may gain an initial tactical advantage but would achieve less asymmetric benefit by directly attacking Army forces. Instead, they may attempt to achieve a strategic advantage by employing WMD against U.S. or allied civilian populations. In doing so, threats hope to use political sensitivity to high civilian casualties to reduce popular support for the United States or its allies. The chance of these attacks occurring in an urban area increases because—

- The area facilitates weapon effects and camouflages delivery means.
- The dense civilian population ensures a high casualty rate.
- The attack (or even the threat of attack) often will receive more publicity and public attention.
- The urban infrastructure is especially vulnerable to WMD, (particularly the systems of the economics and commerce infrastructure located in large urban areas) and may have far-reaching national and global effects.

# THREAT OBJECTIVES

3-5. The threat will seek to achieve several key objectives when opposing Army forces operating in an urban environment (see figure 3-1). These objectives focus more on how a threat might fight in an urban area rather than specifically who the threat might be or in what region of the world the conflict might occur. These objectives are more easily achieved in an urban environment due to—

- The high costs in time, material, and manpower involved in UO.
- The limiting effects of urban areas on many technological advantages.
- The proximity of airfields and ports to urban areas.
- The potential moral dilemmas created by exposing numerous civilians to harm or injury.
- The ability of the media to observe and report the threat's version of events.

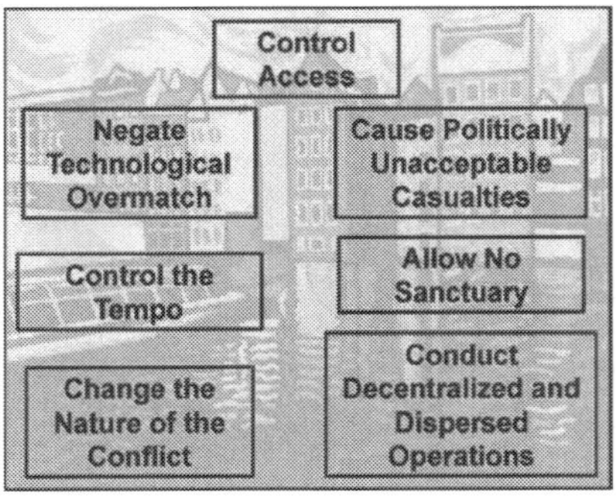


**Figure 3-1. Threat objectives**

3-6. These objectives complement and overlap each other; however, at their core is the threat need to defeat an enemy of superior numbers, technology, or both.

## CONTROL ACCESS

> *... Iraq made no direct effort to impede the buildup in the theater. Planners should consider what might have happened if Iraq had attempted a strategy based on denying access to the region. Planners might also wonder what the outcome would have been if Iraq had attacked US forces in Kuwait before they were ready for the running start. The point is that the conditions in [US Central Command] in 2003 are unlikely to be replicated elsewhere.*
>
> *On Point: The United States Army in Operation IRAQI FREEDOM*

3-7. The Army may not be located where future conflicts are fought. Thus, the Army maintains the ability to rapidly project and sustain combat power over long distances and time spans. This capability demands that Army forces quickly gain and maintain control of seaports or aerial ports of embarkation or debarkation, particularly where the density of U.S. basing and en route infrastructure is low. Commanders gain control of these ports by unopposed (assisted or unassisted) or forcible entry operations. In either case, these phased-entry operations may present potential vulnerabilities, particularly—

- Unsuitable composition of initial or early entry forces lacking necessary combat power for immediate decisive operations.
- Initial command and control difficulties and an immature situational understanding.
- Weak initial port or base defenses without the necessary integration of naval and air component forces.
- Lack of developed logistic support.

3-8. Consequently, threats may attack during initial force projection operations to oppose, delay, disrupt, or otherwise manipulate the build-up of essential combat power into a theater of operations. These attacks may occur anywhere deploying Army forces are located, at overseas bases, at home stations, and even in military communities. Increasingly, deployment facilities such as airfields and ports exist as integral components of urban areas. Threats will invariably use the complex and concealing nature of these urban areas, coupled with the vulnerabilities, to create favorable conditions for their attacks.

## NEGATE TECHNOLOGICAL OVERMATCH

3-9. Threats will always strive to force engagements at a time and place most advantageous to them. They may locate military forces and vital military capabilities in urban areas to achieve sanctuary from the effects of Army capabilities and make Army forces and systems more vulnerable to less-sophisticated weapons.

3-10. The clutter of the physical structures, electromagnetic radiation, and population diminishes Army capabilities. This clutter makes it difficult for Army forces to acquire and effectively engage targets at long ranges. In urban areas, the terrain often allows a threat to operate in closer proximity to friendly forces. Therefore, the threat may "hug" friendly forces to avoid the effects of high-firepower standoff weapon systems and degrade their ability to gain or maintain a thorough common operational picture. Additionally, this threat tactic attempts to inhibit friendly commanders from employing some weapon systems and munitions for fear of fratricide. Threats will also seek to hide and operate in culturally sensitive areas and sites, such as places of worship and schools, to establish temporary safe havens and refuge from Army forces operating within the urban environment.

## CONTROL THE TEMPO

3-11. Threats will try to achieve a decisive advantage by setting and controlling the tempo. To prevent the Army's entry into theater, threats may try to create a high operational tempo to take advantage of the inherent weaknesses in power projection operations outlined earlier. As other efforts deny entry, threats may seize the initiative, achieve surprise, and exploit the tempo differential by attacking with heavy conventional forces potentially possessing greater firepower and more rapid ground mobility than the Army's initial-entry forces.

3-12. If they cannot deny entry or end the conflict quickly, threats may use any preparations made in the initial high-tempo period to prolong the event, aiming to degrade U.S. or allied commitment. (Some threats may conduct operationally planning in terms of decades rather than years.) The complex nature of the urban environment slows operations conducted in, around, and over these areas. Threats may maximize this characteristic by fighting (conventionally or unconventionally) for key urban complexes and infrastructure, forcing friendly forces to operate within these areas. If Army operations focus on one or more urban areas, the overall campaign slows. However, even when UO make up only one component of a much larger campaign, they may consume valuable resources needed for other operations and delay the entire campaign. (See also the discussion of Tempo in Chapter 7.)

**Tempo**

The battle for Aachen, Germany, in the fall of 1944, developed during the U.S. First Army's offensive to breach the Westwall fortifications. Aachen, the ancient capital of Charlemagne, had symbolic political and psychological significance for the Germans and Americans. Furthermore, it was the first city on German soil to face an assault by the Allies. Consequently, the symbolic importance of this first major battle in Germany ensured bitter resistance against American attackers. The Germans surrendered only after the city was destroyed. Expected to take a few days, instead, the battle took weeks. Although the Army had achieved a clear tactical victory, the German defense of Aachen cost the First Army valuable time and resources, and delayed the planned attack to the Rhine River.

## Change the Nature of the Conflict

3-13. Threats may attempt to change the fundamental nature of the urban conflict to exploit ambiguous or tenuous political-military objectives. Many nations must gain and maintain domestic popular support to use their armies for political objectives. The threat may attempt to change the nature of the conflict by modifying its strategy and tactics, the environment, or any combination, ultimately hoping to reduce friendly popular support. For example, introducing an urban terrorist threat to U.S. civilians or Soldiers not directly engaged in operations changes the nature of the conflict. This type of threat may not have been an initial consideration, and this change may reverse public support for the operation. Another example, growing U.S. coalition combat power may cause the threat to switch from open maneuver warfare to urban insurgency operations to avoid decisive combat with superior forces and achieve a stalemate. Originally expecting a quick solution or victory, the political leadership may now envision a longer deployment with less chance of lasting success.

## Cause Politically Unacceptable Casualties

3-14. Threat forces may gain an advantage against superior friendly forces by capitalizing on a perceived weakness of many Western nations: the inability to endure continuous losses or casualties for other than vital national interests or losses for which they are psychologically unprepared. A secondary U.S. interest may equate to national survival on the part of a threat. Therefore, the threat (particularly with fanatical leadership) may willingly sacrifice excessive amounts of money, equipment, and people (soldiers and civilians) to achieve victory. Threats may attempt to weaken U.S. resolve and national will to sustain the deployment or conflict by inflicting highly visible, embarrassing, and if possible, large losses on Army forces, even at the cost of similar losses to themselves. Many threat forces will use UO to inflict mass casualties and destroy premier Army weapon and information systems. The physical characteristics of the urban environment support these ambush techniques. Light infantry or insurgents with readily obtainable, hand-held antiarmor weapons can effectively attack armored vehicles and helicopters, no matter how sophisticated, in an urban area.

### ALLOW NO SANCTUARY

3-15. Threats will attempt to deny Army forces safe haven anytime and anywhere. Terrorism and insurgency may be some of the tactics used to deny sanctuary to Army forces. They will attack Army forces anywhere, particularly while operating in urban areas where the fear from being attacked from any quarter is often greater. Threats may be or employ state-sponsored or independent terrorists, well equipped and motivated to accomplish their assigned missions.

3-16. Military buildings, facilities, and installations in urban areas are particularly vulnerable to high-yield explosive munitions as well as other clever means to create large explosions. The close-in nature of urban areas, large populations, and high volume of vehicle traffic provide a good environment for target reconnaissance, explosives positioning (conventional or high-yield), and cover for an attack. These attacks will likely be preceded by extensive, careful reconnaissance, necessitating a solid friendly counterterrorism, counterinsurgency, and counterintelligence effort.

### CONDUCT DECENTRALIZED AND DISPERSED OPERATIONS

3-17. To a certain extent, decentralized and dispersed operations are an integral part of all threat objectives. However, this concept warrants separate emphasis as an objective since threat forces will likely place great significance on it on future urban battlefields. Both dispersed and decentralized approaches seek to reduce threat vulnerabilities to air power and precision-guided munitions (PGM) while increasing their agility, flexibility, and overall maneuverability in an urban environment.

3-18. Urban terrain tends to fragment and separate forces that operate in it. Threat forces recognize this characteristic, accept it, and make it work to their advantage. They conduct operations from dispersed urban locations to reduce their vulnerability to friendly decisive operations and massed firepower. Although separated, threat forces will attempt to retain the ability to assemble and mass quickly so to strike as opportunities present themselves. Once threat forces complete the operation, they will return to separate locations to avoid potential counterattack. The fluidity and seemingly disjointed appearance of these UO threats will challenge friendly efforts to conduct templating and pattern analysis. Ambushes (air and ground) will be used to deny friendly ground and air reconnaissance of their dispersed locations.

3-19. Dispersed operations normally depend on good command and control to achieve synchronization and massed effects. Threat forces also understand the debilitating effects of the urban terrain on communications and the execution of operations. When they cannot mass their forces or effects, they will depend on decentralized operations to achieve their objectives. They will operate autonomously, guided only by a higher authority's purpose and intent. These operations make them even less vulnerable to massed attacks and PGM as smaller threat forces do not present an objective or target that will allow friendly decisive operations. Again, pattern analysis and templating, although essential, will be extremely difficult. This objective may prolongs the conflict but is central to achieving the other threat objectives.

3-20. When the threat employs decentralized operations, is effectively isolated, or both, Army and coalition forces are much more likely to be able to share lessons learned and emergent TTP than a dispersed threat. Threat isolation, either imposed by Army forces or as a result of the threat's desire to remain concealed from Army intelligence, surveillance, and reconnaissance (ISR) capabilities, reduces the threat's ability to freely and easily exchange information and quickly adapt. As learning organizations, this allows Army forces to develop dominance over a threat's weapons and tactics for longer, albeit temporary, periods of time. This ability to rapidly train and learn is an advantage the Army and coalition forces must maintain while denying the same to future threats.

## URBAN THREAT TACTICS

3-21. Urban areas provide a potentially casualty-producing and stress-inducing environment ideally suited for using specific urban tactics. Moreover, urban areas can provide threats with an unmatched degree of cover and concealment from friendly ISR and firepower systems. Understanding how these tactics will be used often requires a concurrent understanding of the threat's culture, customs, language, and philosophical ideals. While active urban threats may vary widely, many techniques will be common to all.

Figure 3-2 outlines a set of tactics available to potential threats opposing mission accomplishment in urban areas. Army forces may use many of these tactics, except those that violate the law, ethics, and morals, to defeat urban threats. Moreover, using asymmetric means is not the sole domain of the threat. Army commanders can also leverage capabilities, create conditions, and plan operations to develop their own asymmetric advantages to accomplish the mission.

## USE THE POPULATION TO ADVANTAGE

3-22. Many urban areas may be too large to evacuate completely (if at all). Even if desirable, a military force may have no place to safeguard and secure the inhabitants. Therefore, future UO may see large segments of the populace remain. Offensive and defensive operations may be constrained not only by the terrain, but also by the presence of many civilians. Army forces involved in urban stability and civil support operations will certainly conduct missions in and amongst the residents. These residents may restrict operations and, when gathered in large numbers, may (even without initial hostile intent) present a critical force protection issue for the commander.

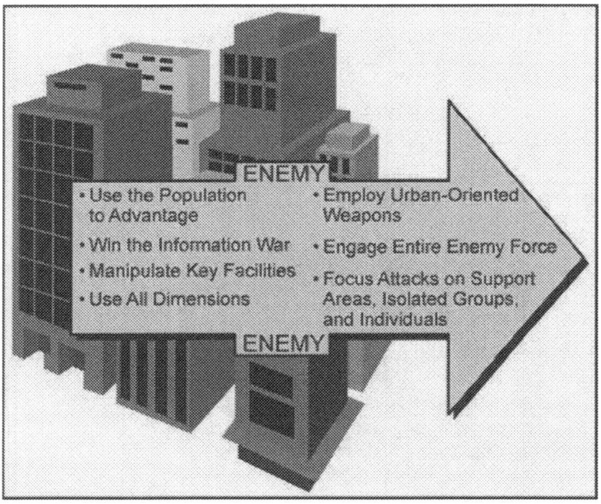


**Figure 3-2. Urban threat tactics**

### Use as Key Terrain and Concealment

3-23. From the threat standpoint, the populace is similar to key terrain: the side that manages it best has an advantage. Threat forces may gain this advantage by using civilians as camouflage, concealment, and a means of deception. Guerrilla and terrorist elements may look no different from any other member of the community. Many foreign conventional and paramilitary troops—in addition to terrorists and insurgents—often have a "civilian" look. Western military forces originally adopted the clean-shaven and close-cut hair standards to combat disease and infection, but future opponents may not adhere to those standards. Instead, they may adopt grooming standards, civilian-looking clothing, and other "nonmilitary" characteristics to make themselves indistinguishable from the civilians. To avoid defeat or capture, threats will often throw down their weapons and attempt to blend into the urban population. Therefore, commanders must adopt effective populace and resource control measures to limit, identify, and track threat movements through the urban area of operation.

**Identifying Threats from Noncombatants**

During Russia's 1994-95 conflict with Chechnya, Russian forces had difficulty identifying Chechen guerrilla forces from Grozny's noncombatant population. Because their dress and ethnic appearance was identical to that of the urban populace, Chechen soldiers could freely walk around the city, suddenly disappear, and then abruptly reappear firing their weapons from basements, windows, or dark alleyways. To distinguish fighters from peaceful city dwellers, Russian forces began looking at men's shoulders to see if they were bruised (from firing weapons) and their forearms to see if there was burned hair or flesh (from the extraction of cartridges). They closely examined their clothing and smelled for gunpowder residue. To identify a Chechen artilleryman, Russian soldiers checked for glossy spots left by artillery and mortar rounds on the bends and cuffs of sleeves. They also turned pockets inside out to check for a shiny, silvery-leaden hue indicating the former presence of small arms ammunition. Russian forces also recognized a grenade launcher operator or mortar man from fibers and crumpled pieces of gun cotton on their clothing. U.S. Army commanders may need to develop similar, imaginative techniques and rapidly integrate technological means to identify and track threat forces such as—

- Military working dogs, ion spray, and electronic vapor tracers to locate explosives (or their former presence).
- Biometrics—measurable biological characteristics—such as fingerprints, voiceprints, facial scans, and retina scans.
- Electronic and chemical tagging devices.
- Monitoring hospitals for injuries that indicate participation in combat activities. For example, improper use of rocket-propelled grenade launchers may result in significant burns to the lower part of the body due to the back blast.

## Gain Cover, Protection, and Increased Mobility

3-24. Threat forces may attempt to gain cover by using the urban inhabitants as human shields. With this increase in protection, they simultaneously increase their mobility. They recognize the Army's focus on developing and applying rules of engagement (ROE). They will take advantage of the restraining effects of international law and the Army ethical values to enhance their mobility in proximity to friendly positions. Knowing the Army's reluctance to cause noncombatant casualties and collateral damage, threats may operate in areas containing civilians and essential facilities to restrict the Army's use of massed or nonprecision firepower. For example, they may use school buses and ambulances to transport fighters and equipment about the urban battlefield. They may also employ "rent-a-crowds"—civilians paid or incited to demonstrate against military forces—armed only with sticks, stones, and Molotov cocktails (a potential asymmetric challenge). Threats can also hide military equipment within sensitive infrastructure—for example, caching weapons in places of worship, schools, or medical facilities. As Army and coalition forces potentially respond by targeting and firing upon these locations, threats will make use of these reactions for their possible propaganda value.

## Make Moral Responsibilities a Weakness

3-25. Depending on their successes, threats may use these tactics and skillful information operations that attack national will and coalition sensitivities in an attempt to force the Army to establish more restrictive ROE. For example, a teenaged fighter coerced or enticed to fire a rifle or rocket-propelled grenade launcher can be quickly stripped of his weapon once dead and displayed at the local hospital as a child victim of coalition forces. Threat forces may also take advantage of the Army's moral responsibilities. By herding refugees into friendly controlled areas, threat forces try to make the civilians a burden on the Army's logistic and security resources. Similarly, some threats may attack to kill NGOs and force the remainder of them out of the area to achieve the same burden effect. At the same time, these threats hope to cause the population to lose confidence in the Army's (and potentially the government's) ability to protect and provide them essential services.

3-26. Threat forces, on the other hand, may not abide by international agreements, such as the Geneva conventions. Threats may not take prisoners unless they can be ransomed or made part of a local prisoner exchange. They may even execute friendly prisoners in front of the media to show their "strength" and, more importantly, to cause friendly forces to overreact and lose their legitimacy. Threat forces can then use such an overreaction to unite others with their cause. Pretending to be civilian noncombatants, they may also feign injury or pretend to need assistance (such as a disabled taxi) with the primary motive of luring Army forces into a deliberate ambush. While the threat may gain a short-term advantage by using ethical responsibilities as a weapon, the Army commander's strict adherence to moral principles will, in the long run, contribute greatly to threat defeat—particularly during long-term urban stability operations.

### Acquire Intelligence and Logistic Support

3-27. Indigenous threat forces can normally use the local population for intelligence and logistic support far more effectively than can an alien army. Threat forces may manipulate local hires serving among U.S. Soldiers, such as those contracted by the Army for base operation purposes or translator duties. In addition, refugees moving through friendly controlled sectors may provide the threat with information on friendly dispositions, readiness, and intent. (Women and children may be specifically used for this purpose since U.S. Soldiers' cultural bias can create an incorrect perception that these noncombatants pose no threat to security.) Even friendly residents may become unwitting or unwilling informants, providing an enemy or a hostile with vital information on friendly activities, dispositions, and capabilities. However, a threat employing particularly cruel, abusive, or repressive measures may easily turn certain groups in the urban area against them, even when they share a common history, culture, and ethnicity with the civilians. This is more likely in those areas with higher population densities.

3-28. Threat forces may also seek to use some nongovernmental organizations (NGOs). They may try to obtain relief supplies either through the organizations' legitimate relief operations or as a target for theft. Some organizations may even be fronts for weapons, food, ammunition, money, and fighters. For example, during Russia's second conflict in Chechnya (1999-2000), documents purportedly found in Grozny by the Russians listed nations such as Sudan, Nigeria, Niger, and Ivory Coast as sending fighters to Chechnya under the guise of the International Islamic Relief Organization. (Chechen fighters also disguised themselves as Red Cross workers.) This deception increases the need for strict security and force protection measures, close coordination with NGOs operating in urban areas, and closer monitoring of suspect organizations' activities by civil affairs personnel.

## WIN THE INFORMATION WAR

3-29. Threat forces will try to win the information war as much as they will directly oppose Army UO. Threat urban campaigns need not be tactical military successes. They need only to weaken legitimacy and make the opposition's campaign appear unpalatable to domestic, U.S., and world support. As a critical part of a threat's overall information operations, they will use the ever-present media to tell their story. Portable video cameras, commercial radios, and cellular telephones, available and easily concealed, will be as important to many threat actors as weapons and ammunition. Internet access, already firmly established in many urban areas, provides the means to easily disseminate threat propaganda, misinformation, and disinformation through web sites and electronic mail. The internet also allows threats to conduct their own distance learning and education. Threats will make good use of the instantaneity and global reach of television and radio broadcasts, particularly local and international media, to obtain information of immediate tactical value and broadcast their own operational and tactical guidance. Hackers, covered and concealed in the interior spaces of the urban area, may gain access to U.S. sites to manipulate and obtain information to the threat's advantage.

**Information and the Media**

The media coverage of the urban battle for Hue, South Vietnam, although only one of hundreds of different attacks of the Tet Offensive, affected the will of both the American people and their political leadership. On January 31, 1968, two North Vietnamese Army (NVA)/Vietcong (VC) regiments and two sapper battalions, moving rapidly and with the element of surprise, attacked and seized part of the walled city (Citadel) of Hue. It was the third largest city in South Vietnam, the former capital of a united Vietnam, the capital of Thua Thien province, and a spiritual and cultural center. Initially intending to hold the city for seven days, the NVA/VC retained portions of the city for approximately three weeks against determined U.S. and South Vietnamese attempts to retake it.

Hue marked a revolution in the coverage of war by modern media. It was the first time Americans could sit at home and watch an ongoing battle on the evening news. One of the most intense and savage battles of the Vietnam conflict, it was televised every evening for almost a month. Although the battle for Hue was a tactical victory for the United States, the North Vietnamese clearly achieved strategic success by searing the American consciousness with the high costs of urban warfare. Had U.S. leaders made winning the information war a central part of the overall campaign plan—for example, exposing the American people to the NVA's brutality by publicizing the civilian executions in Hue—civilian support for the war may have been bolstered and a different outcome achieved. See Chapter 7 for a more detailed account of the battle for Hue.

## MANIPULATE KEY FACILITIES

3-30. Threat forces will attempt to identify and quickly seize control of critical components of the urban area to help shape the operational environment. Urban telephone exchanges can provide threats with simple and reliable communications that can be easily secured with off-the-shelf technologies. Existing public address and loudspeaker systems can be used to incite riots or openly command and control dispersed threat forces. (Local language may provide threats with a measure of communications security.) Sewage treatment plants and flood control machinery can be used to implement WMD strategies or to make sections of the urban area uninhabitable. Media stations significantly improve the information operations abilities of the controlling force. Power generation and transmission sites provide means to control significant aspects of civilian society over a large area. Insurgents will attempt to destroy key facilities and critical infrastructure in order to shape the populace's perception that the existing government is incapable of providing for their safety and meeting their basic needs.

## USE ALL DIMENSIONS

3-31. Threats will think and operate throughout the depth, breadth, and height (including supersurface and subsurface areas) of the urban environment. However, threats will often choose to fight from within interior spaces rather than out in the streets and other open areas. Conventional lateral boundaries will often not apply as threat forces control some stories of the same building while friendly forces control others. The use of all three dimensions of an urban area makes identification, reporting, and targeting of enemy locations more difficult for forces accustomed and trained to acquire and engage targets in primarily two dimensions. Because of these target identification and acquisition difficulties, threats will likely use decoys in the urban area to cause erroneous assessments of their combat capabilities, strength, and disposition. They will also attempt to use these decoys to absorb expensive and limited precision-guided munitions as well as cause misallocation of other critical Army resources.

3-32. Roofs and other supersurface areas provide urban threats with excellent observation points and battle positions above the maximum elevation of many weapons. Shots from upper floors strike armored vehicles in vulnerable points. Basements and other subsurface areas also provide firing points below many weapons' minimum depressions and strike at another weakness in most armor. Sewers and subways may provide covered and concealed access throughout the area of operations.

## EMPLOY URBAN-ORIENTED WEAPONS

3-33. Whether purpose-built or adapted, many weapons are more useful in an urban environment while others may have significant disadvantages. Urban threat weapons are much like the nature of urbanization and the urban environment: inventive and varied. Many threats will integrate widely available off-the-shelf technologies into their weapon systems and armed forces. However, sniper rifles and small, man-portable, fire-and-forget weapons and demolitions and other improvised explosive devices (IEDs), to include suicide and car bombs, will likely dominate the urban environment. Figure 3-3 lists examples of threat weapons favored in UO.

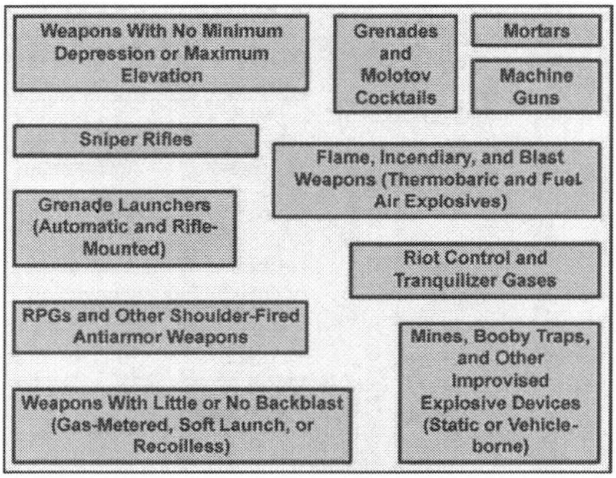


**Figure 3-3. Favored threat weapons**

## ENGAGE ENTIRE ENEMY FORCE

3-34. Threats may attempt to keep all or significant portions of Army forces engaged in continuous operations to increase their susceptibility to stress induced illnesses. UO, by their nature, produce an inordinate number of combat-stress casualties. Continuous operations exacerbate this problem. Threat forces that employ this tactic will often maintain a large reserve or achieve respite by hiding among the civilian population to minimize the psychological impacts on their own forces.

3-35. To accomplish this key tactic, threat UO will likely involve decentralized maneuver, precision fires, and simultaneous operations involving unconventional and special purpose forces and guerilla, insurgent, and terrorist tactics. Threats may also attempt to bait Army forces by ambush or other means with a primary focus of attacking response or reaction forces. Overall, threat forces will take advantage of any exposed weakness in weapons, equipment, and TTP and they will cultivate the flexibility to engage in battles as favorable opportunities present themselves.

### Focus Attacks on Support Areas, Isolated Groups, and Individuals

3-36. To supplement the previous tactic, threat forces will seek to target support areas, small groups, leaders and their headquarters, and unaware Soldiers. Their focus on resupply and convoy operations, casualty evacuation, and other sustainment activities, coupled with the compartmented terrain, navigational challenges, and multiple three-dimensional avenues of approach often makes these locations and Soldiers more susceptible to surprise raids and ambushes. Attacks on these areas and groups are conducted to erode the Army's ability to sustain UO, to inflict maximum casualties, and to induce psychological stress. These attacks can be mitigated by careful, regular evaluation of choke points and other restrictive terrain, regular awareness training for units and individuals operating in or transiting through potential incident-prone areas, well-planned and protected convoys, and thorough after-action analysis of incidents.

## NEGATIVE EFFECTS OF URBANIZATION

3-37. Many urban areas are the engines for increased industrialism and economic growth as an expanding population provides the labor for manufacturing and service needs. However, rapid and inadequately planned growth can result in undesirable consequences. Uncontrolled urbanization may result in an infrastructure and economic base unable to support the growing population. A large transient, ill-housed, and idle population in a close geographic space may produce strife. Classes, cultures, ethnic groups, and races that might otherwise peacefully coexist can clash under the stress of survival. Uncontrolled urban growth has resulted in the negative effects listed in figure 3-4. In many urban stability and civil support operations, these may be the primary "threats" to mission accomplishment.

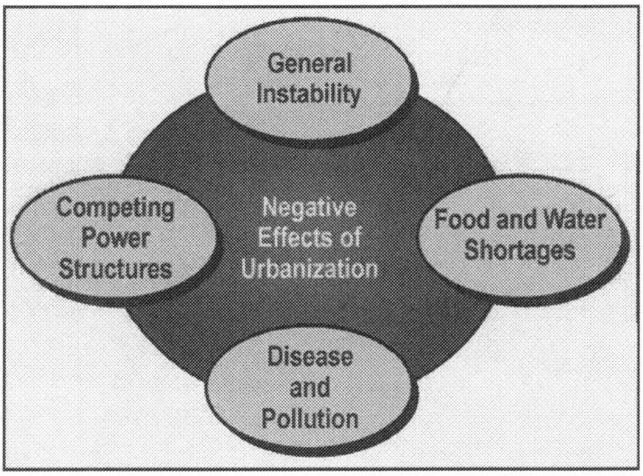


**Figure 3-4. Negative effects of urbanization**

3-38. Not all urban areas prevail as inherently unstable or hotbeds for unrest. Urban growth due to migration may remove sources of conflict, or it may provide the catalyst for violence. Commanders should recognize the possible effects of uncontrolled urbanization. During their intelligence preparation of the battlefield (IPB), they should determine if these conditions exist. Throughout mission analysis and the development of courses of action, commanders should consider the impact (if any) on their operations. At the same time, they must recognize that UO may create similar problems that may affect the current operation as well as the overall campaign.

## General Instability

3-39. Urbanization can enhance stability by generating industrialization and economic growth resulting in more jobs, a higher overall standard of living, and an educated, relatively satisfied populace. However, the population dynamics associated with urbanization can also have an opposite, destabilizing effect. Radical population growth may create overcrowding and generate or aggravate resource and quality of life issues. Intense and destructive competition for employment, housing, and social status may develop in this climate of economic deprivation. The inability of some governments to handle these problems—

- Makes their urban areas potential sources of unrest.
- Increases the likelihood of the Army's involvement in stability operations.
- Complicates all operations conducted in such an urban environment.

Weak civil administrations—perhaps weakened by the effects of previous UO—have difficulty controlling their society, safeguarding their military armaments, and preventing their urban areas from serving as sanctuaries to terrorists, insurgents, and criminal organizations.

3-40. Urbanization in developing countries warrants more concern. Intense migration and growth, coupled with the forced closeness of people once separated by the rural countryside, may stress already struggling institutions, hasten conflict, and lead to overall instability. Commanders must understand that UO, depending on the operation, may either cause massive population movement out of or into urban areas.

3-41. Urban areas with a large youth population may also help to generate conditions for instability. Rural-to-urban migrants tend to be relatively young. In 1999, Cairo, for example, had more than 40 percent of its population younger than 15 years. Young urban populations generate enormous demands for social resources, primarily education and jobs. Even a strong urban economy may fold under the economic expectations of a tremendous influx of young migrants. Disorder and violence may result as hostiles (many nonstate actors) easily mobilize and manipulate the idle young to act politically and criminally. Urbanization and population growth are more dangerous when they combine to produce a cohort of young urban dwellers separated from traditional social controls, such as village elders and clan leaders.

**Cultural and Religious Instability**

The 1992 bombing of the Babri Masid Mosque in Ayodya, India, enflamed an already intense cultural and religious rivalry between Hindus and Muslims and led to rioting throughout many Indian urban areas. Of the 1,500 who died in conflicts and riots, almost 95 percent died in urban areas. The violence struck Ahmedabab and Bombay most seriously, with related acts of murder, gang rapes, and arson occurring months after the destruction of the mosque.

3-42. Ethnic, religious, and other social issues may become the vents for anger and frustration produced by the high tension of urban life. Major acts of violence and destruction, such as occurred in 1992 in India, can directly threaten a nation's security. Army forces may have to conduct large-scale, stability operations to promote peace and protect national interests. In these cases, all levels of command will be particularly concerned with maintaining impartiality and perceived legitimacy.

## Food and Water Shortages

3-43. Rapid urbanization, primarily in developing nations, may lead to severe food shortages that could influence Army forces (or lead to their use). Such shortages may cause instability, massive migration, revolts, or increased support of armed opposition groups. Armed factions may target NGOs that supply aid as a means of furthering dissatisfaction among the populace. In effect, food may become a weapon. Deployed troops may need to provide or support humanitarian food aid networks to keep the humanitarian situation from escalating.

3-44. However, planners must be wary of the unintended consequences of well-intentioned humanitarian assistance operations. For example, providing free, safe food may alleviate starvation, but could also undercut the local agricultural system by reducing demand in the market. If the food is distributed through urban centers, urbanization could increase, further reducing the food supply and adding to the existing strains on the infrastructure. Unless they are trying to use them as a means to separate combatants from noncombatants, commanders should normally use centralized feeding centers as a last resort. Instead, Army forces should bring the food closest to the population to encourage civilians to stay in their homes. If safe areas or camps are created, they should be designed for use over as short a time as is feasible. The general rule should be to return the urban population to their homes as soon as possible. Army forces conducting domestic support or foreign humanitarian assistance operations that cannot maintain the safe food supplies may find the frustrations and hostility of the local population focused on them.

3-45. Water shortages (and quality) are becoming a serious problem in many regions. Sewage, industrial waste, other forms of pollution and deliberate contamination pose threats to existing water supplies. Commanders operating in an urban environment need to know the water supply origins and its treatment, purification, distribution, and vulnerabilities. Before beginning operations, commanders must know if they are providing water for the noncombatants as well as their own forces. Across the spectrum of operations, controlling and protecting a limited water supply is, or may become, an essential operational consideration during UO.

**Food and Water Shortages**

Countries as varied as Indonesia and Algeria exported their food surpluses only two generations ago but now import up to two-thirds of their basic staples. This cycle has resulted in many countries, which once exported agricultural products, facing the growing cost of imports to feed their urban populations. Estimates predict that in the 2010 timeframe, at least 65 countries (including 30 of Africa's 51 countries) may depend completely on food imports. For some countries, the outlook is even worse. The Democratic Republic of Congo (Zaire), once a net food exporter, faces severe malnutrition and utter starvation for a large portion of its population.

Other estimates predict that by 2025, 2.7 to 3.5 billion people will live in water-deficient countries. Even now in urban areas such as Jakarta, Indonesia, for example, urban authorities cannot provide the necessary water to support its growing population. Less than half of Jakarta's population is supplied by the urban area's water infrastructure; the remainder of the area's residents rely on water from wells and other groundwater supplies that lie only a few feet below the surface. These alternate sources need to be sufficiently boiled or residents risk increased disease. In the northern part of Jakarta, land is actually sinking as the urban area's groundwater supply is overused by its residents. Houses on low-lying areas have to be rebuilt every few years to keep them above sea level. Meanwhile, saltwater from the Java Sea is seeping into the land polluting the remaining groundwater and further exacerbating the situation.

## DISEASE AND POLLUTION

3-46. Urban areas frequently spawn epidemics; therefore, widespread disease may pose a significant threat to Army forces that operate there. In many developing nations, rapid urbanization has occurred without a corresponding upgrade, expansion, or even development of adequate sewage and water systems. Some urban areas have only one toilet for every 750 people or more. In these areas, hundreds of thousands live much as they would in poor villages, yet so confined as to ensure high transmission rates for airborne, waterborne, sexually transmitted, and contact-transmitted microbes.

3-47. In urban areas lacking adequate trash and waste management infrastructure, insect-spread diseases proliferate. Mosquitoes that breed in polluted water, open water tanks, and irrigated urban gardens carry malaria and dengue fever—the leading causes of sickness and death from infectious disease in Latin America and Africa. The problem compounds with growing numbers of bacteria resistant to various antibiotics, a shortage of trained medical personnel, inadequate or insufficient medical facilities and supplies, and unclean agricultural and food-processing practices.

3-48. Pollution also creates critical health problems in developing areas and a potential health risk for intervening Army forces. For example, urban areas in China have recorded five to ten times the levels of sulfur dioxide found in the air of urban areas in the developed world. In parts of Poland, toxic waste has so polluted the land and water that ten percent of the babies have birth defects. Pollution may cause immediate health problems but more often, the insidious effects appear months or years after exposure. As discussed earlier, UO may contribute, either intentionally or unintentionally, to an increase in pollution. Destruction of industrial complexes that use, produce, and store hazardous material may produce toxic gas and smoke pollutants and create significant health concerns to exposed Soldiers and civilians. Soldiers forced to fight within the confines of subsurface areas such as sewers and drainage systems further increases their exposure to contaminants.

3-49. Commanders should initiate force health protection planning early, including analysis of the medical threat and other critical medical information requirements during the IPB process. A medical surveillance system should monitor the daily status of Army personnel throughout the operation. In preparation, all personnel must receive a predeployment medical examination. This exam establishes an accurate baseline health status of the force and ensures that Army forces do not introduce new diseases to an urban area, possibly exacerbating the situation. Conversely, Soldiers not immune to native viruses or possessing a weakened immune system due to continuous operations and the stress associated with UO may put Army forces at a significant disadvantage. An outbreak of plague during an operation would have an effect similar to a chemical or biological attack. The closer that Army forces operate to civilians (the humanitarian assistance operations conducted in Port-au-Prince, Haiti, and Mogadishu, Somalia, for example), the more probable that these situations may occur. (See Chapter 9.)

## COMPETING POWER STRUCTURES

3-50. Many groups can exist that become strong enough to rival the power of the governing officials and eventually turn the urban area into a system of divergent and competing power structures. These groups can consist of insurgent forces, a merchant class or an economic elite, criminal organizations, or some other significant source of power such as religious organizations (see Chapter 10), clans, or tribes (figure 3-5). In the absence of a legitimate authority, armed factions headed by "warlords" may vie to fill the power void. Sometimes these groups or organizations, normally at odds with each other, may form alliances to achieve specific goals. Commanders must recognize, identify, and understand these alternate urban power bases and, if necessary, develop engagement strategies to neutralize or harness them to accomplish the Army mission. Four of these groups, not previously covered or covered in a subsequent chapter, are discussed in greater detail below.

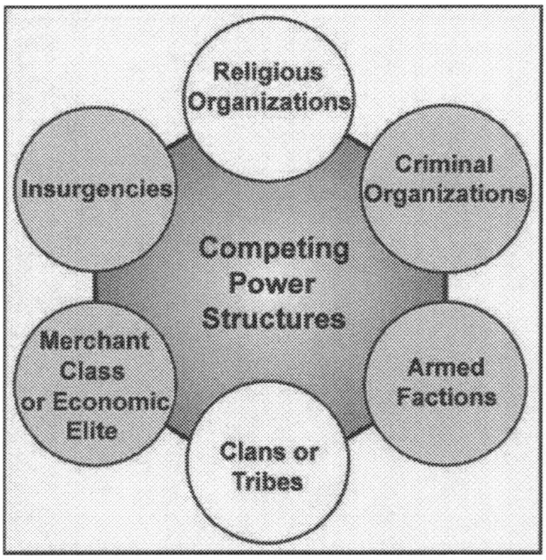


**Figure 3-5. Competing power structures**

### Urban Insurgencies

3-51. As urban migration increases in the developing world, rural guerrillas appear to follow. This transition of insurgencies from rural to urban areas occurs because urban areas offer a rich field of targets for insurgent attacks. People immediately notice any disruption of urban infrastructure, thus having great propaganda value. A concentrated urban population whose allegiance and support are critical to the success of an insurgency is often more susceptible to propaganda, political organization, and terrorism. Insurgents can easily arrange mass demonstrations using available communications facilities, both overt and covert. Travel is effortless and large urban populations provide cover and concealment. On the whole, urban areas, particularly those suffering the effects of general instability outlined earlier, may provide a fertile environment for guerrillas to apply formerly rural insurgent strategies. However, even with a primarily rural-based insurgency, operations in urban areas will often be essential to the insurgents' strategy as they offer them distinct opportunities to disrupt, discredit, and demoralize the government (see FM 3-05.20 and FMI 3-07.22).

## Merchant Class or Economic Elite

3-52. Urban areas normally possess a merchant class or economic elite as part of their social structure. In some urban areas, they may carry more power than the local or central state government. They may isolate themselves physically and socially from the sprawling poor yet wield enormous power over the country's political and economic activities. The degree of economic separation between the merchant class and the poor may be small but still socially or politically significant.

3-53. In a vastly impoverished area where the economy of the urban area is severely disrupted, the merchant class will often continue to operate and function and, as a result, achieve a measure of influence. To continue to operate under acute economic turmoil, they may form alliances in criminal organizations and secure loyalties within the government. Outside resources introduced into a crisis area (such as food, water, fuel, and pharmaceuticals) take on increased value, may replace currency as the medium for exchange, and often become the means to amass and hold wealth. One primary way to obtain wealth is to steal it.

3-54. In some turbulent situations that lead to the need for stability operations, commanders may harness the power of the merchant class as a force for peace and stability instead of one that uses crime to achieve economic goals. For example, in a relief situation, instead of competing with the merchant class by distributing food directly to the needy and possibly creating an environment of looting and black marketeering, it may be possible to monetize food. Food assistance from donor governments could be sold to merchants at an attractive price so they have a reliable source of supply. This could, in turn, create a healthy economic system and separate merchants from criminals and gangs.

## Criminal Organizations

3-55. Organized criminal groups have grown common in urban areas; have also become an important part of the urban social structure (gangs for example); and can exert considerable influence on governments, people, and military forces conducting UO. These organizations can threaten the successful completion of urban operations both directly and indirectly. Criminals and criminal organizations may directly target Army forces, stealing supplies and extorting money or contracts. Criminals may also violently confront Army forces during mission execution. During many UO, particularly during or following combat, civil disturbances, or large natural disasters, looting (organized or unorganized) may become of critical concern.

For a portion of the population, crime may be the only reliable source of income and resources to support their families. Therefore, UO will often require a combined law enforcement and military response.

3-56. Some large criminal organizations relying on international connections often have better resources and equipment than their insurgent counterparts. (A prime example is drug traffickers.) Their large financial resources, long-reaching connections, and ruthlessness provide them the means to corrupt or intimidate local officials and government institutions. Their tactics parallel those of insurgents. They have developed an intuitive cultural understanding of slum neighborhoods and the ability to lure civilians into criminal activities. They have also mastered the management of mobs. They recruit teenagers and young adults in their efforts against rivals and authorities, just as insurgents muster armies from the youth of rural villages. In many developing nations, there exists an alliance between insurgents and organized criminal groups. In these alliances, the insurgents defend the criminals and the criminals fund the insurgents.

3-57. Excessive criminal activity can undermine the commander's efforts to establish a sense of security among the remainder of the urban populace. Commanders will rely on the ability of their military police (MP) units to assess and understand the potential influence that crime and criminal organizations may have on UO. MPs—in coordination with local police—will need to quickly identify law enforcement gaps to include dysfunctional police organizations and stations and inadequate prison systems. Urban police records may be an important intelligence database to assist in this assessment and may be particularly useful in counterterrorism and counterinsurgency operations. Success in ensuring that law and order is maintained or reestablishing law and order will directly affect the population's perceptions and support of Army operations. Also, MP intelligence operations will contribute immeasurably to the overall IPB effort and aid in understanding of the complexities of the urban society (see FM 3-19.50).

**Crime and Criminal Organizations**

Crime and poverty plague urban areas such as Rio de Janeiro, Brazil's second largest urban area, and would affect military operations conducted in their limits. Rio has some of the nation's highest negative urban indicators: the largest number of slum dwellers (1 million), the highest murder rate (1 of 700 residents per year), and the highest kidnapping rate (4 per week). In 1989, the homicide rate of the urban area was three times higher than New York City's, and the rate of urban violence has continued to rise. Therefore, law enforcement management may be a critical issue for Army forces operating in urban environments similar to that of Rio de Janeiro. However, criminal elements or organizations may not always work against Army commanders. They can be co-opted or influenced to serve friendly objectives. For example, during World War II the U.S. Navy worked covertly with the Mafia in New York City to secure the New York harbor from German U-boats believed to be torpedoing ships there. The Mafia controlled most dock activities in New York harbor and was perfectly positioned to monitor other subversive waterfront activity. This capability provided needed information to the Navy for its counterintelligence and security tasks. New York civil authorities, therefore, agreed to permit a Navy-Mafia alliance to operate at the port for the greater good of the country. Although the Mafia was not the preferred ally of the Navy, it had the capability to protect U.S. ships and the interest (patriotism) to help in the war effort. In those circumstances, the temporary alliance worked (see also the civilian threat discussion in Appendix B).

## Armed Factions

3-58. A characteristic of some stability operations has been the deterioration or complete collapse of political authority in the country or urban area in crisis. In some cases, "warlords" or the leaders of various armed factions have attempted to fill the power vacuum (see Appendix C for an example). These individuals often have no particular claim to legitimacy. Their power issues from their weapons, not necessarily from their political skills, human services provided, or popular consent (although they have some popular support to remain in their relative position of authority). Others, however, may already have a substantial base of support (for example, a religious base) and then seek to achieve their goals through violent means. In dealing with the leaders of these armed factions during stability operations, it may appear that there are three general courses of action. The first is to aggressively locate and target the leaders and their forces militarily. The second is to ignore them completely, and the third is to meet and work with them diplomatically. Depending on METT-TC, commanders may reduce some of the risks involved in these first three courses of action by combining the last two to create a fourth approach. Nevertheless, the technique chosen must clearly support political and military objectives and lead to the desired end state.

3-59. In some cases, the leaders of armed factions pose a direct threat to Army forces, the urban populace, and the overall accomplishment of the mission. Therefore, it may become necessary to target them and their forces for capture or destruction. Yet, commanders should generally avoid framing the success of a major operation on the death or capture of one or more individuals. Even with increased ISR capabilities, commanders may find an individual manhunt difficult or impossible to achieve. Instead, it may be better to seek to isolate warlords and their forces militarily, while simultaneously seeking to marginalize their influence through a combination of political, economic, and informational means. This does not necessarily mean that commanders must abandon the search if the death or capture the faction leader will contribute to achieving the mission. However, commanders must maintain the capability to strike with speed and precision if the threat leader's location is discovered in the clutter of the complex urban environment.

3-60. In some situations, commanders may attempt to marginalize warlords by ignoring them. However, refusal to acknowledge warlords may increase the threat to Army forces and NGOs. Their militias may attack Army forces to achieve recognition or simply due to misunderstanding or inherent friction between armed forces. On the other hand, dealing with them may provide legitimacy to the exclusion of other elements of the urban population such as professional groups (for example, doctors or teachers), religious leaders, and traditional clan or tribal chiefs—which may have a greater claim to legitimacy and better form the foundation for a reconstituted urban society.

3-61. A compromise between the last two options may offer the best chance for success. In this fourth course of action, commanders can generally recognize these warlords and work with them diplomatically to decrease the risk of armed conflict and civilian casualties; however, this recognition can be kept at staff levels to avoid bestowing any perceived legitimacy on them. Instead, commanders themselves may visibly meet the other elements of society that have a more legitimate claim to political, social, or economic leadership. Inevitably, commanders may need to meet with warlords. In those circumstances, clan or tribal elders, and others who represent traditional authority, should attend the meetings. Moreover, commanders should ask for and give deliberate consideration to their opinions above those of the leaders of the armed factions.

This page intentionally left blank.

## Chapter 4

# Understanding the Urban Environment's Effects on Warfighting Functions and Tactics

*War is, above all things, an art, employing science in all its branches as its servant, but depending first and chiefly upon the skill of the artisan. It has its own rules, but not one of them is rigid and invariable. As new implements are devised, new methods result in its mechanical execution; but over and above all its mechanical appliances, it rests upon the complex factors of human nature, which cannot be reduced to formulas and rules.*

Captain Francis V. Greene, 1883

Commanders of major operations should understand the potential effects that the urban environment may have on warfighting functions. They should also understand the possible effects that the urban environment may have on lower-level tactics to properly plan, prepare, and execute major operations that may include UO. Otherwise, commanders may ask their subordinates to achieve effects, accomplish objectives, or adhere to a timetable that is unsupportable due to the constraints imposed by the urban environment. Commanders and their staffs must do more than simply understand the impossible; rather, they must apply the art and science of warfighting to the urban environment and determine what it will take to make it possible.

## WARFIGHTING FUNCTIONS

4-1. Understanding the potential effects of the urban environment on warfighting functions permits the urban commander to better visualize his operational environment. With this appreciation, he can conduct a more thorough assessment and thereby determine the most efficient and effective means of employing Army forces. The staff should be intimately familiar with effects in their area of expertise and use that knowledge to understand the problem and develop creative and innovative solutions to achieve their commander's intent.

**WARFIGHTING FUNCTIONS**

- Intelligence
- Movement and maneuver
- Fire support
- Protection
- Sustainment
- Command and control

### INTELLIGENCE

4-2. The intelligence function facilitates understanding of the threat and the environment. The urban environment affects this critical function in many ways. Impacts of the environment on the intelligence function include degraded reconnaissance capability; more difficult IPB process; and an increased importance of credible HUMINT (including the contribution of local civilian liaisons), and an established intelligence reach capability. The Army forces' response to these effects can result in timely, accurate, and actionable intelligence that permits the effective application of other warfighting functions to the mission within the urban environment (see FM 2-0).

## Degraded Reconnaissance and Surveillance Capability

4-3. The physical environment creates a major challenge to the intelligence function. The man-made construction in the urban areas provides nearly complete cover and concealment for threats. Although improving many sensor capabilities cannot penetrate the subsurface facilities and much of the space within supersurface areas. The mass of buildings can also defuse electronic signatures. Tall buildings shield movement within urban canyons from aerial observation except from directly overhead. Urban threats may be less technology dependent and may thwart some signals intelligence efforts simply by turning off their radios and using messengers. Threat forces will likely use elements of the civilian telecommunications infrastructure for C2. These systems may include traditional landline phones, cellular telephones, and computer-to-computer or Internet data communications. Most urban telecommunications systems use buried fiber or cables or employ modern digital signaling technology. Such systems are difficult to intercept and exploit at the tactical level.

4-4. From the above, it is evident that these characteristics make it more difficult for the intelligence function to use electronic means to determine threat dispositions and, in offensive and defensive UO, identify decisive points leading to centers of gravity. While the environment limits some typical collection methods, all enemy electronic and human activity creates some form of observable signature and exposes the enemy to potential collection. Seeking ways to take advantage of these vulnerabilities will provide the commander an information advantage over his opponent.

## Challenging IPB Process

4-5. The sheer complexity of the environment also challenges the intelligence function. The intelligence function applies the IPB process to the urban environment in accordance with Army doctrine (see Appendix B). With more data points for the IPB process to identify, evaluate, and monitor, this application becomes more demanding. The human and societal aspects of the environment and the physical complexity primarily cause this difference. Relationships between aspects of the environment, built on an immense infrastructure of formal and informal systems connecting the population to the urban area, are usually less familiar to analysts. Thus, the urban environment often requires more specifically-focused intelligence resources to plan, prepare for, execute, and assess operations than in other environments.

4-6. Compounding the challenges is the relative incongruity of all urban environments. No two urban areas are alike physically, in population, or in infrastructure. Thus, experience in one urban area with a particular population and pattern of infrastructure does not readily transfer to another urban area. Any experience in UO is valuable and normally serves as a starting point for analysis, but the intelligence function cannot assume (and treat as fact) that patterns of behavior and the relationships in one urban area mirror another urban area. The opposite is as likely to hold true. The intelligence function will have to study each urban area individually to determine how it works and understand its complex relationships.

4-7. Each characteristic of the urban environment—terrain, society, and infrastructure—is dynamic and can change radically in response to UO or external influences. Civilian populations pose a special challenge to commanders conducting UO. Civilians react to, interact with, and influence to varying degrees Army forces. Commanders must know and account for the potential influence these populations may have on their operations. Intelligence analysts must revisit or continuously monitor the critical points looking for changes, relationships, and patterns.

4-8. The actions of Army forces will affect, positively or negatively, their relationship with the urban population and, hence, mission success. NGOs may deliberately or inadvertently influence civilians. The intelligence function can monitor and predict the reactions of the civil population. However, accurate predictive analysis of a large population requires specific training and extensive cultural and regional expertise.

## Increased Importance of Human Intelligence

***HUMINT*** is the collection of foreign information—by a trained HUMINT collector—from people and multimedia to identify elements, intentions, composition, strength, dispositions, tactics, equipment, personnel, and capabilities. It uses human sources as a tool, and a variety of collection methods, both passively and actively, to collect information. FM 2-0

4-9. The intelligence function adjusts to the degradation of its technical intelligence gathering systems by increasing emphasis on HUMINT in UO. HUMINT operations may be the primary and most productive intelligence source in UO. In urban offensive and defensive operations, HUMINT gathers information from refugees, immigrants and former citizens (especially previous civil administrators), civilian contractors, and military personnel who have operated in the area. Credible intelligence of this type can help meet requirements, provide more detail, and alleviate some of the need to physically penetrate the urban area with reconnaissance forces. In many urban operations where HUMINT is the primary source of intelligence, acting on single-source reporting is a constant pitfall. Yet, situations may arise where commanders must weigh the consequences of inaction against any potential negative consequences resulting from acting on uncorroborated, single-source information. (See also the Human Capabilities discussion under the Urban ISR portion of Chapter 5.)

4-10. In urban stability operations, HUMINT identifies threats and monitors the intentions and attitudes of the population. A chief source of information contributing to the development of accurate HUMINT, particularly at the tactical level, is reconnaissance forces—especially small-unit dismounted patrols. Urban reconnaissance forces and patrols should be thoroughly and routinely debriefed by unit intelligence personnel to obtain information that aids in developing a clearer picture of the threat and the urban environment. Reliable and trustworthy HUMINT is particularly important in foreign internal defense, counterterrorism, and support to counterdrug operations. Leaders must organize intelligence resources appropriately, and learn and apply valuable techniques, such as pattern and link analysis (see FM 34-3). Additionally, Soldiers, as part of reconnaissance and patrolling training, should be taught to handle captured documents, weapons, material, and equipment as legal evidence much like military and civilian police. Proper "evidence" handling is often a critical intelligence concern in counterterrorism and counterinsurgency operations.

## Developing Local Liaisons

4-11. Whenever Soldiers encounter the urban populace, the resulting interaction may become an important source of information the commander can use to answer questions about the threat and the urban environment. While military intelligence units are the primary collectors and processors of HUMINT, commanders are not likely to have enough trained HUMINT Soldiers to satisfy their requirements—particularly in a larger urban environment and during longer-term stability operations. Therefore, commanders may need to cultivate and establish local civilian associations to provide relevant information for decision making and to support the overall HUMINT effort.

4-12. Urban liaisons can be developed through positive civil-military interaction with the urban populace. Critical information may be acquired through interface with the urban leadership (both formal and informal), administration officials, business owners, host-nation support workers, inhabitants along a unit's patrol route, pedestrians at a checkpoint, civilian detainees, or any other human source willing to volunteer information to Army forces or who respond positively to tactical questioning. (Noncombatants are never coerced to provide information.) Commanders may also direct unit leaders to conduct liaisons with specific local leaders and key members of the community to obtain command directed information. Critical information may also come from other U.S. and coalition forces and intelligence organizations operating near or within the commander's AO. To this end, commanders should ensure that collectors operating in an urban area coordinate and deconflict activities and, if possible, outbrief subordinate, geographically-responsible commanders with any relevant information that may affect their current operations. Any relevant information obtained incident to civilian liaison activities should be routinely provided to intelligence staffs not only to gain assistance in verifying the credibility of the information but to share the information with all affected echelons and units.

4-13. NGOs operating in urban areas can also be especially beneficial resources for credible and relevant information about the urban environment. (However, they are generally not a good source for information about the threat since providing such information can violate their neutrality thereby making it difficult for them to achieve their humanitarian aid objectives.) During the 1999 fighting in Kosovo, for example, the Red Cross provided the most accurate figures regarding the number of Kosovar refugees, helping U.S. and other coalition forces to estimate the appropriate level of support required to handle their needs. In addition to a developed understanding of the current needs of the local urban populace, NGOs may also have—

- A network of influential associations.
- Historical archives.
- Extensive understanding of the urban infrastructure.
- Key knowledge of political and economic influences.
- A keen awareness of significant changes in the urban environment.
- Insight into the current security situation.
- Up-to-date web sites and maps.

4-14. While productive civilian associations may become long term, they should not be confused with HUMINT source operations. Only trained HUMINT personnel can recruit and task sources to seek out threat information. Information obtained from these societal connections is normally incidental to other civil-military relationships. For example, as part of infrastructure repair in an urban stability operation, a commander may be instrumental in obtaining a generator for a local hospital. Within the context of this relationship, the commander may develop a rapport with one or more of the hospital's administrators or health practitioners. These civilians may be inclined to provide valuable information about the threat and the urban environment—often on a continuing basis. In any civil-military relationship, however, commanders ensure that the information provided is not tied to promises of assistance or that such assistance is in any way perceived as a means to purchase civilian loyalty.

4-15. Commanders also understand that repeated interaction with any one individual may put that individual and his family in danger from threat forces. Before this potential danger becomes a reality, they should refer their civilian connections to trained HUMINT personnel for protection and continued exploitation. In addition to civilian protection considerations, commanders may also deem it necessary to turn their civilian associations over to trained HUMINT collectors anytime during the relationship if they consider the information that the contact is providing (or may provide) is credible, relevant, and—

- Provides essential threat information on a repetitive basis.
- Helps answer higher-level CCIR.
- Affects operations in another AO.
- Requires interrogation or monetary compensation to obtain.

However, turning a liaison over to trained HUMINT teams does not necessarily preclude maintaining a continued, albeit a more guarded, relationship with the individual.

4-16. In developing these civilian liaisons essential to understanding the urban environment, commanders must avoid the distinct possibility of conducting unofficial source operations by non-HUMINT Soldiers. While prohibited by regulatory guidance (see Defense Intelligence Agency Manual 58-11 for requirements and restrictions for source operations), such actions also run the additional risks of—

- Obtaining unevaluated information that cannot be crosschecked with and verified by other sources of information.
- Creating inequities that result from illegally rewarding contracts which can undermine HUMINT Soldiers who are constrained by intelligence contingency fund regulations.
- Disrupting ongoing HUMINT operations when different sources are seen to be treated differently by non-HUMINT Soldiers vice HUMINT Soldiers.
- Providing non-HUMINT and HUMINT Soldiers with the same information potentially leading to a false confirmation of information.

- Increasing the likelihood that untrained Soldiers may fall victim to a threat deception and misinformation.

### Established Intelligence Reach

4-17. Understanding the complex urban environment, particularly the infrastructure and the society, will require more sources of information beyond a unit's organic intelligence capabilities. Therefore, commanders will have to make extensive use of intelligence reach to access information and conduct collaboration and information sharing with other units, organizations, and individual subject matter experts. Before deployment (and throughout the operation), units should establish a comprehensive directory of intelligence reach resources. These resources may include national, joint, Army, foreign, commercial, and university research programs. (Prior to deployment for OPERATION IRAQI FREEDOM, some units established contacts within the local community outside their bases such as police, fire department, and government officials that expanded their reach once in theater—particularly for information regarding civilian infrastructure and urban administration.) Once deployed, intelligence reach includes effective information sharing and collaboration among adjacent units, sister services, coalition partners, and other governmental and nongovernmental agencies operating in the area. Effective information sharing and collaboration requires common network analysis software and databases to be used among all Army forces and, if possible, other governmental agencies.

## MOVEMENT AND MANEUVER

4-18. Army maneuver forces—infantry, armor, cavalry, and attack aviation—move to achieve a position of advantage. Entire urban areas or specific threat forces located within may be isolated from affecting other operations and then bypassed. However, when the situation requires entering the urban area to accomplish the mission—such as when bypassed urban threat forces interdict critical lines of communications (LOCs), the environment will significantly affect the Army's ability to move and maneuver (see figure 4-1). These negative effects include canalization, compartmentalization, and increased vulnerability. However, tactics and techniques equip Army forces to overcome these challenges and maneuver successfully. One tactic, effective combined arms task organization, includes an increased dismounted maneuver capability combined with armor and combat engineers, continuous operations, and technological enhancements.

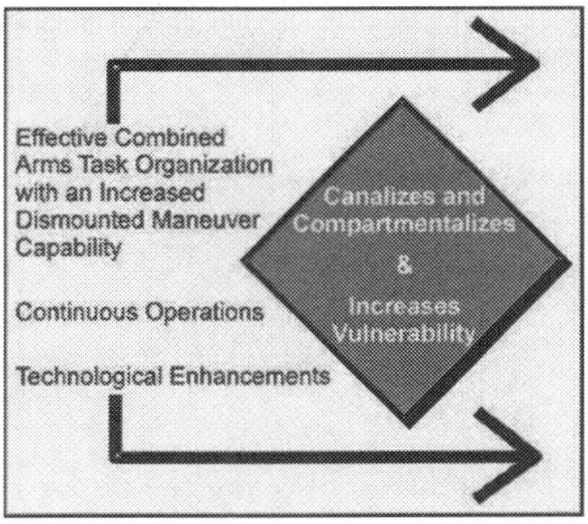


**Figure 4-1. Urban maneuver challenges and means to overcome them**

4-19. In all UO, mobility operations may allow civilian traffic and commerce to resume, letting the urban area return to some semblance of normalcy (often a critical objective). In stability operations, mobility often focuses on keeping lines of communications open and reducing the threat of explosive hazards to Soldiers and civilians. In civil support operations, mobility may focus on removing storm debris or reducing obstacles caused by destroyed property.

### Canalization and Compartmentalization

4-20. The urban terrain will often canalize and compartment forces—and their fires—moving and maneuvering through it. Buildings pose obstacles to both mounted and dismounted movement, forcing units to be canalized along streets. The buildings also block movement between streets, thus compartmenting units. Fires are canalized into open and unmasked areas where vision is unobstructed, producing concentrated fire zones and areas, especially at road intersections and in front of defended positions. Hence, changing directions, repositioning committed forces, reinforcing forces in contact, bypassing threats, and maneuvering to the threat flank become extremely difficult. Units often breach obstacles to help solve this problem. Using helicopters to quickly move forces, both forward into contact and to rear areas as part of repositioning, also permits Army forces to overcome some terrain constraints.

4-21. The canalized and compartmented effects can slow movement and maneuver considerably. However, slowed movement also results from the physically demanding soldier tasks required in an urban environment. Soldiers operate dismounted across rubble and hard surfaces. Operating in three dimensions, they constantly move up the supersurface areas of building interiors and down into basements, cellars, and other subsurface areas. They breach many obstacles and use upper-body strength, ropes, and ladders to scale heights. The inability to see into the next room, floor, or building magnifies stress. The resulting fatigue slows the overall rate of Army force movement and maneuver.

## Increased Vulnerability

4-22. The urban environment increases the vulnerability of Army forces executing movement and maneuver in offensive, defensive, and stability or civil support operations. Both the physical terrain and the urban population provide threat cover and concealment. Air movement and maneuver is vulnerable for many of the same reasons. In offensive or defensive operations, enemy forces can remain undetected in buildings and in position to ambush Army forces. Consistent with METT-TC, Army forces should clear buildings along maneuver routes prior to mounted movement along those axes. Failure to clear routes (and effectively mark cleared portions) may expose mounted movement to ambush at close range. Movement back across streets and obstacles may be difficult particularly if the element of surprise was essential in the initial crossing or breach. The same buildings also provide cover and concealment to enemy air defense capabilities—particularly man-portable air defense systems that can be fired from multiple positions hidden amongst the clutter of fires, lights, smoke and dust, and easily concealed and transported in civilian vehicles throughout the urban area. In all operations, but especially stability operations, civilians can conceal threat elements. The threat can then initiate offensive operations against Army forces from close range and where ROE will hamper applying combat power. Thus, maneuver through a dense population can be a high-risk operation.

## Combined Arms Task Organization

4-23. Effective combined arms task organization ensures that forces are task organized with infantry—the essential building block for all organizations conducting UO. Infantry protects mounted elements as the combined arms unit moves and maneuvers through the urban area. (In some urban situations, mechanized infantry may not be able to provide dismounted support beyond support to its own vehicles—tanks may require the support of additional light infantry). The infantry destroys the enemy in buildings, bunkers, and subsurface areas where they cannot be defeated by mounted forces and prevents infiltration of threat forces back into hard-won urban terrain. Field artillery aids in dismounted and mounted (to include air) maneuver by suppressing known and suspected enemy positions with precision fires. Attack aviation make best use of their standoff capabilities (see later fire support discussion) and aircraft speed to conduct running and diving fires; in UO, hovering fire is generally avoided. Armored elements protect Soldiers from small arms fire and destroy or suppress enemy positions with precise, direct fire. Carefully protected artillery may also be used in this direct fire role. Armored forces and attack helicopters also can facilitate maneuver through shock action that can have a psychological effect, particularly against less well-trained threats and, in discrete instances, hostile crowds. (Although, commanders consider that the "intimidation" value of any method erodes quickly with its repetitive use.)

4-24. Combined arms also ensure that combat engineers support dismounted maneuver by assisting in covered and concealed maneuver through buildings and off exposed streets. In addition to combat engineers, explosive ordnance disposal teams, military police, chemical personnel, and other with essential expertise to conduct mobility missions significantly reduces mobility and maneuver challenges (see FM 3-34.2). Urban buildings are often obstacles to movement and mobility. Combat engineers, trained and equipped for UO, can turn these obstacles into an advantage by breaching them with "mouse holes" made by explosives, sledgehammers, bulldozers or armored vehicles, or high-strength (diamond or carbide-tipped) cutting devices. These breaches permit dismounted movement through buildings under both cover and concealment.

4-25. Combat engineers must also be trained and equipped to facilitate mounted mobility in the urban environment. Generally, buildings restrict mounted movement to the compartmented and canalized streets. Threats can block streets with roadblocks ranging from sophisticated log and concrete cribs reinforced with antitank and antipersonnel mines to the expedient use of cars, buses, and trucks to create obstacles. Combat engineers must be capable of breaching these obstacles to maintain the coherence of the combined arms team (mounted and dismounted). Combat engineers should be forward, often task organized down to platoon level, and have the expertise and equipment to rapidly reduce point obstacles. It even may be necessary that every armored vehicle (or section of two vehicles) be task organized with an associated engineer squad and vehicle. Because of the increased density and hardness of many urban building and construction materials, heavy engineer equipment (such as the D9 bulldozer) will be in great demand to accomplish mobility (and countermobility, and survivability) functions in an urban environment. However, commanders will need to consider increased protection requirements and the availability of equipment transport to move these slower-moving engineer assets around the urban battlefield.

4-26. A major difference of UO combined arms is in proportion and organization. Although based an accurate METT-TC assessment, UO will often require an increased proportion of dismounted infantry and engineer capabilities; armor may not be required in the same high numbers. As significant, genuine combined arms urban operations are required at lower tactical levels where small, well trained and led units will dominate. Commonly, company level will require true combined arms capability and may include combat engineers, military intelligence, reconnaissance, and artillery. Combined arms teams can then form at platoon and squad levels. Because of this, larger units will need more CA, military intelligence, and combat engineers than habitually attached for combat in more open or less restrictive terrain.

4-27. In determining the appropriate task organization, commanders consider a suitable span of control for subordinate commanders. They also consider the potential of dissipating a unit's combat power, capabilities, and synergy by breaking a unit up into smaller units in an attempt to ensure subordinate maneuver units have a complete combined arms capability. For example, an additional engineer battalion may be task organized to a brigade combat team. In turn, the brigade combat team may task organize this battalion into engineer companies under the control of their subordinate maneuver battalions. If this type of organization continues, maneuver companies may end up with an engineer platoon and, in the end; maneuver platoons each having an engineer squad. Ultimately, a combined arms capability may have been established at lower tactical levels but the parent maneuver unit (in this example, the brigade combat team) may have lost the ability to conduct larger engineer operations without having to re-task organize and potentially disrupt current operations and established relationships. As a guide, urban commanders may consider task organizing to create combined arms organizations at lower tactical levels when operations are predominately offensive or defensive, and bringing those assets back under their own control when the operation transitions to predominately stability or civil support operations.

## Continuous Operations and Technology Enhancements

4-28. Two other means to improve Army forces' ability to move and maneuver in urban terrain is through continuous operations and the leveraging of technology, such as the Army's night operations capability. Historically, urban battles have been fought primarily during daylight because of technological limitations and fatigue. By utilizing night vision technologies, accurate situational understanding, a common operational picture (COP), training, and rotated units, Army forces can defeat threats who use the same soldiers in day and night operations and who are less well-equipped and adept at night operations. Night operations are also a means of mitigating the air defense threat against air maneuver. Continuous operations through night maneuver with fresh forces are challenging, but it can overcome many advantages that a stationary force has against maneuver in the urban environment. However, commanders should also consider that streetlights, fires, background illumination (as well as dark building interiors without ambient light), the increased heat absorption of many urban structures, and the skillful use of searchlights by threat forces may limit the effectiveness of night vision devices and make thermal imagery identification difficult.

### Countermobility

4-29. Countermobility capabilities in urban terrain are also an essential consideration in all UO. In defensive operations, commanders use countermobility capability to control where the enemy moves in the urban area. Repositioning defensive forces in the urban area can be difficult and obstacles are essential to limiting the enemy's maneuver options. During offensive operations, countermobility protects exposed flanks and air assaulting forces from counterattack. In stability and civil support operations, countermobility operations may take the form of constructing barriers to assist in populace and resources control at critical urban locations.

## FIRE SUPPORT

4-30. The fire support function includes the collective and coordinated use of several means to attack targets in the urban area (see Appendix D for joint capabilities). These means include target acquisition data, indirect fire weapons, rotary- and fixed-wing aircraft, offensive information operations (IO), and other lethal and nonlethal means. The urban environment affects these components of the fires function and their employment.

### Target Acquisition

4-31. Target acquisition in an urban environment faces several challenges. First, forces have difficulty penetrating the urban environment's increased cover and concealment using sensors and reconnaissance. Acquiring targeting information and tracking targets throughout the depth of the urban area may prove challenging. Moving personnel or vehicular targets are normally easiest to acquire. However, the cover and concealment provided by urban terrain gives moving targets short exposure times requiring firing systems to act rapidly on targeting data. Targeting of opposing indirect fire units by acquisition radar may work more effectively in urban terrain because of the necessary high angles of indirect fire. The urban environment presents similar difficulties for battle damage assessment.

4-32. Targeting challenges are met by innovatively integrating reconnaissance capabilities. These capabilities include SOF, cavalry, unmanned aircraft systems, and aerial observers as well as the standard reconnaissance assets. More artillery systems may need to be used to ensure the responsiveness (rather than the weight) of fires. Positioning numerous artillery systems reduces the dead space (as discussed below) and permits units to establish more direct sensor-to-shooter links.

### The Targeting Process

4-33. Heightened concerns for collateral damage will require that commanders pay particular attention to their targeting process. This process ensures that all available combat power, both lethal and nonlethal (including offensive IO), is effectively integrated and synchronized to accomplish the mission. Commanders ensure that techniques and procedures are in place, rehearsed, and understood by all members of their staffs. Additionally, the C2 system must be responsive and agile; otherwise, an elusive and adaptable threat will likely disappear before units can employ the appropriate weapon systems. In an urban area, even 10-digit grid coordinates may not be sufficient to accurately identify targets as buildings may be connected to each other—often throughout the entire block. Target locations, in addition to grid coordinates, may need to routinely include the street address, number of stories, shape, color, or any other distinguishing characteristic essential for ground and air forces to achieve targeting precision. A common urban reference system with graphics, reference points, and other control measures adequate for both ground and air forces may also help facilitate identification of targets and facilitate the rapid clearance of fires.

4-34. Greater concerns exist for the safety and health (environmental matters) of the urban populace and the protection of critical infrastructure and cultural structures. Hence, CA and staff judge advocates (see Chapters 5 and 10 respectively) will play a greater role for the expert advice they can provide regarding these elements of the urban environment. Nonetheless, all members of the staff ensure that operations minimize collateral damage. That responsibility does not end with identifying potential collateral damage; the goal, as always, is successful mission accomplishment. Again, staffs are guided by the commander's intent and work to develop courses of action that incorporate collateral damage concerns (short- and long-term) yet accomplish the mission. This requires a keen understanding of the legal issues and both friendly and enemy weapon systems' effects in an urban environment.

### Urban Effects on Fire Support Systems

4-35. Both the physical and human components of the urban area affect how units use fire support weapon systems. The physical aspects of the urban environment, such as the heights and concentration of buildings, may cause significant masking and dead space (see figure 4-2). Intervening buildings that stand three or more stories tall hinder close indirect fire support. Tall buildings can potentially mask several blocks of area along the gun-target line of artillery. (For low-angle artillery fire, dead space is about five times the height of the building behind which the target sits.) The potential for collateral damage to adjacent buildings may also prevent engagement with artillery. Such damage might cause noncombatant and friendly troop casualties and unintentional (and unwanted) rubbling. Commanders can offset these effects by carefully placing artillery positions, repositioning artillery as targets change, and using mortars. Mortars have a steep angle of fall and short minimum ranges as a high-angle alternative to field artillery fire. (In comparison to artillery, dead space for mortar fire is only about one-half the height of the building.) For fixed-winged aircraft, precision munitions and weapons with low explosive yields and near-vertical impact angles resulting in bomb burial can significantly reduce collateral damage. Collateral damage concerns may also cause commanders to—

- Maintain approval authority for some sensitive or protected targets (churches or mosques, for example) at higher echelons of command.
- Restrict attacks to certain times of day.
- Give warning prior to an attack so that noncombatants can evacuate the area.
- Incorporate indigenous forces into the operation.
- Abort an attack unless the required level of precision effects can be achieved.
- Prepare specific branches and sequels to the IO plan to inform the populace why the collateral damage was justified. These plans may include filming or otherwise documenting the operation to thwart threat propaganda and claims of excessive collateral damage.
- Develop and rehearse detailed staff battle drills that address clearance of fires in an urban environment.

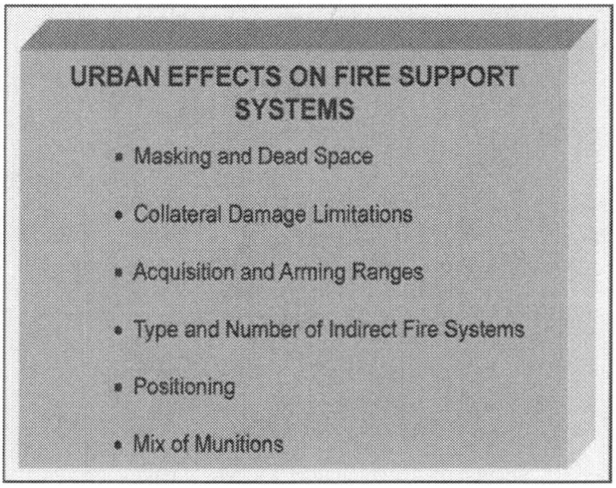


**Figure 4-2. Urban effects on fire support systems**

4-36. Vertical structures interrupt line of sight (LOS) and create corridors of visibility along street axes. The result is thereby shortened acquisition and arming ranges for supporting fires from attack helicopters and subsequently affected engagement techniques and delivery options. Pilots maintain LOS long enough to acquire targets, achieve weapons delivery solutions, and fly to those parameters. For example, tube-launched, optically tracked, wire-guided heavy antitank missile systems require 65 meters to arm. Similarly, the Hellfire missile requires at least 500 meters to reliably arm and stabilize on the intended target. Thus, attack helicopters firing from longer ranges actually improve the probability of a hit. Poor weather and heavy smoke and dust rising from urban fires and explosions may hinder target identification, laser designation, and guidance for rotary- and fixed-winged aircraft. Poor air-to-ground communications may also hinder effective use of airpower. The close proximity of friendly units and noncombatants requires units to agree on, thoroughly disseminate, and rehearse clear techniques and procedures for marking target and friendly locations. The ability for ground units to "talk-on" aircraft using a common reference system described earlier helps expedite aerial target acquisition (and helps mitigate potential fratricide). FM 3-06.1 details other aviation TTP in an urban environment.

4-37. The urban environment also affects the type and number of indirect fire weapon systems employed. Commanders may prefer high-angle fire because of its ability to fire in close proximity to friendly occupied buildings. Tactically, commanders may consider reinforcing units in UO with mortar platoons from reserve units. This will increase the number of systems available to support maneuver units. Unguided Multiple Launch Rocket Systems (MLRSs) may be of limited use in urban areas due to their exceptional destructive capabilities and the potential for collateral damage. However, commanders may use unguided MLRSs to effectively isolate the urban area from outside influence. Commanders may also employ field artillery systems as independent sections, particularly self-propelled systems, in the direct-fire role; decreasing volume and increasing precision of artillery fire helps minimize collateral damage. While discretely applying the effects of high-explosive and concrete-piercing munitions, these self-propelled systems take advantage of the mobility and limited protection of their armored vehicles.

4-38. The urban area may also affect the positioning of artillery. Sufficient space may not exist to place battery or platoon positions with the proper unmasked gun line. This may mandate moving and positioning artillery in sections while still massing fires on specific targets. Commanders must protect artillery systems, particularly when organized into small sections. Threats to artillery include raids and snipers. Therefore, maneuver and firing units will have to place increased emphasis on securing their positions and other appropriate force protection measures.

4-39. The mix of munitions used by indirect fire systems will change somewhat in urban areas. Units will likely request more precision-guided munitions (PGM) for artillery systems to target small enemy positions, such as snipers or machine guns, while limiting collateral damage. Currently, only conventional tube artillery, not mortars, has this capability. However, large expanses of polished, flat reflective surfaces common in urban areas may degrade laser designation for these munitions (as well as attack helicopter PGM). The vertical nature amplifies the geometrical constraints of many precision munitions. Remote designators need to be close enough to accurately designate but far enough away not to be acquired by the PGM during its flight path. PGMs based on the global positioning system (for instance, guided MLRS or the Air Force's joint direct attack munitions) or other optically guided PGMs may be more effective if urban terrain hinders laser designation.

4-40. The urban environment greatly affects the use of nonprecision munitions. Building height may cause variable time fuses to arm prematurely. Tall buildings may also mask the effects of illumination rounds. Units may choose not to use dual-purpose conventional munitions if (similar considerations apply to Air Force cluster bombs)—

- The enemy has several building floors for overhead protection.
- Dismounted friendly units need rapid access to the area being fired on.
- Large numbers of civilians will operate in the target areas soon after combat operations have ceased.

4-41. Depending on the building construction, commanders may prohibit or limit illumination, smoke, and other munitions because of fire hazards. In particular instances, they may specifically use them for that effect. Structure fires in an urban area are difficult to control and may affect friendly units. Conventional high-explosive munitions may work best against concrete, steel, stone, and other reinforced structures. When not used in the direct-fire role, a greater mass of indirect fire is often required to achieve desired effects. Commanders balance firepower and collateral damage since the rubbling caused by massive indirect fires may adversely affect a unit's ability to maneuver and provide a threat with additional cover and concealment.

4-42. Nonlethal weapons, munitions, and devices can help commanders maintain the desired balance of force protection, mission accomplishment, and safety of noncombatants by expanding the number of options available when deadly force may be problematic. As additional nonlethal capabilities are developed, they are routinely considered for their applicability to UO. In determining their use and employment, commanders, in addition to any previous experience at using these weapons, munitions, and devices, consider—

- **Risk**. The use of nonlethal weapons in situations where lethal force is more appropriate may drastically increase the risk to Army forces.
- **Threat Perspective**. A threat may interpret the use of nonlethal weapons as a reluctance to use force and embolden him to adopt courses of action that he would not otherwise use.
- **Legal Concerns**. Laws or international agreements may restrict or prohibit their use (see Chapter 10).
- **Environmental Concerns**. Environmental interests may also limit their use.
- **Public Opinion**. The apparent suffering caused by nonlethal weapons, especially when there are no combat casualties with which to contrast it, may arouse adverse public opinion.

## PROTECTION

4-43. The protection function includes those tasks and systems that preserve the force so that commanders can apply maximum combat power. Preserving the force includes enhancing survivability and properly planned and executed air and missile defense as well as defensive IO (see IO discussion in Chapter 5) and chemical, biological, radiological, nuclear, and high-yield explosive (CBRNE) counterproliferation and consequence management activities (see CBRNE discussion in Chapter 9).

### Survivability

4-44. Survivability in the urban environment is a significant force multiplier. Properly positioned Army forces can take advantage of the increased survivability afforded by the physical terrain. Even a limited engineer effort can significantly enhance the combat power of small Army forces. In stability operations, properly planned and constructed survivability positions can enable small groups of Soldiers to withstand the assaults of large mobs, sniping, and indirect fire. Well-protected support bases are often critically essential to minimizing casualties during long-term stability operations and can become a key engineer task.

4-45. While executing major combat operations or campaigns, in particular defensive operations, well planned and resourced engineer efforts can enhance the survivability characteristics of the urban area. These efforts, though still requiring significant time and materiel, can establish defensive strong points more quickly and with greater protection than can be done in more open terrain. Skillfully integrating the strong point into the urban defense greatly increases the overall effectiveness of the defense disproportionately to the number of forces actually occupying the strong point (see Chapter 8).

4-46. Commanders increase survivability by ensuring that all Soldiers have necessary protective equipment and are trained and disciplined in their use. In addition to standard equipment such as helmets, gloves, boots, and chemical protective overgarments, commanders should ensure, as necessary, availability of other protective equipment and material such as—

- Body armor.
- Goggles or ballistic eye protection.
- Knee and elbow protectors.
- Riot control equipment such as batons, face masks, and shields.

- Barrier material such as preformed concrete barriers, wire, sandbags, and fencing material.
- Up-armored or hardened vehicles.
- Fire extinguishers and other fire-fighting equipment.
- Immunizations.

4-47. The Army's urban survivability operations can become complex if the Army is tasked to support survivability operations for civilians. Such operations can range from constructing civil defense shelters or evacuating the population to assisting the population in preparing for or reacting to the use of weapons of mass destruction. However, Army forces are not organized or equipped to support a major urban area's requirements as well as its own mission needs. Normally, Army forces can render this type of support only as a focused mission using a unique, specially equipped task organization.

## Air and Missile Defense

4-48. The air and missile defense protects the force from air surveillance and air and missile attack. This system uses—

- The careful massing of air and missile defense combat power at points critical to the urban operation.
- The proper mix of air defense weapon and sensor systems.
- Matched (or greater) mobility to the supported force.
- The integration of the air defense plan into the overall urban operation.
- The integration of Army systems with those of joint and multinational forces.

4-49. Properly planned and executed air and missile defense prevents air threats from interdicting friendly forces and frees the commander to synchronize maneuver and other elements of firepower. Even in a major combat operation or campaign, the enemy will likely have limited air and missile capabilities and so seek to achieve the greatest payoff for the use of these systems. Attacking Army forces and facilities promises the greatest likelihood of achieving results, making urban areas the most likely targets for air and missile attack.

### *Rotary- and Fixed-Winged Aircraft*

4-50. Enemy rotary-wing aircraft can be used in various roles to include air assault, fire support, and combat service support. Some threats may use unmanned aircraft systems to obtain intelligence and target acquisition data on friendly forces. Increased air mobility limitations and targeting difficulties may cause enemy fixed-wing aircraft to target key logistics, C2 nodes, and troop concentrations outside the urban area, simultaneously attacking key infrastructure both in and out of the urban area.

### *Increased Missile Threat*

4-51. The intermediate range missile capability of potential threats has increased to be the most likely air threat to an urban area. Urban areas, particularly friendly or allied, make the most attractive targets because of the sometimes-limited accuracy of these systems. By firing missiles at an urban area, a threat seeks three possible objectives:

- Inflict casualties and materiel damage on military forces.
- Inflict casualties and materiel damage on the urban population.
- Undermine the confidence or trust of the civil population (particularly if allied) in the ability of Army forces to protect them.

4-52. If facing a missile threat, commanders conducting UO work closely with civil authorities (as well as joint and multinational forces) to integrate the Army warning system with civil defense mechanisms. Similarly, Army forces may support urban agencies reacting to a missile attack with medical and medical evacuation support, survivor recovery and assistance in damaged areas, and crowd control augmentation of local police forces. Before such an attack, Army engineers might assist and advise urban officials on how to construct shelters.

### *Increased Security of Assets*

4-53. When defending against an air or missile threat in a neutral or hostile urban environment, air defense assets are concerned with security. Separating air defense locations from high population and traffic centers, as well as augmenting these positions with defending forces, can prevent or defeat threat efforts to neutralize them. Additionally, increased density of UO means increased concentration of all friendly and enemy systems engaged in air and counter-air operations. This density may increase friend and foe identification challenges, air space management challenges, and the overall risk in the conduct of air operations. Finally, limited air defense assets, difficulties in providing mutual support between systems, potential mobility limitations, and other effects of the urban environment increase the need for (and effectiveness of) a combined arms approach to air defense (see FM 44-8).

## SUSTAINMENT

4-54. The sustainment function incorporates support activities and technical service specialties, to include maximizing available urban infrastructure and contracted logistics support. It provides the physical means with which forces operate. Properly conducted, the sustainment function ensures freedom of action, extends operational reach, and prolongs endurance. Commanders conducting sustainment to support full spectrum operations must understand the diverse logistic requirements of units conducting UO. They must also understand how the environment (to include the population) can impact sustainment support. These requirements range from minimal to extensive, requiring Army forces to potentially provide or coordinate all life support essentials to a large urban population.

4-55. Commanders and staffs consider and plan for Army sustainment operations that are based in a major urban area. These operations are located in major urban areas to exploit air- and seaports, maintenance and storage facilities, transportation networks, host-nation contracting opportunities, and labor support. These operations are also UO. The commander gains additional factors to consider from basing the sustainment operation in an urban environment. See Chapter 10 for a detailed discussion of urban sustainment considerations.

## COMMAND AND CONTROL

> *Fighting in a city is much more involved than fighting in the field. Here the "big chiefs" have practically no influence on the officers and squad leaders commanding the units and subunits.*
>
> Soviet General Vasili Chuikov
> during the 1942-43 Battle for Stalingrad

4-56. The command and control function is the related tasks and systems that support the commander in exercising authority and direction. The urban environment influences both the commander and his C2 system (which includes INFOSYS). The leader's ability to physically see the battlefield, his interaction with the human component of the environment, and his intellectual flexibility in the face of change all impact the mission. The C2 system faces difficulties placed on the tactical Internet and system hardware by the urban environment, by the increased volume of information, and by requirements to support the dynamic decision making necessary to execute successful UO.

### Unity of Command

4-57. Although severely challenged, the principle of unity of command remains essential to UO. However, the number of tasks and the size of the urban area often require that Army forces operate noncontiguously. Noncontiguous operations stress the C2 system and challenge the commander's ability to unify the actions of his subordinates, apply the full force of his combat power, and achieve success. To apply this crucial principle in an urban environment requires centralized planning, mission orders, and highly decentralized execution. The method of C2 that best supports UO is mission command (see FM 6-0). Mission command permits subordinates to be innovative and operate independently according to clear orders and intent as well as clearly articulated ROE. These orders and ROE guide subordinates to make the right decision when facing—

- A determined, resolute, and adaptive threat.
- A complex, multidimensional battlefield.
- Intermittent or complete loss of communications.
- Numerous potentially hostile civilians close to military operations.
- The constant critique of the media and military pundits.

4-58. Decentralized execution allows commanders to focus on the overall situation—a situation that requires constant assessment and coordination with other forces and agencies—instead of the numerous details of lower-level tactical situations. Fundamentally, this concept of C2 requires commanders who can accept risk and trust in the initiative, judgment, and tactical and technical competence of their subordinate leaders. Many times, it requires commanders to exercise a degree of patience as subordinate commanders and leaders apply mental agility to novel situations.

### Political and Media Impact

4-59. Commanders of a major operation consider how the need to maintain a heightened awareness of the political situation may affect their exercise of C2. A magnified political awareness and media sensitivity may create a desire to micromanage and rely solely on detailed command. Reliance on this method may create tactical leaders afraid to act decisively and with speed and determination—waiting instead for expected guidance from a higher-level commander. Threats may capitalize on this hesitation by conducting operations faster than Army forces can react. Mission orders that express the overarching political objectives and the impact of inappropriate actions, combined with training and trust, will decrease the need for detailed command. Leaders must reduce a complex political concept to its simplest form, particularly at the small-unit level. Even a basic understanding will help curtail potentially damaging political actions and enable subordinates to make the often instantaneous decisions required in UO—decisions that support military and political objectives.

### Commander's Visualization

> *I heard small-arms fire and RPG explosions and felt shrapnel hit the vehicle.... Land navigation at this time was impossible; every time I tried to look out, I was thrown in a different direction.... At this time, I was totally disoriented and had not realized we were on our own.*
>
> Captain Mark Hollis
> "Platoon Under Fire"

4-60. Leaders at all levels need to see the battlefield to lead Soldiers, make effective decisions, and give direction. Sensors and other surveillance and reconnaissance assets alone cannot provide all the information regarding the urban environment that commanders will need. The focus of lead elements narrows rapidly once in contact with a hostile force limiting their assessment to the local area. Therefore, tactical commanders will not be able to observe operations from long, stand-off ranges. Their personal observation remains as critical in urban areas as elsewhere and helps to preclude commanders from demanding their subordinates accomplish a task or advance at a rate inconsistent with the immediate situation. In urban offensive and defensive operations, seeing the battlefield requires that commanders move themselves and their command posts forward to positions that may be more exposed to risk. Thus, commanders modify their C2 system capabilities to make them smaller, reduce their signature, and increase their mobility. Because of the greater threat to C2, security efforts may be more intense.

4-61. In stability operations, commanders often intervene personally to reassure the urban population and community and faction leaders about the intentions of Army forces. In these type operations, threats may attack leaders to gain the greatest payoff with the least expenditure of resources. Commanders carefully evaluate risk and potential benefits of such exposure. These risks however, cannot stop them from seeing the battlefield, personally intervening in situations as appropriate, and leading their Soldiers.

4-62. Commander's visualization also requires having an accurate understanding of friendly and enemy locations, detailed maps, other appropriate intelligence products, and INFOSYS that accurately depict the urban environment and help establish a COP. The reliability of these items is as important to planning major operations as it is to tactical-level operations. The commander of the major operation ensure that subordinate tactical-level commanders have the necessary products to achieve accurate situational understanding and dominate the urban environment as subordinate commands often lack the personnel or assets to develop these products. Frequently, satellite or aerial imagery is requested to compensate for the drastic changes that can occur due to UO, natural disasters, and outdated or imprecise maps. (Even maps developed and maintained by urban area's administrative activities may not be up-to-date. Extensive and continually expanding shantytowns, for example, may not be mapped at all. Maps may have even been purposefully distorted or critical detail intentionally omitted. The systems used to transliterate some languages such as Arabic and Chinese to Anglicized alphabets often result in the same location being spelled several—and frequently considerably different—ways. Maps may also assign names to features that are completely different than those used by locals to refer to them.)

4-63. Other critical intelligence products needed in the COP include overlays or gridded reference graphics. (Whenever possible, gridded reference graphics should conform to standard military grid reference system formats to reduce the probability of error when entering target coordinates into targeting systems that use global positioning systems.) These products should be developed and distributed to all participants prior to the UO. Overall, their focus should be on ease of reference and usefulness for all forces—ground and air (see Appendix B). Overlays and graphics can also portray important societal information or urban infrastructure, such as—

- Religious, ethnic, racial, or other significant and identifiable social divisions.
- Locations of police, fire, and emergency medical services and their boundaries or zones of coverage.
- Protected structures such as places of worship, hospitals, or other historical and culturally significant buildings or locations.
- Underground subway, tunnel, sewer, or water systems.
- Bridges, elevated roadways, and rail lines.
- Electrical generation (to include nuclear) and gas storage and production facilities and their distribution lines.
- Water and sewage treatment facilities.
- Telephone exchanges and television and radio stations.
- Toxic industrial material locations.

## Mental Flexibility

4-64. Commanders conducting UO must remain mentally flexible. Situations can change rapidly because of the complexity of the human dimension. Typical of the change is a stability operation that suddenly requires the use of force. Commanders must be capable of quickly adjusting their mental focus from a noncombat to combat situation. Equally important is dealing with populations during combat operations. Consequently, commanders must also be capable of rapidly adjusting plans and orders for sudden stability or civil support tasks that emerge during or soon after a combat mission. In developing their vision, commanders must consider the second- and third-order effects of UO.

## Information Systems

4-65. The urban environment also challenges INFOSYS that support the commander, especially communications. Urban structures, materials, densities, and configurations (such as urban canyons) and power constraints associated with man-portable radios significantly degrade frequency modulation (FM) communications. This causes problems at brigade-level and below where commanders rely heavily on constant FM radio contact with subordinates. Tactical communication problems might also cause an inability to maintain a COP, to give orders and guidance, to request support, or to coordinate and synchronize elements of the combined arms team. Communication problems in urban areas can prevent the achievement of information superiority and contribute directly to mission failure. In UO, allocating critical or high-value communication assets will be significant and essential to the main effort.

4-66. In an urban environment, units and staffs properly prepare for and mitigate the communication problems in urban areas (see figure 4-3). Adequate communications, in most cases, are ensured by—

- Training in and use of retransmission and relay sites and equipment, which may include unmanned aircraft systems (UAS).
- Airborne command posts, satellite communications, high-frequency radios, and other redundant communication platforms and systems.
- Careful positioning of commanders, command posts, and antennas to take advantage of urban terrain characteristics.
- Detailed communications analysis for movement from one AO to another due to the likely density of units operating in the urban environment.

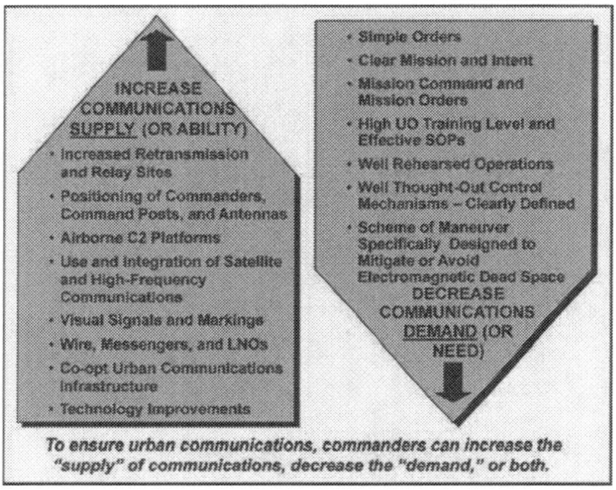


**Figure 4-3. Methods to overcome urban communications challenges**

4-67. Standing operating procedures (SOPs) for visual markings (both day and night) may assist in command and control. These SOPs indicate unit locations and other essential information. They coordinate with units across common boundaries. Given adequate consideration to limitations on multinational capabilities, these SOPs may assist in command and control and preclude fratricide incidents resulting from loss of FM communications. However, visual signals, including pyrotechnics, are less effective in buildings and enclosed spaces.

4-68. In defensive, stability, or civil support operations, positions often do not change as frequently as in offensive operations. Urban commanders then rely more on military wire (properly camouflaged amongst the civilian communications infrastructure), commercial communications, and messengers. Even in combat, some if not all of the urban area's organic communications structure remains intact for Army use. For example, every building may have one or more telephone distribution boxes that can control hundreds of individual telephone lines. Setting up wire communications using these points is relatively simple but, like all wire communications, is susceptible to wire-tapping. Cellular telephones can usually work well in urban areas; however, locating and destroying the repeater stations or other land-based elements of the cellular telephone system (or the effects of natural disasters) can easily disable them. (Cell phones may be a critical and only means to rapidly communicate with key civilian organizations and important community leaders.) Consequently, the C2 system may use these alternatives to FM communications but with proper operations and physical security procedures in place.

**Example of Simple Communications Innovation: Israel's Six-Day War – 1967**

In the 1967 battle for Jerusalem, the Israeli Defense Force placed flags on top of cleared buildings so that aircraft providing close air support could monitor the Israeli forward line of troops. They also used a spotlight during the night to mark specific buildings as close air support targets.

4-69. Units will likely use multiple means to communicate throughout the urban area. Hence, commanders emphasize proper operations security procedures (OPSEC) despite the level of security provided by the communications system. This emphasis helps to lessen the probability that Soldiers will inadvertently compromise essential information as they switch from one mode of communications to another (for example, from secure FM radio to unsecured cellular telephones or from classified to unclassified Internet domains).

4-70. Command posts above brigade-level ensure that they can communicate in an urban area without significant disruption. In stability and civil support operations, immediate and reliable communications between tactical and strategic levels may be necessary. Higher commanders anticipate that although the urban area does not significantly challenge their INFOSYS, the area may severely challenge systems at the lower tactical levels. For this reason, information flow from lower to higher may take longer. If the situation is not acceptable, the higher headquarters takes steps to mitigate it, such as increasing the number of liaison officers operating with units engaged in decisive operations. In some instances, the scheme of maneuver may be specifically designed to account for communications interference, propagation characteristics, and electromagnetic dead space. (However, this will require more time, resources, and a detailed urban communications IPB.)

4-71. Finally, urban areas can overload the INFOSYS with information. UO across the spectrum of operations can generate large volumes of information when crises threaten. This sheer volume can easily overwhelm UO commanders and command posts, and the information conduit connecting the two. Training prepares command posts to handle this volume of information and to filter the critical from the merely informative. Staffs must create products (visual or textual) that help their commanders understand the urban environment, not just present them information to know.

# KEY TACTICAL CONSIDERATIONS

4-72. Commanders and planners of major UO must thoroughly understand the tactical urban battle. Especially, they must understand the effects of the environment on men, equipment, and systems. The complexity of urban environment changes and often compresses many tactical factors typically considered in the planning process. Figure 4-4 shows some of these compressed factors, however, commanders and their staffs should carefully review FM 3-06.11 for doctrine to support tactical urban operations.

**Figure 4-4. Compressed tactical factors**

## Time

4-73. The time available to think and act is compressed in urban combat operations. The tactical engagements that comprise battles and major UO are often quick and decisive; therefore, higher-level commanders require the ability to conduct battle command on the move so that their decision making remains correspondingly fast. The impact of decisions (or lack of) and the outcome of battle can occur in mere minutes. Often the amount of information and the number of decisions can overwhelm the overall ability of INFOSYS to respond. Commanders have little time to influence tactical actions with resources kept in reserve. Reserves and fire support assets are close to the point of decision so that they can respond in time to make a difference. The terrain causes C2 challenges that further inhibit commanders from responding quickly to changes in the situation. Small unit leaders receive training that emphasizes understanding the commander's intent so that they can recognize tactical opportunities and can act quickly to take advantage of them.

## Distances and Density

4-74. Distances in UO are compressed to correspond to the density of threat forces and noncombatants. In open terrain, squads, platoons, and companies may be able to control or influence thousands of meters of space. In UO, large buildings can absorb the efforts of several companies or battalions. Crowds of thousands can assemble in areas of a few hundred meters requiring correspondingly large forces for control. Maximum engagement ranges, as influenced by the urban terrain, are usually closer. Units may require field artillery for direct fire at targets ranging fewer than a hundred meters. Commanders and staffs understand the telescoping nature of the battlefield, the density of threat forces, and the density of noncombatants. In addition to the actual conduct of urban tactical operations, these factors will directly affect training, planning, force deployment, and strength.

4-75. Time-distance considerations are especially important throughout planning cycles. Though distances may be short, the physical nature of the environment can drastically change the planning factors for unit movements. The advance of a battalion may be measured in hundreds of meters per day. Thus, all time and distance calculations that relate to sequencing of forces, synchronizing combat power and other capacities, and making decisions require reevaluation based on the urban conditions.

4-76. Although it may appear counterintuitive, airspace above the urban area may be also be severely compressed as multiple fixed-wing, rotary-wing, and indirect fires (including multinational assets) compete for the same space. Increasing this density is the use of airspace by civilian aircraft and the proliferation of tactical UAS. Due to the potential for a high volume of air traffic, commanders and planers pay close attention to the integration and deconfliction of airspace over urban areas. Commanders should consider specific techniques and procedures for UAS which may include—

- Defining airspace control areas.
- Regulating flight times.
- Creating altitude restrictions.
- Including UAS on the air tasking order.

## Combat Power

4-77. The urban terrain can also compress combat power. This terrain increases the utility and effects of some weapons and systems, increasing overall combat power. One system that dramatically demonstrates this effect is the sniper. In open terrain, snipers slightly influence operations. In UO, snipers—well concealed, positioned, and protected—can take on significance disproportionate to their combat capability in other situations (see FM 23-10).

4-78. The density of ground combat power in a given size area is also increased because of the effect of the terrain on ranges. The complex terrain precludes standoff engagement from extended ranges by dispersed forces. Commanders often position weapon systems closer together and at shorter ranges to mass effects on the same target. Thus, commanders may position armored vehicles, which typically position themselves Hundreds of meters from friendly troops and other vehicles, within a few meters of each other to provide mutual support. Targets, which in open terrain are engaged at thousands of meters, are engaged in tens of meters on the urban battlefield.

4-79. The dense clutter of the urban environment also affects target acquisition. Systems, such as radar optimized for open terrain, will not be able to acquire targets as effectively. Decreased acquisition capability equates to diminished combat power. It may also require increasing the density of acquisition systems to compensate for reduced capability.

4-80. Finally, the density of combat power may also increase the vulnerability of Army forces. Many Army systems are protected from enemy systems at longer ranges. The number of enemy systems that can threaten Army forces at a short range increases dramatically. Lack of dispersal will make it more likely that multiple Army systems can be targeted by a single enemy threat.

## LEVELS OF WAR

4-81. The levels of war are also compressed in the urban area. The tactical actions of individuals and small units can directly influence operational and even national and strategic objectives. Conversely, the decisions of the President can directly affect the conduct of tactical operations. UO have short cause and effect links between the tactical, operational, and strategic levels of operations. Because of the close media scrutiny of UO, the President can sometimes observe the actions of platoons in real time. For example, the media may film a platoon applying nonlethal force for crowd control. The President can view that film on the nightly news before the platoon even disengages from the action, much less reports formally through the various levels of command. If appropriate, the President can decide and direct the strategic and operational commanders to adjust ROE before the platoon has reported. Therefore, commanders at all levels must understand the urban environment's potential compressive effects on the levels of war. A major impact of these effects can be a lower tolerance for tactical errors and a greater need for detailed planning and precision in execution and weapons' effects (lethal and nonlethal).

## DECISION MAKING

4-82. The nature of the urban environment compresses the time available to make decisions and increases the number of decisions to make. This is particularly true at the lower tactical levels. Units observing an urban AO face more potential unknowns than in other situations. A large structure presents many more potential firing positions that are observed than simpler terrain. Movement in one of those windows forces the soldier or unit to quickly make a decision regarding the nature of the target—deciding whether it is a threat or a noncombatant. Incorporating combatant and noncombatant discriminatory considerations into all live-fire training will improve Soldiers' ability to make these critical judgment decisions. Overall, lack of understanding regarding the urban environment requires commanders to rely more on analytic decision making, while greater understanding facilitates adaptability and allows commanders to make rapid and intuitive decisions while still accounting for second, third, and higher-order effects (see Figure 4-5 and FM 6-0). Realistically, however, all decisions made during urban operations will likely require a combination of analytic and intuitive decision-making abilities.

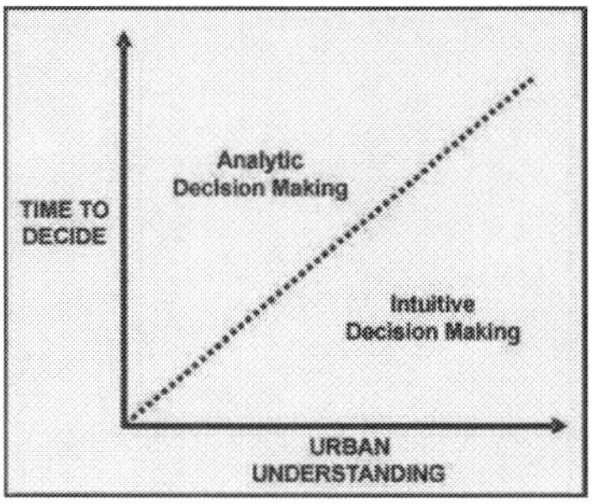


**Figure 4-5. Urban understanding and decision making**

This page intentionally left blank.

# Chapter 5
# Contemplating Urban Operations

*We based all our further calculations on the most unfavorable assumptions: the inevitability of heavy and prolonged fighting in the streets of Berlin, the possibility of German counter-attacks from outside the ring of encirclement from the west and south-west, restoration of the enemy's defence to the west of Berlin and the consequent need to continue the offensive.*

General of the Army, S. M. Shtemenko
describing operational level planning for taking Berlin
*The Soviet General Staff at War*

In any potential situation and in any area, Army commanders will likely need to assess and understand the relevance and impact of one or more urban areas on their operations. They will also need to determine whether full spectrum urban operations (UO) will be essential to mission accomplishment. UO may be the commander's sole focus or only one of several tasks nested in an even larger operation. Although UO potentially can be conducted as a single battle, engagement, or strike, they will more often be conducted as a major operation requiring joint resources. Such actions result from the increasing sizes of urban areas. Army commanders of a major urban operation must then ensure that UO clearly support the operational objectives of the joint force commander (JFC), requesting and appropriately integrating critical joint resources. Whether the urban operation is the major operation itself or one of many tasks in a larger operation, Army commanders must understand and thoroughly shape the conditions so subordinate tactical commanders can engage and dominate in this complex environment.

A ***major operation*** is a series of tactical actions (battles, engagements, strikes) conducted by various combat forces of a single or several services, coordinated in time and place, to accomplish operational, and sometimes strategic objectives in an operational area.

## DETERMINING THE NECESSITY OF URBAN OPERATIONS

5-1. Early in planning, commanders of a major operation must address the necessity and feasibility of conducting operations in urban areas located throughout their areas of operations (AOs). Chapter 1 discussed strategic and operational considerations that compel forces to operate in urban areas. These reasons include the location and intent of the threat force; critical infrastructure or capabilities that are operationally or strategically valuable; the geographic location of an urban area; and the area's political, economic, or cultural significance. Additionally, humanitarian concerns may require the control of an urban area or necessitate operations within it. Several considerations exist, however, that may make UO unnecessary, unwarranted, or prohibitively inefficient. When determining whether to operate in an urban environment, commanders must consider the operational (and accidental) risks and balance them with mission benefits. The factors shown in Figure 5-1 highlight some measures to evaluate the risks associated with UO.

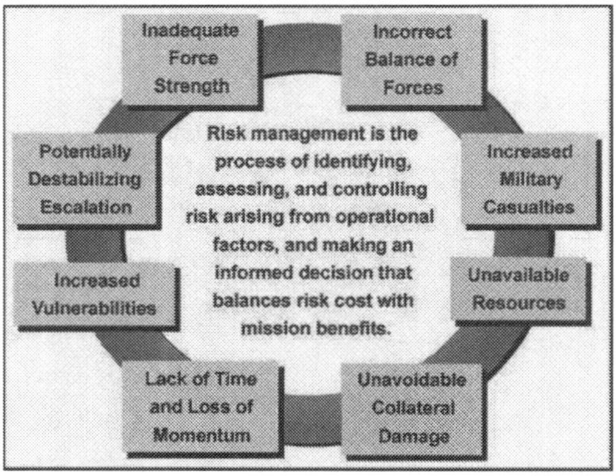


**Figure 5-1. Risk management and the associated risks with urban operations**

## INADEQUATE FORCE STRENGTH

5-2. When facing prospective UO, commanders must consider if they have the necessary troops available to conduct the operation properly and within acceptable risk. Under normal circumstances, large urban areas require many forces merely to establish control. The New York City police department has over forty thousand officers and hundreds of support staff simply to conduct peacetime law enforcement. Major UO, particularly those that are opposed, will often require a significant number of forces. Offensive missions, for example, can require three to five times greater troop density than for similar missions in open terrain. If commanders lack sufficient force to conduct effective operations, they may postpone or consider not initiating those operations until they have the necessary strength. Commanders should also add to their analysis the requirements for troop strength elsewhere in the AO. Additionally, commanders must consider the number (and type) of forces required to make the transition from urban offensive and defensive to stability or civil support operations when determining overall force requirements—not just the number of forces required to realize objectives for major combat operations.

## INCORRECT BALANCE OF FORCES

5-3. Along with force strength, commanders must consider the type and balance of forces available. This consideration includes an assessment of their level of training in urban operations. Generally, UO put a premium on well-trained, dismounted infantry units. The operational or tactical necessity to clear threat forces from the dense urban environment, hold hard-won terrain, and interact with the urban population greatly increase dismounted requirements. Therefore, Army forces conducting UO are often force tailored to include a larger infantry component. In addition, special operations forces (SOF) are invaluable in UO. SOF include psychological operations (PSYOP) and civil affairs (CA) forces. They should always be considered as part of the task organization.

5-4. UO must include combined arms to ensure tactical success in combat. Although masses of heavy forces may not be required, successful UO require the complementary effects of all Army forces. Even if an urban operation is unlikely to involve offensive and defensive operations, armor, combat engineers, and field artillery may be essential to mobility and force protection. In urban stability and civil support operations, successful mission accomplishment requires more robust CA organizations. They are also valuable in urban offensive and defensive operations. While commanders may have sufficient combat and combat support forces, they may lack enough sustainment forces to provide the support necessary to maintain the tempo. Again, commanders without balanced types of forces, to include their proficiency in operating in urban environments, should consider alternatives to UO or delaying UO until proper force types are trained and available in sufficient numbers.

## INCREASED MILITARY CASUALTIES

5-5. Casualties in UO are more likely than in operations in other environments. In urban offense and defense, friendly and threat forces often engage at close range with little space to maneuver. The urban terrain provides numerous advantages to the urban defender; higher casualties occur among troops on the offensive, where frontal assaults may be the only tactical option. Conversely, defenders with limited ability to withdraw can also suffer high casualties when isolated and attacked. Casualties can be more difficult to prevent in urban stability operations because of the dense complex terrain, the close proximity of the urban population, and the possible difficulty in distinguishing friend from foe. The potential for high casualties and the subsequent need for casualty evacuation under difficult circumstances make the positioning and availability of adequate medical resources another important consideration. Additionally, high-intensity urban combat and the potential for increased stress casualties may require additional units to allow for adequate unit rotations so that Soldiers receive the rest they require.

5-6. Though casualties occur in all operations, commanders should recognize the likelihood of more casualties during large-scale or high-intensity UO. During the battle for Hue in 1968, for example, many company-size units suffered more than 60 percent casualties in only a few days of offensive operations. Commanders conducting urban stability operations must know the casualty risk and how it relates to national and strategic objectives. While a lower risk normally exists in stability operations than in offensive and defensive operations, just one casualty may adversely impact the success of the stability mission. A realistic understanding of the risk and the nature of casualties resulting from UO critically affect the decision-making process. If commanders assess the casualty risk as high, they must ensure that their higher headquarters understands their assessment and that the objectives sought within the urban area are commensurate with the anticipated risk.

## UNAVAILABLE RESOURCES

5-7. Offensive and defensive operations in an urban environment put a premium on certain types of munitions and equipment. Forces may want to use vast amounts of precision munitions in the urban environment. At the tactical level, they will likely use more munitions than during operations in other environments. These munitions include—

- Grenades (fragmentation, concussion, stun, riot control, and smoke).
- Mortar ammunition (due to its rate of fire, responsiveness, and high-angle fire characteristic).
- Explosives.
- Small arms.

5-8. Soldiers need access to special equipment necessary to execute small-unit tactics effectively. In urban stability and civil support operations, this equipment may include antiriot gear, such as batons, protective clothing, and other nonlethal crowd-control devices. In urban offensive and defensive operations, special equipment can include sniper rifles, scaling ladders, knee and elbow pads, and breaching equipment such as door busters, bolt cutters, and sledgehammers. Soldiers can conduct UO with standard clothing and military equipment. However, failure to equip them with the right types and quantities of munitions and special equipment will make mission success more difficult and costly. When commanders consider whether to conduct UO, they must evaluate the ability of logistics to provide the necessary resources (see Chapter 10). Considerations must include the ability to supply all Soldiers regardless of branch or military occupational specialty with urban-specific equipment.

## UNAVOIDABLE COLLATERAL DAMAGE

5-9. UO require an expanded view of risk assessment. When considering risk to Army, joint, and multinational forces, commanders must also analyze the risk to the area's population and infrastructure. This comprehensive analysis includes the second- and third-order effects of significant civil casualties and infrastructure damage. Collateral damage can influence world and domestic opinion of military operations and thus directly affect ongoing operations. It also influences the postconflict physical environment and attitudes of the population. Negative impressions of the civilian population caused by collateral damage can take generations to overcome. Destroying an urban area to save it is not a viable course of action for Army commanders. The density of civilian populations in urban areas and the multidimensional nature of the environment make it more likely that even accurate attacks with precision weapons will injure noncombatants. Unavoidable collateral damage of sufficient magnitude may justify avoiding UO, which, though it may be tactically successful, may run counter to national and strategic objectives.

## LACK OF TIME AND LOSS OF MOMENTUM

5-10. Commanders conducting major operations must analyze the time required to conduct UO successfully. UO can be time consuming and can require larger quantities of resources. The density of the environment, the need for additional time to conduct a thorough reconnaissance, the additional stress and physical exertion imposed on Army forces operating in urban areas, and the potential requirements to care for the needs of the urban population consume time and slow momentum. Commanders cannot permit UO conducted as a shaping operation to divert resources from the decisive operation. Nor can they allow UO to interrupt critical time lines, unnecessarily slow tempo, or delay the overall operation. Threat forces may conduct UO with the primary purpose of causing these effects. Commanders must recognize that time generally works against political and military objectives and, hence, they must develop plans and operations to avoid or minimize UO that might unacceptably delay or disrupt the larger, decisive operation. Once commanders achieve major combat objectives, however, they will likely need to shift resources and focus on the urban areas that they previously isolated and bypassed.

## INCREASED VULNERABILITIES

5-11. Commanders must weigh the potential for increased vulnerabilities when executing UO. The density of the environment makes protection (safety, field discipline, force protection, and especially fratricide avoidance) much more difficult. Forces operating in a large urban area increase their risk of isolation and defeat in detail. Joint capabilities, such as air power, may work less effectively to support a close urban battle than in some other environments. Thus, responding to unexpected situations or augmenting disadvantageous force ratios when applying joint capabilities is significantly more difficult. Although organized, trained, and equipped for success in any environment, the Army vulnerability to weapons of mass destruction (WMD) increases when forces concentrate to conduct UO. Commanders may consider not committing forces or limiting the size of a force committed to an urban area because of increased vulnerability to (and likelihood of) attack by WMD.

5-12. Fratricide avoidance is a matter of concern for commanders in all operations. The complex urban terrain and density of participating forces coupled with typical battlefield effects—smoke, dust, burning fires—and weather effects—fog, snow, rain, and clouds—immensely increase the potential for urban fratricide. Additionally, safety requirements make it difficult to fully replicate training conditions that allow leaders to become more aware of the conditions that contribute to fratricide. The effects of fratricide can be devastating to UO and spread deeply within the Army force. Critical effects include—

- Needless loss of combat power.
- Decreased confidence in leadership, weapons, and equipment. These lead to a loss in initiative and aggressiveness, failure to use supporting combat systems, and hesitation to conduct limited visibility operations.
- Disrupted operations and decreased tempo.
- General degradation of cohesion and morale.

5-13. Therefore, commanders should increase fratricide awareness and emphasis on prevention measures during UO. Causes can be procedural, technical, or a combination of the two and include—

- Combat identification failures due to poor situational understanding, lack of communication, failure to effectively coordinate, and short engagement ranges coupled with the need for quick reaction.

- Location errors involving either the target or enemy forces due to poor situational understanding.
- Inappropriate command and control and fire support coordinating measures; a failure to receive, understand, or adhere to these measures.
- Imprecise weapons and munitions effects such as, an antitank round that penetrates several walls before exploding near friendly forces.

Exacerbating these difficulties will be the likelihood of Army forces conducting operations with (or within proximity of) SOF, coalition forces, and indigenous security forces including local police. (During OPERATION IRAQI FREEDOM, the head of the Australian Defence Forces SOF in Iraq made it a point to carefully coordinate with U.S. Army forces—down to the company level—whenever his personnel would be in an American AO. He felt the danger of fratricide was a much greater risk than the loss of operational security.)

## POTENTIALLY DESTABILIZING ESCALATION

5-14. In the urban environment, Army forces cannot avoid close contact with enemy forces and civilians that may potentially become hostiles. In urban stability and civil support operations, commanders should consider the chance of this contact escalating into confrontation and violence, which may become destabilizing. This consideration may delay, limit, or altogether preclude UO using Army forces.

## CONSIDER ALTERNATIVES AND RISK REDUCTION MEASURES

5-15. Since UO are often high risk, commanders should consider courses of action that provide alternatives. When the objective of an urban operation is a facility, commanders should consider replicating that facility outside of the urban area. For example, a critical requirement for an airfield to sustain operations may lead commanders to consider UO to seize or secure one located in an urban area. However, if adequate resources exist (especially time and adequate general engineering support), Army forces may build an airfield outside of the urban area and eliminate the need to conduct the urban operation. Similarly, logistics over-the-shore operations may be an alternative to seizing a port facility. In some situations, the objective of UO may be to protect a political organization such as a government. Relocating the government, its institutions, and its personnel to a safer area may be possible. Commanders can also design an operation to avoid an urban area. For example, if an urban area dominates a particular avenue of approach, use a different avenue of approach. Using a different avenue of approach differs from isolating and bypassing because the entire operation specifically makes the urban area irrelevant.

5-16. If commanders execute UO, they must assess potential hazards, and then they develop controls to either eliminate or reduce the risks to Army forces. The first means to offset risk is always to ensure a thorough understanding of the urban environment and its effects on operations by all members of the force. Other measures to mitigate risk may include—

- Detailed planning to include thorough intelligence preparation of the battlefield and the development of appropriate branches and sequels.
- Integrated, accurate, and timely intelligence, surveillance, and reconnaissance (ISR).
- Clear missions and intent, which includes a well-articulated end state that looks beyond the cessation of combat operations.
- Sufficient reserves and rotation of forces.
- Vigilant physical security precautions to include increased use of barriers and other defenses, particularly when urban areas are used as support areas.
- Operative communications and other information systems (INFOSYS).
- Effective populace and resources control measures.
- Comprehensive and flexible rules of engagement (ROE) continuously reviewed to ensure they remain adequate for the situation.
- Sufficient control measures (which often include a common urban reference system) and standard marking and identification techniques that adequately consider limited visibility concerns for both air and ground forces. Measures should allow commanders to satisfactorily

control UO and minimize fratricide without unreasonably restricting subordinate commanders' ability to accomplish assigned missions. Commanders must ensure that all subordinate units thoroughly disseminate any approved nonstandard reference systems.

- Proper targeting procedures (including effective fire support coordinating measures and a streamlined legal review of targets), positive identification of targets, and controlled clearance of fires. The goal is achievement of precise (yet rapid) effects with both lethal and nonlethal means. In close air support, positive air-to-ground communications are essential to coordinate and authenticate markings.
- Well-synchronized information operations (IO) that begin before introducing Army forces into the urban environment and well through transition. Commanders should emphasize vigilant operations security (OPSEC) particularly when operating closely with the media, nongovernmental organizations (NGOs), and elements of the civilian population.
- Active and effective integrating, synchronizing, and coordinating among all forces, agencies, and organizations involved in the operation. Commanders should allow adequate planning and rehearsal time for subordinates.
- Responsive, sustainable, and flexible urban sustainment.
- Forces well trained in joint, interagency, multinational, and combined arms UO.
- The creation of adaptable, learning organizations. This requires thorough after-action analyses conducted during actual operations as well as after training exercises. In addition to official Army sites (such as the Center for Army Lessons Learned and the Battle Command Knowledge System), commanders must create a unit-level system to allow hard-won, lessons learned and tactics developed to be immediately passed on to other units and soldiers—even in the midst of an operation. This system may be technology-based, procedural, or both. (During OPERATION IRAQI FREEDOM, the 1st Cavalry Division developed an effective web-based knowledge network—called CAVNET—that allowed them to actively capture and share lessons learned among subordinate units.)

# INTEGRATION INTO LAND OPERATIONS

5-17. The commander of the major operation, after determining that urban operations are required, will then integrate the urban operation into his overall operation. He does this by articulating his intent and concept for the urban operation to his subordinates. The commander of the major operation must also set the conditions for successful tactical urban operations by his subordinates. He should define ROE, focus ISR efforts, task organize his capabilities, ensure information superiority, design the operational framework, and initiate and sustain effective coordination with other agencies and organizations (see FM 6-0).

## CONCEPT OF THE OPERATION

5-18. The commander's concept of the operation should address all operationally important urban areas in his AO. It should also articulate his vision of the urban operation through directions to his staff and subordinates. Subordinate commanders should address urban areas that the higher commander does not specifically address. The commander's concept should discuss each urban area in terms of task and purpose (see FM 5-0). The commander should also describes his vision of the situation's end state in terms of—

- The threat.
- The urban environment (terrain, society, and infrastructure).
- Friendly forces.
- The conditions necessary to transition control of urban areas within his AO to another agency or back to legitimate civilian control.

## RULES OF ENGAGEMENT

5-19. National- or joint-level command authorities may develop urban-specific ROE. If not, Army commanders, as part of their assessment, must determine if urban-specific ROE are required for their situation and provide supplemental ROE. However, commanders must forward any conflicts or incongruities to their higher headquarters for immediate resolution.

5-20. Developing effective ROE relies on thoroughly understanding the national and strategic environment and objectives. It also relies on understanding how to conduct urban operations at the tactical level including weapons effects. For example, broad ROE may result in significant collateral damage and civilian casualties. Even in a major operation or campaign, significant collateral damage caused during UO can make postcombat operations difficult. Such damage may even change national and international public opinion or threaten the achievement of national and strategic objectives. In contrast, restrictive ROE can hamper tactical operations causing mission failure, higher friendly casualties, or both. ROE are often part of essential elements of friendly information (EEFI), protected to reduce the potential for threat exploitation. Even in a limited urban operation, ROE will frequently need to change as circumstances warrant. Therefore, commanders should plan ROE "branches" for anticipated changes in the operational environment.

5-21. In urban operations, ROE are flexible, detailed, and understandable. They should preclude the indiscriminate use of deadly force while allowing soldiers latitude to finish the mission and defend themselves. ROE should recognize that the urban area is not homogenous and may vary according to the key elements of the threat and environment: terrain, society, and infrastructure. To be effective, ROE are consistent throughout the force (an increased challenge in multinational urban operations), and soldiers are thoroughly trained and familiar with them.

### Enemy Considerations

5-22. The nature of an urban enemy affects ROE as well. Commanders must consider the type of enemy weapon systems, the degree of defensive preparation, the ability to target enemy vulnerabilities with precision systems, and the ability to distinguish combatant from noncombatant.

### Terrain Considerations

5-23. ROE may vary according to the terrain or physical attributes of an urban area. Physical factors may drive the ROE to preclude certain types of munitions. For example, if the construction of a portion of the area is sensitive to fire, then ROE may preclude using incendiary munitions in that area. The ROE may lift this prohibition when units move into areas of masonry construction. Toxic industrial chemicals or radiological contaminants in an industrial area may also affect ROE.

### Societal Considerations

5-24. The societal or human dimension of the urban environment will often affect ROE the most. Commanders must base the ROE development on a thorough understanding of the civilian population and threat. They evaluate the loyalty of the population, its dynamic involvement in activities that affects the operation, and its size and physical location. A population that is present and supports Army forces will likely elicit more restrictive ROE than a hostile population actively supporting forces opposing the Army forces. A neutral population, not actively involved in behavior affecting Army forces, supports consideration of more restrictive ROE. In all cases, ROE conforms to the law of war. However, ROE may be much more restrictive than the law of war requires.

5-25. The location of the population also affects ROE. The evacuation or consolidation of noncombatants into controlled, safe areas may result in less restrictive ROE. A U.S. or allied population that remains in the urban area conducting routine business in and amongst Army forces during noncombat UO will normally require the most stringent ROE.

### Infrastructure Considerations

5-26. Commanders must consider the urban infrastructure when developing ROE. Urban infrastructure vital to current or future Army operations may dictate that commanders adjust ROE to ensure that critical elements of the infrastructure remain intact during the conduct of operations. If Army forces conduct an urban operation to capture port facilities, for example, the ROE should address damage to the key facilities that are the objective of the operation.

## RESOURCE ALLOCATION

5-27. Commanders of a major operation must ensure that subordinate tactical commanders have the resources necessary to conduct UO effectively. They assign appropriate forces to subordinate commanders tasked to conduct UO; support them with Army forces at the operational level; and request and coordinate their support by joint resources.

### Task Organization

5-28. Task organizing subordinate units for urban operations will depend largely on the nature of the operation. Some units, however, are always part of the task organization to ensure the success of UO. Infantry, CA, aviation, military police, PSYOP, military intelligence, and engineers are units required for all urban operations across full spectrum operations. Other type forces—such as armor, artillery, and chemical—have essential roles in specific types of urban operations but may be less applicable across the spectrum of operations. Commanders and staffs of a major operation must understand their mission, the particular urban environment in which they operate, and the general effects of the environment across the warfighting functions to allocate the appropriate forces to their tactical commanders. See Chapter 4 for details.

### Operational-Level Support

5-29. Commanders of a major operation should also support the tactical commander with forces remaining under their direct control. These forces can include Army SOF, such as CA, PSYOP, and Special Forces, ground and air cavalry, aviation, logistics, engineers, and communications support. These forces may not be under operational control of the supported command, but their efforts are synchronized and coordinated.

### Coordinating and Requesting Joint Support

5-30. Commanders of a major operation will often provide forces to the JFC as well as receive assets. They will also coordinate for and integrate joint assets to support the tactical battle. These assets will usually include air support, such as close air support, tactical airlift, and aerial reconnaissance and surveillance. Intelligence support comes in the form of reach to strategic and national intelligence capabilities and to space-based systems. This reach to space assets provides reliable, robust long-range communications, environmental monitoring, ISR, positioning and navigation, and warning of enemy missile launch. Special operations capabilities can assist the tactical mission with special reconnaissance and direct action (including special operations aviation) against high-payoff targets. Joint resources also provide the Army forces augmentation by Marine ground forces. In coastal areas, Navy forces and Coast Guard elements assist Army forces with security, sealift, and fire support. Additionally, the potential agility, reach, and power projection provided by Navy forces and platforms, combined with their self-sustaining capability, can play a vital role in supporting UO. This is especially true when UO are conducted in the littorals, require forcible entry operations, and when host nation infrastructure is severely taxed or destroyed. Commanders of a major operation will coordinate with the JFC regarding available joint resources and their allocation. They then ensure that their efforts coordinate with and complement those of tactical Army forces in the urban area. Appendix D discusses the potential contribution of joint capabilities to Army UO.

## URBAN ISR

5-31. Commanders at all levels require accurate and timely information to conduct assessments for successful urban operations. This is critical to planning and execution. Senior commanders have a large role in coordinating the urban ISR effort. National strategic sources (as well as open sources) provide most of the information that commanders and staffs require on the characteristics of the human dimension, the physical properties of the terrain, and the infrastructure. The general characteristics of these aspects of the urban environment do not change drastically over time, with one exception. Military operations or natural disasters can change physical characteristics radically. Analysts can obtain crucial information through diligent research of intelligence databases and open sources. However, the disposition and composition of the urban threat is time sensitive and not likely to be discovered through this type of investigation. Due to the effects of the urban environment, deceptive efforts may influence the threat more easily. The urban population is dynamic and critical aspects of this dimension must be updated or confirmed as a prelude to urban operations. Surveillance and reconnaissance (including geospatial assets) provide relevant information regarding threat dispositions, composition and the state of the population, and the specifics of the urban terrain. Successful urban operations depend on the successful conduct of urban reconnaissance. (See also the earlier discussion of effects on the intelligence and command and control warfighting functions in Chapter 4. While counterintelligence (CI) is an essential ISR component, it will be covered later in this chapter under IO.)

### Challenges

5-32. The most significant challenge to urban ISR is physical. The physical organization and complexity of the urban terrain, both man-made and natural, challenges national strategic, operational, and tactical ISR capabilities. Commanders should understand the challenges when planning and allocating time and resources to their ISR efforts. They acknowledge that subordinate commanders will face similar challenges. Therefore, commanders consider subordinate capabilities, limitations, and needs when planning, requesting, allocating, and prioritizing ISR assets and capabilities.

#### *Imagery Capabilities*

5-33. A significant national, strategic, and operational ISR capability is imagery. Even at the tactical level, units employ unmanned aircraft systems (UAS) and vehicle and dismounted electro-optics, commercial cameras (still and video), and co-opt existing civilian surveillance systems to gain situational understanding. However, the structures of the urban area significantly degrade the information that imagery acquires and may make it susceptible to physical deception measures.

5-34. Imagery, however, is an excellent source regarding the arrangement and nature of many other physical aspects. It can be used to create updated mapping products and provide significant detail of major portions of the infrastructure. Imagery can also reveal what may be happening in structures through detailed study of patterns and other exterior indicators. Yet, the bulk of a skillful threat's forces, well positioned and concealed inside or underneath structures in the urban area, are largely immune from rapid detection by overhead imaging systems. The volume of civilian movement in an urban area will itself provide a degree of camouflage and increase the difficulty of employing pattern analysis. The success in 1999 of the Yugoslavian army concealing heavy forces when confronting NATO indicates the limits of these assets to penetrate an urban area. This leaves a requirement to supplement overhead imagery in the urban environment with a reliable vehicle-mounted or handheld imaging system. For example, thermal sights on a vehicle patrolling an urban street during limited visibility may note the hot engine of a vehicle on the side of the road, possibly indicating suspicious activity. Simply providing patrols with digital cameras or video cameras can also help provide needed information and fill in information gaps.

5-35. The requirement for high altitude and low oblique imagery in more open terrain has been supplemented by a growing need for detailed, close-up, and nearly vertical overhead shots of specific portions of the urban area. This forces collection managers and system operators to move expensive and scarce imagery collection systems closer to the target or to employ flight profiles that place them at greater risk from threat air defense systems. While gun-camera images from aircraft can provide a stand-off reconnaissance platform, UAS imagery may be one of the fastest, least risky methods by which commanders can conduct reconnaissance to update and verify current maps of the urban area and determine routes, obstacles such as damaged and destroyed buildings, and intact and destroyed bridges and other significant infrastructure. Increased UAS proliferation will require greater attention to airspace command and control procedures as well as the movement and protection of ground control stations.

5-36. As the media will likely be present in the urban environment, commanders may consider devoting intelligence resources (to include linguists) to monitoring the various local, national and international sources, particularly the medium of television. The media coverage of threat forces may provide important intelligence and relevant information of immediate tactical value. Commanders, however, should always be cognizant that threats may recognize that Army forces will be using this as a source of intelligence and, feeling no compulsion to be truthful, may use it for their own deception means.

### *Electronic Capabilities*

5-37. The physical attributes of the urban area also diminish the effectiveness of electronic ISR capabilities. Buildings and other structures significantly disrupt radio communications in an urban area. Buildings not only make tactical radio communications difficult for the user, they also make them difficult to locate, intercept, and jam. The range and clarity of frequency modulation (FM) signals significantly diminish when antennas are located inside buildings or when buildings block line of sight between the source and receiving station. To mitigate this effect on signals intelligence means, detection capabilities often move closer to the transmission source. Thus, the threat's vulnerability to compromise by means of his FM and other wireless communications in an urban environment is much less than in many other environments.

5-38. Without losing tactical surprise and increasing risk, units cannot effectively use many electronic detection and surveillance capabilities until urban combat is imminent or perhaps already begun. Ground surveillance radars will have limited uses in urban areas because of the lack of wide-open spaces in which they best operate, but can be used along the periphery of urban areas to detect infiltration and exfiltration of threat forces. However, remote systems such as the remotely monitored battlefield sensor system and the platoon early warning device may have great utility in monitoring many of the numerous avenues of approach that cannot be kept under human surveillance due to manpower constraints. Among many urban applications, remote sensors and early warning devices can be used to monitor subsurface passageways, the entrances and exits to buildings, fire escapes, the perimeters of sustainment bases, and covert foot trails between urban and rural areas.

### *Human Capabilities*

5-39. The limits on imagery and electronic ISR capabilities place a premium on human-based visual reconnaissance. Commanders have three types of human reconnaissance assets to augment electronic reconnaissance resources: special reconnaissance, conventional combat reconnaissance, and human intelligence (HUMINT) gathered by military intelligence from individuals and multimedia. The urban environment poses several challenges to these capabilities.

5-40. The urban area challenges special reconnaissance in several ways. First is the access to the urban area. Although avenues of approach may be numerous, concealed avenues of approach into a defended urban area may be limited and thoroughly covered. Air access is also more difficult because aircraft are detected more easily, airspace is smaller, drop and landing zones are limited or not secure, and more air defense systems probably exist. Still, special reconnaissance efforts to penetrate the urban area can be successful using unconventional techniques including high-altitude low-opening parachutes or underwater penetration.

5-41. Special reconnaissance then faces a second challenge: moving in and identifying targets in the urban area. Stealth movement in an occupied urban area is exceptionally difficult. Repositioning to new or alternate positions is also dangerous. The soldiers' ability to conceal themselves among the civil population can mitigate some of these challenges but includes inherent risks of a different nature. Also difficult is establishing observation positions that provide a field of view of several targets.

5-42. Lastly, special reconnaissance may face navigational and reporting challenges. Special reconnaissance's ability to locate themselves and communicate critical locations and routes are challenged by—

- Differences in language and numbering systems. Road signs, if they exist at all, may be removed or deliberately altered to confuse.
- Irregular street patterns.
- Outdated maps.
- Intervening structures that impede communications and global positioning systems.
- Changes to the landscape due to the effects of UO or natural disasters.
- Featureless shantytowns.

5-43. Conventional reconnaissance faces many of the same challenges as special reconnaissance. Conventional reconnaissance also may lack the advantage of surprise and the special equipment and training that provides special reconnaissance stealth capability. Conventional reconnaissance is not likely to operate undetected by the civilian population. Given the constraints discussed above on other sources, conventional reconnaissance units will likely begin their mission with much less information than they would have on threat dispositions in a less complex environment. Commanders may choose to have their reconnaissance elements fight for information in the urban area. While this high-risk option is more favorable under fluid conditions, it can be used at any time. It requires careful planning, rehearsal, and formulation of information requirements.

5-44. As previously indicated, human intelligence will be one of the most valuable sources for information regarding the situation inside an urban area. HUMINT may take advantage of the proximity and large numbers of potential informants to gather information about threat activities and capabilities. It is especially valuable because it can address all elements of the environment. HUMINT sources can describe political and religious nuances that may be significant to commanders. Such information is useful for insights regarding the human dimension but extremely difficult to obtain from other means. This intelligence also can describe the infrastructure uncovering essential details of how the infrastructure functions. Obtaining good HUMINT requires skilled interrogators and linguists—often in large numbers. Commanders must know and account for some of the possible shortcomings of HUMINT:

- It is susceptible to the influence of the threat; the threat can threaten and influence the source as well as local interpreters. Commander may need to consider that people may not want to talk through a local interpreter because they are providing information they might not want others to know; anonymity may be essential. On the other hand, a local interpreter may provide community access. Yet, those trusted by the locals may not be truthful with Army forces. Conversely, those truthful with Army forces may not be trusted by the indigenous population particularly if these interpreters come from another country, region, or demographic group. In such cases, commanders may need more than one interpreter.
- It is limited by the accuracy of the source and interpreter's perceptions. A local interpreter may have difficulty understanding and translating military topics or the intricacies of urban infrastructure. Some local translators may want to please both participants, so they may shade the conversation, telling each side what they want to hear. Finally, accuracy may be affected by the collector's own perceptions if not firmly grounded in an understanding of social, cultural, and other local conditions.
- Multiple echelons collecting from a given area may inadvertently collect from the same sources and thereby give the false impression of independent sources confirming the same data. To help mitigate this possibility, commanders should ensure that all HUMINT sources are properly recorded on intelligence source registers.
- It may not be timely. The process of identifying and cultivating a source (particularly in an environment where most civilians support threat forces), gathering information, analyzing the information, and providing the intelligence to commanders can be extremely time consuming.
- Some informants may come from unscrupulous or sordid elements of the urban society and may also have their own agenda. They may attempt to use protection afforded them by their relationship with Army forces to conduct activities (even atrocities) that will compromise

political and military objectives. In some cases, a local interpreter's own agenda may influence his translation. (For example, a local schoolteacher used as a translator may report to the collector that the source believes that repairing schools in their urban area is critical, when actually the source may have said that potable water is most needed.)

## Conducting Urban ISR

5-45. To be successful, ISR efforts (national to tactical level) are exceptionally comprehensive and synchronized. Moreover, success necessitates the integration of all ISR sources into operational and tactical planning. This requires that ISR assets deploy and execute early, diversify, properly focus, and integrate under a comprehensive ISR plan. Successful ISR also requires flexibility to adapt to the operational and tactical needs of the commander (see Figure 5-2). Commanders must ensure that the appropriate intelligence architecture (including robust links to joint and multinational elements) is established and tested prior to execution of ISR effort.

### *Early Deployment*

5-46. One of the first requirements for effective urban ISR is the early deployment and employment of assets. The complex urban terrain presents a significant challenge. Commanders should consider that ISR assets will normally take longer to gather data amid the complexity.

5-47. Limited national, strategic, and operational imagery intelligence (IMINT) and signals intelligence (SIGINT) capabilities should be considered and quickly requested. If they are approved, they are tasked and deployed or repositioned to begin urban ISR operations. This takes time. Fortunately, spacing the ISR effort over time permits the analysis of the information or data as it is received. Such time also permits subsequently refining the ISR effort before all assets are committed.

5-48. SOF or conventional units will require significantly more time to execute reconnaissance missions and maintain an acceptable survivability rate. Urban reconnaissance operations require additional time for stealthy insertion into the urban area. IMINT and SIGINT capabilities are used to identify possible locations of high-value targets and corresponding observation positions; this helps minimize time-consuming and high-risk repositioning in the urban area. Again, reconnaissance units may require extensive time to observe from observation positions for indicators of threat activity and disposition and identify patterns.

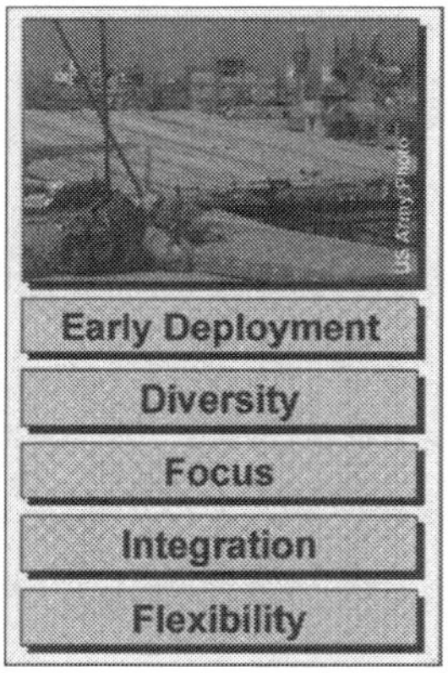


**Figure 5-2. Urban ISR considerations**

5-49. As conventional combat forces prepare to commit to the urban area, conventional reconnaissance precedes their actions. Conventional reconnaissance will often be a slow and methodical effort. Such forces may need time to reconnoiter the interior of structures for snipers and other small threat teams. They may also need time to deploy and destroy snipers and small delaying elements and to breach harassing obstacles. If necessary, they may need time to mass the combat power necessary to fight through enemy security and continue the reconnaissance.

### *Diversity*

5-50. No single ISR capability can solve the riddle of the urban defense. In order to gain an accurate common operational picture of the complex urban terrain necessary to focus combat power on decisive points, commanders must employ diverse ISR capabilities. These capabilities will each contribute pieces of relevant information to permit identifying operational objectives and leveraging tactical combat power to achieve those objectives quickly. Higher-level commanders should understand that tactical reconnaissance capabilities alone often cannot provide all the tactical information required for success at lower echelons.

5-51. Using diverse capabilities challenges the threat's ability to conduct deception and otherwise defeat the friendly ISR effort. A threat that focuses on minimizing his vulnerability to overhead imagery (for example, satellites) may increase his reliance on communications and thus his vulnerability to SIGINT. At the same time, he may decrease his ability to detect the actions of ground reconnaissance units. Conversely, a threat that actively campaigns to detect ground reconnaissance may make himself more vulnerable to SIGINT and IMINT. Moreover, as Army forces enter urban areas, threat activity and susceptibility to SIGNINT often increases.

5-52. Diverse capabilities also facilitate the tactical ISR effort. Tactical reconnaissance units often consist of small dismounted teams and small combined arms teams with a dismounted element and an armor-protected mounted element. Combat engineers with their technical expertise and breaching capability are essential to the combined arms reconnaissance effort. The teams' movements are synchronized and coordinated with other assets, such as UAS and air cavalry reconnaissance. These teams use several movement techniques including infiltration, with the primary objective of conducting zone reconnaissance along key axes that support brigade and battalion actions against decisive points. To accomplish this mission, reconnaissance reconnoiters the proposed routes and alternate approaches. This supports deception and contingency planning. (Reconnaissance of alternate routes and objectives also applies to aerial reconnaissance; helicopters and UAS are not invisible. They can alert a threat to impending operations.) Infiltration of dismounted reconnaissance is made easier when a threat focuses on combined arms reconnaissance teams. Aerial reconnaissance, such as air cavalry and UAS, provides early warning of threat elements to ground reconnaissance, identifies obstacles and ambush sites, and helps select the routes for ground reconnaissance. Air elements may also reduce the mobility of counterreconnaissance forces.

### *Focus*

5-53. Another key to successful ISR is the ability to focus the assets on commander's critical information requirements (CCIR). This focus begins with mission analysis and the commander's initial planning guidance. It is incrementally refined throughout planning and execution as each ISR effort provides information and permits more specific focus in subsequent efforts. The size and complexity of the urban environment require that the ISR effort center strictly on decisive points or centers of gravity (COGs). Therefore, the overall ISR effort will have two major focuses. The first is to uncover and confirm information on the decisive points and COG. The second is to determine the approaches (physical, psychological, or both) leading to the decisive points. The first focus will likely drive ISR in support of major operations. The second focus will likely provide the impetus for tactical ISR efforts. For example, special operations reconnaissance might focus on a major command center that controls the entire urban area and that is one of the higher-echelon command's CCIR. Tactical reconnaissance might focus on the nature of the defense along a particular avenue of approach to the objective.

### *Integration*

5-54. Another important aspect of urban ISR is integration. All reconnaissance capabilities provide both distinctive information as well as information that confirms and adds to that coming from other sources. Essential to urban ISR is the link between all of these sources, either directly or through an integrating headquarters.

5-55. ISR operations must be vertically and horizontally linked. Vertical links ensure that ISR operations among the various levels of command are complementary and that the information flow between these levels is rapid. Horizontal links ensure that forces operating in close proximity (particularly adjacent units), where areas of interest overlap, can rapidly share results of their individual ISR efforts. Together, this helps ensure that all Army forces share a common operational picture and permits the greatest flexibility and survivability of ISR resources.

5-56. ISR operations also are integrated into the planning system, especially the targeting process. As part of targeting, positioned reconnaissance and surveillance elements may become the trigger and terminal control for applying precision fires when appropriate and after considering the risks of compromise of the position or platform. ISR operations and plans must also be synchronized with the air tasking order, as many surveillance platforms are coordinated on the same timeline.

### *Flexibility*

5-57. The urban ISR effort must be more flexible than in other operations. This flexibility permits the ISR effort to meet unforeseen circumstances and to deal with the challenges of the urban environment. As indicated previously, the urban environment is particularly difficult to penetrate. The practical effects of this characteristic are that—

- The initial ISR effort may not be as successful as in other operations.
- More intelligence requirements may be discovered later while executing ISR operations than otherwise.
- The threat may be more successful in active counterreconnaissance because of the concealment advantages of the urban environment (hiding in structures as well as among the urban population).

Therefore, tactical and operational commanders should consider requesting greater than usual ISR support from higher headquarters. Higher headquarters should be proactive in augmenting units conducting urban operations with additional ISR assets. Additionally, ISR assets remaining under the control of the higher headquarters must respond more quickly to the CCIR of supported commanders. Sequencing reconnaissance missions over time provides flexibility by creating uncommitted reconnaissance assets.

5-58. Time sequencing of ISR assets is essential to flexibility. It makes ISR assets more survivable and allows the intelligence cycle to mature the CCIR. It also creates a ready ISR capability to augment committed forces in critical areas if required or diverts them around centers of threat resistance. If not required, the original ISR tasks can be executed as envisioned in planning. Cueing allows a high-value ISR asset to be able to respond to multiple targets based on an ongoing assessment of the overall reconnaissance effort and the changing CCIR. Redundancy permits the effort to overcome line of sight restrictions, the destruction of an ISR asset, and the ability to combine ISR resources to create combat power if required. Maximizing the ISR effort requires applying all available ISR assets to support the urban operation. Additionally, assets—such as air defense artillery and field artillery radars and engineer reconnaissance teams—are integrated into the ISR effort. In urban operations, units will also commit infantry and armor elements (plus their organic reconnaissance elements) into the tactical reconnaissance effort. These units increase the dismount capability and the ability of reconnaissance elements to fight for information and fight through security zones.

## INFORMATION OPERATIONS

5-59. Information operations are an integral part of all Army operations and a critical component in creating and maintaining information superiority. The information environment is the sum of individuals, organizations, or systems that collect, process, and disseminate information; it also includes the information itself. In UO, the information environment is extremely dense due to the proliferation of INFOSYS and widespread access to those systems. In urban operations, commanders must consider how the urban environment, particularly the human component, uniquely relates to executing IO. They must also anticipate the threat's information operation campaign and preclude successful operations on its part. Overall in regard to the urban society, commanders must convince the urban inhabitants of the inevitable success of UO, the legitimacy of Army actions, and the beneficial effects that Army success will eventually generate.

5-60. Because urban operations are likely to be joint, interagency, and multinational, commanders must guard against various forms of information fratricide as a multitude of actors plan and conduct IO in the urban area. For example, the lead governmental agency's key IO objective may be to establish the legitimacy of newly formed civilian authorities. In an attempt to develop close relationships with the civilian populace, Army commanders may continue to work closely with traditional, informal leaders to the exclusion of the new authority. These actions, while they are often conducted out of practical and immediate necessity, may run counter to the lead agency's goal. Overall, Army commanders must nest their IO campaign objectives and themes within those of the lead agency, aggressively coordinate with other governmental agencies and coalition partners, and synchronize activities down to the tactical level to prevent working at odds and avoid information fratricide. However, the process must not be so centralized and rigid that subordinate units lose the flexibility necessary to develop their own products and themes essential to addressing their own AO and the particular situation that confronts them.

5-61. IO are executed using core and supporting elements and related activities (see Figure 5-3 and FM 3-13). The elements of IO are employed in either an offensive or defensive role. Many aspects of IO are not affected differently in an urban environment from any other environment. The following sections outline some IO considerations unique to urban operations.

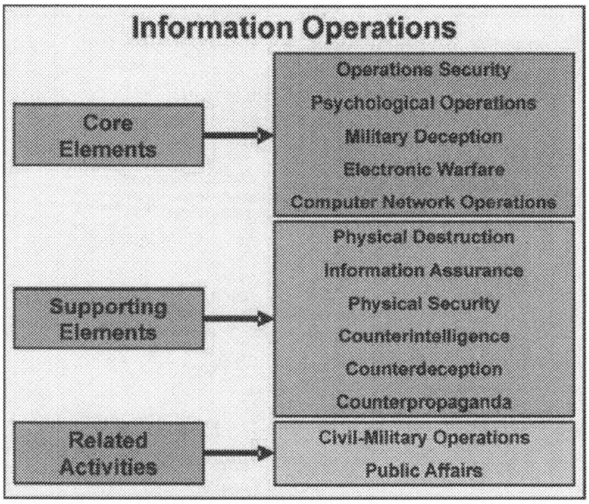


**Figure 5-3. IO elements and related activities**

## Operations Security

5-62. In the urban environment, Army forces can leverage existing urban infrastructure, including the communications and information infrastructure, to enhance Army operations. The danger in integrating these systems is violating OPSEC. Commands ensure that Army forces use only approved systems and proper safeguards exist. Commands also supervise subordinate units for inadvertent breaches of OPSEC policies when using existing urban systems. Finally, established OPSEC procedures must be reasonable and practical. Overly complicated and stringent procedures increase the likelihood that unintentional OPSEC violations may occur as Army forces seek to rapidly accomplish the mission.

5-63. Of particular concern are computers (see also the discussion of Computer Network Operations below). During longer-term UO, commanders will constantly be upgrading and improving the quality of life for their soldiers through the establishment of well-resourced forward operating and sustainment bases. This will likely include the provision of unclassified access to the Internet and e-mail allowing Soldiers to communicate with friends and families around the world. E-mail addresses, facsimile numbers, cell phone numbers, photographs, and other sensitive (but unclassified) information can be valuable sources of threat information. Soldiers must be periodically trained and reminded of everyday procedures to combat potential threat exploitation of these unclassified sources (such as not opening e-mails from unknown sources as they may harbor Trojan programs and viruses).

5-64. To compound concerns, commanders will also establish protected systems to disseminate tactical lessons learned to individual soldiers. This creates the potential that Soldiers can inadvertently release classified information about ongoing operations to threat intelligence systems. In addition to OPSEC training and persistent awareness, commanders must establish procedures to shutdown unclassified Internet and e-mail access immediately, both before and after critical events, to mitigate potential OPSEC violations. (Commanders may also need to randomly shutdown Internet and e-mail access as part of their overall OPSEC measures to preclude such information blackouts serving as a potential warning of impending operations.) Technical surveillance countermeasures—the identification of technical collection activities conducted by threat intelligence entities—will be an important CI service to identify this and other technical OPSEC concerns.

5-65. The close proximity of Army operations to a civil population, particularly in stability and civil support operations, makes Army activities themselves an additional OPSEC concern. Hostile civilians or other threats integrated into the urban population may have more chances to observe Army activities closely. Such observations can provide insight into tactics, techniques, and procedures (TTP) and expose operational vulnerabilities. However, threats may coerce even friendly civilians to provide a threat with EEFI. Therefore, commanders in an urban environment must ensure that civilians cannot observe critical TTP. Any observable patterns and TTP vary and are supplemented with deception efforts. Physical security is increasingly important in urban areas to control civilians' access. Although many urban operations require close coordination with NGOs, commanders should screen information provided to these organizations to protect EEFI. Release of EEFI to NGOs is controlled and done with full recognition and understanding of potential consequences—the benefits must far outweigh the risks involved. (Even the best-intentioned NGO might inadvertently compromise security. For example, by revealing that a commander has closed a particular route the next day, the NGO may unwittingly reveal the objective of a pending offensive operation.)

## Psychological Operations

5-66. PSYOP aim to influence the behavior and attitude of foreign audiences, both military and noncombatant, in the urban environment. PSYOP are a force multiplier and contribute in many ways to mission success (see FM 3-05.30). Effective PSYOP focuses on transmitting selected messages to specific individuals and groups in order to influence their actions. Their ability to influence the attitudes and disposition of the urban population cannot be overstated. While the complexity of the societal component of the urban environment can make PSYOP challenging, the urban society also offers many options and resources. Potentially, PSYOP (with other political and economic actions) may help limit or preclude the use of military force in urban areas. In some circumstances, UO may be relevant to the major operation only in terms of their psychological effect.

5-67. The positive influence created by PSYOP is often useful in developing an effective HUMINT capability particularly in an urban area where many civilians actively or passively support the threat. Persuading and influencing a few to support friendly forces may pay great dividends. These few supporters may allow Army forces to penetrate the urban area and obtain essential information. Such information can apply to threat capabilities, threat intentions, and even the urban environment itself.

5-68. PSYOP, combined with other elements of offensive IO, aid in isolation of a threat—a critical shaping action for any urban operation. For example, commanders may use PSYOP to inform civilians about new food distribution points located away from urban combat operations. This action supports the UO fundamental of separating combatants from noncombatants and helps to further isolate the threat (both physically and psychologically) from the civilian populace. Aside from projecting a positive image of friendly forces over threat forces, PSYOP can also isolate the threat by identifying and exploiting ethnic, cultural, religious, and economic differences between the elements of the civilian populace and threat forces as well as the differences among supportive and unsupportive civilian factions. The complexity of the urban environment enables quick changes in opinion or attitude. Commanders must continually evaluate the results of PSYOP for mission relevance.

## Military Deception

5-69. Urban operations present numerous challenges to tactical commanders; however, higher-level commanders may help to mitigate some challenges through effective military deception Commanders can use military deception efforts designed to mislead threat decision makers as to friendly force disposition, capabilities, vulnerabilities, and intentions. Military deception actions may allow commanders to achieve tactical surprise or improve relative combat power at a selected location. For example, allowing the threat to observe certain activities on a selected avenue of approach may cause the threat to shift his forces (and effort) to the area perceived to be threatened. (This movement may also aid in determining the overall disposition of threat forces and intentions.) Repositioned forces or effort to activities or locations that are not decisive to the achievement of friendly objectives, combined with other IO designed to overwhelm threat information and intelligence systems, may create the force and tempo differential necessary to achieve success. Commanders should tailor urban deception plans to the specific urban area, paying close attention to the societal characteristics of the target population.

## Electronic Warfare

5-70. Electronic warfare (EW) includes all actions that use electromagnetic or directed energy weapons to control the electromagnetic spectrum or to attack a threat. Conducting EW in urban areas seeks to achieve much the same results as in other environments. A major consideration in urban areas is collateral effects on portions of the urban infrastructure that rely on the electromagnetic spectrum for service. Thus, precision is a major factor in planning for EW operations. For example, EW attacking a threat's television broadcasts avoids affecting the television broadcasts of neutral or friendly television. Likewise, EW attacking military communications in a large urban area avoids adversely affecting the area's police and other emergency service communications. Urban offensive and defensive operations will have the least restrictions on EW operations while urban stability may have significant constraints on using EW capabilities.

## Computer Network Operations

5-71. Computer network operations (CNO) include computer network attack (CNA), computer network defense (CND), and computer network exploitation (CNE). CNO are not applicable to units at corps and below. Echelons above corps (EAC) units will conduct CNA and CNE. If tactical units require either of these network supports, they will request it of EAC units.

### *Computer Network Defense*

5-72. In urban operations, CND will require extreme measures to protect and defend the computers and networks from disruption, denial, degradation, or destruction. The nature of the urban environment and configuration of computer networks provides the threat with many opportunities to interdict local area networks (LANs) unless monitored by military forces. LANs controlled by military forces are normally more secure than the civilian infrastructure. Commanders should prepare for opportunities by the threat to insert misinformation.

### *Computer Network Attack*

5-73. Considerations regarding the execution of CNA in urban operations are similar to those of EW: CNAs that do not discriminate can disrupt vital civilian systems. However, possible adverse effects on the civilian infrastructure can be much larger—potentially on a global scale. In the short term, CNAs may serve to enhance immediate combat operations but have a debilitating effect on the efficiency of follow-on urban stability operations. Because of these far-reaching effects, tactical units do not execute CNA. CNA is requested of EAC units. EAC units will receive all requests from lower echelons, carefully consider second- and third-order effects of CNA, and work to ensure its precise application.

### *Computer Network Exploitation*

5-74. CNE consists of enabling operations and intelligence collection to gather data from target or adversary automated INFOSYS or networks. Tactical units do not have the capability for CNE. CNE contributes to intelligence collection at EAC. In UO, CNE will be centrally controlled.

## Physical Destruction

5-75. Physical destruction includes those actions—including direct and indirect fires from air, land, sea, space, and Special Forces—taken with, to augment, or supplement IO actions. Like many other IO elements, major concerns with employing physical destruction in UO are precision and follow-on effects. Thus, commanders using physical destruction to support IO must adhere to the same constraints as all other urban fires.

## Information Assurance

5-76. Information assurance in UO takes on an added dimension. As with other operations, availability of information means timely, reliable access to data and services by authorized users. In UO, the timeliness of information may be restricted because structures block the transmission waves. The need for retransmission facilities may overwhelm the signal community. The reliability can be questioned because of the blockage between units and communications nodes. Unauthorized users may intercept the communications and input misinformation or disinformation. Commanders must protect the integrity of all information from unauthorized changes, including destruction. INFOSYS with integrity operate correctly, consistently, and accurately. The authentication of information may be accomplished by sophisticated electronic means. However, it is more likely that communications-electronics operating instructions authentication tables will authenticate the information. Commanders should consider the confidential nature of all information in UO and establish procedures to protect the information from unauthorized disclosure. Of additional concern, the density of the infrastructure in urban areas may inhibit receipt by the intended individual or unit. Therefore, commanders must develop alternative means to determine if the message was received.

## Physical Security

5-77. Physical security consists of those actions and measures to safeguard personnel; to prevent unauthorized access to equipment, installations, material, and documents; and to safeguard them against espionage, sabotage, damage, and theft. The joint, interagency, and multinational nature of urban operations increases the need for physical security measures, especially with organizations that do not emphasize physical and operational security within their own organizations. The potential for locating sustainment bases in or near urban areas and the use of indigenous personnel on these facilities to accomplish a multitude of tasks mandates that commanders carefully consider the effectiveness of their physical security measures. In some stability operations where Army forces must recruit and train security or police forces, and then conduct joint operations and patrols with them, commanders must examine how best to establish close working relations while maintaining essential physical (and operations) security.

## Counterintelligence

5-78. The urban environment, particularly in stability operations, is ideal for espionage, other intelligence activities, sabotage, or assassination. Threats can approach, conduct reconnaissance, and escape under the concealment offered by the urban environment. The dense environment and electromagnetic clutter serves to hide technological collection means. To counter threat intelligence collection activities, the Army employs CI forces. They provide analysis of the threat's human, imagery, signals, and measurement and signature intelligence capabilities to develop countermeasures against them (see FM 34-60).

5-79. In conjunction with military police, engineers, medical personnel, and other as required, CI help create urban threat and vulnerability assessments necessary in developing comprehensive force protection measures and evaluating their effectiveness. CI personnel can also supplement the screening of linguists and other critical host-nation personnel employed by Army forces. As Soldiers operate in close proximity to the urban society, threats will attempt to use three, time-tested tools to influence them: money, sex, and drugs. Periodic CI refresher training remains crucial to decrease the effectiveness of these threat tactics.

## Counterdeception

5-80. In UO, threat forces can easily accomplish deception operations. The force that controls the area above and below ground will have freedom of movement. Deception aimed at friendly commanders will cause them to deploy combat power at the wrong place and the wrong time. Counterdeception by friendly commanders will identify and exploit threat attempts to mislead friendly forces. Counterdeception is difficult. Cultures of certain rhetoric and actions are more predisposed to deception than others. Knowing a threat's previous deception methods is important. Dismissing tactical indicators because they conflict with preconceptions may allow a hostile deception operation that plays on the preconception to succeed.

## Counterpropaganda

*Four hostile newspapers are more to be feared than a thousand bayonets.*

Napoleon Bonaparte

5-81. Because propaganda is aimed at both combatants and noncombatants, UO are especially concerned with its use. Propaganda can rapidly and dramatically affect the attitudes of the urban population and will probably occur (or increase significantly) after urban operations have begun. Thus, it can create situations in the human dimension of the environment quite different from those discovered in the pre-operations assessment. Counterpropaganda is, therefore, essential to urban operations. To negate, deflect, or destroy the threat's propaganda capability, counterpropaganda requires—

- Monitoring and evaluating the effectiveness the threat's propaganda efforts.
- Determining innovative methods using all Army force capabilities, especially PSYOP and PA units.

5-82. In particular, humanitarian assistance, when coupled with an effective information operations campaign, can be a powerful counterpropaganda tool to sway public support. For example, media coverage of Army forces directly providing or coordinating NGO humanitarian efforts to feed, shelter, and provide medical care to the urban population can help counter threat efforts to discredit Army forces as an evil occupying force. This type of coverage can be used to turn opinion against the threat and could also prove an effective tool in gaining favorable international support.

## Public Affairs

5-83. A closely related activity to IO is public affairs (PA). PA influences urban operations by transmitting information through the media to internal (urban Army forces as well as the urban civilian populace) and external audiences. At higher levels of command, PA can help maintain popular national support for the urban operation by clarifying the links between strategic goals and operational objectives. At both the operational and tactical levels, it links Army units, the urban inhabitants, the U.S. and international public, and the media. The media will often focus on the negative actions that are ongoing in an AO; hence, the urban commander's challenge will be to publicize his unit's good news stories. Commanders should not wait for media inquiries and interview requests. Instead, they should aggressively plan public affairs to get the Army's story out to the media and, ultimately, the public.

5-84. PA can help determine potential media issues that may influence planned UO. It can also aid commanders in understanding the impact of UO on the environment (particularly its citizens) and other agencies and organizations operating in the urban area. PA also helps to counter rumors, uncertainty, fear, loneliness, confusion, and other factors that cause stress (to both soldiers and civilians) and undermine effective UO. If the populace does not understand the mission or the Army forces' capabilities, false expectations may be created that Army forces may not be able to meet. PA can help prepare the American public for the possibility of high casualty rates. Overall, PA supports urban commanders in their goals to achieve information superiority and preserve public support.

5-85. The density of information sources and reporters in UO ensures that all Army activities will be subject to media and public scrutiny. Many reporters will congregate in cities for their own comfort and take advantage of established communications networks. Urban areas are densely populated and, together with Army forces and NGOs operating there, will present the greatest number of human-interest stories.

The local urban or host-nation media, however, will often have their own agendas developed over a longer period of time. This local media may also have a greater influence over the urban population than the international media. The indigenous media may not follow international norms. Commanders are responsible for understanding the media (particularly the local media), its role, and its potential for negative or positive influence.

5-86. Commanders cannot allow themselves to be intimidated by the media. Media sources are going to get and report their story one way or another. Therefore, commanders should support open and independent reporting and grant access to their units as early and as far forward as the situation permits. Successful units often go beyond simply accommodating the media to actually integrating them into their units. In the future, embedding of media may become the standard for reporting Army operations. Other considerations for improving urban media relations include:

- Encouraging the media to report on all ongoing aspects of the unit's urban operations—not exclusively on combat operations. This may create a unit responsibility to protect journalists that participate in covering suggested activities.
- Hosting media luncheons or other events for key journalists. This provides commanders the ability to get the whole story out to the press from their unit's perspective. To gain media interest and unit credibility, this will often require key leadership participation.
- Identifying Soldiers that speak the local and international journalists' native language. Even if journalists speak English as a second language, they may overlook or misunderstand important points due to language subtleties. Additionally, people tend to be friendlier to those that speak their own language.
- Identifying selected Soldiers at lower tactical levels to act as PA representatives for their unit as an additional duty. However, commanders must provide them with appropriate training and resources to accomplish specific PA tasks.
- Writing articles for local newspapers that emphasize the unit's good news. Depending on the situation, subordinate units may be tasked to write stories about the good accomplishments in their AO. Commanders, with appropriate OPSEC considerations, may hire locals to assist in this effort. However, commanders must take extra precautions to ensure that these efforts are not perceived as covert propaganda or as undermining free and independent news reporting.

5-87. PA does not distort, direct, or manipulate information. Its effectiveness stems directly from establishing and maintaining credibility with the urban population and media. PA protects the overall integrity of the Army as an institution by ensuring that the Army is recognized as the most reliable source for official information among all other competing sources. Commanders must synchronize PA with the core and supporting elements of IO (particularly PSYOP and counterpropaganda) to ensure that all Army sources send only one message. Urban commanders must plan for the media and integrate PA into their decision-making and (through IO) targeting processes. Similar to lethal targeting, commanders should consider target audiences and coverage areas for the various new agencies and mediums. They also pay attention to print media circulation statistics and television and radio broadcast times. For example, an urban populace may tend to read morning newspapers vice late editions and watch or listen to nightly news instead of morning or daytime programming. Additionally, low literacy rates among the urban population may indicate that television and radio are better mediums to use to influence the urban population than print media such as newspapers, magazines, leaflets, and posters.

5-88. The PA principles listed in Figure 5-4 and addressed in FM 46-1 summarize PA. They serve as useful guides toward planning and executing PA operations regardless of the environment. However, the principles of "practice security at the source," and "truth is paramount" particularly apply to the urban environment. The compartmented nature of most UO impede commanders' and their PA officers' ability to be at all places where the media will likely be. Friendly inaccuracies, intentional or not, will help fuel the threat information campaign. Therefore, all soldiers are trained, provided with clear and understandable PA guidance, and prepared to accurately communicate to the civilian media. The keys at all levels are understanding, prepared acceptance, and truthfulness tempered with an essential concern for OPSEC.

- Truth is Paramount
- If News is Out, It is Out
- Public Affairs Must be Deployed Early
- Not All News is Good News
- Practice Security at the Source
- Media are Not the Enemy
- Telling Our Story is Good for the Army
- Soldiers and Families Come First

**Figure 5-4. Public affairs principles**

## Civil-Military Operations

5-89. Civil-military operations (CMO) are critical and will be an essential, implied task of any urban operation. They are included here as a closely related activity of IO. Commanders use CMO to establish, maintain, influence, or exploit relations to achieve operational objectives. These relations are among military forces, civilian authorities and organizations (both governmental and nongovernmental), and the civilian population. CMO range from support to combat operations to assisting in establishing political, economic, and social stability. In some instances, commanders may need to be prepared to assume temporary responsibility for the functions and capacities of the urban government. Like PA, effective CMO is based on establishing and maintaining credibility and trust with the urban populace and civilian organizations operating in the urban environment. Because of the potential importance of CMO, commanders may consider placing an assistant commander, executive officer, or some other senior leader to monitor, guide, liaise with other units and organizations, and, overall, ensure that the appropriate command emphasis is placed on these activities.

### *Civil Affairs*

5-90. Civil affairs (CA) units and teams will be critical during UO. While any military force can conduct CMO, CA units are specifically organized, trained, and equipped to conduct activities in support of CMO. They have experience in planning and conducting CMO, a regional focus (which includes enhanced cultural awareness and language training), civilian technical expertise that parallel common urban government and administration functions, and experience in negotiation and mediation. Such experience ensures relevant support to commanders conducting urban operations. (As it is likely that there will not be enough CA forces to cover all requirements, commanders should survey their Soldiers to determine the whether any possess civilian expertise that may be beneficial to CMO.) CA units organize their skills into six functional specialty areas (see Figure 5-5). Commanders should use these skills, often unfamiliar to most military personnel, to—

- Develop their situational understanding of the urban environment (particularly the infrastructure and society).
- Plan CMO to support UO.
- Achieve many of the fundamentals of UO.

---

***Note***: To ensure optimal integration, commanders must establish clear command and support relationships between CA and maneuver forces.

---

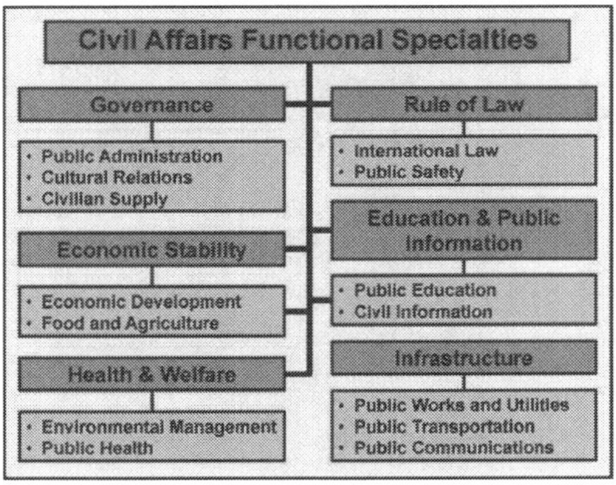


**Figure 5-5. Civil affairs functional specialties**

5-91. In addition to providing essential information for understanding the urban environment, CA personnel and activities help shape the battlefield, engage a civil problem, consolidate gains, and transition to a legitimate civil authority. Specifically, CA teams and CMO help urban commanders—

- Minimize civilian interference with UO and the impact of urban operations on the populace and infrastructure. CA personnel can help establish and run a civil-military operations center (CMOC), discussed later in this chapter, to coordinate UO with civilian agencies (both governmental and nongovernmental), other services, and multinational partners.
- Provide advice and assistance to restore or rehabilitate portions of the infrastructure.
- Plan, supervise, and execute necessary populace and resources control measures (in close coordination with military police units) until no longer required or the urban operation is completed.
- When requested or when military necessity or legitimate directives require, establish all or portions of the urban civil administration.
- Determine available supplies and services in the urban area and if necessary assist in negotiating their acquisition. CA also help commanders assess the capability, dependability, and willingness of urban sources to provide and sustain identified needs as well as to calculate the impact of using them on other aspects of the urban environment.
- In conjunction with the SJA, fulfill the Army's responsibilities toward the urban population under international, host-nation, and U.S. law.
- Plan and conduct the transition of control for the urban area or operation to another military or civilian governmental or nongovernmental organization or agency.

5-92. As part of CMO, stability or civil support operations will be critical as commanders begin to restore essential services to meet the emergency needs of a potentially devastated urban population. Besides repairing or restoring vital services and other critical infrastructure, these and other projects may also stimulate the urban area's economy and provide work for the unemployed. (Employed workers are less likely to be co-opted by terrorists, insurgents, and criminal gangs.) As such, they can be a critical stabilizing tool for commanders as they attempt to transition control and responsibility to other agencies or to legitimate civilian control. These projects may include—

- Providing clean water.
- Restoring waste management functions (less-extensive sewer repairs and trash collection and disposal activities).
- Improving medical care.
- Rehabilitating schools.
- Assisting in the establishment of local governments.
- Improving security and emergency services (police, fire, and rescue).
- Providing electrical generators for hospitals, clinics, and factories.
- Restoring irrigation to surrounding farmland.
- Repairing important cultural infrastructure.

5-93. As part of the initial planning process, CA units conduct an *area assessment*, which can provide commanders with essential information about the environment (see FM 41-10). Commanders should integrate this initial assessment into the overall urban-focused IPB process. To help analyze civil considerations in any environment, commanders and staffs can consider many characteristics such as areas, structures, capabilities, organizations, people, and events (ASCOPE). These characteristics easily align with an urban area's three main characteristics of terrain, society, and infrastructure; and, like them, they are overlapped and interdependent (see FM 6-0). Of these, the category of events is a critical addition to urban characteristics previously categorized and discussed. Significant spikes (and lulls) to threat and civilian activity can be expected to occur in conjunction with important religious, historic, and political events (for example elections). Therefore, commanders should determine, track, and pay particular attention to these events when planning operations and managing transitions (see also the Key Event Chart discussion in Appendix B, paragraph B-43).

5-94. Similar to PA operations, CMO are related to information operations. The nature of CMO and the need for CA personnel to develop and maintain close relationships with the urban population put CA personnel in a favorable position to collect information. CA personnel work daily with civilians, their equipment, and their records that may be prime sources of information. If used correctly, CA personnel can complement the intelligence collection process necessary to understand the dynamic societal component of the urban environment and detect significant changes. However, CA personnel are not, and cannot appear as, intelligence agents; otherwise, it will undermine their ability to interact with the civilian community. Examples of information available to CA units include government documents, libraries, and archives; files of newspapers and periodicals; industrial and commercial records; and technical equipment, blueprints, and plans.

5-95. Overall, CA expertise is critical to a commander's understanding of the complexities of the infrastructure and societal components of the urban area. These components (together with the terrain or physical component of the urban area) interconnect. CA forces help identify and understand the relationships and interactions between these urban components. From this understanding, commanders can anticipate how specific military actions affect the urban environment and the subsequent reactions. CA personnel help commanders predict and consider the second- and third-order effects and reactions as well as the long-term consequences. Understanding these long-term consequences helps in achieving stability and ensuring a smoother and speedier transition of the urban area back to civilian responsibility and control. Oppositely, an unconsidered short-term solution may, in the end, turn out to be the solution requiring the greatest amount of time. Poor solutions may exacerbate the situation. Others may contribute little to conflict resolution and waste precious resources (particularly time) until a more reasoned approach is formulated—one that should have been the first solution.

### *Civil-Military Reconstruction Efforts*

5-96. Reconstruction efforts and activities, an integral part of CMO, should complement NGO efforts and are conducted in concert with larger urban area and nationwide projects often headed by the U.S. Army Corps of Engineers (USACE) and the U.S. Agency for International Development (USAID). For example, extensive, large-scale sewer projects or major repairs to the urban area's electrical infrastructure would be left to the USACE and USAID. When planning and conducting urban reconstruction, commanders should—

- **Keep expectations realistic.** With so much to do, it is easy for Army forces to become overcommitted. Commanders must focus on the achievable, build off successes, and create constructive forward process.
- **Seek opportunities.** While commanders think long-term, they should look for catalytic investment of resources. Minimal resources provided at the right time can generate significant activities resulting in increased momentum toward the desired end state.
- **Align resources with local needs.** In analyzing the merit of a project, commanders must check to determine if it is supported and desired by the urban populace. Local participation in the process reinforces ownership and acceptance and eventual transition of responsibility.
- **Balance speed and quality.** Commanders must balance the achievement of rapid response with quality construction and repairs.
- **Ensure efforts do not undermine locals.** Army reconstruction efforts should not undermine the growth and legitimacy of local institutions and authorities.
- **Integrate in the targeting process.** Commanders should routinely include approval of civil-military reconstruction products as part of the information operation's contribution to the overall targeting process.

## INTEGRATION OF CONVENTIONAL AND SPECIAL OPERATIONS FORCES

5-97. One important Army and joint resource that commanders of a major operation can use to influence urban operations is SOF. Several types of these forces exist (including the CA forces discussed above), each with unique and complementary capabilities. They can be extremely valuable in UO for their ability to execute discrete missions with a higher degree of precision than conventional forces, to provide information, and to enhance cultural understanding. However, the challenges of using SOF include command and control; integration; coordination with conventional forces that will normally command, control, and conduct the bulk of UO tasks; and the imprudent inclination to use SOF forces for conventional purposes. The density and complexity of UO make close coordination and synchronization of conventional forces and SOF essential to mission success. The nature of the environment dictates that both forces will work in close proximity to each other; the separation in space and time between SOF and conventional forces will often be much less in urban areas than in other environments. Overall, the nature of the environment demands a synergistic combination of capabilities to achieve effects on the threat and mission success.

5-98. Successfully integrating SOF occurs with proper integration into, or coordination with, the command structure of the force conducting the UO. SOF within a theater (less CA and tactical PSYOP) ordinarily falls under joint command and control. Therefore, the commander of the major operation responsible for an urban area, if he is not a JFC, will have to coordinate through the JFC to integrate SOF capabilities into the UO. Examples of critical coordination elements include:

- Boundaries.
- No-fire areas. (If assigned, SOF liaison officers (LNOs) must be routinely included in targeting meetings.)
- Coordination points.
- The exchange of key intelligence and information.
- Requirements to support personnel recovery contingencies.

> **Conventional and Special Forces Integration**
>
> On 20 April 2003 during OPERATION IRAQI FREEDOM (OIF), an Army Special Forces (SF) operational detachment alpha (ODA) was patrolling through villages and secondary roads east-southeast of An Najaf. A local Iraqi approached the team and volunteered the location of a senior Ba'ath Party official in the nearby town of Ghamas. Unfortunately, he did not provide an exact location but rather an approximate location and the name of the family dwelling. The ODA coordinated with their company commander and nearby elements of the 101st Airborne Division to conduct a raid on the residence at dawn the next day. Joint planning, preparation, and rehearsals among the units occurred throughout the night.
>
> At 0430, the raid convoy departed its base en route to Ghamas. The small task force consisted of an assault team, a security team, a command and control element, and three blocking forces composed of Soldiers from a scout platoon and antiarmor company of the 101st. At 0515, the convoy hit the release point outside Ghamas. The 101st vehicles moved to their blocking positions at three bridges surrounding the town. The remainder of the task force moved into the town. An interpreter quickly found a local guide who knew where the house was located. He took the SF team to a walled two-story dwelling with a courtyard. The security team isolated the objective and provided overwatch while the assault team forcibly seized the house and apprehended three adult males. Among them was Abd Hamden, the target of the raid and a senior Ba'ath Party official from Baghdad.
>
> After a tactical interrogation, one of the men provided the location of a Fedayeen major nearby. The SF and conventional force raided this house minutes later but only found the major's relatives. As the Soldiers left the town with their captives, they were cheered by the locals throughout the town. Although only one small example, the extensive coordination between SOF and conventional forces during urban operations was a tremendous success during OIF.

# SYNCHRONIZATION OF ACTIVITIES WITH OTHER AGENCIES

5-99. The population density of the urban environment, its economic and political importance, and its life-supporting infrastructure attracts many types of organizations. These organizations include—

- Other U.S. governmental agencies.
- International governmental organizations.
- Allied and neutral national governments.
- Allied and coalition forces.
- Local governmental agencies and politicians.
- NGOs.

5-100. Even in a major operation or campaign, many organizations operate in the area as long as possible before combat or as soon as possible after combat. Therefore, coordination with these organizations sharing the urban AO will be essential to achieve synchronization; however, effective synchronization is challenging, time consuming, and manpower intensive. The staffs of larger headquarters (divisions or higher) normally have the breadth of resources and experience to best conduct the necessary synchronization. They can effectively use or manage the organizations interested in the urban area and mitigate their potential adverse effects on UO. By taking on as much of the synchronization requirements as possible, the operational headquarters permits its tactical subordinates to remain focused on accomplishing their tactical missions. The higher headquarters should assume as much of the burden of synchronization as possible. However, the density of the urban environment will often require that smaller tactical units coordinate and synchronize their activities with other agencies and the local civilian leadership (formal and informal) simply because of their physical presence in the units' AOs. In urban stability and civil support operations, mission accomplishment will require effective civil-military coordination and synchronization activities and measures at all levels as either a specified or implied task.

## Civil-Military Operations Centers

5-101. To coordinate activities among the varied agencies and organizations operating in an urban area and the local population, urban commanders can establish a CMOC. The CMOC synchronizes Army activities and resources with the efforts and resources of all others involved (see FM 41-10). This can be particularly important in stability and civil support operations where combat operations are not the dominant characteristic of the major operation. CMOCs can be established at all levels of command. Hence, more than one CMOC may exist in an AO, particularly large urban areas. CMOCs may be organized in various ways and include representatives from as many agencies as required to facilitate the flow of information among all concerned parties. Commanders must still ensure that force protection and OPSEC requirements are not compromised. Effective CMOCs can serve as clearinghouses for the receipt and validation of all civilian requests for support, can aid in prioritizing efforts and eliminating redundancy, can decrease the potential for inappropriate displays of wealth by one or more of the participating organizations, and, most importantly, can reduce wasting the urban commander's scarce resources.

## Liaison Officers

5-102. LNOs—sufficiently experienced and adequately trained in liaison duties and functions—are necessary to deal with the other agencies that have interests in the urban area. Army LNOs work with the lead agency or other organizations (including local civilian agencies such as police) that the commander has identified as critical to mission success. Together they work to rapidly establish unity of effort and maintain coordination, often before a CMOC is established. The additional coordination afforded by the physical presence of LNOs within these organizations may be required even after the CMOC is fully functional. When commanders lack enough LNOs to meet requirements, they should prioritize and often assign a single LNO to several organizations. That LNO will then share his time and presence to those organizations based on the situation and his commander's guidance.

## Commander's Personal Involvement

5-103. Overall, establishing a close relationship with other agencies and the urban civilian population will often be a major, positive factor in successful mission accomplishment, particularly in urban stability operations. Despite internally established command and control relationships, commanders that develop a direct and personal relationship with the leaders and staff of other agencies can often avoid conflict, win support, foster trust, and help eliminate the "us versus them" mentality that can frustrate cooperation among Army forces and civilian organizations.

# Chapter 6

# Foundations for Urban Operations

*Utilities such as electricity and water are as much weapons of war as rifles, artillery pieces or fighter aircraft. . . . In the case of Manila, where there was a noncombatant, civilian population of one million in place, it was the attacker's aim to capture the utilities which the defender planned to destroy.*

*The Battle for Manila*

Commanders conducting major urban operations (UO) use their ability to visualize how doctrine and military capabilities are applied within the context of the urban environment. An operational framework is the basic foundation for this visualization. In turn, this visualization forms the basis of operational design and decision making. To accurately visualize, describe, and direct the conduct of UO, commanders and their staffs must understand the basic fundamentals applicable to most UO.

## URBAN OPERATIONAL FRAMEWORK

6-1. Army leaders who have an urban area in their area of operations (AO) or are assigned missions in an urban area follow an urban operational framework. They identify the portion of the urban area essential to mission success, shape the area, precisely mass the effects of combat power to rapidly dominate the area, protect and strengthen initial gains without losing momentum, and then transition control of the area to another agency. This framework divides into five essential components: understand, shape, engage, consolidate, and transition. These five components provide a means for conceptualizing the application of Army combat power and capabilities in the urban environment.

6-2. The urban operational framework assists commanders in visualizing urban operations. This framework is simply an aid to the commander. It is not sequential, nor is it a planner's tool for phasing an operation. Commanders should combine the urban operational framework with—

- The principles of war.
- The tenets of Army operations.
- The components of operational design.
- Considerations for stability operations and civil support operations.
- Sustainment characteristics.
- Running estimates.
- Commander's critical information requirements (CCIR).
- Each commander's experience.

The framework contributes to the visualizing, describing, and directing aspects of leadership that make commanders the catalysts of the operations process (see Figure 6-1). In the same manner, the urban operational framework contributes to the overall operations process (see FM 3-0).

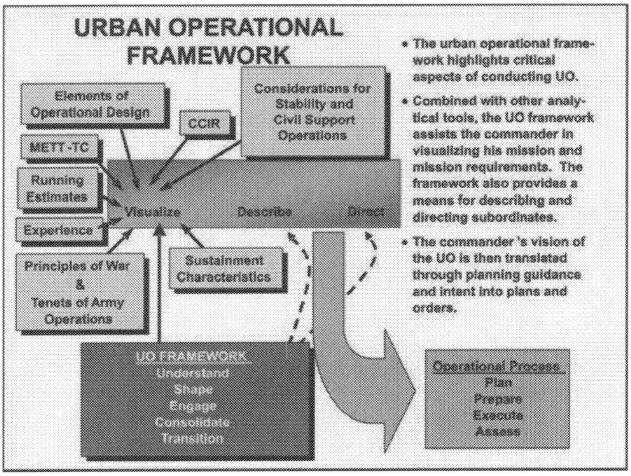


**Figure 6-1. The urban operational framework and battle command**

## UNDERSTAND

6-3. Understanding requires continuous assessment—throughout planning, preparation, and execution—of the current situation and progress of an operation, and the evaluation of it against measures of effectiveness to make decisions and adjustments. Commanders use visualization, staff officers use running estimates, and all use the intelligence preparation of the battlefield (IPB) process to assess and understand the urban environment. Commanders and staffs begin to understand by observing and then collecting information about the situation. They observe and learn about the urban environment (terrain, society, and infrastructure), and other factors of METT-TC—mission, enemy, weather, troops and support available, and time available. They use intelligence, surveillance, and reconnaissance means; information systems (INFOSYS); and reports from other headquarters, services, organizations, and agencies. Then they orient themselves to the situation and achieve situational understanding based on a common operational picture (COP) and continuously updated CCIR. Largely, the ability to rapidly and accurately achieve a holistic understanding of the urban environment contributes to the commanders' abilities to seize, retain, and exploit the initiative during UO.

### Disproportionately Critical

6-4. The Army operations process requires continuous assessment and understanding; it precedes and guides every activity. In UO, however, in-depth understanding is disproportionately critical for several reasons. First, each urban environment is unique. Other environments can be studied and their characteristics quantified in a general manner with accuracy. This is fundamentally not true of different urban areas. The characteristics and experience in one urban area often have limited value and application to an urban area elsewhere. This characteristic sets UO apart from operations in other environments.

### Extremely Dynamic

6-5. The urban environment is also extremely dynamic. Either deliberate destruction or collateral damage can quickly alter physical aspects of the urban environment. The human aspect is even more dynamic and potentially volatile. A friendly civil population, for example, can become hostile almost instantaneously.

These dynamics (combined with initial difficulty of understanding and describing this unique environment) make it difficult for commanders and staffs to initially develop and maintain a COP and establish situational understanding. Furthermore, public reaction to media coverage of the urban operation and political changes influence national security strategy and objectives. Such changes can affect the basic nature of an operation, especially after it has commenced. Anticipating these potential effects and developing appropriate branches and sequels based on an accurate understanding often determines how quickly commanders can achieve the desired end state.

### Risk Assessment

6-6. As in any environment, UO pose both tactical and accident risks. However, the level of uncertainty, ambiguity, and friction can often be higher than that of many other environments. Such challenges increase the probability and severity of a potential loss due to the presence of the enemy, a hostile civilian group, or some other hazardous condition within the urban environment (see Determining the Necessity of Urban Operations in Chapter 5). Therefore, commanders must—

- Identify, assess, and understand hazards that may be encountered in executing their missions.
- Develop and implement clear and practical control measures to eliminate unnecessary risk.
- Continuously supervise and assess to ensure measures are properly understood, executed, and remain appropriate as the situation changes.

6-7. Risk decisions are commanders' business. However, staffs, subordinate leaders, and even individual Soldiers must also understand the risk management process and continuously look for hazards at their level or within their area of expertise. Any risks identified (with recommended risk reduction measures) must be quickly elevated to the appropriate level within the chain of command (see FM 3-100.12 and FM 5-19).

### Complex and Resource Intensive

6-8. The urban environment is the most complex of all the environments in which the Army conducts operations. It is often comprised of a diverse civil population and complex, ill-defined physical components. A sophisticated structure of functional, social, cultural, economic, political, and informational institutions unites it. Thus, the analysis to understand the environment is also complex and time and resource intensive. The nuances of the urban environment can take years to uncover. Hence, constant analysis of the environment requires greater command attention and resources. Accurately understanding the environment is a prerequisite to shaping it, and both understanding and shaping activities are crucial to effectively engage the elements of the urban environment critical to success.

## SHAPE

6-9. Shaping operations, part of all Army operations, are essential to successful UO. They set the conditions for decisive operations at the tactical level in the urban area. Rapid action, minimum friendly casualties, and acceptable collateral damage distinguish this success when the AO is properly shaped. Failure to adequately shape the urban AO creates unacceptable risk. The commander of a major urban operation has several resources with which to begin shaping the AO. Important capabilities include—

- Fires.
- Information operations.
- Special operations capabilities.
- The maneuver of major subordinate units.

6-10. Critical urban shaping operations may include actions taken to achieve or prevent isolation, understand the environment, maintain freedom of action, protect the force, develop cooperative relationships with the urban population, and train Army forces for sustained UO.

### Isolation

6-11. Isolation of an urban environment is often the most critical component of shaping operations. Commanders who's AO includes operationally significant urban areas often conduct many shaping operations to isolate, or prevent isolation of, those areas from other parts of the AO. Likewise, commanders operating in the urban area focus on isolating decisive points and objectives in the urban area or averting isolation of points that are critical to maintaining their own freedom of action. Isolation is usually the key shaping action that affects UO. It applies across full spectrum operations. Most successful UO have effectively isolated the urban area. Failure to do so often contributed to difficult or failed UO. In fact, the relationship between successful isolation and successful UO is so great that the threat often opposes external isolation actions more strongly than operations executed in the urban area (or critical areas within). In some situations, the success of isolation efforts has been decisive. This occurs when the isolation of the urban area compels a defending enemy to withdraw or to surrender before beginning or completing decisive operations. In UO that are opposed, Army forces attempt to isolate the threat three ways: physically, electronically, and, as a resultant combination of these first two, psychologically (see Figure 6-2).

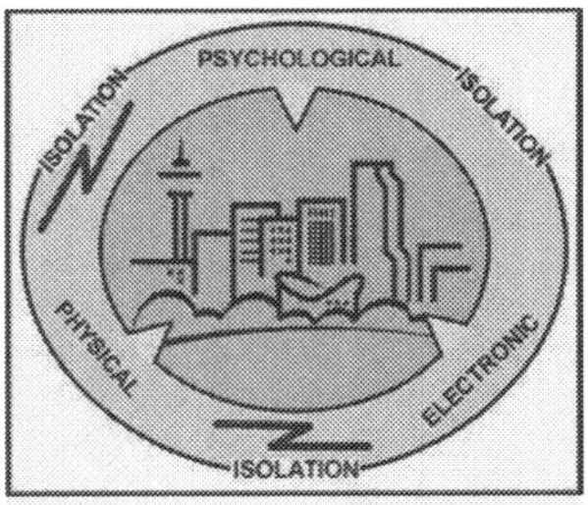


**Figure 6-2. Urban isolation**

### *Physical Isolation*

6-12. In offensive UO, physical isolation keeps the threat from receiving information, supplies, and reinforcement while preventing him from withdrawing or breaking out. Conversely, a defending Army force attempts to avoid its own physical isolation. Simultaneously, this force conducts operations to isolate the threat outside, as they enter, or at selected locations in the urban area. Physical isolation can occur at all levels. In many situations, the commander of a major combat operation may attempt to isolate the entire urban area and all enemy forces defending or attacking it. At the tactical level, forces isolate and attack individual decisive points often using a cordon technique. In stability operations, physical isolation may be more subtly focused on isolating less obvious decisive points, such as a hostile civilian group's individual leaders. In many operations, isolation may be temporary and synchronized to facilitate a decisive operation elsewhere. To effectively isolate an urban area, air, space, and sea forces are often necessary additions to the capabilities of ground forces.

### *Electronic Isolation*

6-13. Electronic isolation is achieved through offensive information operations (IO). Electronic warfare (particularly two of its components: electronic warfare support and electronic attack) and computer network attack are critical to electronic isolation (see FM 3-13 and the Information Operations discussion in Chapter 5). At the operational level, offensive IO aims to quickly and effectively control the information flow into and out of an urban area. This isolation separates the threat's command and control (C2) system in the urban area from its operational and strategic leadership outside the urban area. Offensive IO also focuses on preventing the threat from communicating with civilians through television, radio, telephone, and computer systems. At the tactical level, IO aim to isolate the threat's combat capability from its C2 and leadership within the urban area, thus preventing unity of effort within the urban area. Defensive IO can prevent isolation of friendly forces defending in an urban area.

### *Psychological Isolation*

6-14. Psychological isolation is a function of physical actions, electronic warfare, and other forms of IO, especially military deception and psychological operations. Psychological isolation denies the threat political and military allies. It separates the enemy or hostile civilian group from the friendly population, nongovernmental organizations (NGOs) operating in the urban area, and from political leaders who may consider supporting Army forces. Psychological isolation destroys the morale of individual enemy soldiers or hostile civilians. It creates a feeling of isolation and hopelessness in the mind of the threat. It undermines the threat leadership's ability to think, plan, decide, and act. As such, it undermines the confidence of the threat in their leadership. On the other hand, IO, as well as the disciplined conduct of Army personnel, can help to forge legitimacy for Army operations. In stability operations, psychologically isolating the threat can result in the friendly urban population and NGOs positively supporting Army operations.

### Other Shaping Actions

6-15. Other shaping actions can include the proper sequencing and deployment of forces, reconnaissance operations, and force protection. These actions contribute equally to the success of any urban operation. Commanders must understand how the urban environment affects their ability to accomplish these shaping actions. However, civil-military operations (CMO), another closely related activity of IO, are important to shaping the urban operational environment for decisive operations. The specific civil-military task can vary greatly and may include affecting a cooperative relationship with the civil political system, protecting portions of the civil population or infrastructure, or establishing refugee camps or safe areas for noncombatants. This is most true in stability and civil support operations. Successful CMO also can contribute to the psychological isolation of the threat. (See Civil-Military Operations in Chapter 5 for a more detailed discussion.)

### Training and Education

6-16. Finally, Army commanders must consider that critical shaping actions often occur prior to the urban operation in the form of professional education, and home-station and in-theater training. Commanders can enhance training through joint, interagency, multinational, and combined arms exercises and effective rehearsals.

## ENGAGE

6-17. In UO, Army forces engage by appropriately applying their full range of capabilities against decisive points leading to centers of gravity. Successful engagement takes advantage of the Army force's superior training, leadership, and, within the constraints of the environment, equipment and technology. Successful engagement also requires the establishment of necessary levels of control and influence over all or portions of the urban environment until responsibilities can be transferred to other legitimate military or civilian control. Engagement may range from the overwhelming, yet precise, application of combat power to defeat a threat, to large-scale humanitarian assistance operations, to unobtrusive advice and assistance to the urban security forces. It may require brief engagement in a relatively small portion of an urban area or with small segments of the population to long-term engagement throughout multiple, large urban areas and with vast sectors of the population. The commander's ability to engage in a specific urban operation depends, as always, on the situation and the assigned mission.

### Offense: Attack Decisive Points

6-18. In offensive UO, forces successfully engage by striking at the enemy's center of gravity using multiple offensive actions from unexpected directions and throughout all dimensions. Army forces aim to engage identifiable decisive points. Successful efforts against decisive points lead to effects on the center of gravity. The center of gravity will differ in each offensive situation. It may be an individual enemy leader, the enemy's combat power, the enemy's communications capability, or a physical structure of cultural, political, or economic significance.

### Defense: Deny Vital Functions and Critical Infrastructure

6-19. In defensive UO, engage translates into denying the enemy control of the vital functions and critical infrastructure of the urban area. Forces achieve this by leveraging the defensive advantages of the urban terrain, defending essential areas in depth, using economy of force in nonessential areas, controlling the enemy direction of attack with natural and man-made obstacles, and retaining the initiative through counterattacks.

### Stability and Civil Support: Apply Innovation and Imagination

6-20. The ability to engage in urban stability operations hinges on the type of stability operation commanders execute. In a noncombatant evacuation operation, forces limit engagement to finite geographic areas and times. In contrast, a peace operation may require engagement of a large urban area for an extended time. In this operation, engage is defined as using the array of Army capabilities to create specific conditions among the belligerents. Thus, the techniques used for engagement in stability operations vary according to the situation and as situations mature during long-term operations.

6-21. In urban civil support operations, engagement is accomplished by innovative and subtle application of Army capabilities. Since Army forces usually support other agencies that lead the operation, effective engagement results from carefully and discretely applying Army capabilities to the tasks assigned by the lead agency. In a humanitarian relief situation, Army forces may be tasked to transport supplies in the urban area. Successful engagement in this activity then becomes the goal of Army forces and may be achieved by providing, managing, and protecting transportation assets and clearing and maintaining sufficient lines of communications (LOCs).

## CONSOLIDATE

6-22. Army forces consolidate to protect and strengthen initial gains and ensure retention of the initiative. Consolidation includes actions taken to eliminate or neutralize isolated or bypassed threat forces (including the processing of enemy prisoners and civilian detainees) to increase security and protect LOCs. It includes the rapid repositioning and, as necessary, reorganization of forces and intelligence, reconnaissance, and surveillance (ISR) assets. Consolidation may also include activities in support of the civilian population such as relocating displaced civilians, reestablishing law and order, humanitarian assistance and relief operations, and restoration of key urban infrastructure. A significant consolidation effort may be the reduction or elimination numerous explosive hazards resulting from previous urban combat operations. During consolidation, commanders often coordinate with NGOs to establish unity of effort in the execution of stability or civil support operations. Consolidation begins with the initiation of the urban operation and follows each critical activity.

## TRANSITION

6-23. When planning UO, commanders ensure that they plan, prepare for, and manage transitions. Transitions are movements from one phase of an operation to another or a change in responsibility for all or portions of the urban environment from one unit, organization, or agency to another. They involve significant changes in the type of operation, concept of the operation, mission, situation, task organization, forces, resource allocation and support arrangements, or C2. Transitions occur in all operations, but in UO they occur with greater frequency and intensity, are more complex, and often involve agencies other than U.S. military organizations. For example, a successful attack may transition to a defend mission that includes not only defense tasks but also stability tasks. All operations will likely include a transition of responsibility for some aspect of the urban environment to (or back to) a legitimate civilian authority. Unless planned and executed effectively, transitions can reduce the tempo of the urban operation, slow its momentum, and cede the initiative to the threat.

### Mental and Physical Preparation

6-24. Transitions occur as conditions warrant. They can be carefully planned and controlled, or they can be quick and dramatic, such as the swift transformation of a stability operation into offense or defense—and back again. Units prepare mentally and physically to address rapid transitions. Accordingly, plans include branches and sequels that address anticipated or possible transition points. When the dominant type of operation changes from an offense to stability or civil support, the types of units originally conducting the UO may no longer be appropriate. A large mobile reserve may permit increased flexibility to react to unplanned transition requirements. Operations or units in one part of an urban area may transition before operations or units in a different part of the same urban area. This will require commanders to exercise careful command and control and execute various types of operations and associated tasks simultaneously.

## Transition to Legitimate Civilian Authorities or Agencies

6-25. Combat operations are not an end to themselves. They are inevitably a means of transition, an activity for moving from an unsatisfactory state of affairs to an improved end (as determined by national objectives). In UO, a distinct aspect of transition is the requirement to quickly and efficiently transition the major portions of Army responsibilities to civil agencies. Some tasks to which units will transition are not traditional combat tasks but rather stability tasks more closely associated with CMO. In stability or civil support operations this is often a near-term critical mission objective. In these operations, commanders aim to alleviate the circumstances requiring Army forces and ensure that other civilian agencies assume the functions provided by Army forces. Following combat operations, civilian agencies are assisted and encouraged to quickly resume specific support activities—such as providing sanitary services, food services, law enforcement, and health services—because of their high demand on Army resources.

6-26. Transitioning responsibility for the urban environment (or selected portions) to legitimate civilian agencies or authorities requires commanders to make an accurate assessment and develop an in-depth understanding of the civilian organization's ability to accept and handle the responsibility. Commanders consider not only resources available but whether the organization has the appropriate leadership and technical expertise as well as a vision in consonance with U.S. objectives. Commanders should avoid forcing a transition only to alleviate the burden on Army forces. On the other hand, they may need to encourage acceptance of responsibility while continuing to provide "behind-the-scenes" advice, assistance, and resources. Transitioning responsibility prematurely may alleviate the commander's immediate resource concerns but may, in the longer term, negatively impact the achievement of strategic objectives.

## Clearly Visualize and Describe the End State

6-27. Army UO conclude when Army forces depart and have no further mission requirements in the urban area. At the outset, commanders visualize and describe the intended end state of a unit's execution of UO. Commanders then clarify and update this visualization as the political or strategic situation is refined or changes. Importantly, commanders must create a command climate and procedures that allow their staffs and subordinate commanders the ability to ask questions that will enable all to share in an accurate perception of the commander's visualization. This enables subordinate units to identify likely transitions and ensures that current operational planning takes into account second- and third-order effects. As long as an active Army AO contains an urban area, some type of urban operation will exist. After urban combat successfully ends, combat forces may move on. Support forces conducting sustaining operations may then occupy the area and continue to conduct a different form of UO.

### Applying the Urban Operational Framework: Panama – 1989

The U.S. conducted OPERATION JUST CAUSE in December 1989 to remove the illegal ruler of Panama, Manuel Noriega, and to restore that country to a democracy. It also conducted the operation to ensure the safety of a substantial number of U.S. personnel as well as the security of U.S. interests in Panama. The major focus of JUST CAUSE was in Panama City, the country's capital. Most operations occurred in this large urban area, one of the numerous smaller urban areas, or the urban-like military bases. These bases increased the AO and were directly linked to operations in the capital city. This successful operation illustrates how commanders can apply the urban operational framework to visualize, describe, and direct the critical aspects of urban operations.

Understand

The synchronization achieved during the operation may have obscured the challenges faced in the initial assessment process in Panama. However, it was not as simple as it may have seemed. Using the framework of the urban environment, U.S. forces required details of the physical characteristics of the environment, the infrastructure, and the human dimension including the capabilities of the Panamanian military.

Because Army forces had a long history in Panama, commanders clearly understood the physical challenges and layout of critical urban areas (see Figure 6-3), particularly Panama City. They also understood how the infrastructure in each urban area functioned and which parts was a key element to success. Examples of key portions of the infrastructure included the Madden Dam, which controlled the water flow through the Panama Canal, and the Cerro Azul television tower, which was the main Panamanian broadcast tower.

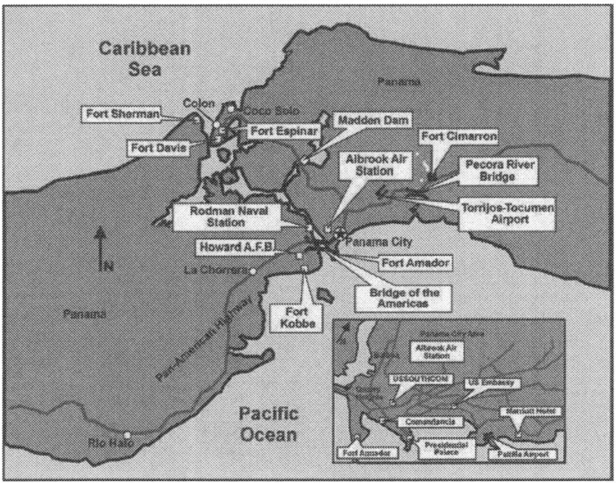


Figure 6-3. Panama

Collecting information and developing intelligence on the human elements of the urban environment was critical to operational success and a challenge. Because the Panamanian Defense Force (PDF) had traditionally been an ally of the United States, Army forces did not have a systemic database that adequately depicted their order of battle and their true capabilities. Additionally, much of the situation in Panama was colored in political terms making it more difficult for traditional military sources to evaluate the status of PDF forces. For example, Army planners needed to know if PDF military units (when faced with a formidable U.S. force) would fight at all for Noriega and if they did fight, how hard and long would they fight. The answers depended largely on their political loyalty to Noriega and on the individual loyalty of the unit officers to the Panamanian president. Thus, Army commanders needed to understand the military characteristics of PDF units and their political affiliations and tendencies.

Because transition from combat to noncombat tasks would be critical to achieving all objectives, particularly the restoration of democracy, Army forces also needed an accurate assessment of the political opposition to Noriega—including that opposition's capabilities and vulnerabilities. Again, Army forces were required to make assessments outside those needed solely for combat operations. Ultimately, assessing the political opposition's vulnerabilities led to assigning Army units to protect them throughout the operation so that they could serve as a foundation for a new democratic government.

Finally, the commander's assessment included an evaluation (often subjective) of the attitudes and disposition of the Panamanian people. Human intelligence (HUMINT) was the primary source of information on the population. Army forces had good access to the population because of their close proximity and historical ties to Panama. Many Soldiers were married to Panamanians, and the Army had total access to local media and to prominent individuals.

National imagery and special operations forces (SOF) also contributed to the ability of Army forces to assess the urban environments of JUST CAUSE. All units executing operations had detailed satellite photos of objective areas. Additionally, key objectives were placed under SOF surveillance well in advance. This surveillance revealed unit readiness, vulnerabilities, detailed disposition, and other patterns critical to mission success. The combination of the two capabilities allowed units to plan and achieve the synchronization necessary for such a complex urban operation.

Shape

During OPERATION JUST CAUSE, commanders conducted numerous shaping operations to establish the conditions for the decisive operations. Many operations were designed to control information, such as an assault on the Azul television tower identified during the assessment of the infrastructure. Planners designed many shaping operations to isolate various garrisons and PDF units. An example of tactical isolation was the plan for the Pacora River Bridge to prevent reinforcements from reaching the garrison at Torrijos-Tocumen Airport.

Operational isolation was achieved through the Ranger Regiment's and 82nd Airborne Division's assault on targets at Rio Hato in the west and Fort Cimarron in the east. These actions in conjunction with the securing of Maddam Dam had the primary objective of isolating Panama City. They were also the largest of the major actions occurring during OPERATION JUST CAUSE. The airborne assault was also the largest airborne operation conducted by U.S. forces since World War II. This large-scale shaping operation demonstrates that shaping operations are critical to mission success and can be more resource intensive than the actual operations that achieve domination.

Engage

U.S. Army forces achieved successful engagement in OPERATION JUST CAUSE by establishing unchallenged military control over Panama City and eliminating Noriega's capability to challenge that control. Toward this end, the operation attacked two decisive points. The first was the assault on the PDF headquarters located in Panama City: the Comandancia. The second was the operation undertaken to locate and seize Noriega himself.

Three battalions of task force (TF) Bayonet (5-87th Infantry, 1-508 Infantry [Airborne], and 4-6th Infantry [Mechanized]) executed the attack on the Comandancia and Fort Amador. They were also tasked to protect the American Embassy in downtown Panama City. To execute these missions, they moved from various staging areas located throughout the city to their assigned objectives using air assault, mounted, and dismounted approaches. The ground movement through the city proved to be the most difficult and hazardous part of the mission due to the vulnerability of the troops in their armored personnel carriers and trucks. The dismounted movement was slower than the mounted movement but allowed the Soldiers greater cover and concealment.

The strongest opposition to TF Bayonet occurred at the Comandancia. Elements of three PDF companies and two public order companies held out for three hours. The troops moving to Comandancia were subject to a large volume of sniper fire, and in the assault, unidentified indirect fire caused significant casualties among the mechanized forces. TF Bayonet forces, supported by airborne armored reconnaissance vehicles and Hellfire missiles from Apache helicopters, captured the

Comandancia. Commanders noted in particular the precision of the supporting fires from attack helicopters. The assault by fire from supporting AC 130 gunships severely damaged much of the reinforced Comandancia building.

Simultaneously, SOF attacked several targets where Noriega might be located. These initial attacks were unsuccessful. However, many subsequent actions neutralized Noriega's influence and eventually resulted in his apprehension on 3 January 1990. These actions included the well-organized and relentless manhunt conducted by SOF units, the isolation of Panama City itself, population control efforts, sophisticated IO, and cooperation with other U.S. agencies.

Consolidate and Transition

OPERATION JUST CAUSE demonstrated the vital need for a thought-out plan that adequately addresses consolidation and the transition from combat to noncombat before commanders initiate operations. Normally in complex UO, commanders cannot leave the details of transition until after the operation has begun without unacceptable risk to overall mission accomplishment. The follow-on stability operation, OPERATION BLIND LOGIC (later renamed OPERATION PROMOTE LIBERTY) began 24 hours after the initial assault and thus both operations were occurring simultaneously. This simultaneity of different types of operations is typical in major operations conducted in a large urban area. The stability operation involved more time than the combat operation and continued well after the close of OPERATION JUST CAUSE and after most of the major combat units had redeployed. It involved significant resources without the same level of risk to U.S. forces as the combat operations.

Civil affairs (CA) were a dominant part of the transition from combat to stability operations. The 96th Civil Affairs battalion was central to this operation. CA forces established a civil police force, emergency food distribution, property protection, production and distribution of newspapers, cleanup of the city, and building support for a new civil government. Most tasks were coordinated through Army CA forces and executed by other Army forces under the supervision of CA.

IO were also a major aspect of affecting a stable transition and successful postcombat operations. These operations built support for the U.S. operation among the population. They emphasized that the U.S. conflict was with Noriega and not the Panamanian people and that the U.S. forces would depart as soon as a new Panamanian government could take over.

Other U.S. agencies played critical roles in stability operations in Panama. The U.S. Drug Enforcement Agency and Justice Department were important to the negotiations that led to Noriega's capture. The U.S. State Department helped to negotiate for Noriega and develop military policies and plans during stability operations. The American Embassy advised commanders regarding the large diplomatic community that existed in Panama City.

# FUNDAMENTALS OF URBAN OPERATIONS

6-28. UO often differ from one operation to the next. However, some fundamentals apply to UO regardless of the mission, geographical location, or level of command. Some of these fundamentals are not exclusive to urban environments. Yet, they are particularly relevant to an environment dominated by man-made structures and a dense noncombatant population (see figure 6-4). Appendix A provides an historical example of how many of these fundamentals applied to an actual conflict situation. Vitally, these fundamentals help to ensure that every action taken by a commander operating in an urban environment contributes to the desired end-state of the major operation.

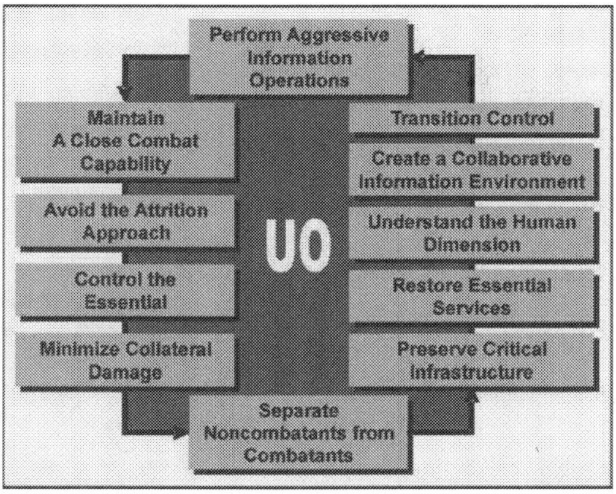


**Figure 6-4. Fundamentals of urban operations**

## Perform Aggressive Information Operations

6-29. Information operations aimed at influencing non-Army sources of information are critical in UO across the spectrum of operations. Because of the density of noncombatants and information sources, the media, the public, allies, coalition partners, neutral nations, and strategic leadership will likely scrutinize how Army forces participate in UO. The proliferation of cell phones, video cameras, Internet capability, and media outlets ensure close observation of the activities of Army forces. With information sources rapidly expanding, public information of Army operations will often be available faster than the internal military INFOSYS can process it. Army forces should aggressively integrate IO into every facet and at all levels of the operation to manage perceptions and mitigate unintended consequences. Under media scrutiny, the actions of one Soldier may have significant strategic implications. IO aim to make the information accurate; placed in the proper context of the Army's mission; and available to all interested parties: the public, the media, and other agencies. Ensuring that the urban population understands how they and their urban area fit within the commander's vision, intent, and end state will be a critical task.

## Maintain a Close Combat Capability

6-30. Close combat is required in all offensive and defensive UO. This core capability must also be present and visible in urban stability operations and may be required in urban civil support operations. Close combat in any urban operation is resource intensive, requires properly trained and equipped forces, and has the potential for high casualties. However, the ability to decisively close with and destroy enemy forces as a combined arms team remains essential. Potential threats respect strength and, instead, seek to attack ill-prepared or complacent Army forces—not necessarily forces that are better protected or armored, but those less likely or capable of inflicting casualties against them. In stability and civil support operations, a lack of respect and fear of Army forces can hinder recovery as much as the ill-advised use of forces. Hence, all Soldiers, regardless of branch or military occupational specialty, must be properly equipped and trained to fight in an urban environment. This ability allows Army forces to deter aggression, compel compliance, morally and physically dominate a threat and destroy his means to resist, and terminate urban conflicts on the Army commander's terms.

## Avoid the Attrition Approach

6-31. Previous Army doctrine was inclined towards a systematic linear approach to urban combat. This approach emphasized standoff weapons and firepower. Army force structure does not support this approach towards UO. It can result in significant collateral damage, a lengthy operation, and an inconsistency with the political situation and strategic objectives. Enemy forces that defend urban areas want Army forces to adopt this approach because of the likely costs in resources. Commanders should only consider this approach to urban combat as an exception and justified by unique circumstances. Instead, commanders should seek to achieve precise, intended effects against multiple decisive points that overwhelm a threat's ability to react effectively.

## Control the Essential

*I'm talking about attacking those things from which the regime draws power but being very careful about it so that we don't get large bodies of young Americans caught up in house-to-house Berlin, World War II-type scenario.*

Lt. Gen. William C. Wallace
Commander, V Corps
OPERATION IRAQI FREEDOM

6-32. Many modern urban areas are too large to be completely occupied or even effectively controlled without an enormous force. Therefore, Army forces focus their efforts on controlling only the essentials to mission accomplishment. At a minimum, this requires control of key terrain. Key terrain is terrain whose possession or control provides a marked advantage to one side or another. In the urban environment, commanders determine key terrain based on its functional, political, economic, or social significance. A power station or a place of worship may be key terrain.

6-33. All principles of war can apply to UO. The principle of mass and the principle of economy of force (in addition to the principle of unity of command discussed in the previous chapter) are particularly important in guiding UO and providing mission focus. Army forces mass combat power only to control those requirements essential for mission success. This permits conservation of combat power. It also implies economy of force and associated risk in those areas where Army forces choose not to exercise control.

## Minimize Collateral Damage

6-34. Forces should integrate precision fires, IO, and nonlethal tactical systems consistent with mission accomplishment while decreasing the potential for collateral damage. Commanders develop unique rules of engagement (ROE) for each urban operation and provide necessary firepower constraints. IO and nonlethal systems may compensate for some restrictions, especially in stability or civil support operations. Commanders continually assess the short- and long-term effects of operations and firepower on the population, infrastructure, subsequent missions, and national and strategic objectives. They must also consider what, if any, provisions should be made to amend or address potential collateral damage. Overall, commanders must balance restraint and precision with speed and overwhelming combat power. By avoiding unnecessary harm to all elements of the urban environment, commanders retain the moral high ground and help sustain legitimacy for their operations. Minimization of collateral damage allows civilians to continue to provide for their own needs or the rapid return of the urban area to civilian self-sufficiency.

## Separate Noncombatants from Combatants

6-35. Promptly separating noncombatants from combatants (psychologically and physically) may make the operation more efficient and diminish some of the threat's potential asymmetrical advantages. This separation also may reduce restrictions on the use of firepower, enhance force protection, and strip the threat from its popular support base. This important task becomes more difficult when the threat is an unconventional force that can mix with civilians.

6-36. In more recent operations, threats have sought to integrate their military capabilities as closely as possible into the civilian population and infrastructure. In these conditions, commanders must increase their efforts to discriminate between the two. Soldiers managing violence in this setting require the highest level of individual and organizational discipline and judgment. Soldiers will require the mental agility to separate their aggression toward threats from the noncombatant civilian population. The training, effort, and command emphasis in this area is as important as fully successful results. Such efforts strongly impact national and international perceptions of the operation.

## PRESERVE CRITICAL INFRASTRUCTURE

6-37. Commanders should analyze the urban area to identify critical infrastructure. They should attempt to preserve and protect the critical elements for postcombat sustainment operations, stability or civil support operations, or the overall health and well-being of the indigenous population. Urban areas remain in the AO after combat operations have ceased. Postcombat UO are unavoidable. Different from simply avoiding collateral damage, Army forces may have to initiate actions to prevent an enemy or a hostile civilian group from removing or destroying critical infrastructure and assets. This may include cultural infrastructure such as religious and historical places. In some cases, preserving the infrastructure and the urban society's sources of economic and cultural wealth may be the assigned objective of the urban operation.

## RESTORE ESSENTIAL SERVICES

6-38. Army forces plan to restore essential services that may fail to function before or during an operation. Essential services include power, food, water, sewage, medical care, and security and law enforcement. When planning for and conducting Army UO, units can use nonlethal and less destructive munitions and capabilities to keep potentially vital infrastructure intact. Initially, Army forces may be the only force able to restore or provide essential services and commanders must plan accordingly. Failure to do so can result in serious health problems for the civilians, which can affect the health of Army forces and negatively impact overall mission success. Army forces seek to transfer responsibility for providing essential services to other agencies, NGOs, or the local government as quickly and effectively as possible. Despite the potential causes for the failure or destruction of essential services, commanders also seek to ensure that civilians continually perceive Army restoration activities as assistance rather than an Army requirement. Otherwise, the civilian populace may be slow to accept or resume responsibility for their urban area.

## UNDERSTAND THE HUMAN DIMENSION

> *Do not try to do too much with your own hands. Better the Arabs do it tolerably than that you do it perfectly. It is their war, and you are to help them, not to win it for them. Actually, also, under the very odd conditions of Arabia, your practical work will not be as good as, perhaps, you think it is. It may take them longer and it may not be as good as you think, but if it is theirs, it will be better.*
>
> T. E. Lawrence
> "Twenty-Seven Articles"

6-39. Commanders carefully consider and manage the perceptions, allegiance, and morale of the civilians. Their assessment of the environment needs to accurately identify the attitudes of the people toward Army forces. Operational guidance to subordinates—including ROE, protection, sustainment operations, and fraternization—is based on this assessment. Commanders expect and consider the demographic variance in the attitudes of an urban population. They cannot inadvertently apply Western cultural norms to a non-Western urban population. Commanders can only make reliable assessments based on a thorough understanding and appreciation of the local society and their culture. Developing this ability also requires the additional ability to effectively share information among all echelons of command.

6-40. Sound policies, proper discipline, adequate consideration for local culture, and rapid engagement of local urban leaders will positively affect the attitudes of the population toward Army forces. Additionally, well-conceived and executed IO will enhance the position of Army forces relative to the urban population. Even during high-intensity urban combat, heightened awareness of and sensitivity toward the civilians can lead to a better postcombat situation than if civil considerations were unobserved or diminished in importance. An improved postcombat situation enhances transition. As the environment of conflict becomes more complex, the human dimension (and associated moral aspects) takes on greater importance and may have the greatest potential for affecting the successful outcome of UO. Therefore, the human aspect and development of cultural acuity creates a discrete planning factor.

## Create a Collaborative Information Environment

6-41. The complexity of the urban environment, particularly the human dimension, requires rapid information sharing—providing and receiving—from the national level to the tactical level, among Army headquarters at each echelon, with sister services and coalition partners, and with participating governmental and (at appropriate times) nongovernmental agencies. The analysis of urban information into the relevant intelligence necessary to refine and deepen a commander's understanding of the urban environment and its infrastructure of systems also demands collaboration among the various information sources and consumers. Therefore, commanders seek to establish streamlined procedures, develop commonality among databases, and make best use of existing information systems to disseminate and receive the necessary intelligence and relevant information for subordinates and partner organizations and agencies to exercise effective leadership and decision making and establish synergistic unity of effort in this multifaceted environment.

## Transition Control

6-42. Because UO are resource intensive, commanders should plan to end them as quickly as possible, yet consistently with successful mission accomplishment. (Depending on the factors of METT-TC, successful transition may take a few days or many years to achieve.) The end state of all UO transfers control of the urban area to another agency or returns it to legitimate civilian control and responsibility. Rapid transition releases Army resources for use elsewhere and improves the civilian morale and disposition toward Army forces. This requires the successful completion of the Army force mission and a thorough transition plan. The transition plan may include returning control of the urban area to another agency a portion at a time as conditions permit. A successful transition plan should consider early alignment of military capabilities with existing urban governmental and administrative organizations, agencies, structures, and districts. Transition planning must be conducted before the onset of operations and continually adjusted as the situation develops.

# Chapter 7
# Urban Offensive Operations

*... Capture Suez City "provided it does not become a Stalingrad situation."*

Order to the Adan Armored Division
prior to its 1973 attack on Suez City
*On the Banks of the Suez*

Offensive urban operations (UO) are one of the most challenging operations that military forces can undertake. Campaigns and wars have sometimes hinged on their success or failure. Costly in resources, even when successful, they are not lightly entered into. Once engaged, they are executed rapidly and decisively. For reasons already discussed, threat forces defending in UO may gain advantages from the environment while Army force capabilities may diminish. Despite the challenges, Army forces conduct successful urban offensive operations by combining the Army's existing offensive doctrine with a thorough understanding of the environment.

## PURPOSE OF URBAN OFFENSIVE OPERATIONS

7-1. Like all offensive operations, urban offensive operations are designed to impose the will of commanders on the threat. The urban offense often aims to destroy, defeat, or neutralize a threat force. However, the purpose may be to achieve some effect relating to the population or infrastructure of the urban area. Army forces may conduct offensive operations to secure a port or a communications center, to eliminate a threat to a friendly government or the urban population, or to deny the threat use of urban infrastructure. No matter the purpose, commanders must use a combined arms approach for successful urban offensive operations.

## OFFENSIVE CHARACTERISTICS APPLIED TO URBAN OPERATIONS

7-2. All offensive operations, including those in urban areas, contain the characteristics of surprise, concentration, tempo, and audacity (see FM 3-0). Commanders consider and incorporate these characteristics in their offensive UO plans.

### SURPRISE

7-3. Army forces can achieve offensive surprise at two levels: operational and tactical. In urban offensive operations, operational surprise can be decisive. The goal is to attack the urban area before the threat expects it, from a direction he doesn't expect, or in a manner he doesn't expect. In UO, this requires an attack against urban areas that the threat believes will provide sanctuary from the technological advantages of Army forces. Usually, urban areas that meet this criterion are not easily accessible. Army forces launch an attack against these urban complexes in different ways: through a vertical assault using airborne or air assault forces, through an amphibious assault, or through a penetration followed by a rapid and deep advance. All three attacks aim to achieve surprise and to deny the threat time to prepare and establish a defense. Surprise in a major urban operation prevents a threat from falling back to occupy prepared positions in and around an urban area.

7-4. At lower tactical levels, forces achieve surprise by attacking asymmetrically. An asymmetric method attacks the threat so he cannot respond effectively. This may be achieved by using special operations forces

(SOF) against a threat prepared for a conventional attack, by attacking decisively with heavy forces when the threat expects an effort by light forces or SOF, or by leveraging Army forces' extensive information operations (IO) capability. Offensive IO—primarily using IO elements of deception, electronic warfare, and operations security (OPSEC)—can help achieve surprise at all levels (see Chapter 4). Attacking at night surprises the threat and maximizes the Army forces' training, command and control (C2), and technological advantages. Attacking from unexpected or multiple directions achieves surprise by leveraging Army information systems (INFOSYS) and superior synchronization of combat power and capabilities.

## CONCENTRATION

7-5. In UO, the attacking force creates a major advantage by concentrating the effects of combat power at the point and time of its choosing. The area and its compartmented effects naturally disperse and dissipate combat capability. Often commanders must position considerable amounts of stay-behind forces to protect hard-fought gains. The environment also hinders repositioning forces rapidly. Such effects can work equally against defending and attacking forces. However, in a well-prepared defense, the defender often has the advantage of interior lines. The defender can reinforce or reposition forces more quickly using covered and concealed routes (such as, sewers, tunnels, or prepared holes made in walls). Successful UO need synchronized air and ground maneuver with overwhelming effects from fires at decisive points on the urban battlefield. To achieve proper synchronization and precise effects, commanders should consider the unique time and distance relationships set by the environment.

## TEMPO

7-6. Tempo is the rate of military action; it differs between UO and operations in more open terrain. Often, the primary purpose of the threat's urban defense is to take advantage of that differential to disrupt the rapidity and overall tempo of the Army's major operation. Additionally, the complexity and the potential for increased risk in an urban environment may invoke a cautious and methodical response on the part of commanders and their staffs thereby exacerbating tempo differences. While careful and meticulous preparation and planning are required in UO, commanders conducting major operations that include urban areas must strive to maintain an active tempo in offensive operations through synchronization of combat power and anticipation of threat reactions. A high tempo allows Army forces to achieve surprise and quickly gain positions of advantage. Controlling operational tempo and not allowing the different tempo of UO to adversely affect other operations is a challenge that commanders of major operations must overcome.

### Operational Context of Urban Operations: Brittany Ports – 1944

The plan for the invasion of Normandy, France, in June 1944 was meticulously developed. The plan not only addressed the invasion itself, but also contained detailed planning for the campaign to follow. A major concern of the detailed campaign planning was logistics. To address this critical concern, and specifically the problem of ports to supply the allied armies once ashore, the pre-invasion planning called for the major ports of the French province of Brittany—Brest, Lorient, and Saint Nazaire—to be objectives of General Patton's Third Army, once it was activated.

Early August 1944, almost two months after the successful Normandy invasion, the operational situation significantly differed from that envisioned by the D-Day planners. General Montgomery's Twenty-first Army Group was still fighting in the Bocage of Normandy. In contrast, General Bradley's Twelfth Army Group had just achieved a major breakthrough at Saint Lo, secured the Cotentin Peninsula, and reached the city of Avranches. Here was a decision point. Bradley and Eisenhower had to decide whether to adhere to the original plan and turn west with Patton's forces to secure the peninsula or to take advantage of the breakout at Saint Lo and turn east to exploit the disruption of the German defenses.

Ultimately they reached a compromise. General Middleton's VIII Corps was tasked to secure the peninsula, and the bulk of Patton's Army, three Army corps, was turned northeast to exploit the operational collapse of the main German defenses. (See figure 7-1.)

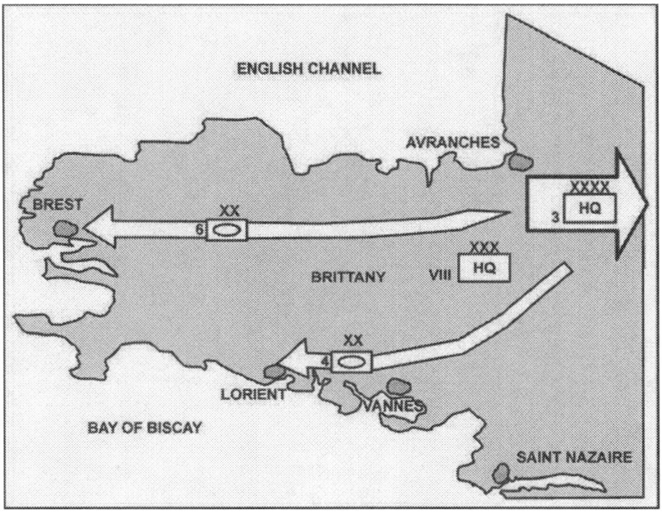


**Figure 7-1. Initial attack in Brittany**

Middleton's corps sprinted into the peninsula with the 4th and 6th Armored Divisions leading the way. However, poor communications, disagreements between commands, and contradictory orders caused the corps to hesitate before pushing the two divisions to continue to exploit toward the ports. The result: the 6th Armored Division missed an opportunity to seize Brest against light resistance by one day. The 4th Armored Division, after capturing the smaller port of Vannes, was also frustrated on the approaches to Lorient. The American reaction to the inability to rapidly seize the ports demonstrated an understanding of changing circumstances. The 6th Armored Division turned the attack at Brest to the 8th Infantry Division and then relieved the 4th Armored Division at Lorient. The 4th Armored was moved to rejoin the rest of Third Army exploiting to the east and north. Ultimately Brest fell to VIII Corps on 19 September after a 43-day siege by three infantry divisions. The victory yielded 36,000 German prisoners of war (POWs). However, the German defense and demolitions of the port left the port without an impact on the logistic situation of the allies. Brest cost the U.S. Army almost 10,000 casualties and the commitment of significant supplies. The experience convinced commanders to surround and bypass the other major Brittany ports. Lorient and Saint Nazaire remained under German control, deep in allied territory, until the war ended ten months later (see figure 7-2).

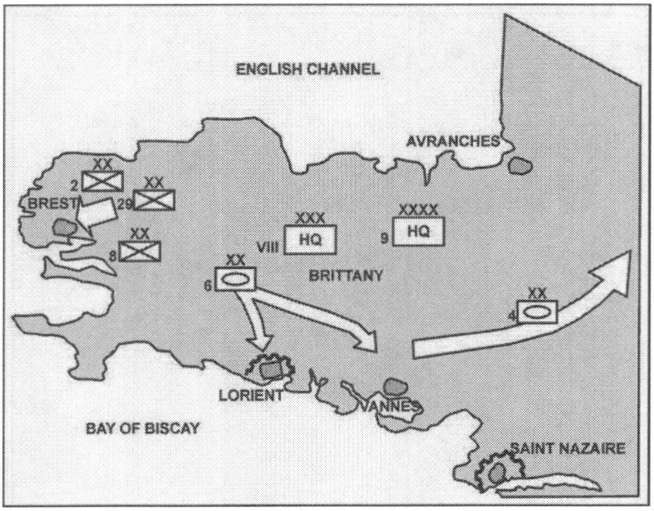


**Figure 7-2. Subsequent disposition of forces in Brittany**

The operational lessons of the Brittany campaign are numerous. First, commanders are responsible to continually assess assumptions and decisions made during planning based on the changing circumstances of the battlefield. This includes the planning decision to conduct urban offensive operations. When the allies arrived at the Brittany Peninsula, the focus of the operational maneuver was no longer securing logistics facilities but exploiting the breakthrough at Saint Lo and the disintegrating the German defense. The bulk of Third Army then was turned to the north and east rather than west into the peninsula.

The Brest experience also demonstrates that the costs of urban offensive operations are continually assessed against the operational value of the objective. This lesson was applied to the cities of Lorient and Saint Nazaire. The cities were never seized from the Germans because their logistic value failed to warrant the required resources to take them. German retention of the ports had no major adverse effect on the overall campaign.

Another lesson is that commanders cannot allow urban operations to disrupt the tempo of other offensive operations. One German goal of defending the ports was to disrupt the rapid tempo of the U.S. exploitation. They failed to achieve this goal because General Bradley continued the exploitation with the bulk of Third Army and executed the original plan with only a single corps.

Finally, commanders cannot allow emotion to color their decision to conduct or continue UO. The failure of 6th Armored Division to seize Brest rapidly caused some commanders to believe that Brest had to be captured because the prestige of the Army was committed to the battle. Costs of the continuing combat operations to seize Brest were significant. These resources might have been better committed elsewhere in the theater.

7-7. Tactical tempo is also important in urban combat. Because of the complex terrain, defending forces can rapidly occupy and defend from a position of strength. Once Army forces initiate tactical offensive operations, they cannot allow the threat to set the tempo of the operation. Instead, attacking forces seek to maintain a high tempo of operations. However, the tactical tempo of urban operations differs from operations in other terrain. Not necessarily slow, it requires a careful balance of preparation, speed, and security. In terms of unit fatigue, resource consumption, and contact with the threat, the tempo of most urban offensive operations may be rated as very high. On the other hand, in distances traveled and time consumed to achieve objectives, the tempo of many urban offensive operations might be rated as slow. The urban battlefield's density concentrates activity and consumes resources in a relatively small area. The lack of terrain seized or secured is not to be construed to mean a low tempo in the battle. Although the tempo may seem excruciatingly slow at higher levels of command and exceeding fast at lower, tactical levels, in reality, the natural tempo of urban operations is not faster or slower than other types of operations, merely different. Creating and operating at a tempo faster than an opponent can maintain, however, can favor forces which are better led, trained, prepared, and resourced.

7-8. A high tactical tempo in urban offensive operations challenges logisticians to provide for the increased consumption of munitions and degrades soldiers' physical capabilities. Commanders must anticipate these challenges and develop the means and abilities to overcome them. In the past, these challenges forced commanders to conduct urban offensives cyclically. They used night and other periods of limited visibility to resupply, rest, and refit forces. The environment influenced the tempo of their operations. This type of "battle rhythm" resulted in the forces spending each new day attacking a rested threat that was in a well-prepared position.

7-9. Army forces must maintain the tempo. Offensive operations continue even during darkness. Moreover, Army forces increase the tempo of operations at night to leverage the limited visibility capabilities, increased situational understanding, training, and INFOSYS that give an advantage to Army forces in all environments. To overcome the physical impact of the environment on soldiers, commanders can retain a large reserve to rotate, continuing offensive operations at night. The force that fights in daylight becomes the reserve, rests, and conducts sustaining operations while another force fights at night. Army forces can then maintain the tempo of operations and leverage technological advantages in urban offensive combat.

7-10. Tempo in UO does not necessarily mean speed. Offensive operations balance speed, security, and adequate firepower. Commanders must plan for the complex tactical environment and the requirements to secure flanks and airspace as the operation progresses. Mission orders allow subordinate units to make the most of tactical advantages and fleeting opportunities.

### AUDACITY

7-11. Audacity is a simple plan of action, boldly executed. Superb execution and calculated risk exemplify it. In an urban attack, a thorough understanding of the physical terrain can mitigate risk. The terrain's complexity can be studied to reveal advantages to the attacker. Audacity can also be embodied in an operation by inventively integrating and coordinating the direct action tasks of SOF throughout the operation. Combining SOF actions with conventional attacks can asymmetrically unhinge a threat's defensive plan. Well trained Soldiers confident in their ability to execute urban offensive operations foster audacity.

## URBAN OFFENSIVE OPERATIONS AND BATTLEFIELD ORGANIZATION

7-12. Urban offensive operations, like all operations, are arranged using the overall battlefield organization of decisive, shaping, and sustaining operations. Each operation is essential to the success of an urban offensive, and usually two or more of these operations occur simultaneously. Decisive operations are attacks that conclusively determine the outcome of UO. These attacks strike at a series of decisive points and directly lead to neutralizing the threat's center of gravity. Shaping operations in urban offensive operations create the conditions for decisive operations. In UO, much of the shaping effort focuses on isolation, which is critical in both major operations and tactical battles and engagements. Sustaining operations in urban offensive operations ensure freedom of action. They occur throughout the area of operations (AO) and for the duration of the operation.

## Decisive Operations

7-13. A tactical commander fights decisive urban combat, whereas commanders conducting a larger major operation influence urban combat by setting the conditions for tactical success. Higher commanders may directly influence urban offensive operations by operational maneuver, by coordinating joint fires, by closely coordinating conventional forces, or with SOF.

7-14. Tactical urban offensive operations quickly devolve into small-unit tactics of squads, platoons, and companies seizing their objectives. The compartmented effect of the terrain and the obstacles to command and control of small units, especially once they enter close combat inside buildings or underground, often restricts the higher commander's ability to influence operations. Commanders can influence the actions of subordinates by clearly identifying the decisive points leading to the center of gravity; using mission orders (as discussed in Chapter 5); developing effective task organizations; synchronizing their decisive, shaping, and sustaining operations; and managing transitions.

7-15. Like all operations, successful decisive operations in UO depend on identifying the decisive points so the forces can destroy or neutralize the threat's center of gravity. Seizing a key structure or system that makes the threat's defense untenable; interdicting a key resupply route that effectively isolates the threat force from his primary source of support; or isolating the threat so that his force can no longer influence friendly activity may be more effective than his outright destruction.

7-16. Commanders must select the right subordinate force for the mission and balance it with appropriate attachments. Higher commanders do not direct how to organize the small tactical combined arms teams, but they must ensure that subordinates have the proper balance of forces from which to form these teams. Successful urban offensive operations require small tactical combined arms teams. Urban offensive operations require abundant infantry as the base of this force. However, successful urban combat requires a combined arms approach adjusted for the conditions of the environment. In urban offensive operations, it will normally not be a question of whether include armor and mechanized elements but, rather, how best to accomplish this task organization. Precision-capable artillery systems generally support urban operations better than unguided rocket artillery.

7-17. Divisions entering urban combat may require additional resources. These resources include military intelligence support in the form of linguists, human intelligence (HUMINT) specialists, and unmanned aircraft systems (UAS). Engineering assets will be at a premium; the task organization of a task force executing the decisive operation may require a one-to-one ratio of engineer units to combat units. Corps and higher engineering support may be necessary to meet these requirements and to repair vital and specialized infrastructure. A tailored and dedicated support battalion or group may need to assist in providing anticipated support to a displaced and stressed civil population. Finally, divisional CA units may require augmentation to deal with NGOs and civilian government issues.

7-18. Successfully conducting decisive operations in the urban environment requires properly synchronizing the application of all available combat power. Army forces have a major advantage in the command and control of operations. Commanders can use this advantage to attack numerous decisive points simultaneously or in rapid succession. They also use it to attack each individual decisive point from as many directions and with as many different complementary capabilities as possible. Commanders must completely understand urban environmental effects on warfighting functions to envision and execute the bold and imaginative operations required. Significantly, these operations require that C2 systems account for the mitigating effects of the environment as execution occurs.

7-19. Properly synchronized actions considerably enhance the relative value of the combat power applied at the decisive points. They present to the threat more requirements than he has resources with which to respond. Synchronized IO and multiple maneuver actions paralyze the threat's decision-making capacity with information overload combined with attacks on his C2 systems. Additionally, well-synchronized actions limit the time the threat has to make decisions and forces him into bad decisions. In the urban environment, these effects are enhanced because C2 systems are already strained, poor decisions are harder to retrieve, and units that do not react effectively are more easily isolated and destroyed.

## SHAPING OPERATIONS

7-20. Shaping operations that support the urban attack separate into those focused on isolating the threat and all others. Army forces isolate the threat to ensure successful urban offensive operations. Depending on the threat reaction to isolation efforts and the nature of the threat center of gravity, this task may become decisive. Other shaping operations include those common to all offensive operations and others unique to urban operations. Unique urban shaping operations may include securing a foothold in a well-fortified defensive sector, securing key infrastructure, or protecting noncombatants. Because of the nature of UO, shaping operations may consume a much larger proportion of the force than during other operations and may take place both inside and outside the urban area (see Applying the Urban Operational Framework: Panama in Chapter 5). By successfully isolating a threat force, the force needed to conduct the decisive operation may be relatively small.

## SUSTAINING OPERATIONS

7-21. Commanders conducting urban offensive operations must ensure security of the sustaining operation and bases; in many situations, protecting lines of communications (LOCs) and sustaining operations may be the greatest vulnerability of the attacking force. Those supporting an urban offensive are tailored to the urban environment and are well forward. Ideally, the supporting forces closely follow the combat forces and move within or just outside the urban area as soon as they secure an area. Operating in the urban area during offensive operations allows the sustaining operation to take advantage of the defensive attributes of the environment for security purposes.

7-22. Counterattacks against sustaining operations may take the form of special operations activities aimed at LOCs leading to or within the urban area. Choke points—such as bridges, tunnels, and mountain passes—are vulnerable to these attacks and may require combat forces to protect them. Threat forces attack the LOC to blunt the Army's combat power advantage in the urban area.

7-23. Attacks against the LOC into the urban area may also attempt to isolate the attacking Army forces from its sustainment base. Isolated forces in an urban area are greatly disadvantaged. Commanders must plan and aggressively execute strong measures to protect their LOC, even if it requires reduced combat power to execute their offensive operation.

7-24. Sustaining operations anticipate the volume and unique logistics requirements of urban operations. Specialized individual equipment—such as grappling hooks, ladders, and pads—is identified and provided to troops in quantity before they are needed. Forces stockpile and distribute their attacking units' special munitions requirements including small arms, explosives, and grenades of all types, precision artillery munitions, and mortar ammunition. Forces also supply transport to move the resources rapidly forward, both to and through the urban environment. Sustaining operations cannot rely on "operational pauses" to execute their tasks. Commanders must plan to continuously supply resources and capabilities to the most forward combatants as offensive operations advance.

7-25. Sustaining operations also anticipate the growth of sustainment requirements as Army forces secure and take responsibility for large portions of the urban area. The success of Army urban offensive operations will often uncover the civil population in former threat occupied areas. It may attract the civil population from sections of the urban area where the Army is not operating to areas occupied by Army forces. Rural populations may migrate to the urban area as the result of successful Army offensive operations.

7-26. Army forces may be required to take initial responsibility to provide for the urban population. This consideration is integrated into logistics planning and organization from the start of the planning process. To be successful and efficient in such a situation, sustainment planning includes Army civil affairs (CA) specialists and local government representatives. It also integrates and consults with the international community and nongovernmental organizations (NGOs) that might augment or supplement Army sustainment capabilities.

# FORMS AND TYPES OF URBAN OFFENSE

7-27. Traditional forms of offensive maneuver include envelopment, turning movement, infiltration, penetration, and frontal attack. Traditional types of offensive operations are movement to contact, attack, exploitation, and pursuit. These traditional forms and types listed apply to urban combat. Some have greater application to an urban environment than others do. Moreover, success will belong to commanders who imaginatively combine and sequence these forms and types throughout the depth, breadth, and height of the urban battlefield. This is true at the lowest tactical level and in major operations.

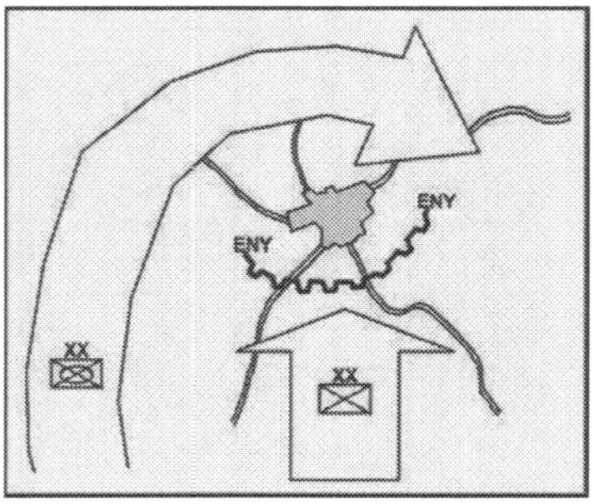


Figure 7-3. Envelopment isolates an urban area

## FORMS OF OFFENSIVE MANEUVER

### Envelopment

7-28. The envelopment is the ideal maneuver for isolating threat elements in the urban area or isolating the area itself. A deep envelopment effectively isolates the defending forces and sets the conditions for attacking the urban area from the flank or rear. Yet, enveloping an objective or threat force in the urban area is often harder since achieving speed of maneuver in the environment is so difficult (see figure 7-3). Vertical envelopment, however, works effectively if Army fires can effectively suppress or neutralize the threat air defense.

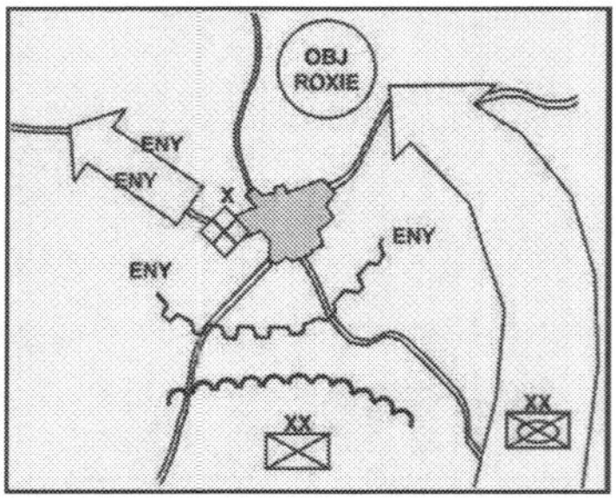


Figure 7-4. Turning movement

### Turning Movement

7-29. Turning movements can also be extremely effective in major operations (see figure 7-4). By controlling key LOCs into the urban area, Army forces can force the threat to abandon the urban area entirely. These movements may also force the threat to fight in the open to regain control of LOCs.

### Infiltration

7-30. Infiltration secures key objectives in the urban area while avoiding unnecessary combat with threat defensive forces on conditions favorable to them (see figure 7-5). This technique seeks to avoid the threat's defense using stealthy, clandestine movement through all dimensions of the urban area to occupy positions of

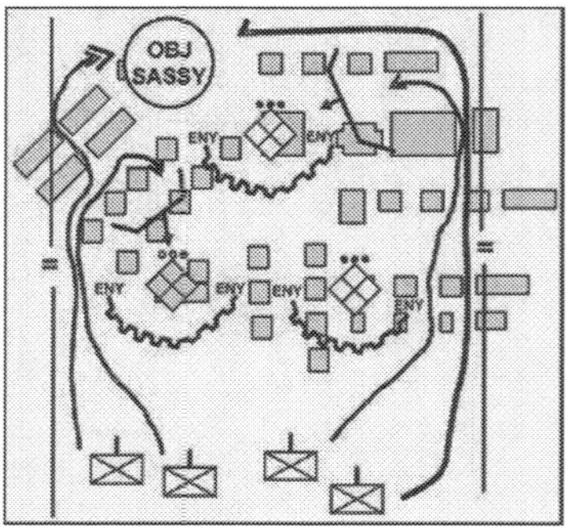


Figure 7-5. Infiltration

advantage in the threat's rear (or elsewhere).It depends on the careful selection of objectives that threaten the integrity of the threat's defense and a superior common operational picture (COP). Well-planned and resourced deception operations may potentially play a critical role in masking the movement of infiltrating forces. The difficulty of infiltration attacks increases with the size and number of units involved. It is also more difficult when Army forces face a hostile civilian population. Under such circumstances, infiltration by conventional forces may be impossible. Armored forces are generally inappropriate for infiltration operations. However, they may infiltrate large urban areas if the threat is not established in strength and had insufficient time to prepare defenses.

## Penetration

7-31. Penetration is often the most useful form of attack against a prepared and comprehensive urban defense (see figure 7-6). It focuses on successfully attacking a decisive point or on segmenting or fragmenting the defense—thereby weakening it and allowing for piecemeal destruction. (The decisive point may be a relatively weak or undefended area that allows Army forces to establish a foothold for attacks on the remainder of the urban area.) Ideally in urban combat, multiple penetrations in all dimensions are focused at the same decisive point or on several decisive points simultaneously. In urban combat, the flanks of a penetration attack are secure, and resources are positioned to exploit the penetration once achieved. Although always a combined arms team, rapid penetrations are enhanced by the potential speed, firepower, and shock action of armored and mechanized forces.

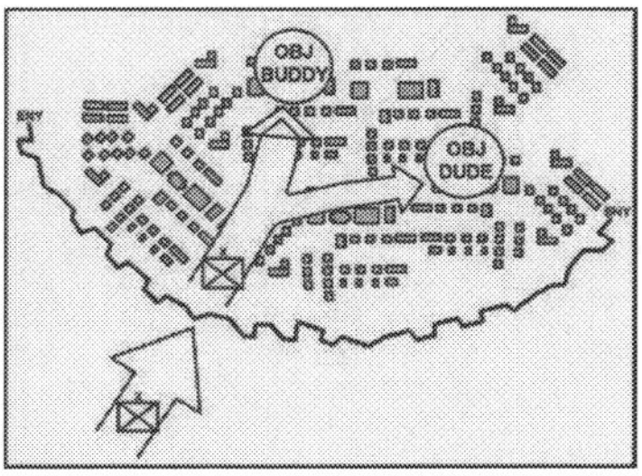


Figure 7-6. Penetration

7-32. Importantly, commanders must consider required actions and resources that must be applied following a successful (or unsuccessful) penetration. A penetration may result in the rapid collapse and defeat of the threat defense and complete capitulation. On the other hand, success may cause threat forces to withdraw but leave significant stay-behind forces (or disperse into the urban population as an insurgent-type force) which may necessitate methodical room-to-room clearance operations by significant dismounted forces. Additionally, securing portions (or all) of an urban area requires occupation by Army forces to prevent re-infiltration of threat forces thereby further increasing manpower requirements. Based on the factors of the situation, commanders may conclude that over time methodical clearance operations conducted from the outset, while frontloading time and resource requirements, will be less costly than a penetration followed by systematic clearing operations.

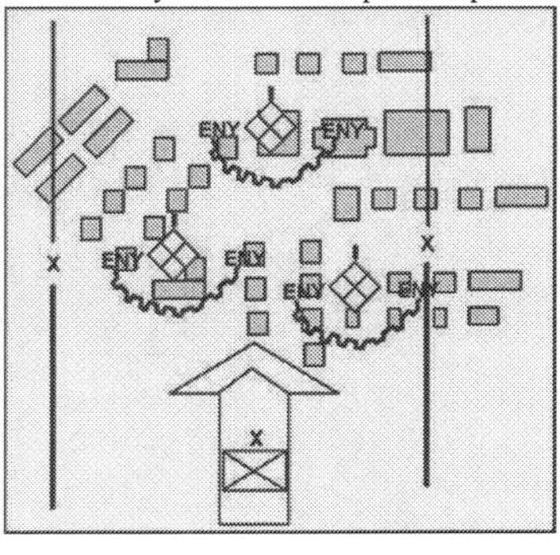


Figure 7-7. Frontal attack

## Frontal Attack

7-33. For the commander of a major operation, the frontal attack is generally the least favorable form of maneuver against an urban area unless the threat is at an obvious disadvantage in organization; numbers, training, weapons capabilities, and overall combat power (see figure 7-7). Frontal attacks require many resources to execute properly, risk dispersing combat power into nonessential portions of the area, and risk exposing more of the force than necessary to threat fires. In urban offensive combat, forces most effectively use the frontal attack at the lowest tactical level once they set conditions to ensure that they have achieved overwhelming combat power. Then the force of the frontal attack overwhelms the threat with speed and coordinated and synchronized combat power at the point of attack. The assigned frontage for units conducting an attack on an urban area depends upon the size and type of the buildings and the anticipated threat disposition. Generally, a company attacks on a one- to two-block front and a battalion on a two- to four-block front (based on city blocks averaging 175 meters in width).

### Forms of Attack in the Urban Offense: Metz – 1944

In November 1944, the U.S. Third Army launched its final effort to take the French city of Metz from the defending Germans. This was the Army's third attempt. The first attempt had been a surprise, mounted attack. This was followed by a series of piecemeal infantry assaults on the surrounding fortresses. Finally, a deliberate effort was made to take the city in a coordinated effort by XX Corps.

The initial failures stemmed from a shortage of resources on the U.S. side, to include fuel and units—especially infantry. This added to the ad-hoc nature of the first two efforts. The third effort, though more deliberate, was still constrained by resources. XX Corps could only muster three nearly full-strength infantry divisions to attack the German's defending with four under-strength divisions.

The third attempt to take Metz demonstrates how a corps operates with multiple divisions using various forms of attack to achieve its objective in urban offensive operations.

The opening phase of the Metz battle had attacks by the 90th and 5th Infantry Divisions to envelop the city from the north and south (see figure 7-8). This isolated the city and ensured the garrison could not escape nor be reinforced. The garrison was under orders from Hitler not to conduct a breakout and to resist to the last man; thus, German forces strongly counterattacked both wings of the envelopment to prevent isolation.

The second phase of the operation was the penetration of the city defenses from multiple directions. The 5th Infantry Division penetrated into the city with the 10th and 11th Infantry Regiments from the south. The 95th Infantry Division penetrated into the city from the north with the 377th Infantry Regiment and TF Bacon. Simultaneously, the 95th Infantry Division infiltrated battalions through the string of fortress positions guarding the western approaches into the city, isolated, and bypassed these positions with its other two regiments.

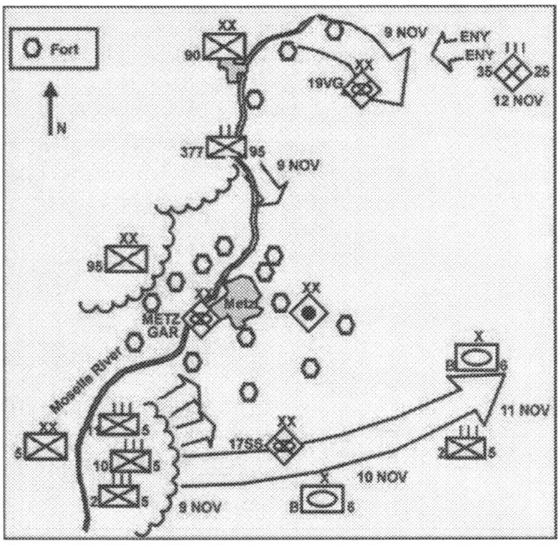


**Figure 7-8. Metz envelopment**

The final reduction of the defense was a series of battalion frontal attacks, which took place against the last remaining strongholds within the city. Even in these final engagements, however, the infantry battalions isolated, bypassed, and then attacked from the flanks and rear whenever possible (see figure 7-9).

The city was declared secured on 19 November. However, at that point more than a half-dozen of the fortresses had yet to be reduced. The 95th Infantry Division, after a four-day rest, quickly moved forward to rejoin the still rapidly advancing corps forward elements. Elements of the 5th Infantry Division remained in siege posture around the remaining strong points, the last of which surrendered on 19 December 1944 when it ran out of food. U.S. forces made no efforts to attempt to assault these bypassed fortresses although extensive psychological operations (PSYOP) were used.

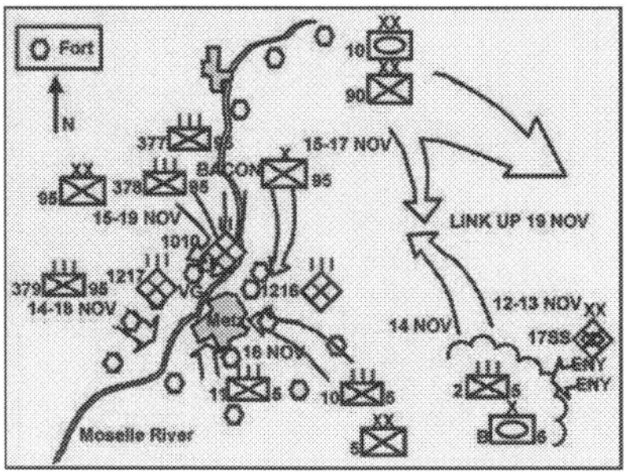


Figure 7-9. Metz final assault

## TYPES OF OFFENSIVE OPERATIONS

### Movement to Contact

7-34. In an urban area where the threat situation is vague, Army forces will often conduct a movement to contact to establish or regain threat contact and develop the situation. A movement to contact in an urban area occurs as both sides try to establish their influence or control over a contested urban area. The situation determines whether the movement to contact or its specific technique, the search and attack, is appropriate. A conventional force-oriented movement to contact will likely take place when friendly and threat conventional forces attempt to establish control simultaneously. Initially, neither side is defensive. The friendly force aims to quickly locate and fix the threat while establishing control of the urban area and its key infrastructure. The search and attack technique works well when a smaller threat has established a noncontiguous defense in an urban area. This operation is characterized by the friendly point defense of key infrastructure, robust reconnaissance, and rapidly concentrated combat power to fix and defeat or destroy threat resistance once located.

7-35. A meeting engagement often results from the movement to contact. It occurs when a moving force that is partially deployed for battle collides with and engages a threat at an unexpected time and place. In a meeting engagement in an urban area, the unit that reacts most quickly and decisively will likely win. Rapid and accurate decision making depends heavily on understanding the nature of the urban area and its impact on operations. Thus, in a meeting engagement, commanders should quickly assess the impact and role of all components of the urban environment (terrain, infrastructure, and society) on the operation. To this end, responsive reconnaissance is important. This permits accurate decision making regarding where to attack, where to defend, and how to allocate resources. Situational understanding enhanced by robust digital INFOSYS that provide an accurate COP also facilitates the rapid reaction of Army units and a synchronized response. This reaction and response allow Army forces to seize the initiative and dominate the threat.

### Attack

7-36. The attack is the most common and likely offensive operation that Army forces will conduct in an urban environment. Commanders conducting major operations and commanders of large tactical units usually execute deliberate attacks. In the urban environment, units larger than battalion-size rarely conduct hasty attacks. Hasty attacks are common below company level as units use their initiative to take advantage of tactical opportunities. However, larger units will conduct hasty attacks when threat defenses are disrupted or unprepared, to take advantage of an unexpected situation, and to prevent the threat from establishing or re-establishing a coherent defense.

### Exploitation

7-37. Exploitation follows a successful attack to disrupt the threat in depth. Commanders of major operations should consider focusing exploitation attacks on urban areas. A threat defeated in an attack will attempt to rally units, reinforce with reserves, and reorganize his defense. With its information and communications capability, transportation network, and defensive attributes, the urban area is the natural focal point to reestablish a disrupted defense. By establishing urban centers as the objectives of the exploitation, commanders deny the threat the sanctuary needed to reorganize and reestablish his defense. The exploitation focuses on the urban area as well as on the remnants of the threat. A successful exploitation to seize an urban area works efficiently because the attack preempts the defense and denies the threat the full advantages of urban terrain.

7-38. Commanders conducting exploitation must acknowledge the vulnerability of their forces to counterattack and ambush in urban areas. An urban area provides ideal cover and concealment to hide threat reserves, reinforcements, or reorganized forces. Constrictions of routes into and through the urban area make exploitation forces a potentially dense target and limit maneuver options. Robust and well-coordinated reconnaissance, tactical dispersal, and use of advance guard security forces protect against this threat (see Defensive Combat Power: Suez City vignette in Chapter 8).

### Pursuit

7-39. The pursuit is designed to destroy threat forces attempting to escape. It focuses on the threat and not on urban areas. When conducting a pursuit, Army forces move through undefended urban areas and, if possible, bypass those in which threat forces successfully take refuge. The threat will likely attempt to use urban areas to disrupt the pursuit and permit the threat main body to escape. Commanders can prevent escape by denying the threat the time to establish forces in urban areas that cannot be bypassed. The agility of Army aviation forces for attack, reconnaissance, and transportation is essential to execute a successful pursuit around and through urban areas.

## URBAN OFFENSIVE CONSIDERATIONS

7-40. The urban operational framework (understand, shape, engage, consolidate, and transition) provides a structure for developing considerations unique to urban offensive operations. The considerations vary depending on the situation and scale of the operation. Some considerations applicable to major operations that include an urban area will also be considerations at the tactical level focused in the urban area. However, no set rules exist. All urban operations are unique. Issues addressed at the operational level in one situation may be addressed in a new situation only at the tactical level. Under the right circumstances, a consideration may become an operational issue, a tactical issue, or a combination of the two. The following identifies some planning and execution issues that commanders conducting major operations should address.

### UNDERSTAND

7-41. The first requirement, and a continuing requirement throughout the conduct of urban operations, is the assessment and understanding of the situation. Commanders should base this understanding on detailed information regarding the particular urban area. Since the threat will likely dominate or control most of the urban area during the planning phase of offensive operations, achieving an accurate understanding of the urban environment will be difficult. A comprehensive intelligence, surveillance, and reconnaissance (ISR) effort in support of a rigorous intelligence preparation of the battlefield (IPB) process overcomes this obstacle.

## Integrated Intelligence, Surveillance, and Reconnaissance

7-42. The commander of a major operation that includes an urban area, unlike his subordinate commanders, can target reconnaissance deep into his AO and area of interest. This begins the application of ISR resources against the urban area that may lead to decisive ground operations. This ISR effort and the understanding it supports continue as long as the urban area remains in the AO. Commanders of major operations initially direct ISR assets on those information requirements that support determining whether or not to conduct urban offensive operations. Once decided, ISR resources shift to support the planning and execution of the operation in the urban area.

7-43. The first resources that a senior commander can use are national and strategic sensors. He requests them through the appropriate joint force commander. The commander aggressively pursues full use of these systems to begin building an initial database for analyzing the significant aspects of the terrain; key infrastructure considerations; the status and disposition of the population; and the size, type, disposition, and intentions of threat forces in the area.

7-44. Simultaneously, multiple intelligence sources contribute to the database. The sources collect, process, store, display, and disseminate the relevant information on large urban areas through open and classified resources. These information sources include—

- Historical research.
- Travel brochures that include cultural information and recent maps.
- Classified debriefings of diplomats, businesses, DOD personnel, and allies.
- Military maps and special geospatial products of the urban area.
- Previous intelligence assessments of the country, government, and population.
- Reachback to appropriate economic, political, cultural, and infrastructure subject matter experts not in the commanders AO.

The gathering and analysis of HUMINT plays a critical part of this process and assists commanders in understanding ethnic, cultural, religious, economic, political, and other societal and infrastructural facets of the environment.

7-45. As the intelligence and the national reconnaissance and surveillance efforts progress, commanders will insert, if available and feasible, Special Forces reconnaissance assets into the urban environment. These elements will seek to confirm or deny the information received from imagery intelligence (IMINT), signals intelligence (SIGINT), and HUMINT sources. Among many factors, using SOF depends on their availability, the particular urban area, the area's ethnic composition, and the relationship between the urban population and the threat. Other joint operational reconnaissance and surveillance assets that higher-echelon commanders may have available might include the Joint Surveillance Target Attack Radar System, Guard Rail targeting aircraft, UAS, and space-based systems.

7-46. The commander's staff will use all sources of information—IMINT and SIGINT sensors, HUMINT, historical research, and reconnaissance—to refine his ability to understand the urban environment. Digitally linking subordinate commanders with information sources helps to develop a COP essential to their situational understanding of the urban environment. The IPB process guides this assessment. As operations progress, additional reconnaissance and surveillance assets may become available. These may include UAS, long-range reconnaissance and surveillance units, counterfire radar, and air and ground cavalry. As these assets are employed, they are linked into the net of sources sharing information and further refine a common situational understanding of the environment.

## Focused Assessment Efforts

7-47. In urban offensive operations, the tactical commander's assessment focuses on defeating the threat in the urban area within the constraints of the environment. Toward this end, identifying and assessing decisive points to attack is a commander's priority assessment task. Some unique aspects of the urban environment also require the focus of the commander's assessment efforts. These include the character of the urban defense, collateral damage considerations, and the effects of the environment on friendly and threat courses of action.

### *Character of the Urban Defense*

7-48. To be both efficient and effective, Army urban offensive operations focus on what is decisive. Decisive points for an urban attack depend primarily on the mission within the urban area. They can vary widely in composition and size. Since commanders only focus on the essential, they may determine the decisive point to be a single building or a limited sector of an urban area. It could be an entire system within the urban infrastructure such as communications and information, or a limited subsystem of the transportation and distribution infrastructure such as a single airfield. Sometimes what is decisive in the urban area is the threat military capability, but even this large an objective, when carefully analyzed, may not require destruction of all threat forces or control of the entire urban area. Decisive points relate directly to the threat's center of gravity and to mission success. Some decisive points related to the urban threat's center of gravity may be physically located outside the urban area.

7-49. To gain specifics on threat dispositions within the urban area requires reconnaissance capability to see into the depths of the area and the intelligence capability to determine the threat's likely defensive course of action. With this information, commanders can determine decisive points and apply Army combat power discretely against them. Effective urban offensive operations require detailed situational understanding of an area of interest that extends well beyond the perimeter of the urban area.

7-50. Commanders see throughout the depth of the urban area using several actions (see figure 7-10). First, they evaluate sensor data and imagery. This guides targeting of special reconnaissance. Simultaneously, HUMINT is conducted using any persons who might know the urban area and threat. This includes civilians (allies, aides, neutrals, obstacles, and hostiles) and POWs. Finally, tactical conventional reconnaissance assets including reconnaissance forces, aviation, artillery radar, signals intelligence, and UAS are directed at the urban area. All these sources and data are linked through digital INFOSYS to provide commanders and their subordinates with improved situational understanding and a COP.

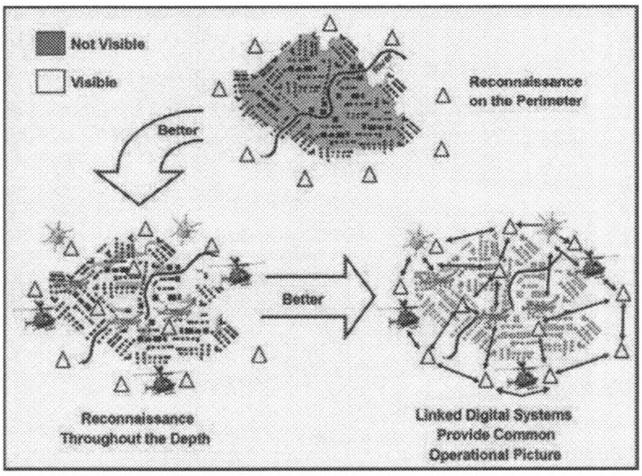


**Figure 7-10. Required urban reconnaissance capabilities**

### *Collateral Damage Considerations*

7-51. Commanders must also assess the collateral damage risks that their operations may include. This assessment helps to initially determine the viability of a course of action. However, commanders should reassess their courses of action at frequent intervals in urban offensive operations based on known information to determine if the original evaluations remain valid. This reassessment minimizes potential collateral damage from a change in mission or a change in a course of action. Many aspects of the environment can change during mission execution.

### *The Environment's Effects on Courses of Action*

7-52. The urban environment's unique aspects can significantly impact the course of action chosen by Army forces and the threat. Commanders must assess these effects in planning, but they must also verify and monitor these effects as forces execute offensive missions. In particular, commanders will want to confirm the civilian population's locations, beliefs, and actions and to monitor any changes. They will need to validate terrain considerations and monitor the effects of any changes due to rubble and other damage to structures. In urban terrain, dead space, cover, and concealment can only be identified physically and may change considerably as operations are executed.

## SHAPE

7-53. Commanders of major operations have a primary contribution to urban operations: the planning and conduct of effective shaping operations that set the conditions for subordinate tactical success. In urban operations, isolation will be a critical condition. Effective isolation will require persistent, continuous surveillance and reconnaissance, innovative use of fires and maneuver (including effective force allocation decisions), and well-established sensor-to-shooter links. These efforts—combined and synchronized with SOF direct actions, IO that minimize noncombatant influences, and necessary shaping attacks (particularly the seizure of a foothold)—establish the conditions necessary for the subsequent offensive domination of the area.

### Isolation is Essential

7-54. One key to success in the history of urban operations has been the effective isolation of the threat force (see figure 7-11). This applies today and equally well to major urban offensive operations as it does to smaller-unit attacks. This isolation not only denies access to the urban area from outside but also contains threat forces within. In a modern metropolis or megalopolis, this can appear a daunting task. Operational isolation requires dominating all physical and electronic contact between the threat in the urban area and supporting threat forces outside the urban area. This does not necessarily require physically encircling the urban area, but it does require that Army forces be able to exert control over the area's entire perimeter, as well as decisive points within. For a sprawling urban area, successful isolation may require the commitment of a large amount of resources.

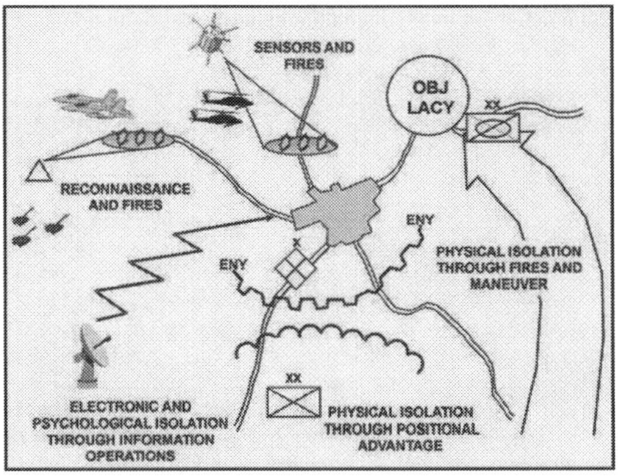


**Figure 7-11. Shaping through isolation**

7-55. Successful isolation of the urban area depends as much on the nature of the threat as it does on any other factor. A conventional threat in a large urban area may be much easier to isolate than an insurgent threat in a much smaller urban area. The forces needed in the former situation may be less than those needed in the latter. The more the characteristics of the threat are conventional in nature generally the easier it will be to isolate him using standard combat methods and equipment. Isolating a more unconventional force requires many of the same techniques as used against conventional forces. However, it also requires a much greater ability to simultaneously conduct offensive IO, to integrate CA units and civil-military operations (CMO), and to work with allies, nongovernmental organizations, and local authorities. Fundamentally, isolating a less conventional threat puts increased emphasis on separating combatants from noncombatants.

#### *Offensive Isolation Objectives*

7-56. Isolation seeks to achieve two primary objectives with respect to defeating a threat's urban defense:

- Weaken the overall coherence of his defense.
- Limit or manipulate his maneuver options.

Isolating the threat in the urban area from external support, as well as isolating him from sources of support within the urban area, weakens his overall defense. The defense is weakened through a combination of attrition (the threat cannot replace his losses) and the diversion of his combat power from the defense to operations to counter the isolation effort. Isolation can also prevent the threat from shifting his forces to reinforce decisive points in the urban area or to conduct counterattacks.

7-57. Commanders may choose not to isolate the urban area completely—or at least make it appear so to threat forces. Instead, they may afford the threat an apparent means of escape, create the conditions for its use through effective fire and maneuver against the defenders, and then destroy the threat through various ambush methods. While friendly forces may be able to move undetected to appropriate ambush sites, it is more likely that this technique will necessitate rapidly mobile air and ground forces moving along carefully chosen routes through the urban area. Commanders must consider maintaining the ability to complete the isolation of the urban area to prevent reinforcement and escape of urban threat forces—particularly if the ambush attack does not achieve desired effects.

#### *Persistent Surveillance*

7-58. Persistent surveillance of the urban area is essential to all types of actions used to isolate an urban area and as complete as resources will allow. Surveillance of the urban area relies on either reconnaissance forces or sensors continuously observing or monitoring urban avenues of approach. This network of ISR assets updates the commander's situational understanding and provides the means to quickly identify and, if necessary, attack threat elements as they move. However, particularly with sensors, commanders must consider that each detection is not necessarily an enemy to be attacked. Noncombatant activity clutters the environment making it easier for threats to disguise themselves and increases the burden (and the number of resources required) on Army forces to distinguish friend from foe.

#### *Fires and Maneuver*

7-59. Fires and maneuver may be used to achieve isolation, either singly or in combination. (As always, effective obstacles, monitored by sensors or observation, are integral to any isolation technique.) First, attacking forces can pre-position themselves along avenues of approach to deny entry and exit through positional advantage. Relying primarily on this method of isolation, particularly around a large urban area with multiple avenues of approach, can be overly resource intensive. Instead, the pairing of fires and maneuver provides attacking commanders more flexibility and allows them to isolate several avenues of approach with fewer resources. Highly mobile attack helicopters, operating outside threat-controlled portions of the urban area, are ideal for this purpose. Inside threat-controlled areas, it is more difficult to identify, eliminate, or effectively suppress the air defense threat. The threats may have numerous man-portable air defense weapons and enhanced effects of small arms used for air defense. Therefore, the risk to using this equipment may outweigh the potential benefits. However, mobile ground units—such as an air assault (subject to the same air defense threat considerations as attack aviation), armored, or mechanized forces—can also rapidly move to attack and destroy a threat moving in or out of an urban area. Potential disadvantages of the combined, fires and maneuver, option are that—

- Critical assets, on standby and dedicated to isolation efforts, may be unavailable for other missions.
- The attacking force may not locate the threat in time to complete its mission (an inherent risk to any attack).

7-60. Another alternative relies on indirect or joint fires alone to destroy the threat force. Its disadvantage is that fires alone often cannot completely destroy or stop a determined force from moving into or out of an urban area. Although targets and avenues of approach will require continual surveillance, it is usually a less resource-intensive option than those that include maneuver. It also does not normally require fires assets to remain on standby to accomplish the mission. However, fires must be able to reliably and quickly respond. For Army field artillery units and naval gunfire, the units must also be in range, which requires careful positioning. A skilled threat can avoid interdiction fires by using the geometry of the area to identify gaps due to obstructing terrain or the firing unit's range limitations. It can also use concealment and weather to avoid observation. However, effective sensor-to-shooter links throughout the urban battlefield will reduce the threat's ability to hide (see figure 7-12). A resolute threat may risk significant losses to fires to prevent isolation or may attempt to use noncombatants as a shield. Ultimately, commanders should use innovative combinations of all techniques discussed. Some units will physically block key avenues of approach. Surveillance will monitor less important routes and avenues. Artillery fires, joint fires, and maneuver units will then respond to the results of surveillance depending on the circumstances.

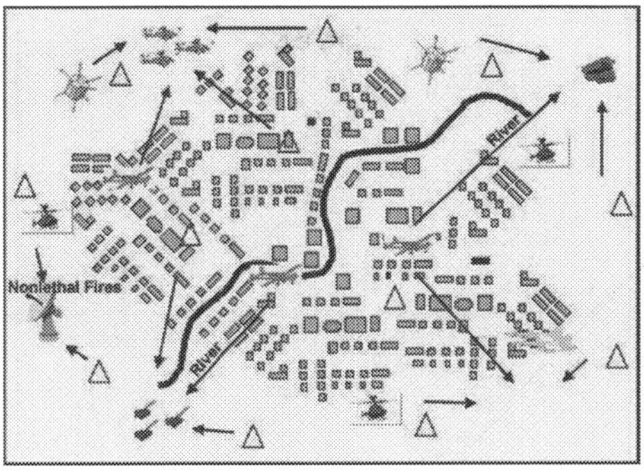


**Figure 7-12. Critical sensor-to-shooter links**

### *Threat Reactions*

7-61. The reaction of the threat to the effects of isolation will depend on his mission, morale, force structure, and overall campaign plan. The threat may recognize isolation actions early and withdraw from the urban area before isolation is completed instead of risking destruction. On the other hand, the threat, based on a different or flawed assessment (perhaps a perception shaped by the Army force commander), may choose to—

- Continue to defend (or hide) and conduct local ambushes and counterattacks.
- Attack to break into the urban area or infiltrate forces and supplies in.
- Attack to break out of the urban area or infiltrate forces out.
- Execute any combination of the above (see figure 7-13).

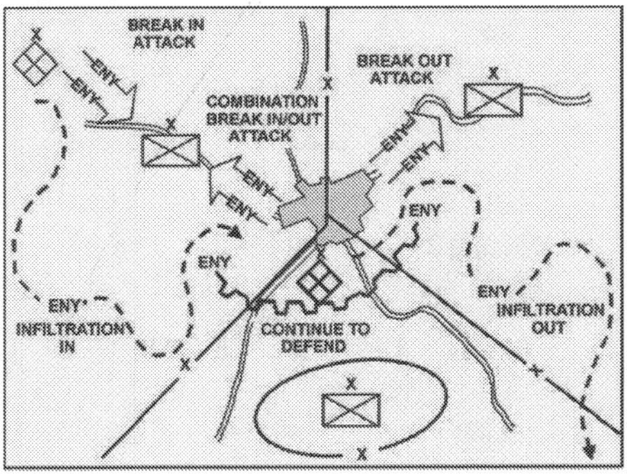


**Figure 7-13. Reactions to isolation**

7-62. Attacking commanders must consider how the threat leadership's subsequent actions will affect the continuance of overall offensive operations. They deliberate many considerations, to include—

- The allocation of more forces to the shaping operations to isolate the urban area.
- The allocation of more combat power to achieve rapid penetration and seizure of objectives to take advantage of developing threat dispositions in the urban area.

**Isolating an Urban Area: Hue, Vietnam – 1968**

On 31 January 1968, the 4th and 6th North Vietnamese Army (NVA) regiments and attached NVA and Vietcong (VC) sapper battalions attempted to seize control of Hue from the South Vietnamese Army's (ARVN) 1st Division as part of the North Vietnamese Tet Offensive. The attack, which was launched with complete surprise, successfully established temporary control over most of the city and facilitated the occupation of strong defensive positions within the city's ancient fortress known as the Citadel. However, it did not lead to the capture of the headquarters of the ARVN 1st Division located in the Citadel nor the U.S. military assistance command—Vietnam compound located in the southeast part of the city.

Both ARVN and U.S. forces swiftly reacted to the NVA attack. Within 24 hours, ARVN infantry and cavalry units counterattacked to recapture the Citadel. South of the Perfume River, the 1st Marine Division began attacking to clear the southern portion of the city (see figure 7-14).

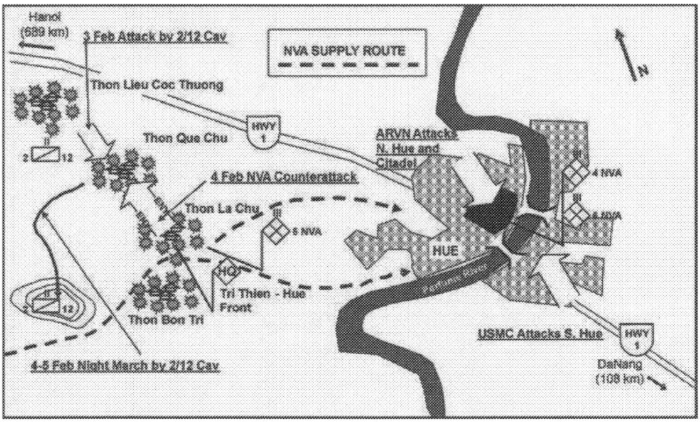


Figure 7-14. Initial attack to isolate Hue

On 2 February, the first Army unit was committed to Hue. The 2-12th Cavalry (an infantry battalion) was ordered to attack southeast along Highway 1 and clear enemy interdicting that route. On 3 February, 2-12th Cavalry began their attack moving along the south side of the highway. The plan was to clear a series of small hamlets in succession en route to the city itself. The first hamlet, Thon Lieu Coc Thuong, was cleared easily. Moving toward the second hamlet, Thon Que Chu, the battalion encountered a strong enemy force that was well entrenched. Air and artillery supported the battalion as it attacked and captured the hamlet against strong resistance. The battalion dug in for the night and prepared to resume the attack against the third hamlet, Thon La Chu, on 4 February.

Unknown to the 2-12th Cavalry, Thon La Chu was the headquarters of the Tri Thien-Hue Front, and it was defended by the NVA's 5th Infantry Regiment. Also important, the hamlet sat astride the NVA's primary supply route to the regiments fighting in Hue. The NVA was determined that 2-12th Cavalry's attack would not succeed; the NVA was determined to destroy the threat to its LOCs.

As the 2-12th Cavalry prepared to resume the attack on 4 February, the NVA fiercely counterattacked with all three battalions of the 5th NVA Regiment. It soon became apparent that 2-12th Cavalry could not continue to attack. As the day continued and the NVA pressure steadily increased, the battalion began to lose its position in Thon Que Chu. To avoid destruction, the battalion broke contact.

As darkness fell on 4 February, 2-12th Cavalry finally broke contact with the NVA. However, instead of retreating north back to its start line, the battalion moved west and then south to good defensive high ground deeper in the NVA AO. Eleven hours later the battalion was set in its new defensive position.

At dawn on 5 February, 2-12th Cavalry was established 5 kilometers west of Hue. The battalion soon observed enemy forces and supplies moving toward Hue. From its high ground position, the battalion directed artillery and air strikes against the NVA forces. By its bold move to bypass the 5th NVA Regiment, the battalion now held excellent position to direct fires on the primary NVA supply line into Hue. These fires were the first step toward isolating the NVA in Hue.

The fires controlled by the 2-12th Cavalry shut down the NVA LOCs into Hue during the daytime. However, under the cover of darkness supplies and reinforcements were still entering the city (see figure 7-15). It became apparent that the isolation of the NVA in Hue would require the capture of Thong La Chu. The problem facing American forces was concentrating combat power against the NVA. All U.S. units at this time were actively engaged against the numerous NVA attacks that constituted the NVA's 31 January Tet Offensive.

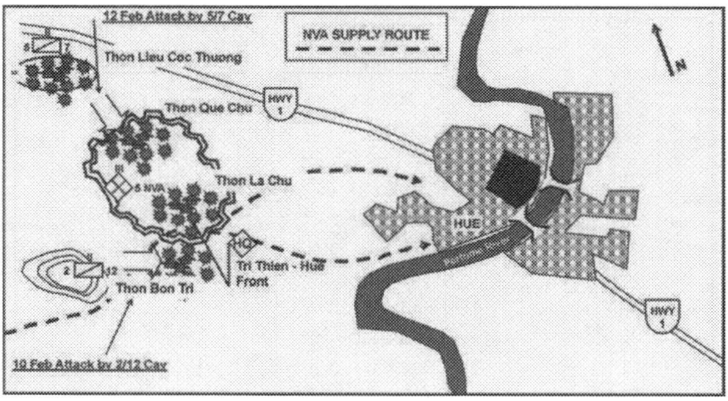


**Figure 7-15. Subsequent attack to isolate Hue**

The first additional American unit was not available until 12 February when the 5 7th Cavalry attacked Thong Que Chu much like the 2-12th Cavalry had attacked previously. The 5-7th Cavalry had even less success against the alert 5th NVA. The 5-7th Cavalry was forced to occupy defensive positions in Thon Lieu Coc Thuong and await the build up of combat power before it could continue to attack. In the interval, 2-12th Cavalry had moved off the high ground and captured the hamlet of Thon Bon Tri, south of the 5th NVA Regiment.

On 21 February, the 1st Cavalry Division had moved enough resources to the area to launch an effective attack to isolate Hue (see figure 7-16). In addition to the 5-7th and 2-12th Cavalry, the 1-7th Cavalry arrived in the AO and the 2-501st Airborne Infantry of the 101st Airborne Division was attached. On 21 February, after a combined artillery, air, and naval gunfire bombardment, the four battalions attacked the Thon La Chu stronghold. Elements of the 5th NVA Regiment were either destroyed in place or fled northeast. The next day resistance in Hue was noticeably lighter. U.S. Marine and ARVN units began the last phase of fighting to recapture the Imperial Palace. On 26 February, the North Vietnamese flag was removed from the Citadel and the ARVN I Corps declared the city secured.

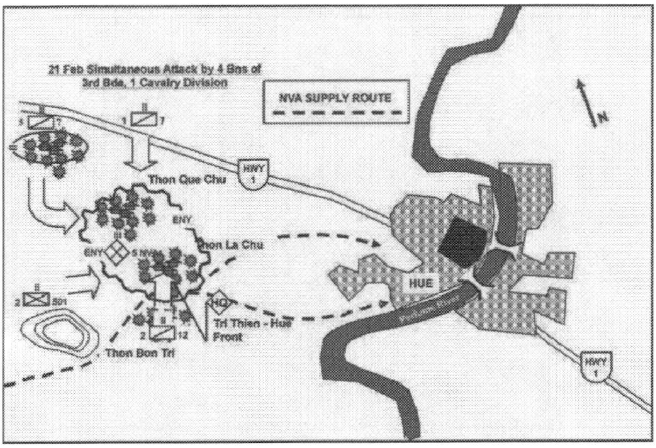


**Figure 7-16. Final attack to isolate Hue**

> The actions of the 1st Cavalry Division forces northwest of Hue demonstrated the importance and the difficulty of isolating an enemy fighting in an urban area. Isolating Hue was difficult not only because of the dispersion and surprise with which the Tet Offensive caught U.S. forces, but also because of the tenacity of the NVA. At least one-third of the combat power of the NVA in the Hue AO was focused on maintaining access to the city.
>
> Although only one aspect of this urban battle, Hue's isolation had an immediate and important, if not decisive, impact on operations. It not only resulted in restriction and then elimination of supplies and reinforcements, but it also immediately impacted the conduct of the defending NVA forces. Isolation caused an immediate drop in NVA morale and changed the nature of the defense. Once the enemy was isolated from external support and retreat, the objective of the NVA in the city changed from defending to avoiding destruction and attempting to infiltrate out of the city.

### *Civilian Reactions*

7-63. Commanders must also consider the potential effects on and reactions and perceptions of the population living in the urban areas that they choose to isolate and bypass—either as a direct effect or as a response of the threat force being isolated. Isolation to reduce the threat's ability to sustain itself will likely have similar (and worse) effects on the civilian population remaining in the isolated area. (If food and water are in short supply, threat forces may take from noncombatants to satisfy their needs, leaving civilians to starve.) Isolation may also create a collapse of civilian authority within an urban area as it becomes apparent that the military arm of their government is suffering defeat. Due to their isolation, elements of the population may completely usurp the governmental and administrative functions of the former regime and establish their own local control, or the population may lapse into lawlessness. Returning later, Army commanders may find that these self-governing residents are proud of their accomplishments and, in some instances, less willing to allow Army forces to assume control since they may be perceived as having done nothing to earn that privilege. Alternatively, as witnessed in some urban areas during OPERATION IRAQI FREEDOM in 2003, a power vacuum may lead to intra-urban conflicts among rival factions coupled with general public disorder, looting, and destruction of the infrastructure (see also the discussion of Competing Power Structures in Chapter 3).

## Direct Action by Special Operations Forces

7-64. Although SOF in urban offensive operations will likely conduct essential reconnaissance, they also have a direct action capability to shape the offensive operation (see figure 7-17). Special Forces and Rangers can use direct action capabilities to attack targets to help isolate the urban area or to directly support decisive actions subsequently or simultaneously executed by conventional forces. Successful attacks against urban infrastructure, such as transportation or communications centers, further the area's physical and electronic isolation. Direct action against command centers, logistics bases, and air defense assets can contribute to the success of conventional attacks by destroying or disrupting key threat capabilities. Direct action can also secure key targets such as airports, power stations, and television stations necessary for subsequent operations. Direct action by Special Forces and Rangers in these operations can help achieve precision and reduce potential damage to the target or noncombatant casualties.

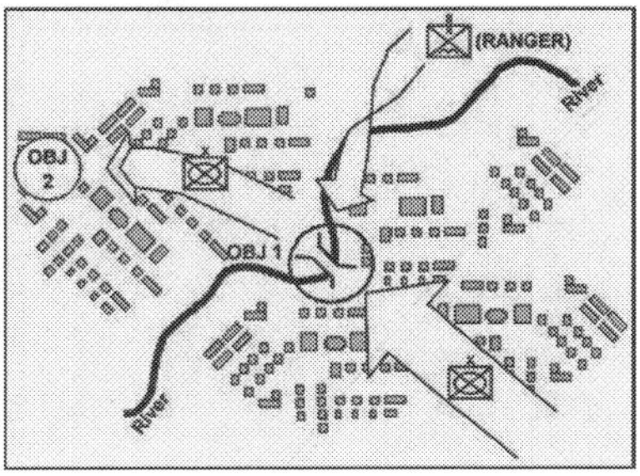


**Figure 7-17. Coordination of SOF and conventional capabilities**

## Information Operations

7-65. Regardless of how Army forces physically isolate the urban area, they combine physical isolation with offensive IO to electronically isolate the threat and undermine his morale. Electronic isolation will cut off communications between forces in the urban area from their higher command to deny both from knowing the other's status. IO combined with isolation may persuade the threat's higher command or leadership that its forces located in the urban area are defeated. Thus, the command or leadership's intentions to break through to the besieged threat forces may be affected. PSYOP can undermine the morale of the threat in the area and reinforce electronic isolation and perceptions of abandonment. IO can be used to reduce any loyalty the civil population may have to the threat. IO can also ensure that civilians have the information that minimizes their exposure to combat and, as a result, overall noncombatant casualties. In addition, IO aim to deceive the threat regarding the time and place of Army force operations and intentions.

## Detailed Leader Reconnaissance

7-66. Army commanders must clearly see the urban environment to understand the challenges facing their brigades, battalions, companies, platoons, and squads. Urban terrain can be deceptive until viewed from the soldier's perspective. Commanders are responsible to intimately know the conditions to allocate resources effectively to subordinate units. Often, particularly at battalion level and above, commanders will not be able to command and control dispersed forces from positions forward, but be forced by the terrain to rely on semifixed command posts. Detailed leader reconnaissance of the AO by commanders, their staff, and their subordinates before the mission can compensate for this challenge. This reconnaissance will give commanders a personal feel for the challenges of the terrain and will facilitate more accurate planning and better decision making during operations.

## Mission Orders

7-67. Before contact, commanders can mitigate some terrain challenges to effective C2 using mission orders. Subordinates have mission orders to take advantage of opportunities before C2 systems can adversely impact the environment. To see the battle and provide effective and timely direction, tactical leaders will follow closely behind units as they assault buildings, floors, and rooms. Thus, only the most mobile INFOSYS can accompany tactical leaders into combat, and they will suffer the degrading effects of the environment. Mission orders permit rapid and decisive execution without commanders intervening at battalion level and above. Higher-level commanders can facilitate mission orders through their subordinates by articulating their desired end state, clearly stating their intent, and building flexibility into the overall plan.

## Effective Task Organization

7-68. Commanders can shape urban offensive operations through effective and innovative task organization. Combined arms, often starting with an infantry base, are essential to success and may be the Army's asymmetric means of defeating an urban threat. Urban attacks will quickly break down into noncontiguous firefights between small units. To achieve the tactical agility for mission success in this nonlinear environment, many Army capabilities are task organized down to the company, platoon, and squad levels. Infantry provides the decisive capability to enter buildings and other structures to ensure threat destruction. Tanks, gun systems, and fighting vehicles provide additional speed and mobility, direct firepower, and protection. Combat engineers provide specialized breaching and reconnaissance capability. Field artillery provides the indirect (and if necessary, direct) firepower. Such mobility and firepower create the conditions necessary for the dismounted infantry to close with and destroy a covered threat in an urban defense. When a threat skillfully uses the urban area to limit ground maneuver, vertical envelopment or aerial attack using precision-guided munitions from Army aviation may circumvent his defenses and achieve necessary effects. Generally, ground systems used within the urban area will not be able to operate independently from dismounted infantry. The infantry will be required to protect armor and mechanized systems from close antiarmor weapons, particularly when those weapons are in well-prepared positions located throughout the urban area but especially on rooftops and in basements.

7-69. In urban offensive operations, direct fire support can be critical. Armor vehicle munitions types do not always achieve decisive effects against some urban structures. In some cases, field artillery high explosive munitions work better than armor for direct fire support of infantry. Large caliber (105 or 155mm) high explosives directly fired at a structure often produce a more severe shock effect than tank and fighting vehicle cannon and machine guns produce. Artillery is also able to achieve higher elevation than armor and engage threats located at greater heights.

7-70. However, commanders must view artillery as not just a weapon but a weapon system. As such, artillery should normally be placed under tactical control (TACON) of maneuver commanders, such as a platoon of three guns TACON to a company or a battery to a battalion, not just one gun to a company or other maneuver unit. Self-propelled artillery has some of the mobility characteristics of armor; however, it provides minimal ballistic protection from fragmentation for the crew. Although these systems seem formidable, they provide less crew protection than a Bradley fighting vehicle, for example, and contain large amounts of onboard ammunition and propellant. They are susceptible to catastrophic destruction by heavy automatic weapons, light cannon, and antitank fire. Therefore, infantry units must carefully secure and protect these systems (even more so than armored vehicles) when employed in urban offensive operations, particularly when forward in the direct fire role.

### Creative Task Organization: Using Artillery in the Direct Fire Role

Task organizing artillery to permit its use in a direct fire role demonstrates the innovative task organization required for urban operations. The following provides three historical examples of task organizing and using field artillery for a direct fire role.

---

In 1944, U.S. Army units of the 1st Infantry Division were assigned to attack and seize the German city of Aachen. The city's internal defense included bunkers designed to serve as air raid shelters. These positions, buildings of stone, seemed impervious to direct fire tank weapons, demolitions, and small arms. To reduce the positions, the 1st Infantry Division relied on the artillery's direct fire.

Field artillery used this way had physical and psychological effects on the defenders. The 26th Regimental Combat Team's history of the battle describes the German reaction to the artillery pieces:

*The chief shock to the defenders, Colonel Wilck (Aachen defense commander) said, came from the self-propelled 155s and tanks. The colonel spoke with considerable consternation of the 155mm self-propelled rifles. A shell from one of the guns, he said, pierced three houses completely before exploding and wrecking a fourth.*

The 26th Infantry Regiment also described how the artillery, one piece attached to each assaulting infantry battalion, helped the infantry to penetrate buildings.

*With solid blocks of buildings comprising most of the city, there wasn't any easy way to get at the Germans in the buildings. The eight-inch gun solved the problem. Beginning on the eastern outskirts the gun would plow a round into the side of the built up block of buildings at about ground level. One shell would usually open an entrance into the first tier of floors, i.e. the first building. Then several more shells were fired through the first hole. Thus a tunnel would be rapidly made all the way to the next cross street. Soldiers could then rush the newly formed entrance, clear the upper floors with hand grenades and rifles and then move on to the next building to repeat the process. When a block or square, was thus completely cleared of Germans—soldiers, skulkers, or even snipers—the next square was treated in the same way, working forward square by square, right and left, thereby avoiding nearly all exposure in the streets.*

---

In 1982, Israeli forces invaded southern Lebanon to destroy base camps of the Palestine Liberation Organization (PLO). This operation involved significant fighting in urban areas including major operations in Beirut. Artillery, firing in a direct fire role, played a major part of the tactical solution. Artillery was particularly effective in the 33-day siege of Beirut. During this siege, Israeli forces used artillery in its traditional role as well as in the direct fire role.

The Israeli army was committed to a policy of disproportionate response during the Beirut siege. When fired on with small arms, crew-served weapons, tanks, or indirect artillery, the Israeli forces responded with intense, high-caliber direct and indirect fire from tanks and artillery positioned around the city. Many firing positions were on heights to the south and southwest that dominated much of the city. These positions had almost unrestricted fields of view. Israeli artillery fired from these positions directly into high-rise buildings concealing PLO gunners and snipers. The artillery, using direct fire, destroyed entire floors, collapsed floors on top of each other, and completely removed some upper floors. Such a response, as in Aachen in 1944, had as much a psychological impact as it did a physical impact on the PLO defenders.

---

In the early hours of 20 December 1989, the United States launched OPERATION JUST CAUSE. One of this operation's objectives was removing the Panamanian dictator, Manuel Noriega. U.S. forces carefully planned using all fires before the operation to minimize casualties and collateral damage. Part of this detailed fire planning called for applying artillery in a direct fire role.

The Panamanian Defense Force's (PDF) 5th Rifle Company based at Fort Amador was one of the key objectives of U.S. forces at the start of hostilities. This unit was high priority because it was the closest PDF unit to Noriega's headquarters, the Comandancia. U.S. forces expected the 500-man company to react first to

> OPERATION JUST CAUSE by reinforcing the defense of the Comandancia. It also posed a threat to U.S. military dependents housed at Fort Amador.
>
> To quickly neutralize this force, the United States assembled a three-company force composed of A, B, and headquarters elements of 1-508th Infantry (Airborne), supported by 105mm towed howitzers of 320th Field Artillery and M113 armored personnel carriers. The howitzers and the personnel carriers were covertly pre-positioned at the fort. At approximately 0100, helicopters transported the two airborne rifle companies into position. The howitzers then suppressed any personnel in the PDF-controlled buildings on Fort Amador while demonstrating the firepower of the U.S. task force. They used direct fire into the PDF barracks. The impact of the 105mm high explosives and .50-caliber fire from the M113s convinced the PDF infantry to give up after token resistance. Following the direct fire, U.S. infantry assaulted and cleared the dozen PDF buildings, finding that most occupants had fled or surrendered. For more details of OPERATION JUST CAUSE, see Applying the Urban Operational Framework: Panama in Chapter 5.
>
> ---
>
> The three examples cited indicate the importance of the situation-dependent and innovative task organization of artillery and its use in the direct fire role. Using artillery helps overcome some challenges of offensive operations in the urban environment, and it has an important psychological effect on a defending threat. Such task organization takes a traditional tool of a higher-level tactical commander and uses it to directly influence the company-level battle. This philosophy of task organization can be applied to other types of forces—not just artillery. PSYOP teams, interpreters, CA specialists, armor, engineers, and reconnaissance teams may require task organization different from traditional organization. The compartmented urban environment drives the requirements for these assets lower in the tactical scheme than in open operations. Consequently, commanders must understand and account for more of these assets for UO than for operations in less restrictive environments.

7-71. Army aviation may also be inventively task organized. It can support urban operations with lift, attack, and reconnaissance capabilities. Tactical commanders down to company may use all these capabilities to positively influence ground close combat. Army attack and reconnaissance aircraft can provide flank security for attacking ground forces. Attack aircraft may also provide direct fire support to individual platoons or squads. Lift may move entire battalions as part of brigade operations, or it may move single squads to a position of advantage (such as a roof) as part of a small unit assault. Army aviation can assist with C2 by providing airborne retransmission capability, airborne command posts, and the confirmed status and position of friendly forces. However, Army aviation is a limited and high-value asset; commanders should review its use in innovative task organizations. It is particularly vulnerable to urban air defense threats unless used over terrain secured by ground forces. From these positions, aircraft can use enhanced sensors to conduct reconnaissance and use precision weapons with standoff capability.

## Shaping Attacks

7-72. In a large urban area, the defending threat cannot be strong everywhere. Shaping operations can also take the form of attacks against vulnerable positions to force the threat to maneuver and redeploy his forces in the urban area. This prevents him from merely defending from prepared positions. Forcing the threat to move negates many of the defensive advantages of urban terrain, confirms his dispositions, exposes vulnerable flanks, and permits target acquisition and engagement with precision standoff fires.

7-73. A critical shaping operation in urban offensive operations is usually an initial attack to seize a foothold. Once Army forces establish this foothold, they accrue some of the defensive advantages of urban terrain. From this protected location, Army forces continue offensive operations and have a position of advantage against neighboring threat defensive positions.

## ENGAGE

7-74. Commanders may employ several methods to decisively engage elements of the urban area during offensive operations. These include—

- Rapid maneuver.
- Appropriate use of SOF.
- Precise application of fires and effects.
- Proper balance of speed and security.

7-75. None is unique to UO. Their effective execution, however, allows Army commanders to dominate in this challenging environment by effectively using resources with the least amount of collateral damage. Overall, decisive engagement results from urban offensive operations when forces achieve the objective of the assigned mission and establish preeminent control over the necessary terrain, population, and infrastructure. Largely, the Army commander's ability to engage is based on superior situational understanding and the correct application of unit strengths to the challenges found in the urban environment.

### Rapid and Bold Maneuver

7-76. Commanders of major operations may have or create the opportunity to seize an urban area with rapid and bold maneuver. Such maneuver requires striking while the area remains relatively undefended—essentially preempting an effective defense. This opportunity occurs when the urban area is well to the rear of defending threat forces or before the onset of hostilities. Under such conditions, an attack requires striking deep behind threat forces or striking quickly with little time for the threat to make deliberate preparations. Attacks under such conditions may entail significant risk; however, the potential benefit of audacious offensive operations may be well worth possible losses. Three potential ways to accomplish such attacks (and their combinations) are:

- Airborne or air assault.
- Amphibious assault.
- Rapid penetration followed by an exceptionally aggressive exploitation, for example, a heavy force using shock, armor protection, and mobility.

7-77. Commanders should analyze all potential urban operations to seek an opportunity or advantage to apply rapid and bold operational maneuver to the task. Using operational maneuver to avoid urban combat against an established threat defense potentially marks a significant operational achievement and can have decisive strategic consequences. Just influencing the threat's morale can positively affect all future operations. However, commanders must evaluate the challenges of such a course of action. These challenges may include—

- Sustaining the operation.
- Avoiding isolation and piecemeal destruction.
- Successfully conducting shaping attacks.
- Achieving the necessary tactical, operational, and strategic surprise.

7-78. Commanders can also build on the shaping effects of isolating the urban area internally and externally by attacking urban decisive points from multiple directions. They can attack multiple decisive points either simultaneously or in a systematic, synchronized manner. This complicates the threat's situational understanding of the urban environment, further impedes his decision making, and allows Army commanders to dictate the tempo.

## Bold Operational Maneuver to Seize an Urban Area: Inchon and Seoul, Korea – 1950

In August 1950, UN forces in Korea were desperately attempting to stave off defeat by establishing a viable defense in southern Korea. This defense, known as the Pusan Perimeter, was the focus of the world's attention. The commander of U.S. Forces Far East, General Douglas MacArthur, was not focused on the U.S. Eighth Army in Pusan. Instead, he focused on how to dramatically and decisively transition to the offense—the recapture of the South Korean capital of Seoul.

Seoul was more than just the South Korean political capital, although that was an important consideration. Seoul was the key to the Korean transportation system, the hub of the national communications system, and the center of the country's economy and culture. It was also strategically placed astride maneuver corridors to the north, south, and northeast. It had been Korea's capital since 1394 when King Taejo Yi Sung-Gye established it as the center of the Chosun dynasty. It had been captured by the surprise communist invasion of the south in June 1950, and the key to success in Korea was its recapture. The question was how.

The obvious military solution to the Korean situation in August 1950 was to build military strength within the Pusan Perimeter and then, when sufficiently strong, to breakout and counterattack north and recapture Seoul. The disadvantage of this course of action was that the counterattack would be through hundreds of miles of mountainous terrain, through several other major cities, and across numerous rivers. The North Korean People's Army (NKPA) would be able to fall back from one mountain defense and one river line to another and would have weeks or months to prepare the defenses of Seoul.

MacArthur's solution was to delay a conventional counterattack, boldly maneuver by sea deep into the flanks of the NKPA, land on the Korean west coast at Inchon, and rapidly seize Seoul before the enemy could react (see figure 7-18). This course of action had numerous disadvantages: achieving surprise; assembling a landing force trained in amphibious operations; few good landing sites; supplying a large force once landed; and needing to simultaneously continue to defend and prepare a counterattack with the Eighth Army from Pusan. MacArthur understood the challenges of the operation but believed that the potential value compensated for the risk.

The potential value of a rapid assault on Seoul through a deep amphibious operation was immense. The operation would trap the bulk of the NKPA in South Korea and facilitate its destruction. It would also capture Seoul before the NKPA could prepare a defense. The quick recapture would immeasurably impact the morale of UN forces and South Korea. MacArthur felt the payoff of success justified the risks and challenges of the operation.

On 15 September, elements of the 1st Marine Division landed in Inchon taking the NKPA completely by surprise. On D+1 they were joined by elements of the 7th Infantry Division, both divisions under the control of U.S. X Corps. Naval gunfire and Marine close air support flying from Navy aircraft carriers supported the landings. Seven days later, the 5th Marine Regiment had battled 25 miles to the outskirts of Seoul and began attacking the city from the north. Three days later, on 25 September, the 1st Marine Division and elements of the 7th Infantry Division had defeated the 10,000 defenders of the NKPA's 18th Rifle Division.

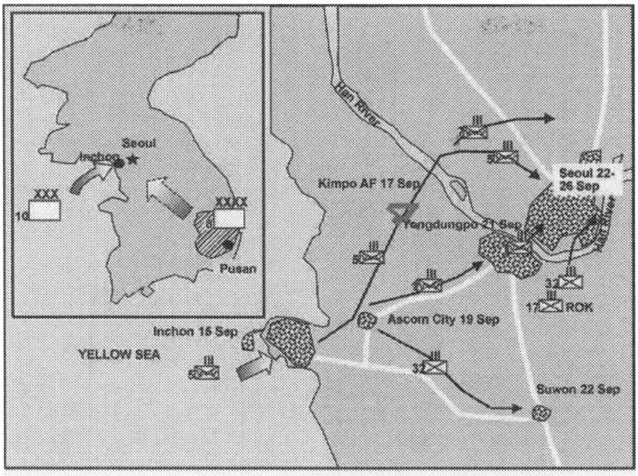


**Figure 7-18. Inchon-Seoul campaign, September 1950**

Most of MacArthur's predictions for the Inchon-Seoul campaign were accurate. Superb staff work, excellent logistics operations, and unmatched training permitted the UN forces to assemble a trained landing force, land it over one of the most difficult shores in the history of military amphibious operations, and keep it supported. More importantly, the operation achieved total strategic and operational surprise. Seoul was recaptured quickly and, although resistance was fierce, the NKPA could not react fast enough to influence the outcome. In conjunction with the Eighth Army counterattack from Pusan, 75 percent of the NKPA was destroyed. The Inchon-Seoul campaign was an important turning point in the war, and had not the Chinese People's Army intervened two months later, it would have been the decisive campaign of the war.

## Appropriate Use of Special Operations Forces

7-79. Sometimes Army forces can decisively engage (instead of simply shape) the urban area using the direct action capability of SOF. When the threat fails to develop a comprehensive defense and does not possess large, capable conventional forces, then Army forces can achieve operational surprise. Commanders, by synchronizing conventional and SOF effects, may actively control offensive operations to dominate the area. Importantly, however, conventional ground forces must be available to quickly assume the mission because SOF units acting as the primary striking force have limited logistic capability to sustain long-term operations. OPERATION JUST CAUSE offers several examples of this type of synchronization (see Applying the Urban Operational Framework: Panama in Chapter 6).

## Precise Application of Fires and Effects

7-80. Precisely applied fires and the massed effects of combat power characterize successful urban attacks. The fires can be direct fire from combined heavy or light ground teams; direct or indirect fires from supporting Army aviation standing off from the target and any possible air defense threat; precision indirect fires from conventional tube artillery; or direct and indirect fires from supporting joint assets including Air Force and naval assets. All efforts strive to reduce collateral damage around the point of attack, consistent with mission success. Forces use fires to deny the threat the ability to maneuver in the urban area and to destroy the threat when he attempts to maneuver. When the threat exposes himself by moving, the environment no longer protects him, and fires can effectively engage him. Overall, precise fires and effects demonstrate the power of Army forces and help the urban population understand that only legitimate military targets are the focus of attacks, potentially building public support of Army UO.

### Proper Balance of Speed and Security

7-81. Attacking units balance speed and security. Forces secure flanks as units advance, control dominating terrain (buildings), evacuate civilians, and keep the integrity and synchronization of the combined arms team. Obstacles are anticipated and rapidly breached. Commanders can choose avenues of approach to—

- Provide cover and concealment for following aviation and support units.
- Permit travel by all classes of vehicles.
- Easily defend from counterattack.
- Avoid nonessential centers of threat resistance.
- Avoid population concentrations.

7-82. Army aviation is a critical resource to protect flanks. Another important resource is engineers who seal off surface and subsurface entries and avenues along the flanks of the attack. Finally, as in all offensive operations, ground and air cavalry are ideal mobile forces to perform security in an economy of force role along flanks allowing decisive forces more freedom of maneuver.

## CONSOLIDATE

7-83. Commanders, at all levels, must consolidate to strengthen their position during urban offensive operations without loss of momentum. They take the steps necessary to make permanent any temporary battlefield successes while maintaining relentless pressure on threat forces. Consolidation provides security and protection, facilitates reorganization, and allows forces to prepare for counterattack.

### Repositioning of Forces

7-84. Following seizure of the objective, commanders normally consolidate by adjusting and repositioning forces. While UO are likely to be noncontiguous, commanders reposition ISR assets (including observation posts and reconnaissance patrols) to maintain contact with the threat, to establish contact with nearby friendly units and ensure that there are no exploitable gaps or seams, and to help maintain freedom of action. Physical occupation of the terrain (as well as continued reconnaissance) will provide commanders with a fuller understanding of the urban environment. With this enhanced understanding, commanders may need to adjust boundaries and other control measures to better adapt to the effects of urban terrain features such as canals, subway tunnels, raised roadways, and tall buildings. As necessary, commanders may also need to reposition communications assets and C2 facilities to enable subsequent operations.

7-85. Immediately after the conduct of successful urban operations, units must also be alert to the potential for rapid and violent counterattacks. Defenders may launch a quick counterattack to regain terrain before the offensive forces have consolidated and fully assumed the defensive advantages of the urban terrain. Delaying a counterattack in UO, even for a few minutes, permits the environment's advantages to shift to the successful attacker. Thus, attacking units anticipate this reaction and reposition forces such that they are prepared to defeat it.

7-86. As necessary, commanders should reposition armor and artillery (and other fire support assets) to account for the changing situation and battlefield geometry. Commanders should also consider consolidating and repositioning armored and artillery forces in positions—either inside or outside the urban area—to add significant combat power to a hasty defense, to defeat threat counterattacks, or to allow for quick resumption of the attack. If integrated into a hasty defense inside the urban area, these forces will require continued infantry protection. As a mobile counterattack force positioned inside the urban area, armored forces will also require careful selection of attack positions and counterattack routes. Damage to buildings and infrastructure may limit maneuver and the use of direct and indirect weapon systems.

Repositioning forces outside the urban area may contribute to strengthening or reestablishing the isolation of the urban area and will make better use of range and standoff capability, allow them to take full advantage of their speed and mobility, and decrease their need for additional protective support.

### Eliminating Pockets of Resistance

7-87. Focusing on controlling the essential and attacking decisive points means that attacking units will have bypassed some elements of the threat's defense. Within the bounds of the initial plan and current situation, commanders should consider whether it is necessary to devote the time and resources to immediately clear and eliminate remaining threat forces and pockets of resistance with all or parts of the attacking force or leave the task to follow-on forces. In their deliberations, commanders should determine if remaining threat forces are capable of consolidating and mounting effective counterattacks before follow-on units can engage the remaining threat and take advantage of their dispersion and disunity. As another part of their deliberations, commanders should also consider if remaining threat forces will likely (and immediately) endanger—

- The urban inhabitants.
- Lines of communications.
- Critical resources within the urban area necessary to support the inhabitants or subsequent operations.

### Reorganization

7-88. Reorganization includes all measures taken by the commander to maintain combat effectiveness or return to a specified level of combat capability. As necessary, these actions include—

- Redistributing or cross-leveling supplies and equipment until resupply can be accomplished.
- Replacing key personnel and combining units or crews to form mission-capable ones.
- Integrating replacement soldiers and systems into the unit and matching operational systems with available crews.
- Recovering, treating, and evacuating casualties, enemy prisoners of war, and damaged equipment.
- Conducting training and disseminating critical lessons learned.
- Other actions to reestablish unit cohesion.

## TRANSITION

7-89. Effective transitioning allows commanders to continue full spectrum operations in the urban area and elsewhere in the AO without unnecessary delays. Commanders can accomplish this by thorough planning—including appropriate branches and sequels (revised as the situation changes)—that gives adequate consideration to post-offensive organizational, training, psychological, and civilian requirements. If properly prepared, commanders can anticipate rather than react to potential mission changes.

### Early and Concurrent Transition Planning

7-90. Commanders can ensure smooth transitions of urban offensive operations by planning for post-offensive operations early. Based on the mission envisioned, they determine which subordinates and what type of force structure to use. Post-offensive missions, like all urban operations, encompass full spectrum operations. At the successful conclusion of offensive operations, Army forces transition to some type of stability or civil support operation. Commanders may leave the subordinate unit in place to execute the new mission, may reorganize the subordinate unit for the mission, or relieve the unit that just completed offensive operations with a new unit.

#### *Changes to Task Organization*

7-91. Commanders should consider the organization of forces following offensive operations. Hostile civilians may require significant combat forces or military police forces to maintain stability. On the other hand, friendly civilians may require a minimum of military police or combat forces, but significant logistic support. Commanders must carefully consider the urban situation before deciding how to use combat forces that recently participated in a high-intensity offensive operation.

#### *Training and Psychological Considerations*

7-92. Many Army combat tasks may not support follow-on stability or civil support operations without considerable modification (if at all). Often, noncombat skills—not normally part of a unit's mission essential task list such as negotiating or mediating skills—will be required. However, the greatest modification required applies to each soldier's mental outlook. Forces that transition directly from combat to stability operations may not be psychologically prepared for a rapid and drastic change of mission. Commanders cannot expect troops who have just completed high-intensity offensive operations to rapidly adjust and exercise the sensitivity and judgment required in most stability operations. This especially applies if the population is hostile to Army forces. If possible, combat forces assisting in stability operations, particularly in hostile civilian situations, should not have had recent experience in high-intensity urban operations and they should have trained for the mission.

#### *Return to Civil Agencies*

7-93. Commanders of major operations also have the critical role of transitioning aspects of the urban offensive operation to civil, allied, NGOs, and other agencies as appropriate. Transition planning is detailed and aims to quickly return as much civil control of the area as is feasible after the attack. Beyond local civil control, outside civil agencies and NGOs assume tasks as completely and as rapidly as possible. These organizations are consulted and integrated into the planning process as early as possible. Commanders should begin planning for transition simultaneously with planning for offensive operations. They consider the feasibility of relinquishing control of urban areas to civil government, law enforcement, or NGOs even before completing offensive operations. During the conduct of urban operations, these transition operations are closely synchronized with the execution of the attack.

### Transition to a New Mission

7-94. In urban offensive operations, like other offensive missions, the change in mission after a successful urban attack may be to a hasty defense or a continuation of offensive operations outside the area. However, in urban offensive operations the mission will just as likely rapidly change to a civil support or stability mission. This is particularly true if the unit has had special training and is task organized for urban operations. Transition to stability or civil support operations is often accompanied by a transition in roles from supported to supporting.

7-95. Even more challenging than transition at the end of the mission is transition during the conduct of the mission. Soldiers may have a difficultly transitioning from stability or civil support to offense and defense, and back again multiple times during an urban offensive operation. Soldiers may apply the tactics, techniques, and procedures of urban offensive operations directly to the stability or civil support missions with potentially disastrous results. Commanders need to segregate missions in time and space, and, if sufficient forces exist, by unit. To this end, commanders should consider permanently designating specific units to conduct civil-military and humanitarian support tasks and avoid rapid mission changes that rotate units (particularly at company level and below) between violent and nonviolent tasks. However, commanders may not have that luxury and may need to rely heavily on preparatory training (including the inculcation of Army values) and strong unit leadership to mitigate potential difficulties.

## Chapter 8

# Urban Defensive Operations

*Generally, a modern city magnifies the power of the defender and robs the attacker of his advantages in firepower and mobility. A city can ingest an invading army, paralyze it for weeks on end, and grind it down to a state of ineffectiveness.*

"Military Operations on Urbanized Terrain:
The 2d Battalion, 26th Infantry, at Aachen, October 1944"

The skillful defense of an urban area can decisively affect a campaign. The urban area offers many advantages to defending forces. An adroit defender can use the advantages of the urban environment to negate combat power disparities, blunt the tempo of an attack, attrit threat forces, and sap the morale of attacking troops. The defender gains an opportunity to concentrate resources, reconstitute attrited units, and transition to the offense. A successful defense of an urban area can also deny the threat vital resources. Defense in the urban environment is an essential Army capability and can significantly affect the outcome of entire campaigns and the achievement of national objectives.

## PURPOSE OF URBAN DEFENSIVE OPERATIONS

8-1. Army forces defend urban areas for various reasons: defeating a threat attack, buying time, economizing forces, protecting an ally's political institutions and economic infrastructure, protecting an urban population, shaping conditions for decisive offensive operations, and shaping conditions for executing stability or civil support operations. During force projection operations, urban areas may be used as initial lodgment areas that Army commanders may need to defend at the outset until they build sufficient combat power. Usually two or more of these purposes apply to the urban defense. Urban defensive operations provide commanders great opportunities to turn the environment's characteristics to the advantage of Army forces. Urban areas are ideal for defensive operations and greatly enhance the combat power of defending units.

## DEFENSIVE CHARACTERISTICS APPLIED TO URBAN OPERATIONS

8-2. There are five general characteristics of the successful defense: preparation, security, disruption, massing effects, and flexibility. All apply to the successful urban defense and to the higher commander supporting a subordinate defending in the urban area.

### PREPARATION

8-3. The urban area suits the defense since the area's physical characteristics naturally enhance the combat power of defending units. These characteristics include protection, obstacles, and concealment. Urban terrain provides superb defensive positions with minimum preparation. With deliberate preparation, urban defensive positions can rapidly become strong points.

8-4. One primary characteristic of urban terrain that enhances the defense is protection. With little or no advance preparation, buildings, subsurface structures, and walls protect Soldiers from direct and indirect fire, interdict indirect fire, limit observation, and limit engagement ranges (requiring Soldiers to be skilled at combat in close quarters and quick-fire techniques). Nearly all buildings provide some ballistic protection from direct and indirect fire. Mason and stone buildings with basements and cellars can protect Soldiers from most fires except the largest caliber or tonnage bomb. Minimal additional preparation turns them into formidable, defensive strong points.

8-5. Buildings in urban areas, because of their height and close proximity, also can protect Soldiers by masking them from indirect fire. The height of a building may interdict the flight path of an artillery round, rocket, missile, or bomb at a point short of the intended target. Masking protects static defending forces and protects forces moving along routes bordered with tall buildings that form urban "canyons". These protected routes can be used for sustainment, counterattacks, and maneuver.

8-6. Structurally significant buildings in an urban area can create major obstacles to maneuver. These obstacles immediately canalize maneuver into existing streets and routes without any preparation by the defense. These obstacles then become kill zones for well-positioned and sited defensive forces. Minimal obstacle construction as point obstacles blocking streets and routes can further restrict the maneuver options of the attacking force. Rubble from structures collapsing into streets after fires (intentional or unintentional) can also block routes.

8-7. Buildings also conceal the location, disposition, and intent of the defense. They limit visual observation to the external perimeter of the urban area. They degrade radar and electronic position identifiers and decrease the utility of overhead imagery. The physical aspect of the urban environment greatly enhances the defense by degrading the opposition's intelligence, surveillance, and reconnaissance (ISR) capabilities. Buildings can conceal static defensive positions and the maneuver of defensive forces in the urban area. Although the environment constrains defensive mobility in much the same manner as offensive mobility, the defender has the time and opportunity to conduct careful reconnaissance and select and prepare routes. This gives the defender the ability to move reserves, maneuver counterattack forces, and plan sustainment without observation. Careful preparation provides the defender a mobility advantage over attacking forces.

## SECURITY

8-8. The urban area can be an advantage or a disadvantage to the security of defending forces. This largely depends on the nature of the human dimension of the environment. If the population is evacuated or allied with Army forces, then the environment may assist in the security of defending Army forces. However, if the population is present and hostile, then the environment may make security difficult.

8-9. The physical aspects of the urban environment, uninfluenced by the human dimension, may assist in the security of defending Army forces. The combat power of small security forces manning observation posts is greatly enhanced. Forces can more easily restrict and monitor avenues of approach for threat reconnaissance. Defending forces positioned mostly in structures are difficult to locate.

8-10. The physical aspects of the environment may also present some security challenges, primarily with observation. The compartmented terrain limits the field of observation from any one point. The defense may require more security forces to adequately observe the mounted and dismounted avenues to prevent infiltration. Threat forces that successfully infiltrate will be more difficult to locate. These forces will gain numerous hide positions for small reconnaissance units in complex terrain and the effect the terrain has in masking electronic signatures.

8-11. Friendly civilians in the urban area can help identify threat forces attempting to conduct reconnaissance. Civilian activity will also help to mask defense preparations. However, a hostile element of the population may pass intelligence information to the threat. They may assist threat reconnaissance to infiltrate the urban area or provide guides, manpower, or resource support for threat forces. Commanders should take measures to ensure strict control of hostile populations. If resources permit, commanders may consider removing potentially hostile civilians from the area.

## DISRUPTION

8-12. The urban environment's attributes assist defending Army forces to disrupt the attacker. It does this through compartmentalization, inhibiting command and control (C2), and facilitating counterattacks.

8-13. The physical aspects of the urban area force the attacking threat into compartmented urban canyons that make mutual support between attacking threat columns difficult. Shifting resources from one portion of the threat attack to another is also difficult. Physically, the urban area disrupts tactical communications making synchronization of combat power difficult.

8-14. The urban terrain hinders the mobility capabilities of the defense. However, careful planning, preparation, and rehearsals can facilitate more rapid movement of larger forces. Defending forces can assemble counterattacks undetected, move them along covered and concealed routes, and achieve surprise at the point of the counterattack. Attacking forces, using the compartmented terrain, often leave forward elements in position to be isolated or expose long and vulnerable flanks to friendly counterattack and interdiction.

### MASSING EFFECTS

8-15. The urban environment allows defenders to better protect their centers of gravity and decisive points. The restrictive terrain reduces the attacker's maneuver options. Defenders can position forces in protected and mutually supportive positions oriented on deadly engagement areas. Relatively few well-positioned defenders can generate significant combat power. Without the positional advantage and the corresponding protective effects of the terrain, attacking forces often mass numbers to achieve the necessary combat power.

8-16. Knowing the complex terrain permits defending forces to plan engagement areas that maximize the effects of their combat power. Defending forces can remove fences, walls, rooftops, and even entire buildings to facilitate fields of fire and unmask indirect fire flight paths. Forces carefully choose firing positions for indirect fire systems so that flight paths travel between buildings into engagement areas. By leveraging this knowledge of the terrain, numerically inferior defenders can synchronize devastating fires on offensive forces that are forced by terrain and reinforcing obstacles to mass in confined spaces where fires can have the greatest effect.

### FLEXIBILITY

8-17. Defensive flexibility results from detailed planning and mental agility. Commanders develop defensive flexibility by ensuring that plans adequately address branches and sequels that include alternate and subsequent positions and emphasize counterattack options. The urban area facilitates defensive flexibility because the urban terrain can be quickly adapted for defensive operations with little or no preparation. The effect is similar to having multiple, prepared positions on nearly every possible approach. The urban area can also permit rapid, covered movement on interior lines. This permits swift movement to and occupation of strong defensive positions with little or no preparation. The defense also has more flexibility since defenders often know and better understand the urban terrain's effects on operations. Normally, defenders will not get lost as easily, will know complex lines of sight and masking effects, and will best understand the ballistic characteristics of individual structures.

8-18. Mental agility allows commanders to see that the best urban defense may actually be to defend outside of the area. Such a defense mitigates the danger to the urban population and potentially reduces collateral damage. It takes advantage of Army long-range engagement capabilities and denies the threat the opportunity to "hug" Army forces or noncombatants as protection from fires. This defense may be appropriate when Army forces have enough resources to defend more open terrain, when time permits deploying extensive obstacles and constructing protected positions, and when natural terrain such as river obstacles aids the defense.

## URBAN DEFENSIVE OPERATIONS AND BATTLEFIELD ORGANIZATION

8-19. Urban defensive operations are organized within the overall battlefield organization of sustaining, shaping, and decisive operations. The success of urban defense depends on each operation, but commanders must synchronize these simultaneous operations as one action. Sustaining operations in defensive urban operations (UO) ensure freedom of action. Critically, urban sustaining operations ensure security of the lines of communications and establish effective movement control. Shaping operations in defensive UO create the conditions for decisive operations. Shaping operations vary greatly depending on the type of defense. For example, in a mobile defense the shaping operation may be the fixing force. In contrast, in an area defense the fixed defense may be the decisive operation. In the urban defense, decisive operations focus on accomplishing the commander's mission. The decisive operation may not defeat the threat's main effort, and it may not prevent threat occupation of large portions of the urban area if those tasks are not essential to mission accomplishment. For example, if the defense's objective is to protect a critical communications node, then, depending on the commander's overall intent, threat actions to secure an airfield elsewhere may not be important.

# TYPES OF URBAN DEFENSE

8-20. Commanders can view urban area defensive operations two ways: as conducting a major defensive operation with an urban area in their area of operations (AO) and as defending entirely in an urban area.

## AREA DEFENSE

8-21. At the operational level, an area defense may include both urban areas and open maneuver areas. The most common defense in an urban area and the most suitable for the characteristics of this distinct environment is the area defense. As a defensive operation, the area defense concentrates on denying threat forces access to designated terrain for a specific time rather than destroying the threat outright. Although an area defense in an urban area does not directly seek to destroy or defeat attacking threat forces, as an objective it does aim to force culmination of the threat's attack. The urban area defense often works effectively to exhaust threat resources and shape conditions for a transition to offensive operations. The urban area may also be used as a strong point to force threat movement in a different direction or to fix threat forces as part of a large, mobile defense taking place in the AO outside the urban area.

## MOBILE DEFENSE

8-22. A mobile defense can operate in an urban area but only under specific conditions. It focuses on destroying or defeating the threat through a decisive attack by a striking force. It requires the defender to have greater mobility than the attacker. To shape a mobility advantage, the urban defender effectively uses the terrain and correctly task organizes his forces' mobility. The principles of applying the mobile defense in the urban area remain the same: a small fixing force stops the threat and limits any ability to maneuver while a striking force quickly maneuvers and counterattacks to destroy him.

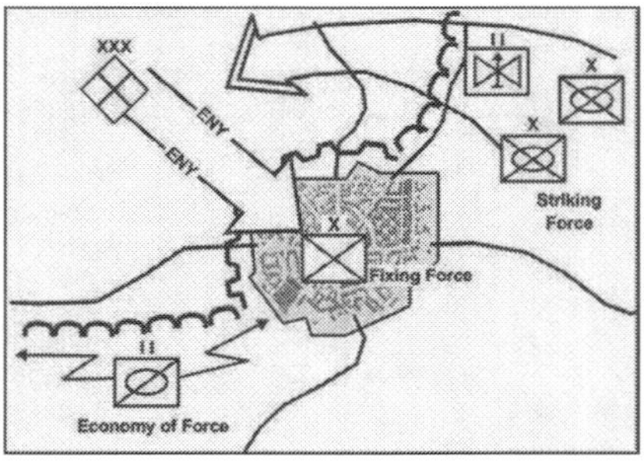


**Figure 8-1. An urban area incorporated into a larger mobile defense**

8-23. One key to executing a mobile defense in the urban area is to entice a threat force into the depths of the urban area where it begins to lose mobility options. A well-placed fixing force augmented with man-made obstacles and taking advantage of the naturally constrictive terrain can stop a much larger force. If the attacking force is largely mounted and armored, its mobility in the urban area may be reduced to less than that of dismounted infantry. In addition, if the attacking force's movement into the urban area is mounted and rapid, the commander's situational understanding also diminishes. Then the striking force, consisting of dismounted infantry forces, can execute the counterattack with surprise from multiple directions and dimensions (subsurface, surface, supersurface, and airspace). Man-portable antiarmor weapons—firing from flanks and top down and supported by precision indirect fires from both organic and joint systems—can rapidly destroy the threat.

8-24. From the perspective of commanders of the major operation, the urban environment can help defending forces achieve a mobility advantage over an attacker in a broader sense. Defending commanders can attempt to shape the battlefield so that the attacker commits significant resources into an urban area, where his maneuver capabilities are reduced (see figure 8-1). A disproportionately small defending force, which relies on the defensive combat power advantages of the urban environment, reduces and fixes the attacker's maneuver capabilities. Other defending forces mass outside the urban area then strike the threat with a combined mobility and firepower advantage.

## Urban Defense in a Major Operation: Stalingrad – 1942 to 1943

The German and Soviet battle for Stalingrad in late 1942 and into early 1943 illustrates how a tactical urban area defense integrates into a larger mobile defense. Stalingrad was a battle fought on a huge scale: it involved army groups on both sides and thousands of square kilometers. Though the city was relatively small, it remained the focus of both German and Soviet offensive and defensive operations during the six-month battle.

In the summer of 1942, the Germans launched a strategic offensive in southern Russia. Its goal was the valuable oil fields of the Caucasus. German forces turning south into the Caucasus exposed a vulnerable flank to Soviet forces positioned between the Don and Volga Rivers. For the German Caucasus operation to succeed, it had to destroy Soviet forces between the Don and Volga, establish a good defensive line, and capture Stalingrad. This city would anchor the German defense and interdict the critical flow of supplies from the Caspian Sea via the Volga River into central Russia. Stalingrad, by virtue of its name, also had important political and cultural value to the Germans and Soviets.

The opening phases of the German offensive were successful; German forces—the 6th Army and 4th Panzer Army—entered the outskirts of Stalingrad in late August 1942 (see figure 8-2).

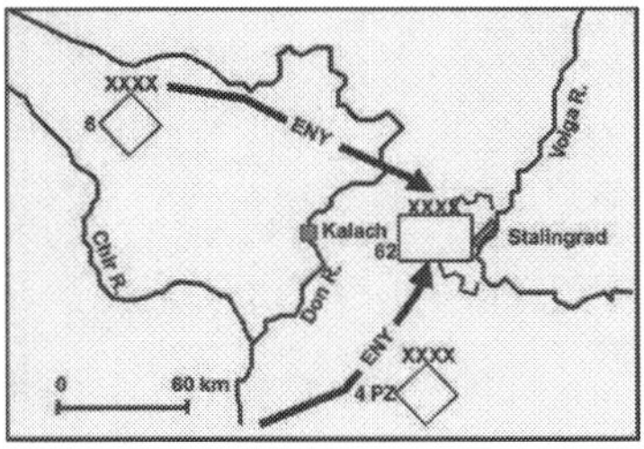


**Figure 8-2. German attacks to seize Stalingrad**

After a month of intense fighting, the Germans possessed nearly 90 percent of the city. At this point, the 6th Army commanded all German forces in and around Stalingrad. The Soviet 62nd Army's defense was reduced to a front only a few hundred meters deep and a couple of kilometers long on the banks of the Volga. The Soviet defenses hinged on fortress-like concrete industrial buildings and the fanatical bravery and tenacity of soldiers and civilians fighting in the city's remains (see figure 8-3 on page 8-6).

Beginning in mid-September, the Soviet command began looking at how to convert the defense of Stalingrad into an operational opportunity. During October and November, the 62nd Army held on to its toehold in Stalingrad. While maintaining the defense of the 62nd Army, the Soviets secretly began to build up strength on both flanks of the German 6th Army. The Germans increased their vulnerability by securing the German 6th Army's flanks with less capable Romanian, Hungarian, and Italian armies. Also, the 6th Army moved powerful German divisions into the city and rotated with German divisions that were exhausted by urban combat.

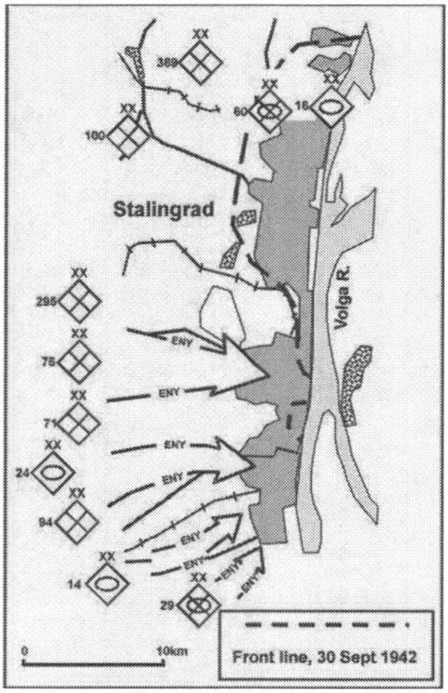


**Figure 8-3. German attacks to seize Stalingrad, September 1942**

On 19 November, the Soviets launched Operation Uranus that attacked two Romanian armies with seven Soviet armies. Simultaneously, the 8th Russian Army attacked to aid the 62nd Army in further fixing the German 6th Army. Within five days, the Soviet armies of the Don Front, Southwest Front, and Stalingrad Front met near the city of Kalach and sealed the fate of the German 6th Army's 300,000 troops in Stalingrad (see figure 8-4).

On the third day of the Soviet offensive, when encirclement seemed inevitable but not yet complete, the 6th Army commander asked permission to withdraw from the trap. The German high command denied permission believing that the Army could be supplied by air and then a renewed offensive could break through to the city. On 12 December, the German LVII Panzer Corps launched an offensive north to break through to Stalingrad. This offensive made progress until another Soviet offensive on 16 December forced its cancellation. This ended any hope of recovering Stalingrad and the 6th Army. On 31 January 1943, the 6th Army surrendered after sustaining losses of almost two-thirds of its strength. The Soviets took over 100,000 prisoners.

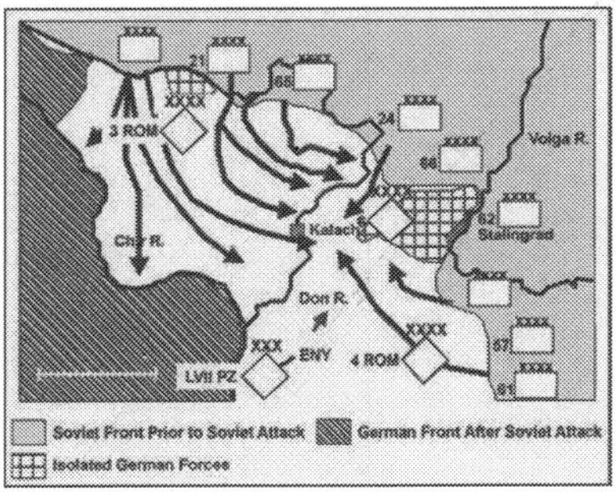


**Figure 8-4. Soviet attacks trap German 6th Army**

Many lessons emerge from the successful defense of Stalingrad. Tactically, the defense showed how using the terrain of a modern industrial city wisely could increase the combat power of an inferior, defending force and reduce the maneuver options of a mobile, modern attacking force. Another element in the Soviet's tactical success was the Germans' inability to isolate the defenders. The Germans never threatened the Soviet supply bases east of the Volga and, despite German air superiority, the Soviets continuously supplied and reinforced the 62nd Army across the Volga River. Also, Soviet artillery west of the river was able to fire in support of Soviet forces and was never threatened with ground attack.

At the operational level, the Soviets demonstrated a keen understanding of using an urban area within the context of a mobile defense. The 62nd Army's stubborn area defense of Stalingrad drew the bulk of the German combat power into the urban area where they were fixed by a smaller and quantitatively inferior defending force. This allowed the Soviets to build combat power outside the urban area. The Soviets set the conditions for a mobile defense by positioning powerful Soviet armor forces in open terrain outside the urban area against quantitatively inferior German allied forces. In OPERATION URANUS, the mobile defense's strike force destroyed the enemy outside the urban area and trapped the greater part of the best enemy formations inside the urban area. The trapped units were then subjected to dwindling resources and extensive psychological operations, further isolated into pockets, and defeated in detail.

## RETROGRADE

8-25. A retrograde involves organized movement away from the threat. Retrograde operations include withdrawals, delays, and retirements. These defensive operations often occur in an urban environment. The urban environment enhances the defending force's ability to conduct retrograde operations successfully (see figure 8-5).

8-26. The cover and concealment afforded by the urban environment facilitates withdrawals where friendly forces attempt to break contact with the threat and move away. The environment also restricts threat reconnaissance, which is less able to detect friendly forces moving out of position, and presents excellent opportunities for deception actions. Finally, a small security force's ability to remain concealed until contact in the urban environment significantly slows threat attempts to regain contact once Army forces have broken contact and begun to move.

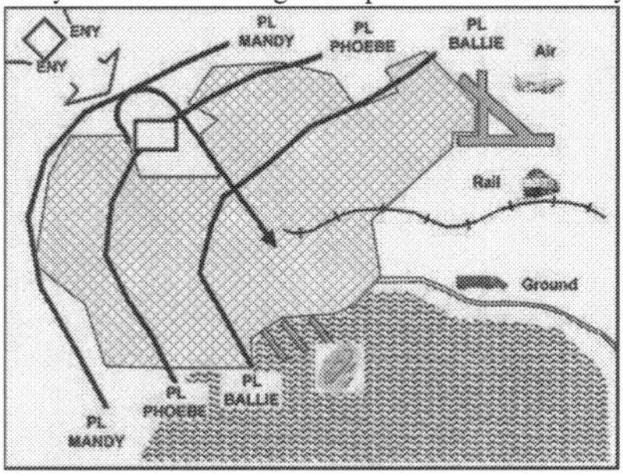


**Figure 8-5. Retrograde through an urban area**

8-27. The urban environment's natural cover and concealment, as well as the compartmented effects, facilitates delays. Delays can effectively draw the threat into the urban area for subsequent counterattack or as an integral part of a withdrawal under threat pressure. Delaying units can quickly displace from one covered and concealed position to another; the repositioning options are vast. Compartmented effects force the attacking threat to move on well-defined and easily interdicted routes and limit the threat's ability to flank or bypass delaying positions.

8-28. The urban area's transportation and distribution network facilitates retiring forces that are not in contact. Properly used, the urban transportation system can quickly move large forces and associated resources, using port facilities, airfields, railheads, and well-developed road networks.

## URBAN DEFENSIVE CONSIDERATIONS

*What is the position about London? I have a very clear view that we should fight every inch of it, and that it would devour quite a large invading army.*

Winston Churchill
*War in the Streets*

8-29. The urban operational framework—understand, shape, engage, consolidate, and transition—provides structure to developing considerations for defensive operations. The considerations can vary depending on the level of war at which the operation is conducted, the type of defense, and the situation. Most issues discussed may, in the right circumstances, apply to both commanders conducting major UO and commanders at lower tactical levels of command.

### UNDERSTAND

8-30. The commander defending in the urban area must assess many factors. His mission statement and guidance from higher commanders help him focus his assessment. If the mission is to deny a threat access to port facilities in an urban area, the commander's assessment will be focused much differently than if the mission is to deny the threat control over the entire urban area. The mission, enemy, terrain and weather, troops and support available, time available, civil considerations (METT TC) structure guides the commander's assessment. Of these, the impacts of the threat and environment—to include the terrain, weather, and civil considerations—are significant to the commander's understanding of urban defensive operations.

#### The Threat

8-31. In the urban defense, a key element is the commander's understanding of the threat. One of his primary concerns is to determine the attacker's general scheme, methodology, or concept. Overall, the attacker may take one of two approaches. The most obvious would be a direct approach aimed at seizing the objectives in the area by a frontal attack. A more sophisticated approach would be indirect and begin by isolating Army forces defending the urban area. Innumerable combinations of these two extremes exist, but the threat's intentions toward the urban area will favor one approach over another. The defending Army commander (whose AO includes but is not limited to the urban area) conducts defensive planning, particularly his allocation of forces, based on this initial assessment of threat intentions. This assessment determines whether the commander's primary concern is preventing isolation by defeating threat efforts outside the area or defeating a threat attacking the urban area directly. For the higher commander, this assessment determines how he allocates forces in and outside the urban area. For the commander in the urban area, this assessment clarifies threats to sustainment operations and helps shape how he arrays his forces.

#### The Environment's Defensive Characteristics

8-32. A second key assessment is the defensive qualities of the urban environment. This understanding, as in any defensive scenario, is based on mission requirements and on a systemic analysis of the terrain in terms of observation and fields of fire, avenues of approach, key terrain, obstacles, and cover and concealment (OAKOC). It is also based on potential chemical, biological, radiological, nuclear and fire hazards that may be present in the urban area. This understanding accounts for the unique characteristics of urban terrain, population, and infrastructure as discussed in Chapter 2.

8-33. Generally, units occupy less terrain in urban areas than more open areas. For example, an infantry company, which might occupy 1,500 to 2,000 meters in open terrain, is usually restricted to a frontage of 300 to 800 meters in urban areas. The density of building in the urban area, building sizes and heights, construction materials, rubble, and street patterns will dictate the actual frontage of units; however, for initial planning purposes, figure 8-6 provides approximate frontages and depths for units defending in an urban area.

| UNIT | Frontage (Blocks*) | Depth (Blocks*) |
|---|---|---|
| Battalion | 4 – 8 | 3 – 6 |
| Company | 3 – 4 | 2 – 3 |
| Platoon | 1 – 2 | 1 |
| *Average block is 175 meters | | |

**Figure 8-6. Approximate defensive frontages and depths**

## SHAPE

8-34. Commanders of a major operation shape the urban battle according to the type of defense they are attempting to conduct. If conducting an area defense or retrograde, they use shaping actions like those for any defensive action. Important shaping actions that apply to all defensive UO include—

- Preventing or defeating isolation.
- Separating attacking forces from supporting resources.
- Creating a mobility advantage.
- Applying economy of force measures.
- Effectively managing the urban population.
- Planning counterattacks.

### Preventing or Defeating Isolation

8-35. Failure to prevent isolation of the urban area can rapidly lead to the failure of the entire urban defense. Its importance cannot be overstated. In planning the defense, commanders must anticipate that the threat will attempt to isolate the urban area. Defensive planning addresses in detail defeating threat attacks aimed at isolation of the urban area. Commanders may defeat this effort by allocating sufficient defending forces outside the urban area to prevent its isolation. Defensive information operations (IO) based on deception can also be used to mislead the threat regarding the defensive array in and outside the urban area. Such information can convince the threat that a direct attack against the urban area is the most favorable approach.

8-36. If the threat has successfully isolated the urban area, commanders of a major operation have several courses of actions. Three options are an exfiltration, a breakout attack by forces defending the urban area, or an attack by forces outside the urban area to relieve the siege. A fourth option combines the last two: counterattacks from both inside and outside the urban area to rupture the isolation (see breakout operations in FM 3-90). Time is critical to the success of either operation. Commanders should plan for both contingencies to ensure rapid execution if necessary. Delay permits threat forces surrounding the urban area to prepare defenses, reorganize their attacking force, retain the initiative, and continue offensive operations. The passage of time also reduces the resources of defending forces and their ability to breakout. Therefore, commanders and staff of a major operation must vigilantly avoid isolation when Army forces are defending urban areas in their AO.

### Separating Attacking Forces from Supporting Resources

8-37. Commanders of the major operation primarily use fires and IO for separating in space and time threat forces attacking the urban area from echelons and resources in support. The purpose of this shaping action is the same as for any conventional area defense. It aims to allow the defending forces to defeat the threat piecemeal as they arrive in the urban area without support and already disrupted by deep fires and IO against information systems. This separation and disruption of the threat also sets the conditions for a mobile defense if commanders choose to execute that type of defense. These operations also prevent the threat commander from synchronizing and massing his combat power at the decisive point in the close battle.

8-38. If the urban area is part of a major mobile defense operation, the urban defense becomes the fixing force. Commanders can shape the defense to encourage the threat to attack into the urban area. They can lure the threat using a combination of techniques depending on the situation. They may make the urban area appear only lightly defended while other alternative courses of action appear strongly defended by friendly forces. Placing the bulk of the defending forces in concealed positions well within the urban area and positioning security forces on the periphery of the urban area portray a weak defense. In other situations, the opposite is true. If the urban area is an important objective to the threat, friendly forces can make the urban area appear heavily defended, thus ensuring that he commits sufficient combat power to the urban area to overwhelm the defense. Both cases have the same objective: to cause a major commitment of threat forces in the urban area. Once this commitment is made, the mobile defense striking force attacks and defeats the threat outside the urban area. This isolates the threat in the urban area and facilitates its destruction.

8-39. In the urban tactical battle, many shaping actions mirror those in all defensive operations. The size and complexity of the urban area prevent defending forces from being strong everywhere; shaping operations designed to engage the threat on terms advantageous to the defense have particular importance. Shaping actions include reconnaissance and security operations, passages of lines, and movement of reserve forces prior to their commitment. In addition, shaping operations critical to urban defense include mobility and countermobility operations, offensive IO, economy of force operations, and population management operations.

## Creating a Mobility Advantage

8-40. In urban terrain, countermobility operations can greatly influence bringing the threat into the engagement areas of defending forces. Countermobility operations—based on understanding the urban transportation system, design, and construction characteristics—can be unusually effective (see Chapter 2). Demolitions can have important implications for creating impassable obstacles in urban canyons as well as for clearing fields of fire where necessary. Careful engineer planning can make the already constrictive terrain virtually impassable to mounted forces where appropriate, thus denying the threat combined arms capabilities. Countermobility operations in urban terrain drastically increase the defense's ability to shape the attacker's approach and to increase the combat power ratios in favor of the defense. As with all aspects of UO, countermobility considers collateral damage and the second- and third-order effects of obstacle construction.

8-41. Well-conceived mobility operations in urban terrain can provide defending forces mobility superiority over attacking forces. This is achieved by carefully selecting routes between primary, alternate, and subsequent positions, and for moving reserves and counterattack forces. These routes are reconnoitered, cleared, and marked before the operation. They maximize the cover and concealment characteristics of the terrain. Using demolitions, lanes, and innovative obstacles denies the defense of these same routes.

## Applying Economy of Force Measures

8-42. Economy of force is extremely important to effective tactical urban defense. A megalopolis is too large and too easily accessible for defending forces to be strong everywhere. Economy of force enables the defending force to mass effects at decisive points. Forces used in an economy of force role execute security missions and take advantage of obstacles, mobility, and firepower to portray greater combat power than they actually possess. They prevent the threat from determining the actual disposition and strength of the friendly defense. If, contrary to expectations, they are strongly attacked, their mobility—stemming from a mounted maneuver capability, planning, and an intimate knowledge of the terrain—allows them to delay until reserves can meet the threat. Security forces in an economy of force role take position in parts of the urban area where the threat is less likely to attack.

## Defensive Combat Power: Suez – 1973

At the end of October, the Israeli Army was in the midst of effective counterattack against the Egyptian Army. The Israelis had success attacking west across the Suez Canal. Their armored divisions were attempting to achieve several objectives, to include destroying Egyptian air defense sites and completing the encirclement of the Egyptian 3rd Army, which was trapped on the canal's east side.

To completely encircle the Egyptian 3rd Army, the Israelis had to seize all possible crossing sites to it from the canal's west bank and the Red Sea. Also, as international negotiations towards a cease-fire progressed, the Israeli government wanted to capture as much Egyptian territory as possible to improve their negotiating position after hostilities.

Consequently, the Israeli Adan Armored Division was tasked to seize the Egyptian Red Sea port of Suez on the morning of 24 October. A cease-fire was to begin at 0700, and the Israeli intent was to be decisively engaged in the city by that time and then consolidate their position as part of the cease-fire compliance.

The Adan Division plan to seize Suez was a two-part operation. Each of the division's armored brigades would have a role. The 460th Brigade would attack west of the city and complete the city's encirclement. Simultaneously, the 217th Brigade would attack in columns of battalions through the city to seize three key intersections in the city. This was in accordance with standard Israeli armored doctrine for fighting in an urban area. The 217th Brigade would seize its objectives through speed, firepower, and shock action. Once the objectives were seized, infantry and armored teams would continue attacking from the secured objectives to mop up and destroy pockets of resistance. The Israeli commanders expected to demoralize the defending Egyptians—two infantry battalions and one antitank company—by this rapid attack. The armored division commander was specifically advised by his commander to avoid a "Stalingrad" situation.

The attack got off to an ominous beginning as mist greatly inhibited a scheduled aerial bombardment in support of the attack. The 217th Brigade began its attack without infantry support and was quickly stopped by antitank missiles and antitank fire. Infantry was quickly integrated into the brigade and the attack resumed.

At the first objective, the Israelis encountered their first problems. A withering barrage of small arms, antitank missiles, and antitank fire hit the lead tank battalion, including direct fire from SU-23 anti-aircraft guns. Virtually all the officers and tank commanders in the tank battalion were killed or wounded, and several tanks were destroyed. Disabled vehicles blocked portions of the road, and vehicles that turned on to secondary roads were ambushed and destroyed. The battalion, however, successfully fought its way through the first brigade objective and on to the final brigade objective.

Hastily attached paratroop infantry in company strength were next in column following the tanks. They were traveling in buses and trucks. As the lead tank battalion took fire, the paratroopers dismounted, and attempted to secure adjacent buildings. The tank battalion's action of fighting through the objective caused the paratroopers to mount up and also attempt to move through the objective. Because of their soft skinned vehicles the paratroopers were unable to remain mounted and again dismounted, assaulted, and secured several buildings that they could defend. Once inside the buildings, the paratroopers found they were cut off, pinned down, and unable to evacuate their considerable casualties, which included the battalion commander. The paratroopers were on the initial brigade objective but were unable to maneuver and were taking casualties.

A second paratroop company also dismounted and quickly became stalled in house-to-house fighting. The brigade reconnaissance company in M113 personnel carriers brought up the rear of the brigade column and lost several vehicles and was also unable to advance.

By 1100 the Israeli attack culminated. Elements of the 217th Brigade were on all three of the brigade's objectives in the city. However, the armored battalion, which had achieved the deepest penetration, was without infantry support and under severe antitank fire. Both paratroop companies were isolated and pinned down. In addition, an attempt to link up with the paratroopers had failed. At the same time, the civilian population of the city began to react. They erected impromptu barriers, ambushed isolated Israeli troops, and carried supplies and information to Egyptian forces.

The Israeli division commander ordered the brigade to break contact and fight its way out of the city. The armored battalion was able to fight its way out in daylight. The paratroop companies were forced to wait until darkness and then infiltrated out of the city carrying their wounded with them. Israeli casualties totaled 88 killed and hundreds wounded in addition to 28 combat vehicles destroyed. Egyptian casualties were unknown but not believed to be significant.

The fight for Suez effectively demonstrates numerous urban defensive techniques. It also vividly demonstrates the significant effect on defensive combat power of the urban environment.

The Egyptian defense demonstrates how the compartmented urban terrain restricts the mobility and the massing of firepower of armored forces. Trapped in column on the road, the Israelis were unable to mass fire on particular targets nor effectively synchronize and coordinate their fires. The short-range engagement, also a characteristic of urban combat, reduced the Israeli armor protection and eliminated the Israeli armor's ability to keep out of small arms range. Thus, hand-held antiarmor weapons were more effective in an urban area. Additionally, Egyptian small arms and sniper fire critically affected Israeli C2 by successfully targeting leaders.

The Egyptian defenders effectively isolated the mounted Israelis by defending and planning engagement areas in depth. The Egyptians synchronized so that they engaged the entire Israeli force simultaneously. This forced the Israelis to fight in multiple directions. It also separated the Israeli infantry from the armor and prevented the formation of combined arms teams necessary for effective urban offensive operations.

Suez also demonstrated how civilians come to the advantage of the defense. After the battle was joined, the population—by threatening isolated pockets of Israelis and building barricades—helped prevent the Israelis from reorganizing while in contact and hindered the Israelis breaking contact. The population was also a valuable source of intelligence for the Egyptians and precluded covert Israeli movement in daytime.

Suez shows the ability of a well-placed defense in depth to fix a superior force in an urban area. Despite the Israeli commander's caution to avoid a "Stalingrad," the Israeli division, brigade, and battalion commanders were quickly trapped and unable to easily break contact. Even a successful defense on the perimeter of the city would not have been nearly as effective, as the Israelis would have easily broken contact once the strength of the defense was recognized.

Another key to the success of the Egyptian defense was the Israelis' inadequate reconnaissance. While the Israelis knew the approximate size of the defending forces, they had no idea of the Egyptian dispositions. In this case, time prevented adequate reconnaissance. Key to a successful defense is adequate security to obscure defense dispositions, which permits surprise and shock effect.

The Suez defense was a decisive defeat of elite Israeli forces by regular infantry units inferior in training, morale, and numbers. Total disaster was averted only because of the professionalism of the Israeli armored forces and paratroopers that permitted them to continue to fight and eventually exfiltrate the urban trap. The Israeli forces thus escaped total destruction. Suez strongly demonstrates how the enhancing effects of the urban environment on defensive combat power are significant enough to permit inferior regular forces to defeat elite formations. Since

> the 1973 Suez battle, U.S. forces in Mogadishu, Somalia, and Russian forces in Grozny, Chechnya have faced similar urban defensive ambushes.

### Effectively Managing the Urban Population

8-43. Another way to shape the urban defensive battle is population management. In most cases, defending force commanders are in the urban area before combat. This time gives them the chance to manage civilians. Consequently, they can better manage and protect the population (a legal requirement) and gain more freedom of action for their forces.

8-44. Managing the civilians during the defense is a function of the size, disposition, and needs of the population and the resources available to the commander. Requesting higher support or coordinating with nongovernmental organizations and the local civil leadership for support may make up shortages of resources. Resources devoted to population management are carefully weighed against availability, military mission requirements, and possible collateral damage affecting tactical, operational, or strategic success. It may prove impractical to evacuate an urban area's population; still, commanders should attempt to create and move most civilians to protected areas. Moving the population allows defending forces to more liberally apply fires, emplace obstacles, and relieve combat units and support units of requirements to continue life support for civilians while executing combat operations. Overall, effective civil-military operations can turn a friendly (or a neutral) population into an effective force multiplier providing support to every warfighting function.

### Planning Counterattacks

8-45. Counterattacks are also an important tool in shaping the battlefield for defensive success. Counterattacks as a shaping tool have two applications: retaining the initiative and separating forces. However, opportunity for effective counterattacks will be brief and, therefore, timing will be critical. If conducted too soon, the counterattack may expend resources required later; if conducted too late, it may not be effective. Commanders should understand the effect of the urban environment on time-distance relationships; otherwise, the timing of the attack may be upset and the operation desynchronized. Additionally, commanders should develop plans beyond the counterattack to exploit potential success.

## ENGAGE

8-46. Engaging the urban area in a defensive operation requires decisively defeating the threat's attacks. Defensive forces use the terrain to their advantage, employ precision supporting fires, and use direct fire from protected positions aligned against carefully selected avenues of approaches and kill zones. The combat power of the defense augmented by shaping actions and the characteristics of urban terrain force culmination of the threat attack. Like urban offensive operations, effective engagement in urban defensive operations typically results from successful actions at the tactical level of war. These actions include—

- Performing aggressive ISR.
- Creating depth.
- Executing an effective obstacle plan.
- Conducting coordinated counterattacks.

### Performing Aggressive ISR

8-47. ISR efforts of the defender are focused initially on identifying relevant information about the location and nature of the threat's main effort. Once identified, the defender's ISR focus shifts to assessing the rate at which the threat attack moves to its culminating point. Indicators of culmination may be physical fatigue of Soldiers, a breakdown in C2 capability, difficulty providing logistics support, or the increasing time required to reorganize small units to attack. When that culmination is achieved, friendly forces counterattack before the threat has a chance to transition to a hasty defense.

### Creating Depth

8-48. Depth in the defense is the key to forcing the threat to culminate. The urban defense cannot allow itself to be penetrated nor permit forward elements to be destroyed. The defense is designed with the greatest depth possible. Defending forces weaken the threat to the fullest extent possible by attack from each position but not permit themselves to be destroyed by fires or close assault. Instead, as threat combat power builds up against individual positions, the use of mission orders permits subordinate leaders to disengage on their own initiative and move on preplanned routes to subsequent positions. Positions are designed to be mutually supporting—withdrawing from one position to a subsequent one while supporting positions cover by fires. The attacker is constantly forced to deploy and reorganize without being able to achieve decisive effects against the defender.

### Executing an Effective Obstacle Plan

8-49. Obstacles in the urban defense are designed to break up the threat's combined arms capability. Separating dismounted forces from mounted forces disrupts the cohesion of the attacker and reduces his combat power. It also exposes his individual elements to the effects of asymmetric counterattack. The leading threat dismounted force can be effectively counterattacked by a friendly combined arms element while the threat armored force remains vulnerable to antiarmor attack by dismounted forces.

### Conducting Coordinated Counterattacks

8-50. The counterattack is one of the key actions of the urban defense. However, commanders do not counterattack unless there is a reasonable chance of success. As the attacker moves into the depth of the urban area, his forces become fatigued, attrited, and increasingly disorganized. He likely also creates an increasingly long and exposed flank. At all levels, forces defending in urban terrain look for opportunities to counterattack. As the offensive force reaches the culmination point where it can no longer continue to attack with the available forces, the defensive commander executes a planned and coordinated counterattack. The counterattack aims to regain the initiative and to make the threat fight in multiple directions. Infiltration using superior knowledge of the terrain (including supersurface and subsurface capabilities) and the skillful use of stay-behind forces permits attacking the threat throughout the depth of his formations. Small-scale counterattacks focus on C2 and sustainment capabilities. These counterattacks can set the conditions for a deliberate attack leading to the ultimate destruction of the attacking threat force.

## CONSOLIDATE

8-51. Consolidation is as important to urban defensive operations as to offensive operations. Many of the same consolidation considerations for the urban offense apply equally well to the defense. Commanders should reinforce or reposition maneuver forces and fire support assets on the urban battlefield based on weaknesses uncovered during rehearsals and opportunities discovered during actual execution. While maximizing the many advantages of the urban defense, the commander of the urban defense aggressively seeks ways of attriting and weakening enemy forces before they enter the urban area and close combat is initiated. Commanders should combine the static and mobile elements of the defense to strengthen their positions in relation to the enemy while seeking every opportunity to transition to urban offensive operations. As in urban offensive operations, commanders should conduct any necessary reorganization actions that they were unable to accomplish during execution.

## TRANSITION

8-52. Transitions in urban defensive operations occur at all levels. As with offensive operations, commanders of major operations should address which units are assigned to continue to operate in the area after defensive operations have ceased. In defensive UO, this task is not as challenging as an occupation mission during urban offensive operations. The psychology of troops defending an urban area differs from those attacking into it. Defending forces become accustomed to the environment, having experience in the environment before combat. In terms of training, it is easier for follow-on missions to be assigned to a unit that has successfully defended the urban area. This course of action takes advantage of the defending unit's experience in the area and its relationships with other agencies—agencies that were operating alongside the units before and possibly during the defense. In defensive operations, regardless of the civilians' attitudes, policies regarding that population are established before the successful defense, and the command likely has experience executing operations with civil authorities and other agencies. Thus, these relationships are neither new nor as significant an issue as in offensive operations. Therefore, commanders must be prepared to execute various stability or civil support operations or use a successful defense to springboard into more decisive offensive operations elsewhere in the commanders' AO.

### Transition Emphasis to Stability or Civil Support Operations

8-53. At the end of a successful urban defense, operational commanders should generally expect civil authority, control, and jurisdiction to increase. Additionally, the civil population may be anxious to return. Defensive combat will require virtually complete military control of the urban area; however, after the successful defense, a rapid transition will occur from military control to civil or joint military and civil control afterward. Although the full spectrum of Army operations will often be conducted simultaneously in UO, commanders will transition emphasis to stability or civil support operations. Important transition tasks will include demilitarizing munitions, clearing obstacles, and searching for isolated threat pockets of resistance. Conclusion of the defensive operations will also require transition to joint civil-military tasks, such as evaluating structures for safety, restoring essential services, and possibly creating joint law enforcement. Commanders of major operations, using a civil-military operations center and G9, should anticipate these requirements and begin early preparations to ensure a smooth, successful transition.

### Transition to Offensive Operations

8-54. Units that have successfully defended the urban area may then transition to offensive operations. A rapid transition to offensive operations will require identification, preparation, and training of units designated to assume missions as the defending units leave the urban area. This preparation emphasizes continuity of policies and relationships already established. A relief in place occurs. The new occupying units provide not only a continuity of policy, but also a continuity of attitude toward the urban area, its population, and its institutions.

# Chapter 9

# Urban Stability and Civil Support Operations

The fundamental shared aims between stability and civil support operations and any operation conducted in an urban environment are the increased significance and influence of the civil population and nonmilitary organizations. Often, no military victory is to be achieved. The center of gravity for these operations normally cannot be attacked through military means alone; the Army (and the military in general) is often but one tool supporting a larger, civil-focused effort. When conducted, military defeat of an armed adversary is only a phase in a larger campaign, the overall objectives of which will be to bring about a more satisfactory political, social, diplomatic, economic, and military situation than existed prior to the conflict. Without a tightly coordinated civil-military effort, overall success will be difficult or impossible to achieve. Commanders who can understand and cope with the complexities of stability or civil support operations gain insights that directly apply to executing any urban operation. Urban stability or civil support operations may complement urban offensive and defensive operations, or may dominate the overall operation. Army forces may need to conduct offensive and defensive operations to defend themselves or destroy urban threats seeking to prevent the decisive stability or civil support mission. During hostilities, urban stability operations may keep armed conflict from spreading, encourage coalition and alliance partners, and secure the civilian population's support in unstable urban areas (and the surrounding rural areas under their influence). Following hostilities, urban stability operations may provide a secure environment for civil authorities to rebuild. Urban civil support operations can range from transporting, feeding, and sheltering the population made homeless as a result of man-made or natural disasters to conducting civil search and rescue and providing urgent medical care.

## PURPOSE OF URBAN STABILITY AND CIVIL SUPPORT OPERATIONS

9-1. Army forces conduct stability and civil support operations to deter war, resolve conflict, promote peace, strengthen democratic processes, retain U.S. influence or access abroad, assist U.S. civil authorities, and support moral and legal imperatives. Stability operations promote and sustain regional and global stability. Civil support operations meet the urgent needs of designated U.S. civilians until civil authorities can accomplish these tasks without military assistance. Nearly every urban operation will involve some type or form of stability operation or civil support operation combined and sequenced with offensive and defensive operations.

# CHARACTERISTICS OF URBAN STABILITY AND CIVIL SUPPORT OPERATIONS

9-2. Worldwide urbanization, migration trends from rural to urban areas, and more centralized populations in urban areas increase the chance that Army forces will conduct stability or civil support operations in or near urban areas. Simply put, many people live in urban areas, and their welfare will be the primary reason for conducting these operations. Urban areas that serve as economic and government centers (the ideal location for U.S. and allied embassies) are often the focal point for many threat activities. Therefore, Army forces may need to conduct stability operations in these foreign cities to counter those threats. Additionally, urban areas may contain the resources and infrastructure to support both types of operations, regardless of whether the overall focus is in urban or rural areas. Repairing or restoring the infrastructure may be a critical task in accomplishing either type of operation. Supported governmental and nongovernmental agencies are not as logistically self-sufficient as the Army. As such, these agencies will often need to center their operations in and around urban areas to use the area's infrastructure to support themselves and their objectives. These agencies may require military protection to accomplish their missions. Figure 9-1 lists some defining characteristics of these wide-ranging operations.

- Long or Short Duration
- Joint and Interagency
- Unilateral or Multinational
- Increased Civil-Military and Legal Considerations
- Greater Potential for Ambiguity
- Increased Constraints Necessitating More Restrictive ROE
- Amplified Need for Cultural and Political Sensitivity

**Figure 9-1. Characteristics of stability and reconstruction and support operations**

9-3. Stability and civil support operations are diverse, varied in duration, and unilateral or multinational. Stability operations are conducted overseas as part of a campaign or major operation, while civil support operations are conducted within the United States as part of homeland security. Like all urban operations (UO), they are usually joint. Unlike urban offensive and defensive operations, they are more often interagency operations and require more restrictive rules of engagement (ROE). The multiplicity of actors involved usually increases the scope and scale of required coordination and communication. In urban stability or civil support operations, adverse conditions arising from natural or man-made disasters or other endemic conditions—such as human suffering, disease, violations of human rights, or privation—will significantly modify the urban environment. Unresolved political issues and tenuous agreements, difficulties discriminating combatants from noncombatants or between parties of a dispute, and the absence of basic law and order all serve to complicate an already complex and uncertain environment. Civil-military and legal considerations take on added significance in all urban operations (see Civil-Military Operations and Legal Support in Chapters 5 and 10 respectively), but even more so in urban stability and civil support operations. Finally, recognizing, defining, and achieving the desired end state is often more difficult than in offensive and defensive operations.

9-4. Overall, commanders of major operations involving urban stability and civil support operations should not expect clear guidance. They must learn, adapt, and live with ambiguity. They cannot expect to operate in a political vacuum (even commanders at the tactical level); rather, they should expect to work alongside both governmental and nongovernmental leaders and organizations. Commanders conducting UO should not expect an easily identifiable enemy located across a clearly demarcated line. In fact in many peace operations, they and their Soldiers must resist the need to have an enemy—difficult at best when one side or another (or both) may be sniping at them. Commanders should also expect changing and additional missions and tasks, without being allowed to use every means at their disposal to carry out those missions. Many tasks required may be ones for which their units have never, or rarely, trained. Finally, commanders should expect to show restraint with a keen sensitivity to political considerations and to alien cultures, either or both of which they might find confusing or even repugnant.

# URBAN STABILITY AND CIVIL SUPPORT OPERATIONS AND BATTLEFIELD ORGANIZATION

9-5. Each urban stability operation or civil support operation will be distinct from one another. These operations differ even more when applied to a specific urban area. Due to the complexity of the environment, commanders must carefully arrange their forces and operations according to purpose, time, and space to accomplish the mission. In most UO the terrain, the dense population (military and civilian), and the participating organizations will further complicate this arrangement.

9-6. The support and assistance that Army forces will provide during these operations is only temporary (although it may be of long duration) and this facet requires particular emphasis. Commanders should plan and execute both of these types of operations with that essential consideration always in mind. Eventually, the government and administration—either foreign or domestic—must secure and support their population by themselves. Therefore, commanders must envision and set the conditions that allow for the transition of control and responsibility to shift to legitimate civilian authorities. While commanders may only provide assistance and support based on specific and well planned civilian requests, more often commanders will need to determine requirements in collaboration with competent civilian authorities and agencies or, in some cases, with little or no initial civilian assistance at all. Ultimately, transition planning should occur as an integral part of the overall operational planning and include collaboration with appropriate civilian agencies and organization as early as possible to ensure a seamless transition to civilian control without major setbacks and loss of forward momentum.

## STABILITY OPERATIONS

9-7. Stability operations establish, sustain, and exploit security and control over foreign areas, populations, and resources. Urban areas will be decisive to accomplishing many types of stability operations because urban areas are the centers of population, culture, economy, and government. Much of the support provided by Army forces will aim to assist local, regional, or national governments to restore essential services and infrastructure and reestablish civil order and authority. The location of civilian authorities in urban areas will, by necessity, be a dominating factor in accomplishing the mission. As importantly, many stability operations—enforcing peace in Bosnia for example—will require interacting with, influencing, controlling, or protecting all or parts of the civilian population. Assessing, understanding, and gaining the support of civilians in key economic, cultural, or political urban areas may influence surrounding regions (smaller urban areas and the rural countryside) and may be decisive to achieving overall stability objectives.

9-8. Stability operations may involve both coercive and cooperative actions. They are conducted in situations in which legitimate civil authority is unable to provide the necessary security and control for the urban population as a result of—

- Deliberate operations to influence a regime change.
- Offensive or defensive operations or natural disasters resulting in ineffective civil authorities, conditions beyond the capabilities of the foreign urban government, or both.

## CIVIL SUPPORT OPERATIONS

9-9. Civil support operations are an integral part of homeland security which aim to protect U.S. citizens and infrastructure from conventional and unconventional threats and mitigate the effects of man-made and natural disasters. (The other part to homeland security is homeland defense.) Army forces conduct civil support operations by providing Army resources, expertise, and capabilities in support of a lead agency—which may be to Department of Defense—to address the consequences of man-made or natural accidents beyond the capabilities of local or regional U.S. civilian authorities. Such a situation could result from a chemical, biological, radiological, nuclear, and high-yield explosive (CBRNE) incident, hurricane, flood, or civil disturbance in a domestic urban area.

### CBRNE INCIDENTS

9-10. CBRNE incidents can be disastrous and are of particular concern for both stability and civil support UO. In urban areas, the potential for catastrophic loss of life and property is enormous. The Army categorizes CBRNE incidents separately from other natural and man-made disasters because it has specific expertise with these weapons. CBRNE incidents usually result from a military or terrorist threat (adding a law enforcement dimension to the disaster). Other urban CBRNE considerations include the following:

- Subways and other subsurface areas offer ideal areas for limited chemical or biological attacks. Nuclear attack (and high-yield explosives) can produce tragic results due to the effects of collapsing structures, flying debris, and fires. Dispersion patterns are affected by the urban terrain and are more difficult to predict and monitor. Large-scale incidents may produce hundreds of thousands of casualties, but even a limited attack may require evacuating and screening large numbers of civilians. Requirements for medical support, basic life support, and, if necessary, decontamination may quickly overwhelm the Army force's capabilities even with augmentation.
- Similar to natural disasters, panic and disorder may accompany the CBRNE event. Fleeing civilians may clog elements of the transportation and distribution infrastructure. Physical destruction may also affect other components of the infrastructure of critical and immediate concern, such as energy and administration and human services (water, sanitation, medical, fire fighting, and law enforcement). Because all elements of the infrastructure may be affected, the overall recovery time may be lengthened and the effects broadened to include much of the surrounding area. The effects of a single urban CBRNE event potentially could be felt nationally or globally.

### DECISIVE OPERATIONS

9-11. In urban stability operations, decisive operations may take many years and include multiple actions before achieving the desired end state. This particularly applies to the strategic and operational levels. Oppositely, decisive operations involved in an urban civil support operation for mitigating or reducing disease, hunger, privation, and the effects of disasters normally achieve faster results. However, any operation that attacks the underlying cause and seeks to prevent or relieve such conditions will usually take longer. In urban areas, establishing law and order to protect critical infrastructure and the inhabitants from lawlessness and violence is often critical and also the decisive operation.

### SHAPING OPERATIONS

9-12. Shaping operations establish and maintain the conditions for executing decisive operations. In urban stability or civil support operations, shaping operations always include information operations (IO) that influence perceptions and maintain legitimacy. Often, various participants, and their potentially divergent missions and methods, are involved. Army commanders must coordinate their planning and efforts (early and continuously) to ensure that their decisive, shaping, or sustaining operations are not working against other agencies' efforts and operations—agencies that may have the lead role in the operation. Thus, a critical shaping operation may be to establish the coordination to help develop a common purpose and direction among agencies, particularly those that may experience continuous personnel turnover during the conduct of a lengthy operation. In some instances and with some organizations and agencies, particularly nongovernmental organizations (NGOs), genuine unity of effort may not be achievable; however, recognizing the differences in aims and goals will allow Army commanders to conduct operations with less friction. Commanders should actively request and include NGOs and appropriate governmental agencies in mission readiness exercises or any other training for stability or civil support operations.

### SUSTAINING OPERATIONS

9-13. Sustaining operations enable decisive and shaping operations and include logistics, rear area and base security, movement control, terrain management, and infrastructure development. Sustainment bases, especially those located in urban areas, become an attractive target for hostile civilians; therefore, commanders must actively and aggressively protect these bases as well as lines of communications (see Chapter 10).

## CONSIDERATIONS FOR URBAN STABILITY AND CIVIL SUPPORT OPERATIONS

*In wars of intervention the essentials are to secure a general who is both a statesman and a soldier; to have clear stipulations with the allies as to the part to be taken by each in the principal operations; finally, to agree upon an objective point which shall be in harmony with the common interests.*

Lieutenant General Antoine-Henri,
Baron de Jomini

9-14. The urban operational framework (understand, shape, engage, consolidate, and transition) provides a structure for developing considerations unique to urban stability and civil support operations. Many considerations presented in urban offensive and defensive operations apply to urban stability and civil support operations, particularly those that address how to understand the urban and overall operational environment. Because the situations in which stability and civil support operations normally occur share strong similarities with any urban environment, many of these considerations are closely linked to the urban fundamentals presented in Chapter 5. Taken together, commanders will often find them useful in conducting UO throughout full spectrum operations. Appendix C has a historical example of how to apply the urban operational framework to a stability operation.

### UNDERSTAND

9-15. In urban stability or civil support operations, commanders must carefully assess and understand the political dimension of the operational environment, as well as their role and the media's part in managing information. These operations are inherently tied to the exercise of diplomatic power. All operations in urban areas are often the focus of the media and thus gain considerable public and political attention. Therefore, military objectives in urban stability or civil support operations are more directly linked with political objectives. The relationship between the levels of war—strategic, operational, and tactical—is often closer than in urban offensive and defensive operations. Military objectives must be carefully nested within political objectives. Commanders must ensure that the ways and means to accomplish their objectives, to include security and force protection measures, will hold up to media scrutiny and are appropriate for the situation and environment. All levels of command must understand the link between political and military objectives, to include a basic understanding at the Soldier level. One uncoordinated, undisciplined, or inappropriate action, even at the lowest level, could negate months or years of previous, disciplined effort. Commanders must balance security and force protection measures with mission accomplishment. Ineffective measures can put Soldiers at too great a risk and jeopardize the mission. Conversely, overly stringent measures may make it difficult for forces to interact with the population closely—essential in many of these operations. Finally, commanders will need a thorough assessment of the governmental and nongovernmental organizations and agencies that will be operating in or near urban areas that fall within their area of operations.

#### Political and Military Objectives

9-16. Commanders must translate political objectives into military objectives that are clear and achievable (clear tasks and purposes) and can lead to the desired end state. (In most stability operations, defeat of a threat is not the ultimate desired end state.) Political objectives may be vague making it difficult for commanders to conduct their mission analysis. This applies to tactical- and even operational-level commanders, unskilled at higher level, strategic political-military assessments. Each type of stability operation or civil support operation is distinct, often unfamiliar to the executing unit, and unique to the specific situation. These factors often make it difficult to confidently determine the specific tasks that will lead to mission success. Therefore, commanders must also establish measures of effectiveness that aid in understanding and evaluating progress and help gauge mission accomplishment.

> ***Note***: Commanders should consult the US Agency for International Development's (USAID) Field Operations Guide for Disaster Assessment and Response when conducting their assessments and developing measures of effectiveness for many urban relief operations.

9-17. These measures should be quantifiable (in some circumstances, a qualitative assessment may be most appropriate) and link cause with effect. They help determine the changes required and are essential to the assessment cycle required for urban stability or civil support operations. In a humanitarian relief operation to aid the starving, commanders could determine that the decisive effort is delivering safe food to the urban area. To judge success or effectiveness, they could determine that the appropriate measure is the number of food trucks dispatched daily to each distribution site; the more trucks, the more effective the efforts. However, this measure must correlate with the overarching measure of effectiveness: decline in the mortality rate. If no significant decrease in deaths due to starvation occurs, they may need to reassess and modify the tasks or measures of effectiveness. A better measure may be to track the amount of food consumed by those in need instead of simply counting the number of trucks dispatched. Areas around which measures of effectiveness can be formed for many stability or civil support operations (including the example above) and which will help return most societies to some degree of normalcy and self-sufficiency include:

- Restoring law and order.
- Decreasing morbidity and mortality rates.
- Securing safe food and water.
- Restoring critical infrastructure.
- Resettling the population.
- Reestablishing economic activity.

Commanders can often develop measures of effectiveness to address these areas in terms of establishing or restoring security and providing logistics. (Figure 9-2 provides example measures of effectiveness from the strategic to tactical levels for a possible stability operation.)

9-18. Political objectives are fluid and modified in response to new domestic and international events or circumstances. Thus, assessment is continuous, and commanders must adjust their own objectives and subsequent missions accordingly. In urban stability or civil support operations, commanders often develop military objectives that support or align with the objectives of another agency that has overall responsibility for the urban operation. In this supporting role, commanders may receive numerous requests for Soldier and materiel assistance from the supported agency and other supporting agencies operating in the urban area (to include elements of the urban population). With such unclear lines of authority and areas of responsibility, they must ensure that the tasks, missions, or requested Army resources fall clearly in the intended scope and purpose of the Army's participation in the operation. They must not develop or execute missions based on inadequate or false assumptions, misinterpreted intent, or well-meaning but erroneously interpreted laws or regulations by any organization, to include even the lead agency. When missions appear outside their scope, commanders must quickly relay their assessment to their higher headquarters for immediate resolution. The commander's goal is not to limit or slow military participation but to contribute as intended and in consonance with political objectives and the law.

## Security and Force Protection Measures

9-19. Commanders must plan for and continually assess the security of their forces operating in an urban area as well as constantly review protection measures. Establishing a robust intelligence—particularly human intelligence (HUMINT)—network that can determine the intentions and capabilities of the threat and the urban populace is the basis for establishing force protection for Army forces operating in the urban environment. However, many such operations, particularly stability operations, require extra time to forge a lasting change. Over time, and particularly in peacetime when objectives center on helping others and avoiding violence, even the complex urban environment may seem benign. Without continued, aggressive command emphasis, Soldiers may become lulled into complacency. Also, during periods of transition or transfer of authority from one unit or organization to another, departing Soldiers' focus often shifts to redeployment activities and away from force protection concerns. It is usually then that Army forces are most vulnerable to terrorist and insurgent tactics, such as bombings, kidnappings, ambushes, raids, and other forms of urban violence.

**Measures of Effectiveness: The Need to Measure Progress**

| Strategic Level Criteria | Operational Level Criteria | Tactical Level Criteria |
|---|---|---|
| Accountable to the American people for defining and measuring progress towards defeating terrorism and meeting national security goals.<br>Examples:<br>• Prevention of the insurgency from receiving aid or resources from other international groups.<br>• A functioning national government.<br>• Amount of international support and aid to reconstruction.<br>• Number of nations contributing manpower to coalition forces. | Accountable to the strategic level for measuring operational success and providing linkage to strategic goals.<br>Examples:<br>• Host-nation security forces trained and equipped.<br>• Denial of the merging of insurgent forces with terrorist groups.<br>• Amount and distribution of—<br>• Electricity.<br>• Liquid propane gas.<br>• Gasoline.<br>• Functioning provincial governments. | Responsible to the operational level for measuring tactical success and providing linkage to operational goals.<br>Examples:<br>• Number of insurgent forces in the AO neutralized.<br>• Reduced attacks on coalition forces in the AO.<br>• Reduced civilian-on-civilian violence in the AO.<br>• Host-nation security force recruitment goals met.<br>• Host-nation security force training goals met.<br>• Number of reliable human intelligence walk-ins.<br>• Amount of unexploded explosive ordnance cleared.<br>• Functioning neighborhood and district advisory councils. |

**Figure 9-2. Example measures of effectiveness**

9-20. Although force protection will not ensure successful urban stability or civil support operations, improper assessment and inadequate force protection measures can cause the operation to fail. In either operation, keeping a neutral attitude toward all elements of the urban population, while maintaining the appropriate defensive posture, enhances security. One threat principle discussed in Chapter 3 was that threats would seek to cause politically unacceptable casualties. An improper threat assessment and a lapse in security at the tactical level could result in casualties. That result could affect strategy by influencing domestic popular support and subsequently national leadership decisions and policy.

9-21. Emphasizing security and force protection measures does not mean isolating Soldiers from contact with the urban population. On the contrary, commanders should balance survivability with mobility according to the factors of METT-TC—mission, enemy, terrain and weather, troops and support available, time available, civil considerations. Survivability measures—such as sandbagging, hardening, or fortifying buildings and installations, particularly where large numbers of Soldiers are fed and billeted—will be necessary and require considerable command attention. On the other hand, mobility operations are essential in preserving freedom of action and denying a threat the opportunity to observe, plan, and attack urban forces. Continual Army presence in the urban area may provide the urban population a sense of security and allow Soldiers to develop a detailed knowledge of the "patterns of life" in their assigned area of operation (AO). Armed with this knowledge, they can detect the absence of the normal or the presence of the abnormal that might indicate a potential threat. Overall, mission degradation and increased risk to the force can result if force protection measures prevent Army forces from conducting prudent missions and establishing an active and capable presence.

**Assessment of Security and Force Protection: Belfast, Northern Ireland**

Since 1969, Belfast has significantly affected the British military campaign for stabilizing the area. British operations in Belfast illustrate the difficulty of balancing the security and protection of forces with maintaining the stabilizing presence necessary to uphold law and order, minimize violence, and control the urban population.

British successes in protecting Belfast's infrastructure and government facilities from terrorist attacks compelled various terrorist cells, especially the Irish Republican Army and the Provisional Irish Republican Army, to attack more military targets.

At the time, British soldiers and bases presented relatively unprotected targets to these factions, and attacks against them solidified their legitimacy as an "army." In response, British commanders implemented extreme security and force protection measures—from ballistic protection vests and helmets to fortress-like operational bases and large unit patrols. These protection measures successfully decreased the violence against British soldiers in Belfast. However, they also decreased the soldiers' interaction with the population and their ability to stabilize the city. The large patrols protected the soldiers, but also inhibited effective saturation of neighborhoods. These patrols, coupled with fortress-like bases and bulky protective clothing, created an "us-versus-them" mentality among civilians and soldiers.

As force protection increased and stabilizing effects decreased, the terrorists were provided more targets of opportunity among the civilians and infrastructure. British commanders reassessed the situation, identified this "see-saw" effect, and adapted to strike a better balance between force protection and effective presence patrols. For example, British forces switched to four-man patrols to enable greater mobility and wore berets instead of helmets to appear less aggressive.

While the above historical example illustrates a critical aspect of stability and civil support operations, it subtly reveals another important lesson for Army commanders. Many of our future coalition partners and assisting governmental and nongovernmental organizations will bring key insights to UO (particularly dealing with urban societies) based on their historical and operational experience. Army commanders and their staffs should seek out and remain open to others' unique understandings that are based on hard-earned and well-analyzed lessons learned.

## Participating Organizations and Agencies

9-22. Across the spectrum of urban operations, but more so in these operations, numerous NGOs may be involved in relieving adverse humanitarian conditions. Dense populations and infrastructure make an urban area a likely headquarters location for them. (In 1994 during OPERATION UPHOLD DEMOCRACY, for example, over 400 civilian agencies and relief organizations were operating in Haiti.) Therefore, commanders should assess all significant NGOs and governmental agencies operating (or likely to operate) in or near the urban area to include their—

- Functions, purposes, or agendas.
- Known headquarters and operating locations.
- Leadership or senior points of contact (including telephone numbers).
- Communications capabilities.
- Potential as a source for critical information.
- Financial abilities and constraints.
- Logistic resources: transportation, energy and fuel, food and water, clothing and shelter, and emergency medical and health care services.
- Law enforcement, fire fighting, and search and rescue capabilities.
- Refugee services.
- Engineering and construction capabilities.
- Other unique capabilities or expertise.
- Previous military, multinational, and interagency coordination experience and training.
- Rapport with the urban population.
- Relationship with the media.
- Biases or prejudices (especially towards participating U.S. or coalition forces, other civilian organizations, or elements of the urban society).

Commanders can then seek to determine the resources and capabilities that these organizations may bring and the possible problem areas to include resources or assistance they will likely need or request from Army forces. These organizations will be critical to meeting the population's immediate needs and minimizing the effects of collateral damage or disaster. However, commanders should consider whether a close relationship with any of these organizations will compromise the organization's appearance of neutrality (particularly threat perceptions) and adversely affect their ability to assist the population.

# SHAPE

9-23. Commanders conduct many activities to shape the conditions for successful decisive operations. In urban stability and civil support operations, two activities rise to the forefront of importance: aggressive IO and security operations.

## Vigorous Information Operations

9-24. IO, particularly psychological operations (PSYOP) and the related activities of civil affairs (CA) and public affairs, are essential to shape the urban environment for the successful conduct of stability or civil support operations. Vigorous IO can influence the perceptions, decisions, and will of the threat, the urban population, and other groups in support of the commander's mission. IO objectives are translated to IO tasks that are then executed to create the commander's desired effects in shaping the battlefield. These operations can isolate an urban threat from his sources of support; neutralize hostile urban populations or gain the support of neutral populations; and mitigate the effects of threat IO, misinformation, rumors, confusion, and apprehension. Developing effective measures of effectiveness is essential to a good IO campaign strategy. One of the most valuable methods for obtaining data for use in this process is face-to-face encounters with targeted audiences by unit patrols and HUMINT, PSYOP, and CA teams. A valuable technique may be to conduct periodic, unbiased surveys or opinion polls of the civilian population to determine changes in their perceptions and attitudes.

## Security Operations

### *Protecting Civilians and Critical Infrastructure*

9-25. Security for NGOs and civilians may also be an important shaping operation, particularly for stability and civil support operations. Commanders may need to provide security to civil agencies and NGOs located near or operating in the urban area so that these agencies can focus their relief efforts directly to the emergency. Commanders may also need to protect the urban population and critical infrastructure to maintain law and order if the urban area's security or police forces are nonexistent or incapacitated, or the urban area's security situation has undergone drastic change (such as the result of a natural disaster) and security or police forces require additional augmentation. (See also the discussion of the legal aspects of Civilians Accompanying the Force in Chapter 10.)

### *Preserving Resources*

9-26. Just as forces may be at risk during urban stability or civil support operations, so may their resources. In urban areas of great need, supplies and equipment are extremely valuable. Criminal elements, insurgent forces, and people in need may try to steal weapons, ammunition, food, construction material, medical supplies, and fuel. Protecting these resources may become a critical shaping operation. Otherwise, Army forces and supporting agencies may lack the resources to accomplish their primary objectives or overall mission.

## Prioritize Resources and Efforts

9-27. During UO, commanders will always face limited resources with which to shape the battlefield, conduct their decisive operations, and accomplish their objectives. They must continually prioritize, allocate, and apply those resources to achieve the desired end state. To this end, they may develop an order of merit list for proposed projects and constantly update it over time. To some degree, the local urban population will usually need to be protected and sustained by Army forces. Hence, commanders must tailor their objectives and shape their operations to achieve the greatest good for the largest number. Commanders first apply the urban fundamental of preserving critical infrastructure to reduce the disruption to the residents' health and welfare. Second, they apply the urban fundamental of restoring essential services, which includes prioritizing their efforts to provide vital services for the greatest number of inhabitants possible. In operations that include efforts to alleviate human suffering, the criticism for any participating organization is likely to be there is not enough being done or Army forces are not being responsive enough. Therefore, commanders must develop clear measures of effectiveness not only to determine necessary improvements to operational plans but also to demonstrate their Soldiers' hard work and sacrifice and U.S. commitment to the operation.

## Improve the Urban Economy

9-28. When conducting reconstruction and infrastructure repair, commanders should consider using such activities to simultaneously improve the urban economy. Hiring local civilians and organizations to do reconstruction work helps satisfies urban job requirements, may inspire critical elements of the urban society to assume responsibility for the success or failure of urban restoration efforts, and may potentially reduce threat influence by diminishing their civilian sources of aid. Hiring indigenous personnel for short-term projects does not replace the need for long-term economic planning and the development of stable jobs. The overall reconstruction effort must be guided by the commander's vision of the end state. However, the commander's guidance and intent should be broad and expansive enough to allow responsive decision making by subordinate commanders based on their analysis of the urban population's needs in their assigned AO.

## ENGAGE

*If there is any lesson to be derived from the work of the regular troops in San Francisco, it is that nothing can take the place of training and discipline, and that self-control and patience are as important as courage.*

Brigadier General Frederick Funston
commenting on the Army's assistance following
the 1906 San Francisco earthquake and fire

9-29. The focus of the Army is warfighting. Therefore, when Army commanders conduct many urban stability and civil support operations, they must adjust their concept of success. Commanders will most often find themselves in a supporting role and less often responsible for conducting decisive operations. They must accept this supporting function and capitalize on the professional values they have instilled in each Soldier, particularly the sense of duty to do what needs to be done despite difficulty, danger, and personal hardship. Commanders must also put accomplishing the overall mission ahead of individual desires to take the lead—desire often fulfilled by being the supported rather than supporting commander. In many stability and civil support operations, success may be described as settlement and compromise rather than victory. Yet, the Army's professionalism and values—combined with inherent adaptability, aggressive coordination, perseverance, and reasonable restraint—will allow Army forces to engage purposefully and dominate during complex urban stability or civil support operations.

### Adaptability

9-30. Adaptability is particularly critical to urban stability or civil support operations because these operations relentlessly present complex challenges to commanders for which no prescribed solutions exist. Commanders often lack the experience and training that provide the basis for creating the unique solutions required for these operations. Since the primary purpose for the Army is to fight and win the nation's wars, the challenge then is to adapt urban warfighting skills to the unique stability or civil support situation.

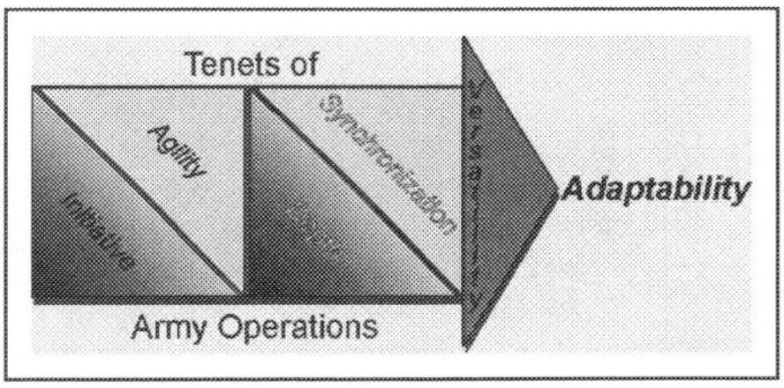


**Figure 9-3. Adaptability**

9-31. Doctrine (joint and Army) provides an inherent cohesion among the leaders of the Army and other services. Still, Army commanders conducting urban stability or civil support operations will often work with and support other agencies that have dissimilar purposes, methods, and professional languages. Army commanders must then capitalize on three of the five doctrinal tenets of Army operations: initiative, agility, and versatility (see figure 9-3 and FM 3-0). Commanders must bend as each situation and the urban environment demand without losing their orientation. They must thoroughly embrace the mission command philosophy of command and control addressed in FM 6-0 and Chapter 5 to encourage and allow subordinates to exercise creative and critical thinking required for planning and executing these UO. Commanders must also be alert to and recognize "good ideas," and effective tactics, techniques, and procedures, regardless of their source—other services, coalition partners, governmental and nongovernmental organizations, and even the threat—and adapt them for their own purposes in UO.

9-32. Adaptability also springs from detailed planning that carefully considers and realistically accounts for the extent of stability and civil support operations. Although no plan can account for every contingency and completely eliminate the unexpected, good plans—which include detailed civil considerations—provide platforms from which to adjust course more readily. Adequate planning allows commanders not only to react more effectively, but also to be forward-thinking and take actions that favorably steer the course of events.

## Aggressive Coordination and Synchronization

9-33. In urban stability and civil support operations, the increased number of participants (both military and nonmilitary) and divergent missions and methods create a significant coordination and synchronization challenge. Significant potential for duplicated effort and working at cross-purposes exists. The success of UO often depends on establishing a successful working relationship with all groups operating in the urban area. The absence of unity of command among civil and military organizations does not prevent commanders from influencing other participants not under his direct command through persuasion, good leadership, and innovative ideas.

9-34. Commanders may consider establishing, as necessary, separate organizations for combat operations and for stability operations or civil support operations to increase coordination and enhance local, NGO, and international support. Further, aligning the unit or subordinate units with NGOs may contribute to establishing popular legitimacy for the operation and place greater pressure on threat forces. In some instances, commanders may consider organizing part of their staff around government, administrative, and infrastructure functions that mirror the urban area in which their forces are operating. Development of a mirror urban area organization may give greater legitimacy to the urban government or administration and ease transition of responsibility once the end state is achieved. Commanders must be mindful, however, that local groups seen as allying themselves with Army or coalition authorities will likely experience pressure to demonstrate their independence as established dates for redeployment or other critical events approach. In some instances that demonstration of independence may be violent.

### Civil Support and Coordination with Civilian Authorities: Los Angeles – 1992

During the spring of 1992, Soldiers from the 40th Infantry Division, California National Guard were among the forces deployed to Los Angeles County to assist the California Highway Patrol, Los Angeles County Sheriffs, and civilian law enforcement. They worked to quell the riots that were sparked by the "not guilty" verdicts concerning four police officers who, following a lengthy high-speed chase through Los Angeles, were accused of brutally beating Rodney King.

Successful accomplishment of this civil support operation was attributed to the exercise of strong Army leadership and judgment at lower tactical levels, particularly among the unit's noncommissioned officers. An essential component of combat power, it was especially critical in executing noncontiguous and decentralized operations in the compartmented terrain of Los Angeles. As important, however, was the clear understanding that Army forces were to support civilian law enforcement—and not the other way around. The 40th Infantry Division aligned its area of operations with local law enforcement boundaries and relied heavily on police recommendations for the level at which Soldiers be armed (the need for magazines to be locked in weapons or rounds chambered).

One incident emphasized the need for coordination of command and control measures with civilian agencies even at the lowest tactical levels. To civilian law enforcement and Army forces, the command "Cover me" was interpreted the same: be prepared to shoot if necessary. However, when a police officer responding to a complaint of domestic abuse issued that command to an accompanying squad of Marines, they responded by immediately providing a supporting base of fire that narrowly missed some children at home. However, the Marines responded as they had been trained. This command meant something entirely different to them than for Army Soldiers and civilian law enforcement. Again, coordination at all levels is critical to the success of the operation (see also the vignette in Appendix B).

9-35. In the constraints imposed by METT-TC and operations security (OPSEC), commanders should seek to coordinate all tactical stability operations with other agencies and forces that share the urban environment. Importantly, they should seek to coordinate appropriate information and intelligence sharing with participating organizations. Commanders must strive to overcome difficulties, such as mutual suspicion, different values and motivations, and varying methods of organization and execution. Frequently, they must initiate cooperative efforts with participating civilian agencies and determine where their objectives and plans complement or conflict with those agencies. Commanders can then match Army force capabilities to the needs of the supported agencies. In situations leading to many urban civil support operations, confusion may initially make it difficult to ascertain specific priority requirements. Reconnaissance and liaison elements—heavily weighted with CA, engineers, and health support personnel—may need to be deployed immediately to determine what type of support Army forces provide. Overall, aggressive coordination will foster trust and make unity of effort possible in urban stability or civil support operations where unity of command is difficult or impossible to achieve.

## Perseverance

9-36. The society is a major factor responsible for increasing the overall duration of urban operations. This particularly applies to urban stability operations where success often depends on changing people's fundamental beliefs and subsequent actions. Modifying behavior requires influence, sometimes with coercion or control, and perseverance. The urban population often must be convinced or persuaded to accept change. Necessary change may take as long as or longer than the evolution of the conflict. Decades of problems and their consequences cannot be immediately corrected. Frequently, the affected segments of the urban society must see that change is lasting and basic problems are being effectively addressed.

9-37. In most stability operations, success will not occur unless the host nation, not Army forces, ultimately prevails. The host urban administration must address the underlying problem or revise its policies toward the disaffected portions of the urban population. Otherwise, apparent successes will be short lived. The UO fundamental of understanding the human dimension is of paramount importance in applying this consideration. After all Army forces, particularly commanders and staff of major operations, understand the society's history and culture, they can begin to accurately identify the problem, understand root causes, quickly engage and assist key civilian leadership, and, overall, plan and execute successful Army UO.

## Reasonable Restraint

9-38. Unlike offensive and defensive operations where commanders seek to apply overwhelming combat power at decisive points, restraint is more essential to success in urban stability and civil support operations. It involves employing combat power selectively, discriminately, and precisely (yet still at decisive points) in accordance with assigned missions and prescribed legal and policy limitations. Similar to the UO fundamentals of minimizing collateral damage and preserving critical infrastructure, restraint entails restrictions on using force. Commanders of major operations should issue or supplement ROE to guide the tactical application of combat power. Excessively or arbitrarily using force is never justified or tolerated by Army forces. Even unintentionally injuring or killing inhabitants and inadvertently destroying their property and infrastructure lessens legitimacy and the urban population's sympathy and support. Collateral damage may even cause some inhabitants to become hostile. In urban stability or civil support operations, even force against a violent opponent is minimized. Undue force often leads to commanders applying ever-increasing force to achieve the same results.

9-39. Although restraint is essential, Army forces, primarily during urban stability operations, must always be capable of decisive (albeit in some circumstances, limited) combat operations. This is in accordance with the UO fundamental of maintaining a close combat capability. This capability must be present and visible, yet displayed in a nonthreatening manner. A commander's intent normally includes demonstrating strength and resolve without provoking an unintended response. Army forces must be capable of moving quickly through the urban area and available on short notice. When necessary, Army forces must be prepared to apply combat power rapidly, forcefully, and decisively to prevent, end, or deter urban confrontations. Keeping this deterrent viable requires readiness, constant training, and rehearsals. It also requires active reconnaissance, superb OPSEC, a combined arms team, and timely and accurate intelligence.

## Restraint: An Najaf, Iraq – 2003

On 31 March 2003, during OPERATION IRAQI FREEDOM, The 101st Airborne (Air Assault) Division transitioned from isolating An Najaf—among the holiest places for Shia Muslims in Iraq and the home of one of their leading holy men, the Grand Ayatollah Ali Hussein Sistani—to an attack to clear Iraqi fighters from the town. The decision to attack was made in response to increasing attacks against 101st forces situated near An Najaf and due to alarming reports that the Fedayeen were killing Iraqi family members to force adult males to fight coalition forces. Since An Najaf had an airfield, control of the town would also allow the 101st to obtain needed hardstands for their aircraft.

Ultimately, the division used two brigades to attack and clear the town. The 2nd Brigade Combat Team (BCT) attacked from the north and the 1st BCT—responsible for the initial penetration—attacked from the southwest. Using tanks, infantry fighting vehicles, and light infantry, the commander of the 1st BCT formed effective combined arms teams supported by artillery and air that successfully fought their way into the city.

On 3 April 2003, Soldiers of TF 2-327 Infantry turned a street corner to face a group of civilian men blocking their way and shouting in Arabic, "God is great." The crowd quickly grew into hundreds as they mistakenly thought the Soldiers were trying to capture Sistani and seize the Imam Ali Mosque. Someone in the crowd threw a rock at the Soldiers, which started a hailstorm of rocks; even the battalion commander was hit on the head, chest, and the corner of his sunglasses.

Based on information from a Free-Iraqi Fighter accompanying the battalion and his own observations during the attack, the commander of TF 2-327 believed that most of the people in An Najaf neither wanted to fight nor obstruct his unit's efforts. Consequently, he ordered his Soldiers to position themselves on one knee, smile, and point their weapons toward the ground. At this gesture, many of the Iraqis backed off and sat down, enabling the commander to identify the true troublemakers. He identified eight. In case these agitators were to produce weapons and start to shoot, the commander wanted to make sure that the remainder of the crowd would know from where the shooting would originate. Next, he ordered his Soldiers to withdraw to allow the tension to subside. With his own rifle pointed toward the ground, the battalion commander bowed to the crowd and led his Soldiers away. When tempers had calmed, the Grand Ayatollah Sistani issued a decree (fatwa) calling on the people of Najaf to welcome the American Soldiers.

The attack on An Najaf by the 101st provides an excellent example of well-led Soldiers capable of understanding cultural differences and possessing the discipline and mental flexibility to transition from the aggressive mindset required for high-intensity urban combat to the restraint essential for the initiation of stability operations. In his weekly radio address, President George W. Bush commented: "This gesture of respect helped defuse a dangerous situation and made our peaceful intentions clear."

## CONSOLIDATE

9-40. As urban operations will often be full spectrum, many of the consolidation activities necessary to secure gains in urban offensive and defensive operations will be applicable to urban stability and civil support operations. However, the greatest obstacles to attaining strategic objectives will come after major urban combat operations. Therefore, emphasis will appropriately shift from actions to ensure the defeat of threat forces to those measures that address the needs of the urban population, manage their perceptions, and allow responsibility to shift from Army forces to legitimate indigenous civilian control (or the intermediate step to other military forces, governmental agencies, and organizations).

### Continued Civilian and Infrastructure Protection

9-41. Following urban offensive or defensive operations, there will likely be a need to secure and protect the civilian population and much of the civilian infrastructure from the civilians themselves. After having minimized collateral damage and preserved critical infrastructure, commanders must implement measures to preclude looting and destruction of critical and essential infrastructure by the urban population and civilian-on-civilian violence. This may be as relatively simple as allowing the urban police force to return to work or may be as difficult as hiring, vetting, and training an indigenous police force. In the latter case commanders may need to determine—

- Number and operability of police stations.
- Responsibility for recruiting, hiring, training, and equipping the urban security or police force. The accountable unit, organization, or agency will need to consider a vetting process, suitable salaries and wages, and appropriate training standards.
- The appropriate responses toward those civilians who threaten, oppose, or harm the new police force.

9-42. Alternatively, civilian security firms (from inside or outside the urban area or country) can provide supplemental protection until indigenous police forces are fully functional. Further, the commander of the major urban operation may plan to manage expected instability primarily with Army forces. Often, this will require larger numbers of infantry, military police, and dismounted forces. Other populace and resource control measures such as curfews may also assist in civilian (including NGOs) and infrastructure protection. Previous shaping operations aimed at improving the local economy can also assist in this regard.

### Resolute Legitimacy

9-43. Closely linked to the restraint described previously is legitimacy or the proper exercise of authority for reasonable purposes. Achieving or maintaining legitimacy during urban stability or civil support operations is essential in gaining and maintaining the support of the urban population. Commanders can ensure legitimacy by building consent among the population, projecting a credible force, and appropriately using that force. Perceptions play a key role in legitimacy, and skillful IO can shape perceptions. Commanders must send messages that are consistent with the actions of their forces. Generally, the urban population will accept violence for proper purposes if that force is used impartially. Perceptions that force is excessive or that certain groups are being favored over others can erode legitimacy and generate resentment, resistance, and, in some situations, violent acts of revenge.

9-44. In stability operations against an elusive insurgent threat, commanders may need to explain to urban residents why damage was necessary, apologize, or make near-instant restitution for some unsuccessful Army operations that may have been planned based on inaccurate or incomplete intelligence. (Soldiers may even make it a point to thank homeowners for allowing the search of their homes during cordon and search operations.) During urban operations, a single Soldier's misbehavior can significantly degrade a commander's ability to project an image of impartiality and legitimacy. Fortunately, disciplined Soldiers can contribute immeasurably to gaining and maintaining legitimacy, mitigating ill will, or otherwise winning the urban population's trust and confidence. In stability operations, the greater fight will often be to win the battle of perceptions and ideas instead of one to seize terrain and triumph over an enemy. Inconsistencies in message and behavior will provide threats with raw material for their propaganda and precipitate doubt in the minds of the urban populace who might otherwise support Army objectives.

## TRANSITION

9-45. Commanders of major operations are the focal point for synchronizing tactical stability operations or civil support operations with strategic diplomatic and political issues. They are also the critical links between national intelligence resources and the tactical commander. Because strategic, diplomatic, and political changes can quickly transition the type of urban operation, they must keep subordinate tactical commanders abreast of changes in intelligence, policy, and higher decisions. The potential to rapidly transition to urban combat operations emphasizes the need to maintain the capability to conduct close, urban combat. Failure to recognize changes and transition points may lead to UO that do not support the attainment of the overall objective and needlessly use resources, particularly Soldiers' lives. Therefore, Army forces on the ground in an urban stability operation must be more aware of the strategic environment than the threat and the civilian population, each of whom will have their own means of monitoring the national and international situation.

## Legitimate and Capable Civilian Control

> *I met immediately with the [special operations forces liaison officer] to discuss how we would establish a new government that would be able to make instant progress. We decided to select a city council with the following departments: commerce, public works, social services, health, emergency services, education, public relations, and agriculture. A city council director would lead the council, and each department would also have a director in charge of its respective services. A critical step in selecting the city council was ensuring that there was a balance on the council between Altun Kupri's three major ethnic groups. An ethnic imbalance on the council or perceived favoritism would have decreased the legitimacy of the Coalition forces' efforts in the city and served as fuel for the fire of competition for power between the Turkomen and Kurds.*
>
> Captain Jeffrey B. Van Sickle
> "Stability Operations in Northern Iraq"

9-46. Commanders must maintain or enhance the credibility and legitimacy of the government and police of the urban area and, in the case of stability operations, of the host nation's military forces operating there. In accordance with the urban fundamental of transitioning control, urban commanders must conclude UO quickly and successfully, often to use assets elsewhere in their area of operations. This entails returning the control of the urban area back to civilian responsibility as soon as feasible. The host nation's military and the urban area's leadership and police are integrated into all aspects of the urban stability or civil support operations to maintain their legitimacy. They must be allowed (or influenced) to take the lead in developing and implementing solutions to their own problems. This requires commanders to transition from "leading from the front" to "leading from behind" in an advisory and assistance role. Effective transition to civilian control and responsibility requires commanders at all echelons who understand the basic operation of civil governments and the administration and management of key urban infrastructure.

9-47. If the host nation's leadership, military, and police are not up to the task, commanders can take steps to increase its capabilities through training, advice, and assistance by CA units or by other nongovernmental or governmental organizations and agencies. Sometimes, new leadership and a restructured police force may be required, particularly when corrupt and no longer trusted by the population. This candid assessment of the urban leadership's ability to govern, protect, and support itself is made early in the planning process. Only then can commanders ensure that resources and a well thought-out and coordinated plan (particularly with civilian organizations and agencies) are available for a speedy transition. IO will be paramount in these instances to ensure that the urban population sees the training and rebuilding process itself as legitimate. Throughout urban stability or civil support operations, commanders shape the conditions to successfully hand over all activities to urban civilian authorities.

## Longer-Term Commitment

### *Maintain the Focus*

9-48. Many stability operations often require perseverance and a longer-term U.S. commitment requiring operational endurance and the establishment of a sustainable battle rhythm. Lengthy operations will also require a transitional rotation of Army units into the area of operations to continue the mission. Considerations for these transitions are similar to a relief in place and battle handover (see FM 3-90) combined with considerations for deployment and redeployment. FM 41-10 contains comprehensive guidance on transition planning and coordination activities applicable to UO. In addition to any threat considerations, planning for urban transitions between units will often include emphasis and understanding of—

- Formal and informal civilian leadership and relationships.
- Government institutions and administrative functions.
- On-going reconstruction projects.
- The urban economy.
- Participating governmental and nongovernmental organizations and established relationships and cooperation activities, particularly information sharing.
- Significant key events affecting (or likely to affect) operations.
- Significant cultural lessons learned as a result of the outgoing unit's operations.

9-49. The commander of the major operation must ensure that the incoming unit also understands the political and strategic objectives behind the tasks that they will be required to accomplish. Otherwise, the new unit may begin to plan operations that are similar to those conducted by the previous unit without achieving the desired end state or accomplishing the mission.

#### *Transition Trust*

9-50. If units are successful in establishing and cultivating relationships with the urban populace and their leadership, and have built an effective level of trust, their anticipated departure from the AO as part of a unit rotation schedule may have detrimental effects on working relationships. As a result, local leaders may develop deep anxieties about the replacement force and their ability to "be as good" as the current command. Therefore, incumbent forces must take steps to identify successful and ongoing efforts and activities in which the new unit can continue to achieve success. In some instances, the current unit may delay a project or activity whose scheduled completion is close to the date of relief or the transfer of authority so as to allow the new unit immediate success. In other instances, projects or activities can be backward planned to ensure that the completion date is scheduled soon after the new unit arrives. The goal for the outgoing unit is to ensure that the new unit is readily accepted and that both units execute a seamless relief in place.

#### *Modify Objectives to Match the Current Environment*

9-51. While it is important that incoming commanders create a seamless transition with their outgoing counterparts, they must make full use of the opportunity to review the current political, strategic, and local urban environment to determine potential modifications to logical lines of operations and unit objectives. Otherwise, units may fall into the trap of executing a series of six-month to one-year rotations that do not significantly contribute to solidifying the conditions required to ultimately transition the urban environment back to legitimate civilian control. In operations that involve longer-term commitments, the potential exists for the same unit to conduct two (or more) rotations back into the same AO. Commanders must resist the desire to simply continue with previous lines of operations that, at the time, were deemed successful. The urban environment will change in response to Army actions and this change will necessitate subsequent operational adaptations.

#### *Provide an End State Not an End Date*

9-52. Though it may not be feasible, due to political constraints, commanders of major urban operations should fight the public desire to establish a fixed transition timetable. Instead, they should plan urban stability operations based on an attainable end state and not on a specified withdrawal or end date. Progress toward strategic (and operational) goals is susceptible to many changes and delay, particularly in coalitions which will likely be a part of most future urban operations. As such, an end date will only serve to unduly constrain commanders in the attainment of their urban operational objectives. Having set and announced a firm timetable, American and coalition publics will expect withdrawal to commence as scheduled and will not easily be convinced of the need to stay longer even if the situation changes drastically. In general, the public can understand and accept a logical and clearly articulated end state as long as commanders can demonstrate adequate progress based on well developed measures of effectiveness. As important, a timetable allows threats to adjust their plans to the anticipated withdrawal date. Instead of yielding to relentless diplomatic, informational, military, and economic pressure backed by firm political resolve, threats will likely exercise strategic and operational patience.

This page intentionally left blank.

Chapter 10

# Urban Sustainment

Sustainment activities exist to enable the Army to initiate and sustain full spectrum operations. Inseparable from decisive and shaping operations, logistics is the central component of sustaining operations and provides the means for commanders to generate and maintain combat power and extend the operational reach of the force. In offensive and defensive operations, sustainment activities are not by themselves likely to be decisive or shaping, yet significantly contribute to those operations. In stability and civil support operations, when the critical objectives may be restoring the infrastructure and the welfare of civilians, sustainment activities will often be decisive. Their success will allow Army forces to dominate this complex environment. However, like all urban operations (UO), sustainment activities affect and are affected by the environment. The urban terrain, infrastructure, and existing resources, coupled with supportive civilians, may facilitate sustainment activities. In contrast, a poorly designed or damaged infrastructure and a hostile population may severely hamper sustainment activities. In the latter case, critical Army resources required elsewhere in the area of operations (AO) may be diverted to repair facilities and control and support the inhabitants of the urban areas.

## URBAN SUSTAINMENT CHARACTERISTICS

10-1. Fundamental sustainment characteristics (see figure 10-1) guide prudent sustainment planning regardless of the environment. They provide commanders an excellent framework to analyze and help develop urban sustainment requirements, assess the impact of the environment on the provision of logistics, and gauge the effectiveness of urban sustainment activities.

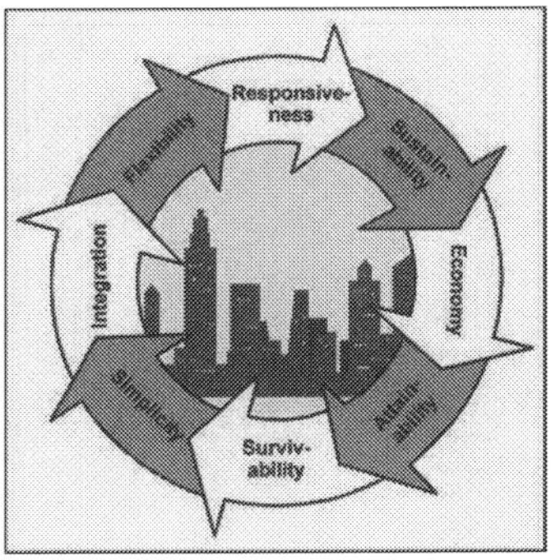


**Figure 10-1. Sustainment characteristics**

## Responsiveness and Sustainability

10-2. UO require responsiveness and sustainability to establish and maintain the tempo necessary for success. Responsiveness—providing the right support in the right place at the right time—is the essential sustainment characteristic. It requires that sustainment commanders and planners accurately forecast urban operational requirements. Continuous urban operations will drain personnel, equipment, and supplies at rates vastly different than in other environments. Therefore, sustainability—the ability to maintain continuous support throughout all phases of the operation—will be a significant concern. Anticipation is critical to both responsiveness and sustainability. It requires that sustainment commanders and planners comprehend the potential effects that components of the urban environment (terrain, infrastructure, and society) may have on operations and sustainment, either benefiting or impeding UO. Effective urban operational and sustainment planning cannot be accomplished separately. Operational and sustainment planners, as well as sustainment operators, must be closely linked to aid in synchronizing and attaining responsiveness and sustainability. The key is often the right balance between efficiency and effectiveness.

## Economy and Attainability

10-3. A thoughtful assessment and understanding of the urban environment can also help determine how specific urban areas can contribute to or frustrate the achievement of economy and attainability. Economy is providing the most efficient support at the least cost to accomplish the mission. Attainability means generating the minimum essential supplies and services necessary to begin operations. If available, obtaining support in the AO costs less than purchasing the supplies outside the area and then transporting them there. Critical resources may be available in urban areas to support the operation. However, relying on sources outside the established military sustainment system may create conflict with other sustainment characteristics. A strike by longshoremen, for example, may shut down port operations (at least temporarily) lowering responsiveness and sustainability.

## Survivability

10-4. Survivability is being able to protect support functions from destruction or degradation. Commanders often choose to locate sustainment functions in an urban area because the buildings may better protect and conceal equipment, supplies, and people. Urban industrial areas and airports are frequently chosen as support areas because they offer this protection as well as sizeable warehouses, large parking areas, and materials handling equipment (MHE). Such areas facilitate the storage and movement of equipment and supplies. They may also provide readily available water, electricity, and other potentially useful urban resources and infrastructure. However, these areas may also contain toxic industrial materials (TIM) (see Chapter 2). These materials and chemicals in close proximity to support areas may unjustifiably increase the risk to survivability. Sustainment activities in any environment will always be targeted by threat forces. Furthermore, sustainment activities located in any type of confined urban area can offer lucrative targets for terrorists, insurgents, or even angry crowds and mobs. Therefore, no sustainment activity should be considered safe from attack during UO. (During OPERATION JUST CAUSE, Panamanian paramilitary forces and deserters even attacked marked ambulances). Although host-nation support may include assets to assist in defending sustainment units, bases, and lines of communications (LOCs), sustainment commanders must carefully consider if adequate protection measures can ensure survivability. The sustainment commander's greatest challenge to force protection may be complacency born of routine.

**Base Security: Tan Son Nhut, Vietnam – 1968**

Colonel Nam Truyen was the commander of the 9th Vietcong Division who planned and conducted the attack on the U.S. airbase at Tan Son Nhut during the 1968 Tet Offensive. He had previously entered the airbase during the 1967 Christmas cease-fire using forged identity papers to conduct his own personal and highly thorough reconnaissance.

UO increase the likelihood that the indigenous population will be utilized as a workforce to supplement Army sustainment activities. Hence, strict operations security will be paramount. Commanders will need to carefully screen potential workers to determine their relative trustworthiness. Human intelligence (HUMINT) and counterintelligence (CI) assets will likely be required to assist in this effort. However, higher-level commanders should seek to take on as much of this burden as possible so that lower-level, tactical commanders can use their limited intelligence assets to accomplish other critical intelligence collection and analysis activities.

All commanders with base protection responsibilities should implement random, changing force protection measures that focus on civilians—even those civilians who have, over time, earned a perceived measure of trust. (Civilians can be co-opted and coerced by threat forces.) As such, all civilian activity that seems out of the ordinary should be promptly investigated regardless of whether a specific force protection measure is in effect. Although leaders and Soldiers seek to establish mutual trust and respect between themselves and the local populace, commanders should normally err on the side of protecting their Soldiers.

## SIMPLICITY

10-5. Simplicity is required in both planning and executing sustainment activities in this complex environment. Developing standard procedures among the Army, other services, and especially civilian governmental and nongovernmental agencies, liaison and open channels of communication, simple plans and orders, and extensive rehearsals contribute immeasurably to attaining this necessary characteristic.

## INTEGRATION

10-6. The need for sustainment integration increases in UO due to its joint and multinational nature and greater numbers of other governmental and nongovernmental organizations (NGOs) and agencies operating in or near urban areas. More NGOs will likely exist because urban areas often contain most of a region's population. Most NGOs focus on people. Army forces and other military and nonmilitary groups must cooperate and coordinate their actions. Much of their coordination will revolve around sustainment activities. Cooperation and coordination will take advantage of each group's sustainment capabilities, help to avoid duplicated effort (contributing to economy), and create sustainment synergy. It will also help to curtail competition for the same urban resources and assist in developing a unified list of priorities. Such coordination will help ensure that other operations by one force or agency will not disrupt or destroy portions of the urban infrastructure critical to another's sustainment operations and the overall mission.

10-7. Success in UO requires combined arms and innovative task organizations. As such, UO will often result in the integration of heavy and light units necessitating a simultaneous integrated approach to sustainment. Gaining units must integrate forces into their maintenance systems and carefully consider potentially increased resource requirements. (For example, armor units consume greater amounts of Class III—bulk and petroleum—than a similar-sized light unit.) When integrating into a light unit, armored and mechanized units should bring as much of their critical parts and supplies as possible and immediately integrate into the gaining unit's logistical reporting system to ensure adequate resupply.

## FLEXIBILITY

10-8. Lastly, commanders must develop flexibility. Although they and their staffs must thoroughly understand the urban environment essential to planning sustainment operations, they cannot possibly anticipate every eventuality. Urban commanders must possess the ability to exploit fleeting opportunities. Knowledge of the environment, particularly its infrastructure, can aid in developing innovative solutions to sustainment acquisition and distribution problems. Flexibility enables sustainment personnel to remain adaptive and responsive to the force commander's needs. Key to maintaining flexibility is the constant assessment of the sustainment situation and the readiness to modify or change procedures to adapt to the current conditions confronting Army forces operating within the urban environment.

10-9. Maneuver and sustainment commanders should consider and prioritize these characteristics as they visualize UO. Each characteristic does not affect every operation and urban area in the same way. The sustainment characteristics seldom exert equal influence, and their importance varies according to mission, enemy, terrain and weather, troops and support available, time available, civil considerations (METT-TC). Like the principles of war, commanders must not ignore the potential impact of sustainment characteristics and how their influence changes as the operation evolves (see FM 4-0).

# LOGISTICS PREPARATION OF THE THEATER

> *[Sustainment] planning accounts for increased consumption, increased threats to lines of communications, and anticipated support to noncombatants...Urban operations place a premium on closely coordinated, combined arms teams and carefully protected [sustainment activities]. Urban operations are [logistic]-intensive, demanding large quantities of material and support for military forces and noncombatants displaced by operations.*
>
> FM 3-0

10-10. A thorough logistics preparation of the theater (LPT) is critical for an adaptable UO support plan. Sustainment planners conduct the LPT to understand the situation from a sustainment perspective and determine how best to support the maneuver commander's plan. However, decisions that impact the urban environment and the political situation require a combined operational and sustainment perspective. Sustainment planners must understand the urban environment, the fundamentals of UO, the proposed course of action, and the urban environment's effects on sustainment (as well as the other warfighting functions). Such knowledge allows the planners to develop a detailed estimate of support requirements. A thorough LPT helps commanders determine the most effective method of providing adequate, responsive support to meet support estimates while minimizing the sustainment footprint. Overall, it helps tie together UO requirements with acquisition and distribution. As with all operations, but particularly in a dynamic urban environment, this assessment process is continuous since requirements will change as the urban operation unfolds and matures.

## SUPPORT TO INTELLIGENCE

10-11. The LPT resembles and runs parallel to the intelligence preparation of the battlefield (IPB). Products generated under IPB may be useful in the sustainment analysis. Conversely, the LPT may contribute to the IPB by identifying critical resources and infrastructure and assessing their potential to influence (positively or negatively) the operational plan. This information may warrant a course of action that includes offensive or defensive operations to seize, secure, or destroy those critical resources. In UO initially planned for other than sustainment reasons, the information may require altering the plan or imposing additional constraints to protect the identified resources. While these resources may or may not be critical to current operations, they are usually important to set or shape the conditions necessary for Army forces to consolidate and transition to subsequent missions or redeploy. This close relationship between IPB and LPT underscores the need to quickly and continuously involve sustainment personnel for their sustainment expertise and perspective in planning UO.

10-12. As sustainment units are among the most dispersed and omnipresent in any AO, sustainment Soldiers are potentially valuable intelligence collectors. Soldiers working in and around urban sustainment bases and executing frequent convoys through urban areas quickly become very familiar with the urban terrain and the routine associated with that terrain. As with all Soldiers, sustainment personnel must be incorporated into a comprehensive human intelligence collection system.

## URBAN SUSTAINMENT INFORMATION

10-13. Figure 10-2 illustrates that a thorough analysis of the key components of urban areas in the commander's AO provides the data for an accurate LPT and subsequent UO sustainment plan (see Chapter 2 and Appendix B). Analyzing the urban terrain and infrastructure helps to determine—

- Geographic influences on consumption factors and on the provision of support (weather, climate, and topography).
- The availability of supplies, such as safe food, potable water, petroleum, electrical energy, barrier material, and compatible repair parts.
- The location of facilities, such as warehouses, cold-storage sites, water treatment facilities, manufacturing plants, hospitals, hotels for billeting, and waste treatment facilities.
- Transportation information, such as seaport and harbor facilities, airfields, rail and road networks, traffic flow patterns, choke points, and control problems.
- Locations and accessibility of maintenance facilities and equipment, and machine works for the possible fabrication of parts.
- The available general skills among the urban population, such as linguists, drivers, mechanics, MHE operators, longshoremen, and other vital trade skills.

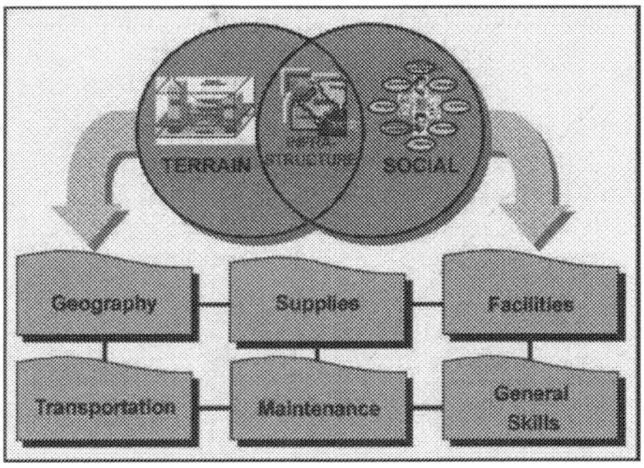


**Figure 10-2. Urban environment and essential elements of sustainment information**

## POTENTIAL RESTRICTIONS

10-14. Commanders must be aware of restrictions that apply to the use of some non-U.S. resources. Security and requirements for U.S. national control dictate that only U.S. assets may perform certain services and functions. Therefore, some foreign urban-area capabilities, even if abundantly available, may not be used. These might include—

- Command and control of medical supply, service, maintenance, replacements, and communications.
- Triage of casualties for evacuation.
- Treatment of chemical, biological, radiological, nuclear, and nuclear (CBRN) casualties, as well as the decontamination of U.S. equipment, personnel, and remains.
- Identification and burial of deceased U.S. personnel.
- Veterinary subsistence inspection.
- Law and order operations over U.S. forces and U.S. military prisoner confinement operations, as well as accountability and security of enemy prisoners of war in U.S. custody.

## URBAN SOCIETAL CONSIDERATIONS

10-15. As in all aspects of UO, the urban society is a critical element of the LPT analysis. Sustainment planners must assess whether they can acquire and use urban resources and infrastructure without overly disrupting the urban society and their environment. If the resources are only sufficient for the urban area's inhabitants (and dependent populations in outlying areas), and the facilities cannot increase production to accommodate the needs of Army forces, then commanders may not be able to rely on those resources to support their operations. In fact, the opposite may be true. The effects of UO on the inhabitants, particularly during offensive and defensive operations, may place increased burdens on the Army's resources. For example, commanders may require civilians to evacuate all or parts of an urban area to safeguard them from the effects of planned combat operations. This precautionary measure may require Army forces to support civilian evacuees with temporary housing, food, water, clothing, and medical and sanitary facilities. It will also require carefully planned evacuation routes to ensure that the movement does not interfere with sustainment of the fight.

10-16. Sustainment planners must take into account the potential effects that the provision of supplies to the urban population may have on the operational environment (provided by Army forces or not). For example, an embargo on petroleum products was lifted after Army forces entered Haiti in the 1990s and a less repressive regime assumed governmental control. Few civilian vehicles had moved on the streets of the Port-au-Prince, the capital city, when fuel was scarce and the movement of mounted forces through the urban area was virtually unimpeded. This changed almost overnight once fuel supplies became available, initially making it very difficult for forces to move where required (unless transported by helicopter).

10-17. Sustainment planners must also consider the urban society's ability to restore their own facilities and provide for themselves (if necessary with assistance from Army forces). Throughout this analysis, civil affairs (CA) units can advise and assist in identifying and assessing urban supply systems, services, personnel, resources, and facilities. Critically, commanders must understand that purchasing local goods and services may have the unintended negative consequence of financially sustaining the most disruptive and violent factions in the area. Army forces must seek to purchase urban products and services that will not contribute to prolonging the conflict or crisis. In many stability or civil support operations, they must also attempt to distribute the contracts for goods and services purchased locally as fairly as possible among urban factions and ethnic groups to maintain impartiality and legitimacy. As part of their coordination efforts, commanders should attempt to achieve the cooperation of relief agencies and other NGOs in this endeavor.

10-18. Finally, sustainment planners must also identify potential threats and increased protection requirements that the urban society (criminals, gangs, and riotous mobs) may present, particularly when support units and activities are positioned in urban areas. The disposition or allegiance of the urban population is also important to consider. The infrastructure of an urban area may exhibit great potential to support the sustainment efforts of Army forces, but if the population is hostile or unreliable, use of the facilities and resources may be unfeasible.

## Urban Support Areas and Bases

10-19. A major influence on the operation plan and its subsequent execution is often the proper identification and preparation of support areas and bases. The LPT helps commanders determine the need, advantages, and disadvantages of using urban areas in the AO as bases from which to provide support and conduct distribution operations. Ideally, these areas should support reception, staging, onward movement, and integration operations. They should also allow easy sea and air access, offer adequate protection and storage space, increase the throughput of supplies and equipment, and be accessible to multiple LOCs. Consequently, commanders often establish support areas and bases near seaports and airports that are part of a larger urban area. However, threats recognize the Army's need for ports and airfields and may devote substantial resources and combat power to defend them. Army commanders also face the challenge of integrating airfield and port defensive operations with air and naval component commanders to deny threats the ability to conduct stand-off attacks against aircraft and ships. Only an integrated, joint approach can ensure persistent air and sea support– essential not only to sustainment but to firepower as well—is not degraded by the threat. Therefore, planners may determine during the LPT that the risks of seizing or establishing urban lodgment areas may be too high (see Chapter 4). Instead, they may recommend building an airfield, conducting logistics over-the-shore operations, or constructing sustainment bases in more isolated—but more easily secured—locations.

## Overall Assessment

10-20. As shown above, the LPT process and analysis help to determine if urban areas in the AO—

- Are suitable as areas and bases for support.
- Can contribute sufficient quantities of and are a dependable source for resources for the overall operation.
- May additionally drain the supported commander's resources.

10-21. The results of this process often serve as a basis for reviewing requirements for civilian contract support and host-nation support and for developing sustainment input into time-phased force and deployment data. This chapter focuses on the effects urban areas may have on accomplishing sustainment functions and related activities, particularly when sustainment units and activities are in urban areas.

# ACHIEVING FORCE AGILITY

10-22. To maintain the responsiveness necessary to support full spectrum operations against a wide array of threats and operational environments, sustainment forces must remain agile, possessing the mental and physical agility to transition within or between types of operation with minimal augmentation, no break in contact, and no significant additional training. Responsiveness, flexibility, and economy are key characteristics that enable sustainment forces to support a dynamic combat force; they represent the capability of sustainment forces to provide the resources to initiate, sustain, and extend UO. Agile sustainment forces are characterized by modular force designs, the ability to tailor sustainment organizations for the supporting mission, and the ability to conduct split-based operations. These sustainment aspects are critical to achieving the agility necessary for Army forces to succeed in UO.

## MODULAR DESIGN

10-23. Sustainment units are structured as modular organizations, or must possess the ability to modularize in order to conduct decentralized sustainment operations required in UO. At the company level, modularity enables each major company sub-element to assume a cross-section of the company's total capabilities, allowing commanders to employ individual modules to provide a support function, while the rest of the unit remains operational. Modularity enhances responsiveness while enabling support forces to project capabilities as far forward as possible into urban areas, yet with the minimum forward footprint necessary to sustain operations.

## FORCE TAILORING

10-24. Sustainment force tailoring refers to the process of determining and deploying the right mix of capabilities to support the force or mission. Sustainment commanders may task organize specific capabilities into functional elements in order to maximize effectiveness and efficiency, and to minimize the sustainment footprint. During UO, the sustainment commander can tailor the support element required to accomplish a specific mission or task, thereby mitigating the risk associated with deploying a larger, more robust capability package forward into the urban area.

## SPLIT-BASED OPERATIONS

10-25. Split-based operations refer to performing certain sustainment administrative and management functions outside the area of operations. During UO, Soldiers and civilians can perform personnel, materiel, and distribution management functions without deploying forward into the urban area if the information systems are adequate. This is essential to minimizing risk to support forces and the sustainment footprint in the urban area, and still ensures that all support requirements are met.

**SUSTAINMENT FUNCTIONS**

- Supply and Field Services
- Maintenance Support
- Transportation Support
- Force Health Protection
- Explosive Ordnance Disposal Support
- Human Resources Support
- Financial Management Support
- Religious Support
- Legal Support to Operations
- General Engineering Support

**Figure 10-3. Sustainment functions**

# SUSTAINMENT FUNCTIONS

10-26. Sustainment consists of eleven functions that must be carefully planned, managed, and synchronized to provide responsive and efficient support to maneuver commanders (see figure 10-3).

Similar to the components of the urban environment, particularly its infrastructure, they overlap and are interdependent. The success of one function depends on the success of several others. Like urban infrastructure, they have two components: a physical component (supplies, equipment, and facilities) and a human component (the Soldiers and civilians who execute these functions). Like city mayors, commanders must plan, manage, and synchronize these functions to provide responsive and efficient sustainment to UO.

10-27. Commanders and planners should consider two essential aspects when addressing these sustainment functions. One aspect looks outward and one looks inward. The first aspect is how these functions can best support full-spectrum UO—the outward analysis. The second aspect is how the urban area affects the conduct of sustainment functions, particularly when those functions are located or performed in an urban area—the inward analysis.

## SUPPLY AND FIELD SERVICES

10-28. The supply function involves acquiring, managing, receiving, storing, protecting, maintaining, salvaging, and distributing all classes of supply (except Class VIII) required to equip and sustain Army (and joint) forces. For UO, commanders may need to decide early in the planning cycle whether to stockpile resources forward or to rely on distribution-based sustainment management to satisfy requirements. Some specialized items identified below may not be available through the normal military supply system and may take sustainment personnel much longer to obtain or fabricate. Operation planners must quickly identify the special equipment and increased supply requirements for UO to give sustainment personnel the time necessary to acquire them.

10-29. Field services are those essential services required to enhance a Soldier's quality of life. They consist of clothing exchange, laundry and shower support, textile repair, mortuary affairs, preparation for aerial delivery, food services, billeting, and sanitation. The urban commander determines the need and priority of each service after careful METT-TC analysis. Some facilities such as shower, laundry, and cold storage may be available in the urban area through theater support contracting or provided by external support contracts such as the logistics civilian augmentation program (LOGCAP). Additionally, requirements to care for the urban population will increase requirements for field services immensely. In some circumstances, most notably urban stability or civil support operations, adequately protected field service units or activities will be critical and may be the only support provided.

### Increased Urban Supply Requirements

#### *Water*

10-30. Water production and distribution will become more difficult for sustainment forces during UO. (Many parts of an urban area can become virtual deserts if public water supplies are cut off or contaminated due to deliberate military action or natural disasters.) Consequently, sustainment managers will be called upon to find more efficient and effective means of moving and distributing potable water, especially when bulk storage of water within the urban area may not be possible or feasible. In sustainment bases that do not have access to urban or other water sources, the transportation and storage of non-potable water may be as great a concern as potable water. Unless adequate storage is available for both potable and non-potable water, units may be forced to rely solely on potable water for all bulk requirements to mitigate health risk. Due to the compartmented nature of many urban operations, planners may need to ensure that additional containers for water (and fuel) are available to support dispersed stock at small-unit level, and that man-portable systems are available in sufficient quantities to ensure timely replenishment.

### *Class II*

10-31. Clothing and individual equipment exposed to the urban environment (concrete, glass, and steel) experience wear and damage at accelerated rates. Units will also require items such as rope, grappling hooks, crowbars, ladders, chain saws, elbow- and kneepads, special vehicle and personnel body armor, fire-fighting equipment, packboards, carpentry tools, and other specialized items to conduct UO. Planners must recognize this and plan accordingly to ensure operations are not interrupted due to equipment shortfalls.

### *Class III (Bulk)*

10-32. With the potential exception of aviation fuel, bulk Class III requirements for UO generally decrease at the maneuver unit level. However, increased fuel requirements for engineer and power-generation equipment attached to or operating with forward units may offset these decreases. Generally, fuel sources within the urban area (refinement facilities, gas stations, garages, and airfields) should not be utilized unless tested and approved for use by a fuel laboratory.

### *Class IV*

10-33. Requirements for construction and barrier material (and the power tools to cut it) during UO will likely far exceed the capability of the sustainment force to provide. Commanders must prioritize distribution, defensive projects, and construction and repair efforts in order to maximize the efficient use of available Class IV. Every effort should be made to obtain supplemental Class IV material from the local economy, as long as that effort does not adversely affect operations or the restoration and improvement of critical urban infrastructure. Of note, specialized, prefabricated road barriers will also be in great demand to block roads, create checkpoints, reinforce defensive positions, and protect headquarters and sustainment activities.

### *Class V*

10-34. Urban combat operations will often increase requirements for Class V. Ammunition consumption rates for urban environments have been as much as five to ten times greater than operations in other environments. Urban offensive and defensive operations will often increase overall requirements for certain types of ammunition (crew-served and small arms ammunition, mortars, antitank rounds, grenades, mines, and demolitions) while others may typically decrease (tank, artillery, attack helicopter). Since forces operating on urban terrain may not be capable of storing ammunition, an established system of distribution is vital to ensuring a constant flow of ammunition necessary to sustain continuous operations and retain the initiative.

### *Class IX*

10-35. Requirements for repair parts generally decrease in UO, with the notable exception of those systems that experience increased ammunition consumption. The heavy use of crew-served and small arms, mortar, and antitank systems in the urban area causes wear and failure at significantly increased rates. Additionally, wheeled-vehicle tires and armored-vehicle track pads and road arms, road wheels, center guides, wedge bolts, and track blocks may wear rapidly on urban streets, rubble, and debris. Based on the tempo of operations, it may be more feasible to replace whole sets of tracks rather than have crews replace individual track pads. Hence, commanders may consider resourcing all wheeled vehicles with spare tires and creating containerized and rapidly transportable sets of track and track parts for speedy delivery to urban sustainment bases. On the whole, planners should consider the use of repair parts packages specifically configured for UO that can be transported by tailored, modular sustainment elements that directly support operations within the urban environment. During long-term stability operations, outgoing units may leave large portions of their prescribed load list in theater to ensure that relieving units have adequate supplies to maintain the tempo of operations.

## Storage and Distribution

10-36. UO present challenges to storage and distribution that may not be experienced outside of the urban environment. Forces operating within these areas may not be static and, therefore, cannot store equipment and resources beyond what they can transport organically. Even forces conducting operations from secure locations within the urban area may not be able to store materiel, necessitating a continuous, coordinated distribution system to ensure forces are adequately resourced. During UO, sustainment forces may need to utilize unit or throughput distribution techniques to provide the level of support required to sustain the force.

10-37. Urban commanders must also consider the benefits of stockpiling resources forward against relying on the Army's distribution-based sustainment system. Stockpiling brings supplies close to urban forces and helps ensure supplies are immediately available to support potentially increased consumption rates. However, this method may burden the support structure that moves, handles, and protects large quantities of resources often on a repetitive basis. The Army's normal distribution-based sustainment system reduces this burden significantly; however, available transportation assets (sea, air, and ground) impact delivery response times. These transportation assets often combine military assets overlaid on the host nation's (and urban area's) transportation and distribution infrastructure. Although the civilian infrastructure may initially support the Army's distribution system, later effects of UO, such as destruction of equipment and facilities or loss of civilian workers, may degrade the system. Moreover, Army forces may have to share these assets with other military, multinational force, or civilian organizations participating in the urban operation, as well as with the indigenous civilian population at large. This shared system may put at risk the timely delivery of critical supplies to Army forces.

10-38. Sustainment planners must understand the urban environment and its effects on the proposed method of distribution. They must also understand how urban operations (to include sustainment UO) may affect the urban environment. Storing bulk fuel or ammunition in or near an urban area, for example, may increase the risk of fire and explosive hazard to civilians and Army forces. With this awareness, planners present the commander with an estimate that considers both risks and benefits. Depending on the particular area and other METT-TC factors, they may recommend one method of distribution then transition later to another or a combination of methods. Preconfigured resupply loads delivered as close as possible to where they are needed is vital to responsiveness. Only an agile distribution-based sustainment system will allow Army forces to be responsive and operationally effective across full spectrum operations in an urban environment.

## Food Preparation

10-39. The daily feeding standard for Soldiers operating in urban areas remains the same: three quality meals per Soldier. UO creates an increased energy demand on Soldiers, requiring a caloric intake of about 5,000 calories each day. Since producing and delivering prepared meals to forward elements may be difficult or even impossible due to the urban environment, the innovative use of meal supplements can be critical to meeting the dietary requirements of the force. Energy drinks and bars, rich in electrolytes and carbohydrates, are an essential supplement to prepared meals and should be provided in quantities that support or exceed the requirements of the forces operating in the urban area.

10-40. Food (and water) may be available in the urban area; however, local sources must be tested, carefully monitored, and medically approved before consumption. Garbage disposal may be an important consideration in the urban area. Improper trash disposal may leave a signature trail (particularly during urban defensive operations) that may degrade sanitary conditions leading to increased disease and nonbattle injuries (DNBIs). Commanders must understand that food operations, if not properly positioned and secured, can become a focal point for the urban population. Strict policy regarding distribution and control of any Class I supplies (including waste products) will be enforced. Black marketers will be attracted to Army food service activities as well (even during offensive and defensive operations).

## Water Purification

### *Vigilant Monitoring*

10-41. Urban areas will often have a ready source of water to support the urban inhabitants and its infrastructure. However, this water may not be suitable for U.S. and allied forces (though the urban population may have developed immunity toward its microorganisms). The higher concentration of TIM in urban areas compounds this problem, as ground water is highly susceptible to chemical contamination, even supplies located miles away from the source of contamination. U.S. Soldiers should be trained and cautioned against using water from an urban area (to include ice and bottled water) until preventive medicine and veterinary personnel can determine its quality. When water quality is unknown, commanders ensure use of tactical water purification equipment that will upgrade it to the Army's water quality standards. Even if initial testing indicates that urban water sources are safe for Army forces, personnel must continuously monitor the water quality. However, Army water purification, storage, and transportation requirements for UO can be greatly reduced if the existing urban water supply can be integrated into sustainment operations. An early assessment of the feasibility of this course of action is critical to sustainment planning.

### *Potentially Greater Requirements*

10-42. Offensive and defensive UO are often intense and can produce more casualties, including civilians. Consequently, medical facilities, already consumers of large volumes of water, may require even more water. Water purification, particularly in the urban areas of developing nations and during urban disaster relief operations, will be a critical and constant concern for Army forces.

### *Possible Key Terrain*

10-43. Sites that can control the water of the urban area may be key terrain, providing not only a resource for Army forces, but also a means to control the threat, the civilians, or both. These sites may be the sources of the water—the river, lake, reservoir, or storage tanks—or the means that process and transport the water—pipelines, pumping stations, or treatment facilities. Many sites may be outside the urban area, as many large urban areas draw water from distant sources. The seizure of a pumping station or pipeline may make it possible for commanders to control water supplies without expending resources required to enter the urban area. To preserve critical infrastructure, commanders may increase security to protect these locations from contamination or destruction. Engineers also may need to restore, maintain, or operate existing water facilities damaged by the threat or disaster and to drill new wells and construct new water facilities.

## Mortuary Affairs

10-44. Mortuary affairs provide the necessary care and disposition of deceased personnel. It supports the Army across the spectrum of operations. It may directly and suddenly impact (positively or negatively) the morale of the Soldiers and the American public and may influence relations with the civilian population in the AO. It can also affect the health of Soldiers and the urban populace. Overall, commanders must carefully plan evacuation routes and temporary collection and internment sites, adhering to local customs and traditions to lessen potentially negative consequences.

10-45. Units are responsible for recovering the remains of their own fatalities and evacuating them to the closest mortuary affairs collection point, usually located at the nearest support area. Urban governmental authorities, family, and representatives of appropriate nongovernmental organization are responsible (and should always be relied on to the greatest extent possible) for recovering and handling indigenous civilian remains. However, because of the sheer density of noncombatants in UO, commanders may also find themselves responsible for civilian remains. Deaths of civilians under Army control, such as urban evacuees and refugees at Army-operated sites, often obligate the Army to care for their remains including medical certification and records of death. High-intensity urban combat may result in civilian deaths, and health concerns will require Army forces to deal with civilian remains expeditiously. Commanders should consult local religious leaders, the Staff Judge Advocate (SJA), CA personnel, and chaplains to verify that they are abiding by law and customs. Overall, commanders must ensure that forces treat all deceased, including civilians, with dignity and respect (another important aspect of adhering to the urban fundamental of understanding the human dimension).

### Aerial Delivery

10-46. Aerial delivery is the movement by fixed- or rotary-wing aircraft and delivery by the use of parachute or sling load of Soldiers, supplies, and equipment. As a vital link in the distribution system, it contributes to flexibility and provides the capability of supplying the force even when urban ground LOCs are disrupted. Forces use aerial delivery to deliver supplies and equipment when not possible or timely by other means. During urban stability or civil support operations, it is used extensively to move supplies to meet the urgent needs of a population in crisis. In all UO, delivery aircraft are highly vulnerable to small arms, rockets, and air defense systems. A threat may further decrease an already limited number of urban drop zones (aircraft may be able to avoid air defense systems, but ground forces may not be able to secure the drop zone and retrieve the cargo). Equipment and supplies transported by helicopter sling-load lessen the latter disadvantage. There are usually more available sites to deposit sling loads, such as rooftops (engineers may need to determine if they have sufficient structural integrity), parking lots, and athletic fields. However, load instability during flight may restrict a helicopter's airspeed and maneuver capabilities making it more vulnerable to small arms and man-portable air defense systems. For these reasons, aerial delivery of supplies in UO may be much less efficient than in many other environments.

## MAINTENANCE SUPPORT

10-47. Maintenance entails keeping materiel in operating condition, returning it to service, or updating and upgrading its capability. It includes recovering and evacuating disabled equipment; replacing forward; performing preventive maintenance checks and services (PMCS); increasing battle damage assessment and repair (BDAR); and determining potential maintenance resources in the urban area.

### Rapid Recovery Essential

10-48. Disabled vehicles produce obstacles to movement during UO, blocking narrow thoroughfares and creating vulnerable targets for threat forces. This makes rapid recovery operations essential to avoid clogging limited LOCs and mounted avenues with vehicle evacuation operations. Hence, maintenance personnel should be proficient in hastily securing unit maintenance collection sites near the damaged equipment, ensuring route security to and from the disabled vehicle, and methods of egress under fire. The increased potential for the task organization of armored units into smaller attachments during UO (often platoon-sized or smaller) will strain limited recovery assets and increase the need for sections to be well trained and resourced (to include adequate tow-bars and chains) for self-recovery operations. Units must maintain centralized and responsive control over these potentially critical recovery assets and position them as far forward as the situation allows.

### Replace Forward

10-49. One of the guiding maintenance principles is to replace forward and fix rear. Maintenance activities, with a forward focus on system replacement, task and use the distribution and evacuation channels to push components and end items to the sustainment level for repair. However, the urban environment may make distribution and evacuation difficult or even impossible. Repairing equipment as far forward as possible will often be vital to maintaining the tempo necessary to succeed in UO. When recovery is required, equipment is moved only as far back as the point where repairs can be made. When selecting a maintenance site, commanders consider:

- Security and force protection measures.
- Sufficient area around equipment for lift or recovery vehicles.
- Availability of existing maintenance facilities or garages.

10-50. The unforgiving urban terrain will invariably increase damage to man-portable weapons and equipment, particularly electronic equipment sensitive to jarring. More frequent jarring can also translate into more frequent calibration requirements. Although by definition man-portable weapons and equipment are easier to transport, evacuating these systems may prove as difficult as evacuating vehicles and larger, heavier equipment. Therefore, field support maintenance support teams (MSTs) will frequently need to repair equipment at (or as near as possible to) the point where it was damaged. Equipment operators are responsible to properly diagnose the fault or damage. Such action ensures that the correct repair parts and maintenance personnel are sent forward to complete necessary repairs. In UO, particularly offensive and defensive operations, units may need to replace rather than repair equipment. Subsequently, sustainment personnel may need to plan for increased replacement of what might normally be repairable equipment, as well as plan for increased repair parts for man-portable items.

## Add-On Protection Increases Wear

10-51. To increase protection against small arms, mines, rocket-propelled grenades, and lightweight antiarmor weapons, units may need to attach additional armor and improvised wire-mesh screens and cages to both wheeled and tracked vehicles operating in an urban environment. (Units can also sandbag vehicles to achieve a degree of increased protection.) These modifications, however, increase wear on brakes, springs, suspension, and tires, all of which are already vulnerable to the increased amount of debris present in the urban environment. Over time, this wear will begin to increase failure rates in major assemblies, such as engines, transmissions, and transfers. To offset the increase in wear and, subsequently repair parts requirements, commanders conducting UO should consider emphasizing these aspects during daily operator PMCS and increasing the frequency of scheduled services.

## Increased Battle Damage Assessment and Repair

10-52. In UO, operators, crews, MSTs, and recovery teams execute BDAR far more than in other environments. BDAR quickly restores minimum essential combat capabilities for a specific mission (normally of short duration) or allows the equipment to self-recover by expediently fixing or bypassing components. Commanders may need to authorize supervised battlefield cannibalization and controlled exchange when units lack critical parts or cannot bring them forward. Sustainment commanders should consider the use of BDAR kits, specifically designed for the systems they support. The use of these kits, though providing only temporary repair, will extend the availability of combat systems until full repair capability is accessible.

## Potential Urban Maintenance Resources

10-53. Although urban areas can complicate maintenance, they may contribute to this sustainment function. Analyzing the urban area in the commander's AO may reveal potential sources of parts, tools, equipment, and facilities, and the expertise necessary to fix equipment, repair components, and fabricate critical parts. Sites of potential value may include automotive repair shops, foundries, car dealerships, parts stores, junkyards, scrap-metal dealers, and machine shops. Urban areas may also serve as key sources for parts and facilities (and theater support contractor personnel) to repair automation and network communication equipment.

## Host Nation Repair Operations: OPERATION IRAQI FREEDOM

During operations in Mosul, Iraq in the summer months of 2003, elements of the 2nd Brigade Combat Team (BCT), 101st Airborne Division (Air Assault) suffered from a decline in combat power that outpaced the ability of the distribution-based sustainment system to provide repair parts. As readiness across the BCT threatened to degrade operational capability, sustainment personnel from the brigade's forward support battalion (FSB)—drawing on lessons learned from the 1990-1991 Gulf War—scoured the local populace for repair parts, facilities, and expertise in an attempt to offset the losses suffered due to supply shortfalls.

Initially, local retailers provided limited repair parts and facilities to the FSB, but the effect was significant and readiness rates began to stabilize. As time passed, however, the continued wear on equipment operating in the urban environment, coupled with the harsh summer weather, began to take a toll on reparable components and major assemblies. With limited availability in theater, and order-ship times as long as 60 days, requisitions would not be filled before the effect on combat fleet readiness reached a critical point.

Recognizing a trend that threatened the success of the BCT's urban operations in Mosul, the FSB established a contract maintenance facility in the brigade support area, used its contingency contracting officer to hire a group of local mechanics certified through General Motors, and opened the first host nation repair facility in Iraq. The group repaired major assemblies and reparable components and systems, maintaining a failure rate consistent with major CONUS rebuild facilities. The effect was immediate: during the first four months of the program, a full two-thirds of the combat platforms repaired and returned to service operated with components or assemblies rebuilt in the host nation facility.

Over time, the program was expanded to support other divisional units in northern Iraq. This effort provided maneuver commanders with responsive maintenance support and demonstrated the effect of sustainment flexibility and sustainability in urban operations. Furthermore, this action met the division commander's intent of infusing much-needed funds into the local economy while strengthening American-Iraqi relations in the region.

### TRANSPORTATION SUPPORT

10-54. Transportation supports the concept of the urban operation by moving and transferring units, Soldiers, equipment, and supplies. Transportation incorporates military, commercial, host-nation, and urban area capabilities to establish a flexible system that expands to meet the needs of the force. Transportation includes movement control, terminal operations, and mode operations.

10-55. Urban areas are often critical to transportation operations. These areas may serve as a lodgment or support area for entry of Army forces and sustainment supplies. The existing transportation and distribution infrastructure may be essential to reception, staging, and onward movement. Theater support contracts, as well as host-nation support agreements, may greatly increase the ability of Army forces to use the urban area's facilities, which may include docks, airfields, warehouses, and cargo handling equipment. Urban support may also include skilled urban workers, such as longshoremen, truck drivers, and MHE operators.

## Urban Terminals

10-56. In addition to serving as major seaports and aerial ports of debarkation, urban areas may provide additional terminals in the AO. Forces may use these terminals for further staging, loading, discharging, and transferring the handling between various inland transportation modes and carriers (motor, air, rail, and water). These urban terminals—with synchronized movement control—permit commanders to rapidly shift transportation modes and carriers. This increases flexibility and ensures the continued forward movement of equipment and supplies to support ongoing operations. Movement control, particularly in urban areas, relies heavily on support from military police in their maneuver and mobility support role. Without this support, urban LOCs may become congested, hinder movement and maneuver, and degrade force effectiveness (see FM 3-19.4). Urban commanders may need to establish multiple checkpoints, roadblocks, and traffic control points; restrict selected roads to military traffic; and reroute movement to unaffected road networks when civil support and refugee control operations compete for available routes. Again, military police operations are critical in this regard and will require continuous, close coordination with urban civilian police (if available).

## Obstacles to Ground Transportation

10-57. Although urban areas can contribute to transportation operations, rubble and other damage can become obstacles to ground movement. Even in an undamaged urban environment, road and bridge weight restrictions, barriers and medians, steep embankments, large drainage systems, and other urban structures may limit transportation operations. Urban route maintenance, to include reinforcing bridges and constructing bypasses, may become a priority task for engineer units. Bypassed pockets of resistance and ambushes pose a constant threat along urban supply routes and routes in close proximity to urban areas. Urban convoy and resupply operations will often require increased security in the form of continuous route security operations, regular (daily, if necessary) mine and improvised explosive device (IED) clearance operations, numerous observation posts, attack helicopter support, a larger, more mobile tactical combat force, and adequate communications capabilities among convoy vehicles and with supporting forces and higher headquarters. While the level of security and force protection may vary, convoy operations are never considered an "administrative" operation (see FM 4-01.45). Security needs may increase manpower requirements for sustaining operations and potentially reduce resources from decisive operations. Moving critical supplies may require heavily armed convoys and armored vehicles. Drivers must be well trained in convoy operations, rehearsed, and alert. They must recognize and avoid potential mines and IEDs (stationary and vehicle-borne) and be able to react rapidly to ambushes. Additionally, commanders may need to institute deception measures such as randomly closing and opening supply routes to confuse threat pattern analysis. As before, aerial resupply can alleviate problems due to ground obstacles, but the air defense threat and proximity of threat forces may preclude their routine use.

## Population Effects

10-58. The ability of Army forces to use vital urban transportation facilities depends largely on the civilians and the threat. The civilian population can affect the transportation system if they do not support the goals of Army operations. Urban transportation systems—such as ports, railroads, and rivers—require many specialists to operate. Without these specialists, the system's utility is degraded and may not function at all. In urban stability or civil support operations, Army forces will share the system with civilians and other agencies. Civilian authorities may refuse to allow Army forces to use any portions of an urban area's transportation system. Negotiating for access to that system under these circumstances then becomes a command priority.

## Threat Effects

10-59. The threat can significantly affect urban transportation systems. Many are composed of smaller subsystems. Each subsystem is vulnerable to attack, which in turn often shuts down the whole system. A large canal system, for example, may have entrance and exit facilities, the canal itself, a means to pull the vessel along such as a locomotive engine, and the civilians that run each of these subsystems. Both an attacker and defender understand the components of the particular transportation system. If important to current or subsequent operations, defending forces then develop plans and allocate forces to protect these subsystems. Attacking forces, on the other hand, often avoid collateral damage to the system, while simultaneously preventing enemy destruction of the facilities.

## FORCE HEALTH PROTECTION

10-60. Force health protection (FHP) consists of all services performed, provided, or arranged to promote, improve, conserve, or restore the mental or physical well-being of personnel in the Army and, as directed, for other services, agencies, and organizations. FHP conserves the force by—

- Preventing DNBIs and controlling combat stress.
- Clearing the urban area of casualties.
- Providing forward medical treatment and hospitalization and en route care during medical evacuation.
- Providing required veterinary, dental, and laboratory services.
- Ensuring that adequate Class VIII supplies, medical equipment, and blood are available.

10-61. FHP operations minimize the effects of wounds, injuries, disease, urban environmental hazards, and psychological stresses on unit effectiveness, readiness, and morale. Effective UO require acclimated Soldiers trained in specific urban tactics, techniques, and procedures. FHP helps maintain the health of urban forces, thereby conserving that trained manpower. Historically, urban combat operations have generated three- to six-times greater casualty rates than operations in any other type environment. (Greater enemy wounded and a need to provide care for them will also exacerbate this situation.) FHP operations that maintain the health and readiness of Army forces reduce the strain on the replacement and evacuation systems. Such care allows Soldiers to concentrate on the task at hand instead of the increased risks associated with UO.

10-62. As part of the overall LPT, commanders and medical planners must analyze and continuously assess the urban area. They must identify medical threats, required medical resources, and the quality and availability of medical facilities and resources (to include civilian medical personnel). This assessment prevents duplicated services and permits more effectively and efficiently organized medical resources.

10-63. FHP analysis also identifies available hospitals, clinics, medical treatment facilities, and medical supplies and equipment (including production facilities) in the urban area. It may also identify NGOs capable of providing medical services and supplies. (However, stringent federal regulations, standards of medical care, and a need for unavailable advanced technologies may limit their use by Army forces. But, this analysis will still be crucial in determining whether urban medical facilities can support civilian requirements.) FHP personnel keep abreast of the operational situation and its impact on FHP and the urban population. Peace operations, for example, may rapidly transition to high-intensity offensive and defensive operations requiring medical support able to handle potential mass casualty scenarios which may include civilians.

### Care of Civilians

10-64. During combat operations, the military normally does not provide injured civilians with medical care, instead relying on civilian medical personnel to provide this service. However, based on METT-TC and requirements under Geneva Conventions, commanders may need to recover, evacuate, and treat numerous civilians (particularly in urban civil support or stability operations) until local civilian medical personnel and facilities can be reconstituted and supplied. In urban civil support operations involving weapons of mass destruction, the primary focus of Army support may be FHP.

10-65. If commanders provide civilian medical support, they adhere to the UO fundamental of transitioning control and transferring responsibilities for medical care of civilians to another agency or into civilian hands as soon as is practical. In UO, commanders consider and address the medical treatment of civilians (enemy and friendly) early in the planning process. Any initial assessment or survey teams therefore contain medical representatives. Of significant concern, commanders may need to ensure that adequate supplies and equipment are on hand to treat the urban population's children as combat health service support is normally focused on the treatment of adult Soldiers. Additionally, follow-on military care for civilians after treatment at a combat support hospital is often nonexistent or extremely limited to a case-by-case basis necessitating the quick restoration of civilian hospital capabilities. Finally, military health practitioners must also assess and consider cultural factors related to civilian medical treatment. (For example, male doctors may not be able to examine or treat indigenous women.) To alleviate confusion and preclude avoidable negative repercussions and unrealistic expectations, commanders may need to develop specific medical policies, directives, and standing operating procedures. These "medical rules of engagement" (ROE) ensure that subordinates know how much medical care they may provide to—

- The urban population.
- Other host-nation and third-country civilians.
- Coalition and host-nation forces.
- Contractor personnel.

10-66. Commanders should also consider that medical aid to the local urban population and other civilians can be a powerful influential tool. Properly planned and executed, its potential benefits for Army forces include enhanced security and intelligence collection created by increasing the willingness of civilians to interact with Soldiers. These effects cannot be left to happenstance. They should be a conscious part of the information operations campaign nested within the overall UO plan. Consideration must also be given, however, to mitigating the potential negative consequences involved in the inevitable reduction or curtailing of medical assistance by Army FHP personnel. As with any aspect of UO, providing care without considering the longer-term consequences could result in far greater disadvantages than benefits.

## Disease and Nonbattle Injury

10-67. DNBI is a major medical threat during all operations and presents additional risk during UO. Urban pollution hazards and potential exposure to TIM may increase the risk to soldier health. Some urban areas, particularly those in developing countries, are already large sources of communicable diseases, such as tuberculosis, cholera, typhus, hepatitis, malaria, dengue, and acquired immune deficiency syndrome (AIDS). Other factors that increase the number of disease vectors and the potential for the transmission of infectious diseases might include—

- Physical damage or deterioration of urban infrastructure—such as electricity, water, and sewage services.
- Industries that use or produce hazardous materials.
- Increased habitats for rodents and other pests due to rubble, debris, and uncollected trash.
- Inability for the threat or local authorities to recover and dispose of human remains and animal carcasses.
- Greater threat of animal diseases transmittable to humans due to larger numbers of animals wounded or injured by the effects of UO.
- Greater susceptibility of urban civilians to infectious diseases—increasing the potential of person-to-person disease transmission—due to declining public health capabilities and nutrition levels.

10-68. The density of the environment extends these urban medical risks to Army forces. Commanders should establish a medical epidemiological surveillance system early to assess the health of the force and promptly identify unusual or local occurrences that may signal preventive medicine problems or the influence of biological or chemical agents. These potential hazards, particularly the release of TIM, will influence the type of medical supplies needed by medical personnel and will also necessitate critical planning and preparation for potential mass casualties (civilian as well as military).

10-69. Adhering to the UO fundamental of preserving critical infrastructure will likely prevent surges in DNBI. Preventive medicine personnel must identify the diseases and recommend control and preventive measures. In urban areas, they may also conduct civilian health screening, health education, and immunization programs. Medical screening of military personnel, particularly multinational forces, may be required to prevent introducing new diseases (especially drug-resistant strains) into an urban area. A new disease may tax the medical system and introduce a new medical problem into an area already in crisis. Lastly, field sanitation training (to include a general emphasis on hand washing; training in the use of barrier protection such as latex gloves when rendering care to any person; and animal, rodent, and pest control), equipment, and supplies are part of overall preventive medicine measures and considerations.

## Combat Stress

10-70. Stress occurs in every operation and type of environment; while limited stress levels are beneficial, too much is harmful. Controlled combat stress can invoke positive, adaptive reactions such as loyalty, selflessness, and heroism. On the other hand, uncontrolled combat stress can result in negative, harmful behavior and combat stress casualties. Such behaviors and casualties—battle fatigue, misconduct stress behaviors, and post-traumatic stress disorder—can interfere with the unit's mission. Physical and mental factors leading to combat stress result from the environment and the deliberate enemy actions aimed at killing, wounding, or demoralizing Soldiers. However, many stressors are also generated from the Soldier's own leaders and mission demands.

### Combat Stress: Chechnya – 1994 to 1996

Russia's 1994-1996 conflict with Chechnya, a republic in the southwestern part of the Russian Federation, produced an increased number of psychological trauma and combat stress casualties. One medical survey found 72 percent of the soldiers screened had some sort of psychological disorder symptoms. Of that, 46 percent exhibited asthenic depression (a weak, apathetic, or retarded motor state). The other 26 percent exhibited psychotic reactions such as high states of anxiety, excitement, or aggressiveness and a deterioration of moral values or interpersonal relations. The statistics showed more troops experienced combat stress disorders than during their 1980s war in Afghanistan. One primary difference was that in Chechnya, Russian forces conducted combat mostly in cities rather than in mountains, valleys, and other rural areas.

Combat always invokes fear in soldiers. However, poor training and planning, uncertainty in their cause, and urban populations that resented their presence exacerbated the psychological climate for the Russian forces in Chechnya. Acts of subversion and terrorism by Chechen guerrillas kept the Russians in a constant, high state of readiness and anxiety; the soldiers viewed every civilian—young or old, male or female—as a potential enemy. This psychological pressure was not simply a by-product, but an objective of information operations and a prime reason for taking the fight into the close confines of Chechnya's urban areas. Torture and mutilation of prisoners; immediate execution of captured pilots; imitative electromagnetic deception (Chechens mimicked Russian radio transmissions and directed Russian close air support against their own forces); and propaganda to convince civilians that Russia's actions had a religious bias against Muslims and Islam were conducted to exert intense, unremitting psychological pressure on Russian forces—with great success.

The characteristics of urban areas combined with Chechen insurgent activities and information operations, civilians that did not welcome foreigners, and an unpopular and poorly supported conflict with an open-ended mission reinforced the need for a national will to strengthen and support the urban fight. These characteristics also reinforce the need for clear objectives, proper leadership and training, and available medical assets able to properly diagnose and treat combat stress casualties.

10-71. During UO, compartmented urban combat leads to physical isolation while difficulties in transmitting radio signals lead to communications isolation; together, they create an overwhelming sense of loneliness. Snipers, mines, and IEDs combined with the closeness and high intensity of urban combat contribute to an unremitting fear of attack from any quarter that further increases stress casualties. Additionally, seeing and perhaps accidentally inflicting casualties on civilians (especially women and children) increase battle fatigue. If civilians are hostile or a threat uses the population as cover and concealment, then the potential for misconduct stress behaviors often increases. Urban areas may provide temptations for looting, alcohol and substance abuse, black marketeering, and harmful social interactions; these temptations may increase misconduct stress behaviors. Leaders can prevent or rapidly identify, successfully manage, and treat stress-related casualties (see FM 22-51 and FM 6-22.5) and prevent misconduct stress behaviors as well as potential violations of the law of war. To reduce the number of stress casualties, commanders must—

- Provide tough, realistic training in urban combat (and combat stress management).
- Plan for rotation of forces and adequate rest and relaxation opportunities.
- Establish effective ROE.
- Create unit cohesion.
- Ensure adequate mental health support is available.
- Provide strong, capable leadership. Leaders, forward on the battlefield, will determine whether the necessary discipline perseveres.

## Evacuation

10-72. Transportation restrictions may preclude the medical evacuation (MEDEVAC) of urban casualties. MEDEVAC by air ambulance remains the preferred method; however, it may not be feasible. The threat from small arms fire, rocket-propelled grenades, and other shoulder-fired weapons may prohibit the employment of helicopters. Aeromedical evacuation can also be hampered by the lack of landing zones due to narrow streets, heavily damaged buildings, and barricaded or booby-trapped roofs. If the employment of air ambulances is not feasible, a ground evacuation operation will be necessary. In UO, ground ambulances should often be—

- Hardened to improve the safety of the crew and patients as well as to increase the survivability of the evacuation platform.
- Field-sited as close to the operation as is tactically and physically possible.
- Maneuverable in tight spaces and narrow streets.

10-73. Evacuation by ground ambulance may have their own set of obstacles to surmount such as rubble; debris; barricades; mines, booby-traps, and IEDs; damaged or destroyed vehicles, and flooded or snow-filled streets. If enough ground ambulances are unavailable due to increased casualties or the tactical situation does not allow, units will need to execute casualty evacuation (CASEVAC) operations.

10-74. In UO, CASEVAC may entail evacuation by litter and the expedient use of a vehicle of opportunity or other military vehicle not dedicated for medical evacuation. Based on the threat situation, CASEVAC may also be limited to armored vehicles, conducted only during limited visibility, or both. For evacuation by litter, units may require more litter bearers to move the injured to a point where they can be further evacuated by ground or air ambulance. Lengthy evacuation routes require more litter bearers, as multiple litter relay teams will be necessary to conserve energy and expedite evacuation. However, unless augmented or relieved of this responsibility by another unit, these litter bearers will come from the casualties' own unit thereby diminishing the unit's strength levels necessary to accomplish its primary mission. Instead, tactical commanders may need to wait until a lull in or a cessation of the battle occurs. Depending on projected casualty rates, higher-level commanders may augment units with additional personnel to perform evacuation or may assign maneuver units this mission. Evacuation personnel will require specific training in urban evacuation techniques (moving casualties from subsurface and supersurface levels to and along the surface level).

10-75. Overall, urban commanders at all echelons must develop detailed casualty evacuation plans that include the establishment of treatment facilities as far forward as the situation allows. Engineers are critical to clear routes for medical resupply and evacuation. Based on the threat situation, additional armored vehicles may need to accompany ambulances in a multi-vehicle convoy not only to and from the patient pickup point but also through to a higher level of care. Army forces may also need to plan for and acquire unique capabilities, equipment, and skills now typically seen in civilian urban search-and-rescue teams to clear debris and search for urban casualties. For example, specially trained dogs may play a vital role in locating victims among urban rubble.

### Treatment

10-76. First aid training will have increased significance in UO. The compartmented nature of UO, transportation restrictions, communications difficulties, and the low density of trauma specialists (combat medics) may limit the urban casualty's initial treatment administered by nonmedical personnel or to self-treatment measures. To mitigate this risk, units identify and train combat lifesavers (CLS) to perform in the absence of trauma specialists. Since the likelihood of Army forces performing UO continues to increase, commanders strive to meet or exceed Army standards for the number of fully trained CLS required for their specific unit. CLS (and trauma specialist) training must be realistic, hands-on, and conducted at appropriate intervals to ensure CLS can perform lifesaving tasks to include starting intravenous lines. This increase in self, buddy, and CLS care, as well as longer delays in evacuation, may also increase requirements for additional first aid, CLS, and trauma specialist-carried supplies. (Ultimately, commanders may have to consider weighting the main effort with greater overall FHP support for units conducting UO than for units conducting operations in other environments.)

10-77. The increased potential for delayed evacuation during UO mandates that Army trauma specialists be skilled in prolonged casualty care. During the 3-4 October 1993 battle in Somalia, for example, seven trauma specialists managed 39 casualties for more than 14 hours before they could be evacuated. Evacuation delays significantly increase the potential for infection; therefore, trauma specialists should also be trained to administer antibiotics on the battlefield.

10-78. FHP personnel must be capable of recognizing and treating injuries due to IEDs and incendiary or thermobaric weapons (fuel-air explosives)—favored, urban-oriented threat weapons (see Chapter 3). Thermobaric weapons explode; create a cloud of volatile gases, liquids, or powders; and then ignite, creating an immense fireball consuming oxygen and creating enormous overpressure. When employed in an urban structure, the blast wave or overpressure of any explosive is greatly amplified. Injuries resulting from these weapons are massive burns, broken or crushed bones, concussions and traumatic brain injuries, missile injuries, and internal injuries. To counter this threat, medical personnel must be aware of the capabilities of these blast weapons, wounding patterns, and treatment methods.

10-79. The increased use of body armor during UO will help prevent penetrating chest and abdomen wounds. As such, FHP personnel should expect more groin, pelvis, head, and extremity injuries. Furthermore, when fighting soldiers that are known to use body armor, a threat (particularly snipers) can be expected to target the head and face more often than other anatomic areas resulting in more head injuries. Medical personnel must also anticipate and be prepared to treat an increase in hearing loss due to the firing of weapons (particularly recoilless) in enclosed spaces with little ventilation. Medical personnel (and engineers) should also be trained on how to properly remove rubble that has fallen on a Soldier or civilian so as not to increase the extent of his injuries. Lastly, FHP personnel will need to research and develop unique treatments for injuries due to an ever-increasing arsenal of nonlethal weapons.

## Explosive Ordnance Disposal Support

10-80. Explosive ordnance disposal (EOD) support provides the capability to neutralize domestic or foreign conventional explosive hazards which include unexploded explosive ordnance (UXO), booby traps, IEDs, captured enemy ammunition (CEA), and bulk explosives. Such material and devices threaten military operations and military and civilian facilities, equipment, and personnel. Explosive hazards create a much greater risk during UO than operations in any other environment. Confined spaces, hard surfaces, and more personnel (both Soldiers and civilians) in the vicinity may magnify the detonating effects of explosives. Dense terrain makes all types of explosive hazards more difficult to locate. Moreover, terrorists select urban settings for their booby traps and IEDs to potentially kill and destroy more, thus gaining greater visibility for their message.

10-81. 10-81. EOD units detect, mark, identify, render safe, and dispose of explosive hazards. Importantly, EOD specialists will work closely with unit intelligence personnel to conduct explosive forensics—technical intelligence collection and exploitation—to help identify the makers of the devices as well as their ever-evolving methods of construction, placement, concealment, and detonation. Chemical reconnaissance with their array of chemical sensors will also work closely with EOD units to reduce the uncertainty of the chemical threat. This integrated forensic approach is a proactive method to take the advantage away from a terrorist or insurgent threat that seeks to employ these devices as a primary tactic.

10-82. Urban operations will increase EOD specialists' role as advisors and instructors on explosive hazards, protection measures, and disposal techniques. They will advise and train other Army forces, other services, multinational partners, and civilian authorities. EOD specialists will also work closely with public affairs and psychological operations personnel to increase public awareness of the horrific effects of these munitions and devices. This education process is designed not only to teach the urban population to identify and avoid explosive hazards, especially UXO, booby traps, and IEDs, but to gain civilian assistance in reporting their locations. (In this regard, rewards for information leading to the discovery of unexploded ordnance may be beneficial. However, the largest reward should normally be reserved for information leading to the arrest or capture of bomb makers.)

10-83. In an urban environment where there are large amounts of UXO, IEDs, and CEA, the sheer amount of explosives may overwhelm the capabilities of available EOD personnel. Because of their training in demolitions and explosives, commanders will often use combat engineers to augment the destruction and clearance effort. Engineers are normally given responsibility for UXO clearance and disposal while EOD teams use their technical skills for IEDs. Commanders may also consider embedding EOD expertise within engineer units to provide the knowledge necessary to determine the correct amount and type of explosive necessary to detonate UXO and to understand potential risks and effects of detonation of various munitions. For example, EOD specialists would better understand the second-order effects and risk reduction measures necessary to detonate a large missile containing numerous submunitions.

10-84. Unfortunately, there is no single device or technique that can eliminate every explosive hazard encountered. Clearance operations are dangerous and time-consuming. Based on METT-TC, commanders may consider the following options when dealing with these explosive hazards:

- Report, mark, and bypass the hazard.
- Employ tactical breaching procedures.
- Self-extract to alternate routes or positions.
- Restrict routes or positions from further use.
- Clear specific areas or positions, or conduct an extensive route clearance.
- Lastly, accept risk of casualties and continue with the assigned mission.

10-85. Effective tactics, techniques, and procedures (TTP) for addressing explosive hazards must be captured and shared with friendly forces and coalition partners. Commanders recognize, however, that adaptive threats will continually modify their own techniques to counter the unit's response. See FM 3-34.119, FM 20-32, FM 3-100.38, FM 4-30.51, and FM 4-30.16.

## Human Resources Support

10-86. Human resources support (HRS) encompasses the following functions: manning the force, personnel support, and personnel services. These activities include personnel readiness; replacement operations; personnel accounting; personnel information management; casualty operations; essential personnel services; postal operations; morale, welfare, and recreation; and band support provided to Soldiers, their families, Department of Army civilians, and contractors.

10-87. Successful UO require HRS functions. Clearly, HRS focuses on caring for the needs of people: Soldiers and civilians who support them. Since a critical component of the urban environment is the population, these activities, when required, may also support them. For example, HRS elements, in conjunction with mortuary affairs, military police, and CA, may provide administrative support to assist in the accountability of displaced persons and civilian internees and the establishment of other populace and resources control measures. This support may include the provision of identification and registration cards, travel permits, and other documents, and administrative support to casualty operations. Human resources managers may also assist with community activities and coordination with the Red Cross. If necessary, a postal platoon may assist urban officials in training and reestablishing civilian postal operations. Overall, this type of assistance and support helps to care for the immediate needs of the urban population and, ultimately, speed transition back to legitimate civilian responsibility and control.

10-88. In offensive and defensive UO, HRS personnel will need to account for more casualties and more frequent reconstitution. In all UO, success relies on training individual replacements. Urban combat requires Soldiers skilled in specific urban combat TTP. They understand the societal aspects of the urban population and have training in crowd control to avoid escalating potentially volatile situations. Training these replacements prior to their actual deployment or while in the replacement system frees urban commanders from having to do so. It also helps reduce Soldier (and supporting civilian) isolation, anxiety, and fear associated with urban areas and operations.

## FINANCIAL MANAGEMENT SUPPORT

10-89. Money and its management are often critical to the success of UO. As such, financial management support makes resources available when and where they are needed and assist the urban commander in maintaining fiscal responsibilities. Financial management operations are necessary to conduct the contracting necessary to restore and rebuild infrastructure and provide real-time information, accounting, and services. Financial management operations ensure that urban operational policies and procedures adhere to laws and regulations, develop command resource requirements, and leverage appropriate fund sources to meet them.

10-90. UO will likely include other U.S. military services, governmental agencies, and contractors; multinational forces; NGOs; and various elements of the host nation. So many actors complicate financial accounting, resource management, and cost capturing. Yet, these activities are vital to accomplishing UO and maintaining legal requirements. Army forces will receive support from and provide support to these participants. Financial managers provide crucial expertise to the urban commander on the obligation authority. They also furnish the documentation necessary to obtain and pay for local goods and services using contract or commercial vendor services. However, commanders (with financial management assistance) must assess the economic impact of UO on the economy. An appropriate analysis includes how well the urban economics and commerce infrastructure can support the deployed force and how an influx of U.S. currency may affect the overall economy either positively or negatively.

### Labor, Reward, and Solatia Programs

10-91. Financial management elements can also provide joint pay support and non-U.S. pay support for host-nation employees and day laborers supporting UO, as well as for civilian internees. These units can support reward programs and solatia (financial compensation). Commanders can use reward programs to purchase weapons, radios, information, and other items from the urban inhabitants. However, they must use caution when paying for critical human intelligence. First, they must ensure that multiple collection agencies in the command avoid paying the same source for the same information and interpreting these repeated inputs as validation that information is accurate and reliable. Second, they must establish a price or pay scale so subordinate units (particularly in multinational UO) equally compensate their sources for information. Commanders may also need to make solatia to alleviate grief, suffering, and anxiety resulting from injuries and property or personal loss resulting from some aspect of UO. At other times, commanders make solatia to meet cultural expectations of the urban population. These payments are nominal in amount and made without an admission of liability by the Army. Significantly, they may help maintain the legitimacy of the operation and preserve the support of the urban populace.

### Financial Support to Civil-Military Projects

> *...[I]n an endeavor like that in Iraq, money is ammunition. In fact, depending on the situation, money can be more important than real ammunition—and that has often been the case in Iraq since early April 2003 when Saddam's regime collapsed and the focus rapidly shifted to reconstruction, economic revival, and restoration of basic services. Once money is available, the challenge is to spend it effectively and quickly to rapidly achieve measurable results. This leads to a related observation that the money needs to*

> *be provided as soon as possible to the organizations that have the capability and capacity to spend it in such a manner.*
>
> Lieutenant General David H. Petraeus
> Commander, 101st Airborne Division (Air Assault) and, later,
> Commander, Multi-National Security Transition Command-Iraq
> during OPERATION IRAQI FREEDOM

10-92. Other financial resources separate from operational funds may be appropriated to support civil-military operations (CMO) projects during stability operations. (During OPERATION IRAQI FREEDOM, these funds were referred to as commander's emergency response program [CERP] funds.) These funds are typically disbursed as bulk currency to tactical units, where the funds are allocated against lower-cost, high-impact projects that support the immediate needs of civil authorities and their citizens. As such, commanders must be prepared to receive, account, secure, and properly disburse these resources in accordance with established procedures. Commanders have an inherent responsibility for ensuring the proper use and disbursement of these funds and must execute appropriate judgment, fiscal responsibility, and supervisory oversight of any resources appropriated in support of CMO. While commanders establish appropriate guidelines to accomplish this fiscal responsibility, they must provide their subordinate commanders—who have the need, capacity, and capability to put these funds to use—reasonable flexibility in how they spend at least a portion of the money, so that they can address emerging needs within their AO.

### Contracting Expertise

10-93. Due to the nature of UO, tactical-level units will be likely required to plan and initiate contracts not only for reconstruction projects but to supplement their own sustainment needs. Therefore, higher-level commands must be prepared to assist tactical-level units with contracting expertise. This expertise is required to overcome difficulties in delineating the scope of work, defining construction standards, and establishing viable delivery dates. For small projects and purchases, a standard contractual format may be sufficient. However, for large projects, commanders must provide their subordinates with knowledgeable contracting personnel, who can write contracts with clearly-understood specifications, create purchase orders that specify quality and delivery times, and help monitor work for quality control and assurance. Payment guidelines and standards among all services and members of a coalition are essential to eliminate competition and potential overpayment. Translators will be also essential to accurately translate requirements and standards in addition to trusted local agents to assist in selecting quality engineering and construction firms and workers. Overall, subordinate commanders need adequate contracting support and expertise to meet civilian needs as well as their expectations for timeliness and quality. Without it, commanders may risk losing critical opportunities to positively influence civilian perceptions. (See also Legal Support to Operations below.)

## RELIGIOUS SUPPORT

10-94. Religious support entails providing and performing religious support operations for commanders to protect the free exercise of religion for Soldiers, family members, and authorized civilians. This includes personal delivery of rites, sacraments, ordinances, pastoral and spiritual care, and religious education. Such support also consists of:

- Advising the commander on matters of religion, morals, and morale as affected by religion.
- The impact of local religions on threat courses of action, the urban society, and, subsequently, the mission.
- The ethical impact of command decisions.

### Moral and Ethical Climate

10-95. Chaplains can help commanders sense the moral and ethical climate in Army units and understand potential moral dilemmas associated with planned UO. The urban environment affects Soldiers' mental health and can increase combat stress casualties, especially misconduct stress behaviors. Chaplains are an important part of combat stress support. The chaplain's presence and faith can sustain Soldiers throughout periods of great trauma. Religious support contributes to the total well being of Soldiers and aids in their return to combat readiness. Chaplains help bolster Soldiers' moral and ethical behaviors through spiritual fitness training so that Soldiers may better cope with ambiguous moral and ethical situations. Chaplains can help identify ethical concerns before they become critical command problems. To this end, their observations help develop and modify ROE, which—due to ambiguity or overly restraining rules—may be part of the problem.

### Influence of Local Religions

10-96. Chaplains advise commanders on matters of religion as it affects the Soldiers within their units. They also explain the influences of local religions on threat courses of action and the urban populace and, subsequently, their potential effects on Army forces and UO. Religion is a crucial aspect of assessing the societal component of the urban environment. Understanding the major tenets and concepts of the religions and the impact of faith on civilians' lives may help commanders understand what motivates the threat and the urban populace. (Historically, when religion was a critical factor in armed conflict, the level of lethality has been significantly higher.) To understand threat strategy and tactics, commanders understand their ideology, which, in many cases, involves strong religious beliefs. This understanding also helps commanders to appreciate the inhabitants' attitudes toward other races, religions, and cultures and to identify unacceptable kinds of social interaction (particularly between Soldiers and civilians). Failure to recognize and respect religious beliefs can rapidly erode the perceived legitimacy of the mission and increase the length of the operation. In addition to its influence on threat courses of action, a thorough analysis of the urban environment also includes the degree of influence religion and religious leaders have on the area's government and economy.

## Legal Support to Operations

10-97. Legal support provides operational law support in all legal disciplines (including military justice, international law, civil law—composed of contract, fiscal, and environmental law, claims, and legal assistance). This support assists in command and control, sustainment, and HRS of UO. Legal considerations are important in any operation; they take on added significance during UO. They form the foundation for establishing ROE and are critical in the targeting process. (For example, with the increased likelihood that the threat will use the media in an attempt to categorize Army attacks as unlawful, legal support will be critical in determining the protected status of targets.) Adequate legal instruction and training may help subordinate commanders and their Soldiers better understand some aspects ROE and, with this understanding, help them make the rapid decisions often required in a complex urban environment.

10-98. Legal considerations also affect how units acquire goods and services from urban areas and provide support to other agencies and organizations operating in an area. The environment's complex nature requires commanders and their staffs to review and closely consider applicable legal constraints when developing and executing friendly courses of action. Most urban areas have a highly developed legal system nested within county, provincial, state, and national systems. The SJA support to commanders addresses this urban system in conjunction with other legal systems and their potential to affect, positively or negatively, the success of UO.

10-99. International, host-nation, and U.S. law and other regulatory guidelines may vary in their applicability by time and place; actions permissible in one situation may be prohibited in another. These exceptions and complexities increase requirements for SJA, often working with CA personnel, to identify and resolve technical legal issues. Therefore, the SJA must actively advise and participate in all aspects of UO from predeployment training and initial planning through transition and redeployment. FM 27-100 contains detailed legal guidelines affecting UO.

### International and Host-Nation Law

10-100. International law consists primarily of agreements, treaties, and customary law to include the law known as the law of war (see FM 27-10). The law of war consists of four general principles applicable when conducting any operation but requiring particular attention during UO. Figure 10-4 lists and describes the four principles: military necessity, discrimination (or distinction), unnecessary suffering (or humanity), and proportionality.

10-101. International law may affect urban operational issues, such as the right of entry, base operations, use of urban infrastructure, and overflight and landing rights. Status-of-forces agreements (SOFAs) exist or can be negotiated to resolve legal issues, such as the status of Soldiers (and, as necessary, contractors) operating in foreign areas to include criminal and civil jurisdictions, taxation concerns, and claims for damages and injuries. Unless a SOFA or other convention exists, Soldiers operating in foreign urban areas have the same legal status as tourists; they are subject to the laws and judicial process of the host nation. (During armed conflict, however, Soldiers always have the rights afforded to them by the Geneva Conventions). Commanders are responsible for understanding the international and host-nation agreements and laws that influence foreign UO. If local law hinders the operation, commanders may be able to inform the local U.S. diplomatic mission and request that it negotiate a solution.

10-102. Commanders may encounter civilian resistance groups whose actions may range from providing enemies with sustainment support to actively fighting Army and coalition forces. Members of such resistance groups must be dealt with in accordance with applicable provisions of the law of war. Commanders should seek legal guidance concerning detention and disposition of persons participating in various acts harmful to friendly forces and detrimental to the mission.

## U.S. Law

10-103. Commanders conducting UO must also comply with U.S. law whether it is in the form of a statute, executive order, regulation, or other directive from a federal branch or agency. U.S. law influences UO by governing the acquisition of supplies and services for Army forces, regulating the assistance that can be rendered to foreign nations, and controlling intelligence activities. The Posse Comitatus Act, for example, makes it a crime for Army forces to enforce civil law. Similarly, portions of the Foreign Assistance Act prohibit Soldiers from performing law enforcement activities in foreign urban areas. However, circumstances—expressly authorized by the Constitution, acts of Congress, and other exceptions to these statutes—exist that allow the Army to support civilian law enforcement. Although not nearly all-inclusive, the above demonstrates how U.S. law further complicates urban operations, particularly stability operations, and increases the need for proactive SJA advice and counsel in all facets of UO.

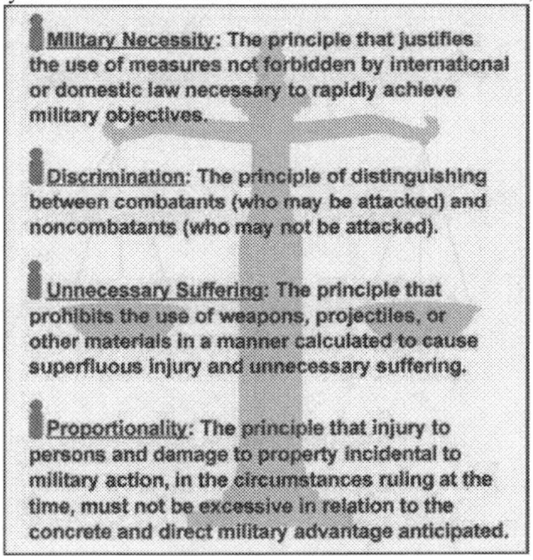


**Figure 10-4. General principles of the law of war**

## Legal Aspects of Nonlethal Force

10-104. Nonlethal capabilities can augment the means of deadly force and extend urban firepower options. They enhance the commander's ability to apply force in proportion to the threat and to allow discrimination in its use. The range of nonlethal capabilities includes offensive information operations, smoke and obscurants, irritants (such as chemical riot control agents), nonpenetrating projectiles, high-pressure water devices, and directed-energy weapons. These continually expanding capabilities give commanders more options to confront situations that do not warrant deadly force but require Soldiers to employ overwhelming decisive power. However, nonlethal capabilities are subject to the same legal constraints as lethal force (in fact, some nonlethal capabilities can cause serious injury and death, particularly if not employed properly) and undergo the same legal review. Like lethal force, nonlethal capabilities show military necessity, distinguish between combatants and noncombatants, distinguish between military objectives and protected property, are used proportionally, and do not result in unnecessary suffering.

10-105. Of special note, commanders cannot employ chemical herbicides or riot control agents as a method of warfare without prior Presidential approval. However, the Secretary of Defense and regional combatant commanders have authority to use herbicides and riot control agents in limited circumstances during armed conflict and in peacetime operations. Considering the complexity of this issue, commanders of major operations must streamline the approval process so that subordinate commanders do not discount use of these agents simply because of the high-level approval authority. Subordinate commanders must, however, be able to anticipate and clearly justify their potential use in UO. Also, any approval does not mean that these agents must be employed.

### Civilians Accompanying the Force

10-106. LOGCAP personnel and other civilian support contractors provide a variety of sustainment functions for the Army. These functions may range from providing unskilled labor, transportation support, and health care to technical support of sophisticated equipment and weapons systems. Commanders must ensure that civilians providing support in their AO are not placed in positions of jeopardy, but if they are, these civilians must understand the risks that they assume when they engage in activities that might be misconstrued as direct or active participation in hostilities.

10-107. This last caution indicates an important legal consideration for commanders. They must try to ensure that threat forces can distinguish civilian support contractors from combatants and thereby remove any justification for them to be targeted or attacked. (As Army forces seek to work closely with the members of governmental and nongovernmental agencies and organizations in urban stability operations, commanders may need to take steps necessary to mitigate the risk of attack caused by their involvement with Army forces.) In the event that threat forces capture civilian support contractors, this same concern helps to ensure that, as lawful civilians accompanying the force, they will be afforded prisoner of war status. As the role of contractor support to Army operations continues to expand, commanders must stay abreast of changes in the law (to include evolving interpretative distinctions) as it affects the use of contractors on the urban battlefield. Other important considerations for commanders with civilian contractors supporting their force are—

- **Minimize Exposure to Harm**. Civilian contractors should be assigned duties at locations that minimize their exposure to harm. Commanders will need to periodically review these locations since they are likely to change as the urban operation evolves and transitions.
- **Prohibit the Performance of Combat Functions**. Civilian support personnel should not be working in situations that involve combat operations where they might be conceived as combatants. As an example, they should not perform force protection for Army forces such as fortification construction and checkpoint security.
- **Prohibit the Use of Arms**. Civilians should not be armed, unless approved by the combatant or service component commander on a limited, by-exception basis for personal defense.
- **Provide Identification Cards**. Civilians should be provided, and carry on their person, a Geneva Convention identification card identifying them as civilians authorized to accompany military forces and entitling them to be treated, if captured, as prisoners of war.
- **Restrict the Wear of Military Uniforms**. Civilians should not normally wear distinctive U.S. military uniforms unless the combatant or service component commander authorizes them to do so. Regardless of their clothing, however, civilians will wear a symbol that establishes their civilian status.

## General Engineering Support

10-108. General engineering support will be essential during UO. This support helps assess, construct, maintain, and restore essential LOCs and urban facilities to sustain Army forces, the urban population, or both. Using civilian resources and investing Army general engineering resources requires careful consideration by commanders and staff planners. Since all elements of the urban infrastructure interconnect, general engineering support touches each category to some degree. Figure 10-5 illustrates how urban-specific, general engineering tasks align primarily with the transportation and distribution, energy, and administration and human services components of the urban infrastructure. These engineering tasks are significant and readily apply to UO. The last two, providing fire fighting support and waste management, have not been previously addressed and require more specific consideration.

**Construct, Maintain, or Restore**

- Roads and Highways
- Over-the-Shore Facilities
- Ports
- Railroad Facilities
- Airports and Heliports
- Fixed Bridges
- Electric Power Facilities
- Petroleum Pipelines and Storage Facilities
- Water Facilities

**and Provide**

- Mobile Electric Power
- Construction Support
- Fire Fighting Support
- Waste Management

**Figure 10-5. General engineering support tasks**

## The Use or Investment of Resources

10-109. During urban offensive and defensive operations, Army engineer units accomplish tasks to sustain or improve movement and mobility, protection, and sustainment of U.S. and allied forces. These units should maximize the existing urban facilities, host-nation support, civilian contractors, and joint engineer assets. However, commanders should consider how using urban facilities to support military forces may negatively affect the population. On the other hand, construction and repair may benefit both Army units and the urban inhabitants. Restoring the urban transportation network not only improves military LOCs, but may also allow needed commerce to resume. Repairing urban airfields or ports increases throughput capabilities for military supplies, facilitates medical evacuation operations to the support base, accelerates needed relief efforts, and allows international commerce to proceed. Commanders may first invest resources and conduct general engineering tasks to restore facilities for civilian use. Such actions stem future drains on operational resources or facilitate later transition of control back to civilian authorities. For example, repairing police stations, detention facilities, and marksmanship ranges may help urban governments reestablish law and order after completing urban offensive or defensive operations. During stability or civil support operations, the focus of general engineering will often be to support and assist the urban population rather than Army forces.

## Fire Fighting Support

10-110. Fire protection and prevention, as well as fire fighting, takes on added significance during UO, particularly offensive and defensive operations. Most ordnance produces heat and flame. This, coupled with an abundance of combustible material (buildings, furniture, gasoline, oil, and propane), poses a serious risk to Soldiers, civilians, and the urban operation. Large shantytowns can exacerbate this problem. In highly combustible areas, commanders may even need to limit or preclude the use of small-arms tracer ammunition. Fire threats to urban areas can be categorized as—

- **Isolated Fires**. These are restricted to a single structure or a specific area within a structure.
- **Area Fires**. These fires consume two or more structures and may extend to encompass an entire block. Generally, streets will serve as firebreaks and help contain the fire within a single block.
- **Firestorms**. These are the most violent and dangerous fires capable of rapidly consuming large areas by creating windstorms and intense heat. Firestorms are often inextinguishable until they have consumed all available combustible materials.

- **Explosive Hazards**. These are present in areas containing fuels and chemicals as well as military explosive hazards.

10-111. When analyzing the administration and human services component of the infrastructure, commanders must determine the adequacy of existing civilian fire fighting support. A deteriorated or nonexistent infrastructure that cannot support the urban area will likely fail to handle the increased risk due to military operations. Commanders may need to provide fire fighting teams to support their own forces and civilians.

10-112. A military force task organized with multiple fire fighting teams (even with maximum use of available civilian fire-fighting assets) will only be able to fight some fires in the AO. Water distribution systems damaged during operations, chemical and other TIM, and hostile activities will further complicate and limit fire fighting capabilities. Commanders must develop priorities for equipment, facility, and infrastructure protection. All Soldiers need training in fire prevention and initial or immediate response fire fighting. Such training includes planning covered and concealed movement, withdrawal, and evacuation routes. Soldiers should be trained to identify and remove ignition and fuel sources and provided additional fire fighting material such as extinguishers, sand, and blankets (see FM 5-415 and the U.S. Department of Transportation's current version of the Emergency Response Guidebook).

## Waste Management

10-113. Management of all forms of waste, particularly human, putrescible (such as food), and medical, may become a critical planning consideration for Army forces. This particularly applies if the urban waste management infrastructure was previously inadequate or damaged during UO or natural disasters, the Army force is operating in the urban area for an extended period of time, and a significant number of the urban population remains. Failure to adequately consider this aspect, possibly coupled with the decaying remains of humans and animals and an inadequate or tainted water supply (such as may be caused by flooding), may create unacceptable sanitary and hygiene conditions and subsequently increase DNBIs as well as civilian casualties.

# Appendix A

# Siege of Beirut: An Illustration of the Fundamentals of Urban Operations

*The [Israeli Defense Forces] IDF had neither the strategy nor the experience nor the configuration of forces to fight and sustain a house-to-house campaign in Beirut.*

Richard A. Gabriel
*Operation Peace for Galilee: The Israeli-PLO War in Lebanon*

## OVERALL STRATEGIC SITUATION

A-1. In 1982, Israel launched OPERATION PEACE FOR GALILEE designed to destroy the Palestine Liberation Organization (PLO) presence in southern Lebanon. On 1 June, the IDF launched a massive assault across the border into southern Lebanon. The Israeli attack focused on the PLO, but the operations quickly involved major ground and air combat between Israel and Syrian forces.

A-2. In the first few weeks, Israeli forces quickly pushed back both the Syrians and the PLO. However, except for some PLO forces isolated in bypassed urban areas, such as Tyre and Sidon, most of the PLO fell back into Beirut (see Figure A-1). By 30 June, Israeli forces had reached the outskirts of southern Beirut, occupied East Beirut, isolated the city from Syria and the rest of Lebanon, and blockaded the sea approaches to the city. Even so, with most of the PLO intact inside and with significant military and political capability, the Israelis had yet to achieve the objective of OPERATION PEACE FOR GALILEE. The Israeli command had to make a decision. It had three choices: permit the PLO to operate in Beirut; execute a potentially costly assault of the PLO in the city; or lay siege to the city and use the siege to successfully achieve the objective. The Israelis opted for the latter.

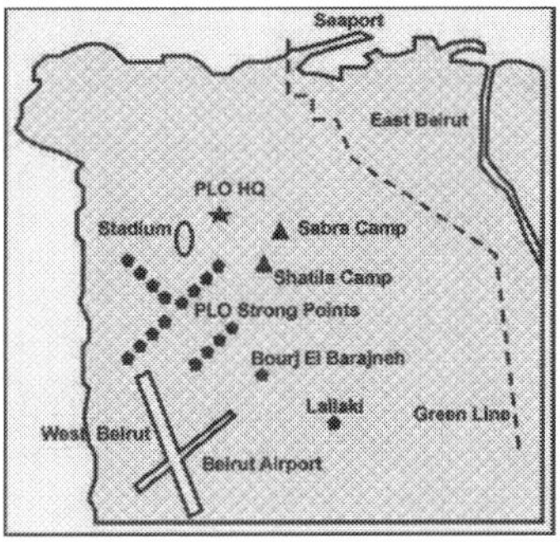


Figure A-1. The city of Beirut

### Israeli Military Position

A-3. The Israelis had an excellent position around Beirut. They occupied high ground to the south and west, virtually dominating the entire city. Israeli naval forces controlled the seaward approaches to Beirut. The Israelis' position was also strong defensively, capable of defeating any attempt to break out of or into the city from northern Lebanon or Syria. The Israeli air force had total and complete air superiority. The Israelis controlled the water, fuel, and food sources of West Beirut. Although the PLO forces had stockpiles of food and supplies, the Israelis regulated the food, water, and generating power for the civil population.

A-4. Despite the superior positioning of Israeli forces, the IDF faced significant challenges to include the combat power of the PLO, Syria, and other threats in Beirut. Israeli doctrine and training did not emphasize urban operations. Additionally, Israel was constrained by its desire to limit collateral damage and friendly and noncombatant casualties. Organizationally, the Israeli army was not optimized to fight in urban terrain.

Armor and self-propelled artillery formations dominated the Israeli forces, and most Israeli infantry was mechanized. The Israeli forces had only a few elite formations of traditional dismounted infantry.

# PLO MILITARY POSITION

A-5. Despite being surrounded and cut off from support, the PLO position in Beirut offered numerous advantages in addition to the characteristic advantages of urban defense. The PLO had long anticipated an Israeli invasion of southern Lebanon; it had had months to prepare bunkers, obstacles, and the defensive plan of Beirut and other urban areas. Approximately 14,000 Arab combatants in West Beirut readied to withstand the Israeli siege. This was done with the advice of Soviet, Syrian, and east European advisors. The preparation included stockpiling essential supplies in quantities sufficient to withstand a six-month siege. Also, the PLO fighters integrated into the civil populations of the urban areas. Often their families lived with them. The civil population itself was friendly and provided both information and concealment for PLO forces. PLO fighters were experienced in urban combat and knew the urban terrain intimately. PLO forces had been involved in urban fighting against Syrian conventional forces and Christian militias in Beirut several years prior to the Israeli invasion. Finally, the organization of the PLO—centered on small teams of fighters armed with machine-guns and antitank weapons, and trained in insurgent, hit-and-run tactics—was ideally suited to take maximum advantage of the urban environment.

# ROLE OF CIVILIANS

A-6. Various ethnic and religious groups make up the civil population of southern Lebanon. However, West Beirut's population was heavily Palestinian and Lebanese. The civil population of West Beirut was between 350,000 and 500,000. The Palestinian population supported the PLO. The Lebanese population may be described as friendly neutral to the Israelis. Although unhappy under Palestinian dominance, this population was unwilling to actively support Israel. The civilian population was a logistic constraint on the PLO, which would have become significant had the siege lasted longer. The civilians in West Beirut were an even larger constraint on the Israelis. The presence of civilians significantly limited the ability of the Israelis to employ firepower. However, the Palestine refugee camps located in West Beirut were both civilian centers and military bases. The Israeli constraints on artillery and other systems against these parts of the city were much less restrictive than in other parts of West Beirut where the population was mostly Lebanese and where fewer key military targets existed.

A-7. The PLO knew of the Israeli aversion to causing civil casualties and purposely located key military centers, troop concentrations, and logistics and weapons systems in and amongst the population—particularly the refugee Palestinian population in the southern part of West Beirut. Tactically, they used the civilians to hide their forces and infiltrate Israeli positions.

A-8. The friendly Palestinian population provided intelligence to the PLO while the friendly Lebanese population provided intelligence for the IDF. Throughout the siege, the IDF maintained a policy of free passage out of Beirut for all civilians. This policy was strictly enforced and permitted no weapons to leave the city. Some estimates are that as many as 100,000 refugees took advantage of this policy.

# INFORMATION OPERATIONS

A-9. The siege of Beirut involved using information operations (IO) to influence the media. PLO information operations were aimed at controlling the media and hence the international perception of the operation. This was done by carefully cultivating a select group of pro-PLO media years before hostilities even began. Once hostilities started, only these media sources were permitted to report from the besieged portions of the city, and they were only shown activities that portrayed the IDF negatively. The IDF did not vigorously counter the PLO plan. In fact, the IDF contributed to it by limiting media access to their activities. The PLO information operations had a successful impact. The international community was constantly pressuring the Israeli government to end hostilities. This put pressure on the IDF to conduct operations rapidly and to limit firepower and casualties.

# CONDUCT OF THE URBAN OPERATIONS

A-10. The siege of Beirut began 1 July (see Figure A-2). By 4 July, Israeli forces occupied East Beirut, the Green Line separating East and West Beirut, and dominating positions south of the airport. IDF naval forces also controlled the sea west and north of Beirut. On 3 and 4 July, IDF artillery and naval fire began a regular campaign of firing on military targets throughout West Beirut. On 4 July, the IDF cut power and water to the city.

A-11. From 5 to 13 July, the Israeli fires continued to pound PLO targets in West Beirut. The PLO gave one significant response, firing on an Israeli position south of the city and causing several casualties. On 7 July, reacting to international pressure, the IDF returned power and water to West Beirut's civil population. On 11 July, the IDF launched its first attack, probing the southern portion of the airport with an armored task force (see Figure A-3). The PLO repulsed this attack and destroyed several IDF armored vehicles.

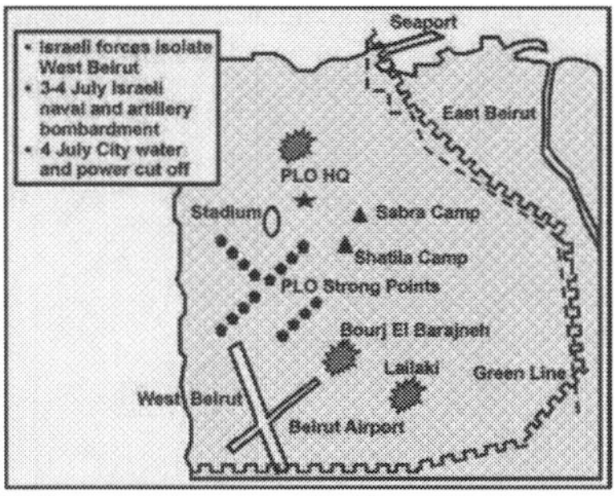


**Figure A-2. Initial conduct of the urban operation**

A-12. On 13 July, both sides entered into a cease-fire that lasted until 21 July. They began negotiations, mediated by international community, to end the siege. The PLO used this period to continue to fortify Beirut. The Israelis used the time to train their infantry and other arms in urban small unit tactics in Damour, a town the Israeli paratroopers had captured.

A-13. The cease-fire ended on 21 July as PLO forces launched three attacks on IDF rear areas. The Israelis responded with renewed and even more vigorous artillery, naval, and air bombardment of PLO positions in the city. The IDF attacks went on without respite until 30 July. On 28 July, the IDF renewed its ground attack in the south around the airport (see Figure A-4). This time IDF forces methodically advanced and captured a few hundred meters of ground establishing a toehold.

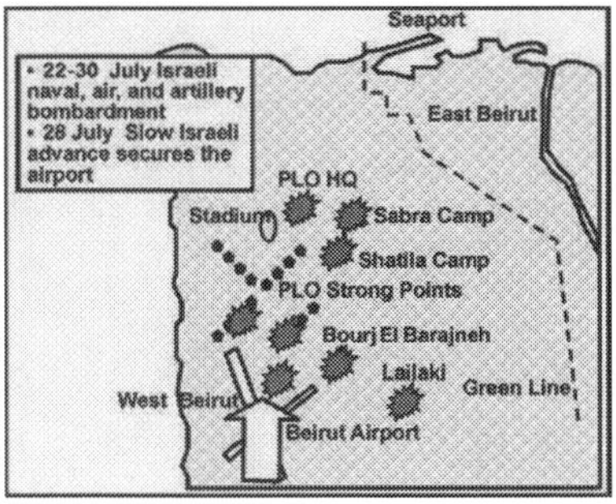


**Figure A-3. Israeli probe of PLO defenses**

A-14. The Israeli bombardment stopped on 31 July. However, on 1 August the IDF launched its first major ground attack, successfully seizing Beirut airport in the south (see Figure A-5). Israeli armored forces began massing on 2 August along the green line, simultaneously continuing the attack from the south to the outskirts of the Palestinian positions at Ouzai. On 3 August, the Israeli forces continued to reinforce both their southern attack forces and forces along the green line to prepare for continuing offensive operations. On 4 August, the IDF attacked at four different places. This was the much-anticipated major Israeli offensive.

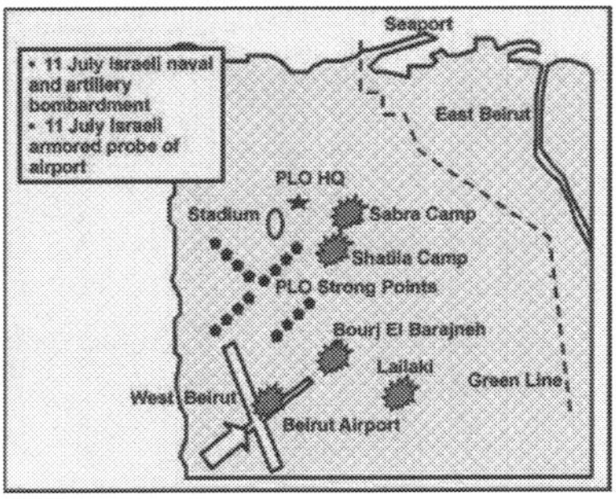


**Figure A-4. Initial Israeli attack**

A-15. The Israeli attack successfully disrupted the coherence of the PLO defense. The southern attack was the most successful: it pushed PLO forces back to their camps of Sabra and Shatila and threatened to overrun PLO headquarters. Along the green line the IDF attacked across three crossing points. All three attacks made modest gains against stiff resistance. For this day's offensive, the Israelis suffered 19 killed and 84 wounded, the highest single day total of the siege, bringing the total to 318 killed. Following the major attacks on 4 August, Israeli forces paused and, for four days, consolidated their gains and prepared to renew the offensive. Skirmishes and sniping continued, but without significant offensive action. On 9 August, the IDF renewed air and artillery attacks for four days. This activity culminated on 12 August with a massive aerial attack that killed over a hundred and wounded over 400—mostly civilians. A cease-fire started the next day and lasted until the PLO evacuated Beirut on 22 August.

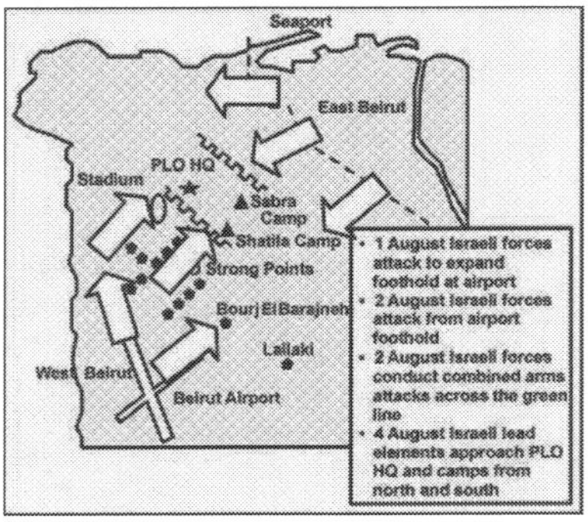


**Figure A-5. Final Israeli attack**

# LESSONS

A-16. The Israeli siege of West Beirut was both a military and a political victory. However, the issue was in doubt until the last week of the siege. Military victory was never in question; the issue in doubt was whether the Israeli government could sustain military operations politically in the face of international and domestic opposition. On the other side, the PLO faced whether they could last militarily until a favorable political end could be negotiated. The answer was that the PLO's military situation became untenable before the Israeli political situation did.

A-17. This favorable military and political outcome stemmed from the careful balance of applying military force with political negotiation. The Israelis also balanced the type of tactics they employed against the domestic aversion to major friendly casualties and international concern with collateral damage.

## PERFORM AGGRESSIVE INFORMATION OPERATIONS

A-18. The PLO devoted considerable resources and much planning on how to use IO to their best advantage. They chose to focus on media information sources as a means of influencing international and domestic opinion.

A-19. The PLO's carefully orchestrated misinformation and control of the media manipulated international sentiment. The major goal of this effort was to grossly exaggerate the claims of civilian casualties, damage, and number of refugees—and this was successfully accomplished. Actual casualties among the civilians were likely half of what the press reported during the battle. The failure of the IDF to present a believable and accurate account of operations to balance PLO efforts put tremendous pressure on the Israeli government to break off the siege. It was the PLO's primary hope for political victory.

A-20. In contrast to the overall weak performance in IO, the IDF excelled in psychological operations. IDF psychological operations attacked the morale of the PLO fighter and the Palestinian population. They were designed to wear down the will of the PLO to fight while convincing the PLO that the IDF would go to any extreme to win. Thus defeat was inevitable. The IDF used passive measures, such as leaflet drops and loudspeaker broadcasts. They used naval bombardment to emphasize the totality of the isolation of Beirut. To maintain high levels of stress, to deny sleep, and to emphasize their combat power, the IDF used constant naval, air, and artillery bombardment. They even employed sonic booms from low-flying aircraft to emphasize the IDF's dominance. These efforts helped to convince the PLO that the only alternative to negotiation on Israeli terms was complete destruction.

## Maintain a Close Combat Capability

A-21. The ground combat during the siege of Beirut demonstrated that the lessons of tactical ground combat learned earlier in the twentieth century were still valid. Small combined arms teams built around infantry, but including armor and engineers, were the key to successful tactical combat. Artillery firing in direct fire support of infantry worked effectively as did the Vulcan air defense system. The Israeli tactical plan was sound. The Israelis attacked from multiple directions, segmented West Beirut into pieces, and then destroyed each individually. The plan's success strongly influenced the PLO willingness to negotiate. Tactical patience based on steady though slow progress toward decisive points limited both friendly and noncombatant casualties. In this case, the decisive points were PLO camps, strong points, and the PLO headquarters.

A-22. The Israeli willingness to execute close combat demonstrated throughout the siege, but especially in the attacks of 4 August, was decisive. Decisive ground combat was used sparingly, was successful and aimed at decisive points, and was timed carefully to impact on achieving the political objectives in negotiations. The PLO had hoped that their elaborate defensive preparations would have made Israeli assaults so costly as to convince the Israelis not to attack. That the Israelis could successfully attack the urban area convinced the PLO leadership that destruction of their forces was inevitable. For this reason they negotiated a cease-fire and a withdrawal on Israeli terms.

## Avoid the Attrition Approach

A-23. The Israelis carefully focused their attacks on objects that were decisive and would have the greatest impact on the PLO: the known PLO headquarters and refugee centers. Other areas of West Beirut were essentially ignored. For example, the significant Syrian forces in West Beirut were not the focus of Israeli attention even though they had significant combat power. Selectively ignoring portions of the urban area allowed the Israelis to focus their combat power on the PLO and limit both friendly casualties and collateral damage.

## Control the Essential and Preserve Critical Infrastructure

A-24. The Israeli siege assured Israeli control of the essential infrastructure of Beirut. The initial Israeli actions secured East Beirut and the city's water, power, and food supplies. The Israelis also dominated Beirut's international airport, closed all the sea access, and controlled all routes into and out of the city. They controlled and preserved all that was critical to operating the city and this put them in a commanding position when negotiating with the PLO.

## Minimize Collateral Damage

A-25. The Israeli army took extraordinary steps to limit collateral damage, preserve critical infrastructure, and put in place stringent rules of engagement (ROE). They avoided randomly using grenades in house clearing, limited the use of massed artillery fires, and maximized the use of precision weapons. With this effort, the Israelis extensively used Maverick missiles because of their precise laser guidance and small warheads.

A-26. The strict ROE, however, conflicted with operational guidance that mandated that Israeli commanders minimize their own casualties and adhere to a rapid timetable. The nature of the environment made fighting slow. The concern for civilian casualties and damage to infrastructure declined as IDF casualties rose. They began to bring more field artillery to bear on Palestinian strong points and increasingly employed close air support. This tension underscores the delicate balance that Army commanders will face between minimizing collateral damage and protecting infrastructure while accomplishing the military objective with the least expenditure of resources—particularly soldiers. ROE is but one tool among many that a commander may employ to adhere to this UO fundamental.

## Understand the Human Dimension

A-27. The Israelis had a noteworthy (although imperfect and at times flawed) ability to understand the human dimension during their operations against the PLO in Beirut. This was the result of two circumstances. First, the PLO was a threat with which the Israeli forces were familiar after literally decades of conflict. Second, through a close alliance and cooperation with Lebanese militia, the Israelis understood a great deal regarding the attitudes and disposition of the civil population both within and outside Beirut.

## Separate Noncombatants from Combatants

A-28. Separating combatants from noncombatants was a difficult but important aspect of the Beirut operation. The Israelis made every effort to positively identify the military nature of all targets. They also operated a free passage system that permitted the passage of all civilians out of the city through Israeli lines. The need to impose cease-fires and open lanes for civilians to escape the fighting slowed IDF operations considerably. Additionally, Israeli assumptions that civilians in urban combat zones would abandon areas where fighting was taking place were incorrect. In many cases, civilians would try to stay in their homes, leaving only after the battle had begun. In contrast, the PLO tied their military operations closely to the civilian community to make targeting difficult. They also abstained from donning uniforms to make individual targeting difficult.

A-29. Earlier in OPERATION PEACE FOR GALILEE when the IDF attacked PLO forces located in Tyre, Israeli psychological operations convinced 30,000 Lebanese noncombatants to abandon their homes and move to beach locations outside the city. However, the IDF was subsequently unable to provide food, water, clothing, shelter, and sanitation for these displaced civilians. IDF commanders compounded the situation by interfering with the efforts by outside relief agencies to aid the displaced population (for fear that the PLO would somehow benefit). Predictably, many civilians tried to return to the city complicating IDF maneuver and targeting—that which the separation was designed to avoid. IDF commanders learned that, while separation is important, they must also adequately plan and prepare for the subsequent control, health, and welfare of the noncombatants they displace.

## Restore Essential Services

A-30. Since essential services were under Israeli command, and had been since the beginning of the siege, the Israelis had the ability to easily restore these resources to West Beirut as soon as they adopted the cease-fire.

## Transition Control

A-31. In the rear areas of the Israeli siege positions, the Israeli army immediately handed over civic and police responsibility to civil authorities. This policy of rapid transition to civil control within Israeli lines elevated the requirement for the Israeli army to act as an army of occupation. The Israeli army believed the efficient administration of local government and police and the resulting good will of the population more than compensated for the slightly increased force protection issues and the increased risk of PLO infiltration.

A-32. Upon the cease-fire agreement, Israeli forces withdrew to predetermined positions. International forces under United Nations (UN) control supervised the evacuation of the PLO and Syrian forces from Beirut. These actions were executed according to a meticulous plan developed by the Israeli negotiators and agreed to by the PLO. Israeli forces did not take over and occupy Beirut as a result of the 1982 siege (an occupation did occur later but as a result of changing situations).

# SUMMARY

A-33. The Israeli siege of West Beirut demonstrates many of the most demanding challenges of urban combat. Apart from Israel's poor understanding of the strategic influence of the media on operations (a lesson that they learned and subsequently applied to future operations), the IDF's successful siege of Beirut emerged from their clear understanding of national strategic objectives and close coordination of diplomatic efforts with urban military operations. A key part of that synchronization of capabilities was the understanding that the efforts of the IDF would be enhanced if they left an escape option open to the PLO. This option was the PLO's supervised evacuation that occurred after the siege. Although the PLO was not physically destroyed, the evacuation without arms and to different host countries effectively shattered the PLO's military capability. Had Israel insisted on the physical destruction of the PLO in Beirut, the overall operation might have failed because that goal may not have been politically obtainable in view of the costs in casualties, collateral damage, and international opinion.

This page intentionally left blank.

# Appendix B

# An Urban Focus to the Intelligence Preparation of the Battlefield

*Maneuvers that are possible and dispositions that are essential are indelibly written on the ground. Badly off, indeed, is the leader who is unable to read this writing. His lot must inevitably be one of blunder, defeat, and disaster.*

*Infantry in Battle*

The complexity of the urban environment and increased number of variables (and their infinite combinations) increases the difficulty of providing timely, relevant, and effective intelligence support to urban operations (UO). Conducted effectively, however, the intelligence preparation of the battlefield (IPB) allows commanders to develop the situational understanding necessary to visualize, describe, and direct subordinates in successfully accomplishing the mission.

IPB is the systematic process of analyzing the threat and environment in a specific geographic area—the area of operations (AO) and its associated area of interest (see figure B-1). It provides the basis for intelligence support to current and future UO, drives the military decision-making process, and supports targeting and battle damage assessment. The procedure (as well as each of its four steps) is performed continuously throughout the planning, preparation, and execution of an urban operation.

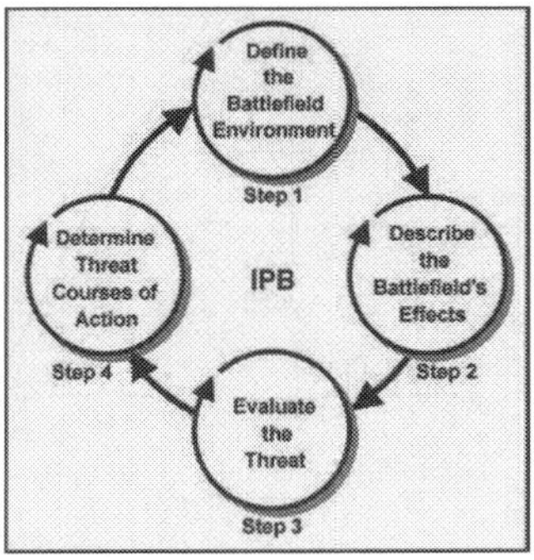


**Figure B-1. The steps of IPB**

## UNAFFECTED PROCESS

B-1. The IPB process is useful at all echelons and remains constant regardless of the operation or environment. However, an urban focus to IPB stresses some aspects not normally emphasized for IPBs conducted for operations elsewhere. The complex urban mosaic is composed of the societal, cultural, or civil dimension of the urban environment; the overlapping and interdependent nature of the urban infrastructure; and the multidimensional terrain. This mosaic challenges the conduct of an urban-focused IPB. There is potential for full spectrum operations to be executed near-simultaneously as part of a single major operation occurring in one urban area with multiple transitions. Such multiplicity stresses the importance of a thorough, non-stop IPB cycle aggressively led by the commander and executed by the entire staff. Overall, the art of applying IPB to UO is in properly employing the steps to the specific operational environment. In UO, this translates to analyzing the significant characteristics of the environment, the role that its populace has in threat evaluation, and understanding how these affect the planning and execution of UO. FM 2-01.3 details how to conduct IPB; FM 34-3 has the processes and procedures for producing all-source intelligence. This appendix supplements the information found there; it does not replace it.

## Increased Complexity

B-2. Uncovering intricate relationships takes time, careful analysis, and constant refinement to determine actual effects on friendly and threat courses of action (COAs). These relationships exist among—

- Urban population groups.
- The technical aspects of the infrastructure.
- The historical, cultural, political, or economic significance of the urban area in relation to surrounding urban and rural areas or the nation as a whole.
- The physical effects of the natural and man-made terrain.

B-3. A primary goal of any IPB is to accurately predict the threat's likely COA (step four—which may include political, social, religious, informational, economic, and military actions). Commanders can then develop their own COAs that maximize and apply combat power at decisive points. Understanding the decisive points in the urban operation allows commanders to select objectives that are clearly defined, decisive, and attainable.

### Reducing Uncertainty and Its Effects

B-4. Commanders and their staffs may be unfamiliar with the intricacies of the urban environment and more adept at thinking and planning in other environments. Therefore, without detailed situational understanding, commanders may assign missions that their subordinate forces may not be able to achieve. As importantly, commanders and their staffs may miss critical opportunities because they appear overwhelming or impossible (and concede the initiative to the threat). They also may fail to anticipate potential threat COAs afforded by the distinctive urban environment. Commanders may fail to recognize that the least likely threat COA may be the one adopted precisely because it is least likely and, therefore, may be intended to maximize surprise. Misunderstanding the urban environment's effect on potential friendly and threat COAs may rapidly lead to mission failure and the unnecessary loss of Soldiers' lives and other resources. A thorough IPB of the urban environment can greatly reduce uncertainty and contribute to mission success.

### Training, Experience, and Functional Area Expertise

B-5. Not all information about the urban environment is relevant to the situation and mission—hence the difficulty and the reason for IPB and intelligence analysis. Although it may appear daunting, institutional education, unit training, and experience at conducting intelligence support to UO will improve the ability to rapidly sort through all the potential information to separate the relevant from merely informative. (This applies to any new or difficult task.) The involvement and functional expertise of the entire staff will allow commanders to quickly identify the important elements of the environment affecting their operations. Fortunately, IPB as part of the entire intelligence process (plan, prepare, collect, process, and produce) is comprehensive enough to manage the seemingly overwhelming amounts of information coming from many sources. Accomplished properly, it allows commanders to recognize opportunities often without complete information.

B-6. As in any operational environment, tension exists between the desire to be methodical and the need to create the tempo necessary to seize, retain, and exploit the initiative necessary for decisive UO. Quickly defining the significant characteristics of the urban environment requiring in-depth evaluation (not only what we need to know but what is possible to know) allows rapid identification of intelligence gaps (what we know versus what we don't know). Such identification leads to information requirements and will drive the intelligence, surveillance, and reconnaissance (ISR) plan (how will we get the information we need). Commanders must carefully consider how to develop focused priority information requirements (PIR) to enable collectors to more easily weed relevant information from the plethora of information. Commanders can then make better decisions and implement them faster than urban threats can react.

## AMPLIFIED IMPORTANCE OF CIVIL (SOCIETAL) CONSIDERATIONS

B-7. The Army focuses on warfighting. The experiences in urban operations gained at lower echelons often center on the tactics of urban offensive and defensive operations where the influences of terrain and enemy frequently dominate. At higher echelons, the terrain and enemy are still essential considerations, but the societal component of the urban environment is considered more closely and throughout the operational process. Moreover, the human or civil considerations gain importance in civil support or stability operations regardless of the echelon or level of command. In addition to the echelon and the type of operation, a similar relationship exists between the key elements of the urban environment and other situational factors. These factors can include where the operation lies within the range of operations or the level of war and the conventional or unconventional nature of the opposing threat. Figure B-2 graphically represents the varying significance of these elements to urban IPB. Overall, population effects are significant in how they impact the threat, Army forces, and overall accomplishment of strategic and operational goals.

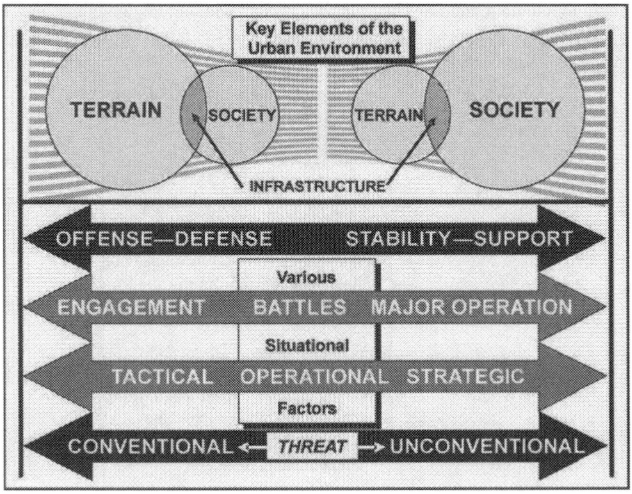


**Figure B-2. Relativity of key urban environment elements**

B-8. Describing the battlefield's effects—step two in IPB—ascribes meaning to the characteristics analyzed. It helps commanders understand how the environment enhances or degrades friendly and threat forces and capabilities. It also helps commanders understand how the environment supports the population. It also explains how changes in the "normal" urban environment (intentional or unintentional and because of threat or friendly activities) may affect the population. Included in this assessment are matters of perception. At each step of the IPB process, commanders must try to determine the urban society's perceptions of ongoing activities to ensure Army operations are viewed as intended. Throughout this process, commanders, staffs, and analysts cannot allow their biases—cultural, organizational, personal, or cognitive—to markedly influence or alter their assessment (see FM 34-3). This particularly applies when they analyze the societal aspect of the urban environment. With so many potential groups and varied interests in such a limited area, misperception is always a risk.

# SIGNIFICANT CHARACTERISTICS

B-9. Urban intelligence analysis must include consideration of the urban environment's distinguishing attributes—man-made terrain, society, and infrastructure—as well as the underlying natural terrain (to include weather) and the threat. Because the urban environment is so complex, it is often useful to break it into categories. Then commanders can understand the intricacies of the environment that may affect their operations and assimilate this information into clear mental images. Commanders can then synthesize these images of the environment with the current status of friendly and threat forces and develop a desired end state. Then they can determine the most decisive sequence of activities that will move their forces from the current state to the end state. Identifying and understanding the environment's characteristics (from a friendly, threat, and noncombatant perspective) allows commanders to establish and maintain situational understanding. Then they can develop appropriate COAs and rules of engagement that will lead to decisive mission accomplishment.

B-10. Figures B-3, B-4, and B-5 are not intended to be all-encompassing lists of urban characteristics. Instead, they provide a starting point or outline useful for conducting an urban-focused IPB and analysis that can be modified to fit the specific operational environment and meet the commander's requirements. Commanders and staffs can compare the categories presented with those in the civil affairs area study and assessment format found in FM 41-10 and the IPB considerations for stability and civil support operations found in ST 2-91.1.

## INTERCONNECTED SYSTEMS

B-11. Since the urban environment comprises an interconnected "system of systems," considerations among the key elements of the environment will overlap during an urban intelligence analysis. For example, boundaries, regions, or areas relate to a physical location on the ground. Hence, they have urban terrain implications. These boundaries, regions, or areas often stem from some historical, religious, political, administrative, or social aspect that could also be considered a characteristic of the urban society. Overlaps can also occur in a specific category, such as infrastructure. For instance, dams are a consideration for their potential effects on transportation and distribution (mobility), administration and human services (water supply), and energy (hydroelectric).

B-12. This overlap recognition is a critical concern for commanders and their staffs. In "taking apart" the urban environment and analyzing the pieces, commanders and staffs cannot lose perspective of how each piece interacts with any other and as part of the whole. Otherwise, their vision will be shortsighted, and they will fail to recognize the second- and third-order effects of their proposed COAs; the actual end state differing dramatically from the one envisioned by the commander. The increased density of combatants and noncombatants, infrastructure, and complex terrain means that a given action will likely have unintended consequences—positive or negative. Those consequences will be more widely felt and their impact will spread in less time than in other environments. These unintended results may have important strategic and operational consequences. The multiple ways these dynamic urban elements and characteristics combine make it necessary to approach each urban environment as a unique challenge for intelligence analysis.

## URBAN TERRAIN AND WEATHER

### Terrain

B-13. Earlier admonitions that civil considerations are more closely considered in UO do not necessarily mean that consideration for urban terrain is de-emphasized. In every urban operation, terrain and its effects on both threat and friendly forces must be assessed and understood. Then commanders can quickly choose and exploit the terrain (and weather conditions) that best supports their missions. Effective terrain analysis thoroughly assesses urban structures as well as the ground on which they stand (see figure B-3 and FM 5-33). An analysis of urban terrain first considers broader urban characteristics and effects and then progresses to a more detailed examination.

### *Natural Terrain*

B-14. The natural terrain features beyond the urban area and beneath urban structures significantly influence unit operations. They dictate where buildings can be constructed, the slope and pattern of streets, and even the broad urban patterns that develop over longer periods of time, thereby influencing a unit's scheme of maneuver. The military aspects of terrain—observation and fields of fire, avenues of approach, key terrain, obstacles, and cover and concealment (OAKOC)—remain critical to the analysis of natural terrain in, under, and around urban areas. Fortunately, commanders and their staffs are normally accustomed to this type of analysis.

**Forms and Functions**

- Cores
- Industrial areas
  - Toxic industrial material production and storage facilities
  - Standard signs and markings for toxic chemicals
- Outlying high-rise areas
- Residential areas and shantytowns
- Commercial ribbon areas
- Forts and military bases

**Broad Urban Patterns**

- Types
  - Satellite
  - Network
  - Linear
  - Segment
- Dominant or central hub (if any)
- Area covered (square miles)

**Street Patterns**

- Basic Types
  - Radial
  - Grid
  - Irregular (planned and unplanned)
- Variations
  - Rayed
  - Radial-Ring
  - Contour-forming
  - Combined
- Widths

**Construction and Placement**

- Construction
  - Mass or framed
  - Light or heavy clad
  - Material (dirt, wood, stone, brick, cinder block, concrete, steel, and glass)
  - Density and thickness (roofs, floors, and interior and exterior walls)
  - Load bearing walls and columns
  - Height (floors)
  - Doors, windows, fire escapes, and other openings
  - Interior floor plan (including crawl spaces, elevators, and stairs)
- Placement
  - Random
  - Close-orderly block
  - Dispersed
- Ownership

**Military Aspects of Terrain: OAKOC**

- Observation and fields of fire
  - Smoke (fire), dust (explosions), and flying debris
  - Rubble
  - Engagement ranges (including minimum safe distances and backblast factors) and obliquity/angle (ricochets)
  - Elevation and depression considerations
  - Lasers and reflective concerns
- Avenues of approach (mobility corridors)
  - Airspace
  - Surface
  - Supersurface
  - Subsurface
- Key terrain
  - Landmarks
  - Buildings of significant cultural, social, political, historical, or economic significance
- Obstacles
  - Rubble and vehicles
  - Steep embankments
  - Medians
  - Inadequate bridges and overpasses (destroyed, weight-restricted, or narrow)
  - Tunnels and underpasses (destroyed or narrow)
  - Mines and roadside improvised explosive devices
  - Masking of fires
  - Burning buildings or other fire hazards
  - Rivers and lakes
- Cover and concealment
  - Building protection
  - Weapon penetration (single shot and multiple rounds) considerations
  - Rubble and vehicles

**Figure B-3. Significant urban terrain characteristics**

## Analysis of an Urban Area's Underlying Terrain: Mitrovica, Kosovo – 1999

An urban area's underlying terrain provides many clues into its history, economy, society, and current situation. Mitrovica, Kosovo is an illustrative example. The Ibar River creates a natural line of communications through the middle of the city as well as an obstacle that bisects the urban area. This bisection naturally divides the two resident ethnic groups: Albanians and Serbs. The separation became significant at both the strategic and tactical levels during 1999 deployments to Kosovo. Army forces had to ensure that the Orthodox Church located south of the Ibar was accessible to Serbs residing in the north. North Atlantic Treaty Organization (NATO)

> peacekeepers built a footbridge across the river that allowed reliable, safe passage. The natural feature separating the two groups assisted NATO troops in maintaining stability in the region.

### *Man-Made Terrain*

B-15. Building composition, frontages, placement, forms and functions, size, floor plans, and window and door locations affect maneuver, force positioning, and weapons deployment considerations. Angles, displacement, surface reflection, and antenna locations influence command and control. Structures also influence ISR operations. The increased density and volume created by man-made structures increases how much information commanders and their staffs collect and assess as well as the number of forces required. Building materials and construction will also influence force structures to include weapons and equipment required. Heavily constructed buildings combined with hot and cold extremes may affect target identification for thermal sights. Thick walls, for example, may make combat vehicle identification difficult by distorting hotspots. Additionally, the increased use of heaters and warming fires may clutter thermal sights with numerous hotspots. The ability to maneuver through the urban dimensions—airspace, supersurface, surface, and subsurface—and shoot through walls, ceilings, and floors also creates increased psychological stress. The physical characteristics of man-made terrain can also be analyzed using OAKOC.

## Weather

B-16. Weather and its effects are often considered when examining the military aspects of terrain. Military aspects of weather include temperature (heat and cold), light conditions, precipitation (cloud cover, rain, snow, fog, and smog), and wind. Their military effects during UO are similar to any operational environment (see FM 34-81 and FM 34-81-1). Extremes of heat and cold affect weapon systems and the Soldiers that operate them. The extra luminescence provided by the ambient light of an urban area, unless controlled, may affect night vision capabilities and the ability of the Army to "own the night." Precipitation affects mobility and visibility. Smog inversion layers are common over cities. An inversion layer may trap smoke and chemicals in the air and increase ground and air temperatures to the detriment of Soldiers' health. (If the conditions are severe enough, it might require the use of protective masks.) Changes in temperature as a result of air inversions can also affect thermal sights during crossover periods of warm to cold and vice versa. Winds, which may increase as they are funneled through urban canyons, may—

- Increase other weather effects (for example, wind chill).
- Decrease visibility (blowing debris, sand, rain, and snow).
- Spread radiation, biological, and chemical hazards.
- Adversely affect low-altitude air mobility and airborne ISR platforms.

B-17. Commanders also analyze weather for its potential effect on civilians and civilian infrastructure as well as Soldiers and military equipment. Rain might create sewage overflow problems in urban areas with collapsed sewage infrastructure and in refugee camps, increasing disease and even creating panic. Rain and flooding may also make some subsurface areas impassable or extremely hazardous to civilian and military forces alike. Chemical agents and TIM may be washed into underground systems creating toxic hotspots. These effects become more pronounced as chemical agents or TIM are absorbed by brick and unsealed concrete sewers. Other weather effects on UO can include—

- Heavy snowfall in an urban area that may paralyze area transportation and distribution infrastructure, hindering the urban administration's ability to provide vital human services (police, fire fighting, medical, and rescue). Heavy rains and flooding may have similar effects especially on poorly designed and constructed roads or roads that have been damaged by tracked vehicles.
- Extreme hot and cold weather climates that increase the dependence (and military significance) of many elements of the infrastructure. For example, the energy infrastructure may be critical; without it, civilians may not be adequately cooled or heated or they may not be able to cook their food.

- In urban areas located in tropical regions, it can rain at the same time each day during the wet season. Threat forces may attack during these periods knowing aircraft will have difficulty responding. Bad weather also reduces the effectiveness of surveillance, direct and indirect fire, and logistic support.
- Inclement weather may preclude demonstrations or rallies by threats. Good weather may mean a maximum turnout of civilians for events such as festivals, sporting events, and other social, cultural, or religious activities.
- Severe weather may affect psychological, civil-military and humanitarian assistance operations. Heavy rains and severe dust storms may disrupt leaflet drops, construction projects, food and water distribution, and medical and veterinary assistance programs.

## URBAN SOCIETY

B-18. This manual shows that societal considerations take on added importance in UO. Critical to operational success is knowing and understanding which groups live in an urban area, what relationships exist among them, and how each population group will respond to friendly and threat activities. Often determining any of this is very difficult. Cultural acuity is also essential in helping commanders and their staffs to view the urban area as the residents view it. The demographics presented depict what conditions exist, while the other categories help to explain the root causes or why conditions exist (see figure B-4 on page B-8). Other categories besides basic demographics that are important to gain this understanding include health, history, leadership, ethnicity and culture, religion, and government and politics.

### From Cultural Awareness to Understanding and Competence

B-19. This societal understanding must go beyond a superficial awareness of gestures and taboos to developing a thorough understanding of the organizing principles that make the urban society hold together and function as a coherent entity—or not. In UO, particularly longer-term stability operations, commanders use this knowledge and insight to design and plan operations that work effectively within that society. For example in OPERATION IRAQI FREEDOM, Army units, as they shifted to long-term stability operations, were required to recruit, organize, equip, and train indigenous security forces in order to be able to transition security responsibility from U.S. forces to the burgeoning Iraqi government. Army commanders were successful—to a point. The first units that they built and trained attained a remarkable level of unity and tactical ability in a relatively short period of time. The newly formed units effectively fought alongside Army forces. However when prodded into conducting operations with minimal Army supervision, many Iraqi units lost unity and coherence. Army commanders initially failed to account for a basic Iraqi societal consideration: their society is patriarchal—each member or organization in Iraqi society always seeks a patron or sponsor to support them and provide for their needs. When Iraqi units trained and operated with Army forces, this relationship was satisfied. Operating on their own, however, Iraqi commanders and units lost this patronage without an equivalent Iraqi replacement. In hindsight, the establishment of an effective and more robust ministry of defense, either as a precursor or simultaneously with the establishment of tactical units, may have provided the Iraqi security forces the patronage that their culture required.

### Resource the Societal Analysis

B-20. Aside from friendly and threat forces, the society is the only thinking component of the urban environment able to rapidly impact the urban operation. (Even people going about their daily routines can unwittingly hamper the mission.) Urban residents create conditions for restrictive rules of engagement, increase stress on Soldiers and logistic capabilities, and confuse threat identification (see Threat Considerations in this appendix). Demographic, health, safety, ethnic, and cultural concerns will be essential considerations in most UO. Other situational factors—the mission, enemy, and time available—dictate the balance between the level of detail and analysis to support the overall urban operation with the level of detail that commanders and their staffs can achieve. However, an analysis that fails to devote enough time and resources to understanding the urban society can find large elements of the population turned against the Army force. Detailed analysis of the urban society will help save time and ISR resources. While not necessarily a resource saver in the short term, a thorough understanding of the urban society and its culture ultimately contributes to faster achievement of strategic goals.

**Population Demographics**

- General population size
  - Village
  - Town
  - City
  - Metropolis
  - Megalopolis
- Group size based on race age, sex, political affiliation, economics, religion, tribe, clan, gang, criminal activities, or other significant grouping
  - Significant US or allied populations
  - Distribution, densities, and physical boundaries and overlaps
  - Majority, minority, and dominant groups
- Increasing or decreasing migration trends
  - Dislocated civilians
- Nongovernmental organizations (NGOs)
  - Local
  - National
  - International
- Languages (distribution, dialects, relationship to social structure)
- Educational levels and literacy rates
- Crime rates
- Birth and death rates
- Labor statistics and considerations
  - Skilled and unskilled
  - Imported and exported
  - Unemployment
  - Standard wages and per capita income
  - Workday and workweek norms

**Health**

- Diseases
- Nutritional deficiencies
- Local standards of care
- Pollution and environmental hazards (air, water, food, and soil)
- Health workers (types, numbers, and degree of skill)

**History**

- General and for a specific group
  - Internal or external
  - Recent conflicts
- Relationship with US, allies, and other participating multinational forces
- Applicable international treaties
- Status-of-forces agreements
- Antagonists/protagonists
- Heroes
- Events, facts, and dates considered important or celebrated
- Urban area's historical importance

**Leadership and Prominent Personalities**

- Identification, location, and prioritization of influential leaders (exploitation, evacuation, protection, etc.)
- Affiliation (ethnic, religion, military, government, industry, criminal, or entertainment)
- Education attained
- Organization and distribution of power
- Associations among different leaders and groups

**Ethnicity and Culture**

- Values, moral codes, taboos, and insults (verbal and non-verbal)
- Attitudes towards age, sex, and race (including same-sex interaction)
- Role of the clan, tribe, or family
- Biases between ethnic groups
- Privacy and individuality
- Recreation, entertainment, and humor
- Fatalism or self-determination
- Exchanges of gifts
- Displays of emotion
- Lines of authority
- Dating and marriage
- Greetings, leave-takings, and gestures
- Visiting practices
- Alcohol and drug use
- Important holidays, festivals, sporting, or entertainment events
- Eating and dietary practices
- Significance of animals and pets
- Urban-rural similarities and differences
- Driving habits
- Clothing

**Religion**

- Sects, divisions, and overlaps
- Religious biases and problems
- Relationship and influence on government, politics, economics, and education
- Impact on ethnic and cultural beliefs
- Key events or celebrations (daily, weekly, monthly, or annually)
- Funeral and burial practices

**Government and Politics**

- Present and past forms
- Organization and powers (executive, legislative, judicial, and administrative divisions)
- Scheduled elections and historical turnouts
- Degree of control over the population
  - Identification required
  - Border-crossing procedures
- Relationship with US or multinational governments, national governments, and criminal elements
- Political factions and boundaries
- Political traditions
- Grievances
- Censorship
- Nepotism and other clan, tribal, or social ties
- Civil defense and disaster preparedness (organization, plans, training, equipment, and resources)
  - Evacuation routes
- Legal system
  - System of laws
  - Applicable treaties
  - Courts and tribunals
  - Procedures
  - Records (birth and deeds)
- Property control
- Monetary system (formal and informal)
- Domestic and foreign trade
  - Taxation and tariffs
  - Customs requirements
  - Rationing and price controls
  - Economic performance and contribution to gross national product
- Economic aid
- Perception of relative deprivation
- Trade unions
- Competition with the black market and organized crime

**Figure B-4. Significant urban societal characteristics**

**Transportation and Distribution**

- Water
  - Shipyards and other port and harbor facilities
  - Inland waterways, canals, and locks
  - Offshore pipeline berths
  - Cargo storage and handling
  - Types and number of ships, boats, and ferries
  - Dams
- Streets and roads
  - Bridges and fords
  - Over- and underpasses
  - Raised embankments, tunnels, culverts, and other subterranean features (widths and clearances)
  - Parking areas (surface, subsurface, and supersurface)
  - Weight restrictions
  - Traffic light operations
  - Traffic patterns
  - Widths
  - Surface materials
- Rail
  - Lines
  - Terminals
  - Switchyards and junctions
  - Subways, bridges, elevated rail lines, and underpasses (clearances)
  - Track gauges
  - Types and number of rolling stock
  - Electrification
- Air
  - Airfields and runways (including capabilities)
  - Heliports and helipads (including rooftop)
  - Types and number of aircraft
  - Cargo storage and handling
- Trucking companies and delivery services
- Available material-handling equipment
- Rush hour and market time considerations
- Seasonal (weather) effects
- Rubble effects
- Impact of dislocated civilians and migration patterns
- Likely population congregation points
- Identifiable primary and alternate lines of communication

**Economics and Commerce**

- Industries
  - Types and locations
  - Important companies (including US or allied)
  - Military production facilities
- Sources of raw materials
- Use of toxic industrial materials and biological agents
  - Agriculture (insecticides, herbicides, and fertilizers)
  - Manufacturing
  - Cleaning
  - Research
- Food types, quantities, and sources
  - Requirements and availability
  - Storage and processing
  - Cleanliness standards
- Stores, shops, restaurants, hotels, and strip malls
- Recreation facilities
  - Outdoor and amusement parks
  - Stadiums and other sports facilities
- Machine shops
- Brick and lumber yards
- Banking and investment institutions

**Administration and Human Services**

- Police and fire protection
  - Headquarters, station, and key facilities locations
  - Organization and strengths
  - Equipment
  - Functions, authority, and jurisdictional boundaries
  - Contract guard services
- Welfare and public assistance
  - Monetary assistance
  - Orphanages
  - Elderly care facilities
- Water supply systems
  - Water sources and storage (lakes, reservoirs, cisterns, pools, and public baths)
  - Water treatment and quality
  - Pumping stations and other distribution methods (trucks, bottles)
  - Hydrant locations
- Snow removal capabilities
- Street light operations
- Health facilities
  - Hospitals
  - Emergency medical services
  - Mental institutions
  - Medical supplies and equipment
  - Research and pharmaceutical buildings
  - Blood banks
- Governmental buildings
  - Embassies
  - Capitol building
  - Legislative, judicial, and ministry buildings
  - Hall of records
- NGOs providing assistance
- Effects of military control measures on providing vital human services
- Waste and sanitation
  - Types (solid, sewage, and toxic)
  - Collection, processing, and disposal
  - Dumps or landfills
  - Drainage systems

**Cultural**

- Religious buildings and places of worship
- Shrines, monuments and other historical structures
- Schools
- Museums
- Theaters
- Libraries

**Energy**

- Types
  - Electric
  - Oil
  - Coal
  - Natural gas
  - Nuclear
  - Solar
  - Hydroelectric
  - Geothermal
- Facilities
  - Production and processing
  - Storage
- Distribution
  - Pipelines (above and below ground)
  - Power lines (overhead and underground)
  - Water, rail, and road
  - Potential hazards

**Communication and Information**

- Print media
  - Newspapers, periodicals, and pamphlets
  - Billboards and posters
  - Postal facilities
- Telephone facilities
  - Wire or wireless
  - Facsimile machines
- Telegraph facilities
- Radio facilities
- Police, fire, and rescue systems
- Security systems
- Television facilities
- Computers and the Internet
- Antennas, towers, relay stations, and lines (surface and subsurface)
- Integration of space-based capabilities
- Public forums and speech
- Low-technology media (car horns, drums, graffiti, and burning fires)
- Key media organizations and reporters
  - Local
  - International
  - US

**Figure B-5. Significant urban infrastructure characteristics**

### URBAN INFRASTRUCTURE

B-21. Functional and analytical interdependence is readily apparent when examining urban infrastructures (see figure B-5 on page B-9). They are composed of physical structures or facilities and people. Hence, much of the analysis conducted for terrain and society can apply when assessing the urban infrastructure. For example, commanders, staffs, and analysts could not effectively assess the urban economic and commercial infrastructure without simultaneously considering labor. All aspects of the society relate and can be used to further analyze the urban work force since they are a sub-element of the urban society. Similarly, the OAKOC aspects used to evaluate terrain may also apply to the urban infrastructure, especially considerations of key terrain. Depending on the mission, elements of the urban infrastructure may become key terrain. While infrastructure analysis is not an engineer-specific task, Army engineers may provide the general and geospatial engineering expertise necessary to better understand the urban terrain and infrastructure.

## THREAT CONSIDERATIONS

> *... [T]he adversaries of freedom ... send arms, agitators, aid, technicians and propaganda to every troubled area.... [S]ubversives and saboteurs and insurrectionists . ... [possess] the power to conscript talent and manpower for any purpose. and long experience in the techniques of violence and subversion....It is a contest of will and purpose as well as force and violence—a battle for minds and souls as well as lives and territory.*
>
> John F. Kennedy
> Message to Congress, 5 May 1961

B-22. Chapter 3 outlines the instability and uncertainty of the strategic environment. Commanders, staffs, and analysts identify and analyze the threat's composition, strength, disposition, leadership, training, morale, weapons and capabilities, vulnerabilities, internal logistics and external support, doctrine (if any), strategy or modus operandi, and tactics. The threat can take a variety of forms:

- Conventional military forces.
- Paramilitary forces.
- Guerrillas and insurgents.
- Terrorists.
- Militia or special police organizations.

B-23. A general study of guerrilla and insurgent tactics, techniques, and procedures may prove beneficial to many types of operations regardless of the actual composition or type of threat forces. Insurgent strategies and tactics may work especially well in this complex environment and will likely be a part of any threat COA. Particularly, commanders should understand how a threat might restrict itself by the laws of land warfare and similar conventions, or exploit the use of these conventions to its own gain. Commanders can refer to FM 31-20-3 and FMI 3-07.22 for more information. For many of the above threats, no doctrinal templates may exist. Intelligence staffs will need to identify and track common threat patterns and tactics, techniques, and procedures as they develop and change during the course of the operation. Understanding culture will be critical to determining threat motives that drive decisions. Comman415081ders, staffs, and analysts must evaluate, update (or create), and manage threat databases early (and continuously) during the operational process.

### URBAN ENVIRONMENTAL THREATS

B-24. While threats vary, they share a common characteristic: the capability and intent to conduct violence against Army forces to negatively influence mission accomplishment. These threats are often the most recognizable for forces trained for warfighting—these are often the enemy. In order for Army units to be able to conduct full spectrum urban operations, commanders broaden their concept of the threat when analyzing the urban environment's terrain, societal, and infrastructure characteristics. This analysis includes many environmental dangers (potentially affecting both sides of a conflict as well as noncombatants) such as—

- Natural disasters (earthquakes, tornados, hurricanes, volcanic eruptions, fires, floods, tidal waves or tsunamis, and heavy snows).
- Food shortages.
- Water shortages (both quantity and quality).
- Rampant disease.
- Pollution and toxic industrial materials.

B-25. A critical difference between the latter forms of threat and the former is the lack of intent to do harm. The latter may stand alone as threats, or these conditions may be created, initiated, or used by the enemy or a hostile civilian group as a weapon or tool. Threat analysis includes identifying and describing how each relevant characteristic of the area of operations can hinder mission accomplishment. This analysis, particularly during stability or civil support operations, will require extensive coordination and cooperation with urban civil authorities, law enforcement, and numerous governmental and nongovernmental organizations.

## CIVILIANS

B-26. In a major combat operation or campaign where offensive and defensive operations are conducted against a conventional enemy, threat identification is more readily accomplished. However, adaptation may be necessary to even further broaden the threat concept to include specific elements of the urban society and, in some instances, nongovernmental organizations (NGOs) and other civilian agencies working in the urban area. In many stability operations, this modification can account for opposing armed forces that are not an enemy but are a threat to the mission. As discussed in Chapter 9, Army forces in many stability or civil support operations must avoid classifying or thinking of these threats as the enemy. Further, commanders ensure that Soldiers do not shift the animosity that they feel toward urban threats onto the civilian population. In OPERATION IRAQI FREEDOM, for example, some Soldiers began to refer to Iraqi civilians as "hajjis" in reference to the annual pilgrimage—the Hajj—that forms one of the tenets or pillars of Islamic faith. Those who have completed the pilgrimage are entitled to add the phrase al-Hajj or hajji (pilgrim) to their name. However, the Soldiers were not using the term in its proper context but rather to denigrate. In response, commanders prohibited the inappropriate use of this term as one of many measures to stress and develop cultural understanding and respect for the people that Army forces were there to assist.

### Need for More Accurate Categories

B-27. Urban commanders recognize that threat evaluation is not solely a straightforward assessment of the capabilities of a known, armed enemy. Instead, they develop a system of categories that mitigates potential situational uncertainty and better indicates the level of threat (or utility) that civilian groups pose for Army forces conducting UO. Categories can range from a simple construct—threat, neutral, or friendly—to a more refined one (see figure B-6 on page B-12). In most UO, a more intricate approach is necessary to detect and monitor shifts in key relevant relationships. While this analysis is important in any urban operations, it is especially critical when Army forces are opposed by urban insurgent and guerrilla forces.

B-28. In any system, commanders consider that the classification of a group is relative to the perspective from which it is viewed. This is an especially important consideration in multinational and interagency UO. During operations in Somalia, for example, U.S. Army forces may have viewed a particular clan as a hostile element. The United Nations' Italian contingent, with their colonial background in the region, may have considered the same clan as neutral or even an ally.

### Shifting and Overlapping Groups

B-29. Commanders recognize that no system of categorization will precisely classify any given group; no system can reflect the overall nature and complexity of the urban society. A single group may fit in a particular category. It may also have components in two or more categories simultaneously. Often, it can shift among categories during an operation. A given group may have individuals in it who have interests identical to or different from that group and these individual interests may change over time.

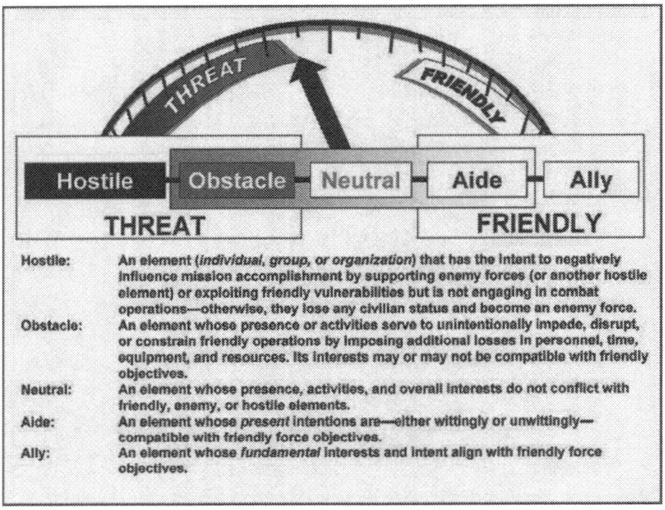


**Figure B-6. A refined approach to civilian allegiance**

B-30. A peace enforcement operation illustrates the varying nature of groups. An identified criminal group might be classified as an obstacle to the commander's mission because its illegal activities impede unit progress. Its compelling interest, however, is to make money rather than resist the actions of friendly forces. In the same operation, one of the armed belligerents may be intent on disrupting the peace process and would be, therefore, classified as a hostile. (Again, they are not an enemy unless the belligerents are engaged in planned combat operations against the peacekeeping force. Commanders should consider, however, that under international law, if any civilian takes a direct or active role in hostilities, that civilian can be subject to attack.) The belligerent force may finance the criminal organization to assist in further obstructing the peace mission. In this case, the criminal organization moves from being an obstacle to being a hostile.

B-31. This classification effort, therefore, is not a one-time undertaking; commanders must constantly review and update it. Groups or individuals can be influenced into assisting either the friendly or opposing force. People will also act opportunistically, shifting support and alliances as perceived advantages arise. Even seemingly passive and law abiding members of the urban society may conduct themselves in unexpected ways given the right conditions (mob violence, for example).

**Shifting Civilian Interests and Intent**

Among other applicable lessons (see also the vignette in Chapter 9), the Los Angeles riots of 1992 illustrate how urban population groups can shift their relative positions due to changing conditions in an urban AO. Several gangs exist in the Los Angeles area. Usually, these gangs are hostile to one another. During the riots, however, several rival gangs formed a "united front" against what was seen as a larger obstacle to their own interests: law enforcement. As a result, the hostile gangs became one another's aides during that time.

### Similarities, Differences, Capabilities, and Vulnerabilities

B-32. Similarities and differences in interests and interdependencies between groups are often focal points for analysis (and the allocation of ISR assets). They may indicate how commanders may influence, coerce, or align civilian interests and intentions with mission objectives. Simultaneously, commanders should consider an analysis of the civilian element's (individual, group, or organization) capability to influence the accomplishment of friendly objectives. They should also consider civilians' vulnerabilities and dependencies. If a civilian group's fundamental interests align with friendly objectives and this group has the intent to assist friendly forces, it is clearly an ally. However, with limited or no capability, a specific group will not help move the commander any closer to achieving his desired end state and accomplishing the mission. Then the commander would normally limit the resources expended on this group to those necessary for maintaining their commitment to common goals and objectives.

B-33. In contrast, commanders may provide a group with resources to enhance or create the capability to assist in mission accomplishment. They may do this if they felt that the assistance gained (or reduction to threat support) exceeds the potential diminishment of their own force's capabilities from losing those same resources. Commanders would also consider the group's dependencies (such as food, infrastructure, and protection) and overall vulnerability to outside influence. If vulnerable to friendly influence or control (understanding urban societal considerations and matters of perception are critical in this regard), then forces are likely susceptible to enemy or hostile manipulation. Even if commanders can generate extra resources (and not significantly affect their own force's capabilities), they should still conduct this same cost-benefit analysis to determine which civilian group (if any) should receive resources.

### Greatest Potential Nearer the Center

B-34. The most critical population sectors often are those nearest the middle of the spectrum shown in figure B-6, particularly if their capabilities (or potential) significantly degrade or enhance mission accomplishment. These are the obstacle, neutral, or aide categories. If their beliefs and interests can be adequately understood, commanders may have great chance to influence the population segment and significantly contribute to mission accomplishment. In the fight for information superiority, persuading a neutral population to become an aide may provide critical information required about the urban environment and threat forces operating within.

### Political and Strategic Concerns

B-35. The aide category may be of significant political or strategic concern. An aide group may be invaluable for accomplishing intermediate objectives but become a vulnerability to accomplishing a larger stability operation. (Even an urban offensive or defensive operation is likely to transition to a stability operation.) For example, commanders may provide resources to a criminal organization to assist in defeating insurgent forces during urban combat operations. Once these forces have been subdued, the interests (monetary gain and defeat of the threat) linking friendly forces with this criminal element disappear. What may remain is a criminal organization with more power than a reconstituted or newly established law enforcement agency and a truly destabilizing force. This also illustrates the second- and third-order thinking that will be required of commanders and their staffs during UO.

# URBAN INTELLIGENCE TOOLS AND PRODUCTS

B-36. Developing tools and products to assist with situational understanding is not the task of intelligence sections alone. Commanders and all elements of their staffs must develop and adapt products and tools suited to their particular requirements. Listed below are some of the analytic tools and products that may help meet those requirements in a complex environment (see figure B-7 on page B-14 and FMI 2-91.4). Standard tools and products include: modified combined obstacle overlays and doctrinal, situation, event, and decision support templates or matrices. In addition to these standards aids, staffs and analysts may develop or produce other innovative tools to assist commanders in their situational understanding of the complex urban environment. These tools may contain overlapping information as different types of information are compared to determine patterns among them. Staffs and analysts may also initiate requests for products (or information) from their higher headquarters or other agencies with the technical means or control over assets when the capability lies outside the Army force's means. The tools that may be developed or requested include—

- Imagery.
- Three-dimensional representations.
- Infrastructure blueprints.
- Hydrographic surveys.
- Psychological profiles.
- Matrices, diagrams, or charts.
- Various urban overlays.

B-37. There are many software applications available to the Army that can be used to conduct intelligence analysis and create many of the relevant products described above. These applications currently range from such programs as the Analyst Notebook and Crimelink which have link analysis, association matrix, and pattern analysis software tools to the Urban Tactical Planner developed by the Topographic Engineering Center as an operational planning tool. The focus of the following information, however, is the various types of tools that can be developed and used to understand the urban environment rather than the constantly evolving and changing software or hardware that may be used to create them.

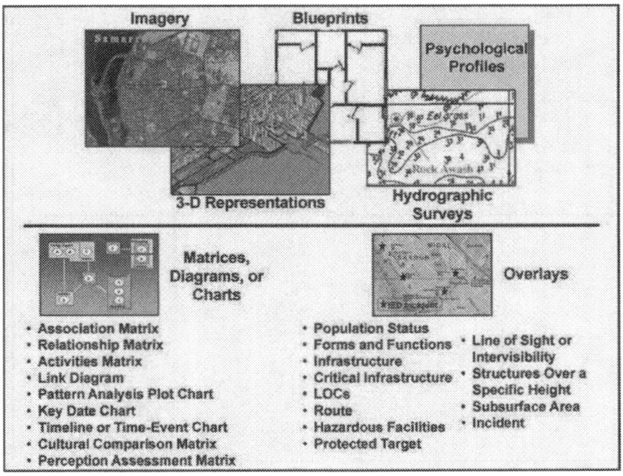


**Figure B-7. Urban intelligence tools and products**

## IMAGERY

B-38. Recent satellite imagery or aerial photography will be required for most types of UO. Such images clarify vague and inaccurate maps and other graphic representations. Satellite assets provide responsive data input into the geographic information systems (GIS). (The National Geospatial Intelligence Agency [NGA] is one important source of imagery and map data.) GIS will often form the basis for creating the three-dimensional representations and the various overlays described below. Frequently updated (or continuous real-time) satellite or aerial imagery may be required for detailed pattern analysis and maintaining accurate situational understanding. For example, imagery taken during an area's rainy season may appear significantly altered during the summer months. Finally, digital, hand-held still and video cameras, particularly at the tactical level, will also be critical in developing and attaining situational understanding in a complex urban environment.

## THREE-DIMENSIONAL REPRESENTATIONS

B-39. Often, physical or computer-generated (virtual) three-dimensional representations may be required to achieve situational understanding. These models or computer representations include specific sections of the urban area or specific buildings or structures. Such detail is particularly important for special operating forces and tactical-level units. These units require detail to achieve precision, increase the speed of the operation, and lessen friendly casualties and collateral damage.

## INFRASTRUCTURE BLUEPRINTS

B-40. Urban police, fire, health, public utilities, city engineer, realty, and tourist agencies, and other urban organizations often maintain current blueprints and detailed maps. Such documents may prove useful to update or supplement military maps or to clarify the intricacies of a specific infrastructure. They may prove critical in operations that require detailed information to achieve the speed and precision required for success. Without such detail, analysts determine interior configurations based on a building's outward appearance. That task generally becomes more difficult as the building size increases.

## HYDROGRAPHIC SURVEYS

B-41. Many urban areas are located along the world's littoral regions and major rivers. Therefore, commanders may need current hydrographic surveys to support amphibious, river crossing, riverine, and sustainment operations.

## PSYCHOLOGICAL PROFILES

B-42. Psychological profiles analyze how key groups, leaders, or decision makers think or act—their attitudes, opinions, and views. They include an analysis of doctrine and strategy, culture, and historical patterns of behavior. The degree to which the attitudes, beliefs, and backgrounds of the military either reflect or conflict with the urban populace's (or civilian leadership's) core values is extremely important in this analysis. Psychological profiles help to assess the relative probability of a threat (or noncombatant group) adopting various COAs as well as evaluating a threat's vulnerability to deception. These profiles are derived from open-source intelligence as well as signals and human intelligence.

## MATRICES, DIAGRAMS, AND CHARTS

B-43. Matrices, diagrams, and charts help to identify and understand key relationships among friendly and threat forces and other significant elements of the urban environment. While similar, each looks at information in a different way to uncover hidden patterns and connections useful in understanding the complex urban environment, particularly the threat and the urban society. These tools and products might include—

- **Association Matrix**. The association matrix helps identify the nature and relationship between individuals. Association matrices also help to identify those personalities and associations needing a more in-depth analysis necessary to establish the degree of relationship, contacts, or knowledge between individuals. Threat organizational structure and functions are uncovered as connections between personalities are made.

- **Relationship Matrix**. Relationship matrices are an extension of association matrices described above. They are used to examine the relationship between groups and organizations—threat forces, friendly forces, NGOs, the media, and the various elements of the urban population. Urban terrain and infrastructure may also be included as analytical factors to help expose relationships. These matrices graphically depict how human elements of the urban area interact with the physical elements to promote or degrade mission success. A significant relationship matrix may be a comparison of cultural perspectives—ideology, politics, religion, acceptable standards of living, and mores—between urban population groups and Army (and multinational) forces to help understand and accurately predict a civilian element's actions. The relationships discovered may suggest ways that commanders can shape the environment. For example, conflicting ideals or issues between two disparate threat forces may be emphasized to cause threats to focus their resources against each other rather than Army forces. Conversely, stressing common ideals identified between opposing noncombatant organizations may cause these groups to unite and work harmoniously toward a goal reflecting their commonalities.
- **Activities Matrix**. An activities matrix connects individuals to organizations, professions, events, activities, or addresses. Information from this matrix (combined with information from association and relationship matrices) also helps to link personalities, uncover the structure of an organization (threat or otherwise), and recognize differences, similarities, and dependencies for possible exploitation.
- **Link Diagram**. This tool graphically depicts many of the relationships and associations described above—people, groups, events, locations, or other factors deemed significant in any given situation. It helps commanders and analysts visualize how people and other factors are interrelated to determine key links.
- **Pattern Analysis Plot Chart**. This chart depicts the times and dates of a selected activity (such as ambushes, sniper and mortar attacks, bombings, and demonstrations) to search for patterns of activity for predictive purposes as well as to discern intent. This analysis can be conducted using a time-event wheel as well as by plotting events on maps using multiple historical overlays (analog or digital). (Compare with Incidents Overlay covered later in this chapter.) During pattern analysis, commanders and analysts consider not only what is occurring but also what is not occurring (or ceases to occur).
- **Key Event Chart**. In many urban operations, particularly stability operations, key holidays, historic events, and other significant cultural or political events can be extremely important for commanders to understand. More than just dates, these charts depict what can be expected to happen on each particular event. In Bosnia, for example, weddings are often held on Fridays and celebratory fire is a common occurrence. Understanding this cultural phenomenon could reduce collateral damage and accompanying civilian deaths.
- **Timeline or Time-Event Chart**. Timelines are a list of significant historical dates with relevant information and analysis that provide the commander with a record of past activities necessary to understand current operational conditions. A timeline may highlight a specific feature of the present situation such as population movements or political shifts or outline the general chronological record of the urban area, perhaps highlighting the activities of a certain population sector. Understanding past events also helps commanders predict future reactions to proposed COAs. Similarly to link diagrams, time-event charts display large amounts of information in a small space. These charts may help commanders discover larger-scale patterns of activity and relationships.
- **Cultural Comparison Matrix**. Commanders must avoid ethnocentricity—assuming that only one cultural perspective exists (often their own) and using that as the single lens with which to view the situation. To avoid this common obstacle to an accurate understanding of the urban society, commanders use this matrix to compare local ideology, attitudes, beliefs, and acceptable standards of living with associated U.S. and Army norms and values. This matrix not only highlights differences but also similarities.

- **Perception Assessment Matrix**. All of the above tools contribute to developing the perception assessment matrix; they allow commanders to see the urban environment from the perspective of its inhabitants. The perception assessment matrix is another tool that uses this refined perspective and the framework outlined in the matrix to predict the urban population's perception of a proposed COA. Although perceptions are not actions, they—more than reality—drive decision-making and resultant civilian activity. (As an example, legitimacy is a critical factor in developing many COAs that affect the civilian population, particularly during stability operations. This legitimacy must be viewed from the perspective of the inhabitants and not based on the Army commander's own culturally-shaped perceptions. Moreover, the legitimacy of an operation may need to be considered from other perspectives as well. These might include those of the American public, coalition partners and their publics, and those of neighboring states in the region.) This same matrix can later be used as a foundation to track and gauge the effectiveness of the chosen COA with favorable perception as one criterion of success. Commanders must be wary that perceptions are not fixed. They may change based on factors that the commander may not be able to control or influence. Therefore, commanders must continually monitor perceptions for deviation.

## Various Urban Overlays

B-44. Staffs can produce various map overlays. These overlays depict physical locations of some aspect critical to the planning and conduct of the urban operation. Given adequate lead time, NGA can produce many overlays as an integrated map product (including satellite imagery). These overlays can include the—

- **Population Status Overlay**. This tool depicts the physical location of various groups identified by any significant social category such as religion, tribe, or language. During offensive and defensive operations, it may simply be where significant numbers of people are "huddled" or located throughout the battlefield. Population dispersal can vary significantly through the day, particularly at night, and must be considered as part of the overall analysis leading to the development of this tool.
- **Forms and Functions Overlay**. Based on the urban model, this overlay depicts the urban core or central business district, industrial areas, outlying high-rise areas, commercial ribbon areas, and residential areas, to include shantytowns.
- **Infrastructure Overlay**. This overlay is actually a series of overlays. It depicts identifiable subsystems in each form of urban infrastructure: communications and information, transportation and distribution, energy, economics and commerce, administration and human services, and cultural. Each subsystem can be broken down into more detail. Infrastructure data may be used to develop three other overlays—
  - **Critical Infrastructure Overlay**. This tool displays specific elements of the urban infrastructure that, if harmed, will adversely affect the living conditions of the urban society to the detriment of the mission. These elements may include power generation plants, water purification plants and pumping stations, and sewage treatment plants. This information could be coded as part of the overall infrastructure overlay.
  - **Lines of Communications (LOCs) Overlay**. The LOCs overlay highlights transportation systems and nodes, such as railways, road, trails, navigable waterways, airfields, and open areas for drop zones and landing zones. It also includes subsurface areas and routes such as sewage, drainage, and tunnels and considers movement between supersurface areas. The LOCs overlay and the route overlay (below) consider traffic conditions, times, and locations, to include potential points where significant portions of the urban population may congregate.
  - **Route Overlay**. This overlay emphasizes mobility information to assist commanders and planners in determining what forces and equipment can move along the urban area's mobility corridors. Pertinent data includes street names, patterns, and widths; bridge, underpass, and overpass locations; load capacities; potential sniper and ambush locations (which may be its own overlay); and key navigational landmarks such as major roads and highways, rivers and canals, cemeteries, bridges, stadiums, and churches. The structures over a specific height

overlay and subsurface overlay may assist in its development. As with the LOCs overlay, commanders, planners, and analysts think in all dimensions.

- **Line of Sight or Intervisibility Overlay**. This product creates a profile view (optical or electronic) of the terrain from the observer's location to other locations or targets. It can show trajectory or flight-line masking as well as obstructed or unobstructed signal pathways.
- **Structures over a Specific Height Overlay**. This level of detail may also be critical to communications, fires, and Army airspace command and control (air mobility corridors especially low-level flight profiles). Incorporated as part of this overlay, it may include floors or elevations above limitations for particular weapon systems at various distances from the structure.
- **Subsurface Area Overlay**. As an alternate to the building or structure height overlay, this product provides the locations of basements, underground parking garages, sewers, tunnels, subways, naturally-occurring subterranean formations, catacombs, and other subsurface areas. Similar to elevation "dead spaces," this overlay may show areas that exceed depression capabilities of weapon systems and potential threat ambush locations—again, affecting maneuver options.
- **Urban Logistic Resources Overlay**. This product identifies the locations of urban logistic resources that may contribute to mission accomplishment. It may contain specific warehouse sites, hospitals and medical supply locations, viable food stores, building material locations, fuel storage areas, car or truck lots, maintenance garages, and appliance warehouses. (NGO locations, taken from an NGO relationship matrix, may be an essential, overlapping element of this overlay.)
- **Hazardous Facilities Overlay**. This overlay identifies urban structures with known or suspected chemical, biological, or radiological features, such as nuclear power plants, fertilizer plants, oil refineries, pharmaceutical plants, and covert locations for producing weapons of mass destruction. These locations are critical to maneuver and fire planning.
- **Protected Target Overlay**. This overlay depicts terrain that should not be destroyed or attacked based on restrictions due to international, host-nation, or U.S. law and subsequent rules of engagement. These may include schools, hospitals, historical or other culturally significant monuments, and religious sites. This overlay may incorporate no-fire areas, such as special operations forces locations, critical infrastructure, logistic sources, and hazardous sites that must be protected as part of the commander's concept of the operation.
- **Incident Overlay**. Similar to the pattern analysis plot chart, this product depicts the location of different threat actions and types of tactics employed to uncover recurring routines, schemes, methods, tactics, or techniques and overall threat interests, objectives, or the desired end state.

B-45. The above IPB tools and products constitute a small sampling of what staffs and analysts can produce. They are limited only by their imaginations and mission needs (not all tools presented above may be relevant or necessary to every operation). Many products can be combined into a single product or each can generate further products of increasing level of detail. This is similar to transparent overlays positioned one atop another on a map. As discussed earlier, technology—software and hardware—will allow for more urban data to be combined, compared, analyzed, displayed, and shared—ultimately allowing scaleable products that can expand and contract to meet the intelligence requirements of any echelon of command down to the individual Soldier. The challenge will remain, however, to provide timely, accurate, complete, and relevant information in an understandable, usable, and share-capable form—share-capable among different levels of command as well as among multinational partners and various governmental and (as required) nongovernmental organizations—without overloading the commander.

## Appendix C

# Operations in Somalia: Applying the Urban Operational Framework to Stability Operations

*It's impossible for an American mother to believe that a Somali mother would raise children to avenge the clan.*

Major General Thomas M. Montgomery

## GENERAL SITUATION

C-1. Following decades of political unrest and the fall of Somali dictator Siad Barre, a civil war broke out as 14 clans vied for power. The resulting nation composed of hostile social factions was held together by weak political alliances—none strong enough to unite and lead the country to national reconciliation. An ongoing drought led to famine and compounded the ethnic tensions and political instability. This volatile situation rapidly led to a phased U.S. involvement (see figure C-1). Army forces combined, sequenced, and proportionally emphasized the different types of operations to accomplish changing political objectives. Throughout all operations in Somalia, urban areas were critical to achieving mission success.

## SOMALI OPERATIONS

### INITIAL UN RESPONSE

C-2. The United Nations (UN) initially responded to requests for assistance from international relief organizations by sending supplies and other forms of humanitarian aid to Somalia. However, widespread looting, fighting between gangs, and other lawlessness prevented supplies from reaching the hungry and sick. Only 20 percent of the food entering the country reached the people who needed it. An estimated 25 percent of Somalia's 6 million people died of starvation or disease. In April 1992, the UN issued Security Council Resolution (UNSCR) 751 which authorized 50 unarmed observers, but the action had little effect. Under continuing pressure for additional measures to ensure the delivery of supplies and relief, the UN authorized 500 armed peacekeepers (furnished by Pakistan and transported by U.S. sea- and airlift) to protect humanitarian workers. The battalion's limited mission, designated UN Operations in Somalia I (UNOSOM I), encompassed safeguarding the unloading of ships and providing convoy security.

### PROVIDE RELIEF (UNOSOM I)

C-3. In July 1992, the UN requested an increased airlift of supplies and the United States quickly responded. U.S. Central Command (USCENTCOM) activated joint task force (JTF) OPERATION PROVIDE RELIEF. Based on careful mission analysis, USCENTCOM limited the JTF's actions to—

- Deploying a humanitarian assistance survey team to assessing relief requirements.
- Providing an emergency airlift of supplies.
- Using Air Force cargo aircraft for daily relief sorties into Somalia.

C-4. USCENTCOM restricted the sorties to flying during daylight hours and to locations that would provide a permissive and safe environment. In mid-September 1992, the United States prudently expanded its role by stationing the amphibious ready group, USS Tarawa, offshore to provide support to the Pakistani security battalion and to provide security for U.S. airlift operations. The 11th Marine expeditionary unit (MEU) was on board the USS Tarawa to rapidly respond to any change in mission (see Appendix D for a description and the capabilities of a MEU).

| Operation | Dates | UN Security Council Resolution | Relative Proportionality Between Types of Operations |
|---|---|---|---|
| PROVIDE RELIEF (UNOSOM I) | Aug 92 – Dec 92 | UNSCR# 751 24 Apr 92 | 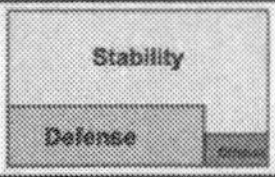 |
| RESTORE HOPE (UNITAF) | Dec 92 – May 93 | UNSCR# 794 3 Dec 92 | 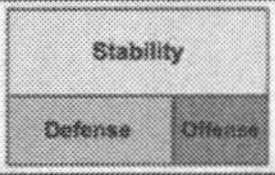 |
| CONTINUED HOPE (UNOSOM II) | May 93 – Mar 94 | UNSCR# 814 26 Mar 93 | 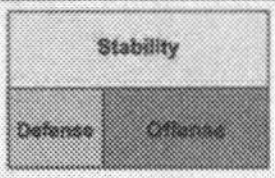 |

**Figure C-1. Relative proportionality between types of operations**

## RESTORE HOPE (UNITAF)

C-5. By November 1992, the magnitude of the task, UN organizational deficiencies, and a continued lack of security precluded delivery of sufficient supplies to the needy. Notably, a ship laden with relief supplies was fired on in the harbor at Mogadishu, forcing its withdrawal before the supplies could be brought ashore, and a Pakistani peacekeeper was shot when his car was hijacked. Subsequently, the United States offered to provide forces and lead a UN-sponsored operation to reopen the flow of food to where it was needed most. In December 1992, the UN issued UNSCR 794, which authorized member states "to use all necessary means to establish a secure environment for humanitarian relief operations in Somalia" and demanded all "factions in Somalia immediately cease hostilities." To allay concerns of colonialism by a number of African countries, the UN Secretary-General was given oversight of the operation. The resolution also required soldiers to be withdrawn once order was restored; however, it provided no exit strategy. As clearly as possible, the USCENTCOM mission statement (below) for OPERATION RESTORE HOPE reflected the UN mandate:

> *When directed by the [President or the Secretary of Defense], [Commander, USCENTCOM] will conduct joint/combined military operations in Somalia to secure the major air and sea ports, key installations and food distribution points, to provide open and free passage of relief supplies, provide security for convoys and relief organization operations, and assist UN/[nongovernmental organizations] in providing humanitarian relief under UN auspices. Upon establishing a secure environment for uninterrupted relief operations, [Commander, USCENTCOM] terminates and transfers relief operations to UN peacekeeping forces.*

C-6. Mogadishu was the largest port in the country and the focal point of previous humanitarian relief activities of nongovernmental organizations (NGOs). It was also the headquarters of the coalition of 20 nations and over 30 active humanitarian relief organizations. As such, Mogadishu became the entry point for the operational buildup of the multinational force known as Unified Task Force (UNITAF) and the key logistic hub for all operations in Somalia. UNITAF immediately gained control over the flow of relief supplies into and through Mogadishu and stabilized the conflict among the clans. In less than a month, UNITAF forces expanded control over additional ports and interior airfields. They secured additional distribution sites in other key urban areas in the famine belt to include Baidoa, Baledogle, Gialalassi, Bardera, Belet Uen, Oddur, Marka, and the southern town of Kismayo (see figure C-2). With minimal force, the U.S.-led UNITAF established a secure environment that allowed relief to reach those in need, successfully fulfilling its limited—yet focused—mandate.

## CONTINUED HOPE (UNOSOM II)

C-7. In March 1993, the UN issued UNSCR 814 establishing a permanent peacekeeping force, UNOSOM II. However, the orderly transition from UNITAF to UNOSOM II was repeatedly delayed until May 1993. (The UN Secretary-General urged the delay so that U.S. forces could effectively disarm bandits and rival clan factions in Somalia.) This resolution was significant in two critical aspects:

- It explicitly endorsed nation building with the specific objectives of rehabilitating the political institutions and economy of Somalia.
- It mandated the first ever UN-directed peace enforcement operation under the Chapter VII enforcement provisions of the Charter, including the requirement for UNOSOM II to disarm the Somali clans. The creation of a peaceful, secure environment included the northern region that had declared independence and had hereto been mostly ignored.

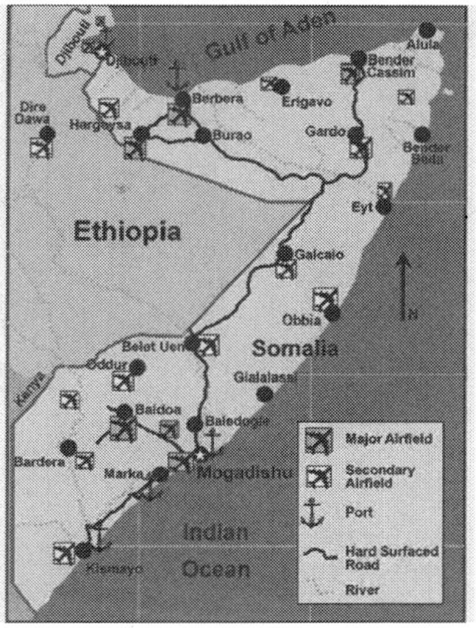


**Figure C-2. Map of Somalia**

C-8. These far-reaching objectives exceeded the limited mandate of UNITAF as well as those of any previous UN operation. Somali clan leaders rejected the shift from a peacekeeping operation to a peace enforcement operation. They perceived the UN as having lost its neutral position among rival factions. A more powerful clan leader, General Mohammed Farah Aideed (leader of the Habr Gidr clan), aggressively turned against the UN operation and began a radio and propaganda campaign. This campaign characterized UN soldiers as an occupation force trying to recolonize Somalia.

C-9. The mounting crisis erupted in June 1993. Aideed supporters killed 24 Pakistani soldiers and wounded 57 in an ambush while the soldiers were conducting a short-warning inspection of one of Aideed's weapons arsenals. UNSCR 837, passed the next day, called for immediately apprehending those responsible and quickly led to a manhunt for Aideed. The United States deployed 400 rangers and other special operations forces (SOF) personnel to aid in capturing Aideed, neutralizing his followers, and assisting the quick reaction force (QRF), composed of 10th Mountain Division units, in maintaining the peace around Mogadishu.

### PHASED WITHDRAWAL

C-10. On 3 October 1993, elements of Task Force (TF) Ranger (a force of nearly 100 Rangers and SOF operators) executed a raid to capture some of Aideed's closest supporters. Although tactically successful, 2 helicopters were shot down, 75 soldiers were wounded, and 18 soldiers were killed accomplishing the mission. The U.S. deaths, as well as vivid scenes of mutilation to some of the soldiers, increased calls to Congress for withdrawing U.S. forces from Somalia. The President then ordered reinforcements to protect U.S. Forces, Somalia (USFORSOM) as they began a phased withdrawal with a 31 March deadline. The last contingent sailed from Mogadishu on 25 March, ending OPERATION CONTINUED HOPE and the overall U.S. mission in Somalia.

C-11. Although U.S. forces did not carry out the more ambitious UN goals of nation building, they executed their missions successfully, relieving untold suffering through humanitarian assistance with military skill and professionalism. Operations in Somalia occurred under unique circumstances, yet commanders may glean lessons applicable to future urban stability operations, as well as civil support operations. In any operations, commanders balance changing mission requirements and conditions.

# UNDERSTAND

C-12. Although accomplished to varying degrees, U.S. forces failed to adequately assess and understand the urban environment, especially the society. Somali culture stresses the unity of the clan; alliances are made with other clans only when necessary to elicit some gain. Weapons, overt aggressiveness, and an unusual willingness to accept casualties are intrinsic parts of the Somali culture. Women and children are considered part of the clan's order of battle.

C-13. Early in the planning for OPERATION RESTORE HOPE, U.S. forces did recognize the limited transportation and distribution infrastructure in Mogadishu. The most notable was the limited or poor airport and harbor facilities and its impact on the ability of military forces and organizations to provide relief. Therefore, a naval construction battalion made major improvements in roads, warehouses, and other facilities that allowed more personnel, supplies, and equipment to join the relief effort faster.

## Understanding the Clan (The Human Dimension)

C-14. During OPERATION RESTORE HOPE, the UNITAF worked with the various clan leaders as the only recognized leadership remaining in the country. The UNITAF was under the leadership of LTG Robert B. Johnston and U.S. Ambassador to Somalia, Robert Oakley. In addition, UNITAF forces also tried to reestablish elements of the Somali National Police—one of the last respected institutions in the country that was not clan-based. This reinstated police force manned checkpoints throughout Mogadishu and provided crowd control at feeding centers. Largely because of this engagement strategy, the UNITAF succeeded in its missions of stabilizing the security situation and facilitating humanitarian relief. Before its termination, the UNITAF also worked with the 14 major Somali factions to agree to a plan for a transitional or transnational government.

C-15. The UN Special Representative of the Secretary-General, retired U.S. Navy Admiral Jonathon Howe, worked with the UNOSOM II commander, Turkish General Cevik Bir. During OPERATION CONTINUED HOPE, Howe and General Bir adopted a philosophy and operational strategy dissimilar to their UNITAF predecessors. Instead of engaging the clan leaders, Howe attempted to marginalize and isolate them. Howe initially attempted to ignore Aideed and other clan leaders in an attempt to decrease the warlord's power. Disregarding the long-established Somali cultural order, the UN felt that, in the interest of creating a representative, democratic Somali government, they would be better served by excluding the clan leadership. This decision ultimately set the stage for strategic failure.

## Threat Strategy and Tactics

C-16. During OPERATION RESTORE HOPE, U.S. forces also failed to properly analyze and understand their identified threat's intent and the impact that the urban environment would have on his strategy, operations, and tactics. The UN began to view eliminating Aideed's influence as a decisive point when creating an environment conducive to long-term conflict resolution. Aideed's objective, however, remained to consolidate control of the Somali nation under his leadership—his own brand of conflict resolution. He viewed the UN's operational center of gravity as the well-trained and technologically advanced American military forces, which he could not attack directly. He identified a potential American vulnerability—the inability to accept casualties for an operation not vital to national interests—since most Americans still viewed Somalia as a humanitarian effort. If he could convince the American public that the price for keeping troops in Somalia would be costly, or that their forces were hurting as many Somalis as they were helping, he believed they would withdraw their forces. If U.S. forces left, the powerless UN would leave soon after, allowing Aideed to consolidate Somalia under his leadership.

## Vulnerability and Risk Assessment

C-17. U.S. forces failed to assess, understand, and anticipate that Aideed would adopt this asymmetric approach and attack the American public's desire to remain involved in Somalia. By drawing U.S. forces into an urban fight on his home turf in Mogadishu, he could employ guerrilla insurgency tactics and use the urban area's noncombatants and its confining nature. Such tactics made it difficult for the U.S. forces to employ their technological superiority. If U.S. forces were unwilling to risk harming civilians, his forces could inflict heavy casualties on them, thereby degrading U.S. public support for operations in Somalia. If, on the other hand, the U.S. forces were willing to risk increased civilian casualties to protect themselves, those casualties would likely have the same effect.

C-18. However, an assessment and understanding of the Somali culture and society should have recognized the potential for Aideed's forces to use women and children as cover and concealment. Accordingly, the plan should have avoided entering the densely populated Bakara market district with such restrictive rules of engagement. As legitimacy is critical to stability operations, TF Ranger should have been prepared and authorized to employ nonlethal weapons, to include riot control gas, as an alternative to killing civilians or dying themselves.

C-19. U.S. forces also failed to assess and understand the critical vulnerability of their helicopters in an urban environment and the potential impact on their operations. TF Ranger underestimated the threat's ability to shoot down its helicopters even though they knew Somalis had attempted to use massed rocket-propelled grenade (RPG) fires during earlier raids. (Aideed brought in fundamentalist Islamic soldiers from Sudan, experienced in downing Russian helicopters in Afghanistan, to train his men in RPG firing techniques). In fact, the Somalis had succeeded in shooting down a UH-60 flying at rooftop level at night just one week prior to the battle. Instead, TF Ranger kept their most vulnerable helicopters, the MH-60 Blackhawks, loitering for forty minutes over the target area in an orbit that was well within Somali RPG range. The more maneuverable AH-6s and MH-6s could have provided the necessary fire support. Planning should have included a ready ground reaction force, properly task organized, for a downed helicopter and personnel recovery contingency.

C-20. Information operations considerations apply throughout the entire urban operational framework; however, operations security (OPSEC) is critical to both assessment and shaping. OPSEC requires continuous assessment throughout the urban operation particularly as it transitions among the spectrum of operations. As offensive operations grew during OPERATION CONTINUED HOPE, U.S. forces did little to protect essential elements of friendly information. Combined with the vulnerability of U.S. helicopters, Aideed's followers used U.S. forces' inattention to OPSEC measures to their advantage. The U.S. base in Mogadishu was open to public view and Somali contractors often moved about freely. Somalis had a clear view both day and night of the soldiers' billets. Whenever TF Ranger would prepare for a mission, the word rapidly spread through the city. On 3 October 1993, Aideed's followers immediately knew that aircraft had taken off and, based on their pattern analysis of TF Ranger's previous raids, RPG teams rushed to the rooftops along the flight paths of the task force's Blackhawks.

## SHAPE

C-21. One of the most critical urban shaping operations is isolation. During OPERATION CONTINUED HOPE, U.S. forces largely discounted other essential elements of friendly information and did not establish significant public affairs and psychological operations (PSYOP) initiatives. In fact, Army forces lacked a public affairs organization altogether. Consequently, Aideed was not isolated from the support of the Somali people. This failure to shape the perceptions of the civilian populace coupled with the increased use of lethal force (discussed below) allowed Aideed to retain or create a sense of legitimacy and popular support. Ironically, the focus on Aideed helped create a popular figure among the warlords through which many noncompliant, hostile, and even neutral Somalis could rally behind.

C-22. During OPERATION RESTORE HOPE, Aideed conducted propaganda efforts through "Radio Aideed"—his own radio station. UNITAF countered these efforts with radio broadcasts. This technique proved so effective that Aideed called MG Anthony C. Zinni, UNITAF's director of operations, over to his house on several occasions to complain about UNITAF radio broadcasts. General Zinni responded, "If he didn't like what we said on the radio station, he ought to think about his radio station and we could mutually agree to lower the rhetoric." This approach worked.

# ENGAGE

C-23. The complexity of urban operations requires unity of command to identify and effectively strike the center of gravity with overwhelming combat power or capabilities. Complex command and control relationships will only add to the complexity and inhibit a commander's ability to dominate and apply available combat power to accomplish assigned objectives. Stability operations as seen in Somalia required commanders to dominate within a supporting role and, throughout, required careful, measured restraint.

## UNITY OF COMMAND (EFFORT)

C-24. During OPERATION RESTORE HOPE, UNITAF successfully met unity of command challenges through three innovations. First, they created a civil-military operations center (CMOC) to facilitate unity of effort between NGOs and military forces. Second, UNITAF divided the country into nine humanitarian relief sectors centered on critical urban areas that facilitated both relief distribution and military areas of responsibility. Third, to establish a reasonable span of control, nations that provided less than platoon-sized contingents were placed under the control of the Army, Marine Corps, and Air Force components.

C-25. On the other hand, during OPERATION CONTINUED HOPE, UNOSOM II command and control relationships made unity of command (effort) nearly impossible. The logistic components of USFORSOM were under UN operational control, while the QRF remained under USCENTCOM's combatant command—as was TF Ranger. However, the USCENTCOM commander was not in theater. He was not actively involved in planning TF Ranger's missions or in coordinating and integrating them with his other subordinate commands. It was left to TF Ranger to coordinate with the QRF as needed. Even in TF Ranger, there were dual chains of command between SOF operators and the Rangers. This underscores the need for close coordination and careful integration of SOF and conventional forces (see Chapter 4). It also emphasizes overall unity of command (or effort when command is not possible) among all forces operating in a single urban environment.

C-26. Following TF Ranger's 3 October mission, the command structure during OPERATION CONTINUED HOPE was further complicated with the new JTF-Somalia. This force was designed to protect U.S. forces during the withdrawal from Somalia. JTF-Somalia came under the operational control of USCENTCOM, but fell under the tactical control of USFORSOM. Neither the JTF nor USFORSOM controlled the naval forces that remained under USCENTCOM's operational control. However unity of effort (force protection and a rapid, orderly withdrawal) galvanized the command and fostered close coordination and cooperation among the semiautonomous units.

## MEASURED RESTRAINT

C-27. During OPERATIONS PROVIDE RELIEF and RESTORE HOPE, U.S. forces dominated within their supporting roles. Their perseverance, adaptability, impartiality, and restraint allowed them to provide a stable, secure environment. Hence, relief organizations could provide the food and medical care necessary to reduce disease, malnourishment, and the overall mortality rate. However, during OPERATION CONTINUED HOPE, U.S. operations became increasingly aggressive under the UN mandate. However, peace enforcement also requires restraint and impartiality to successfully dominate and achieve political objectives. The increased use of force resulted in increased civilian casualties, which in turn reduced the Somalis' perception of U.S. legitimacy. As a result, most moderate Somalis began to side with Aideed and his supporters. Many Somalis, without effective information operations to shape their perceptions, felt that it was fine for foreign military forces to intervene in their country to feed the starving and even help establish a peaceful government, but not to purposefully target specific Somali leaders as criminals.

# CONSOLIDATE AND TRANSITION

C-28. Across the spectrum of conflict, Army forces must be able to execute full spectrum operations not only sequentially but, as in the case of operations in Somalia, simultaneously. OPERATION PROVIDE RELIEF began primarily as a stability operation in the form of foreign humanitarian assistance. Later this operation progressed to include peacekeeping (another stability operation), defensive operations to protect

UN forces and relief supplies, and minimum offensive operations. As operations transitioned to OPERATION RESTORE HOPE, it became apparent that while foreign humanitarian assistance was still the principal operation, other operations were necessary. More forceful peacekeeping, show of force, and arms control (all forms of stability operations), and offensive and defensive operations grew increasingly necessary to establish a secure environment for uninterrupted relief operations. In the final phase of U.S. involvement during OPERATION CONTINUED HOPE, major changes to political objectives caused a major transition in the form of the stability operation to peace enforcement with an even greater increase in the use of force, offensively and defensively, to create a peaceful environment and conduct nation building. Finally, the TF Ranger raid demonstrated the need to maintain a robust, combined arms force capable of rapidly transitioning from stability operations (primarily peace operations) to combat operations.

# SUMMARY

C-29. OPERATIONS PROVIDE RELIEF and RESTORE HOPE were unquestionably successes. Conversely, during OPERATION CONTINUED HOPE, the 3-4 October battle of Mogadishu (also known as the "Battle of the Black Sea") was a tactical success leading to an operational failure. TF Ranger succeeded in capturing 24 suspected Aideed supporters to include two of his key lieutenants. Arguably, given the appropriate response at the strategic level, it had the potential to be an operational success. After accompanying Ambassador Oakley to a meeting with Aideed soon after the battle, MG Zinni described Aideed as visibly shaken by the encounter. MG Zinni believed Aideed and his subordinate leadership were tired of the fighting and prepared to negotiate. Unfortunately, the U.S. strategic leadership failed to conduct the shaping actions necessary to inform and convince the American public (and its elected members of Congress) of the necessity of employing American forces to capture Aideed. The President was left with little recourse after the battle of Mogadishu but to avoid further military confrontation.

C-30. Despite this strategic failing, the operational commanders might have avoided the casualties, and any subsequent public and Congressional backlash, had they better communicated among themselves and worked with unity of effort. Recognizing the separate U.S. and UN chains of command, the UN Special Representative, along with the USCENTCOM, USFORSOM, and TF Ranger commanders, should have established the command and control architecture needed. This architecture would have integrated planning and execution for each urban operation conducted. These commanders failed to "operationalize" their plan. They did not properly link U.S. strategic objectives and concerns to the tactical plan. The TF Ranger mission was a direct operational attempt to obtain a strategic objective in a single tactical action. Yet, they failed to understand the lack of strategic groundwork, the threat's intent and capabilities, and the overall impact of the urban environment, to include the terrain and society, on the operation. Such an understanding may not have led to such a high-risk course of action but instead to one that de-emphasized military operations and emphasized a political solution that adequately considered the clans' influence.

This page intentionally left blank.

## Appendix D

# Joint and Multinational Urban Operations

*[Joint force commanders] synchronize the actions of air, land, sea, space, and special operations forces to achieve strategic and operational objectives through integrated, joint campaigns and major operations. The goal is to increase the total effectiveness of the joint force, not necessarily to involve all forces or to involve all forces equally.*

JP 3-0

As pointed out earlier, Army forces, brigade size and larger, will likely be required to conduct operations in, around, and over large urban areas in support of a joint force commander (JFC). The complexity of many urban environments, particularly those accessible from the sea, requires unique leveraging and integration of all the capabilities of U.S. military forces to successfully conduct the operation. This appendix discusses many of these capabilities; JP 3-06 details joint urban operations.

## PURPOSE

D-1. In some situations, a major urban operation is required in an inland area where only Army forces are operating. Army commanders determine if the unique requirements of the urban environment require forming a joint task force (JTF) or, if not, request support by joint capabilities from the higher joint headquarters. Sometimes the nature of the operation is straightforward enough or the urban operation is on a small enough scale that conventional intraservice support relationships are sufficient to meet the mission requirements.

D-2. Most major urban operations (UO), however, require the close cooperation and application of joint service capabilities. A JTF may be designated to closely synchronize the efforts of all services and functions in an urban area designated as a joint operations area (JOA). If a large urban area falls in the context of an even larger ground force area of operations, a JTF dedicated to the urban operation may not be appropriate. These situations still require joint capabilities. In such cases, the responsible JFC designates support relations between major land units and joint functional commands. The major land units can consist of Army forces, Marine Corps forces, or joint forces land component command. The joint functional commands can consist of the joint special operations task force, joint psychological operations task force, or joint civil-military operations task force.

D-3. This appendix describes the roles of other services and joint combatant commands in UO. It provides an understanding that enables Army commanders to recommend when to form a JTF or to request support from the JFC. It also provides information so commanders can better coordinate their efforts with those of the JFC and the commanders of other services or components conducting UO. Lastly, this appendix describes some considerations when conducting UO with multinational forces.

## SISTER SERVICE URBAN CAPABILITIES

D-4. Army forces conducting UO rely on other services and functional joint commands for specialized support in the urban environment. These capabilities are requested from and provided through the commanding JFC. Army forces request the assets and capabilities described in this annex through their higher headquarters to the joint command. The JFC determines if the assets will be made available, the appropriate command relationship, and the duration of the support. Army forces prepare to coordinate planning and execution with other services and to exchange liaison officers. These capabilities can greatly increase the Army's ability to understand, shape, engage, consolidate, and transition within the context of UO.

## AIR FORCE

D-5. Air Force support is an important aspect of the Army force concept for urban operations. Air Force elements have a role to play in UO across the spectrum of operations.

D-6. Air Force intelligence, surveillance, and reconnaissance (ISR) systems contribute significantly to understanding the urban environment. These ISR systems include the E-8 Joint Surveillance Target Attack Radar System (JSTARS) (see figure D-1), U2 Dragon Lady, RC-135 Rivet Joint, or RQ-4 Global Hawk and RQ/MQ-1 Predator unmanned aerial systems. Air Force ISR systems can provide vital data to help assess threat intentions, threat dispositions, and an understanding of the civilian population. These systems also can downlink raw information in real-time to Army intelligence processing and display systems, such as the common ground station or division tactical exploitation system.

**Figure D-1. USAF E-8 JSTARS platform**

D-7. Air interdiction (AI) can be a vital component of shaping the urban operational environment. Often, AI of the avenues of approach into the urban area isolates the threat by diverting, disrupting, delaying, or destroying portions of the threat force before they can be used effectively against Army forces. AI is especially effective in major combat operations where restrictions on airpower are limited and the threat is more likely to be a conventionally equipped enemy. In 1991 during OPERATION DESERT STORM, AI helped prevent the Iraqi 5th Mechanized Division from reaching Khafji.

D-8. Precision-guided munitions (PGMs) delivered from combat aircraft can positively and directly affect the conduct of Army close combat actions in the urban area. Special munitions designed to penetrate reinforced concrete can provide unique support to land forces executing UO. Although laser-guided bombs are extremely accurate with an effective laser lock-on, environmental factors such as smoke, haze, and weather can adversely affect successful laser designation. If the launching aircraft can achieve a successful laser designation and lock-on, these weapons have devastating effects, penetrating deep into reinforced concrete before exploding with great force. If launched without a lock-on, or if the laser spot is lost, these weapons are unpredictable and can travel long distances before they impact. However, continual advancements in tactics and weapons systems, such as joint direct attack munition (JDAM) weapons, help decrease the adverse effects of these environmental factors. (For example, in 2003 during OPERATION IRAQI FREEDOM, the Air Force supported the Army's attack on An Najaf with JDAM-equipped F 16s and B 1s during a blinding sandstorm. With near zero visibility, JDAM weapons provided significant fire support, allowing Army forces to complete their mission and continue the attack north.) When PGMs are unavailable, general-purpose bombs from 500 to 2,000 pounds can also prove effective. However, there may be risk factors associated with their use such as collateral damage due to decreased accuracy, threat exposure to the delivering aircraft due to lower delivery altitudes, or reduced weapons effectiveness due to shallower penetration angles. (See FM 3-06.1 for a more detailed description of various aircraft weapons and capabilities.)

D-9. In addition to shaping and engaging the UO through firepower, commanders can use Air Force capabilities to improve and augment the urban transportation and distribution infrastructure. Air Force units can repair or improve airfields, revitalize civil aviation maintenance facilities, manage air-delivered cargo, and control civil and military air traffic. These latter capabilities particularly enhance urban stability and civil support operations. These capabilities may even be decisive. Air-delivered cargo and air traffic management, for example, were the decisive factors in U.S. forces' successful resistance of the Soviet blockade of Berlin in 1948. In OPERATION RESTORE HOPE, from December 1992 to May 1993, Air Force operations in Mogadishu were critical to airlifting and staging supplies and forces. The Air Force determined the ultimate success of the humanitarian assistance operation (see Appendix C).

D-10. In unique situations, such as the Berlin Blockade and OPERATION RESTORE HOPE, exercising Air Force urban capabilities may be the decisive action of the operation. Most often, however, Air Force capabilities will play a shaping role, sustaining role, or both in joint urban operations because of the inherent requirement to occupy terrain and interface with the population.

## MARINE CORPS

D-11. The Marine Corps can assault across water obstacles into a defended urban environment. This capability is an invaluable tactical and operational tool. The mere threat of this capability can divert many enemy forces from other avenues of approach and obscure the true nature of an attack. The impact of the threat of amphibious assault was vividly demonstrated during OPERATION DESERT STORM where embarked Marine Forces diverted several Iraqi divisions to defensive positions along the coast and near Kuwait City.

D-12. The presence of Marine amphibious equipment, apart from Marine infantry, provides Army forces unique capabilities. In UO, the amphibious operation is often not an assault from the sea, but rather an assault river crossing. In 1950, the 7th U.S. Infantry Division used amphibious tractor support from the 1st Marine Division to conduct an assault river crossing of the Han River into downtown Seoul. See FM 3-31.1.

D-13. The worldwide deployment of Marine air-ground task forces (MAGTFs) enables a short notice response into any urban areas accessible from the sea. Typically, a deployed MAGTF is a Marine expeditionary unit (special operations capable) (MEU[SOC]). The MEU(SOC) can perform forcible entry operations, seize lodgments, and may execute these tasks anticipating reinforcement by Army or joint forces. They are also well positioned and equipped to rapidly reinforce Army forces already deployed in theater. The special-operations-capable training that these units accomplish before deploying includes urban warfare training and contributes to their value in UO. The MEU(SOC) is relatively small (its core unit is a Marine infantry battalion), is forward deployed, and has a wide spectrum of organic capabilities. It is an important asset in crisis stability and civil support operations.

D-14. The Marine expeditionary brigade is the MAGTF between a MEU and a Marine expeditionary force (the largest MAGTF). All MAGTFs include an aviation combat element, which is particularly skilled at providing accurate and timely close air support. The responsiveness and accuracy of Marine close air support aptly suits it to UO where the lethality of combat and the close range of engagements demand accurate and responsive fires.

D-15. Marine forces that conduct UO work well in littoral urban areas because of their unique relationship with naval forces and thus their capability to closely integrate land and sea operations. A supporting arms liaison team may be attached to Army forces at battalion level to provide ship-to-shore communications and coordination for naval gunfire support.

## NAVY

D-16. Many major urban areas are accessible from the sea. Therefore, Army commanders should understand how sea power can influence and support UO. The Navy brings several major capabilities to UO. These include naval gunfire support, naval air support, and port and coastal security. In addition, the Navy can offer secure command post sites and meeting areas; provide limited emergency shelter and critical care for Soldiers and, as necessary, civilians; and store and transport critical supplies ashore providing flexibility in meeting support requirements. The ability to provide support from "over the horizon" not only enhances force protection but also the legitimacy of urban and host nation authorities during transition.

D-17. Naval surface fire support (which includes naval gunfire support) particularly applies to forcible entry operations in littoral urban areas. It provides an initial indirect fire support capability until Army forces land ashore. However, naval gunfire support lacks a precision munitions capability. In defensive operations, naval gunfire support can also add major fires, especially during retrograde operations through an urban area. Naval gunfire support of the 3rd Infantry Division was the primary fire support for the last two days of the X Corps retrograde operation out of the port of Hungnam, North Korea, in December 1950.

D-18. Destroyers and cruisers, which mount one and two 127mm MK45 lightweight gun systems respectively (see figure D-2), are capable of providing naval gunfire support. This gun system can provide a rate of fire of 20 rounds per minute per gun to a range of approximately 23 kilometers. Destroyers, cruisers, and attack submarines are capable of precision strikes with Tomahawk cruise missiles. Depending on the location of the urban area, naval air can provide responsive aviation support to UO alleviating the need for fixed-wing aviation bases.

**Figure D-2. USN MK45 lightweight gun system**

D-19. Due to its flat trajectory, terrain masking affects naval gunfire more than field artillery. Naval gunfire also results in large range probable errors (the dispersion pattern of the naval gun is roughly elliptical with the long axis in the direction of fire). Hence, coverage of targets such as roads and airfields is most effective when the gun-target line (GTL) coincides with the long axis of the target. Very close supporting fire can be delivered when the GTL is parallel to the front line of troops. Oppositely, a GTL perpendicular to the front trace can endanger friendly forces. Within the limits of hydrographic conditions, the ship can maneuver to achieve a better GTL, but ship movement also makes it difficult to adjust fire. Overall, naval and air threats, bad weather, and large range probable errors make naval gunfire difficult and can cause cancellation of supporting fires.

D-20. Water terminals located in urban areas are usually the debarkation points for the bulk of Army forces as well as a joint force theater logistics requirement. Army forces are responsible for water terminal operations while naval capabilities protect these strategic and operationally vital facilities.

D-21. Naval coastal warfare (NCW) is the responsibility of the JFC and is often exercised through the Navy component commander. He may assign a naval coastal warfare commander for an appropriate geographic area. NCW includes coastal sea control, port security, and harbor defense. While coastal sea control is conducted in the environment of the open seas, port security and harbor security include the urban environment. Port security is the safeguarding of vessels, harbors, ports, waterfront facilities, and cargo from internal threats. It includes destruction, loss, or injury from sabotage or other subversive acts; accidents; thefts; or other causes of similar nature. The Navy's role in protecting essential urban infrastructure is often key. This task is also important when executing stability operations if the threats against urban infrastructure will likely be unconventional.

D-22. Harbor defense protects harbor approaches, harbors, anchorages, and ports from external threats. Harbor defense focuses on the conventional defense of port infrastructure. It is a task appropriate in a major operation or campaign and often includes port security as a subtask. The JFC executes NCW (focused on harbor defense and port security) using combined Navy capabilities including surface warfare, aviation, and naval special operations. See also discussions in this appendix on transportation command and special operations command.

### COAST GUARD

D-23. The Coast Guard, like the Navy, can significantly influence the conduct of UO when the urban area is accessible from the sea. The U.S. Coast Guard (USCG) is the federal authority for port security and harbor defense of domestic facilities. When directed by the President, the USCG can augment the Navy in operations overseas. Historically, the entire USCG was under Navy control during both World Wars I and II. USCG elements deployed overseas and operated under Navy control during the Vietnam War and during OPERATIONS DESERT SHIELD and DESERT STORM.

D-24. The USCG uses surface warfare systems and aerial reconnaissance systems to conduct its missions. Its air systems are unarmed, whereas its surface systems are armed for self-defense and law enforcement operations. The USCG is experienced and adept at supporting other agencies, local governments, and law enforcement. Its capabilities can best support Army UO in civil support operations or the civil support aspects of an operation. The USCG works effectively against an unconventional threat and threats with on-water capability.

## URBAN FUNCTIONAL COMBATANT COMMAND CAPABILITIES

D-25. Three commands provide urban functional combatant command capabilities. Transportation Command manages the global transportation system. Strategic Command integrates plans for and employs space forces ad assets. Special Operations Command is responsible for providing special operations forces.

### TRANSPORTATION COMMAND

D-26. U.S. Transportation Command (USTRANSCOM) provides strategic air, land, and sea transportation for the Department of Defense to deploy, employ, sustain, and redeploy U.S. military forces worldwide. USTRANSCOM provides global transportation management, using an integrated transportation system across the spectrum of operations through its transportation component commands (TCCs). The TCC consists of Air Mobility Command (AMC), Military Sealift Command (MSC), and Surface Deployment and Distribution Command (SDDC).

D-27. During urban operations, USTRANSCOM, through its TCC, can provide common-user terminal services in support of strategic transportation movements to a theater of operations. AMC provides common-user airlift, air refueling, and aeromedical evacuation services. It is the worldwide aerial port manager and, where designated, the operator of common-user aerial ports of embarkation and aerial ports of debarkation. MSC provides common-user sealift services between seaports of embarkation (SPOEs) and seaports of debarkation (SPODs). SDDC provides common-user ocean terminal service and, where designated, serves as the single port manager at SPOEs and SPODs. The urban transport system, in the form of railheads, ports, and airfields, is integral to many urban operations across the spectrum of Army operations and often the objective of Army UO. Army forces planning, preparing, and executing urban operations engage the supported geographic commander of a combatant command or his component commands in all aspects of conducting UO when the objective is transport related.

D-28. In offensive operations, commanders use transportation expertise to identify the urban transportation infrastructure (both the terrain and social [human] aspects) that is secured or that can affect current and future operations. This analysis also includes second- and third-order support systems. USTRANSCOM is consulted regarding the degree of acceptable damage that the system can sustain and still meet mission requirements. During execution, USTRANSCOM units may integrate into the operation so they can begin operating the transportation systems as early as possible. During the Inchon landing of September 1950, Army forces had begun rail operations on D+1 and port operations under way by D+3 of the forcible entry.

D-29. In defensive and stability or civil support operations, USTRANSCOM units safeguard and prevent disruption of the transport system by the conduct of defensive operations or stability tasks.

USTRANSCOM advises Army commanders of the impact of defensive tactics, techniques, and procedures (TTP) on USTRANSCOM operations as well as security requirements for USTRANSCOM facilities. The JFC provides guidance to deconflict any issues that may arise from these potentially divergent missions and tasks.

## STRATEGIC COMMAND

D-30. The United States Strategic Command (USSTRATCOM) is the combatant command that integrates and synchronizes space operations, information operations, integrated missile defense, global command and control (C2); ISR, global strike, and strategic deterrence to provide. A unified source for greater understanding of specific threats around the world and the means to rapidly respond to those threats including weapons of mass destruction. USSTRATCOM operates assigned forces through its service component commands: U.S. Army Space Command/Army Forces Strategic Command; Air Force Space Command, and Naval Net Warfare Command.

D-31. USSTRATCOM provides critical space systems that offer global coverage and potential for real-time and near real-time support to urban operations. Space systems are unconstrained by political boundaries. Commanders can use space systems during peacetime or times of crisis to monitor an urban area before inserting friendly forces. Space systems enhance operations and assist commanders in overcoming some of the physical challenges of the urban environment. These enhancement operations include ISR; positioning and navigation; environmental monitoring; and communications. Space-based imagery and sensors are important ISR capabilities that contribute to situational understanding throughout the depth, breadth, and height of the urban area. ISR systems can provide route and target information for mission planning, locate presurveyed missile launch sites, detect camouflage, assess threat operations and movements, and warn of hostile acts and reconnaissance. Positioning and navigation systems also assist situational understanding through links to digital information systems (INFOSYS) while assisting tactical navigation in ambiguous terrain found in some urban areas. These enhancements can include precise location and position information for urban fires, ingress and egress routes, and rendezvous coordination. Environmental monitoring systems can provide weather and ionospheric information needed to assess weapon selection, air routes, ground and water trafficability, and communications. Communications systems provide secure, survivable links between elements of Army and joint forces to disseminate plans, orders, and warnings. These systems may form a critical link in the INFOSYS that transmit data to assessment centers and intelligence to key decision makers. However, some of the same environmental influences and degradation as ground-based systems may affect space-based systems. Commanders should refer to FM 3-14 for more information on space support to Army operations.

## SPECIAL OPERATIONS COMMAND

D-32. U.S. Special Operations Command (USSOCOM) exercises combatant command of all active and reserve special operations forces stationed in the United States. USSOCOM also provides trained and combat-ready special operations forces (SOF) to the geographical combatant commanders and, when directed by the President or the Secretary of Defense, command-designated special operations.

D-33. SOF provide commanders with capabilities critical to success in the urban environment. The density of this environment in both space and time requires the careful integration of SOF and conventional forces. Army forces conducting UO have a clear and unambiguous C2 relationship with the SOF in the urban area to ensure coordination, massing of effects, and unity of effort.

D-34. Each service has unique special operations (SO) capabilities. For example, SOF can identify and seize or destroy key terrain or infrastructure in denied areas; secure or capture key personnel; identify supportive civilians (aides and allies) who can facilitate future infiltrations and the cache of critical supplies; counter urban insurgencies; and conduct unconventional warfare in enemy-held urban areas. SOF can also emplace covert sensors and surveillance devices, provide clandestine intelligence collection, and provide target acquisition information or emplace beacons (for weapons guidance or navigation) in the highly restrictive terrain of the urban environment. In multinational UO, they can provide coalition support teams with trained, culturally aware, language proficient, military liaison personnel with organic communications connectivity. Army leaders understand the SO capabilities available to ensure that they request the right support for Army UO and to ensure unity of effort within the urban JOA.

## Army SOF

D-35. The Army provides five types of SOF units to USSOCOM. Two of these, psychological operations (PSYOP) and civil affairs, are discussed in Chapter 4.

### *Ranger Forces*

D-36. The U.S. Army 75th Ranger Regiment can rapidly deploy light infantry forces from company through regimental size. Rangers specialize in direct action (DA) SO missions and focus on airfield seizure and raids. Typically, Ranger units turn over their objectives to conventional units upon mission completion. The Ranger DA capability especially applies to UO because of the many critical infrastructures that often prompt Army forces to engage in UO. Ranger capabilities are ideal for seizing critical facilities to preempt their defense or destruction. The size and combat power of Ranger units permit execution of offensive and defensive operations against enemy conventional units for periods of limited duration; austere combat service support capabilities limit the Ranger regiment's ability to sustain combat action without extensive augmentation.

### *Special Forces*

D-37. U.S. Army Special Forces Command (Airborne) trains and prepares Army Special Forces (SF) to deploy and execute operational requirements for geographical combatant commanders. SF units are small and capable of extended operations in remote and hostile locations. SF units execute seven basic missions: foreign internal defense (FID), unconventional warfare (UW), counterproliferation of weapons of mass destruction, special reconnaissance (SR), DA, counterterrorism, and information operations. Important collateral activities consist of coalition support, combat search and rescue, counterdrug activities, humanitarian demining activities, humanitarian assistance, security assistance, and special activities.

D-38. SR, DA, and coalition support are particularly important in UO. SF units may physically penetrate an urban area to conduct SR to determine threat strengths, dispositions, and intentions. In some situations, SF units, due to their language and cultural training, can gather accurate information regarding the disposition and attitudes of the population. SF special reconnaissance can also determine or verify the functional status of urban infrastructure as well as conduct target acquisition, area assessment, and poststrike reconnaissance.

D-39. SF teams can execute terminal guidance and with qualified joint tactical air controllers can control operations for Army and joint precision fires. Although these teams possess limited organic combat power for their short-duration, DA missions, they can bring significant effects to bear against high-payoff targets. Yet, SF units have no capability to conduct a sustained defense of such targets.

D-40. SF units can advise, train, and assist urban indigenous movements already in existence to conduct unconventional warfare and possibly accompany these groups into combat. The upsurge in urban insurgency and terrorism has caused worldwide concern since it is not confined to developing countries. Present day dissident groups are well aware of this situation. These groups realize that to be successful, they must center the insurgent activities on the major cities or political center of their countries.

### *Special Operations Aviation Forces*

D-41. Army special operations aviation provided by the 160th Special Operations Aviation Regiment operates primarily to support SO missions. They execute insertion, extraction, and resupply missions to support SO. Similar to conventional aviation, this capability may be more vulnerable to concealed air defenses when operating over hostile or unsecured urban terrain than in many other environments.

## Navy SOF

### *Sea-Air-Land Teams*

D-42. Navy sea-air-land teams (SEALs) specialize in water approaches to targets. They operate in small, squad-size teams and have many of the same capabilities as Army Special Forces (see above). Navy SEALs do not typically have the cultural and language training of Army Special Forces. Their capability to insert from sea gives them a unique ability to penetrate into urban areas that are accessible from the sea.

### *Special Boat Units*

D-43. Special boat units (SBUs) employ, operate, and maintain various surface combatant craft (see figure D-3) to conduct and support naval and joint special operations, riverine warfare, and coastal patrol and interdiction. The SBU can infiltrate and exfiltrate forces; provide small-caliber gunfire support; and conduct coastal patrol, surveillance, harassment, and interdiction of maritime lines of communications. These units are ideal in UO that include ports and rivers. They can assist in port security, conduct river patrols, and participate in harbor defense. They are well suited for preventing sea infiltration by unconventional threats.

**Figure D-3. USN MK V special**

## Operations CraftAir Force SOF

D-44. The Air Force has three primary SOF elements: special operations air units, battlefield Airmen, and combat aviation advisory units.

### *Special Operations Wings and Groups*

D-45. Air Force special operations aviation elements operate both fixed-wing and rotary-wing aircraft (which include variants of both C-130s and MH-53s). Air Force SOF conduct core missions which include specialized air mobility, precision aerospace fires, aerospace surface interface, personnel recovery operations, specialized refueling, combat aviation advisory, information operations, and PSYOP. As a result, Air Force SOF conduct infiltrations, exfiltrations, resupplies, close air support of SOF ground elements, electronic warfare, and aerial C2 support.

D-46. The AC-130 U and AC-130 H model aircraft (see figure D-4) are designed specifically to provide close air support to ground SOF. However, they will be an invaluable asset to any ground force operating in an urban environment. Hence, Army commanders seek opportunities to train with AC-130 gunships prior to the conduct of UO. AC-130s are armed with one 40mm autocannon and one 105mm howitzer. The AC-130 U is also armed with one 25mm autocannon. This aircraft has night capability and is extremely accurate. Its fires are responsive and can be decisive in close urban combat. Its cannon and howitzer are accurate enough to concentrate fire onto a single spot to create a rooftop breach that allows fire to be directed deeper into the building. The aircraft can provide excellent covert illumination with its infrared spotlight. The AC-130 is very vulnerable to air defense systems, which friendly forces must suppress or destroy to effectively use this system. (During OPERATION DESERT STORM in 1991, an AC-130 was shot down over Kuwait.)

#### *Battlefield Airmen*

D-47. Battlefield Airmen consist of special tactics teams (STTs) and special operations weather teams (SOWTs). Combat controllers and pararescue personnel constitute STTs. STTs are specially tailored to meet mission criteria and may vary from a small three-man team to a larger twenty-five-man element. STTs support the UO Army commanders by—

- Performing air-land-sea personnel recovery operations.
- Providing terminal attack control or guidance.
- Establishing and operating navigational aids and beacons.
- Providing liaison to ground commanders.
- Providing visual flight rules and limited instrument flight rules air traffic control.
- Providing positive control of the terminal objective area aviation environment during SOF operations.

---

*Note*: SOWTs are normally attached to Army SOF to provide weather observation and limited tactical forecasting.

---

#### *Combat Aviation Advisory (CAA) Operations*

D-48. CAA operations are part of an interagency process aimed toward freeing and protecting the host nation's population from subversion, lawlessness, and insurgency. Specifically, CAA operations are tailored to assess, train, advise, and assist foreign aviation forces in air operations employment and sustainability. CAA operations support geographical combatant commanders throughout the spectrum of conflict primarily by facilitating the availability, reliability, safety and interoperability, and integration of friendly and allied aviation forces supporting joint, interagency, and multinational forces and organizations. CAA operations provide assistance in the interrelated areas of FID, coalition support, UW, and humanitarian and disaster relief. CAA operations also include a liaison role in coalition support.

## MULTINATIONAL CONSIDERATIONS

D-49. Army UO in foreign urban areas will often be joint and likely have a multinational component. When properly executed, integrating multinational forces into UO greatly enhances the operation's military (as well as political) effectiveness. Properly integrating multinational forces into UO requires a thorough understanding of both the urban environment and the nature of individual national forces. This understanding includes the political, cultural, and historical characteristics of the other national forces. Such understanding also includes the national force's doctrine and military capabilities, strengths, and weaknesses. In UO, these considerations are critical because these factors will alter the urban population's attitude toward multinational forces and the behavior of such forces as they interact with the urban population. Combining this understanding with effective C2 and an equal understanding of the urban environment results in effective multinational UO (see FM 100-8).

### CULTURAL COMPATIBILITY

D-50. When evaluating and assigning UO tasks, Army commanders also consider the degree of interaction with the civilian population. A national force from a Muslim-majority country may work better with a civilian population that is also Muslim than a force having a different religion. In such a situation, the national force with the same religion as the urban population may be assigned tasks that require close relations with civilians. Army forces, in this situation, may be assigned tasks that are more remote from the population. National forces that have a national history of animosity to the civil population (or that sympathize with antagonists within the civilian population) are not used in tasks requiring diplomacy and close cooperation with the civilians or government.

## Military Capabilities and Restrictions

D-51. Army forces are responsible to understand the military capabilities of national forces with which they work. Some national forces, as part of their normal capabilities, are adept at police functions that enable them to operate with little training in a law enforcement role. Other national forces specialize in small-unit, light infantry patrolling. These forces may be ideal in a stability operation. In contrast, a national force composed of conscripts and trained primarily in conventional warfare techniques may best work as a reinforcing force or may require extensive training before mission execution in an urban environment or a stability situation. Army commanders also consider the type of weapon systems with which participating multinational forces are equipped; they may be more or less effective in an urban environment. For example, some countries may still possess the Vulcan antiaircraft gun system (or similar weapon) that can be very useful in urban offensive and defensive operations.

D-52. Commanders in multinational UO must also understand nation-specific rules of engagement (ROE). Some nations will severely limit their participating military's use of lethal and nonlethal fires. On the other hand, other nations may not be as constrained as Army units. Ideally, commanders attempt to create ROE that are the same across the coalition but recognize that this may not be achievable. Hence, in an environment where minimization of collateral damage will be paramount, Army commanders will need to carefully consider national limitations on the employment of forces when planning and executing UO. Relatedly, commanders take into account each nation's ability to achieve precision effects with their available forces and systems.

## Intelligence Sharing

D-53. Due to classification issues, sharing intelligence with coalition partners may be challenging; the United States may have close intelligence ties with some countries within the coalition and few or none with others. In many cases intelligence personnel from other countries have unique skills (and cultural perspectives) that can significantly contribute to the friendly intelligence effort. Reports from some coalition members may fill intelligence gaps for Army forces and the coalition as a whole. On the other hand, few countries have the sophisticated intelligence collection assets available to Army forces and information that Army units may provide could be critical to a coalition partner's force protection and overall mission success.

## Information Management

D-54. National capabilities to collect, process, store, display, and disseminate information and establish a common operational picture vary considerably between coalition partners. As the capabilities of Army information systems increase, those of our future partners may not keep pace. Even when capabilities are similar, equipment may not be compatible. The complex nature of the urban environment and coalition language differences (and linguist shortages) will only serve to exacerbate incompatibilities and limitations. Therefore commanders must understand capabilities and limitations to ensure that orders, control measures, and information can be appropriately disseminated, displayed, and understood. Significantly, they must understand the coalition's ability to accurately locate and track friendly forces as well as identify friend from foe. Commanders must take the appropriate steps to ensure that essential intelligence and relevant information is provided to coalition partners in a form (and language) that they can accept and use. This will be critical to establishing the common operational picture necessary to facilitate rapid decision making and effective command and control across the coalition.

## Logistic Requirements

D-55. Logistically, agreement among the multinational forces should include support consolidation whenever possible. Creating a multinational logistics office or coordination center works best when accomplished early. This office can coordinate local contracts as well as already agreed upon host-nation support. This coordination among participating nations will reduce the competition for local assets that could otherwise have detrimental effects on one or more participating nations. Additionally, many nations providing forces for a coalition effort may not be able to support them logistically and will rely almost entirely on the United States or another coalition army for logistic support. Finally, pooled resources such as transportation assets and maintenance personnel could potentially reduce deployment requirements for one or more coalition partners. As the coalition matures, the role of this office or center may be expanded to include command activities.

This page intentionally left blank.

# Source Notes

These are sources quoted or paraphrased in this publication. They are listed by page number. Quotations are identified by the first few word of the quote. Where a quote is embedded within a paragraph, the paragraph number is listed. Boldface indicates the title of historical vignettes.

**Chapter 1–The Urban Outlook**

1-1 "Today's security environment demands more...": FM 1, *The Army*, 14 June 2005: 1-19.

1-4 **Rome: A Microcosm of Urban Warfare**: R. Ernest Dupuy and Trevor N. Dupuy, *The Encyclopedia of Military History: From 3500 B.C. to the Present* (San Francisco: Harper & Row, 1977).

1-6 **The Three Block War**: as quoted in Russell W. Glenn, Steven Hartman, and Scott Gerwehr, *Urban Combat Service Support Operations: The Shoulders of Atlas* (Santa Monica, CA: RAND, 2003): 26.

1-8 **Winning the Peace in Iraq: The Requirement for Full Spectrum Operations**: Peter W. Chiarelli and Patrick R. Michaelis, "Winning the Peace: The Requirement for Full-Spectrum Operations," *Military Review* (July/August 2005): 4.

**Chapter 2–Understanding the Urban Environment**

2-1 "From a planning perspective, commanders view cities....": FM 3-0, *Operations*, 14 June 2001: 6-19.

2-16 "Me and Somalia Against the World": cited in Andrew S. Natsios, "Food Through Force: Humanitarian Intervention and U.S. Policy," *The Washington Quarterly* (winter 1994): 136.

2-18 **Understanding the Effects of Unit and Soldier Actions in Iraq**: *Initial Impressions Report: Operation Iraqi Freedom—Stability Operations-Support Operations,* Center for Army Lessons Learned (CALL) Newsletter No. 04-13 (Fort Leavenworth, KS: Center for Army Lessons Learned, U.S. Training and Doctrine Command [TRADOC], 2004): ii.

2-21 "Whoever coined the phrase... ": General Sir Rupert Smith, "Wars in Our Time–A Survey of Recent and Continuing Conflicts," *World Defence Systems*, volume 3:2 (London: Royal United Services Institute for Defence Studies, 2001): 121.

2-25 **Understanding the Urban Environment: Paris – 1944**: Russell W. Glenn, "Urban Combat is Complex," *Proceedings* (February 2002): 65. Russell W. Glenn, Steven Hartman, and Scott Gerwehr, *Urban Combat Service Support Operations: The Shoulders of Atlas* (Santa Monica, CA: RAND, 2003): 10-11.

**Chapter 3–Understanding the Urban Threat**

3-1 "...[T]he United States could be forced to intervene...": *Quadrennial Defense Review Report*, 30 September 2001: 6.

3-3 "...Iraq made no direct effort...": Gregory Fontenot, E. J. Degen, and David Tohn, *On Point: US Army in Operation IRAQI FREEDOM* (Fort Leavenworth, KS: Combat Studies Institute Press, 2004): 388.

3-4 **Tempo**: 26th Infantry Regimental Association, *Aachen. Military Operations in Urban Terrain* (Lititz, PA: 26th Infantry Regimental Association, 1999); Irving Werstein, *The Battle of Aachen* (New York: Thomas Y. Crowell Company, 1962); Charles Whiting, *Bloody Aachen* (New York: Stein and Day, 1976).

3-6 "Chechen fighters sometimes disguised...": Olga Oliker, *Russia's Chechen Wars 1994–2000: Lessons from Urban Combat* (Santa Monica, CA: RAND, 2001): 21.

3-7 **Identifying Threats from Noncombatants**: Olga Oliker, *Russia's Chechen Wars 1994–2000: Lessons from Urban Combat* (Santa Monica, CA: RAND, 2001).

3-9 **Information and the Media**: Eric Hammel, *Fire in the Streets: The Battle for Hue, Tet 1968* (New York: Dell Publishing, 1991).

3-12 **Cultural and Religious Instability**: Sid Heal, "Crowds, Mobs and Nonlethal Weapons," *Military Review* (March/April 2000): 45–50.

3-13 **Food and Water Shortages**: Leif R. Rosenberger, "The Strategic Importance of the World Food Supply," *Parameters* (spring 1997): 84–105.

3-15 **Insurgencies and the Urban Society**: Roger Trinquier, *Modern Warfare: A French View of Counterinsurgency* (New York: Praeger Publisher, 1964): 16-17.

3-16 **Crime and Criminal Organizations**: Eugene Linden, "The Exploding Cities of the Developing World," *Foreign Affairs* (January/February 1996): 52–65; Andrew S. Natsios, "Commander's Guidance: A Challenge of Complex Humanitarian Emergencies," *Parameters* (summer 1996): 50–66.

**Chapter 4–Understanding the Urban Environment's Effects on Warfighting Functions and Tactics**

4-1 "War is, above all things, an art...": Francis V. Greene, "The Important Improvements in the Art of War During the Past Twenty Years and Their Probable Effect on Future Military Operations," *Journal of the Military Service Institution of the United States* 4, no. 13 (1883): 41.

4-13 "Fighting in a city is much...": Vasili I. Chuikov, *The Battle for Stalingrad* (New York: Holt, Rinehart, and Winston, 1964): as cited by JP 3-06, *Doctrine for Urban Operations*, 16 September 2002: 1-8.

4-14 "I heard small-arms fire and...": Mark A. B. Hollis, "Platoon Under Fire," *Infantry* (January/April 1998): 29–30.

4-17 **Example of Simple Communications Innovation: Israel's Six-Day War – 1967**: Trevor N. Dupuy and Paul Martell, *Flawed Victory: The Arab-Israeli Conflict and the 1982 War in Lebanon* (Fairfax, VA: Hero Books, 1986); Richard A. Gabriel, *Operation Peace for Galilee: The Israeli-PLO War in Lebanon* (New York: Hill and Wang, 1984).

**Chapter 5–Contemplating Urban Operations**

5-1 "We based all our further calculations...": S. M. Shtemenko, *The Soviet General Staff at War, 1941-1945: Book One* (Moscow: Progress Publishers, 1981): 317–318.

5-19 "Four hostile newspapers...": Napoleon Bonaparte, Justin Wintle (ed.), *The Dictionary of War Quotations* (New York: The Free Press, 1989): 73.

5-24 **Conventional and Special Forces Integration**: Gregory Fontenot, E. J. Degen, and David Tohn, *On Point: US Army in Operation IRAQI FREEDOM* (Fort Leavenworth, KS: Combat Studies Institute Press, 2004): 403.

**Chapter 6–Foundations for Urban Operations**

6-1 "Utilities such as electricity…": Richard Connaughton, John Pimlott, and Duncan Anderson, *The Battle for Manila: The Most Devastating Untold Story of World War II* (Novata, CA: Presidio Press, 1995): 103.

6-7 **Applying the Urban Operational Framework: Panama – 1989**: Ronald H. Cole, *Operation Just Cause* (Washington DC: Joint History Office, 1995); Thomas Donnelly, Margaret Roth, and Caleb Baker, *Operation Just Cause: The Storming of Panama* (New York: Lexington Books, 1991); John T. Fishel, *The Fog of Peace: Planning and Executing the Restoration of Panama* (Carlisle, PA: Strategic Studies Institute, 1992); Malcolm McConnell, *Just Cause: The Real Story of America's High-Tech Invasion of Panama* (New York: St. Martin's Press, 1991); John Embry Parkerson, Jr., "United States Compliance with Humanitarian Law Respecting Civilians During Operation Just Cause," *Military Law Review* (1991): 31–140; Jennifer Morrison Taw, *Operation Just Cause: Lessons for Operations Other Than War* (Santa Monica, CA: RAND, 1996).

6-12 "I'm talking about attacking...": William S. Wallace quoted in Michael R. Gordon, "Baghdad Targets Picked if Hussein Holes Up There," *New York Times* (7 March 2003): A11.

6-13 "Do not try to..." T. E. Lawrence, "Twenty-Seven Articles," *The Arab Bulletin* (20 August 1917): as cited in David H. Petraeus, "Learning Counterinsurgency: Observations from Soldiering in Iraq," *Military Review* (January/February 2006): 4–5.

**Chapter 7–Urban Offensive Operations**

7-1 "…Capture Suez City...": Abraham Adan, *On the Banks of the Suez* (Novato, CA: Presidio Press, 1980): 409.

7-2 **Operational Context of Urban Operations: Brittany Ports – 1944**: Kent H. Butts, "The Strategic Importance of Water," *Parameters* (spring 1997): 65–83; Paul Carell, *Stalingrad: The Defeat of the German 6th Army* (Atglen, PA: Schieffer Publishing, 1993); R. Ernest Dupuy and Trevor N. Dupuy, *The Encyclopedia of Military History: From 3500 B.C. to the Present* (San Francisco: Harper & Row, 1977); Ralph Peters, "Our Soldiers, Their Cities," *Parameters* (spring 1996): 43–50; John J. Peterson, *Into the Cauldron* (Clinton, MD: Clavier House, 1973).

7-10 **Forms of Attack in the Urban Offense: Metz – 1944**: John Colby, *War From the Ground Up: The 90th Division in WWII* (Austin, TX: Nortex Press, 1991); Anthony Kemp, *The Unknown Battle: Metz, 1944* (New York: Stein and Day, 1981).

7-17 **Isolating an Urban Area: Hue, Vietnam – 1968**: James R. Arnold, *Tet Offensive 1968: Turning Point in Vietnam* (London: Osprey, 1990); Eric Hammel, *Fire in the Streets: The Battle for Hue, Tet 1968* (New York: Dell Publishing, 1991).

7-21 "Often what seems to be...": Russell W. Glenn, *An Attack on Duffer's Downtown* (Santa Monica, CA: RAND, 2001): 28.

7-22 **Creative Task Organization: Using Artillery in the Direct Fire Role**: 26th Infantry Regimental Association, *Aachen. Military Operations in Urban Terrain* (Lititz, PA: 26th Infantry Regimental Association, 1999); Ronald H. Cole, *Operation Just Cause* (Washington DC: Joint History Office, 1995); Thomas Donnelly, Margaret Roth, and Caleb Baker, *Operation Just Cause: The Storming of Panama* (New York: Lexington Books, 1991); David Eshel, *Mid-East Wars: The Lebanon War 1982* (Hod Hasharon, Israel: Eshel-Dramit, 1983); Richard A. Gabriel, *Operation Peace for Galilee: The Israeli-PLO War in Lebanon* (New York: Hill and Wang, 1984); Malcolm McConnell, *Just Cause: The Real Story of America's High-Tech Invasion of Panama* (New York: St. Martin's Press, 1991); Irving Werstein, *The Battle of Aachen* (New York: Thomas Y. Crowell Company, 1962); Charles Whiting, *Bloody Aachen* (New York: Stein and Day, 1976); Bruce Allen Watson, *Sieges: A Comparative Study* (Westport, CT: Praeger, 1993).

7-26 **Bold Operational Maneuver to Seize an Urban Area: Inchon and Seoul, Korea – 1950**: Robert Debs Heinl, Jr., *Victory at High Tide: The Inchon-Seoul Campaign* (Annapolis, MD: Nautical & Aviation Publishing Company of America, 1979); Douglas MacArthur, *Reminiscences* (New York: McGraw-Hill, 1964); Shelby L. Stanton, *Ten Corps in Korea, 1950* (Novato, CA: Presidio Press, 1989).

## Chapter 8–Urban Defensive Operations

8-1 "Generally, a modern city magnifies...": Christopher R. Gabel, "Military Operations on Urbanized Terrain: The 2d Battalion, 26th Infantry, at Aachen, October 1944," *Urban Combat Operations: tactics, Techniques and Procedure.* Center for Army Lessons Learned (CALL) Newsletter No. 99-16 (Fort Leavenworth, KS: Center for Army Lessons Learned, U.S. Training and Doctrine Command [TRADOC], November 1999): 1-9.

8-5 **Urban Defense in a Major Operation: Stalingrad –1942 to 1943**: Anthony Beevor, *Stalingrad* (New York: Penguin Books, 1998); Vasili I. Chuikov, *The Battle for Stalingrad* (New York: Holt, Rinehart, and Winston, 1964); William Craig, *Enemy at the Gates: The Battle for Stalingrad* (New York: Reader's Digest Press, 1973).

8-8 "What is the position...": Michael Dewar, *War in the Streets: The Story of Urban Combat from Calais to Khafji* (New York: Sterling Publishers, 1992): 8.

8-10 **Defensive Combat Power: Suez – 1973**: Abraham Adan, *On the Banks of the Suez* (Novato, CA: Presidio Press, 1980).

## Chapter 9–Urban Stability and Civil Support Operations

9-1 "The Rangers were bound...": Mark Bowden, *Black Hawk Down: A Story of Modern War* (New York: Atlantic Monthly Press, 1999): 18.

9-5 "In wars of intervention...": Antoine-Henri Jomini, "The Art of War," in *Roots of Strategy: Book 2* (Harrisburg, PA: Stackpole Books, 1987): 441.

9-8 **Assessment of Security and Force Protection: Belfast, Northern Ireland**: Michael Dewar, *British Army in Northern Ireland* (New York: Sterling Publishers, 1985); Desmond Hamill, *Pig in the Middle: The Army in Northern Ireland, 1969-1984* (London: Methuen, 1985).

9-11 "If there is any lesson...": Frederick Funston, "How the Army Worked to Save San Francisco," *Cosmopolitan Magazine* (July 1906). Available at http: //www.sfmuseum.org/1906/cosmo.html.

9-12 **Civil Support and Coordination with Civilian Authorities: Los Angeles – 1992**: James D. Delk, *Fires & Furies: The LA Riots* (Palm Springs, CA: Etc. Publications, 1995); William W. Mendel, *Combat in Cities: The LA Riots and Operation Rio* (Fort Leavenworth, KS: Foreign Military Studies Office, 1996).

9-14 **Restraint: An Najaf, Iraq – 2003**: Gregory Fontenot, E. J. Degen, and David Tohn, *On Point: US Army in Operation IRAQI FREEDOM* (Fort Leavenworth, KS: Combat Studies Institute Press, 2004).

9-16 "I met immediately with the...": Jeffrey B. Van Sickle, "Stability Operations in Northern Iraq: Task Force Altun Kupri," *Infantry* (January/February 2005): 26.

**Chapter 10–Urban Sustainment**

10-1 "Even supply is different..." Ralph Peters, "Our Soldiers, Their Cities," *Parameters* (spring 1996): 48.

10-2 **Base Security: Tan Son Nhut, Vietnam – 1968**: James R. Arnold, *Tet Offensive 1968: Turning Point in Vietnam* (London: Osprey, 1990); Keith William Nolan, *The Battle for Saigon: Tet 1968* (New York: Pocket Books, 1996).

10-4 "[Sustainment] planning accounts for...": FM 3-0, *Operations*, 14 June 2001: 6-19.

10-8 "Greater friendly force density would appear...": Russell W. Glenn, Steven Hartman, and Scott Gerwehr, *Urban Combat Service Support Operations: The Shoulders of Atlas* (Santa Monica, CA: RAND, 2003): 4.

10-13 **Host Nation Repair Operations: OPERATION IRAQI FREEDOM**: Steven Leonard, personal experience written specifically for FM 3-06.

10-18 **Combat Stress: Chechnya – 1994 to 1996**: Timothy L. Thomas and Charles P. O'Hara, "Combat Stress in Chechnya: 'The Equal Opportunity Disorder'," *Army Medical Department Journal* (January/March 2000): 46–53.

10-22 "...[I]n an endeavor like that in Iraq...": David H. Petraeus, "Learning Counterinsurgency: Observations from Soldiering in Iraq," *Military Review* (January/February 2006): 4–5.

**Appendix A–Siege of Beirut: An Illustration of the Fundamentals of Urban Operations**

A-1 **Siege of Beirut: An Illustration of the Fundamentals of Urban Operations**: Trevor N. Dupuy and Paul Martell, *Flawed Victory: The Arab-Israeli Conflict and the 1982 War in Lebanon* (Fairfax, VA: Hero Books, 1986); David Eshel, *Mid-East Wars: The Lebanon War 1982* (Hod Hasharon, Israel: Eshel-Dramit, 1983); Richard A. Gabriel, *Operation Peace for Galilee: The Israeli-PLO War in Lebanon* (New York: Hill and Wang, 1984).

A-1 "The [Israeli Defense Forces] IDF had...": Richard A. Gabriel, *Operation Peace for Galilee: The Israeli-PLO War in Lebanon* (New York: Hill and Wang, 1984): 128.

**Appendix B–An Urban Focus to the Intelligence Preparation of the Battlefield**

B-1 "Maneuvers that are possible...": *Infantry in Battle, 2d ed.* (Richmond, VA: Garret & Massie, 1939; Reprint, Fort Leavenworth, KS: U.S. Army Command and General Staff College, 1981): 69.

B-5 **Analysis of an Urban Area's Underlying Terrain: Mitrovica, Kosovo – 1999**: Jamison Jo Medby and Russell W. Glenn, *Street Smart: Intelligence Preparation of the Battlefield for Urban Operations* (Santa Monica, CA: RAND, 2002).

B-10 "...[T]he adversaries of freedom...": John F. Kennedy, Message to Congress, 25 May 1961. Available at <http://www.jfklibrary.org/Historical+Resources/Archives>.

B-12 **Shifting Civilian Interests and Intent**: James D. Delk, *Fires & Furies: The LA Riots* (Palm Springs, CA: Etc. Publications, 1995); Jamison Jo Medby and Russell W. Glenn, *Street Smart: Intelligence Preparation of the Battlefield for Urban Operations* (Santa Monica, CA: RAND, 2002).

**Appendix C–Operations in Somalia: Applying the Urban Operational Framework to Stability Operations**

C-1 **Operations in Somalia: Applying the Urban Operational Framework to Stability and Reconstruction Operations**: Kenneth Allard, *Somalia Operations: Lessons Learned* (Washington, DC: National Defense University Press, 1995).

C-1 "It's impossible for an American mother...": Thomas M. Montgomery, interview, "Ambush in Mogasdishu," *Frontline*, WGBH Boston, 29 September 1998, <http://www.pbs.org/wgbh/pages/frontline/programs/1998.html>.

C-2 C-5 ..." to use all necessary...": *United Nations Security Council Resolution 794* (1992) (New York: United Nations, 1992), <http://daccessdds.un.org/doc/UNDOC/GEN/N92/772/11/PDF>.

C-2 C-5 "When directed by the...": CENTCOM mission statement as cited in Kenneth Allard, *Somalia Operations: Lessons Learned* (Washington, DC: National Defense University Press, 1995): 13-14.

### Appendix D–Joint and Multinational Urban Operations

D-1 "[Joint force commanders] synchronize the actions…" JP 3-0, *Doctrine for Joint Operations*, 10 September 2001: II-4.

# Glossary

## SECTION I – ACRONYMS AND ABBREVIATIONS

| | |
|---|---|
| **AI** | air interdiction |
| **AIDS** | acquired immune deficiency syndrome |
| **AMC** | Air Mobility Command |
| **AO** | area of operations |
| **ARVN** | Army of the Republic of Viet Nam |
| **ASCOPE** | areas, structures, capabilities, organizations, people, and events |
| **BCT** | brigade combat team |
| **BDAR** | battle damage assessment and repair |
| **BENELUX** | Belgium, Netherlands, and Luxembourg |
| **C2** | command and control |
| **CA** | civil affairs |
| **CAA** | combat aviation advisory |
| **CASEVAC** | casualty evacuation |
| **CBRN** | chemical, biological, radiological, nuclear |
| **CBRNE** | chemical, biological, radiological, nuclear, high-yield explosive |
| **CCIR** | commander's critical information requirement |
| **CI** | counterintelligence |
| **CEA** | captured enemy ammunition |
| **CERP** | commander's emergency response program |
| **CLS** | combat lifesaver |
| **CMO** | civil-military operations |
| **CMOC** | civil-military operations center |
| **CNA** | computer network attack |
| **CND** | computer network defense |
| **CNE** | computer network exploitation |
| **CNO** | computer network operations |
| **COA** | course of action |
| **COG** | center of gravity |
| **CONUS** | continental United States |
| **COP** | common operational picture |
| **DA** | direct action |
| **DNBI** | disease and nonbattle injury |
| **DOTMLPF** | doctrine, organization, training, materiel, leadership and education, personnel, and facilities |
| **EAC** | echelons above corps |
| **EEFI** | essential elements of friendly information |

| | |
|---|---|
| **EOD** | explosive ordnance disposal |
| **EW** | electronic warfare |
| **FHP** | force health protection |
| **FID** | foreign internal defense |
| **FM** | field manual; frequency modulation |
| **FMI** | field manual, interim |
| **FSB** | forward support battalion |
| **G-9** | assistant chief of staff, civil affairs |
| **GIS** | geographic information system |
| **GTL** | gun-target line |
| **HRS** | human resources support |
| **HUMINT** | human intelligence |
| **IDF** | Israeli Defense Forces |
| **IED** | improvised explosive device |
| **IMINT** | imagery intelligence |
| **INFOSYS** | information system |
| **IO** | information operations |
| **IPB** | intelligence preparation of the battlefield |
| **ISR** | intelligence, surveillance, and reconnaissance |
| **JDAM** | Joint Direct Attack Munition |
| **JFC** | joint force commander |
| **JOA** | joint operations area |
| **JP** | joint publication |
| **JTF** | joint task force |
| **JUO** | joint urban operations |
| **LAN** | local area network |
| **LNO** | liaison officer |
| **LOC** | line of communications |
| **LOGCAP** | logistics civilian augmentation program |
| **LOS** | line of sight |
| **LPT** | logistics preparation of the theater |
| **MAGTF** | Marine air-ground task force |
| **MANPADS** | man-portable air defense system |
| **MEDEVAC** | medical evacuation |
| **METT-TC** | mission, enemy, terrain and weather, troops and support available, time available, civil considerations |
| **MEU** | Marine expeditionary unit |
| **MEU(SOC)** | Marine expeditionary unit (special operations capable) |
| **MHE** | materials handling equipment |
| **MLRS** | Multiple Launch Rocket System |
| **MP** | military police |

| | |
|---|---|
| **MSC** | Military Sealift Command |
| **MST** | maintenance support team |
| **NATO** | North Atlantic Treaty Organization |
| **NCW** | naval coastal warfare |
| **NGA** | National Geospatial-Intelligence Agency |
| **NGO** | nongovernmental organization |
| **NKPA** | North Korean People's Army |
| **NVA** | North Vietnamese Army |
| **OAKOC** | observation and fields of fire, avenues of approach, key terrain, obstacles, and cover and concealment |
| **ODA** | operational detachment alpha |
| **OIF** | Operation Iraqi Freedom |
| **OPSEC** | operations security |
| **PA** | public affairs |
| **PDF** | Panamanian Defense Force |
| **PGM** | precision-guided munitions |
| **PLO** | Palestinian Liberation Organization |
| **PMCS** | preventive maintenance checks and services |
| **POW** | prisoner of war |
| **PSYOP** | psychological operations |
| **QRF** | quick reaction force |
| **ROE** | rules of engagement |
| **RPG** | rocket-propelled grenade |
| **SBU** | special boat unit |
| **SDDC** | Surface Deployment and Distribution Command |
| **SEAL** | sea-air-land (team) |
| **SF** | special forces |
| **SIGINT** | signals intelligence |
| **SJA** | Staff Judge Advocate |
| **SO** | special operations |
| **SOF** | special operations forces |
| **SOFA** | status-of-forces agreement |
| **SOP** | standing operating procedure |
| **SOWT** | special operations weather team |
| **SPOD** | seaport of debarkation |
| **SPE** | seaport of embarkation |
| **SR** | special reconnaissance |
| **STT** | special tactics team |
| **SWET** | sewer, water, electricity, trash |
| **SWEAT** | sewer, water, electricity, academics trash |
| **SWEAT-MS** | sewer, water, electricity, academics trash-medical, security |

| | |
|---|---|
| **TACON** | tactical control |
| **TC** | training circular |
| **TCC** | transportation component command |
| **TF** | task force |
| **TIM** | toxic industrial material |
| **TTP** | tactics, techniques, and procedures |
| **UAS** | unmanned aircraft system |
| **UN** | United Nations |
| **UNITAF** | Unified Task Force |
| **UNOSOM** | United Nations Operations in Somalia |
| **UNSCR** | UN Security Council resolution |
| **UO** | urban operations |
| **USACE** | United States Army Corps of Engineers |
| **USAID** | United States Agency for International Development |
| **USCENTCOM** | United States Central Command |
| **USFORSOM** | United States Forces, Somalia |
| **USCG** | United States Coast Guard |
| **USTRANSCOM** | United States Transportation Command |
| **UXO** | unexploded ordnance |
| **UW** | unconventional warfare |
| **VC** | Viet Cong |
| **WMD** | weapons of mass destruction |

## SECTION II – Terms and Definitions

**agility** – The ability to move and adjust quickly and easily. (FM 3-0)

**air interdiction** – Air operations conducted to destroy, neutralize, or delay the enemy's military potential before it can be brought to bear effectively against friendly forces at such distance from friendly forces that detailed integration of each air mission with the fire and movement of friendly forces is not required. (JP 1-02)

**area defense** – A type of defensive operation that concentrates on denying enemy forces access to designated terrain for a specific time rather than destroying the enemy outright. (FM 3-0)

**area of interest** – That area of concern to the commander, including the area of influence, areas adjacent thereto, and extending into enemy territory to the objectives of current or planned operations. This area also includes areas occupied by enemy forces who could jeopardize the accomplishment of the mission. (JP 1-02)

**area of operations** – An operational area defined by the joint force commander for land and naval forces. Areas of operations do not typically encompass the entire operational area of the joint force commander, but should be large enough for component commanders to accomplish their missions and protect their forces. (JP 1-02)

**ARFOR** – The senior Army headquarters and all Army forces assigned or attached to a combatant command, subordinate joint force command, joint functional command, or multinational command. (FM 3-0)

**assessment** – (Army) The continuous monitoring of the current situation and progress of an operation.(FMI 5-0.1)

**asymmetry** – Dissimilarities in organization, equipment, doctrine, and values between other armed forces (formally organized or not) and US forces. Engagements are symmetric if forces, technologies, and weapons are similar; they are asymmetric if forces, technologies, and weapons are different, or if a resort to terrorism and rejection of more conventional rules of engagement are the norm. (FM 3-0)

**attack** – An offensive operation that destroys or defeats enemy forces, seizes and secures terrain, or both. (FM 3-0)

**attainability** - One of the eight characteristics of combat service support: generating the minimum essential supplies and services necessary to begin operations. (FM 4-0)

**attrition** - The reduction of the effectiveness of a force caused by loss of personnel and materiel. (JP 1-02)

**avenue of approach** – An air or ground route of an attacking force of a given size leading to its objective or to key terrain in its path. (JP 1-02)

**axis of advance** – (Army) The general area through which the bulk of a unit's combat power must move. (FM 3-90)

**battle** – A set of related engagements that lasts longer and involves larger forces than an engagement. (FM 3-0)

**battlefield organization** – The allocation of forces in the area of operations by purpose. It consists of three all-encompassing categories of operations: decisive, shaping, and sustaining. (FM 3-0)

**battle rhythm** – The sequencing of command and control activities within a headquarters and throughout the force to facilitate effective command and control. (FMI 5-0.1)

**branch** – (Army) A contingency plan or course of action (an option built into the basic plan or course of action) for changing the mission, disposition, orientation, or direction of movement of the force to aid success of the current operation, based on anticipated events, opportunities, or disruptions caused by enemy actions. Army forces prepare branches to exploit success and opportunities, or to counter disruptions caused by enemy actions. (FM 3-0)

**breakout** – An operation conducted by an encircled force to regain freedom of movement or contact with friendly units. It differs from other attacks only in that a simultaneous defense in other areas of the perimeter must be maintained. (FM 3-90)

**buffer zone** – A defined area controlled by a peace operations force from which disputing or belligerent forces have been excluded. A buffer zone is formed to create an area of separation between disputing or belligerent forces and reduce the risk of renewed conflict. Also called **area of separation** in some United Nations operations. (JP 1-02)

**bypass** – A tactical mission task in which the commander directs his unit to maneuver around an obstacle, position, or enemy force to maintain the momentum of the operation while deliberately avoiding combat with an enemy force. (FM 3-90)

**campaign** – A series of related military operations aimed at accomplishing a strategic or operational objective within a given time and space. (JP 1-02)

**canalize** – (Army) A tactical mission task in which the commander restricts enemy movement to a narrow zone by exploiting terrain coupled with the use of obstacles, fires, or friendly maneuver. (FM 3-90)

**capability** - The ability to execute a specified course of action (a capability may or may not be accompanied by an intention) (JP 1-02)

**casualty evacuation** – (Army) A term used by nonmedical units to refer to the movement of casualties aboard nonmedical vehicles or aircraft. (FM 8-10-6)

**centers of gravity** – Those characteristics, capabilities, or sources of power from which a military force derives its freedom of action, physical strength, or will to fight. (JP 1-02)

**civil affairs** – Designated Active and Reserve component forces and units organized, trained, and equipped specifically to conduct civil affairs activities and to support civil-military operations. (JP 1-02)

**civil considerations** – The influence of manmade infrastructure, civilian institutions, and attitudes and activities of the civilian leaders, populations, and organizations within an area of operations on the conduct of military operations. (FM 6-0)

**civil disturbance** – Group acts of violence and disorder prejudicial to public law and order. (JP 1-02)

**civil-military operations** – The activities of a commander that establish, maintain, influence, or exploit relations between military forces, governmental and nongovernmental civilian organizations and authorities, and the civilian populace in a friendly, neutral, or hostile operational area in order to facilitate military operations, to consolidate and achieve US objectives. Civil-military operations may include performance by military forces of activities and functions normally the responsibility of the local, regional, or national government. These activities may occur prior to, during, or subsequent to other military actions. They may also occur, if directed, in the absence of other military operations. Civil-military operations may be performed by designated civil affairs, by other military forces, or by a combination of civil affairs and other forces. (JP 1-02)

**civil-military operations center** – An ad hoc organization, normally established by the geographic combatant commander or subordinate joint force commander, to assist in the coordination of activities of engaged military forces, and other United States Government agencies, nongovernmental organizations, and regional and international organizations. There is no established structure, and its size and composition are situation dependent. (JP 1-02)

**civil support operations** – Domestic operations that address the consequences of man-made or natural accidents and incidents beyond the capabilities of civilian authorities. (FM 1)

**clear** – (Army) 1. A tactical mission task that requires the commander to remove all enemy forces and eliminate organized resistance within an assigned area. (FM 3-90) 2. The total elimination or neutralization of an obstacle that is usually performed by follow-on engineers and is not done under fire. (FM 3-34.2)

**close air support** – Air action by fixed- and rotary-wing aircraft against hostile targets that are in close proximity to friendly forces and that require detailed integration of each air mission with the fire and movement of those forces. (JP 1-02)

**close combat** – Combat carried out with direct fire weapons, supported by indirect fire, air-delivered fires, and nonlethal engagement means. Close combat defeats or destroys enemy forces or seizes and retains ground. (FM 3-0)

**coalition** – An ad hoc arrangement between two or more nations for common action. (JP 1-02)

**collateral damage** – Unintentional or incidental injury or damage to persons or objects that would not be lawful military targets in the circumstances ruling at the time. Such damage is not unlawful so long as it is not excessive in light of the overall military advantage anticipated from the attack. (JP 1-02)

**combatant command** – A unified or specified command with a broad continuing mission under a single commander established and so designated by the President, through the Secretary of Defense and with the advice and assistance of the Chairman of the Joint Chiefs of Staff. Combatant commands typically have geographic or functional responsibilities. (JP 1-02)

**combat power** – The total means of destructive and/or disruptive force which a military unit/formation can apply against the opponent at a given time. (JP 1-02)

**combined arms** – The synchronized or simultaneous application of several arms—such as infantry, armor, field artillery, engineers, air defense, and aviation—to achieve an effect on the enemy that is greater than if each arm were used against the enemy separately or in sequence. (FM 3-0)

**combined arms team** – (Army) Two or more arms mutually supporting one another, usually consisting of infantry, armor, cavalry, aviation, field artillery, air defense artillery, and engineers. (FM 3-90).

**command and control system** – (Army) The arrangement of personnel, information management, procedures, and equipment and facilities essential for the commander to conduct operations. (FM 6-0)

**commander's critical information requirements** – (Army) – Elements of information required by commanders that directly affect decisionmaking and dictate the successful execution of military operations. (FM 3-0)

**commander's intent** – (Army) A clear, concise statement of what the force must do and the conditions the force must meet to succeed with respect to the enemy, terrain, and civil considerations that represent the operation's desired end state. (FMI 5-0.1)

**commander's visualization** – The mental process of developing situational understanding, determining a desired end state, and envisioning how the force will achieve that end state. (FMI 5-0.1)

**common operational picture** – (Army) An operational picture tailored to the user's requirements, based on common data and information shared by more than one command. (FM 3-0)

**computer network attack** – Actions taken through the use of computer networks to disrupt, deny, degrade, or destroy information resident in computers or computer networks, or the computers and networks themselves. (JP 1-02)

**computer network defense** – Actions taken through computer networks to protect, monitor, analyze, detect and respond to unauthorized activity within Department of Defense information systems and computer networks. (JP 1-02)

**concept of operations** – (Army) How commanders see the actions of subordinate units fitting together to accomplish the mission. As a minimum, the description includes the scheme of maneuver and concept of fires. The concept of operations expands the commander's selected course of action and expresses how each element of the force will cooperate to accomplish the mission. (FM 3-0)

**constraint** – A restriction placed on the command by a higher command. A constraint dictates an action or inaction, thus restricting the freedom of action a subordinate commander has for planning. (FM 5-0)

**control** – (Army) 1. The regulation of forces and warfighting functions to accomplish the mission in accordance with the commander's intent. (FMI 5-0.1) 2. A tactical mission task that requires the commander to maintain physical influence over a specified area to prevent its use by an enemy. (FM 3-90)

**controlled exchange** – The removal of serviceable parts, components, or assemblies from unserviceable, economically reparable equipment and their immediate reuse in restoring a like item of equipment to a combat operable or serviceable condition. (FM 4-30.3)

**control measure** – A means of regulating forces or warfighting functions. (FMI 5-0.1)

**counterattack** – (Army) A form of attack by part or all of a defending force against an enemy attacking force with the general objective of denying the enemy his goal in attacking. See FM 3-0.

**counterdeception** – Efforts to negate, neutralize, diminish the effects of, or gain advantage from a foreign deception operation. Counterdeception does not include the intelligence function of identifying foreign deception operations. (JP 1-02)

**counterdrug** – Those active measures taken to detect, monitor, and counter the production, trafficking, and use of illegal drugs. (JP 1-02)

**counterinsurgency** – Those military, paramilitary, political, economic, psychological, and civic actions taken by a government to defeat insurgency. (JP 1-02)

**counterintelligence** – Information gathered and activities conducted to protect against espionage, other intelligence activities, sabotage, or assassinations conducted by or on behalf of foreign governments or elements thereof, foreign organizations, or foreign persons, or international terrorist activities. (JP 1-02)

**countermobility operations** – The construction of obstacles and emplacement of minefields to delay, disrupt, and destroy the enemy by reinforcement of the terrain. The primary purpose of countermobility operations is to slow or divert the enemy, to increase time for target acquisition, and to increase weapon effectiveness. (JP 1-02)

**counterpropaganda** – Programs of products and actions designed to nullify propaganda or mitigate its effects. (FM 3-05.30)

**counterterrorism** – Operations that include the offensive measures taken to prevent, deter, preempt, and respond to terrorism. (JP 1-02)

**course of action** – 1. Any sequence of activities that an individual or a unit may follow. 2. A possible plan open to an individual or a commander that would accomplish or is related to the accomplishment of a mission. 3. The scheme adopted to accomplish a job or mission. 4. A line of conduct in an engagement. 5. A product of the Joint Operation Planning and Execution System concept development phase. (JP 1-02)

**cover** – (Army) 1. Protection from the effects of fires. (FM 6-0) 2. A form of security operation whose primary task is to protect the main body by fighting to gain time while also observing and reporting information and preventing enemy ground observation of and direct fire against the main body. Unlike a screening or guard force, the covering force is a self-contained force capable of operating independently of the main body. (FM 3-90)

**dead space** – 1. An area within the maximum range of a weapon, radar, or observer, which cannot be covered by fire or observation from a particular position because of intervening obstacles, the nature of the ground, the characteristics of the trajectory, or the limitations of the pointing capabilities of the weapon. 2. An area or zone which is within range of a radio transmitter, but in which a signal is not received. 3. The volume of space above and around a gun or guided missile system into which it cannot fire because of mechanical or electronic limitations. (JP 1-02)

**debarkation** – The unloading of troops, equipment, or supplies from a ship or aircraft. (JP 1-02)

**decision point** – (Army) An event, area, or point in the battlespace where and when the friendly commander will make a critical decision. (FM 5-0)

**decision support matrix** – An aid used by the commander and staff to make battlefield decisions. This matrix is a staff product of the wargaming process that lists the decision point, location of the decision point, the criteria to be evaluated at the point of decision, the action or operations to occur at the decision point, and the unit or element that is to act and has responsibility to observe and report the information affecting the criteria for the decision. (FM 5-0)

**decision support template** – (Army) A staff product initially used in the wargaming process that graphically represents the decision points and projected situations, and indicates when, where, and under what conditions a decision is most likely to be required to initiate a specific activity or event. (FM 5-0)

**decisive engagement** – In land and naval warfare, an engagement in which a unit is considered fully committed and cannot maneuver or extricate itself. In the absence of outside assistance, the action must be fought to a conclusion and either won or lost with the forces at hand. (JP 1-02)

**decisive operation** – The operation that directly accomplishes the task assigned by the higher headquarters. Decisive operations conclusively determine the outcome of major operations, battles, and engagements. (FM 3-0)

**decisive point** – A geographic place, specific key event, critical system or function that allows commanders to gain a marked advantage over an enemy and greatly influence the outcome of an attack. (JP 1-02)

**defeat** – A tactical mission task that occurs when an enemy force has temporarily or permanently lost the physical means or the will to fight. The defeated force's commander is unwilling or unable to pursue his adopted course of action, thereby yielding to the friendly commander's will, and can no longer interfere to a significant degree with the actions of friendly forces. Defeat can result from the use of force or the threat of its use. (FM 3-90)

**defeat in detail** – Concentrating overwhelming combat power against separate parts of a force in sequence rather than defeating the entire force at once. (FM 3-90)

**defense in depth** – The siting of mutually supporting defense positions designed to absorb and progressively weaken attack, prevent initial observations of the whole position by the enemy, and to allow the commander to maneuver the reserve. (JP 1-02)

**defensive information operations** – The integration and coordination of policies and procedures, operations, personnel, and technology to protect and defend information and information systems. Defensive information operations ensure timely, accurate, and relevant information access while denying adversaries the opportunity to exploit friendly information and information systems for their own purposes. (FM 3-0)

**defensive operations** – Operations that defeat an enemy attack, buy time, economize forces, or develop conditions favorable for offensive operations. Defensive operations alone normally cannot achieve a decision. Their purpose is to create conditions for a counteroffensive that allows Army forces to regain the initiative. (FM 3-0)

**delay** – A form of retrograde in which a force under pressure trades space for time by slowing the enemy's momentum and inflicting maximum damage on the enemy without, in principle, becoming decisively engaged. (FM 3-90)

**depth** – The extension of operations in time, space, and resources. (FM 3-0)

**destroy** – 1. A tactical mission task that physically renders an enemy force combat-ineffective until it is reconstituted. 2. To damage a combat system so badly that it cannot perform any function or be restored to a usable condition without being entirely rebuilt. (FM 3-90)

**direct action** – Short-duration strikes and other small-scale offensive actions conducted as a special operation in hostile, denied, or other politically sensitive environments and which employ specialized military capabilities to seize, destroy, capture, exploit, recover, or damage designated targets. Direct action differs from conventional offensive actions in the level of physical and political risk, operational techniques, and the degree of discriminate and precise use of force to achieve specific objectives. (JP 1-02)

**direct approach** – The application of combat power directly against the enemy center of gravity or the enemy's principal strength. (FM 3-90)

**direct fire** – Fire delivered on a target using the target itself as a point of aim for either the weapon or the director. (JP 1-02)

**disinformation** – Information disseminated primarily by intelligence organizations or other covert agencies designed to distort information or deceive or influence United States decisionmakers, United States forces, coalition allies, key actors, or individuals via indirect or unconventional means. (FM 3-13)

**displaced person** – (Army) A civilian who is involuntarily outside the national boundaries of his or her country or as an internally displaced person is a civilian involuntarily outside his area or region within his country. (FM 2-0)

**display** – (Army) An information management activity: to represent relevant information in a usable, easily understood audio or visual form tailored to the needs of the user that conveys the common operational picture for decisionmaking and exercising command and control functions (FM 6-0)

**disposition** – 1. Distribution of the elements of a command within an area, usually the exact location of each unit headquarters and the deployment of forces subordinate to it. (JP 1-02)

**disrupt** – 1. A tactical mission task in which a commander integrates direct and indirect fires, terrain, and obstacles to upset an enemy's formation or tempo, interrupt his timetable, or cause his forces to commit prematurely or attack in piecemeal fashion. (FM 3-90) 2. An engineer obstacle effect that focuses fire planning and obstacle effort to cause the enemy to break up his formation and tempo, interrupt his timetable, commit breaching assets prematurely, and attack in a piecemeal effort. (FM 90-7)

**distribution system** – That complex of facilities, installations, methods, and procedures designed to receive, store, maintain, distribute, and control the flow of military materiel between the point of receipt into the military system and the point of issue to using activities and units. (JP 1-02)

**doctrinal template** – A model based on known or postulated adversary doctrine. Doctrinal templates illustrate the disposition and activity of adversary forces and assets conducting a particular operation unconstrained by the effects of the battlespace. They represent the application of adversary doctrine under ideal conditions. Ideally, doctrinal templates depict the threat's normal organization for combat, frontages, depths, boundaries and other control measures, assets available from other commands, objective depths, engagement areas, battle positions, and so forth. Doctrinal templates are usually scaled to allow ready use with geospatial products. (JP 1-02)

**doctrine** – Fundamental principles by which the military forces or elements thereof guide their actions in support of national objectives. It is authoritative but requires judgment in application. (JP 1-02)

**economy** – One of the eight characteristics of combat service support: providing the most efficient support at the least cost to accomplish the mission. (FM 4-0)

**economy of force** – One of the nine principles of war: allocate minimum essential combat power to secondary efforts. (FM 3-0)

**effect** – (Army) A result, outcome, or consequence of an action. (FMI 5-0.1)

**electromagnetic deception** – The deliberate radiation, re-radiation, alteration, suppression, absorption, denial, enhancement, or reflection of electromagnetic energy in a manner intended to convey misleading information to an enemy or enemy electromagnetic-dependent weapons, thereby degrading or neutralizing the enemy's combat capability. Among the types of electromagnetic deception are: a. **manipulative electromagnetic deception**—Actions to eliminate revealing, or convey misleading, electromagnetic telltale indicators that may be used by hostile forces; b. **simulative electromagnetic deception**—Actions to simulate friendly, notional, or actual capabilities to mislead hostile forces; c. **imitative electromagnetic deception**—The introduction of electromagnetic energy into enemy systems that imitates enemy emissions. (JP 1-02)

**electromagnetic spectrum** – The range of frequencies of electromagnetic radiation from zero to infinity. It is divided into 26 alphabetically designated bands. (JP 1-02)

**electronic attack**—That division of electronic warfare involving the use of electromagnetic energy, directed energy, or antiradiation weapons to attack personnel, facilities, or equipment with the intent of degrading, neutralizing, or destroying enemy combat capability and is considered a form of fires. Electronic attack includes: 1. actions taken to prevent or reduce an enemy's effective use of the electromagnetic spectrum, such as jamming and electromagnetic deception, and 2. employment of weapons that use either electromagnetic or directed energy as their primary destructive mechanism (lasers, radio frequency weapons, particle beams), or antiradiation weapons. (JP 1-02)

**electronic warfare** – Any military action involving the use of electromagnetic and directed energy to control the electromagnetic spectrum or to attack the enemy. The three major subdivisions within electronic warfare are: electronic attack, electronic protection, and electronic warfare support. (JP 1-02)

**electronic warfare support**—That division of electronic warfare involving actions tasked by, or under direct control of, an operational commander to search for, intercept, identify, and locate or localize sources of intentional and unintentional radiated electromagnetic energy for the purpose of immediate threat recognition, targeting, planning, and conduct of future operations. Thus, electronic warfare support provides information required for immediate decisions involving electronic warfare operations and other tactical actions such as threat avoidance, targeting, and homing. Electronic warfare support data can be used to produce signals intelligence, provide targeting for electronic or destructive attack, and produce measurement and signature intelligence. (JP 1-02)

**embarkation** – The process of putting personnel and/or vehicles and their associated stores and equipment into ships and/or aircraft. (JP 1-02)

**encirclement** – An operation where one force loses its freedom of maneuver because an opposing force is able to isolate it by controlling all ground lines of communications. (FM 3-0)

**end state** – (Army) At the operational and tactical levels, the conditions that, when achieved, accomplish the mission. At the operational level, these conditions attain the aims set for the campaign or major operation. (FM 3-0)

**engagement** – A tactical conflict, usually between opposing lower echelon maneuver forces. (JP 1-02)

**envelopment** – (Army) – A form of maneuver in which an attacking force seeks to avoid the principal enemy defenses by seizing objectives to the enemy rear to destroy the enemy in his current positions. At the tactical level, envelopments focus on seizing terrain, destroying specific enemy forces, and interdicting enemy withdrawal routes. (FM 3-0).

**essential elements of friendly information** – (Army) The critical aspects of a friendly operation that, if known by the enemy, would subsequently compromise, lead to failure, or limit success of the operation, and, therefore, must be protected from enemy detection. (FM 3-13)

**essential task** – A specified or implied task that must be executed to accomplish the mission. Essential tasks are always included in the unit's mission statement. (FM 5-0)

**exfiltration** – The removal of personnel or units from areas under enemy control by stealth, deception, surprise, or clandestine means. (JP 1-02)

**exploitation** – 1. Taking full advantage of success in military operations, following up initial gains, and making permanent the temporary effects already achieved. 2. An offensive operation that usually follows a successful attack and is designed to disorganize the enemy in depth. (JP 1-02)

**explosive ordnance disposal** – The detection, identification, on-site evaluation, rendering safe, recovery, and final disposal of unexploded explosive ordnance. It may also include explosive ordnance which has become hazardous by damage or deterioration. (JP 1-02)

**firepower** – 1. The amount of fire which may be delivered by a position, unit, or weapon system. 2. Ability to deliver fire. (JP 1-02)

**fires** – The effects of lethal or nonlethal weapons. (JP 1-02)

**fire support coordinating measure** – A measure employed by land or amphibious commanders to facilitate the rapid engagement of targets and simultaneously provide safeguards for friendly forces. (JP 1-02)

**fix** – (Army) 1. A tactical mission task where a commander prevents the enemy from moving any part of his force from a specific location for a specific period of time. 2. An engineer obstacle effect that focuses fire planning and obstacle effort to slow an attacker's movement within a specified area, normally an engagement area. (FM 3-90)

**flexibility** – One of the eight characteristics of combat service support: being able to adapt combat service support structures and procedures to changing situations, missions, and concepts of operations. (FM 4-0)

**force projection** – The ability to project the military element of national power from the continental United States (CONUS) or another theater in response to requirements for military operations. Force projection operations extend from mobilization and deployment of forces to redeployment to CONUS or home theater. (JP 1-02)

**force protection** – Actions taken to prevent or mitigate hostile actions against Department of Defense personnel (to include family members), resources, facilities, and critical information. These actions conserve the force's fighting potential so it can be applied at a decisive time and place and incorporates the coordinated and synchronized offensive and defensive measures to enable the effective employment of the joint force while degrading opportunities for the enemy. Force protection does not include actions to defeat the enemy or protect against accidents, weather, or disease. (JP 1-02)

**force tailoring** – The process of determining the right mix and sequence of units for a mission. (FM 3-0)

**foreign humanitarian assistance** – Programs conducted to relieve or reduce the results of natural or manmade disasters or other endemic conditions such as human pain, disease, hunger, or privation that might present a serious threat to life or that can result in great damage to or loss of property. Foreign humanitarian assistance (FHA) provided by US forces is limited in scope and duration. The foreign assistance provided is designed to supplement or complement the efforts of the host nation civil authorities or agencies that may have primary responsibility for providing FHA. FHA operations are those conducted outside the United States, its territories, and possessions. (JP 1-02)

**foreign internal defense** – Participation by civilian and military agencies of a government in any of the action programs taken by another government to free and protect its society from subversion, lawlessness, and insurgency. (JP 1-02)

**fratricide** – The unintentional killing or wounding of friendly personnel by friendly firepower. (FM 3-0)

**frontal attack** – (Army) A form of maneuver in which the attacking force seeks to destroy a weaker enemy force or fix a larger enemy force in place over a broad front. (FM 3-0)

**full spectrum operations** – The range of operations Army forces conduct in war and military operations other than war. (FM 3-0)

**graphic control measure** – A symbol used on maps and displays to regulate forces and warfighting functions. (FMI 5-0.1)

**gun-target line** – An imaginary straight line from gun to target. (JP 1-02)

**hazardous material** – Any substance which has a human health hazard associated with it. Special storage, use, handling, and shipment safety procedures and protocols must be followed to help protect against accidental exposure. Hazardous materials are specifically identified under federal law. (FM 3-100.4)

**high angle fire** – (Army) Fire delivered to clear an obstacle (such as a hill) that low-angle fire cannot, or fire delivered to attack targets on the reverse side of an obstacle (such as a hill) that cannot be attacked with low-angle or direct fire. (FM 6-30).

**host nation** – A nation that receives the forces and/or supplies of allied nations, coalition partners, and/or NATO organizations to be located on, to operate in, or to transit through its territory. (JP 1-02)

**host-nation support** – Civil and/or military assistance rendered by a nation to foreign forces within its territory during peacetime, crises or emergencies, or war based on agreements mutually concluded between nations. (JP 1-02)

**human intelligence** – A category of intelligence derived from information collected and provided by human sources. (JP 1-02)

**imagery intelligence** – Intelligence derived from the exploitation of collection by visual photography, infrared sensors, lasers, electro-optics, and radar sensors such as synthetic aperture radar wherein images of objects are reproduced optically or electronically on film, electronic display devices, or other media. (JP 1-02)

**implied task** – A task that must be performed to accomplish a specified task or mission, but is not stated in the higher headquarters order. (FM 5-0)

**indirect fire** – Fire delivered on a target that is not itself used as a point of aim for the weapons or the director. (JP 1-02)

**infiltration** – (Army) 1. A form of maneuver in which an attacking force conducts undetected movement through or into an area occupied by enemy forces to occupy a position of advantage in the enemy rear while exposing only small elements to enemy defensive fires. (FM 3-90)

**information** – (Army) 1. In the general sense, the meaning humans assign to data. 2. In the context of the cognitive hierarchy, data that have been processed to provide further meaning. (FM 6-0)

**information fratricide** – The results of employing information operations elements in a way that causes effects in the information environment that impede the conduct of friendly operations or adversely affect friendly forces. (FM 3-13)

**information management** – The provision of relevant information to the right person at the right time in a usable form to facilitate situational understanding and decisionmaking. It uses procedures and information systems to collect, process, store, display, and disseminate information. (FM 3-0)

**information operations** – (Army) The employment of the core capabilities of electronic warfare, computer network operations, psychological operations, military deception, and operations security, in concert with specified supporting and related capabilities, to affect and defend information and information systems and to influence decisionmaking. (FM 3-13)

**information requirements** – (Army) All information elements the commander and staff require to successfully conduct operations, that is, all elements necessary to address the factors of METT-TC. (FM 6-0)

**information superiority** – The operational advantage derived from the ability to collect, process, and disseminate an uninterrupted flow of information while exploiting or denying an adversary's ability to do the same. (JP 1-02)

**information system** – (Army) The equipment and facilities that collect, process, store, display, and disseminate information. This includes computers—hardware and software—and communications, as well as policies and procedures for their use. (FM 3-0)

**insurgency** – An organized movement aimed at the overthrow of a constituted government through the use of subversion and armed conflict. (JP 1-02)

**integration** – One of the eight characteristics of combat service support: the total inclusion of Army combat service support into the operations process (plan, prepare, execute, assess), as well as into other logistics comnponents of the unified force. (FM 4-0).

**intelligence** – 1. The product resulting from the collection, processing, integration, analysis, evaluation, and interpretation of available information concerning foreign countries or areas. 2. Information and knowledge about an adversary obtained through observation, investigation, analysis, or understanding. (JP 1-02)

**intelligence preparation of the battlefield** – The systematic, continuous process of analyzing the threat and environment in a specific geographic area. Intelligence preparation of the battlefield (IPB) is designed to support the staff estimate and military decisionmaking process. Most intelligence requirements are generated as a result of the IPB process and its interrelation with the decisionmaking process. (FM 34-130)

**intelligence, surveillance, and reconnaissance** – An activity that synchronizes and integrates the planning and operation of sensors, assets, and processing, exploitation, and dissemination systems in direct support of current and future operations. This is an integrated intelligence and operations function. (JP 1-02)

**intelligence, surveillance, and reconnaissance plan** – An integrated plan for collection of information from all available sources and analysis of that information to produce intelligence to meet requirements. Specifically, a logical plan for transforming priority intelligence requirements into orders or requests to reconnaissance and surveillance assets to collect pertinent information within a required time limit. (FM 34-3)

**interagency** – Activities or operations conducted by or through coordination with two or more agencies or an agency and one or more Services of the same nation. (FM 3-07)

**interdict** – A tactical mission task where the commander prevents, disrupts, or delays the enemy's use of an area or route. (FM 3-90)

**interior lines** – A force operates on interior lines when its operations diverge from a central point. (FM 3-0)

**isolate** – A tactical mission task that requires a unit to seal off—both physically and psychologically—an enemy from his sources of support, deny an enemy freedom of movement, and prevent an enemy unit from having contact with other enemy forces. (FM 3-90)

**joint force** – A general term applied to a force composed of significant elements, assigned or attached, of two or more Military Departments, operating under a single joint force commander. (JP 1-02)

**joint force commander** – A general term applied to a combatant commander, subunified commander, or joint task force commander authorized to exercise combatant command (command authority) or operational control over a joint force. (JP 1-02)

**joint operations** – A general term to describe military actions conducted by joint forces, or by Service forces in relationships (e.g., support, coordinating authority), which, of themselves, do not create joint forces. (JP 1-02)

**joint task force** – A joint force that is constituted and so designated by the Secretary of Defense, a combatant commander, a subunified commander, or an existing joint task force commander. (JP 1-02)

**key terrain** – Any locality, or area, the seizure or retention of which affords a marked advantage to either combatant. (JP 1-02)

**landing zone** – Any specified zone used for the landing of aircraft. JP 1-02)

**law of war** – That part of international law that regulates the conduct of armed hostilities. (JP 1-02)

**liaison** – That contact or intercommunication maintained between elements of military forces or other agencies to ensure mutual understanding and unity of purpose and action. (JP 1-02)

**line of communications** – A route, either land, water, and/or air, that connects an operating military force with a base of operations and along which supplies and military forces move. (JP 1-02)

**lines of operations** – (Army) A line that defines the orientation of the force in time and space, or purpose, in relation to an enemy or objective. (FMI 5-0.1).

**logistics** – The science of planning and carrying out the movement and maintenance of forces. In its most comprehensive sense, those aspects of military operations which deal with: a. design and development, acquisition, storage, movement, distribution, maintenance, evacuation, and disposition of materiel; b. movement, evacuation, and hospitalization of personnel; c. acquisition or construction, maintenance, operation, and disposition of facilities; and d. acquisition and furnishing of services. (JP 1-02)

**logistics-over-the-shore operations** – The loading and unloading of ships without the benefit of deep draft-capable, fixed port facilities; or as a means of moving forces close to tactical assembly areas dependent on threat force capabilities. (JP 1-02)

**logistics preparation of the theater** – Actions taken by combat service support personnel to optimize means—force structure, resources, and strategic lift—of supporting the joint force commander's plan. (FM 4-0)

**major operation** – A series of tactical actions (battles, engagements, strikes) conducted by various combat forces of a single or several Services, coordinated in time and place, to accomplish operational and, sometimes, strategic objectives in an operational area. These actions are conducted simultaneously or sequentially in accordance with a common plan and are controlled by a single commander. (JP 1-02)

**maneuver** – One of the nine principles of war: place the enemy in a disadvantageous position through the flexible application of combat power. (FM 3-0)

**Marine air-ground task force** – The Marine Corps principal organization for all missions across the range of military operations, composed of forces task-organized under a single commander capable of responding rapidly to a contingency anywhere in the world. The types of forces in the Marine air-ground task force (MAGTF) are functionally grouped into four core elements: a command element, an aviation combat element, a ground combat element, and a combat service support element. The four core elements are categories of forces, not formal commands. The basic structure of the MAGTF never varies, though the number, size, and type of Marine Corps units comprising each of its four elements will always be mission dependent. The flexibility of the organizational structure allows for one or more subordinate MAGTFs to be assigned. In a joint or multinational environment, other Service or multinational forces may be assigned or attached. (FM 1-02)

**Marine expeditionary force** – The largest Marine air-ground task force (MAGTF) and the Marine Corps principal warfighting organization, particularly for larger crises or contingencies. It is task-organized around a permanent command element and normally consists of one or more Marine divisions, Marine aircraft wings, and Marine force service support groups. The Marine expeditionary force is capable of missions across the range of military operations, including amphibious assault and sustained operations ashore in any environment. It can operate from a sea base, a land base, or both. In a joint or multinational environment, it may contain other Service or multinational forces assigned or attached to the MAGTF. (FM 1-02)

**Marine expeditionary unit** – A Marine air-ground task force (MAGTF) that is constructed around an infantry battalion reinforced, a helicopter squadron reinforced, and a task-organized combat service support element. It normally fulfills Marine Corps' forward sea-based deployment requirements. The Marine expeditionary unit provides an immediate reaction capability for crisis response and is capable of limited combat operations. In a joint or multinational environment, it may contain other Service or multinational forces assigned or attached to the MAGTF. (FM 1-02)

**mass** – One of the nine principles of war: the effects of combat power at the decisive place and time. (FM 3-0)

**mass casualty** – Any large number of casualties produced in a relatively short period of time, usually as the result of a single incident such as a military aircraft accident, hurricane, flood, earthquake, or armed attack that exceeds logistical support capabilities. (JP 1-02)

**measure of effectiveness** – (Army) A criterion used to assess changes in system behavior, capability, or operational environment that is tied to measuring the attainment of an end state, achievement of an objective, or creation of an effect. (FMI 5-0.1)

**medical evacuation** – The timely and efficient movement of the wounded, injured, or ill while providing en route medical care to and between medical treatment facilities. (FM 4-02)

**meeting engagement** – A combat action that occurs when a moving force engages an enemy at an unexpected time and place. (FM 3-0)

**METT-TC** – A memory aid used in two contexts: (1) In the context of information management, the major subject categories into which relevant information is grouped for military operations: mission, enemy, terrain and weather, troops and support available, time available, civil considerations. (2) In the context of tactics, the major factors considered during mission analysis. (FM 6-0)

**military deception** – Actions executed to deliberately mislead adversary military decisionmakers as to friendly military capabilities, intentions, and operations, thereby causing the adversary to take specific actions (or inactions) that will contribute to the accomplishment of the friendly mission. (JP 1-02)

**military decisionmaking process** – A process that integrates the activities of the commander, staff and subordinate commanders in developing and operation plan or order. It establishes procedures for analyzing a mission; developing, analyzing, and comparing courses of action; selecting the best course of action; and producing an operation plan or order (FMI 5-0.1).

**misinformation** – Incorrect information from any source that is released for unknown reasons or to solicit a response or interest from a nonpolitical or nonmilitary target. (FM 3-13)

**mission** – 1. The task, together with the purpose, that clearly indicates the action to be taken and the reason therefor. 2. In common usage, especially when applied to lower military units, a duty assigned to an individual or unit; a task. (JP 1-02)

**mission command** – The conduct of military operations through decentralized execution based upon mission orders for effective mission accomplishment. Successful mission command results from subordinate leaders at all echelons exercising disciplined initiative within the commander's intent to accomplish missions. It requires an environment of trust and mutual understanding. (FM 6-0)

**mission orders** – A technique for completing combat orders that allows subordinates maximum freedom of planning and action in accomplishing missions and leaves the "how" of mission accomplishment to the subordinate. (FM 6-0)

**mobile defense** – (Army) A type of defensive operation that concentrates on the destruction or defeat of the enemy through a decisive attack by a striking force. (FM 3-0)

**mobility corridor** – Areas where a force will be canalized due to terrain restrictions. They allow military forces to capitalize on the principles of mass and speed and are therefore relatively free of obstacles. (JP 1-02)

**mobility operations** – Obstacle reduction by maneuver and engineer units to reduce or negate the effects of existing or reinforcing obstacles. The objective is to maintain freedom of movement for maneuver units, weapon systems, and critical supplies. (FM 3-34)

**modified combined obstacle overlay** – A joint intelligence preparation of the battlespace product used to portray the effects of each battlespace dimension on military operations. It normally depicts militarily significant aspects of the battlespace environment, such as obstacles restricting military movement, key geography, and military objectives. (JP 1-02)

**monitoring** – (Army) Continuous observation of the current situation to identify opportunities for the force, threats to the force, gaps in information, and progress according to the plan or order. (FMI 5-0.1)

**movement to contact** – A form of the offensive designed to develop the situation and to establish or regain contact. (JP 1-02)

**naval coastal warfare** – Coastal sea control, harbor defense, and port security, executed both in coastal areas outside the United States in support of national policy and in the United States as part of this Nation's defense. (JP 1-02)

**naval gunfire support** – Fire provided by Navy surface gun systems in support of a unit or units tasked with achieving the commander's objectives. A subset of naval surface fire support. (JP 1-02)

**naval surface fire support** – Fire provided by Navy surface gun and missile systems in support of a unit or units. (JP 1-02)

**neutral** – (Army) 1. An individual, group of individuals, organization, or nation that is not hostile to or in any way supportive of only one belligerent force in a hostile environment. (FM 3-07)

**no-fire area** – A land area, designated by the appropriate commander, into which fires or their effects are prohibited. (JP 1-02)

**noncombatant** – 1. An individual in an area of combat operations who is not armed and is not participating in any activity in support of any of the factions or forces involved in combat. 2. An individual, such as chaplain or medical personnel, whose duties do not involve combat. (FM 3-07)

**noncombatant evacuation operations** – Operations directed by the Department of State, the Department of Defense, or other appropriate authority whereby noncombatants are evacuated from foreign countries when their lives are endangered by war, civil unrest, or natural disaster to safe havens or to the United States. (JP 1-02)

**nongovernmental organization** – A private, self-governing, not-for-profit organization dedicated to alleviating human suffering; and/or promoting education, health care, economic development, environmental protection, human rights, and conflict resolution; and/or encouraging the establishment of democratic institutions and civil society. (JP 1-02)

**nonlethal fires** – Any fires that do not directly seek the physical destruction of the intended target and are designed to impair, disrupt, or delay the performance of enemy operational forces, functions, and facilities. Psychological operations, electronic warfare (jamming), and other command and control countermeasures are all nonlethal fire options. (FM 6-20)

**nonlethal weapons** – Weapons that are explicitly designed and primarily employed so as to incapacitate personnel or materiel, while minimizing fatalities, permanent injury to personnel and undesired damage to property and the environment. a. Unlike conventional lethal weapons that destroy their targets through blast, penetration, and fragmentation, nonlethal weapons employ means other than gross physical destruction to prevent the target from functioning. b. Nonlethal weapons are intended to have one, or both, of the following characteristics: (1) They have relatively reversible effects on personnel or materiel. (2) They affect objects differently within their area of influence. (JP 1-02)

**objective** – (Army) 1. One of the nine principles of war: direct every military operation toward a clearly defined, decisive and attainable objective. (FM 3-0) 2. A location on the ground used to orient operations, phase operations, facilitate changes of direction, and provide for unity of effort. (FM 3-90)

**offensive** – One of the nine principles of war: seize, retain, and exploit the initiative. (FM 3-0)

**offensive information operations** – The integrated use of assigned and supporting capabilities and activities, mutually supported by intelligence, to affect enemy decisionmakers or to influence others to achieve or promote specific objectives. (FM 3-0)

**offensive operations** – Operations which aim at destroying or defeating an enemy. Their purpose is to impose US will on the enemy and achieve decisive victory. (FM 3-0)

**operation** – 1. A military action or the carrying out of a strategic, operational, tactical, service, training, or administrative military mission. 2. The process of carrying on combat, including movement, supply, attack, defense, and maneuvers needed to gain the objectives of any battle or campaign. (JP 1-02)

**operational control** – Command authority that may be exercised by commanders at any echelon at or below the level of combatant command. Operational control is inherent in combatant command (command authority) and may be delegated within the command. When forces are transferred between combatant commands, the command relationship the gaining commander will exercise (and the losing commander will relinquish) over these forces must be specified by the Secretary of Defense. Operational control is the authority to perform those functions of command over subordinate forces involving organizing and employing commands and forces, assigning tasks, designating objectives, and giving authoritative direction necessary to accomplish the mission. Operational control includes authoritative direction over all aspects of military operations and joint training necessary to accomplish the missions assigned to the command. Operational control should be exercised through the commanders of subordinate organizations. Normally this authority is exercised through subordinate joint force commanders and Service and/or functional component commanders. Operational control normally provides full authority to organize commands and forces and to employ those forces as the commander in operational control considers necessary to accomplish assigned missions; it does not, in and of itself, include authoritative direction for logistics or matters of administration, discipline, internal organization, or unit training. (JP 1-02)

**operational framework** – The arrangement of friendly forces and resources in time, space, and purpose with respect to each other and the enemy or situation. It consists of the area of operations, battlespace, and battlefield organization. (FM 3-0)

**operations process** – The major command and control activities performed during operations: planning, preparation, execution with continuous assessment. These activities occur continuously throughout an operation, overlapping and recurring as required. (FMI 5-0.1)

**operations security** – A process of identifying critical information and subsequently analyzing friendly actions attendant to military operations and other activities to: a. identify those actions that can be observed by adversary intelligence systems; b. determine indicators hostile intelligence systems might obtain that could be interpreted or pieced together to derive critical information in time to be useful to adversaries; and c. select and execute measures that eliminate or reduce to an acceptable level the vulnerabilities of friendly actions to adversary exploitation. (JP 1-02)

**paramilitary forces** – Forces or groups distinct from the regular armed forces of any country, but resembling them in organization, equipment, training, or mission. (JP 1-02)

**patrol** – A detachment of ground, sea, or air forces sent out for the purpose of gathering information or carrying out a destructive, harassing, mopping-up, or security mission. (JP 1-02)

**peace enforcement** – Application of military force, or the threat of its use, normally pursuant to international authorization, to compel compliance with resolutions or sanctions designed to maintain or restore peace and order. (JP 1-02)

**peacekeeping** – Military operations undertaken with the consent of all major parties to a dispute, designed to monitor and facilitate implementation of an agreement (ceasefire, truce, or other such agreement) and support diplomatic efforts to reach a long-term political settlement. (JP 1-02)

**peace operations** – A broad term that encompasses peacekeeping operations and peace enforcement operations conducted in support of diplomatic efforts to establish and maintain peace. (JP 1-02)

**penetration** – (Army) A form of maneuver in which an attacking force seeks to rupture enemy defenses on a narrow front to disrupt the defensive system. (FM 3-90).

**perceptions** – Mental images the commander wants the deception target to believe are real. (FM 3-13)

**phase line** – A line utilized for control and coordination of military operations, usually an easily identified feature in the operational area. (JP 1-02)

**physical destruction** – The application of combat power to destroy or degrade adversary forces, sources of information, command and control systems, and installations. It includes direct and indirect forces from ground, sea, and air forces. Also included are direct actions by special operations forces. (FM 3-13)

**physical security** – 1. That part of security concerned with physical measures designed to safeguard personnel; to prevent unauthorized access to equipment, installations, material, and documents; and to safeguard them against espionage, sabotage, damage, and theft. 2. In communications security, the component that results from all physical measures necessary to safeguard classified equipment, material, and documents from access thereto or observation thereof by unauthorized persons. (JP 1-02)

**pickup zone** – A geographic area used to pick up troops or equipment by helicopter. (FM 90-4)

**plan** – A design for a future or anticipated operation. (FM 5-0)

**planning** – The process by which commanders (and staffs, if available) translate the commander's visualization into a specific course of action for preparation and execution, focusing on the expected results. (FMI 5-0.1)

**port of debarkation** – (DOD) The geographic point at which cargo or personnel are discharged. This may be a seaport or aerial port of debarkation; for unit requirements, it may or may not coincide with the destination. (JP 1-02)

**port of embarkation** – (DOD) The geographic point in a routing scheme from which cargo and personnel depart. This may be a seaport or aerial port from which personnel and equipment flow to a port of debarkation; for unit and nonunit requirements, it may or may not coincide with the origin. (JP 1-02)

**principles of war** – Principles that provide general guidance for conducting war and military operations other than war at the strategic, operational, and tactical levels. The nine principles of war are: objective, offensive, mass, economy of force, maneuver, unity of command, security, surprise, and simplicity (FM 3-0)

**propaganda** – Any form of communication in support of national objectives designed to influence the opinions, emotions, attitudes, or behavior of any group in order to benefit the sponsor, either directly or indirectly. (JP 1-02)

**psychological operations** – Planned operations to convey selected information and indicators to foreign audiences to influence their emotions, motives, objective reasoning, and ultimately the behavior of foreign governments, organizations, groups, and individuals. The purpose of psychological operations is to induce or reinforce foreign attitudes and behavior favorable to the originator's objectives. (JP 1-02)

**public affairs** – Those public information, command information, and community relations activities directed toward both the external and internal publics with interest in the Department of Defense. (JP 1-02)

**pursuit** – An offensive operation designed to catch or cut off a hostile force attempting to escape, with the aim of destroying it. (JP 1-02)

**push** – In logistics, the delivery of a predetermined amount of supplies to a user on a scheduled basis without the user requesting them. (FM 4-0)

**railhead** – A point on a railway where loads are transferred between trains and other means of transport. (JP 1-02)

**reachback** – The process of obtaining products, services, and applications, or forces, or equipment, or material from organizations that are not forward deployed. (JP 1-02)

**rear area** – (Army) For any particular command, the area extending forward from its rear boundary to the rear of the area assigned to the next lower level of command. This area is provided primarily for the performance of support functions, and is where the majority of the echelon's sustaining functions occur. (FM 3-0)

**refugee** – A person who, by reason of real or imagined danger, has left his home country or country of nationality and is unwilling or unable to return. (JP 1-02)

**relevant information** – All information of importance to commanders and staffs in the exercise of command and control. (FM 3-0)

**relief in place** – (Army) An tactical enabling operation in which, by direction of higher authority, all or part of a unit is replaced in an area by the incoming unit. (FM 3-90)

**reorganization** – Action taken to shift internal resources within a degraded unit to increase its level of combat effectiveness. (FM 100-9)

**reserve** – Portion of a body of troops which is kept to the rear or withheld from action at the beginning of an engagement, in order to be available for a decisive movement. (JP 1-02)

**responsiveness** – One of the eight principles of combat service support: providing the right support at the right place at the right time. (FM 4-0)

**retrograde** – A type of defensive operation that involves organized movement away from the enemy. (FM 3-0)

**riot control agent** – (Army) A substance that produces temporary irritating or disabling physical effects that disappear within minutes of removal from exposure. There is no significant risk of permanent injury, and medical treatment is rarely required. (FM 3-11.11)

**risk** – (DOD) Probability and severity of loss linked to hazards. (JP 1-02)

**risk management** – The process of identifying, assessing, and controlling risk arising from operational factors, and making decisions that balance risk cost with mission benefits. (JP 1-02)

**rules of engagement** – Directives issued by competent military authority that delineate the circumstances and limitations under which United States forces will initiate and/or continue combat engagement with other forces encountered. (JP 1-02)

**running estimate** – A staff section's continuous assessment of current and future operations to determine if the current operation is proceeding according to the commander's intent and if future operations are supportable. (FMI 5-0.1)

**search and attack** – A technique of conducting a movement to contact that shares many of the characteristics of an area security mission. (FM 3-0)

**secure** – (Army) 1. A tactical mission task that involves preventing a unit, facility, or geographical location from being damaged or destroyed as a result of enemy action. (FM 3-90)

**security** – One of the nine principles of war: never permit the enemy to acquire an unexpected advantage (FM 3-0).

**seize** – (Army) A tactical mission task that involves taking possession of a designated area using overwhelming force. (FM 3-90)

**sequel** – (Army) An operation that follows the current operation. It is a future operation that anticipates possible outcomes – success, failure, or stalemate – of the current operation. (FM 3-0).

**shaping operation** – An operation at any echelon that creates and preserves conditions for the success of decisive operations. (FM 3-0)

**show of force** – An operation designed to demonstrate US resolve that involves increased visibility of US deployed forces in an attempt to defuse a specific situation that, if allowed to continue, may be detrimental to US interests or national objectives. (JP 1-02)

**signals intelligence** – 1. A category of intelligence comprising either individually or in combination all communications intelligence, electronic intelligence, and foreign instrumentation signals intelligence, however transmitted. 2. Intelligence derived from communications, electronics, and foreign instrumentation signals. (JP 1-02)

**simplicity** – One of the nine principles of war: prepare clear, uncomplicated plans, and clear, concise orders to ensure thorough understanding. (FM 3-0) [Note: "simplicity" is also one of the eight characteristics of combat service support.]

**situational understanding** – (Army) The product of applying analysis and judgment to the common operational picture to determine the relationship among the factors of METT-TC. (FM 3-0)

**space operations** – The employment of space system capabilities that provide the means to enhance command and control, facilitate the maneuver of forces, reduce the commander's uncertainty, and improve fire support, air defense, intelligence collection, and combat service support operations which will support strategic, operational, and tactical missions across the operational continuum in the near, mid, and far term. (FM 3-14)

**special operations** – Operations conducted in hostile, denied, or politically sensitive environments to achieve military, political, economic, and/or informational objectives employing military capabilities for which there is no broad conventional force requirement. These operations often require covert, clandestine, or low visibility capabilities. Special operations are applicable across the rnage of military operations. They can be conducted independently or in conjunction with operations of conventional forces or other government agencies and may include operations through, with, or by indigenous or surrogate forces. Special operations differ from conventional operations in degree of physical and political risk, operational techniques, mode of employment, independence from friendly support, and dependence on detailed operational intelligence and indigenous assets. (JP 1-02)

**special operations forces** – Those Active and Reserve Component forces of the Military Services designated by the Secretary of Defense and specifically organized, trained, and equipped to conduct and support special operations. (JP 1-02)

**special reconnaissance** – Reconnaissance and surveillance actions conducted by special operations forces in hostile, denied, or politically sensitive environments to collect or verify information of strategic or operational significance, employing military capabilities not normally found in conventional forces. These forces provide an additive capability for commanders and supplement other conventional reconnaissance and surveillance actions. (JP 1-02)

**stability operations** – Operations that promote and protect US national interests by influencing the threat, political, and information dimensions of the operational environment through a combination of peacetime developmental, cooperative activities and coercive actions in response to crisis. (FM 3-0)

**status-of-forces agreement** – An agreement that defines the legal position of a visiting military force deployed in the territory of a friendly state. Agreements delineating the status of visiting military forces may be bilateral or multilateral. Provisions pertaining to the status of visiting forces may be set forth in a separate agreement, or they may form a part of a more comprehensive agreement. These provisions describe how the authorities of a visiting force may control members of that force and the amenability of the force or its members to the local law or to the authority of local officials. To the extent that agreements delineate matters affecting the relations between a military force and civilian authorities and population, they may be considered as civil affairs agreements. (JP 1-02)

**strategy** – The art and science of developing and employing instruments of national power in a synchronized and integrated fashion to achieve theater, national and/or multinational objectives. (JP 1-02)

**striking force** – A committed force organized to conduct the decisive attack in a mobile defense. It normally comprises the maximum combat power available to the commander at the time of the attack. (FM 3-0)

**strong point** – (Army) A heavily fortified battle position tied to a natural or reinforcing obstacle to create an anchor for the defense or to deny the enemy decisive or key terrain. (FM 3-90)

**surprise** – One of the nine principles of war: strike the enemy at a time or place or in a manner for which he is unprepared. (FM 3-0)

**survivability** – (Army) 1. Concept which includes all aspects of protecting personnel, weapons, and supplies while simultaneously deceiving the enemy. Survivability tactics include building a good defense; employing frequent movement; using concealment, deception, and camouflage; and constructing fighting and protective positions for both individuals and equipment. Encompasses planning and locating position sites, designing adequate overhead cover, analyzing terrain conditions and construction materials, selecting excavation methods, and countering the effects of direct and indirect fire weapons. (FM 5-103) 2. One of the eight characteristics of combat service support: being able to shield support functions from destruction or degradation. (FM 4-0)

**sustainability** - One of the eight characteristics of combat service support: the ability to maintain continuous support, rear area and base security, movement control, terrain management, and infrastructure development. (FM 4-0)

**sustaining operation** – An operation at any echelon that enables shaping and decisive operations by providing combat service support, rear area and base security, movement control, terrain management, and infrastructure development. (FM 3-0)

**synchronization** – 1. The arrangement of military actions in time, space, and purpose to produce maximum relative combat power at a decisive place and time. See FM 3-0. 2. In the intelligence context, application of intelligence sources and methods in concert with the operation plan. (JP 1-02)

**tactical combat force** – A combat unit, with appropriate combat support and combat service support assets, that is assigned the mission of defeating Level III threats. (JP 1-02)

**tactical control** – Command authority over assigned or attached forces or commands, or military capability or forces made available for tasking, that is limited to the detailed direction and control of movements or maneuvers within the operational area necessary to accomplish missions or tasks assigned. Tactical control is inherent in operational control. Tactical control may be delegated to, and exercised at any level at or below the level of combatant command. When forces are transferred between combatant commands, the command relationship the gaining commander will exercise (and the losing commander will relinquish) over these forces must be specified by the Secretary of Defense. Tactical control provides sufficient authority for controlling and directing the application of force or tactical use of combat support assets within the assigned mission or task. (JP 1-02)

**tactical level of war** –The level of war at which battles and engagements are planned and executed to accomplish military objectives assigned to tactical units or task forces. Activities at this level focus on the ordered arrangement and maneuver of combat elements in relation to each other and to the enemy to achieve combat objectives. (JP 1-02)

**target acquisition** – The detection, identification, and location of a target in sufficient detail to permit the effective employment of weapons. (JP 1-02)

**targeting** – The process of selecting and prioritizing targets and matching the appropriate response to them, taking account of operational requirements and capabilities. (JP 1-02)

**target reference point** – An easily recognizable point on the ground (either natural or man-made) used to initiate, distribute, and control fires. Target reference points (TRPs) can also designate the center of an area where the commander plans to distribute or converge the fires of all his weapons rapidly. They are used by task force and below, and can further delineate sectors of fire within an engagement area. TRPs are designated using the standard target symbol and numbers issued by the fire support officer. Once designated, TRPs also constitute indirect fire targets. (FM 3-90)

**task organization** – A temporary grouping of forces designed to accomplish a particular mission. (FM 3-0)

**task-organizing** – (Army) The process of allocating available assets to subordinate commanders and establishing their command and support relationships. (FM 3-0)

**tempo** – The rate of military action. (FM 3-0)

**terminal control** – 1. The authority to direct the maneuver of aircraft which are delivering ordnance, pasengers, or cargo to a specific location or target. Terminal control is a type of air control. 2. Any electronic, mechanical, or visual control given to aircraft to facilitate target acquisition and resolution. (JP 1-02)

**terrorism** – The calculated use of unlawful violence or threat of unlawful violence to inculcate fear; intended to coerce or to intimidate governments or societies in the pursuit of goals that are generally political, religious, or ideological. (JP 1-02)

**terrorist** – An individual who uses violence, terror, and intimidation to achieve a result. (JP 1-02)

**throughput distribution** – The bypassing of one or more intermediate supply echelons in the supply system to avoid multiple handling. (FM 4-0)

**time-phased force and deployment data** – The Joint Operation Planning and Execution System database portion of an operation plan; it contains time-phased force data, non-unit-related cargo and personnel data, and movement data for the operation plan, including the following: a. In-place units; b. Units to be deployed to support the operation plan with a priority indicating the desired sequence for their arrival at the port of debarkation; c. Routing of forces to be deployed; d. Movement data associated with deploying forces; e. Estimates of non-unit related cargo and personnel movements to be conducted concurrently with the deployment of forces; and f. Estimate of transportation requirements that must be fulfilled by common-user lift resources as well as those requirements that can be fulfilled by assigned or attached transportation resources. (JP 1-02)

**toxic industrial material** – A generic term for toxic radioactive compounds in solid, liquid, aerosolized or gaseous form. These may be used, or stored for use, for industrial, commercial, medical, military, or domestic purposes. Toxic industrial materials may be chemical, biological, or radioactive and described as toxic industrial chemical, toxic industrial biological, or toxic industrial radiological. (FM 3-11.4)

**track** – 1. To display or record the successive positions of a moving object. 2. To keep a gun properly aimed, or to point continuously a target-locating instrument at a moving target. (JP 1-02)

**troop leading procedures** – A dynamic process used by small unit leaders to analyze a mission, develop a plan, and prepare for an operation. (FM 5-0)

**turning movement** – (Army) A form of maneuver in which the attacking force seeks to avoid the enemy's principal defensive positions by seizing objectives to the enemy rear and causing the enemy to move out of his current positions or divert major forces to meet the threat. (FM 3-0)

**unconventional warfare** – A broad spectrum of military and paramilitary operations, normally of long duration, predominantly conducted through, with, or by indigenous or surrogate forces who are organized, trained, equipped, supported, and directed in varying degrees by an external source. It includes, but is not limited to, guerrilla warfare, subversion, sabotage, intelligence activities, and unconventional assisted recovery. (JP 1-02)

**unexploded explosive ordnance** – Explosive ordnance which has been primed, fused, armed, or otherwise prepared for action, and which has been fired, dropped, launched, projected, or placed in such a manner as to constitute a hazard to operations, installations, personnel, or material, and remains unexploded either by malfunction or design or for any other cause. (JP 1-02)

**unity of command** – One of the nine principles of war: for every objective, ensure unity of effort under one responsible commander.

**unity of effort** – Coordination and cooperation among all military forces and other organizations toward a commonly recognized objective, even if the forces and nonmilitary organizations are not necessarily part of the same command structure. (FM 6-0)

**unmanned aircraft system** – A powered aerial vehicle that does not carry a human operator, uses aerodynamic forces to provide lift, can fly automomously or is remotely operated, can be expendable or recoverable, and can carry a lethal or nonlethal payload. Ballistic or semi-ballistic vehicles, cruise missiles, and artillery projectiles are not considered unmanned aircraft systems.

**urban area** – A topographical complex where manmade construction or high population density is the dominant feature. See also **urban environment**; **urban operations**. (FM 3-06)

**urban environment** – The physical urban area, as well as the complex and dynamic interaction among its key components—the terrain (natural and manmade), the population, and the supporting infrastructure—as an overlapping and interdependent system of systems. (FM 3-06)

**urban operations** – Offense, defense, stability, and support operations conducted in a topographical complex and adjacent natural terrain where manmade construction and high population density are the dominant features. (FM 3-0)

**versatility** – The ability of Army forces to meet the global, diverse mission requirements of full spectrum operations. (FM 3-0)

**visualize** – To create and think in mental images. (FM 6-0)

**vulnerability** – 1. The susceptibility of a nation or military force to any action by any means through which its war potential or combat effectiveness may be reduced or its will to fight diminished. 2. The characteristics of a system that cause it to suffer a definite degradation (incapability to perform the designated mission) as a result of having been subjected to a certain level of effects in an unnatural (manmade) hostile environment. 3. In information operations, a weakness in information system security design, procedures, implementation, or internal controls that could be exploited to gain unauthorized access to information or an information system. (JP 1-02)

**warfighting function** – A group of tasks and systems (people, organizations, information, and processes), united by a common purpose, that commanders use to accomplish missions and training objectives. The six warfighting functions are: a. **command and control** – the related tasks and systems that support the commander in exercising authority and direction; b. **fire support** – the related tasks and systems that provide collective and coordinated use of Army indirect fires, joint fires, and offensive information operations; c. **intelligence** – the related tasks and systems that facilitate understanding of the enemy, terrain, weather, and civil considerations; d. **movement and maneuver** – the related tasks and systems that move forces to achieve a position of advantage in relation to the enemy; e. **protection** – the related tasks and systems that preserve the force so the commander can apply maximum combat power; f. **sustainment** – the related tasks and systems that provide support and services to ensure freedom of action, extend operational reach, and prolon endurance. (FMI 5-0.1)

**weapons of mass destruction** – Weapons that are capable of a high order of destruction and/or of being used in such a manner as to destroy large numbers of people. Weapons of mass destruction can be high explosives or nuclear, biological, chemical, and radiological weapons, but exclude the means of transporting or propelling the weapon where such means is a separable and divisible part of the weapon. (JP 1-02)

This page intentionally left blank.

# References

Military publications are listed by title. When a field manual has been published under a new number for the first time, the old number is provided in parenthesis after the new number.


## REQUIRED PUBLICATIONS

These documents must be available to intended users of this publication.

### JOINT PUBLICATIONS

Joint publications are available at <http://www.dtic.mil/doctrine/jel/>.

JP 1-02. *Department of Defense Dictionary of Military and Associated Terms*, 12 April 2001.

JP 3-06. *Doctrine for Joint Urban Operations*. 16 September 2002.

### ARMY PUBLICATIONS

Army doctrinal publications are available at <http://www.apd.army.mil/>.

FM 1-02 (101-5-1). *Operational Terms and Graphics*. 21 September 2004.

FM 34-130. *Intelligence Preparation of the Battlefield*. 8 July 1994. FM 34-130 will be republished as FM 2-01.3

FM 3-0 (100-5). *Operations*. 14 June 2001.

FM 3-06.11 (90-10-1). *Combined Arms Operations in Urban Terrain*. 28 February 2002.

FM 3-07 (100-20). *Stability Operations and Support Operations*. 20 February 2003. When updated, FM 3-07 will be retitled *Stability Operations*.

FM 3-90. *Tactics*. 04 July 2001.

FM 5-19 (100-14). *Composite Risk Management*. 21 August 2006.

FM 7-0. *Training the Force*. 22 October 2002.

FM 7-1. *Battle Focused Training*. 15 September 2003.

FM 5-0 (101-5). *Army Planning and Orders Production*. 20 January 2005.

## RELATED PUBLICATIONS

These sources contain relevant supplemental information.

### JOINT PUBLICATIONS

Joint publications are available at <http://www.dtic.mil/doctrine/jel/>.

JP 3-0. *Doctrine for Joint Operations*. 10 September 2001.

JP 3-05. *Doctrine for Joint Special Operations*. 17 December 2003.

JP 3-08. *Interagency Coordination During Joint Operations* (2 Volumes). 9 October 1996.

JP 3-16. *Joint Doctrine for Multinational Operations*. 05 April 2000.

JP 4-0. *Doctrine for Logistic Support of Joint Operations*. 06 April 2000.

JP 5-00.2. *Joint Task Force Planning Guidance and Procedures*. 13 January 1999.

*Joint Military Operations Historical Collection*. 15 July 1997.

### ARMY PUBLICATIONS

Army doctrinal publications are available at <http://www.apd.army.mil/>.

#### Army Regulations

FM 1. *The Army*. 14 June 2005.

FM 2-0 (34-1). *Intelligence*. 17 May 2004.

FM 2-22.3 (34-52). *Human Intelligence Collector Operations*. 6 September 2006.

FM 3-05 (100-25). *Army Special Operations Forces*. 20 September 2006.

FM 3-05.20 (31-20). *Special Forces Operations*. 1 April 2004.

FM 3-05.30 (33-1). *Psychological Operations*. 15 April 2005.

FM 3-05.105. *Multiservice Tactics, Techniques, and Procedures for Special Operations Forces in Nuclear, Biological, and Chemical Environments*. 28 September 2001.

FM 3-05.212 (TC 31-25). *Special Forces Waterborne Operations*. 31 August 2004.

FM 3-05.301 (33-1-1). *Psychological Operations Tactics, Techniques, and Procedures*. 31 December 2003.

FM 3-05.401 (41-10-1). *Civil Affairs Tactics, Techniques, and Procedures*. 23 September 2003.

FM 3-06 (90-10). *Urban Operations*. 01 June 2003.

FM 3-06.1. *Aviation Urban Operations: Multi-service Tactics, Techniques, and Procedures for Aviation Urban Operations*. 09 July 2005.

FM 3-07.2. *Antiterrorism*. (Future Publication).

FM 3-11.3 ( ). *Multiservice Tactics, Techniques, and Procedures for Chemical, Biological, Radiological, and Nuclear Contamination Avoidance*. 2 February 2006.

FM 3-11.9 (3-9). *Potential Military Chemical/Biological Agents and Compounds*. 10 January 2005.

FM 3-11.14 (3-14). *Multiservice Tactics, Techniques, and Procedures for Nuclear, Biological, and Chemical Vulnerability Assessment*. 28 November 2004.

FM 3-11.19 (3-19). *Multiservice Tactics, Techniques, and Procedures for Nuclear, Biological, and Chemical Reconnaissance*. 30 July 2004.

FM 3-11.21 (3-21). *Multiservice Tactics, Techniques, and Procedures for Nuclear, Biological, and Chemical Aspects of Consequence Management*. 12 December 2001.

FM 3-11.22. *Weapons of Mass Destruction Civil Support Team Tactics, Techniques, and Procedures*. 06 June 2003.

FM 3-13 (100-6). *Information Operations: Doctrine, Tactics, Techniques, and Procedures*. 28 November 2003.

FM 3-14 (100-18). *Space Support to Army Operations*. 18 May 2005.

FM 3-19.4 (19-4). *Military Police Leaders' Handbook*. 04 March 2002.

FM 3-19.15 (19-15). *Civil Disturbance Operations*. 18 April 2005.

FM 3-19.17. *Military Working Dogs*. 22 September 2005.

FM 3-19.40 (19-40). *Military Police Internment/Resettlement Operations*. 01 August 2001.

FM 3-19.50. *Police Intelligence Operations*. 21 July 2006.

FM 3-22.40 (90-40). *NLW: Tactical Employment of Nonlethal Weapons*. 15 January 2003.

FM 3-28. *Civil Support Operations*. (Future Publication).

FM 3-31. *Joint Force Land Component Commander Handbook (JFLCC)*. 13 December 2001.

FM 3-31.1 (90-31). *AMCI: Army and Marine Corps Integration in Joint Operations*. 21 November 2001.

FM 3-34 (5-100, 5-114). *Engineer Operations*. 02 January 2004.

FM 3-34.2. *Combined Arms Breaching Operations*. 31 August 2000. FM 3-34.2 will be republished as FM 3-90.11

FM 3-34.230 (5-105). *Topographic Operations*. 03 August 2000.

FM 3-61.1. *Public Affairs Tactics, Techniques and Procedures*. 01 October 2000.

FM 3-90.12 (90-13). *Combined Arms Gap Crossing Operations*. TBP.

FM 3-97.6 (90-6). *Mountain Operations*. 28 November 2000.

FM 3-100.12. *Risk Management: Multiservice Tactics, Techniques, and Procedures for Risk Management*. 15 February 2001.

FM 3-100.38. *UXO: Multi-service Tactics, Techniques and Procedures for Unexploded Ordnance Operations*. 18 August 2005.

FM 4-0 (100-10). *Combat Service Support*. 29 August 2003.

FM 4-01.45. *Tactical Convoy Operations: Multiservice Tactics, Techniques, and Procedures for Tactical Convoy Operations*. 24 March 2005.

FM 4-02.51 (8-51). *Combat and Operational Stress Control*. 6 June 2006.

FM 4-30.16. *EOD: Multiservice Procedures for Explosive Ordnance Disposal in a Joint Environment*. 27 October 2005.

FM 4-30.51. *Unexploded Ordnance (UXO) Procedures*. 13 July 2006.

FM 5-01.12 (90-41). *JTF LNO Integration: Multiservice Tactics, Techniques, and Procedures for Joint Task Force (JTF) Liaison Officer Integration*. 27 January 2003.

FM 5-33. *Terrain Analysis*. 8 September 1992. FM 5-33 will be republished as FM 3 34.330.

FM 5-103. *Survivability*. 10 June 1985. FM 5-103 will be republished as FM 5-34.112.

FM 5-104. *General Engineering*. 12 November 1986. FM 5-104 will be republished as FM 3-34.250.

FM 5-170. *Engineer Reconnaissance*. 05 May 1988. FM 5-170 will be republished as FM 3 34.212.

FM 5-415. *Fire-Fighting Operations*. 09 February 1999. FM 5-415 will be republished as FM 3-34.281.

FM 6-0. *Mission Command: Command and Control of Army Forces*. 11 August 2003.

FM 6-22.5 (22-9). *Combat Stress*. 23 June 2000.

FM 7-15. *The Army Universal Task List (Incl. changes 2)*. 6 July 2006.

FM 20-32. *Mine/Countermine Operations*. 1 April 2005.

FM 22-51. *Leaders' Manual for Combat Stress Control*. 29 September 1994. FM 22-51 will be republished as FM 4-02.22.

FM 22-100. *Army Leadership: Be, Know, Do*. 31 August 1999. FM 22-100 will be republished as FM 6-22.

FM 23-10. *Sniper Training*. 17 August 1994. FM 23-10 will be republished as FM 3-22.10.

FM 27-10. *The Law of Land Warfare*. 18 July 1956. FM 27-10 will be republished as FM 1-04.10.

FM 27-100. *Legal Support to Operations*. 01 March 2000. FM 27-100 will be republished as FM 1-04.

FM 31-20-3. *Foreign Internal Defense Tactics, Techniques, and Procedures for Special Forces*. 20 September 1994. FM 31-20-3 will be republished as FM 3-05.202.

FM 31-50. *Combat in Fortified and Built-Up Areas*. March 1964.

FM 34-3. *Intelligence Analysis*. 15 March 1990. FM 34-3 will be republished as FM 2-33.4.

FM 34-60. *Counterintelligence*. 03 October 1995. FM 34-60 will be republished as FM 2-01.2.

FM 34-81. *Weather Support for Army Tactical Operations*. 31 August 1989. FM 34-81 will be republished as FM 2-33.2.

FM 34-81-1. *Battlefield Weather Effects*. 23 December 1992. FM 34-81-1 will be republished as FM 2-33.201.

FM 41-10. *Civil Affairs Operations*. 14 February 2000. FM 41-10 will be republished as FM 3-05.40.

FM 44-8. *Combined Arms for Air Defense*. 01 June 1999. FM 44-8 will be republished as FM 3-01.8.

FM 46-1. *Public Affairs Operations*. 30 May 1997. FM 46-1 will be republished as FM 3-61.

FM 90-8. *Counterguerrilla Operations*. 29 August 1986. FM 90-8 will be republished as FM 3-07.11.

FM 90-10. *Military Operations on Urbanized Terrain (MOUT) (How to Fight)*. 15 August 1979.

FM 100-5. *Operations*. 1944.

FM 100-8. *The Army in Multinational Operations*. 24 November 1997. FM 100-8 will be republished as FM 3-16.

FM 100-11. *Force Integration*. 15 January 1998. FM 100-11 will be republished as FM 3-100.11.

FMI 2-91.4. *Intelligence Support to Operations in the Urban Environment*. 30 June 2005. FMI 2-91.4 will be republished as FM 2-91.4.

FMI 3-07.22. *Counterinsurgency Operations*. 01 October 2004. FMI 3-07.22 will be republished as FM 3-07.22.

FMI 3-34.119. *Improvised Explosive Device Defeat*. 21 September 2005.

FMI 4-30.50. *Modular Explosive Ordnance Disposal Operations*. 31 July 2006.

FMI 5-0.1. *The Operations Process*. 31 March 2006.

TC 3-15. *Nuclear Accident and Incident Response and Assistance (NAIRA)*. 27 December 1988.

TC 5-230. *Army Geospatial Guide for Commanders and Planners*. 28 November 2003.

TC 7-98-1. *Stability and Support Operations Training Support Package*. 05 June 1997.

TC 9-21-01. *Soldiers Improvised Explosive Device (IED) Awareness Guide: Iraq and Afghanistan Theaters of Operation*. 28 May 2004.

TC 90-1. *Training for Urban Operations*. 01 April 2002.

## DEPARTMENT OF DEFENSE PUBLICATIONS

Defense Intelligence Agency Manual 58-11. *Defense Human Resources Intelligence Collection Management*. 20 August 1968.

*Quadrennial Defense Review Report*. 30 September 2001.

U.S. Army Training and Doctrine Command (TRADOC), Deputy Chief of Staff for Intelligence (DCSINT). *TRADOC DCSINT Handbook No. 1: A Military Guide to Terrorism in the Twenty-First Century* (Version2.0). Fort Leavenworth, KS: U.S. Government Printing Office, 2004.

## NONMILITARY PUBLICATIONS

26th Infantry Regimental Association. *Aachen: Military Operations in Urban Terrain*. Lititz, PA: 26th Infantry Regimental Association, 1999.

Adan, Abraham. *On the Banks of the Suez*. Novato, CA: Presidio Press, 1980.

Akers, Frank H., Jr. and George B. Singleton. "Task Force Ranger: A Case Study Examining the Application of Advanced Technologies in Modern Urban Warfare." Oak Ridge, TN: National Security Program Office, November 2000.

Allard, Kenneth. *Somalia Operations: Lessons Learned*. Washington, DC: National Defense University Press, 1995.

Arnold, James R. *Tet Offensive 1968: Turning Point in Vietnam*. London: Osprey, 1990.

Bass, Bernard M. "Leading in the Army after Next." *Military Review* (March/April 1998): 46–58.

Beevor, Anthony. *Stalingrad*. New York: Penguin Books, 1998.

Bowden, Mark. *Black Hawk Down: A Story of Modern War*. New York: Atlantic Monthly Press, 1999.

Butts, Kent H. "The Strategic Importance of Water." *Parameters* (spring 1997): 65–83.

Carell, Paul. *Stalingrad: The Defeat of the German 6th Army*. Atglen, PA: Schieffer Publishing, 1993.

Celeski, Joseph D. "Joint Urban Operations: Special Forces in Urban Campaign Planning." *Special Warfare* (fall 2001): 2–13.

Chiarelli, Peter W. and Patrick R. Michaelis. "Winning the Peace: The Requirement for Full-Spectrum Operations." *Military Review* (July/August 2005): 4–17.

Chiarelli, Peter W., Patrick R. Michaelis, and Geoffrey A. Norman. "Armor in Urban Terrain: The Critical Enabler." *Armor* (March/April 2005): 7–12.

Chuikov, Vasili I. *The Battle for Stalingrad.* New York: Holt, Rinehart, and Winston, 1964.

Colby, John. *War From the Ground Up: The 90th Division in WWII.* Austin, TX: Nortex Press, 1991.

Cole, Ronald H. *Operation Just Cause.* Washington, DC: Joint History Office, 1995.

Connaughton, Richard, John Pimlott, and Duncan Anderson. *The Battle for Manila: The Most Devastating Untold Story of World War II.* Novata, CA: Presidio, 1995.

Craig, William. *Enemy at the Gates: The Battle for Stalingrad.* New York: Reader's Digest Press, 1973.

Delk, James D. *Fires & Furies: The LA Riots.* Palm Springs, CA: Etc. Publications, 1995.

Dewar, Michael. *British Army in Northern Ireland.* New York: Sterling Publishers, 1985.

______. *War in the Streets: The Story of Urban Combat from Calais to Khafji.* New York: Sterling Publishers, 1992.

Donnelly, Thomas, Margaret Roth, and Caleb Baker. *Operation Just Cause: The Storming of Panama.* New York: Lexington Books, 1991.

Dupuy, R. Ernest and Trevor N. Dupuy. *The Encyclopedia of Military History: From 3500 B.C. to the Present.* San Francisco: Harper & Row, 1977.

Dupuy, Trevor N. and Paul Martell. *Flawed Victory: The Arab-Israeli Conflict and the 1982 War in Lebanon.* Fairfax, VA: Hero Books, 1986.

Edwards, Sean J. A. *Freeing Mercury's Wings: Improving Tactical Communications in Cities.* Santa Monica, CA: RAND, 2001.

Eshel, David. *Mid-East Wars: The Lebanon War 1982.* Hod Hasharon, Israel: Eshel-Dramit, 1983.

Fishel, John T. *The Fog of Peace: Planning and Executing the Restoration of Panama.* Carlisle, PA: Strategic Studies Institute, 1992.

Fontenot, Gregory, E. J. Degen, and David Tohn. *On Point: US Army in Operation IRAQI FREEDOM.* Fort Leavenworth, KS: Combat Studies Institute Press, 2004.

Funston, Frederick. "How the Army Worked to Save San Francisco." *Cosmopolitan Magazine* (July 1906): <http://www.sfmuseum.org/1906/cosmo.html>.

Gabel, Christopher R. "Military Operations on Urbanized Terrain: The 2d Battalion, 26th Infantry, at Aachen, October 1944." *Urban Combat Operations: tactics, Techniques and Procedure.* Center for Army Lessons Learned (CALL) Newsletter No. 99-16. Fort Leavenworth, KS: Center for Army Lessons Learned, U.S. Training and Doctrine Command [TRADOC], November 1999: 1-9–1-13.

Gabriel, Richard A. *Operation Peace for Galilee: The Israeli-PLO War in Lebanon.* New York: Hill and Wang, 1984.

Glenn, Russell W. *An Attack on Duffer's Downtown.* Santa Monica, CA: RAND, 2001.

______. *Heavy Matter: Urban Operations' Density of Challenges.* Santa Monica, CA: RAND, 2000.

______. "Urban Combat is Complex." *Proceedings* (February 2002): 62–65.

______. "…We Band of Brothers": *The Call for Joint Urban Operations Doctrine.* Santa Monica, CA: RAND, 1999.

Glenn, Russell W., Steven Hartman, and Scott Gerwehr. *Urban Combat Service Support Operations: The Shoulders of Atlas.* Santa Monica, CA: RAND, 2003.

Gordon, Michael R. "Baghdad Targets Picked if Hussein Holes Up There." *New York Times* (7 March 2003): 1A2.

Grau, Lester W. "Something Old, Something New: Guerrillas, Terrorists, and Intelligence Analysis." *Military Review* (July/August 2004): 42–49.

Grau, Lester W. and Timothy Smith. *A 'Crushing' Victory: Fuel-Air Explosives and Grozny 2000.* Fort Leavenworth, KS: Foreign Military Studies Office, 2000.

Hamill, Desmond. *Pig in the Middle: The Army in Northern Ireland, 1969–1984.* London: Methuen, 1985.

Hammel, Eric. *Fire in the Streets: The Battle for Hue, Tet 1968*. New York: Dell Publishing, 1991.

Heal, Sid. "Crowds, Mobs and Nonlethal Weapons." *Military Review* (March/April 2000): 45–50.

Heinl, Robert Debs, Jr. *Victory at High Tide: The Inchon-Seoul Campaign*. Annapolis, MD: Nautical & Aviation Publishing Company of America, 1979.

Herzog, Cham. *The War of Atonement, October, 1973*. Boston: Little, Brown and Company, 1975.

Higham, Robin. *Bayonets in the Streets: The Use of Troops in Civil Disturbances*. Lawrence, KS: University Press of Kansas, 1969.

Hoar, Joseph P. "A CINC's Perspective." *JFQ Forum* (autumn 1993): 56–63.

Hollis, Mark A. B. "Platoon Under Fire." *Infantry* (January/April 1998): 27–34.

*Infantry in Battle*. 2d ed. Garret & Massie: Richmond, VA, 1939. Reprint, Fort Leavenworth, KS: U.S. Army Command and General Staff College, 1981.

*Initial Impressions Report: Operation Iraqi Freedom—Stability Operations-Support Operations* Center for Army Lessons Learned (CALL) Newsletter No. 04-13 (Fort Leavenworth, KS: Center for Army Lessons Learned, U.S. Army Training and Doctrine Command [TRADOC], 2004).

*Initial Impressions Report: Operation Iraqi Freedom—Stability Operations-Support Operations*. Center for Army Lessons Learned (CALL) Newsletter No. 03 (Fort Leavenworth, KS: Center for Army Lessons Learned, U.S. Army Training and Doctrine Command [TRADOC], 2003).

Jomini, Antoine-Henri. "The Art of War" in *Roots of Strategy: Book 2*. Harrisburg, PA: Stackpole Books, 1987.

Kemp, Anthony. *The Unknown Battle: Metz, 1944*. New York: Stein and Day, 1981.

Kennedy, John F. Message to Congress, 25 May 1961. Available at <http://www.jfklibrary.org/Historical+Resources/Archives>.

Linden, Eugene. "The Exploding Cities of the Developing World." *Foreign Affairs* (January/February 1996): 52–65.

Mabry, Robert L. et al. "United States Army Rangers in Somalia: An Analysis of Combat Casualties on an Urban Battlefield." *Journal of Special Operations Medicine* (fall 2001): 24–40.

MacArthur, Douglas. *Reminiscences*. New York: McGraw-Hill, 1964.

Manwaring, Max G. *Street Gangs: The New Urban Insurgency*. Carlisle, PA: Strategic Studies Institute, 2005.

Maxwell, Mark D. "The Law of War and Civilians on the Battlefield: Are We Undermining Civilian Protections?" *Military Review* (September/October 2004): 17–25.

McConnell, Malcolm. *Just Cause: The Real Story of America's High-Tech Invasion of Panama*. New York: St. Martin's Press, 1991.

Medby, Jamison Jo and Russell W. Glenn. *Street Smart: Intelligence Preparation of the Battlefield for Urban Operations*. Santa Monica, CA: RAND, 2002.

Mendel, William W. *Combat in Cities: The LA Riots and Operation Rio*. Fort Leavenworth, KS: Foreign Military Studies Office, 1996.

Monmonier, Mark. *How to Lie with Maps*. Chicago: University of Chicago Press, 1996.

Montgomery, Thomas M. Interview. "Ambush in Mogasdishu." *Frontline*, WGBH Boston. 29 September 1998. Available at <http://www.pbs.org/wgbh/pages/frontline/programs/1998.html>.

Moore, John C. "Sadr City: The Armor Pure Assault in Urban Terrain." *Armor* (November/December 2004): 31–37.

Natsios, Andrew S. "Commander's Guidance: A Challenge of Complex Humanitarian Emergencies." *Parameters* (summer 1996): 50–66.

______. "Food Through Force: Humanitarian Intervention and U.S. Policy." *The Washington Quarterly* (winter 1994): 136.

Nolan, Keith William. *The Battle for Saigon: Tet 1968*. New York: Pocket Books, 1996.

Oakley, Robert B. "An Envoy's Perspective." *JFQ Forum* (autumn 1993): 44–55.

Oliker, Olga. *Russia's Chechen Wars 1994–2000: Lessons from Urban Combat*. Santa Monica, CA: RAND, 2001.

*Operation Restore Hope Lessons Learned Report: 3 December 1992 – 4 May 1993, Operations Other than War*. Fort Leavenworth, KS: Center for Army Lessons Learned (CALL), U.S. Army Combined Arms Command (CAC), 1993. Reprint, Washington, DC: U.S. Government Printing Office, 1996.

Parkerson, John Embry, Jr. "United States Compliance with Humanitarian Law Respecting Civilians During Operation Just Cause." *Military Law Review* (1991): 31–140.

Peters, Ralph. "Our Soldiers, Their Cities." *Parameters* (spring 1996): 43–50.

Peterson, John J. *Into the Cauldron*. Clinton, MD: Clavier House, 1973.

Petraeus, David H. "Learning Counterinsurgency: Observations from Soldiering in Iraq." *Military Review* (January/February 2006): 2–12.

Ritchie, Elspeth C. and Robert Mott. "Caring for Civilians During Peace Keeping Missions: Priorities and Decisions." *Military Medicine* (Vol. 167, Supplement 3, 2002): 14–16.

Rosenberger, Leif R. "The Strategic Importance of the World Food Supply." *Parameters* (spring 1997): 84–105.

Schadlow, Nadia. "War and the Art of Governance." *Parameters* (autumn 2003): 85–94.

Shelton, H. Hugh and Timothy D. Vane. "Winning the Information War in Haiti." *Military Review* (November/December 1995): 3–9.

Shtemenko, S. M. *The Soviet General Staff at War, 1941-1945: Book One*. Moscow: Progress Publishers, 1981.

Smith, Rupert. "Wars in Our Time: A Survey of Recent and Continuing Conflicts." *World Defence Systems,* volume 3:2. London: Royal United Services Institute for Defence Studies, 2001.

Stanton, Shelby L. *Ten Corps in Korea, 1950*. Novato, CA: Presidio Press, 1989.

Taw, Jennifer Morrison. *Operation Just Cause: Lessons for Operations Other Than* War. Santa Monica, CA: RAND, 1996.

Thomas, Timothy L. "The Battle for Grozny: Deadly Classroom for Urban Combat." *Parameters* (summer 1999): 87–102.

Thomas, Timothy L. and Charles P. O'Hara. "Combat Stress in Chechnya: 'The Equal Opportunity Disorder'." *Army Medical Department Journal* (January/March 2000): 46–53.

Trinquier, Roger. *Modern Warfare: A French View of Counterinsurgency*. New York: Praeger Publisher, 1964.

*United Nations Security Council Resolution 794* (1992). New York: United Nations, 1992. Available at <http://daccessdds.un.org/doc/UNDOC/GEN/N92/772/11/PDF>.

U.S. Agency for International Development, Bureau for Humanitarian Response, Office of Foreign Disaster Assistance. *Field Operations Guide for Disaster Assessment and Response, Version 3.0*. Washington, DC: U.S. Government Printing Office, 2003. Available at <http://www.usaid.gov/our_work/humanitarian_assistance/>.

U.S. Department of Transportation, Research and Special Programs Administration, Office of Hazardous Materials Initiative and Training. *2004 Emergency Response Guidebook*. Washington, DC: U.S. Government Printing Office, 2004. Available at <http://hazmat.dot.gov/pubs/erg/erg2004.pdf>.

U.S. Environmental Protection Agency, Office of Emergency and Remedial Response. *HAZMAT Team Planning Guidance*. Washington, DC: U.S. Government Printing Office, 1990.

Van Sickle, Jeffrey B. "Stability Operations in Northern Iraq: Task Force Altun Kupri." *Infantry* (January/February 2005): 25–29.

Watson, Bruce Allen. *Sieges: A Comparative Study*. Westport, CT: Praeger, 1993.

Werstein, Irving. *The Battle of Aachen*. New York: Thomas Y. Crowell Company, 1962.

Whiting, Charles. *Bloody Aachen*. New York: Stein and Day, 1976.

Wintle, Justin, ed. *The Dictionary of War Quotations*. New York: The Free Press, 1989.

Yates, Lawrence A. "Military Stability and Support Operations: Analogies, Patterns and Recurring Themes." *Military Review* (July/August 1997): 51–61.

# RECOMMENDED READINGS

The following sources contain relevant supplemental information.

Alexander, John B. *Future War: Non-Lethal Weapons in Twenty-First-Century* Warfare. New York: St. Martin's Press, 1999.

Ashworth, G. J. *The City and War*. London: Routledge, 1991.

Barker, A. J. *Dunkirk: The Great Escape*. New York: David McKay, 1977.

Baumann, Robert F., Lawrence A. Yates, and Versalle F. Washington. "My Clan Against the World": *U.S. and Coalition Forces in Somalia, 1992-1994*. Fort Leavenworth, KS: Combat Studies Institute Press, 2004.

Birtle, Andrew J. *U.S. Army Counterinsurgency and Contingency Operations Doctrine: 1860-1941*. Washington, DC: Center of Military History, 2001.

Blaufarb, Douglas S. *The Counterinsurgency Era: U.S. Doctrine and Performance 1950 to Present*. New York: The Free Press, 1977.

Breuer, William B. *Hitler's Fortress Cherbourg: The Conquest of a Bastion*. New York: Stein and Day, 1984.

Brown, Stephen D. "PSYOP in Operation Uphold Democracy." *Military Review* (September/October 1996): 57–76.

Bulloch, Gavin. "Military Doctrine and Counterinsurgency: A British Perspective." *Parameters* (summer 1996): 4–16.

*Urban Combat Operations: Tactics, Techniques and Procedures*. Center for Army Lessons Learned (CALL) Newsletter No. 99-16 (Fort Leavenworth, KS: Center for Army Lessons Learned, U.S. Army Training and Doctrine Command [TRADOC], November 1999).

Cordesman, Anthony H. and Abraham R. Wagner. *The Lessons of Modern War—Volume II: The Iran-Iraq War*. Boulder, CO: Westview Press, 1990.

Corum, James S. and Wray R. Johnson. *Airpower in Small Wars: Fighting Insurgents and Terrorists*. Lawrence, KS: University Press of Kansas, 2003.

Desch, Michael C., ed. *Soldiers in Cities: Military Operations on Urban Terrain*. Carlisle Barracks, PA: Strategic Studies Institute, 2001.

Duffy, Christopher. *Fire & Stone: The Science of Fortress Warfare, 1660–1860*. Mechanicsburg, PA: Stackpole Books, 1996.

______. *The Fortress in the Age of Vauban and Frederick the Great, 1660–1789: Siege Warfare Volume I*. London: Routledge & Kegan Paul, 1985.

______. *Siege Warfare: The Fortress in the Early Modern World, 1494–1660*. London: Routledge & Kegan Paul, 1979.

Edwards, Sean J. A. *Mars Unmasked: The Changing Face of Urban Operations*. Santa Monica, CA: RAND, 2000.

Eikenberry, Karl W. "Take No Casualties." *Parameters* (summer 1996): 109–118.

Erickson, John. *The Road to Berlin: Continuing the History of Stalin's War with Germany*. Boulder, CO: Westview Press, 1983.

Gerwehr, Scott and Russell W. Glenn. *The Art of Darkness: Deception and Urban Operations*. Santa Monica, CA: RAND, 2000.

Glenn, Russell W. *Combat in Hell: A Consideration of Constrained Urban Warfare*. Santa Monica, CA: RAND, 1996.

______. *Marching Under Darkening Skies: The American Military and Impending Urban Operations Threat.* Santa Monica, CA: RAND, 1998.

Glenn, Russell W., ed. *Capital Preservation: Preparing for Urban Operations in the Twenty-First Century: Proceedings of the RAND Arroyo-TRADOC-MCWL-OSD Urban Operations Conference, March 22-23, 2000.* Santa Monica, CA: RAND, 2001.

Glenn, Russell W. et al. *Denying the Widow-Maker, Summary of Proceedings, RAND-DDBL Conference on Military Operations on Urbanized Terrain.* Santa Monica, CA: RAND, 1999.

______. *Getting the Musicians of Mars on the Same Sheet of Music: Army, Joint, Multinational, and Interagency C4ISR Interoperability.* Santa Monica, CA: RAND, 2000.

Glenn, Russell W. et al., eds. *The City's Many Faces: Proceedings of the RAND Arroyo-MCWL-J8 UWG Urban Operations Conference, April 13-14, 1999.* Santa Monica, CA: RAND, 2000.

Gouré, Leon. *The Siege of Leningrad.* Stanford, CA: Stanford University Press, 1962.

Grau, Lester W. and Jacob W. Kipp. "Urban Combat: Confronting the Specter." *Military Review* (July/August 1999): 9–17.

Groves, John R. "Operations in Urban Environments." *Military Review* (July/August 1998): 30–39.

Habeck, Mary R. Lecture. "Jihadist Strategies in the War on Terrorism." Washington, DC: The Heritage Foundation, 8 November 2004. Available at http://www.heritage.org/research/nationalsecurity/hl855.cfm.

Hahn, Robert F., II and Bonnie Jezior. "Urban Warfare and the Urban Warfighter of 2025." *Parameters* (summer 1999): 74–86.

Harman, Nicholas. *Dunkirk: The Patriotic Myth.* New York: Simon and Schuster, 1980.

Heckstall-Smith, Anthony. *Tobruk: The Story of a Siege.* New York: W. W. Norton, 1959.

Hibbert, Christopher. *The Battle of Arnhem.* London: Batsford, 1962.

House, John M. "The Enemy After Next." *Military Review* (March/April 1998): 22–28.

Jackson, W. G. F. *The Battle for Italy.* New York: Harper & Row, 1967.

Karslake, Basil. *1940 – The Last Act: The Story of the British Forces in France After Dunkirk.* Hamden, CT: Archon Books, 1979.

Kostof, Spiro. *The City Shaped: Urban Patterns and Meanings Through History.* London: Thames and Hudson, 1999.

______. *The City Assembled: The Elements of Urban Form Throughout History.* Boston: Little Brown and Company, 1999.

Kuehl, Daniel T. "Airpower vs. Electricity: Electric Power as a Target for Strategic Air Operations." *Journal of Strategic Studies* (March 1995): 237–266.

Larson, Eric. *Casualties and Consensus: The Historical Role of Casualties in Domestic Support of U.S. Military Operations.* Santa Monica, CA: RAND, 1996.

Lieven, Anatol. *Chechnya: Tombstone of Russian Power.* New Haven: Yale University Press, 1998.

______. "The World Turned Upside Down: Military Lessons of the Chechen War." *Armed Forces Journal International* (August 1998): 40–43.

McKee, Alexander. *Caen: Anvil of Victory.* New York: St. Martin's Press, 1964.

McLaurin, R. D. "Military Forces in Urban Antiterrorism." Aberdeen Proving Ground, MD: U.S. Army Human Engineering Laboratory, 1989.

McLaurin, R. D. and R. Miller. "Urban Counterinsurgency: Case Studies and Implications for U.S. Military Forces." Aberdeen Proving Ground, MD: U.S. Army Human Engineering Laboratory, 1989.

Mercier, Charles E. "Terrorists, WMD, and the U.S. Army Reserve." *Parameters* (autumn 1997): 98–118.

Milton, T. R. "Urban Operations: Future War." *Military Review* (February 1994): 37–46.

Moskin, J. Robert. *Among Lions: The Battle for Jerusalem, June 5–7, 1967*. New York: Ballantine Books, 1982.

Northam, Ray M. *Urban Geography*. New York: Wiley, 1975.

Peters, Ralph. "The Culture of Future Conflict." *Parameters* (winter 1995 96): 18–27.

______. "The Future of Armed Warfare." *Parameters* (autumn 1997): 50–59.

______. "The Human Terrain of Urban Operations." *Parameters* (spring 2000): 4–12.

Robertson, William G., ed. *Block by Block: The Challenges of Urban Operations*. Fort Leavenworth, KS: U.S. Army Command and Staff College Press, 2003.

Rosenau, William G. "'Every Room is a New Battle': The Lessons of Modern Urban Warfare." *Studies in Conflict and Terrorism* (October–December 1997): 371–394.

Scalard, Douglas. "People of Whom We Know Nothing: When Doctrine Isn't Enough." *Military Review* (July/August 1997): 4–11.

Scales, Robert H. "The Indirect Approach: How U.S. Military Forces Can Avoid the Pitfalls of Future Urban Warfare." *Armed Forces Journal International* (October 1998): 68–74.

Siegel, Adam B. *Eastern Exit: The Noncombatant Evacuation Operation (NEO) from Mogadishu, Somalia, in January 1991*. Alexandria, VA: Center for Naval Analyses, 1992.

Spiller, Roger J. *Sharp Corners: Urban Operations at Century's End*. Fort Leavenworth, KS: Combat Studies Institute, 2001.

Stech, Frank J. "Winning CNN Wars." *Parameters* (autumn 1994): 37–56.

Stevenson, Jonathan. *Losing Mogadishu: Testing U.S. Policy in Somalia*. Annapolis, MD: Naval Institute Press, 1995.

Taw, Jennifer Morrison and Bruce Hoffman. *The Urbanization of Insurgency: The Potential Challenge to U.S. Army Operations*. Santa Monica, CA: RAND, 1994.

Tully, Andrew. *Berlin: Story of a Battle*. Westport, CT: Greenwood Press, 1977.

United Nations Department of Economic and Social Affairs, Population Division. *World Urbanization Prospects: The 2003 Revision*. New York: United Nations, 2003.

Vick, Alan et al. *Aerospace Operations in Urban Environments: Exploring New Concepts*. Santa Monica, CA: RAND, 2000.

Wykes, Alan. *The Siege of Leningrad*. New York: Ballentine Books Inc., 1968.

# Index

This page intentionally left blank.

**FM 3-06**
**26 October 2006**

By Order of the Secretary of the Army:

**PETER J. SCHOOMAKER**
*General, United States Army*
*Chief of Staff*

Official:

**JOYCE E. MORROW**
*Administrative Assistant to the*
*Secretary of the Army*
0627803

**DISTRIBUTION:**

*Active Army, Army National Guard, and U. S. Army Reserve:* To be distributed in accordance with initial distribution number 111233, requirements for FM 3-06.

*FM 20-32

Field Manual
No. 20-32

Headquarters
Department of the Army
Washington, DC, 29 May 1998

# MINE/COUNTERMINE OPERATIONS

## Table of Contents

Page



DISTRIBUTION RESTRICTION: Approved for public release; distribution is unlimited.

*This manual supersedes FM 20-32, 30 September 1992.

Page

Page

Page

Page

## Part Three. Special Mining Operations

Page

Page

# LIST OF ILLUSTRATIONS

## Figures

Page

Page

Page

Page

Page

Page



# Tables

Page

# Preface

Field Manual (FM) 20-32 provides United States (US) armed forces with tactical, technical, and procedural guidance for conducting mine and countermine operations. It applies to all elements of the combined arms team for maneuver and engineer staff planning and coordination. The manual is presented in three parts—mine operations, counteroperations, and special-mining operations.

The guidance provided focuses on individual skills of emplacing and removing mines, team and squad tasks, platoon and company organization and planning, and battalion/task force (TF) organization and coordination for successful obstacle reduction and breaching operations.

The provisions of this publication support existing doctrine established by FMs 5-34, 5-100, 90-7, and 90-13-1. It also contains new and improved techniques for emplacing row mines; marking, reporting, and recording minefields; reducing simple and complex obstacles; and emplacing a standard-pattern minefield. This manual reflects new doctrine from FMs 5-10, 5-71-2, and 5-71-3.

This publication implements the following International Standardization Agreements (STANAGs) between North Atlantic Treaty Organization (NATO) forces:

- STANAG 2036. *Land Minefield Laying, Marking, Recording, and Reporting Procedures.* Edition 5.
- STANAG 2889. *Marking of Hazardous Areas and Routes Through Them.* Edition 3.
- STANAG 2990. *Principles and Procedures for the Employment in Land Warfare of Scatterable Mines with a Limited Laid Life.* Edition 1.

**NOTE: US policy regarding the use and employment of antipersonnel land mines (APLs) outlined in this FM is subject to the Convention on Certain Conventional Weapons and Executive Orders. Current US policy limits the use of non-self-destructing APLs to (1) defending the US and its allies from armed aggression across the Korean demilitarized zone and (2) training personnel engaged in demining and countermine operations. The use of the M18A1 claymore in the command-detonation mode is not restricted under international law or Executive Order.**

All references to US employment of non-self-destructing APLs (such as row mining) in this manual are intended to provide doctrine for use in Korea only. This information is provided in bold lettering throughout the manual. Detailed doctrine on APLs is also provided to ensure that US forces recognize how the enemy can employ these weapons.

As the US military seeks to end its reliance on APLs, commanders must consider the increased use of other systems such as the M18A1 claymore, nonlethal barriers (such as wire obstacles), sensors and surveillance platforms, and direct and indirect fires.

This publication includes the following appendixes:

- Appendix A. Installation and Removal of US Mines and Firing Devices.
- Appendix B. Controls and Components of Special-Purpose Munitions.
- Appendix C. Threat Mine/Countermine Operations.
- Appendix D. Air Volcano.
- Appendix E. Safety and Training.
- Appendix F. Mine Awareness.
- Appendix G. Countermine Data.
- Appendix H. Metric Conversion Chart.

The proponent for this publication is Headquarters, US Army Training and Doctrine Command (TRADOC). Forward comments and recommendations on Department of the Army (DA) Form 2028 to Commandant, US Army Engineer School, ATTN: ATSE-DME-MWF, Fort Leonard Wood, Missouri 65473-5000.

Unless this publication states otherwise, nouns and pronouns do not refer exclusively to men.

Chapter 1

# Introduction

This chapter provides the mechanics and characteristics of antitank (AT) mines and munitions, antipersonnel (AP) mines and munitions, and antihandling devices (AHDs). The information contained in this chapter also provides a foundation for the rest of the manual.

Land-based mines and munitions are hand-emplaced, remote-delivered, ground-delivered, or air-delivered:

- Hand-emplaced mines and munitions require manual arming and are labor-, resource-, and transport-intensive.
- Remote- and air-delivered mines and munitions require less time and labor; however, they are not as precisely placed as hand-emplaced mines and munitions.
- Ground-delivered mines are less resource-intensive than hand-emplaced mines. They are not precisely placed; however, the minefield boundaries are.

Soldiers can surface lay or bury mines and munitions and can place AHDs on hand-emplaced AT mines.

**NOTE: Some countries employ AHDs on AP mines, but US forces are not authorized to employ AHDs on any type of AP mine.**

## MECHANICS OF MINES

### CHARACTERISTICS AND FUNCTIONING

A land mine is an explosive device that is designed to destroy or damage equipment or personnel. Equipment targets include ground vehicles, boats, and aircraft. A mine is detonated by the action of its target, the passage of time, or controlled means. There are two types of land-based mines—AT and AP. Mines generally consist of the following parts (Figure 1-1, page 1-2):

- Firing mechanism or other device (sets off the detonator or igniter charge).
- Detonator or igniter (sets off the booster charge).
- Booster charge (may be attached to the fuse or the igniter or be part of the main charge).
- Main charge (in a container; usually forms the body of the mine).
- Casing (contains all the above parts).

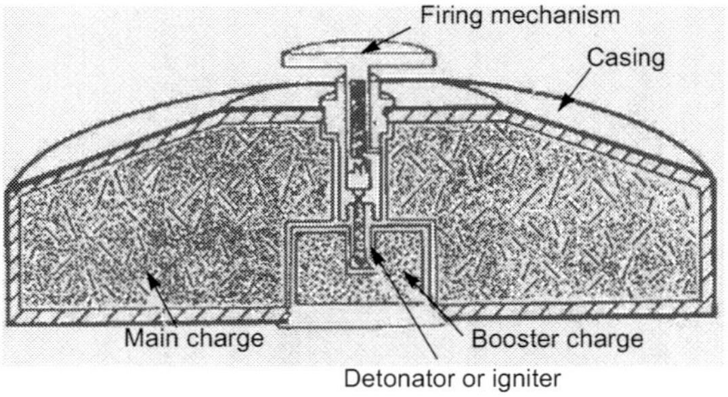


**Figure 1-1. Mine components**

### Components and Initiating Actions

A firing mechanism prevents the mine from exploding until it makes contact with, or is influenced by, its target. Once a mine has been armed, the firing mechanism may be actuated by the following methods (Figure 1-2):

- Applying pressure (including tilt rod).
- Pulling a trip wire.
- Releasing tension or breaking a trip wire.
- Releasing pressure.
- Passage of time (time-delay mechanism).
- Impulses.
  - Electrical.
  - Vibration.
  - Magnetic-influence.
  - Electromagnetic-frequency.
  - Infrared-sensored.
  - Acoustic.

To arm some mines, you must position the igniter, set the mechanism properly, and disengage the safety device (usually by removing a safety pin). The fuse is the initial component in the firing chain; it has a low-explosive (LE) powder but is highly sensitive. The fuse is actuated by an initiating action. Although mines are issued with a standard fuse, alternate fuses are issued separately for some mines.

The four main fuse types are shown in Figure 1-3, page 1-4.

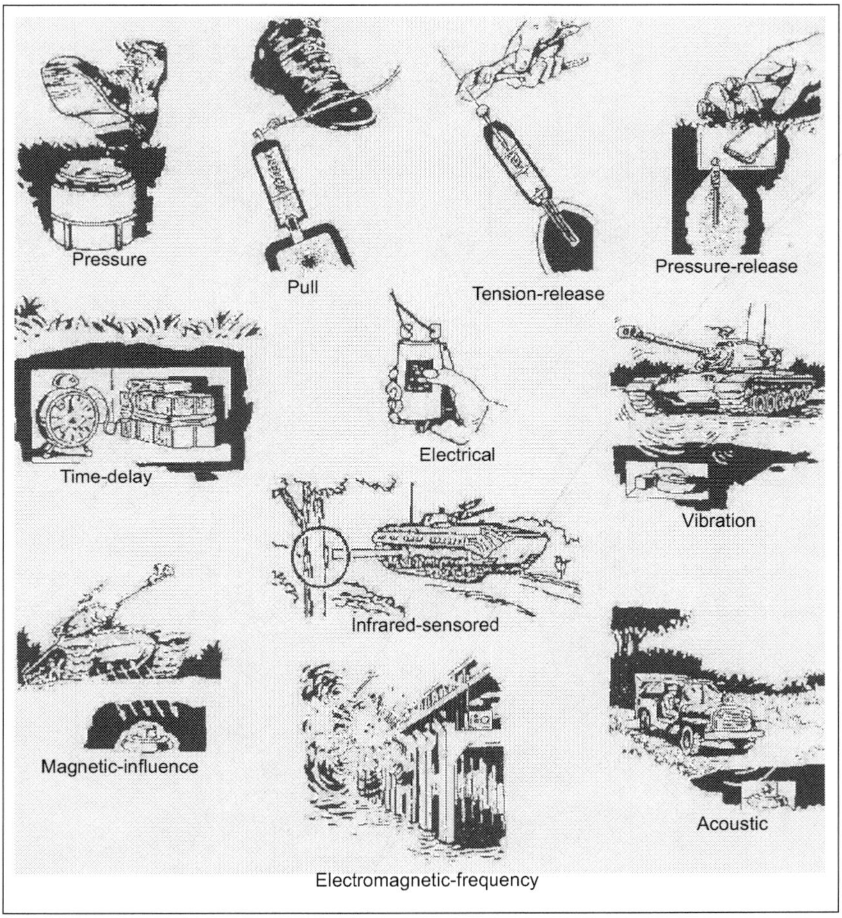


**Figure 1-2. Methods of actuating mines**

| | |
|---|---|
| **Mechanical.** A spring drives a striker against a percussion cap, which fires the detonator. | 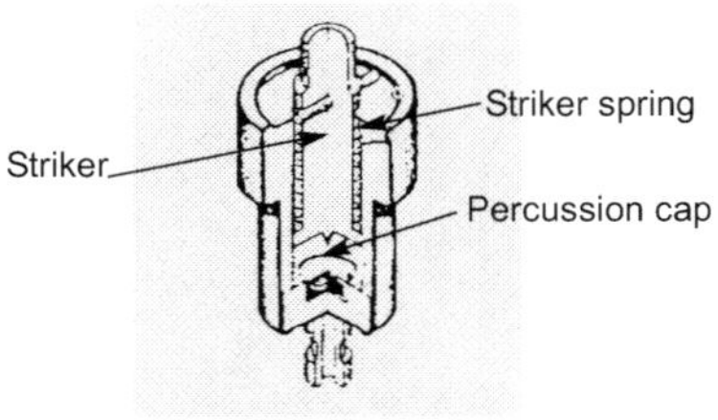  |
| **Chemical.** A small container of a chemical compound is broken by the initiating action. The chemical compound reacts with another substance to generate heat, which ignites the detonator. | 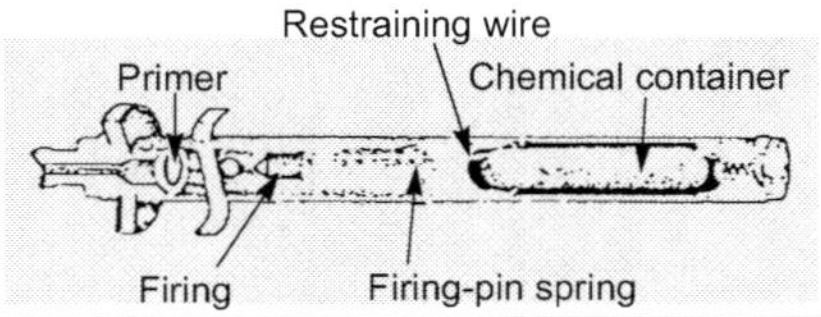  |
| **Friction.** The initiating action ignites substances inside the fuse by friction. The flame fires the detonator. | 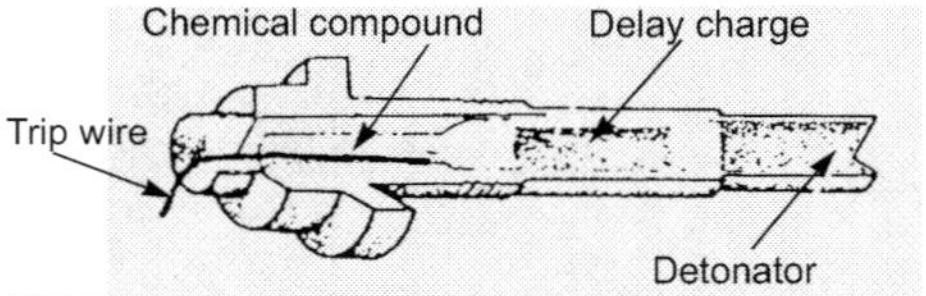  |
| **Electrical.** The initiating action closes an electrical circuit, which functions an electrical detonator. | 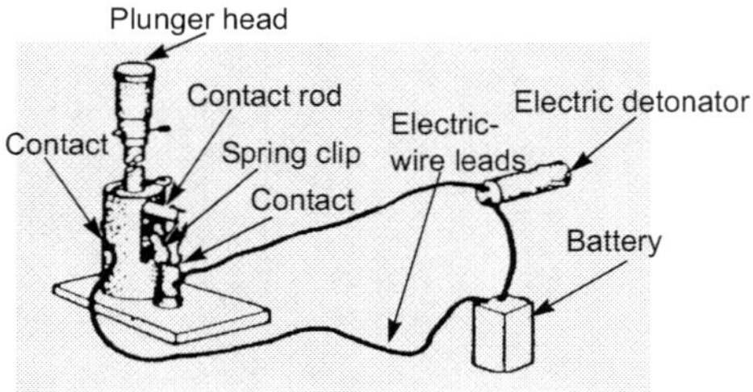  |

**Figure 1-3. Types of fuses**

## ANTITANK MINES

AT mines are designed to immobilize or destroy vehicles and their occupants.

### TYPES OF KILLS

An AT mine produces a mobility kill (M-Kill) or a catastrophic kill (K-Kill). An M-Kill destroys one or more of the vehicle's vital drive components (for example, breaks a track on a tank) and immobilizes the target. An M-Kill does not always destroy the weapon system and the crew; they may continue to function. In a K-Kill, the weapon system and/or the crew is destroyed.

#### TYPES OF SENSING

AT fuses fall into three design categories:

- Track-width. Usually pressure-actuated, requiring contact with the wheels or tracks of a vehicle.
- Full-width. Activated by several methods—acoustics, magnetic-influence, tilt-rod, radio-frequency, infrared-sensored, command, or vibration. Tilt-rod or magnetic-influence fuses are the most common. Full-width fuses are designed to be effective over the entire target width and can cause a K-Kill from penetration and spalling metal or from secondary explosions. When a full-width fuse is activated solely by contact with the wheels or tracks of the target vehicle, it usually causes an M-Kill because most of the energy is absorbed by the wheels or tracks.
- Off-route. Designed to be placed along the side of a route likely to be taken by armored vehicles. It has numerous fuzing possibilities, including infrared, seismic, break wire, and magnetic. It produces an M-Kill or a K-Kill, depending on the location of the target at the time of mine detonation.

#### TYPES OF WARHEADS

AT mines can be identified by their warheads:

- Blast AT mines derive their effectiveness from the force generated by high-explosive (HE) detonation. They usually produce an M-Kill when the blast damages the track or the vehicle, but a K-Kill is also possible.
- Shaped-charge mines use a directed-energy warhead. A shaped charge is formed by detonating an explosive charge behind a cone of dense metal or other material. Upon detonation, the cone collapses and forms a metal slug and a gaseous metal jet that penetrate the target. A K-Kill is probable if the crew or ammunition compartment is hit.
- Explosive-formed penetrating (EFP) mines have an explosive charge with a metal plate in front. Upon detonation, the plate forms into an inverted disk, a slug, or a long rod. A K-Kill is probable if the crew or ammunition compartment is hit.

## ANTIPERSONNEL MINES

#### TYPES OF KILLS

AP mines can kill or incapacitate their victims. The injuries and deaths they cause commit medical resources, degrade unit morale, and damage nonarmored vehicles. Some types of AP mines may break or damage the track on armored vehicles.

### TYPES OF SENSING

AP mines can be fused in many ways, to include pressure, seismic, wire, or command detonation:

- Pressure fuses usually activate an AP mine when a load is placed on the fuse.
- Seismic fuses activate an AP mine when the sensor detects vibrations.
- Trip wires or break wires activate an AP mine when something disturbs barely visible wires.
- Command-detonated mines are activated by a soldier when he detects the enemy in the mines' blast area.

### TYPES OF EFFECTS

AP mines contain five types of effects:

- Blast. Cripples the foot or leg of a soldier who steps on it; can also burst the tires of a wheeled vehicle that passes over it.
- Bounding-fragmentation. Throws a canister into the air; the canister bursts and scatters shrapnel throughout the immediate area.
- Direct-fragmentation. Propels fragments in the general direction of enemy soldiers.
- Stake-fragmentation. Bursts and scatters shrapnel in all general directions.
- Chemical. Disperses a chemical agent to whoever activates it; contaminates the surrounding area.

## ANTIHANDLING DEVICES

AHDs perform the function of a mine fuse if someone attempts to tamper with the mine. They are intended to prevent moving or removing the mine, not to prevent reduction of the minefield by enemy dismounts. An AHD usually consists of an explosive charge that is connected to, placed next to, or manufactured in the mine. The device can be attached to the mine body and activated by a wire that is attached to a firing mechanism. US forces can employ AHDs on conventional AT mines only. Other countries employ AHDs on AT and AP mines.

Some mines have extra fuse wells that make it easier to install AHDs (Figure 1-4). An AHD does not have to be attached to the mine; it can be placed underneath the mine (Figure 1-5). Mines with AHDs are sometimes incorrectly called *booby-trapped* mines.

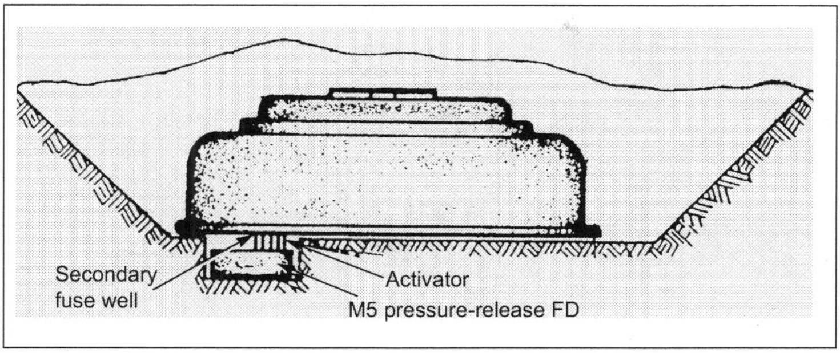


**Figure 1-4. AHD incorporating a release mechanism**

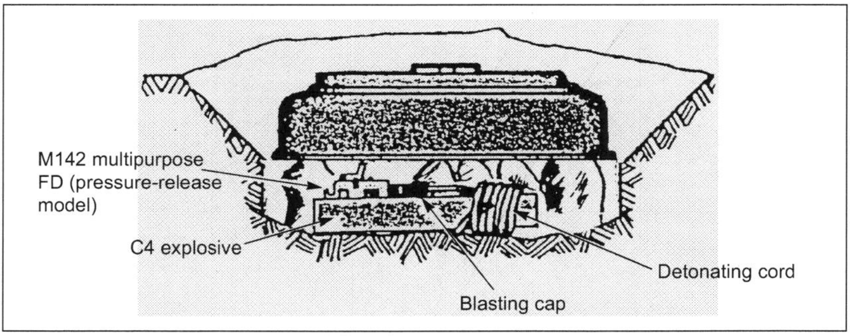


**Figure 1-5. AHD not attached to the mine**

The following hand-emplaced AHDs are used by US forces (Figure 1-6, page 1-8):

- M5 pressure-release firing device (FD).
- M142 multipurpose FD.

These devices use a spring-loaded striker with a standard base, and they function in one or more modes—pressure, pressure-release, tension, and/or tension-release. When an FD is employed as an AHD on certain AT mines, it requires the use of an M1 or M2 activator. FDs and activators are described in Appendix A.

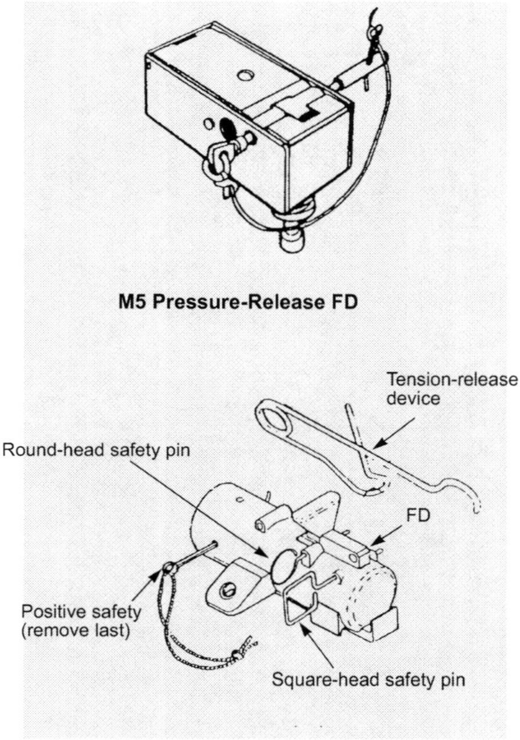


**Figure 1-6. Hand-emplaced US AHDs**

# PART ONE

# Mine Operations

This part of the manual provides tactical and technical information on mines, minefields, and mine-delivery systems; emplacement and employment methods and responsibilities; and reporting and recording procedures for US mine operations. Current US policy for restrictions on AP mines and other devices is also discussed.

## Chapter 2

## Mine-Warfare Principles

This chapter provides guidance to staff personnel who must plan the employment of minefields for tactical operations. It defines the four types of minefields—protective, tactical, nuisance, and phony. The remainder of the chapter provides guidance on the employment of tactical minefields—specifically their functions, designs, and integration principles.

### MINE-WARFARE CONCEPTS

Mines are explosive devices that are emplaced to kill, destroy, or incapacitate enemy personnel and/or equipment. They can be employed in quantity within a specified area to form a minefield, or they can be used individually to reinforce nonexplosive obstacles. They can also be emplaced individually or in groups to demoralize an enemy force. A minefield is an area of ground that contains mines or an area of ground that is **perceived** to contain mines (a phony minefield). Minefields may contain any type, mix, or number of AT and/ or AP mines. Minefields are used to—

- Produce a vulnerability on enemy maneuver that can be exploited by friendly forces.
- Cause the enemy to piecemeal his forces.
- Interfere with enemy command and control ($C^2$).
- Inflict damage to enemy personnel and equipment.
- Exploit the capabilities of other weapon systems by delaying enemy forces in an engagement area (EA).
- Protect friendly forces from enemy maneuver and infiltration.

### TYPES OF MINEFIELDS

There are four general types of minefields—protective, tactical, nuisance, and phony. Each type is determined by its distinct battlefield purpose. Therefore, minefields are employed differently, and they target the enemy in unique ways that support the overall concept of the operation.

- Protective minefields are employed to protect soldiers, equipment, supplies, and facilities from enemy attacks or other threats.
- Tactical minefields directly effect the enemy's maneuver in a way that gives the defending force a positional advantage.
- Nuisance minefields impose caution on enemy forces and disrupt, delay, and sometimes weaken or destroy follow-on echelons.
- Phony minefields deceive the enemy about the exact location of real minefields. They cause the attacker to question his decision to breach and may cause him to expend his reduction assets wastefully. Phony minefields may be employed in conjunction with other minefields, but should be used only after the enemy has become mine-sensitive.

It is important to distinguish the difference between the types of minefield and the means of emplacement. Volcano, Modular Pack Mine System (MOPMS), standard-pattern, and row mining are not types of minefields; they are just some of the means used to emplace tactical, nuisance, and protective minefields. They may also be the method of emplacement that is replicated by a phony minefield.

### PROTECTIVE MINEFIELDS

Protective minefields are employed to protect soldiers, equipment, supplies, and facilities from enemy attacks or other threats. Other threats range from enemy surveillance to theft of supplies and equipment. In tactical operations, protective minefields provide friendly forces with close-in protection and defeat the enemy's ability to maneuver or utilize the tenants of offense. They deny mechanized penetration and dismounted infiltration. In military operations other than war (MOOTW), protective minefields may focus on preventing unauthorized access to facilities and installations, rather than assisting in the destruction of an enemy force.

Protective minefields are usually employed and emplaced at the small-unit level (platoon or company/team). The authority to emplace protective minefields is normally delegated to the company/team commander. In some cases, such as a hasty defense, protective minefields are laid on short notice by units that use mines from their basic load or local stock. More commonly, protective minefields are used as part of a unit's deliberate defense. The mines are laid so that they are easy to detect and recover by the laying unit.

An important aspect of protective minefields is the requirement to recover them before leaving the area. This is often overlooked and is difficult to control because they are emplaced at the small-unit level. When a unit is being relieved in place by an adjacent unit, protective minefields are turned over to the relieving unit (minefield turnover is further defined later in this chapter). The decentralized emplacement of protective obstacles makes consolidating reports and records difficult and requires command involvement.

Much like final protective fires (FPF), protective minefields provide the defender with close-in protection during the enemy's final assault. Protective minefields serve two purposes. First, they impose a delay on an attacker that allows the defender time to break contact as the unit displaces to another battle position. Secondly, they break up the enemy's assault to complete its destruction. The composition of a protective minefield is driven by the vulnerability of the defender:

- Dismounted infantry is the greatest close-combat threat to a defending tank company/team. Protective minefields encountered in this case consist predominantly of AP mines that limit enemy dismounts from closing with the armor defender.
- A tank force is the greatest threat to an infantry defense. Protective minefields in this case consist predominantly of AT mines that reduce the enemy's ability to close quickly onto the infantry's position.

Neither AP nor AT mines are used in isolation. The preponderance of mine composition is designed against the most severe close-combat threat and the likelihood of that threat.

A protective minefield may take many forms. It may be only a few mines in front of a platoon, or it may be a standard-pattern minefield around an airfield. Protective minefields are used in both close and rear operations, and they are classified as either hasty or deliberate:

- Hasty protective minefields are temporary in nature and are used as part of a unit's defense perimeter. They are usually laid by units using mines from their basic load. If time permits, mines should be buried to increase their effectiveness; but they can be laid on top of the ground. AHDs and low-metallic mines are not used so that the minefield can be easily recovered. Mines are employed outside the hand-grenade range but within the range of small-caliber weapons. All mines are picked up by the emplacing unit upon leaving the area, unless enemy pressure prevents mine retrieval or the minefield is being transferred to a relieving commander. The brigade commander has the initial authority to employ hasty protective minefields. This authority may be delegated to a battalion or company commander on a mission basis. Procedures for emplacing a hasty protective row minefield are contained in Chapter 6.
- Deliberate protective minefields are more permanent, require more detailed planning, and usually require more resources. They are commonly used to protect static assets (vital sites)—logistical sites, communication nodes, depots, airfields, missile sites, air-defense artillery (ADA) sites, and permanent-unit locations. A typical deliberate protective minefield is the standard-pattern minefield; however, a row minefield can also be used. Deliberate protective minefields are usually emplaced for extended periods of time and can be transferred to another unit. Techniques for emplacing deliberate protective minefields are discussed in Chapters 6 and 7.

### TACTICAL MINEFIELDS

Tactical minefields are employed to directly attack enemy maneuver and to give the defender a positional advantage over the attacker. Tactical minefields may be employed by themselves or in conjunction with other types of tactical obstacles. They attack the enemy's maneuver by disrupting its combat formations, interfering with its $C^2$, reducing its ability to mass fires, causing him to prematurely commit limited breaching resources, and reducing his ability to reinforce. The defender masses fires and maneuver to exploit the positional advantage created in part by tactical obstacles.

Tactical minefields add an offensive dimension to the defense. They are a commander's tool for recapturing and maintaining the initiative that is normally afforded to an attacker. Combined with fires, tactical obstacles force the attacker to conform to the defender's plan.

Tactical minefields may be emplaced during offensive operations to protect exposed flanks, isolate the objective area, deny enemy counterattack routes, and disrupt enemy retrograde. This chapter further discusses the principles behind designing, integrating, siting, and emplacing tactical minefields.

### NUISANCE MINEFIELDS

Nuisance minefields are a form of tactical minefields. They are mainly used to impose caution on enemy forces and to disrupt, delay, and sometimes destroy follow-on echelons. Once nuisance minefields are emplaced, they do not require cover by observation or direct fire. Nuisance minefields are usually irregular in size and shape; they can be a single group of mines or a series of mined areas. They can be used to reinforce existing obstacles and can also be rapidly emplaced on main avenues of approach (AAs). Conventional mines and scatterable mines (SCATMINEs) may be used in nuisance minefields.

### PHONY MINEFIELDS

Phony minefields are areas of ground that are altered to give the same signature as a real minefield and thereby deceive the enemy. Phony minefields serve two primary functions. First, they confuse an attacker's breach cycle and cause him to question his breach decision. Secondly, they may cause an attacker to wastefully expend reduction assets to reduce mines that are not really there.

The success of phony minefields depends on the enemy's state of mind. The bluff succeeds best when the enemy is mine-conscious and has already suffered the consequences of a mine encounter. A fear of mines can quickly evolve into paranoia and break the momentum of the enemy's attack. Therefore, phony minefields are normally employed in conjunction with real minefields and are seldom employed alone. Once the enemy has become mine-conscious, phony minefields may produce considerable tactical effects with very little investment in time, labor, and material. Phony minefields may also be used to extend the front and depth of live minefields when mines or labor are in short supply or when time is restricted. They may be used to conceal minefield gaps through live minefields. There is **no** guarantee that phony minefields will achieve their purpose.

There are two mission-essential tasks inherent in the employment of a phony minefield:

- The phony minefield must completely replicate a live minefield in every detail, using a specific method of emplacement as a model. This becomes the deception story, and every aspect of the phony minefield must support the deception story. For example, if the deception story is a buried row minefield, the depth, front, and marking must be similar to that of a live buried row minefield. The ground should be disturbed, and tracks should be made on the ground in the same pattern as other minefields to give the ground the same signature. Occasional empty mine crates, discarded fuses, or other mine-laying supplies add to the deception.

- The deception story must never be compromised. Once emplaced, the phony minefield must be regarded by friendly forces as live until the tactical situation no longer warrants maintaining the deception. This can be extremely painful for the friendly unit. There is great temptation to drive through, rather than around, a known phony minefield—particularly if it is intended to be a gap between live minefields. However, one vehicle driving through a phony minefield and observed by enemy reconnaissance compromises the minefield's effectiveness.

Live mines are never laid in a phony minefield. A minefield designated as phony implies that the area contains no live mines. Emplacing even a single live mine within a phony minefield makes it a live minefield. Empty tins and such may be laid in a phony minefield but is seldom worthwhile. Minefield marking and covering fire should be the same as for a live minefield. Employment authority and reporting requirements are the same as for the minefield being simulated.

## PROTECTIVE VERSUS TACTICAL MINEFIELDS

As discussed, minefields can be tactical or protective obstacles (Figure 2-1, page 2-6). Tactical and protective obstacles have different purposes with regard to the enemy's maneuver. This difference causes them to have a particular relative place on the battlefield. Tactical obstacles attack enemy maneuver and are placed on the battlefield where the enemy maneuvers from march, prebattle, and attack formations. Protective obstacles are used to protect the force from the enemy's final assault onto the force's position. Protective obstacles are close to defensive positions and are tied in with the FPF of the defending unit. Additional information on obstacles can be found in FM 90-7.

## TACTICAL MINEFIELDS

Tactical minefields are designed, sited, emplaced, and integrated with fires to produce four specific tactical-obstacle effects—disrupt, turn, fix, and block (Figure 2-2, page 2-6). Each obstacle effect has a specific impact on an enemy's ability to maneuver, mass, and reinforce. Obstacle effects also increase the enemy's vulnerability to friendly fires. They support the friendly scheme of maneuver by manipulating the enemy in a way that is critical to the commander's intent. Minefield design is the means by which an emplacing unit varies minefield width, minefield depth, mine density, mine composition, the use of AHDs, and the irregular outer edge (IOE) to best achieve one of the four tactical-obstacle effects. Modifying these variables is at the heart of tactical minefield employment principles.

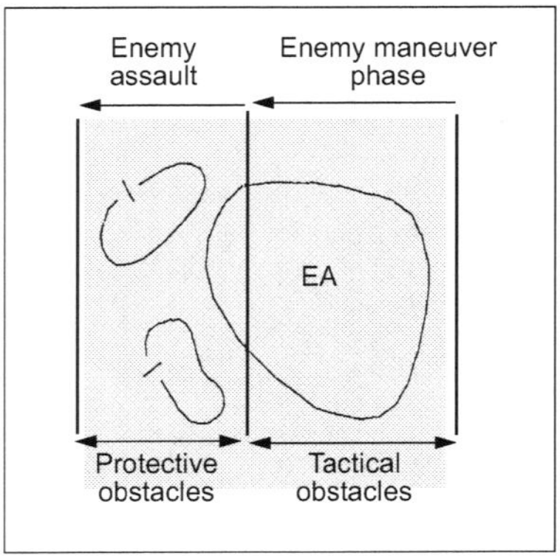


**Figure 2-1. Tactical versus protective obstacles**

| Obstacle Effect | Application | Examples Conveying Intent | Description |
|---|---|---|---|
| Disrupt 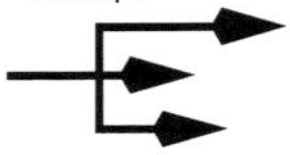 | The short arrow indicates where an enemy is attacked by obstacles. The longer arrows indicate where the bypass is allowed and attacked by fires. | 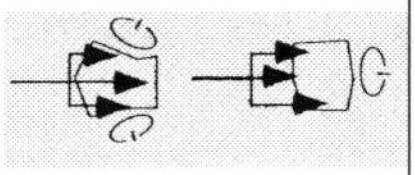 | Causes an enemy to break up its formation and tempo, interrupt its timetable, commit its reduction assets prematurely, and piecemeal the attack. |
| Turn 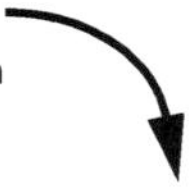 | The heel of the arrow is the anchor point. The direction of the arrow indicates the desired direction of turn. |  | Manipulates an enemy's maneuver in a desired direction. |
| Fix  | The irregular part of the arrow indicates where an enemy advance is slowed by obstacles. | 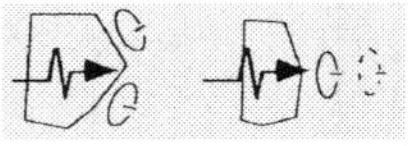 | Slows an enemy within a specific area, normally an EA. Gives the defender time to acquire, target, and destroy the attacking enemy throughout the depth of an EA. |
| Block 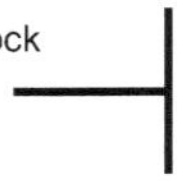 | The ends of the vertical line indicate the limit of an enemy's advance and where obstacles tie in to no-go terrain. | 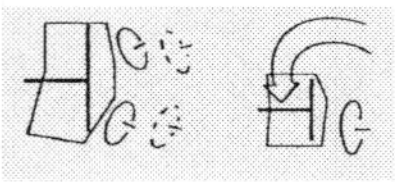 | Stops an enemy along a specific AA or prevents it from passing through an EA. |

**NOTES:**
**1. Arrows indicate the direction of enemy attack.**
**2. Minefields must be integrated with fires to achieve the desired effect.**

**Figure 2-2. Tactical-obstacle effects**

## MINEFIELD VARIABLES

First, it is important to understand how the variables relate to minefield effects. Figure 2-3 clearly defines some of the terms used to discuss minefield variables.

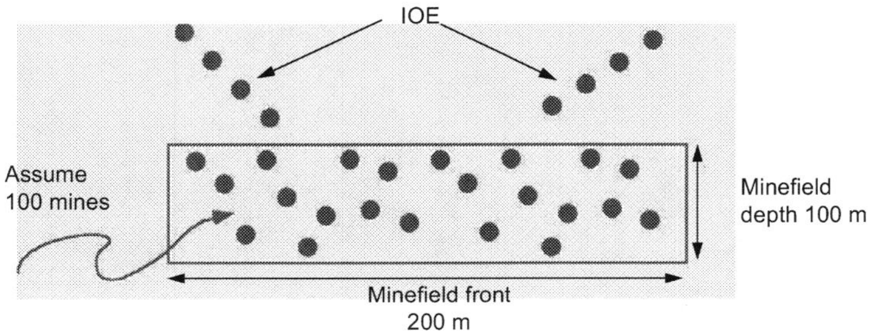


Linear density: 100 mines ÷ 200 meters = 0.5 mine per meter of front
Area density: 100 mines ÷ (100 x 200) = 0.005 mine per square meter

**Figure 2-3. Minefield variables**

### Front

Minefield front is the dimension of a minefield that defines how much of the attacking enemy formation is affected by the minefield. The front of a minefield is based on the desired obstacle effect (disrupt, turn, fix, or block) and the attack front of a company-size enemy force. The front of an attacking enemy depends largely on the type of enemy force (armored, motorized, or dismounted infantry) and norms by which the enemy army fights. For armored warfare, the minefield front is based on effecting a doctrinal company attack front of 500 meters (13 to 18 combat vehicles). For dismounted warfare, the minefield front is based on effecting a company attack front of 150 meters. The front may vary and require a study of enemy force and terrain. Groups of individual minefields are employed to achieve a larger front for battalion and larger enemy formations. For example, a battalion consisting of 52 to 72 combat vehicles has a front of 1,500 meters and requires more minefields.

### Depth

Minefield depth is based on the amount of reduction assets that will be required by an enemy to reduce a lane. The standard should start with 100 meters and increase in depth if denying the enemy the use of a mobility corridor (MC) is the intent (turn or block).

### Density

Minefield density depicts the number of mines in a minefield. It is expressed in linear or area density:

- Linear density is the average number of mines within a 1-meter-wide path through the minefield's depth, anywhere along the front. In Figure 2-3, page 2-7, the minefield contains 100 mines, with a minefield front of 200 meters. The linear density is 0.5 mine per meter of front (100 mines/200 meters of front).
- Area density is the average number of mines within a square meter, anywhere in the minefield. In Figure 2-3, the minefield contains 100 mines within a 20,000-square-meter area. The area density is 0.005 mine per square meter (100 mines/[200 meters x 100 meters]). Area density is normally used to express the density of scatterable minefields.

### Mine Composition

This variable includes the effective use of different types of mines. By using full-width mines, the probability of kill increases for the minefield. AT mines with AHDs **(Korea Only: as well as AP mines)** are used where the enemy is expected to use dismounted reduction techniques.

#### *Antitank Mines*

If the enemy is an armored force, tactical obstacles are predominantly AT mines. Track-width mines (M15s with the M603 fuse) have a lower probability of kill (M-Kill or K-Kill) than full-width mines (M21s and M15s with the M624 fuse). The ratio of full-width versus track-width mines in a minefield depends on the kill required. In general, a track-width minefield does not adequately affect the enemy's maneuver.

#### *Antipersonnel Mines*

AP mines target dismounted soldiers. Their composition in tactical minefields depends on the threat and the enemy's reduction assets. Based on current technology, most breaching operations are accomplished by mechanical or explosive means. **(Korea Only: If the minefield group's intent is to exhaust the enemy's breaching assets, AP mines should be integrated to attack its dismounted reduction ability.)**

### Probabilities of Encounter and Kill

The probabilities of encounter and kill measure a minefield's lethality.

#### *Probability of Encounter*

The probability of encounter is measured by the chance (in percent) that a vehicle, blindly moving through a minefield, will detonate a mine. The probability of encounter is based on mine density, the type of mine, and the type of enemy vehicle. In short, the more dense a minefield, the higher the probability is of encountering a mine. Probability of encounter also depends on the fuse capability of the mines. Tilt-rod and magnetic-influence mines will detonate if they are encountered anywhere along the width of the enemy vehicle. Pressure-fused mines detonate if a vehicle's track or wheel actually runs over them. The probability of encounter is also affected by the type of enemy vehicle. The smaller the width or track signature of the vehicle, the less likely it will encounter and detonate a mine.

Figure 2-4 illustrates the relationship between mine density and the probability of encounter for light versus heavy tracked vehicles and for track-width versus full-width mines. Figure 2-4 also provides general guidance for varying the mine density to yield the necessary probability of encounter when developing disrupt, fix, turn, and block minefields. Varying mine density is further discussed later in this chapter.

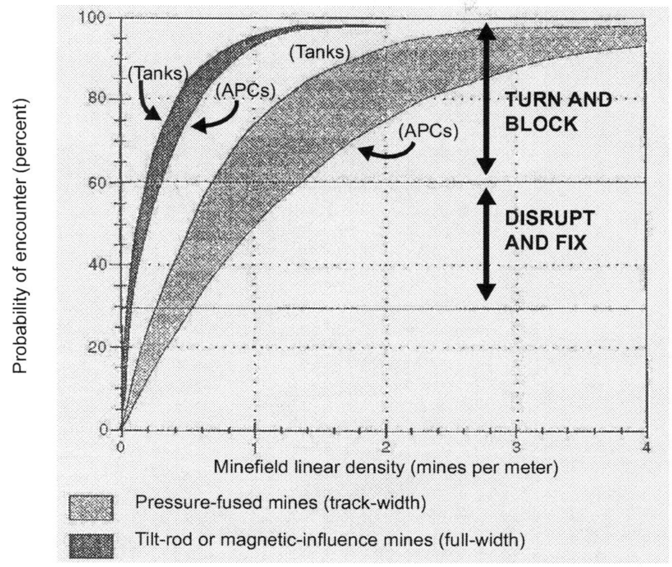


**Figure 2-4. Vehicle mine encounter probability versus minefield density**

***Probability of Kill***

The probability of kill is measured by the chance (in percent) that a vehicle will no longer be mission-capable (M-Kill or K-Kill) because of mine effects. It is a function of the combined probability that a vehicle will encounter a mine and the probability that the mine effect will produce an M-Kill or a K-Kill.

**Antihandling Devices**

Emplacing AHDs on mines is time-intensive. AHDs are added to a minefield to discourage manual removal and reuse of mines by the enemy and to demoralize the enemy who is attempting to reduce the minefield. AHDs do not prevent an enemy from reducing the minefield; they only discourage manual reduction methods.

**Irregular Outer Edge**

An IOE is a strip/row or multiple strips/rows of mines that normally extend toward the enemy from the first (enemy side) row of mines. An IOE is employed to break up the otherwise regular pattern of a minefield. It is used to confuse the enemy about the exact limits of the minefield, particularly its leading edge. An IOE adds an unknown quality to a minefield that makes the enemy's decision of whether to breach or bypass more difficult. The effect an IOE has on enemy actions may increase the overall lethality of a minefield.

## DESIGN

Modifying minefield variables to achieve the desired obstacle effect is a challenge for the engineer, both technically (resourcing and designing) and tactically (supporting the maneuver scheme). Experience will provide the best basis for designing minefields. Figures 2-5 through 2-8, pages 2-10 through 2-13, provide guidelines for varying minefield depth, front, density, and composition to best achieve disrupt, fix, turn, and block effects.

**These are guidelines, not fixed rules. Minefield designs must be based on a threat analysis. The designs are simply considerations or parameters to use when designing tactical minefields, regardless of the emplacement method.** They apply to conventional mine-laying techniques as well as the employment of SCATMINE dispensers. These parameters give the engineer the flexibility to design and emplace tactical minefields based on mission, enemy, terrain, troops, time available, and civilian considerations (METT-TC) (particularly resources and terrain) and still achieve the required effect. These norms are also the basis for developing minefield packages and emplacement procedures outlined throughout this manual. Chapter 3 discusses the characteristics and emplacement procedures for each of the SCATMINE systems, Chapter 6 outlines procedures for row mining using conventional mines, and Chapter 7 is dedicated to the standard-pattern minefield. Each chapter describes standard disrupt, fix, turn, and block minefield packages particular to that method of emplacement or dispensing system.

Each tactical-obstacle effect has a specific resourcing factor. In short, this numeric value helps determine the amount of linear obstacle effort that is needed to achieve the desired effect. The resource factor is multiplied by the width of the AA or MC to get the total amount of linear obstacle effort required. The linear obstacle effort is then divided by the minefield front norm for the specific effect (rounded up) to yield the number of individual minefields required in the obstacle group.

### Disrupt

A disrupt effect (Figure 2-5) focuses fire planning and obstacle effort to cause the enemy to break up its formation and tempo, interrupt its timetable, commit reduction assets prematurely, and piecemeal the attack. It also deceives the enemy about the location of friendly defensive positions, separates combat echelons, or separates combat forces from their logistical support. A disrupt effect should not be time-, manpower-, or resource-intensive. It should not be visible at long range but easily detected as the enemy nears it. Commanders normally use the disrupt effect forward of EAs.

| | |
|---|---|
| Resource factor | 0.5 (3 point obstacles) x AA |
| Group dimensions | W = 0.5 x AA |
| Probability of kill | 50% |
| Minefield front | 250 m |
| Minefield depth | 100 m |
| AT mines | Yes (pressure/tilt) |
| AP mines | No **(Korea Only: optional, based on threat analysis)** |
| AHD | Optional, based on threat analysis |
| IOE | No |

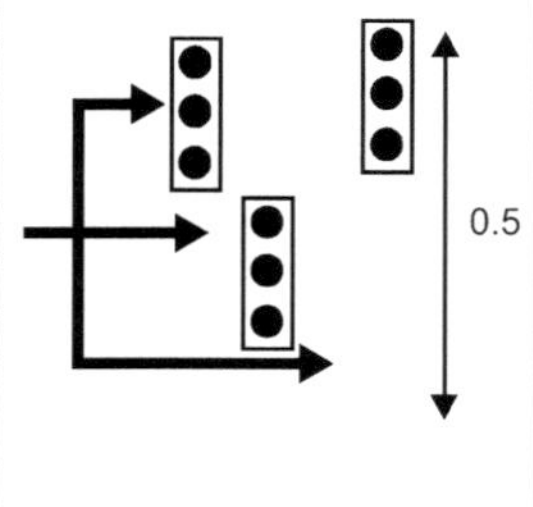


**Figure 2-5. Disrupt-effect group**

Normally, only half of the enemy's AA is attacked with minefields or other tactical obstacles to achieve a disrupt effect. For a minefield to disrupt an enemy company, half of the formation must react to the minefield. The typical width of a disrupt-effect minefield is 250 meters with a standard minefield depth of 100 meters. When designing a disrupt effect to attack an enemy battalion, three disrupt-effect minefields are arrayed in a group to achieve a width that is about half the size of the battalion's attack front (750 meters of minefield). Multiply the resource factor of 0.5 by the width of the AA to provide the amount of linear obstacle effort required for the disrupt effect. When the AA is narrow, an alternative disrupt group is three point obstacles along the AA.

Disrupt-effect minefields should be designed with approximately 50 percent probability of mine encounter to achieve the desired disrupt effect (see Figure 2-4, page 2-9). They should contain predominantly track-width AT mines and include full-width AT mines at the leading edge of the minefield to increase the probability of mine encounter. This should cause the enemy to commit its reduction assets.

AHDs can be added to disrupt-effect minefields to frustrate the enemy's breaching and clearing operations. However, adding AHDs may be too resource-intensive for the return in effect. An IOE is not required.

## Fix

A fix effect (Figure 2-6) focuses fire planning and obstacle effort to slow an attacker within a specific area, normally an EA. The fix effect is primarily used to give the defender time to acquire, target, and destroy the attacking enemy throughout the depth of an EA or AA. A fix effect may be used to generate the time necessary for the force to break contact and disengage as the enemy maneuvers into the area (typically used for delays). Fix-effect minefields in the obstacle group must be employed in depth, causing the enemy formation to react and breach repeatedly. Fix-obstacle groups must span the entire width of the AA. Commanders normally use the fix effect inside the EA.

| | |
|---|---|
| Resource factor | 1.0 x AA |
| Group dimensions | W = 1.0 x AA |
| Probability of kill | 50% |
| Minefield front | 250 m |
| Minefield depth | 120 m |
| AT mines | Yes (pressure/tilt) |
| AP mines | No **(Korea Only: optional, based on threat analysis)** |
| AHD | Optional, based on threat analysis |
| IOE | Yes |

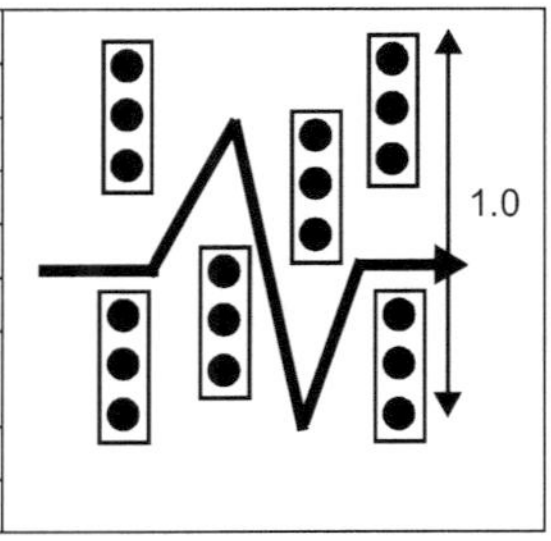


**Figure 2-6. Fix-effect group**

Individual fix-effect minefields must not appear too difficult to reduce. The enemy should be enticed into the area. The concept is to employ multiple minefields that individually attack a portion of a deploying company formation. Therefore, the fix-effect minefield front is 250 meters. It takes on the characteristics of a disrupt-effect minefield with a similar density, composition, and probability of encounter (Figure 2-4, page 2-9), with two exceptions. First, AHDs are not used because the application of massed direct and indirect fires complicate the enemy's breaching effort. Secondly, an IOE is added to further delay the enemy and confuse the attacker on the exact orientation of individual minefields. This also serves to increase the effective lethality of the minefield. The majority of mines are track-width AT, but full-width AT mines are used in the IOE and the leading edge of the minefield because they are the most lethal.

While individual minefields are designed to attack only portions of an enemy company formation, the fix-obstacle group is resourced, arranged, and sited to attack the entire front of an enemy battalion. Figure 2-6, page 2-11, depicts a fix group effect on an attacking enemy battalion. In this case, six fix-effect minefields are arrayed in an area the full width of the battalion AA (1,500 meters) by 1,500 meters deep. Accordingly, the resource factor for a fix-effect minefield group is one; the amount of linear minefield that should be resourced equals the width of the AA.

## Turn

A turn effect (Figure 2-7) manipulates the enemy's maneuver in a desired direction. One technique or a combination of techniques aids in achieving the turn effect. First, in order to entice the enemy to maneuver in the desired direction rather than reduce the obstacle, the obstacle must have a subtle orientation relative to the enemy's approach. Secondly, the obstacle and fires must allow bypass in the direction desired by the friendly scheme of maneuver. Obstacles in the start of the turn are visible and look more complex than those in the direction of the turn. Finally, the obstacle is tied into severely restricted terrain (restricted terrain as a minimum) at the initial point of the turn. The point where the severely restricted terrain feature and the minefield meet is known as the *anchor point*. Commanders normally use the turn effect on the flanks of an EA.

| | |
|---|---|
| Resource factor | 1.2 x AA |
| Group dimensions | W = 1.0 x AA |
| Probability of kill | 75% |
| Minefield front | 500 m |
| Minefield depth | 300 m |
| AT mines | Yes (pressure/tilt) |
| AP mines | No **(Korea Only: optional, based on threat analysis)** |
| AHD | Optional, based on threat analysis |
| IOE | No |

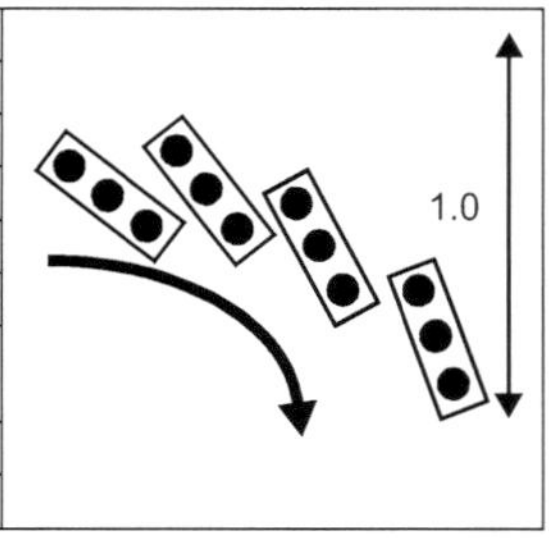


**Figure 2-7. Turn-effect group**

The standard turn-effect minefield has a width of 500 meters and a depth of 300 meters. One turn-effect minefield affects the entire width of an enemy company's front. It must be deep enough to cause multiple applications of line-charge reduction assets. In short, the minefield must discourage any attempts to breach and must entice the enemy to bypass rather than reduce. Figure 2-7 depicts a turn effect on an attacking battalion, with turn-effect minefields arrayed in a group across a 1,500-meter AA. The angle of the minefields should be subtle, encouraging the enemy to bypass individual minefields. Each minefield overlaps another one to tie the minefields together and prevent gaps. This is considered in the resource factor (1.2) for a turn-obstacle group. This factor, multiplied by the width of the AA, equals the amount of linear obstacle effort required for this turn effect.

As shown in Figure 2-4, page 2-9, turn-effect minefields must be extremely lethal and achieve approximately 80 percent probability of encounter. In other words, an enemy vehicle attempting to reduce or pass through the minefield will likely encounter a mine. This forces the small-unit commander to make an immediate decision—breach or bypass. A lethal minefield that is covered by intense fires and has an easily detectable bypass reduces breach decisions to instinct and causes the enemy to choose the bypass (turn). To produce this lethality, the majority of mines should be full-width AT. Full-width mines in the first rows the enemy encounters and in the depth of the minefield either exhaust the enemy's breaching assets or convince him to bypass early. AHDs are not required because the enemy force will seldom commit to dismounted breach when faced with intense direct and indirect fires. An IOE should not be used because the enemy must be able to determine the orientation of the minefield and the bypass.

## Block

A block effect (Figure 2-8) integrates fire planning and obstacle effort to stop an attacker along a specific AA or prevent him from passing through an EA. Block obstacles are complex and are integrated with intense fires; they do not stop an attacker by themselves. Individual block obstacles are employed successively in a relatively shallow area. When the enemy reduces one block obstacle, it is critical that he encounters another, thus denying him to project combat power and maintain momentum. Block obstacles must defeat the enemy's breaching effort (mounted and dismounted) as well as his maneuver. A block effect must span the entire width of an AA and prevent a bypass.

| | |
|---|---|
| Resource factor | 2.4 x AA |
| Group dimensions | W = 1.0 x AA |
| Probability of kill | 80% |
| Minefield front | 500 m |
| Minefield depth | 300+ m |
| AT mines | Yes (pressure/tilt) |
| AP mines | No **(Korea Only: Yes)** |
| AHD | Yes |
| IOE | Yes |

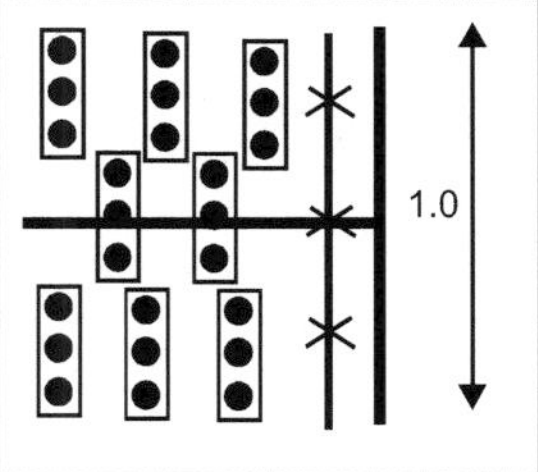


**Figure 2-8. Block-effect group**

The typical block-effect minefield is 500 meters wide and 300+ meters deep (includes an IOE). Figure 2-8 depicts a block effect on an attacking battalion. Note how individual minefields are arrayed to affect the entire width of the

AA but in a relatively shallow depth. Eight block-effect minefields are required in this example to achieve the necessary depth and width. The block group is the most resource-intensive. A resource factor of 2.4 is multiplied by the width of the AA to determine the linear obstacle effort required.

The lethality of a block-effect minefield (80 percent or higher) is similar to that of a turn-effect minefield (see Figure 2-4, page 2-9). The lethality of the group is considerably higher, since there are enough minefields in the group to cover more than twice the width of the AA. This lethality is produced by a density slightly greater than one mine per meter of front and the use of predominantly full-width AT mines.

A block-effect minefield must be capable of defeating mechanical and dismounted breach efforts. Therefore, AHDs **(Korea Only: and AP mines)** are used to target dismounted breaching. An IOE confuses the attacker about the exact minefield limits and complicates his employment of mechanical reduction assets. The depth of the block-effect minefield requires employing multiple line charges.

The above minefields are not standard solutions to every situation. The terrain could dictate a decrease or an increase in the effort required. Incorporating other reinforcing obstacles (AT ditches, road craters, wire, SCATMINEs) aid in attacking the different reduction assets.

## TACTICAL-OBSTACLE INTEGRATION PRINCIPLES

Tactical minefields are considered tactical obstacles and follow the same basic integration principles. Tactical obstacle $C^2$ focuses on obstacle emplacement authority and obstacle control.

### OBSTACLE EMPLACEMENT AUTHORITY

Obstacle emplacement authority is the jurisdiction that a unit commander has to emplace tactical obstacles. In a theater of operations (TO), theater commanders have the authority to emplace obstacles. In most cases, they delegate that authority to corps commanders who further delegate it to division commanders. Division commanders then have obstacle emplacement authority in their area of operations (AO), unless the authority is subsequently withheld or restricted by a higher commander. Commanders subordinate to corp and division do not have the authority to emplace obstacles unless the higher commander delegates authority for a current operation. Commanders use control measures and other specific guidance or orders to grant obstacle emplacement authority to subordinate commanders. Emplacement authority for SCATMINEs is covered in Chapter 3.

### OBSTACLE CONTROL

Commanders exercise obstacle control to ensure that obstacles support current and future operations. Obstacle control ensures that subordinate commanders emplace obstacles to best support the higher commander's scheme of maneuver. It also ensures that subordinate commanders do not interfere with future operations. Commanders maintain obstacle control by focusing or withholding emplacement authority or by restricting the types or locations of obstacles. Commanders use control measures, specific guidance, and orders to maintain obstacle control.

## OBSTACLE CONTROL MEASURES

Obstacle control measures are specific control measures that simplify granting obstacle emplacement authority and providing obstacle control for all levels of command (Table 2-1). Obstacle control measures are classified as—

- Zone.
- Belt.
- Group.
- Restriction.

**Table 2-1. Echelons of obstacle control and effect**

| Obstacle Control Measure | Echelon | Specific Obstacle Effects Assigned | Size of Enemy AA/MC | | Planning Guidance |
|---|---|---|---|---|---|
| | | | Armored | Light vs Armored | |
| Zone | Corps or division | Optional | Division/ brigade | Brigade/ battalion | Requires anticipating belts and intents |
| Belt | Brigade | Optional but normal | Brigade/ battalion | Battalion/ company | Requires anticipating groups and intents |
| Group | Corps, brigade, division, or battalion/TF | Mandatory | Battalion/TF | Company/ platoon | Based on individual obstacle norms |
| Restriction | All | NA | NA | NA | Used only when necessary to support the scheme of maneuver |

A protective obstacle is the only obstacle that can be employed outside a designated obstacle zone, belt, or group.

A specific obstacle effect (disrupt, turn, fix, or block) may be assigned to obstacle control measures. This enables the commander to direct the overall effect of obstacles within a designated zone, belt, or group to support his plan. At corps and division levels, assigning specific effects to obstacle zones is optional. At brigade level, the commander will normally assign a specific effect to the obstacle belts. At TF and battalion levels, obstacle effects for obstacle groups are required. This ensures that subordinate commanders emplace tactical obstacles that support the maneuver and fire plans. Assigning a specific obstacle effect to a control measure becomes obstacle intent, giving the obstacle effect, target, and location. Obstacle intent provides a direct link between the obstacle plan, the scheme of maneuver, the commander's intent, and the fire plan (direct and indirect). Obstacle intent is critical at brigade level and below, and it becomes the foundation for obstacle group development and design at the TF level.

### Obstacle Zones

Obstacle zones (Figure 2-9, page 2-16) are graphic control measures that corps and division commanders use to grant obstacle emplacement authority to brigades (including armored cavalry regiments [ACRs] and other major subordinate units). Corps and division commanders use zones to ensure that subordinates emplace obstacles that support the higher commander's scheme of maneuver, and to ensure that the obstacles do not interfere with future operations.

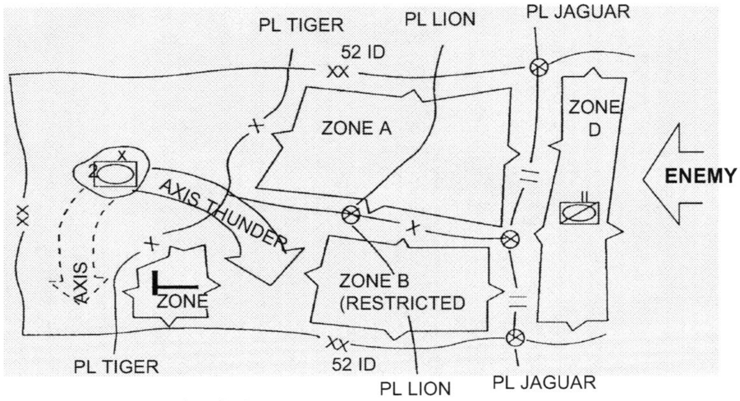


Obstacle number prefix: IO52

| Unit | Zone | Priority | Remarks |
|---|---|---|---|
| 1st Brigade | A | 3 | |
| 3d Brigade | B | 2 | No SCATMINEs with SD after H+2 |
| 3d Brigade | C | 1 | Block intent |
| Cavalry Squadron | D | 4 | |

**Figure 2-9. Obstacle zones**

## Obstacle Belts

Obstacle belts (Figure 2-10) are graphic control measures that brigade commanders use to constrain tactical-obstacle employment, and the belts should not cross unit boundaries. Commanders plan obstacle belts within assigned obstacle zones to grant obstacle emplacement authority to their major subordinate units. This is normally the first level in which the commander assigns an intent to the obstacle plan. It gives TF commanders the necessary guidance on the overall effect of obstacles within a belt. It does not designate that all obstacle groups within the belt must be the same effect. It simply means that the sum effect of groups within the belt must achieve the assigned belt effect. This serves to synchronize the obstacle effort within the brigade, particularly between adjacent TFs. Obstacle belts also focus obstacles in support of the brigade scheme of maneuver and ensure that obstacles do not interfere with the maneuver of any higher headquarters.

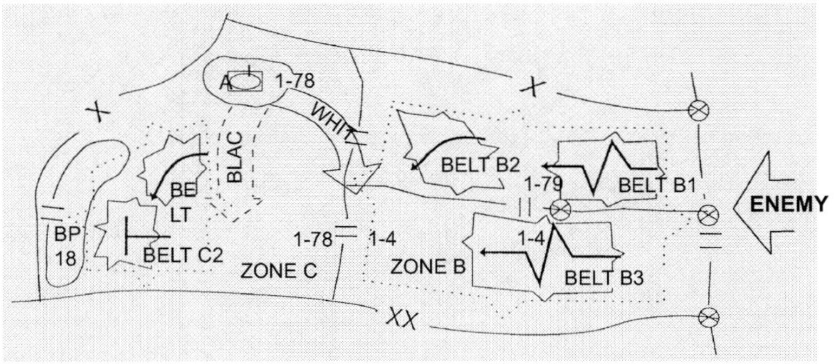


| Unit | Belt | Intent | Priority | Remarks |
|---|---|---|---|---|
| 1-78 | C1 | Turn | 2 | |
| 1-78 | C2 | Block | 1 | |
| 1-79 | B1 | Fix | 5 | No long-duration SCATMINEs; must SD by H+2 |
| 1-79 | B2 | Turn | 3 | No long-duration SCATMINEs; must SD by H+2 |
| 1-4 | B3 | Fix | 4 | No long-duration SCATMINEs; must SD by H+2 |

**Figure 2-10. Obstacle belts**

Belts are planned to attack enemy regiments based on an analysis of enemy battalion MCs. Brigades allocate maneuver companies based on the motorized rifle battalion (MRB) AAs and organize TFs to defeat the motorized rifle regiment (MRR). Obstacle belts and their intents are directed against MRR AAs. This provides the appropriate level of guidance while preserving the TF's need to refine the obstacle intent, based on how the TF will fight its allocated companies.

## Obstacle Groups

Obstacle groups (Figure 2-11, page 2-18) contain one or more individual obstacles that are grouped together to provide a specific obstacle effect. TFs use obstacle groups to ensure that company teams emplace individual obstacles that support the TF scheme of maneuver. In rare cases, brigades, divisions, or even corps may use obstacle groups for specific tactical obstacles.

Also, units perform detailed integration of obstacle groups with direct- and indirect-fire plans.

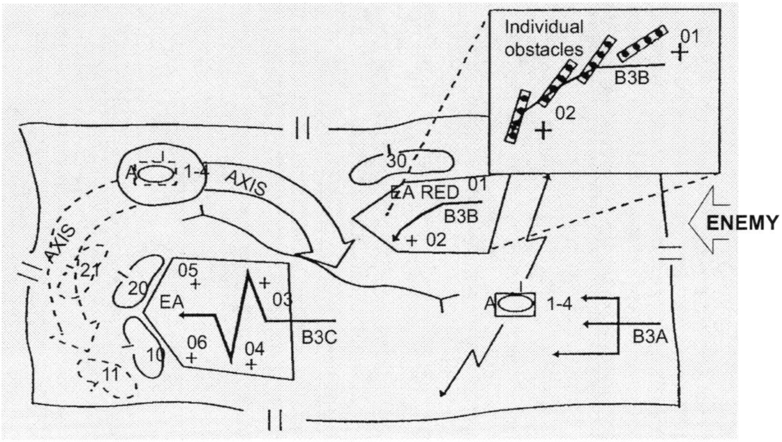


| Unit/BP | Group | Intent | Priority | Remarks |
|---|---|---|---|---|
| Company A/Screen | B3A | Disrupt | 3 | All groups restricted:<br>• Long-duration SCATMINEs are not allowed.<br>• Short-duration SCATMINEs must be emplaced by H-3. |
| Team B/BP 30 | B3B | Turn | 1 | |
| Team D/BP 10 | B3C | Fix | 2 | |

**Figure 2-11. Obstacle groups**

A TF allocates platoons against motorized rifle company (MRC) MCs and task-organize them into companies/teams to defeat MRBs. Likewise, direct-fire plans are designed based on the maneuver of MRBs and independent MRCs. Therefore, obstacle groups are used to attack the maneuver of MRB-size forces. Groups are designed specifically to support the direct-fire plan of the TF. The TF designates groups rather than obstacles because the location of individual obstacles hinges on siting at the company/team level. The group effect or obstacle intent drives obstacle siting and is therefore more important to convey to commanders. There can be more than one type of obstacle group to support the overall intent of an obstacle belt. This is because the belt design is based on the brigade's scheme of maneuver, without knowing the TF's direct-fire plan and scheme of maneuver. Groups are developed once the fire plan is established.

### Obstacle Restrictions

Commanders at all levels may use obstacle restrictions to provide additional obstacle control. They may use obstacle restrictions to limit the specific types of obstacles used (for example, no buried mines and no SCATMINEs that do not self-destruct within 48 hours). These restrictions ensure that subordinates do not use obstacles with characteristics that impair future operations. It also allows commanders to focus the use of limited resources for the main effort by restricting their use elsewhere. Commanders may also use restrictions to prevent subordinates from emplacing obstacles in a certain area. This type of restriction may be shown graphically as an obstacle-restricted area.

### FRATRICIDE PREVENTION

The modern tendency toward maneuver warfare and the disappearance of the linear battlefield places repositioning forces at an increased risk of fratricide by minefields. Obstacle control and the use of graphic obstacle control measures are vital in preventing minefield fratricide at every echelon.

Obstacle control is further facilitated by positive $C^2$ of all sustainment traffic, tactical repositioning, obstacle turnover, well-established and disseminated traffic plans with traffic control, and strict adherence to minefield marking procedures (discussed later in this chapter).

### MANEUVER-PLAN SUPPORT

Commanders include obstacle planning in each level of the decision-making process. This ensures that obstacle integration is effective and that the obstacle plan is flexible enough to allow changes during the planning, preparation, and execution phases of an operation. The following method is used to integrate obstacle planning at the TF level; it uses the decision-making doctrine contained in FM 101-5. Obstacle planning for brigade and higher levels can be found in FM 90-7.

The focus of obstacle planning is to integrate obstacles into the maneuver direct- and indirect-fire plans. This planning is directive and detailed in nature and focuses on the determination of obstacle groups and the type and amount of prepared positions. Actual obstacle siting, emplacement, and position location are the purview of the company/team commander and are normally supported by an engineer platoon.

TF-level defensive planning is part of the military decision-making process. The engineer battlefield assessment (EBA) process provides the basis for integrating engineer issues into the decision-making process.

### Mission Analysis

The key activities during mission analysis are to—

- Determine facts and assumptions.
- Analyze relative combat power.
- Analyze the engineer battalion's/brigade's mission and the commander's intent.
- Issue the commander's guidance.

#### *Determine Facts and Assumptions*

Defensive planning normally starts with the receipt of a warning order (WO) from the higher headquarters to defend. The company executive officer (XO) and the battlefield information control center (BICC) (or the TF engineer and the Intelligence Officer [US Army] [S2]) begin by developing a situation template (SITEMP) that includes a modified combined obstacle overlay (MCOO). The MCOO is a product developed during the intelligence preparation of the battlefield (IPB) process. The MCOO development is a joint effort of the engineer and the intelligence section of the TF's tactical operations center (TOC).

The MCOO should define the AAs and the MCs within the TF's AO. This information is vital to obstacle planning. Obstacles are placed on AAs to attack enemy maneuver. The AA analysis details potential EAs and indicates where forces can defend with limited survivability construction, because a reverse slope or undulating terrain provides natural concealment and cover.

The threat evaluation and the enemy course-of-action (COA) development detail how the enemy will potentially attack. They also provide an insight as to what and where the enemy's objective and routes might be. The SITEMP helps the engineer to understand how the enemy will traverse through the TF's sector and allows the engineer to gain an understanding of how and where he can best attack the enemy's maneuver.

The SITEMP also depicts how the enemy's reconnaissance forces will enter the sector. This is especially important when countering the enemy's ability to reconnoiter obstacle efforts.

The engineer must articulate the current capabilities of the engineer forces, its current combat power, and its ability to support the TF. Assumptions of future capability or potential reinforcement by other engineers should be analyzed. Specific characteristics of special engineer equipment and SCATMINE systems are detailed for the staff. An initial Class IV/V supply-point location and an operation plan should be developed with the TF staff (note that the TF has responsibility for Class IV/V supply-point operation). The XO/first sergeant (1SG) works with the TF and the engineer battalion Supply Officer (US Army) (S4) to ensure that delivery of Class IV/V barrier material supports the TF countermobility plan.

***Analyze Relative Combat Power***

The engineer compares friendly and enemy combat power and identifies possible obstacle requirements that offset potential enemy breaching and direct- and indirect-fire capabilities. The actual inclusion of the obstacles normally occurs after COA development. During this phase, the engineer finishes his EBA to gain an understanding for the engineer company's ability to support the TF.

***Analyze the Engineer Battalion's/Brigade's Mission and the Commander's Intent***

The staff analyzes and identifies information from the maneuver brigade order and the commander's intent that will potentially impact defensive planning. The engineer analyzes the maneuver brigade/engineer battalion commander's intent to determine potential obstacle placement, obstacle intent, and construction priority based on his concept of the operation.

The TF must identify tasks and limitations imposed from the brigade operation order (OPORD). These might include obstacle belts with or without specific intents, obstacle-restricted areas, or restrictions on the type of obstacles. Also, the brigade OPORD might specify reserve, situational, or directed obstacle groups.

The engineer must identify the TF's total obstacle capabilities. Available assets include engineer units, SCATMINE systems, and other units that can provide additional manpower for obstacle construction. Engineer-equipment status, work rates, and the time available must be evaluated in detail.

**NOTE: Work rates should only be used in the absence of unit-developed planning factors.**

Table 2-2 provides planning factors for the mine dump. Tables 2-3 through 2-6, pages 2-21 through 2-23, provide planning factors for obstacles. Special focus to limited visibility work rates and engineer squad strengths must be made when making total-capability estimates. Plan to accomplish obstacle siting and Class IV/V supply-point setup during daylight hours, and plan to emplace mines during limited visibility hours as much as possible.

**Table 2-2. Planning factors for the mine dump**

| Number of Personnel | Quantity of Mines | Required Equipment |
|---|---|---|
| 2-man team (2 minutes per mine) | 25 mines per hour | Shears, metal cutting<br>Grease, automotive and artillery<br>Rags<br>Work gloves<br>Flashlight<br>Night-vision goggles<br>Pliers |
| Squad (7 soldiers and an NCO) | 100 mines per hour | |
| Platoon (with leadership) | 300 mines per hour; 3,600 mines per day | |
| Company | 10,800 mines per day | |
| **NOTE: Soldiers work 50 minutes per hour, 12 hours per day.** | | |

**Table 2-3. Planning factors for work rates**

| Survivability | Time Required to Construct | | |
|---|---|---|---|
| | **With D7F Dozer** | **With ACE** | **With SEE** |
| Hull-defilade position | 1 BTH | 1.5 BTH | NA |
| Turret-defilade position | 2.5 BTH | 3.5 BTH | NA |
| HMMWV TOW position | 1.5 BTH | 2 BTH | NA |
| Vehicle-protective position | 0.75 BTH | 1 BTH | NA |
| Dismount-crew position | NA | NA | 1 SEEH |
| Individual-fighting position | NA | NA | 0.5 SEEH |
| **Countermobility** | **With D7F Dozer** | **With ACE** | **In Man-Hours** |
| Antitank ditch | 1 BTH/70 m | 1 BTH/50 m | NA |
| Standardized disrupt minefield | NA | NA | 1.5 PH |
| Standardized fix minefield | NA | NA | 1.5 PH |
| Standardized turn minefield | NA | NA | 3.5 PH |
| Standardized block minefield | NA | NA | 5 PH |
| Triple-standard concertina | NA | NA | 1 PH/300 m |
| Road crater | NA | NA | 1.5 SH |
| Point minefield | NA | NA | 1 SH |
| Concertina roadblock | NA | NA | 1 SH |
| Bridge demolition (massive) | NA | NA | 2 SH |
| Bridge demolition (steel) | NA | NA | 1 SH |
| Mine preparation at the TF Class IV/V supply point | NA | NA | 1 SH/100 mines |
| **LEGEND:**<br>BTH (blade team hour). One blade team working for one hour. A blade team consists of two engineer blades (two dozers, two ACEs, or one ACE and one dozer). One vehicle digs (cutter) while the other spreads the spoil (striker). A dozer-ACE blade team uses the dozer BTH.<br>SEEH (SEE hour). One SEE working for one hour.<br>PH (platoon hour). One platoon (3 squads) working for one hour.<br>SH (squad hour). One squad working for one hour. | | | |

**Table 2-4. Planning factors for standardized row minefields**

| Effect | Resource Factor | Front | Depth | Full-Width AT Mines | Track-Width AT Mines | Frag AP Mines |
|---|---|---|---|---|---|---|
| Disrupt | 0.5 | 250 m | 100 m | 42 | 84 | NA |
| Fix | 1.0 | 250 m | 120 m | 63 | 84 | NA |
| Turn | 1.2 | 500 m | 300 m | 336 | 168 | NA |
| Block | 2.4 | 500 m | 320 m | 378 | 168 | **84 (Korea Only)** |

**Table 2-5. Planning factors for scatterable minefields**

| System | Minefield Size | SD Time | Arming Time |
|---|---|---|---|
| ADAM | 400 x 400 m<br>200 x 800 m | 4 hr<br>48 hr | Within 1 min after ground impact |
| RAAM | 400 x 400 m<br>200 x 800 m | 4 hr<br>48 hr | 2 min 45 sec |
| Volcano (one load = 160 canisters or 960 mines [800 AT and 160 AP]) | Turn or block (1 per load):<br>• Ground: 555 x 320 m<br>• Air: 557 x 320 m<br>Fix or disrupt (4 per load):<br>• Ground: 277 x 120 m<br>• Air: 278 x 120 m | 4 hr<br>5 days<br>15 days | 2 min |
| MOPMS | 70 x 35 m | 4 hr* | 89 sec |
| *Can be recycled 3 times for a total of 13 hr | | | |

**Table 2-6. Ranges of common weapons**

| Weapon | | Maximum Effective Range | Planning Range* |
|---|---|---|---|
| **FRIENDLY WEAPON SYSTEMS** | | | |
| M16A2 | | 580 m | 400 m |
| M249 SAW | | 1,000 m | 800 m |
| M60 | | 1,100 m | 1,100 m |
| M203 | Area | 350 m | 350 m |
| | Point | 160 m | 160 m |
| M2, .50 Cal | Area | 1,830 m | 1,830 m |
| | Point | 1,200 m | 1,200 m |
| MK19 | Area | 2,200 m | 2,200 m |
| | Point | 1,600 m | 1,600 m |
| AT4 | | 300 m | 300 m |
| M47 Dragon | | 1,000 m | 800 m |
| Javelin | | 2,000 m | 2,000 m |
| M1 Abrams tank | 105 mm | 2,500 m | 2,000 m |
| | 120 mm | 3,000 m | 2,500 m |
| M2 Bradley ITV | 25 mm (APDS) | 3,000 m | 1,700 m |
| | 25 mm (HEI-T) | 3,000 m | 1,700 m |
| | TOW2 | 3,750 m | 3,750 m |
| 60-mm mortar | HE | 3,400 m | 50 m (min) |
| | WP | 4,800 m | 50 m (min) |
| | ILLUM | 931 m | 50 m (min) |
| 81-mm mortar | HE | 4,595 m | 75 m (min) |
| | WP | 4,595 m | 75 m (min) |
| | ILLUM | 3,150 m | 75 m (min) |
| 4.2-in mortar | HE | 6,840 m | 770 m (min) |
| | WP | 5,650 m | 920 m (min) |
| | ILLUM | 5,490 m | 400 m (min) |
| **SOVIET-STYLE WEAPON SYSTEMS** | | | |
| BMP, 73 mm | | 800 m | 800 m |
| AT3 missile | | 3,000 m | 3,000 m |
| AT5 missile | | 4,000 m | 4,000 m |
| BMP-2 | | 2,000 m | 2,000 m |
| BTR, 14.5 mm | | 2,000 m | 1,000 m |
| T-72 tank, 125 mm | | 2,100 m | 2,000 m |
| T-80 tank, 125 mm | | 2,400 m | 2,000 m |
| T-80 AT8 | | 4,000 m | 4,000 m |
| *The planning range is based on ideal weather conditions during daylight. | | | |

### *Issue the Commander's Guidance*

The commander should be as specific as possible with his initial obstacle guidance. If the commander narrows the COA focus, he should also provide obstacle guidance. His guidance is a key factor to an early start and must be solicited if not offered.

## Course-of-Action Development

Detailed planning begins following the COA development (Figure 2-12). The engineer focuses on four specifics of obstacle planning in the scheme of engineer operations (SOEO) for the defensive plan:

- Direct-/indirect-fire analysis.
- Obstacle-intent integration.
- Obstacle priority.
- Mobility requirements.

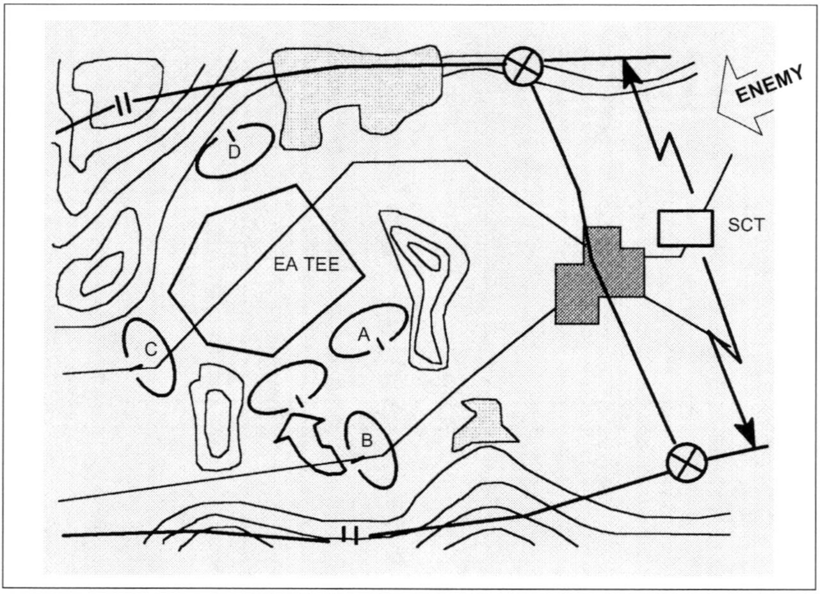


**Figure 2-12. TF defense COA**

### *Direct-/Indirect-Fire Analysis*

The direct-/indirect-fire analysis examines how engineers can best use obstacles (within the commander's intent) to enhance the direct-/indirect-fire plan. Figure 2-13 shows a sample direct-fire analysis. This analysis can be used to formulate obstacle locations with the direct-fire plan. The engineer must have a fundamental understanding of the direct-/indirect-fire and maneuver plans and the TF's organization of the EA to effectively integrate obstacles with the direct-/indirect-fire plan.

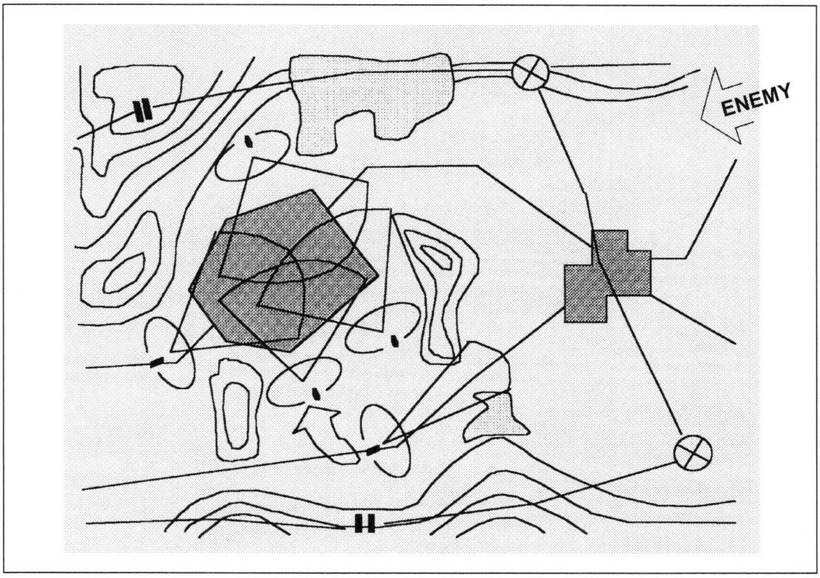


Figure 2-13. TF direct-fire analysis

Synchronization of direct and indirect fires with obstacles multiplies the relative effect on the enemy. An obstacle is an excellent location for preplanned artillery and mortar fires. These fires can eliminate dismounted breaching efforts. Indirect fires contribute to the threat's ability to breach, making the obstacle more effective and providing direct-fire systems a higher probability to kill.

### *Obstacle-Intent Integration*

The engineer determines locations for the directed obstacle groups. Groups are placed on the COA overlay to support the maneuver plan. This location is for planning only and will normally be adjusted after the ground reconnaissance.

Obstacle groups target specific enemy elements based on the SITEMP. The engineer generally allocates an obstacle group against a battalion-size AA. This approach mirrors the staff's placement of a company/team against the same enemy force. The company's/team's fire responsibility drives the placement of the obstacle groups. The engineer advises the commander on which specific effect each directed obstacle group must achieve. He plans obstacle groups to—

- Disrupt the enemy.
- Turn the enemy into an area where friendly units can mass fires.
- Fix the enemy in the EA and enhance his direct-fire destruction.
- Block the enemy from using an AA.

The engineer integrates directed obstacle groups with the COA. The obstacle effects are shown on the COA overlay using obstacle-effect graphics (Figure 2-14). The engineer draws the obstacle-group graphic to reflect the location, the target, and the specific intent of the group as accurately as possible. The engineer should visualize how the terrain naturally effects maneuver. Terrain visualization is vital to proper obstacle-group design.

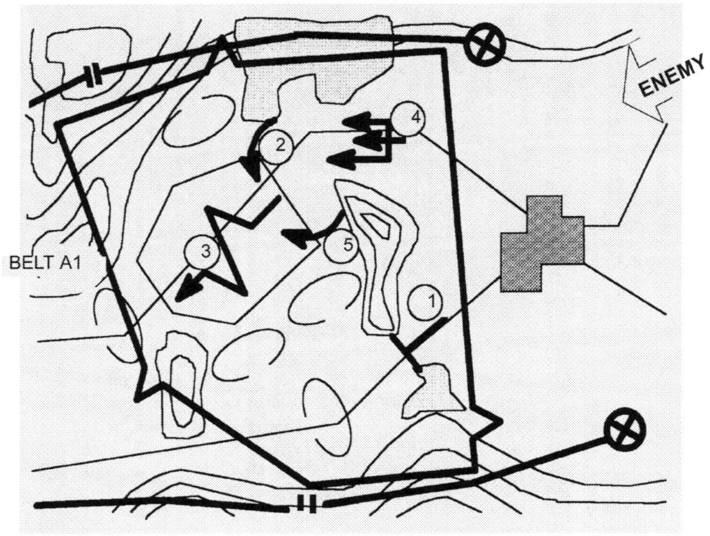


**Figure 2-14. TF obstacle-intent integration and priorities**

Note the placement and the effect of obstacle groups in Figure 2-14. First, the engineer must manipulate the MRB into the EA. The turn groups (2 and 5), combined with a heavy volume of AT fires from a company team at the turning point, achieves this. In the EA, particularly where the TF fires are massed, a fix group (3) slows the enemy and increases the effects of the fires. A block group (1) in the south, along with direct fires from a company team, will stop the advance of any element along the southern AA. A disrupt group (4), with indirect fires, will break up the $C^2$ and the tempo of the attacking force.

Protective obstacles in front of all team positions protect the teams from the enemy's final assault.

### *Obstacle Priority*

The staff determines the priority of each group depicted on the overlay. Priority is established by the commander's intent and the most likely enemy COA. The obstacle priority should reflect the TF's greatest obstacle requirement. The primary obstacle effort can be with an economy of force where the commander needs more obstacles to overcome a shortage of direct-fire systems. The TF engineer should be cognizant of flank protection, weapon types and ranges, and the overall commander's intent for the entire force before placing obstacle priority on the main EA. Priorities assist the engineer in allocating resources and ensuring that the most critical obstacle groups are emplaced first.

In Figure 2-14, the first priority is to turn the enemy where the fires are massed. The second priority is to deny the enemy access (block effect) to the southern AA. The fix effect is the third priority because it enhances the TF fires in the EA but only slows the enemy). The block effect is a higher priority than the fix effect because it stops the enemy from flanking the TF. The disrupt effect is the last priority.

### *Mobility Requirements*

The engineer identifies the TF's mobility requirements. Obstacle groups should not be arrayed along potential counterattack routes or where there is a potential to hamper unit repositioning. Mobility assets should be used to counter potential enemy situational obstacles and friendly obstacles that might hinder friendly maneuver. The TF engineer must consider the commander's mobility requirements and plan for mobility assets so he can be ready when and where he is needed.

## Course-of-Action Analysis

The staff war-games the COAs to determine their viability and recommends the best COA to the commander. The engineer refines the SOEO during this process as well. Obstacles should be considered within the context of the maneuver COA (Figure 2-15, page 2-28).

The engineer staff officer should consider the following:

- Enemy reactions at the obstacle groups (breaching or bypassing capability) versus the desired obstacle effect.
- Enemy breaching capabilities that make one obstacle type preferable to another (such as an AT ditch versus a minefield).
- Obstacle locations that hinder friendly maneuver.
- The compatibility of obstacle effects and weapon-system capabilities.
- Adequate direct-/indirect-fire control measures and targeting that support the obstacle effect. The effects of artillery and obstacles must be synchronized to gain the desired effect on the enemy's maneuver.
- Locations and types of enemy situational obstacles that make one type of breaching asset preferable to another.

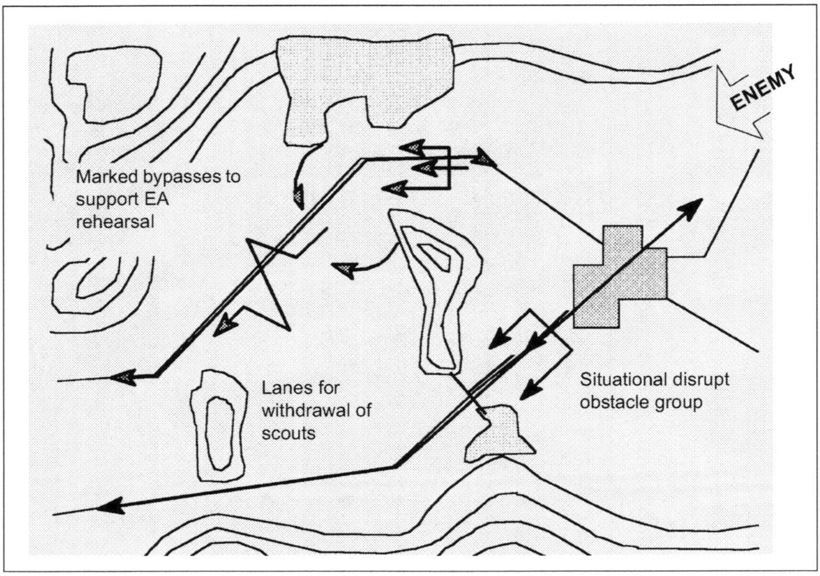


**Figure 2-15. Obstacle-plan refinement**

After war gaming, the staff adjusts the COA (including the obstacle plan) by—

- Changing the location of obstacle groups.
- Changing the obstacle effects.
- Adding more situational obstacles.
- Adding more reserve obstacles.
- Identifying other mobility requirements.
- Refining artillery targets based on obstacle-group changes.

***Mobility Requirements***

The staff determines which obstacles require lanes and determines the closure criteria for the lanes. It also determines obstacle-restricted areas that support the TF's maneuver. Lanes and bypasses are determined using tactical repositioning requirements developed during the COA analysis. Requirements for rehearsal movement, placement of the target reference point (TRP), and logistical support of forward TF elements are also considered in lane development. Mobility requirements identified during COA development are synchronized and refined during COA analysis. Additional mobility requirements identified during war gaming are resourced and planned. (Lane marking is discussed later in this chapter.)

### *Obstacle Design/Resourcing*

After the COA analysis, the engineer conducts a detailed study of the obstacle plan to determine resource requirements. Groups are resourced using the methods previously discussed. The TF engineer resources the obstacle groups based on their assigned priorities. Once the engineer has developed the resource requirements for the obstacle groups, he plans the individual obstacles within the group.

If time permits, a detailed ground reconnaissance of the obstacle-group location can be conducted. This will allow a more detailed analysis of the obstacle requirement for that AA, and then the engineer can plan individual obstacles. The TF engineer usually designates the intent to guide the companies/teams; the company/team commanders and their supporting engineers complete the actual design of the obstacle groups.

## Decision and Execution

The engineer makes adjustments to the SOEO based on the COA that the commander approves. The engineer then provides oral, written, and/or graphical orders with sufficient detail to allow the subordinate units to conduct the operation. The engineer provides critical information using the scheme-of-obstacle overlay and the obstacle-execution matrix.

### *Scheme-of-Obstacle Overlay*

The scheme-of-obstacle overlay (Figure 2-16, page 2-30) depicts the location of the TF's obstacle groups, brigade-directed obstacle groups (if any), and obstacle belts within the TF's sector. The overlay also includes any obstacle restrictions dictated from a higher headquarters. The overlay depicts the obstacle groups using the standardized obstacle-effect symbols. The overlay does not generally show individual obstacles unless the engineer has had sufficient time to conduct a thorough ground reconnaissance where exact obstacle locations have been identified. The engineer must exercise extreme caution if he uses individual obstacles on the overlay. He must ensure that inexperienced leaders do not attempt to emplace obstacles exactly as shown on the overlay, but instead, properly site the obstacle with the company/team commander. The scheme-of-obstacle overlay graphically depicts how the commander seeks to influence enemy maneuver through obstacles.

### *Obstacle-Execution Matrix*

The obstacle-execution matrix includes specific instructions and detailed information concerning the obstacle groups shown on the scheme-of-obstacle overlay. Develop an obstacle-execution matrix for all situational, reserve, and direct obstacles within the TF area. As a minimum, the matrix should include the information shown in Figure 2-17, page 2-31.

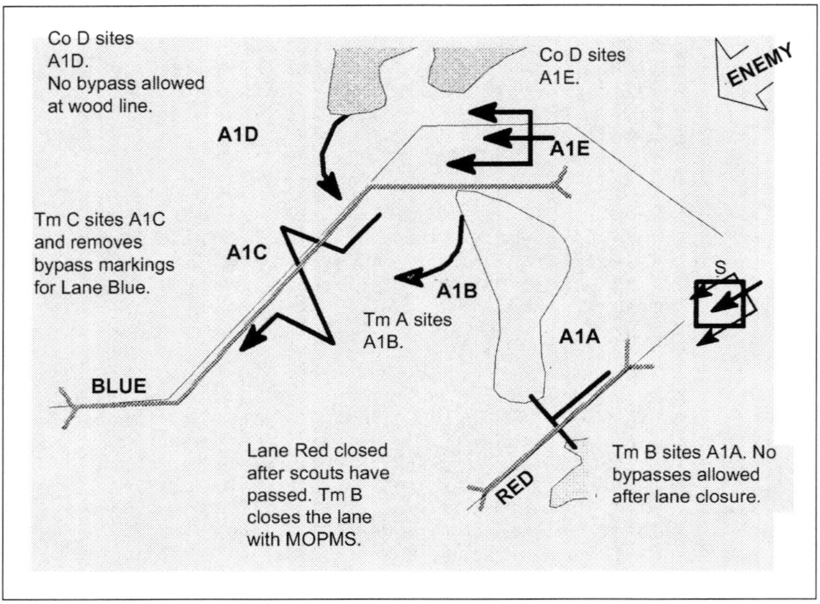


**Figure 2-16. Scheme-of-obstacle overlay**

| Percent Complete | Priority | Standard Obstacle Number | | | | Obstacle Location | Laying Unit | Overwatch Unit | Class IV/V Planned and Pre-position Grid | Start Time | Finish Time | Lane Grid (Open or Closed) | Remarks |
|---|---|---|---|---|---|---|---|---|---|---|---|---|---|
| | | Obstacle Authority | Zone, Belt, or Group | Obstacle Type and Number | Status | | | | | | | | |
| | | | | | | | | | | | | | |
| | | | | | | | | | | | | | |
| | | | | | | | | | | | | | |
| | | | | | | | | | | | | | |
| | | | | | | | | | | | | | |
| | | | | | | | | | | | | | |
| | | | | | | | | | | | | | |
| | | | | | | | | | | | | | |
| | | | | | | | | | | | | | |
| | | | | | | | | | | | | | |
| | | | | | | | | | | | | | |
| | | | | | | | | | | | | | |
| | | | | | | | | | | | | | |
| | | | | | | | | | | | | | |

Figure 2-17. Sample obstacle-execution matrix

## SITING AND EMPLACING TACTICAL MINEFIELDS

This section outlines the principles for siting tactical minefields to support the company/team. These principles apply to all methods of emplacement—standard pattern, row mining, and SCATMINE systems. The focal point of the discussion on siting is the coordination that must occur between the emplacing engineer (normally the engineer platoon leader) and the maneuver company commander. Coordination between the engineer platoon and the maneuver company is perhaps the most vital component of effective obstacle integration, and it is also a vital component in EA development. Obstacles are directly integrated with weapon effects, capability, and the fire plan at this level. The two subcomponents of tactical-obstacle siting are coordinating with the maneuver commander and siting the minefield.

### COORDINATING WITH THE MANEUVER COMMANDER

Effective coordination with the maneuver company/team commander who will fight the obstacle(s) is essential to realizing the full potential of minefields as a combat multiplier. In short, the emplacing engineer becomes the maneuver company/team commander's team engineer for the mission. The engineer, the fire-support team (FIST), and the maneuver commander must work closely to ensure complete integration of the minefield into all aspects of the company plan. The engineer must be integrated into the maneuver company/team EA development process. Throughout each step of the process, the engineer must provide the maneuver commander and the FIST with the engineer expertise necessary to ensure complete and effective obstacle integration.

Before the emplacing engineer can conduct effective coordination, he must have tools and information from the TF order that serves as common ground between the emplacing engineer, the FIST, and the maneuver commander. The order drives the integration of tactical obstacles into the fire plan and ensures that the obstacles affect their intended enemy target in a way that supports the scheme of maneuver.

#### Modified Combined Obstacle Overlay

The MCOO is a product from the IPB process that graphically depicts the maneuverability of the terrain. It depicts slow-go and no-go terrain relative to the type of enemy force. It also defines AAs and MCs that the enemy may use for its attack. Since tactical obstacles attack the enemy's maneuver and must complement the existing terrain, the MCOO is vital to obstacle siting. It helps ensure that obstacles correctly address the enemy AAs and MCs. It also helps select how and what part of the enemy formation will be directly attacked by obstacles, and it shows the effect the obstacles will have on the enemy's maneuver.

#### Situation Template

The SITEMP is developed by the maneuver battalion S2 and the TF engineer during the IPB. It estimates how the enemy will attack, in terms of size and type of units, and the formations it will use. Tactical obstacles are employed to produce specific effects on specific enemy targets. Therefore, the SITEMP helps the engineer and the maneuver commander site and emplace obstacles in a way that attacks the intended target. The SITEMP may also depict the likely routes for enemy reconnaissance elements. This helps the engineer and the maneuver commander analyze requirements for reconnaissance and surveillance (R&S) patrols that defeat enemy attempts to reconnoiter the obstacles and reduce enemy effectiveness before the attack. The type of formations the S2 expects the enemy to use during the entire course of the attack is also vital information. The SITEMP should identify when the enemy is in march, prebattle, and attack formations. The enemy formation may impact on the necessary front of obstacle groups and the obstacle groups' effectiveness in achieving the intended effect on enemy maneuver.

### Commander's Intent

The emplacing engineer, the FIST, and the maneuver commander must have a common understanding of the battalion commander's intent. The battalion commander's intent is his vision of the battle and normally outlines what actions the unit must do to accomplish the mission. The commander's intent may include key aspects of the plan that he wants to emphasize to subordinates to synchronize the actions of subordinates toward a single purpose. The engineer must understand the commander's intent and how it relates to integrating obstacles. The engineer should always ensure that the obstacles he is emplacing support the commander's overall intent.

### Maneuver Graphics and the Fire Plan

In order to fully support the scheme of maneuver, the engineer must have and understand the maneuver graphics on the battalion's operational overlay. The maneuver graphics use symbols to depict the missions of each subunit within the battalion. Maneuver control measures such as battle positions, sectors, phase lines, passage lanes/points, and counterattack axis are vitally important to understanding the plan and integrating tactical obstacles. The maneuver graphics may include direct-fire control measures that direct how and where combat forces will mass, shift, and lift fires to destroy the enemy. Direct-fire control measures include EAs, trigger lines, and TRP and unit boundaries. In short, they dictate the direct-fire responsibilities of each subordinate. Understanding the direct-fire plan and the organization of the engagement is fundamental to integrating obstacles with fires. The maneuver graphics also give the engineer an appreciation of how tactical obstacles supporting one unit must complement the adjacent units. This is particularly true of adjacent EAs or plans requiring any tactical repositioning of forces.

### Obstacle-Execution Matrix

The obstacle-execution matrix includes specific instructions and detailed information concerning the obstacle groups shown on the scheme-of-obstacle overlay. This matrix gives the engineer critical information on minefield groups that will be emplaced within the company's/team's AO. As a minimum, the obstacle-execution matrix should include the information shown in Figure 2-17, page 2-31.

### Scheme-of-Obstacle Overlay

At the maneuver battalion level, the scheme-of-obstacle overlay depicts the location of brigade-directed belts, TF obstacle groups, and any directed obstacles within the battalion sector. Any obstacle restrictions attached to an obstacle control measure (belt or group) that preclude the employment of certain types of obstacles are annotated on the overlay. A scheme-of-obstacle overlay is a graphic control measure that defines the general location of the obstacle groups and the effect to be achieved by them. The scheme-of-obstacle overlay does not normally depict individual obstacle locations. The location of individual obstacles within a group is determined during the siting process between the emplacing engineer and the maneuver company commander. When overlaid on the maneuver graphics and the SITEMP, the scheme-of-obstacle overlay should depict the essential elements of obstacle integration:

- Enemy targeted by the obstacle.
- Location of the obstacle on the battlefield.
- Unit covering the obstacle.
- Directed link between the obstacle effects and the fire plan.

### Fire-Support Plan

The emplacing engineer should be familiar with the key elements of the fire-support plan. He must understand the general scheme of fires and how the fires support the scheme of maneuver, the commander's intent, and the obstacle plan. Normally, the emplacing engineer does not need the entire fire-support overlay depicting the location of all targets. However, he should know the location of fire-support targets directed by the battalion to cover obstacles. The emplacing engineer should know who has the priority of fires for each phase of the battle. The emplacing engineer should also know the location and the type of priority targets or FPF allocated to the maneuver company he is supporting. During coordination with the maneuver company, the emplacing engineer should discuss the fire-support plan with the company FIST and get updates on changes to the plan as well as any company-level fire-support plans that may impact on the integration of obstacles.

### Combat Service Support

The engineer must be familiar with the plan for combat service support (CSS). In particular, the engineer must know the location of major supply routes into and through the battalion area, the location of the battalion logistics release point (LRP), what routes the maneuver company will take from its position to the LRP, and the location of key battalion logistics nodes. The emplacing engineer must always be cognizant of the sustainment traffic flow and the impact obstacle emplacement has on sustainment operations. Of particular concern to the engineer is the location of the Class IV/V supply point and the routes to it.

### Battlefield Operating System

During coordination with the maneuver company commander, a checklist or a framework is a useful tool for organizing thoughts and formulating questions. Below is a list of considerations or points of coordination that should drive the integration of the emplacing engineer and the maneuver company commander. The list is organized using the Battlefield Operating System (BOS) because it provides a logical sequence and a framework that is easily remembered.

- Intelligence.

  - Enemy AAs and MCs (mounted and dismounted).
  - Likely enemy COA and possible reactions to the obstacles.
  - Enemy breach capability.
  - Enemy reconnaissance routes and friendly counterreconnaissance or R&S plans, particularly company-level patrols.
  - Likely enemy formations and transitions between formations.
- Maneuver.
  - Specified, implied, and essential tasks of the maneuver company.
  - Higher commander's intent.
  - Organization of the defensive position, including—
    - Task organization (type of weapons).
    - Decisive point or defeat mechanism.
    - Organization of direct fires in the EA.
    - Location and marking of direct-fire control measures.
    - Position of weapons to cover assigned direct-fire responsibilities.
  - Tactical mobility requirements of the maneuver company and any adjacent units, including—
    - Counterattack axis.
    - Repositioning of forces and their routes to alternate, supplementary, or subsequent battle positions.
    - Employment of reserves.
    - Passage of lines.
  - Obstacle protection measures.
- Mobility/survivability.
  - Intent of tactical obstacles covered by the maneuver company (enemy target, obstacle location, and obstacle effect).
  - Obstacle control measures and restrictions imposed by higher headquarters.
  - Mobility requirements (lanes/gaps), as identified above in maneuver.
  - Mutual support between the obstacle location, the fire plan, obstacle effects, and survivability positions.
  - Security for engineers provided by the maneuver unit supported.
- Fire support.
  - Location of the company FIST and frequency of fire support.
  - Updates on the tentative fire-support plan.

- — Allocation of fires to the company, including—
  - Artillery or mortar targets.
  - Priority targets, types of targets, and the FPF.
- — Covering obstacles and their effects with indirect fires.
- — Indirect-fire control measures to synchronize direct fires, indirect fires, and obstacles.
- — Area-denial artillery munition (ADAM)/remote antiarmor mine (RAAM) use (lane closure and breached obstacle repair).
- — Registering fires. (Deconflict with obstacle emplacement; registration should occur after obstacles are sited but before emplacement.)
- — Company fire-support execution matrix.
- — Means for obtaining fire support, if enemy contact is made during emplacement.

- Air defense.
  - — Enemy air AAs during emplacement.
  - — Update on changes to air-defense warning and weapons status.
  - — Location of air-defense systems that can cover engineers emplacing obstacles.
  - — Method of obtaining early air-defense warning.
- CSS.
  - — Tentative location of the mine dump, if used, within the company position and routes from the mine dump to obstacles.
  - — Routes the company plans on using to conduct logistics package (LOGPAC) operations that must remain open.
  - — Manpower assistance for operations at the Class IV/V supply point and the mine dump.
  - — Casualty evacuation routes for scouts, observation posts (OPs), and ADA systems.
- $C^2$.
  - — Location of the commander during defensive preparation.
  - — Frequency-modulated (FM) net of the supported company and the means of communication.
  - — Unit boundaries affecting obstacle emplacement.
  - — Time and place of the company/team order.
  - — Coordination that must occur with adjacent units.
  - — Obstacle reporting and recording requirements.

- Control and execution of situational and reserve obstacles.
- Lane-closure responsibilities and procedures.
- Time and method of obstacle turnover, including lanes.
- Company/team understanding of the obstacle intent.

## SITING THE MINEFIELD

The emplacing engineer and the company/team commander site individual obstacles to achieve synchronization between the obstacle effect and fires. Siting is a key component to the EA development process, and it represents the final adjustments to the obstacle location and the fire control measures before emplacement.

Certain preconditions are necessary to site individual obstacles. First, the company/team commander decides where he plans to mass fires and marks the necessary fire control measures on the ground. The location of these control measures must be clear, since they are the basis for obstacle siting. The commander then identifies tentative locations for key weapons within the position or the sector. Finally, the commander and the engineer must both understand the intent of the obstacle group.

Obstacle siting concentrates on marking the obstacle group as a whole instead of marking each individual obstacle. In broken terrain, however, it may be easier to site individual obstacles. The company/team commander and the emplacing engineer use vehicles or soldiers from the company/team, the engineer platoon, or both to simulate the enemy force and do the physical marking. The simulated enemy force moves into the EA to the enemy side of the obstacle group. The engineer platoon leader and the company/team commander collocate near the weapons covering the obstacle. As a technique, one or all of the tanks, Bradleys, and other crew-served weapons may occupy their position and contribute to the siting process. All participants in the siting process use a common FM net to communicate during siting.

The simulated enemy force moves into the EA, simulating the enemy's attack. It deploys into a formation of front that is similar to the expected enemy formation. Once it is near the marked fire control measures, it places markers at intervals as it drives the trace of the obstacle-group effect (or individual obstacles in broken terrain). It remains oriented on key fire control measures to ensure that obstacle location and effect are synchronized with fires. During the process, each participant verifies that he can cover the obstacle, notes the location of fire control measures and obstacles, and records the appropriate data on range cards. As the simulated force drives the obstacle trace, siting participants also identify dead space and requirements to refine the location of the obstacle group and fire control measures. Figure 2-18, page 2-38, illustrates how the engineer and the company/team commander work together to site turn- and fix-obstacle groups.

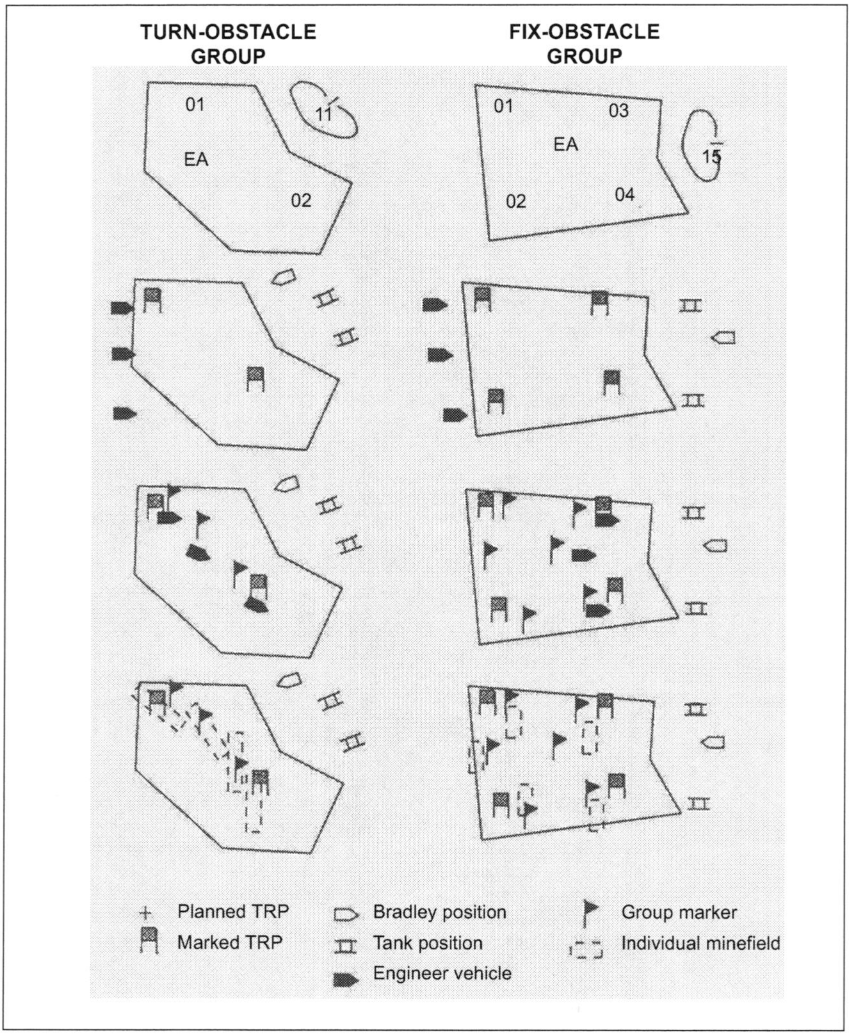


Figure 2-18. Minefield siting

Once the company/team marks the general limits and the orientation of the obstacle group, the engineers can begin marking individual obstacles (if not already done). To mark individual obstacles, the engineer platoon uses the group markers as guides. As shown in Figure 2-18, the group markers may lend themselves well as the start and end points of individual obstacles; however, this is not always the case. As the engineer platoon refines the group limits into the site of individual obstacles, the platoon can then begin the necessary site layout based on the method of obstacle emplacement.

Siting is not the last thing done during preparations. The time and resources involved in emplacing tactical obstacles require that siting begins concurrently with establishing the defensive position. It is imperative that the unit sites the obstacles as soon as the company/team commander has established the EA and identified tentative positions for key weapons. It is not necessary that all weapons are in place and dug in before siting. Normally, well-marked fire control measures and one known position per maneuver platoon (not dug in) are all that is required to site the obstacles effectively.

### EMPLACING MINEFIELDS

Based on the group effect, resources allocated, and the engineer plan, the platoon leader determines the method of emplacement for individual minefields. The procedures for emplacing scatterable, row, and standard-pattern minefields are contained in Chapters 3, 6, and 7, respectively.

### DETERMINING RESOURCE REQUIREMENTS

The engineer must determine the number of individual minefields needed to make up the group and ensure the allocation of required resources. The amount of linear obstacle effort for a group is equal to the width of the AA, multiplied by the resource factor. In Figure 2-19, page 2-40, the AA is 1,500 meters wide, the tactical-obstacle effect is to turn the enemy, and the resource factor is 1.2. The linear minefield requirement is 1,800 meters. One turn-effect minefield has a front of 500 meters (1,800/500 = 4 minefields [round up]). The number of mines and the time required to emplace each minefield depends on the emplacement method.

## MINEFIELD SUPPLY OPERATIONS

Requests for Class IV/V obstacle material that are originated at TF level or below go to the TF S4. The TF S4 processes the requests and sends them to the forward support battalion (FSB). The FSB processes the requests and forwards them to the brigade-level FSB, the division material-management center (DMMC), and the corps material-management center (CMMC).

The issue of Class IV obstacle material normally involves large quantities of material; therefore, corps support elements normally use their transportation assets to deliver the material directly to the emplacement sites or to the Class IV/V supply point.

Units request Class V obstacle material somewhat differently. The TF S4 notifies the brigade S4 of Class V requirements. The brigade S4 notifies the division ammunition officer (DAO) in the DMMC, who authorizes Class V issue by the ammunition transfer point (ATP). Class V obstacle material,

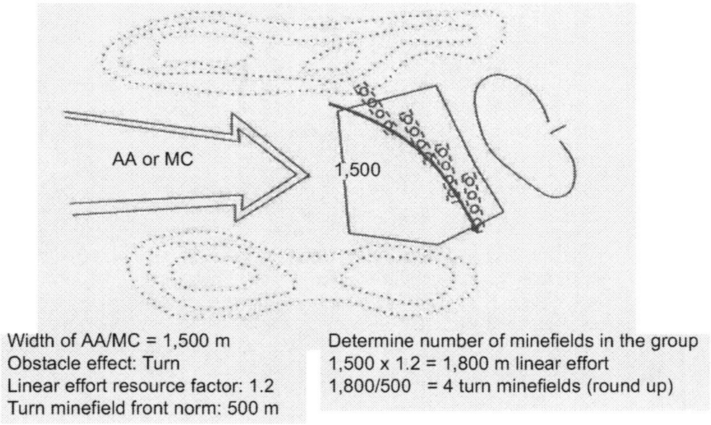


**Figure 2-19. Example of minefield resourcing**

unlike most ammunition, is delivered to the user at the obstacle emplacement site or the Class IV/V supply point.

A supply request includes the quantity, the required delivery time, the transportation responsibilities, and the desired delivery location. The quantity includes the total for each type of obstacle. There may be several Department of Defense identification codes (DODICs) and national stock numbers (NSNs) involved, depending on the types of obstacles required. The required delivery time is very important for ensuring an early start on the preparation of the battlefield because the lack of material could adversely affect the mission. Transportation responsibilities must be clearly understood. Material handling equipment (MHE) is required to ensure a rapid turnaround of haul assets.

In addition, the brigade staff coordinates with the TF staff to identify the location of Class IV/V supply points in the TF sectors. Prompt identification of TF Class IV/V supply points is required if the obstacle material is forwarded from the corps into the TF sector. If the material is not forwarded into the TF sector, the brigade is responsible for delivering it to the TF.

At the TF level, sustaining obstacle operations is an extremely difficult task. Centralized throughput operations by the corps or the division stop at the TF level. Mass quantities of obstacle material, especially mines, are centrally received, broken down into minefield packages, and then distributed throughout the sector based on the obstacle plan. At some point in the distribution plan, the TF turns over control of the obstacle material to engineers who then emplace the obstacles. Obstacle logistics (especially for mine warfare) at the TF level can be complex, require prudent use of scarce haul assets and MHE, and demand positive $C^2$.

The supply of obstacle material may vary slightly for obstacle groups developed at corps, division, and brigade levels. The staff at the level where the obstacle group is planned determines the resources required for the obstacle. It also determines how the emplacing unit will get the material. For example, if the corps staff plans a reserve obstacle group but the detailed planning is done at TF level, the TF resources the obstacle group. However, if the corps staff plans the obstacle group in detail, it determines the resources required. The corps staff also plans the delivery of obstacle material to the emplacing unit. Alternately, the corps staff could direct the emplacing unit to pick up the obstacle material from a location such as the corps storage area.

This section describes some of the underlying principles in mine supply operations. It concentrates on the flow of Class IV/V material (mines) through the battalion sector. The maneuver unit is responsible for the flow of obstacle material within the maneuver battalion sector; however, it is effectively a shared responsibility between the engineer and the maneuver unit.

### RESUPPLY NODES

There are two critical mine resupply nodes within the TF sector—the Class IV/V supply point and the mine dump. The relative locations of the Class IV/V supply point and the mine dump are shown in Figure 2-20.

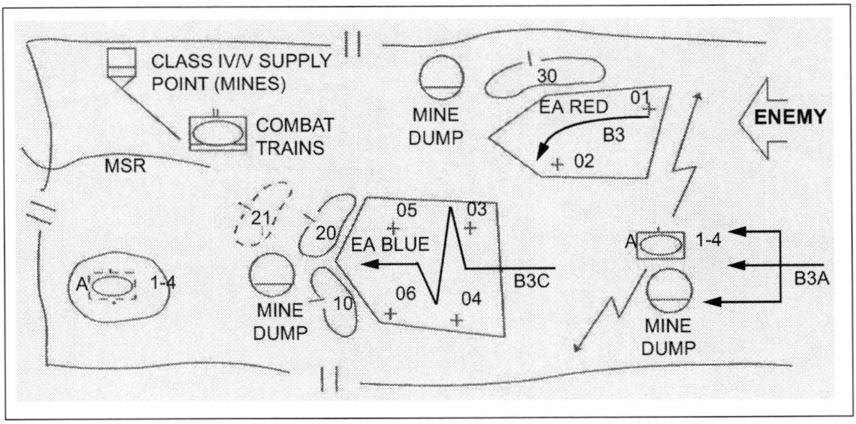


**Figure 2-20. Mine resupply**

### Class IV/V Supply Point

A Class IV/V supply point is a central receiving point for obstacle material in a TF sector. It is where the TF receives and transfers control of obstacle material pushed forward by higher levels. The supply point is established and operated by the TF and is centrally located to support all planned obstacles within the TF sector. Where the tactical-obstacle plan allows, the supply point should be located near the TF combat trains to better facilitate $C^2$ and the availability of equipment. The TF combat trains command post (CTCP) should provide $C^2$ of operations at the Class IV/V supply point. The TF engineer will normally furnish a representative that provides technical assistance to the CTCP.

The main purpose of a Class IV/V supply point operation is to receive obstacle material and reconfigure it based on the requirements for each obstacle group. Each supply point must have a dedicated S4 representative to track the flow of obstacle material in and out of the supply point. The supply point should have dedicated MHE to off-load the bulk quantities of obstacle material and reconfigure them into minefield packages as required. Obstacle material is normally broken down into minefield packages if the material is not already delivered in combat-configured loads. This may require a dedicated engineer representative to ensure that obstacle material is configured properly. Table 2-7 shows personnel requirements for a Class IV/V supply point.

**Table 2-7. Personnel requirements for a Class IV/V supply point**

| Personnel | Responsibilities |
|---|---|
| TF S4/S4 NCOIC | Provides overall $C^2$ and Class IV/V accountability<br>Coordinates for MHE<br>Is assisted by the RTO |
| TF engineer representative | Supervises organization of Class IV/V material into packages to support obstacle groups<br>Provides technical assistance on Class IV/V supply-point setup and mine uncrating and inspection |
| Infantry squad* | Downloads incoming Class V trucks |
| Infantry squad* | Downloads incoming Class IV trucks |
| Infantry squad* | Uploads outgoing Class IV/V trucks |
| Two infantry squads* | Uncrates and inspects mines |
| Support PL/PSG | Provides $C^2$ for cross loading of Class IV/V material |
| Infantry squad* | Provides local security for the Class IV/V supply point |
| * or equivalent-sized element from a supported unit | |

The most labor-intensive task at the Class IV/V supply point is the uncrating of mines. This requires dedicated manpower from the supported maneuver force and the tools needed to break shipping bands and uncrate the mines from their containers. Another important aspect of uncrating mines is tracking fuses and booster charges. As the mines are uncrated, fuses and booster charges are separated. However, the same number and type of fuses and booster charges must be task-organized with minefield packages. This requires strict supervision; mistakes can quickly lead to confusion and a waste of emplacement time.

Because of the assets involved at a Class IV/V supply point, a TF is normally capable of operating only one supply point at any given time. If the TF sector is extremely wide or deep, several supply points may be planned; however, only one can be operated at a time, based on the commander's priorities for obstacle emplacement.

## Mine Dump

A mine dump is the most forward mine resupply node. It is where mines are task-organized into mine-strip packages and then inspected, prepared, and loaded onto emplacing vehicles. A mine dump is not a permanent supply point, and it is not always used. Its use depends on the method of minefield resupply, and these techniques are discussed in more detail later in this chapter. When used, one mine dump supports a single obstacle group. It is activated or deactivated upon initiation and completion of the obstacle group's emplacement. Mine-dump operations are primarily an engineer company or platoon responsibility. However, it is a good technique to augment mine-dump operations with personnel from the company/team overwatching the obstacle group being emplaced. A mine dump may be located in the vicinity of the company/team position or closer to the obstacle group.

There are three critical tasks that must be accomplished at the mine dump:

- The minefield packages are further task-organized into mine-strip packages (complete with the right number, type, and mix of fuses and boosters) as they are transported to the mine dump. For example, if a platoon is emplacing a standardized disrupt row minefield, mines are task-organized into three mine-strip packages. As the engineer platoon moves to the mine dump to resupply, each emplacing vehicle loads a designated mine-strip package.
- The mines are prepared for emplacement; but they are not fused at the mine dump. Preparation includes loosening and greasing fuse and booster wells and ensuring proper functioning of the mine.
- The mines are transloaded onto emplacing vehicles or a mine-delivery system.

Transportation of mines from the Class IV/V supply point to the mine dump is a supported TF responsibility. However, it is normally shared between the engineer company and the TF, since neither one has the haul capability to simultaneously service all active mine dumps.

## Resupply Rules

The following rules govern mine resupply:

- Mines should be uncrated at the Class IV/V supply point to preserve transportation assets going forward.
- Mines are task-organized into minefield packages at the Class IV/V supply point.
- Transportation from the Class IV/V supply point to the mine dump is a shared engineer and maneuver unit responsibility when a mine dump is used.
- Mines are inspected and prepared at the last supply node (Class IV/V supply point or mine dump) before they are loaded onto an emplacing vehicle or a mine-delivery system.

- Class IV/V supply points are set up using authorized ammunition procedures and distance requirements.

### Supply Location

Consider the following when selecting a location for the Class IV/V supply point or the mine dump:

- Carrying capacity. The location of key supply nodes and the type of resupply method used depends in part on the type, amount, and availability of haul assets. The carrying capacity plays a large role. In short, the more material a vehicle can carry, the more turnaround time you can afford. Table 2-8 provides the mine haul capacity for various types of vehicles.
- Traffic circuit. Vehicles must be able to enter, load, unload, and exit without interfering with the loading and unloading of other vehicles.
- Camouflage and cover. Protection from observation and thermal imaging is desired. Protection from artillery and air attack should be considered. Residue must be removed.
- Defense. The site must be organized for defense against enemy patrols and saboteurs.
- Time. Time factors for handling the obstacle material, to include all unloading, uncrating, inspecting, and loading, must be considered.
- Operators. Leaders and soldiers must be specifically allocated to operate Class IV/V supply points and mine dumps. They will probably remain there until the task is complete. The supply node may have to be collocated with or be near a source of manpower. Table 2-7, page 2-42, provides general guidance on how much manpower is required to sustain mine resupply operations.

### Resupply Methods

The methods for obstacle material resupply are—

- Supply point.
- Service station.
- Tailgate.

In each method, corps or division transport delivers Class IV/V supplies forward to a designated Class IV/V supply point in each TF sector. The primary differences in the methods are how the material is delivered from the Class IV/V supply point to the obstacle location and whether or not mine dumps are established.

#### Supply Point

The supply-point method (Figure 2-21, page 2-46) requires that the emplacing engineer platoon return to the Class IV/V supply point each time it resupplies. This method does not require a separate mine dump. In effect, it moves the normal tasks associated with a mine dump to the supply point. Mines are prepared and inspected at the supply point as they are being loaded onto an emplacing vehicle or a mine-delivery system.

**Table 2-8. Class IV/V haul capacity**

| Vehicle | Concertina Wire[1] | M15 AT Mine | M19 AT Mine | M21 AT Mine | M16 AP Mine | M14 AP Mine | MOPMS Mine | Volcano Mine | MICLIC Reload[2] | Hornet |
|---|---|---|---|---|---|---|---|---|---|---|
| HMMWV 1,124 kg, 6 cu m | 2 | 51 | 34 | 27 | 55 | 56 | 15 | 1 | NA | 1 |
| M35 2½-ton truck 2,250 kg, 12.5 cu m | 4 | 102 | 69 | 55 | 111 | 113 | 30 | 2 | 2 | 2 |
| M1078 2½-ton truck 2,250 kg, 13.4 cu m | 4 | 102 | 69 | 55 | 111 | 113 | 30 | 2 | 2 | 2 |
| M54 5-ton truck 4,500 kg, 13.6 cu m | 7 | 204 | 138 | 109 | 222 | 227 | 61 | 5 | 3 | 5 |
| M1083 5-ton truck 4,500 kg, 15.6 cu m | 8 | 204 | 138 | 109 | 222 | 227 | 61 | 5 | 3 | 5 |
| M930 5-ton dump truck (without sideboards) 4,500 kg, 3.8 cu m | 2 | 112 | 64 | 32 | 168 | 71 | 23 | 3 | 2 | 2 |
| M930 5-ton dump truck (with sideboards) 4,500 kg, 8.2 cu m | 4 | 204 | 138 | 70 | 222 | 153 | 51 | 5 | 3 | 4 |
| M1090 5-ton dump truck 4,500 kg, 3.8 cu m | 2 | 112 | 64 | 32 | 168 | 71 | 23 | 3 | 2 | 2 |
| HEMTT truck 9,000 kg, 15 cu m | 8 | 408 | 277 | 128 | 444 | 317 | 94 | 10 | 7 | 8 |
| 12-ton S&T 10,800 kg, 24.5 cu m | 13 | 489 | 333 | 208 | 533 | 514 | 148 | 12 | 9 | 13 |
| 40-ton lowboy 36,000 kg, 49.3 cu m | 27 | 1,466 | 1,035 | 419 | 1,777 | 1,035 | 308 | 30 | 27 | 27 |
| M548 cargo 5,400 kg, 14.9 cu m | 8 | 244 | 166 | 125 | 266 | 272 | 74 | 6 | 4 | 6 |
| M1077 PLS flat rack 14,900 kg, 17.6 cu m | 9 | 440 | 352 | 164 | 586 | 293 | 110 | 11 | 9 | 9 |
| **No of mines per box** | NA | 1 | 2 | 4 | 4 | 90 | 21 | 240 | NA | 30 |
| **Weight per box (kg)** | 531 | 22 | 33 | 41 | 21 | 20 | 73 | 833 | 1,195 | 810 |
| **Size of box (cu m)** | 1.8 | 0.04 | 0.05 | 0.12 | 0.03 | 0.06 | 0.16 | 1.6 | 1.8 | 1.8 |

[1]The number of concertina = bundles; 1 bundle = 40 rolls
[2]Line charge + rocket

Several considerations may drive the use of supply-point resupply. First, if there are no additional haul assets to transport obstacle material forward from the Class IV/V supply point, the supply-point method may be the only viable technique. Secondly, the minefield group may be close enough to the supply point that any other method is less efficient.

- Advantages.
  - — Minimizes unloading and loading of material.
  - — Requires minimal augmentation of haul assets.

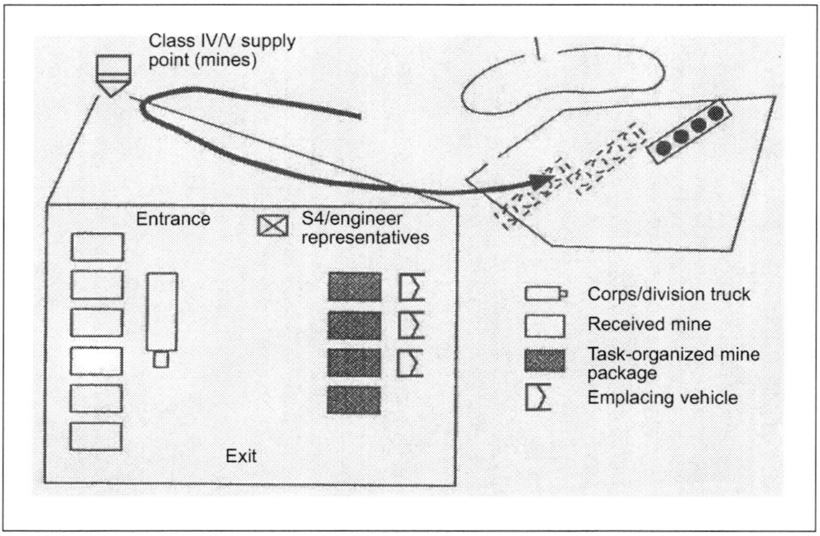


**Figure 2-21. Supply-point resupply method**

— Allows manpower and equipment to be massed at a single supply point.

— Streamlines $C^2$ of material.

- Disadvantages.

— Requires more movement of the platoon, which may take away from emplacement time.

— Requires that the platoon move in and out of the area where the minefields are being emplaced, increasing the risk of fratricide.

— May disrupt the emplacement of individual obstacles when emplacing vehicles cannot carry enough material to start and complete the obstacle. This causes emplacing vehicles to stop work, reload, and pick up where they left off.

— Requires a larger Class IV/V supply point that is capable of receiving mass quantities of obstacle material and multiple loading platoons simultaneously.

## Service Station

The service-station method (Figure 2-22) centers on the activation of a mine dump forward of the Class IV/V supply point. The mines are transported to a mine dump using a combination of engineer and TF haul assets that are normally under the control of the emplacing engineer. At the mine dump, material is stockpiled and prepared by the mine-dump party. Obstacle material is further task-organized into minefield and mine-strip packages. The emplacing platoon moves to a mine dump to resupply emplacing vehicles or dispensers. Once the obstacle group is emplaced, the mine dump is deactivated or moved to support another obstacle group.

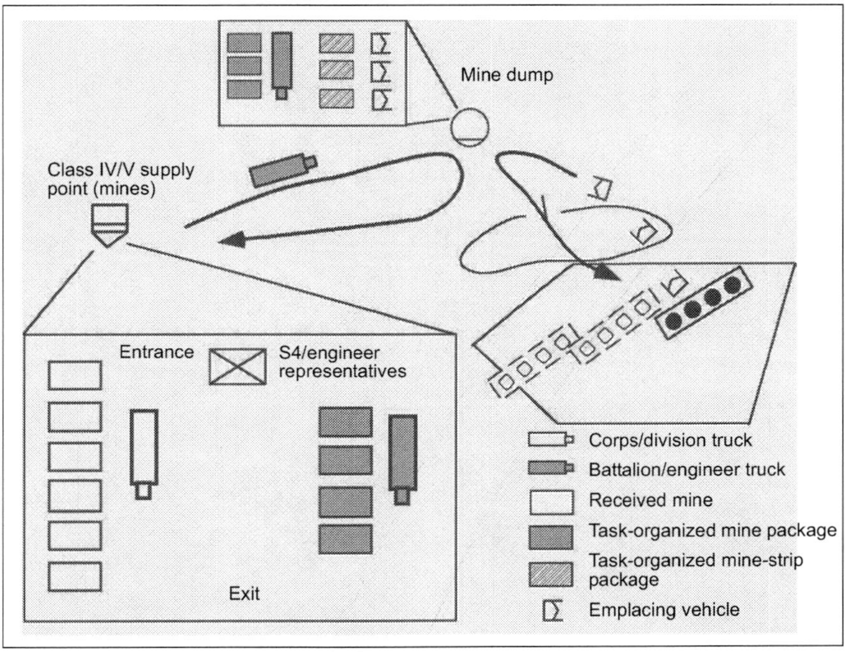


**Figure 2-22. Service-station resupply method**

There are several considerations for using the service-station method. First, it is used when the obstacle group is located too far from the Class IV/V supply point to allow efficient turnaround. Since this method provides for prestocking obstacle material forward, it is used when available haul assets have a relatively small capacity. This requires making frequent, short-duration resupply trips and stocking mines to keep pace with emplacement. It also streamlines emplacement since there is an opportunity to task-organize the mines into mine-strip packages, based on the emplacement method and the type of minefield. Finally, while it still requires the emplacing platoon to stop laying and resupply, it minimizes the distance and the amount of time the platoon must travel to reload. This requires that a small party be left at the minefield to help pick up where emplacement stopped.

- Advantages.

— Allows prestockage of obstacle material to keep pace with emplacement.

— Minimizes the distance and the amount of time the emplacing platoon must travel to reload.

— May provide additional manpower and security if it is located near a company/team.

- Disadvantages.

— Requires additional loading and unloading of obstacle material.

— May require augmentation with haul assets.

— Disrupts emplacement by requiring the emplacing platoon to stop obstacle emplacement, move to the supply point, reload, and return to the minefield.

**Tailgate**

The tailgate resupply method (Figure 2-23) transports obstacle material directly from the Class IV/V supply point to the emplacing platoon at the obstacle site. Obstacle material is transported with TF and engineer haul assets. Obstacle material is transloaded by emplacing engineers to emplacing vehicles or dispensers at the obstacle site.

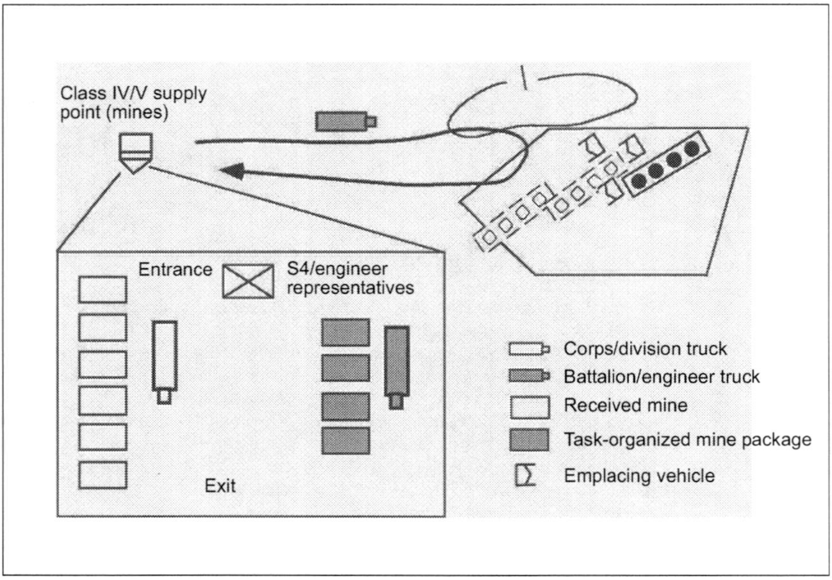


**Figure 2-23. Tailgate resupply method**

Two overriding considerations drive the decision to use the tailgate resupply method. First, if obstacle emplacement is being conducted during limited visibility, the tailgate method is the primary resupply method. It minimizes disruption of emplacement and the chance of fratricide as engineers move back into the work area after reloading. Secondly, tailgate resupply is the primary method used when establishing a hasty defense or when the situation is unclear and an attack can happen at anytime. Since obstacle material remains loaded until transferred to an emplacing vehicle, the tailgate method enables engineers to quickly break contact without risking a loss of obstacle material to the enemy. The tailgate resupply method is the preferred method for light forces.

- Advantages.
  - Minimizes loading and unloading of obstacle material.
  - Allows engineers to break contact rapidly in the event of enemy attack without losing obstacle material to the enemy.
  - Minimizes the movement of platoons in and out of the minefield (suitable for limited visibility).
- Disadvantages.
  - Requires augmentation by high-capacity transportation assets that are capable of offsetting the loss in turnaround time if the vehicle has to wait at the obstacle site.
  - May result in inefficient use of haul assets.
  - Complicates $C^2$ in linking mine transport assets with emplacing engineers since the engineers continue emplacement.
  - Causes task-organizing of minefield packages to be conducted concurrently with loading.

## MINEFIELD MARKING

### CRITERIA

Minefields must be marked to prevent fratricide. Marking ensures that friendly soldiers do not accidentally enter a minefield, and it is a requirement under STANAGs and Geneva Convention agreements. When emplacing minefields behind the forward line of own troops (FLOT) (in the main battle or rear area), mark the minefields on all four sides. This includes air-delivered Volcano minefields that are sited and emplaced before the enemy attacks.

Gator, RAAM, and ADAM minefields are exceptions to the rule. To preserve the system's flexibility and because of the relative inaccuracy of emplacement, these minefields are not normally marked before emplacement unless the tactical situation permits. Marking the area where mines are to be emplaced by artillery or fixed-wing aircraft is not recommended. Mines could likely be emplaced outside the marked area.

Forward of the FLOT, minefields are not generally marked before emplacement. However, commanders must make every attempt to mark these minefields as soon as the tactical situation allows. For scatterable minefields, a commander may choose to remove markings once the self-destruct (SD) time of the mines has expired; but the location of the minefield must still be recorded and forwarded to higher and adjacent units in case some of the mines did not self-destruct.

### Perimeter

Construct a perimeter fence (Figure 2-24) to mark a minefield. Start emplacing the perimeter fence before emplacing mines, preferably during site layout if the tactical situation permits. For conventional minefields, ensure that the perimeter fence is at least 15 meters outside the nearest mine or cluster. For scatterable minefields, the area inside the perimeter fence must include the safety zone. (See Chapter 3 for more details.)

Place warning signs for areas containing emplaced mines. Space the warning signs 10 to 50 meters apart, depending on the terrain. If using pickets and barbwire to mark the minefield, ensure that the wire is waist-high. If using concertina wire, use a one-roll height. Place additional strands of barbwire or rolls of concertina at the discretion of the commander.

### Techniques

A commander may decide to mark individual minefields in an obstacle group or to mark the group as a whole (Figure 2-25, page 2-52). Depending on the size and the location of minefields, either technique may have the advantage of using fewer resources or labor. Normally, marking individual minefields in a fix-obstacle group requires less resources than marking the entire obstacle group. The opposite is usually true for disrupt-, turn-, and block-obstacle groups. The decision to mark individual minefields or obstacle groups should not be based solely on logistical considerations. A commander must consider the amount of tactical and/or sustainment movement required in and around the obstacle groups as well as the capability of the unit's $C^2$ forces.

The following advantages and disadvantages to marking individual minefields versus marking the obstacle group are provided to help commanders make the best decision.

#### Marking Individual Minefields

- Advantages.
  - Returning units forward of the minefields have more routes (tactical lanes or bypasses) through the obstacle group.
  - Tactical lanes need only pass through individual minefields.
  - Mine recovery is easier.
- Disadvantages.
  - Obstacle may not provide the desired effect.
  - Enemy units can more easily bypass individually marked minefields in a fix- or block-obstacle group.

#### Marking Obstacle Groups

- Advantages.

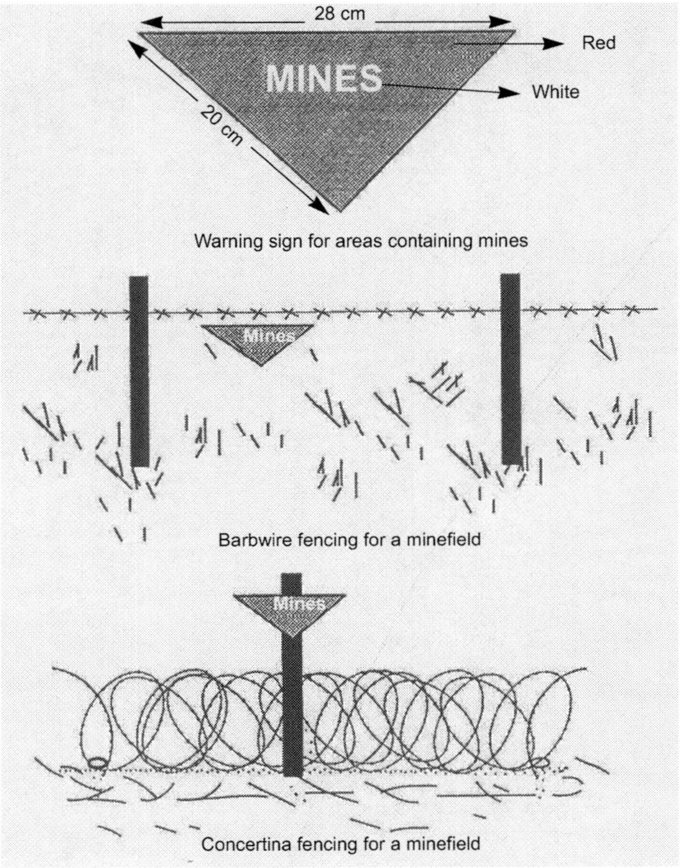


**Figure 2-24. Minefield marking**

- Obstacle is more likely to provide the desired effect.
- The enemy cannot easily discern individual minefields and decide when to employ breach assets in a fix-or block-effect minefield.

- Disadvantages.
  - Friendly patrols cannot clearly see if the minefield is tampered with unless they are within the perimeter fence.

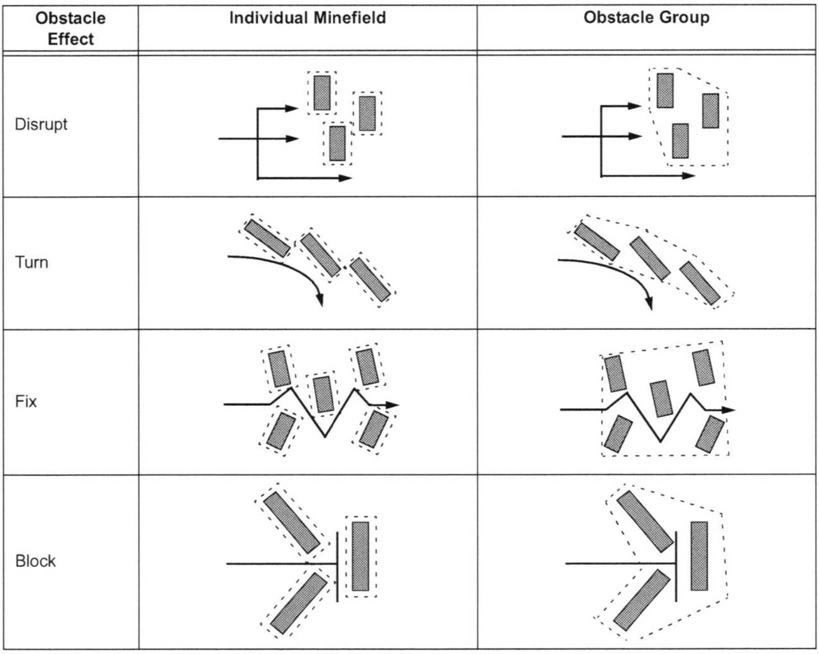


**Figure 2-25. Marking of minefields and obstacle groups**

— Tactical lanes need to pass through the entire obstacle group. Friendly units passing through the lanes will be slowed considerably.

## MINEFIELD TURNOVER

Once an obstacle group is completed, the emplacing unit conducts minefield or obstacle turnover with the owning unit. Occasionally, the owning unit will transfer responsibility for a minefield to another unit. Minefield or obstacle turnover ensures that the commander of the owning unit is familiar with the minefield and understands his responsibilities concerning it. Turnover is conducted whether or not there are lanes/gaps to be closed. Minefield turnover is a must; the time and the location for the turnover is established during the initial siting coordination.

The engineer must verbally address the following items with the overwatching company/team:

- Intelligence.
  - — Provide an update on enemy activity forward of the minefield.

- — Discuss expected enemy reconnaissance efforts.
- — Brief on local, friendly, and enemy situations.

- Maneuver.
  - — Discuss obstacle protection against enemy dismounted patrols. Recommend that the maneuver unit conducts security/patrols to protect the minefield during limited visibility.
  - — Discuss fire control measures.
- Mobility/survivability.
  - — Discuss the obstacle's intended effect on enemy maneuver.
  - — Discuss the minefield front and depth and walk/ride the minefield trace. Provide grid coordinates of the minefield trace.
  - — Discuss minefield composition.
  - — Discuss friendly minefield marking.
  - — Discuss lane/gap closure, if applicable. Confirm the signal or the activity that initiates lane closure.
  - — Train units on how to close lanes. This may mean training the unit on emplacing conventional mines or using the MOPMS.
- Fire support.
  - — Update the company FIST on grid coordinates for the minefield trace.
  - — Discuss indirect fires covering the minefield.
- CSS. Provide mines/material required to close lanes/gaps and ensure that all necessary material is available and prepared.
- $C^2$.
  - — Transfer graphics and documentation (minefield records, demolition-target folders, or other written records).
  - — Report completion of the turnover to the higher engineer and supported unit headquarters.
  - — Complete an obstacle-turnover report (a sample work sheet is shown in Figure 2-26, page 2-54) and submit it to higher headquarters.
  - — Forward the written minefield report and record (DA Form 1355 or 1355-1-R) to the next higher commander common to both units.

| OBSTACLE-TURNOVER/TRANSFER REPORT | | |
|---|---|---|
| **OBSTACLE DATA** | | |
| Obstacle Number | Obstacle Type | Emplacing Authority |
| Obstacle Location *(8-digit UTM grid to center of mass)* | Obstacle Effect | Emplacing Unit |
| **TURNOVER DATA** | | |
| Transferring Unit | Receiving Unit | Date/Time of Transfer |
| Next Higher Command Common to Both | | Date Obstacle Emplaced |

**TURNOVER CHECKLIST**

- ☐ Intelligence.
  - Enemy activity forward of the obstacle *(discuss enemy recon and breaching efforts)*.
  - Friendly activity in the vicinity of the obstacle *(discuss LOGPAC, recon, counterrecon maneuver)*.
- ☐ Maneuver.
  - Obstacle protection against enemy recon/breaching *(discuss overwatch, counterrecon patrols, sunrise sweep)*.
  - Location of TRPs and indirect-fire targets *(point out location on graphics and terrain)*.
  - Rearward/forward passage of lines *(discuss planned and routine activity around the obstacle)*.
- ☐ Mobility/survivability.
  - Obstacle trace *(discuss corner grids on map and point them out on terrain)*.
  - Obstacle marking *(rear side, all four sides, fence, signs, HEMMS poles, or no marking)*.
  - Obstacle composition.
  - Location of gaps/lanes.
  - Method of closing gaps/lane *(MOPMS, ADAM, RAAM, conventional mines)*.
  - Location of material to close gaps/lanes.
- ☐ CSS. Planned casualty evacuation and supply routes in the vicinity of the obstacle.
- ☐ $C^2$.
  - DA Form 1355 or 1355-1-R, dated ________, transferred to receiving unit.
  - Obstacle overlay, dated ________, transferred to receiving unit.
  - Target folder, dated ________, preparing headquarters ______________, transferred to receiving unit.
  - Higher headquarters (transferring, receiving, and emplacing units) notified of transfer.
  - DA Form 1355 or 1355-1-R forwarded. (**NOTE: Three copies are required for transfer—one to transferring unit, one to receiving unit, and one to next higher headquarters common to both.)**

**CERTIFICATION**

The undersigned certify that the receiving unit has assumed full responsibility for the obstacle indicated above. The unit commander understands all the information related to the obstacle, the obstacle intent, and his responsibility to transfer or recover the obstacle upon his departure.

| | |
|---|---|
| ______________________________ | ______________________________ |
| Signature of Transferring Unit Commander | Signature of Receiving Unit Commander |
| ______________________________ | ______________________________ |
| Printed Name and Rank | Printed Name and Rank |

**Figure 2-26. Sample obstacle-turnover work sheet**

## MINEFIELD INSPECTION AND MAINTENANCE

Mines left in the ground for a long time may deteriorate and malfunction for one or more of the following reasons:

- Moisture may have entered the igniter or the body of the mine and either neutralized the explosive or corroded the metal parts. Such actions may be aggravated by local factors (soil acidity or wide temperature swings).
- Frost or heat may have subjected the mine to mechanical strain and caused distortion.
- Insects or vegetation may have caused obstructions.
- Animals may have turned mines over or detonated them.

Technical inspections should only be made by experienced engineers or explosive ordnance disposal (EOD) personnel. When a minefield deteriorates below the operating level, additional mine strips/rows are added to restore its effectiveness. They are sited to the front or the rear of the existing minefield to increase its depth. New mine strips/rows are treated as separate minefields.

Technical inspections of minefields are normally done at three-month intervals. They are done more frequently during extreme weather conditions. Detailed information on mine inspections can be found in Technical Manual (TM) 9-1345-203-12.

This chapter implements STANAG 2990.

Chapter 3

# Scatterable Mines and Mine Delivery Systems

SCATMINEs are laid without regard to a classical pattern. They are designed to be delivered or dispensed remotely by aircraft, artillery, missile, or a ground dispenser. All US SCATMINEs have a limited active life and self-destruct after that life has expired. The duration of the active life varies with the type of mine and the delivery system.

SCATMINE systems enable a tactical commander to emplace minefields rapidly in enemy-held territories, contaminated territories, and in most other areas where it is impossible for engineers to emplace conventional minefields. Some systems allow for rapid emplacement of minefields in friendly areas. As with all minefields and obstacles, scatterable minefields are an engineer responsibility.

Based on the tactical plan, the maneuver commander's staff engineer determines the minefield location, size, density, and emplacement and SD times. With this information and a thorough understanding of the available systems, he can then recommend the type of minefield (conventional or scatterable) to be emplaced. If a scatterable minefield is selected, he recommends the delivery system and coordinates the minefield with appropriate staff officers.

## GENERAL CHARACTERISTICS

Most US SCATMINEs have similar characteristics. SCATMINEs are much smaller in size and weight than conventional mines. For example, a standard AT SCATMINE weighs approximately 1.8 kilograms and has 600 grams of explosive; an M15 conventional mine weighs 13.5 kilograms and has 10 kilograms of explosive. Arming mechanisms, arming times, and SD times of SCATMINEs differ based on the dispensing system.

### ANTIPERSONNEL MINES

There are two general categories of AP SCATMINEs—wedge-shaped and cylindrical (Figure 3-1, page 3-2). Table 3-1, page 3-2, summarizes the characteristics of each AP SCATMINE.

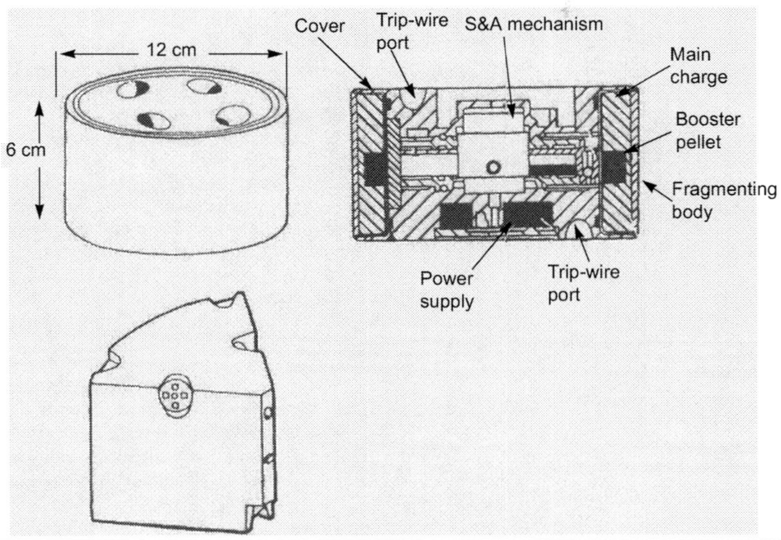


**Figure 3-1. AP SCATMINEs**

**Table 3-1. Characteristics of AP SCATMINEs**

| Mine | Delivery System | DODIC | Arming Time | Fuse | Warhead | AHD | SD Time | Explosive Weight | Mine Weight | Number of Mines |
|---|---|---|---|---|---|---|---|---|---|---|
| M67 | 155-mm artillery (ADAM) | D502 | Within 1 min after ground impact | Trip wire | Bounding frag | 20% | 4 hr | 21 g Comp A5 | 540 g | 36 per M731 projectile |
| M72 | 155-mm artillery (ADAM) | D501 | Within 1 min after ground impact | Trip wire | Bounding frag | 20% | 48 hr | 21 g Comp A5 | 540 g | 36 per M692 projectile |
| BLU 92/B | USAF (Gator) | K291<br>K292<br>K293 | 2 min | Trip wire | Blast frag | 100% | 4 hr<br>48 hr<br>15 days | 540 g Comp B4 | 1.44 kg | 22 per CBU 89/B dispenser |
| M77 | MOPMS | K022 | 2 min | Trip wire | Blast frag | 0% | 4 hr (recycle up to 3 times) | 540 g Comp B4 | 1.44 kg | 4 per M131 dispenser |
| Volcano | Ground/ air | K045 | 2 min | Trip wire | Blast frag | 0% | 4 hr<br>48 hr<br>15 days | 540 g Comp B4 | 1.44 kg | 1 per M87 canister |

The M67 and M72 AP SCATMINEs are wedge-shaped and dispensed from an ADAM projectile, which is a special 155-millimeter artillery munition. Each mine weighs 540 grams and is 7 centimeters high.

The M74, BLU 92/B, M77, and Volcano AP SCATMINEs are all cylindrical in shape. They are 6 centimeters high and 12 centimeters in diameter. Cylindrical AP SCATMINEs kill enemy soldiers through the combined effects of blast and fragmentation. Each mine contains 540 grams of composition B4 as its main charge. The charge detonates upon actuation and shatters the mine's metal casing to produce shrapnel. Shrapnel is propelled upward and outward from the mine and produces fatal casualties to a distance of 15 meters. Each mine has eight trip wires (four on the top and four on the bottom) that deploy after ground impact up to 12 meters from the mine. Trip wires are similar in appearance to very fine thread; they are olive-drab green in color and weighted at the free end. A tension of 405 grams applied to one trip wire is enough to create a break in the electrical circuit and cause the mine to detonate.

### ANTITANK MINES

All AT SCATMINEs (Figure 3-2) have similar functional characteristics. They are cylindrical in shape, weigh approximately 1.8 kilograms, contain 585 grams of cyclonite (RDX) explosive as the main charge, and have a magnetically induced fuse. The characteristics of each AT SCATMINE are summarized in Table 3-2, page 3-4.

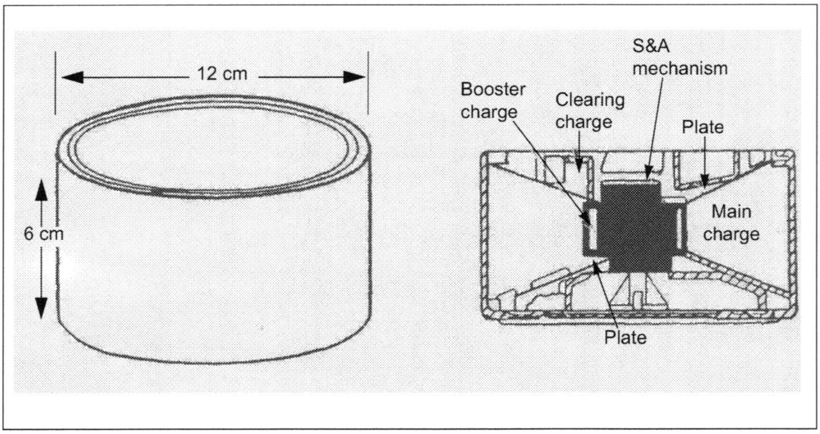


**Figure 3-2. AT SCATMINE**

**Table 3-2. Characteristics of AT SCATMINEs**

| Mine | Delivery System | DODIC | Arming Time | Fuse | Warhead | AHD | SD Time | Explosive Weight | Mine Weight | Number of Mines |
|---|---|---|---|---|---|---|---|---|---|---|
| M73 | 155-mm artillery (RAAM) | D503 | Within 1 min after ground impact | Magnetic | M-S plate | 20% | 48 hr | 585 g RDX | 1.7 kg | 9 per M718 projectile |
| M70 | 155-mm artillery (RAAM) | D509 | Within 1 min after ground impact | Magnetic | M-S plate | 20% | 4 hr | 585 g RDX | 1.7 kg | 9 per M741 projectile |
| BLU 91/B | USAF (Gator) | K291<br>K292<br>K293 | 2 min | Magnetic | M-S plate | NA | 4 hr<br>48 hr<br>15 days | 585 g RDX | 1.7 kg | 72 per CBU 89/B dispenser |
| M76 | MOPMS | K022 | 2 min | Magnetic | M-S plate | NA | 4 hr (recycle up to 3 times) | 585 g RDX | 1.7 kg | 17 per M131 dispenser |
| Volcano | Ground/ air | K045 | 2 min 30 sec | Magnetic | M-S plate | NA | 4 hr<br>48 hr<br>15 days | 585 g RDX | 1.7 kg | 5 per M87 canister; 6 per M87A1 canister |

AT SCATMINEs are designed to produce a K-Kill instead of an M-Kill. They produce a kill by using an SFF warhead (created from an M-S plate). The warhead penetrates the vehicle's belly armor, and spalling metal from the vehicle (caused by the mine blast) kills occupants instantly. Even though the crew is killed, the drive train may be undamaged and the vehicle may continue to move. On enemy tanks with autoloaders, the detonation of rounds in the belly-mounted ammunition carousel is very likely. The mine may not achieve a kill when the track of an armored vehicle runs directly over it.

The magnetic fuse is designed to detonate as the magnetic field changes over the mine. The warhead is bidirectional, meaning that it can fire from the top or the bottom. AHDs are built into 20 percent of M70, M73, and M75 mines. Although Volcano, M76, and BLU 91/B mines do not have AHDs, they may detonate when moved, because the mine may sense a significant change from its original orientation.

Due to their small size, the reduced explosive, and the possibility of landing with an improper orientation (on their side or at an angle), AT SCATMINEs have less chance of destroying a vehicle than a conventional full-width AT mine. An armored vehicle will not always be destroyed after an encounter with an AT SCATMINE. Further, the effectiveness of SCATMINEs in water obstacles is reduced even more, because 5 centimeters of water prevents the formation of the M-S slug. Although the blast wave is accentuated by underwater placement (attacking hatches and covers), mining of banks and approaches is recommended instead.

# CAPABILITIES

## Faster Response

SCATMINEs can be emplaced more rapidly than conventional mines, so they provide a commander with greater flexibility and more time to react to changes in situations. The commander can use SCATMINEs to maintain or regain the initiative by acting faster than the enemy. Using SCATMINEs also helps preserve scarce mine resources.

## Remote Placement

All SCATMINEs are remotely emplaced. This enhances battlefield agility and allows the maneuver commander to emplace mines rapidly to best exploit enemy weaknesses. SCATMINEs can be used as situational obstacles or to attack enemy formations directly through disrupt, fix, turn, and block obstacles. Modern fusing, sensing, and AHDs allow SCATMINEs to better defeat enemy attempts to reduce the minefield.

## Increased Tactical Flexibility

Upon expiration of the SD time, the minefield is cleared and the commander can move through an area that was previously denied to enemy or friendly forces. In many cases, the SD period may be set at only a few hours. This feature allows for effective counterattacks to the enemy's flank and rear areas.

## Efficiency

SCATMINEs can be emplaced by a variety of delivery methods. They can be deployed by fixed-wing aircraft, helicopters, artillery, manpack, or ground vehicles. They satisfy the high mobility requirements of modern warfare. Manpower, equipment, and tonnage are reduced for their emplacement.

## Increased Lethality

AT SCATMINEs utilize an SFF that is created from two M-S plate charges to produce a full-width kill. In simple terms, a metal plate is formed into a high-velocity slug that punches a hole in the belly of a tank. The effect produces an M-Kill against the vehicle's engine, track, or drive train; or it produces a K-Kill when the on-board ammunition is set off and the crew is killed or incapacitated or the vehicle's weapon system is destroyed. AT SCATMINEs are designed to destroy any tank in the world. In order to form an SFF, the mine requires a certain standoff between the vehicle and the target. Mines must also be nearly perpendicular to the target (laying on either side). The M-S plate is actually two plates—one facing the top of the mine and one facing the bottom. This ensures that it will successfully attack the target while lying on either side.

AP SCATMINEs are actuated by a trip wire and utilize a blast-fragmentation warhead.

# LIMITATIONS

## Extensive Coordination

Because SCATMINEs are a very dynamic weapon system, great care must be taken to ensure that proper coordination is made with higher, adjacent, and subordinate units. To prevent friendly casualties, all affected units must be notified of the location and the duration of scatterable minefields. Recording and reporting procedures for SCATMINEs are discussed in detail in Chapter 8, and they were specifically designed to minimize friendly casualties.

### Proliferation of Targets

SCATMINEs may be regarded by some commanders as easy solutions to tactical problems. Target requests must be carefully evaluated, and a priority system must be established because indiscriminate use of weapon systems will result in rapid depletion of a unit's basic load. Controlled supply rates (CSRs) will probably be a constraint in all theaters.

### Visibility

SCATMINEs are highly effective, especially when fires and obscurants strain the enemy's $C^2$. SCATMINEs lay on the surface of the ground, but they are relatively small and have natural coloring.

### Accuracy

SCATMINEs cannot be laid with the same accuracy as conventional mines. Remotely delivered SCATMINE systems are as accurate as conventional artillery-delivered or tactical aircraft-delivered munitions.

### Orientation

Between 5 and 15 percent of SCATMINEs will come to rest on their edges; mines with spring fingers will be in the lower percentile. If there is mud or snow more than 10 centimeters deep, the number will be in the higher percentile. When employing ADAMs or RAAMs in more than 10 centimeters of snow or mud, high-angle fire should be used and the number of mines increased. AP mines may be less effective in snow, because the deployment of trip wires is hindered. Melting of the snow may also cause the mines to change positions and activate AHDs.

## LIFE CYCLE

All SCATMINEs have a similar life cycle, although specific times vary based on the SD time and the dispensing system.

For safety reasons, SCATMINEs must receive two arming signals at launch. One signal is usually physical (spin, acceleration, or unstacking), and the other is electronic. This same electronic signal activates the mine's SD time.

Mines start their safe-separation countdown (arming time) when they receive arming signals. This allows the mines to come to rest after dispensing and allows the mine dispenser to exit the area safely. Table 3-1, page 3-2, and Table 3-2, page 3-4, show arming times for individual SCATMINEs.

Mines are armed after the arming time expires. The first step in arming is a self-test to ensure proper circuitry. Approximately 0.5 percent of mines fail the self-test and self-destruct immediately.

After the self-test, mines remain active until their SD time expires or until they are encountered. Mines actually self-destruct at 80 to 100 percent of their SD time. The time period from when the mines begin to self-destruct and when they finish is called the SD window (Table 3-3). No mines should remain active after the SD time has been reached. Two to five percent of US SCATMINEs fail to self-destruct as intended. Any mines found after the SD time must be treated as unexploded ordnance (UXO). For example, mines with a 4-hour SD time will actually start self-destructing at 3 hours and 12 minutes. When the 4-hour SD time is reached, no unexploded mines should exist.

**Table 3-3. SD windows**

| SD Time | SD Window Begins |
|---|---|
| 4 hours | 3 hours 12 minutes |
| 48 hours | 38 hours 24 minutes |
| 5 days | 4 days |
| 15 days | 12 days |

# LETHALITY AND DENSITY

## LETHALITY AND TACTICAL-OBSTACLE EFFECT

Scatterable minefields are employed to reduce the enemy's ability to maneuver, mass, and reinforce against friendly forces. They increase the enemy's vulnerability to fires by producing specific obstacle effects (disrupt, fix, turn, and block) on the enemy's maneuver. To achieve this aim, individual minefields must be emplaced with varying degrees of lethality. During emplacement, lethality is varied primarily by changing the minefield density. Therefore, there is a direct correlation between the obstacle effect and the minefield density. In order to achieve the tactical-obstacle effect, use the following guidance when selecting minefield density:

- Disrupt.
  - Low density.
  - Probability of encounter: 40 to 50 percent.
  - Linear density: 0.4 to 0.5 mine per meter.
- Fix.
  - Medium density.
  - Probability of encounter: 50 to 60 percent.
  - Linear density: 0.5 to 0.6 mine per meter.
- Turn.
  - High density.
  - Probability of encounter: 75 to 85 percent.

— Linear density: 0.9 to 1.1 mines per meter.

- Block.
  - High density.
  - Probability of encounter: 85+ percent.
  - Linear density: More than 1.1 mines per meter.

**DENSITY**

Density is normally expressed as linear or area. For conventional mines, linear density is normally used and is expressed in the average number of mines per meter of minefield front. For SCATMINE systems, area density is normally used and is expressed as the average number of mines per square meter. Since SCATMINE systems normally employ a preset combination of AT and AP mines, the area density includes both. For example, a scatterable minefield with an area density of 0.006 mine per square meter may have an AT density of 0.004 AT mine per square meter and an AP density of 0.002 AP mine per square meter. Due to the varying dimensions of scatterable minefields that can be created by the different types of employment devices, the exact density of a scatterable minefield cannot be determined. However, an estimate of the average density can be determined by using the following formulas:

- Linear density equals the number of mines divided by the minefield front.

$$\frac{\textit{number of mines}}{\textit{minefield front}} = \textit{mines per meter}$$

- Area density equals the number of mines divided by the minefield area.

$$\frac{\textit{number of mines}}{\textit{front} \times \textit{depth}} = \textit{mines per square meter}$$

- Area density can be converted to linear density by multiplying the area density by the minefield depth. **(NOTE: Converting area density to linear density is not always accurate due to the space between minefield strips.)**

$$\textit{area density} \times \textit{minefield depth} = \textit{linear density}$$

**EXAMPLE:** A 650- by 200-meter Gator minefield contains 564 mines (432 AT and 132 AP).

- Area density: 564 ÷ (200 x 650) = 0.004 mine per square meter.
  - AT area density: 432 ÷ (200 x 650) = 0.003 mine per square meter.

— AP area density: 132 ÷ (200 x 650) = 0.001 mine per square meter.

- Linear density: 564 ÷ 650 = 0.87 mine per meter.
  — AT linear density: 432 ÷ 650 = 0.67 mine per meter.
  — AP linear density: 132 ÷ 650 = 0.2 mine per meter.

## COMMAND AND CONTROL

Due to the delivery means, $C^2$ of SCATMINEs is more complex than conventional mines. SCATMINEs are very dynamic weapon systems because they can be rapidly emplaced and then cleared via their SD capability. Also, the physical boundary of a scatterable minefield is not clearly defined. These characteristics require impeccable communications and coordination to ensure that all friendly units know where mines are located, when they will be effective, and when they will self-destruct.

## AUTHORITY

The corps commander has emplacement authority for all scatterable minefields within the corps AO. He may delegate this authority to lower echelons according to the guidelines contained in Table 3-4.

**Table 3-4. Emplacement authority**

| System Characteristics | Emplacement Authority |
|---|---|
| Ground- or artillery-delivered, with SD time greater than 48 hours (long duration) | The corps commander may delegate emplacement authority to division level, who may further delegate it to brigade level. |
| Ground- or artillery-delivered, with SD time of 48 hours or less (short duration) | The corps commander may delegate emplacement authority to division level, who may further delegate it to brigade level, who may further delegate it to TF level. |
| Aircraft-delivered (Gator), regardless of SD time | Emplacement authority is normally at corps, theater, or army command level, depending on who has air-tasking authority. |
| Helicopter-delivered (Volcano), regardless of SD time | Emplacement authority is normally delegated no lower than the commander who has command authority over the emplacing aircraft. |
| MOPMS, when used strictly for a protective minefield | Emplacement authority is usually granted to the company, team, or base commander. Commanders at higher levels restrict MOPMS use only as necessary to support their operations. |

Based on how the commander wants to shape the battlefield, he must specifically delegate or withhold the authority to employ SCATMINE systems. The commander's guidance concerning SCATMINEs is found in the unit's OPORD/operation plan (OPLAN). Additional information is included in their engineer and fire-support annexes, if used.

Due to the complete control a commander has over the MOPMS, emplacement authority guidelines do not apply to the MOPMS. It is impractical for the corps or brigade commander to authorize every MOPMS protective minefield. Therefore, authority to emplace MOPMS minefields is specifically delegated. In general, units can emplace MOPMS protective minefields as required for their own self-defense and report them as they do any protective obstacle. Any MOPMS minefield used as part of an obstacle plan must be reported as a scatterable minefield.

## COORDINATION

Basic responsibilities of key commands, staff elements, and units are outlined in Table 3-5. The fire-support coordinator (FSCOORD) is involved in planning artillery-delivered (ADAM and RAAM) SCATMINEs, and the air liaison officer (ALO) is involved in planning air-delivered (Gator and Volcano) SCATMINEs. The engineer has primary responsibility for planning and employing SCATMINE systems. It is vital that coordination be conducted with all units and subunits that will be effected by the employment of SCATMINEs. A scatterable minefield warning (SCATMINWARN) will be sent to all effected units before the emplacement of the minefield (see Chapter 8 for more details).

**Table 3-5. Coordination responsibilities**

| Element | Responsibilities |
|---|---|
| G3/S3 with Engineer FSCOORD/ALO | Plan and coordinate the minefield location, size, composition, density, SD time, safety zone, and emplacement time |
| | Designate and brief the emplacing unit |
| | Incorporate the minefield and the safety zone into the obstacle plan |
| | Receive and forward the scatterable minefield report and record |
| | Disseminate information concerning the minefield in the SCATMINWARN to adjacent and subordinate units prior to laying |
| | Post operation maps with the minefield location, safety zone, and DTG of the SD time; and disseminate the SCATMINWARN 1 hour prior to initiation of the SD sequence |
| Emplacing Unit | Calculate the logistical requirements |
| | Calculate the safety zone |
| | Emplace the minefield |
| | Report the amount of ammunition expended |
| | Prepare and forward the scatterable minefield report and record to the authorizing commander via appropriate channels |
| Maneuver Units | Be aware of the calculated safety-zone boundary and advise subunits of its location |

## EMPLOYMENT AND EMPLACEMENT

Employment considerations and emplacement techniques and procedures differ for each type of SCATMINE and delivery system. This section discusses the characteristics of each delivery system and provides tactical considerations for the employment of each system on the battlefield. Techniques and procedures for emplacing minefields intended to disrupt, fix, turn, and block are also discussed; and they build on tactical-obstacle design principles discussed in Chapter 2.

## AREA-DENIAL ARTILLERY MUNITIONS AND REMOTE ANTIARMOR MINES

ADAMs and RAAMs are delivered by a 155-millimeter howitzer (Figure 3-3). There are no special modifications or adaptations necessary for the firing system. Mines are contained within a projectile and are dispensed while the projectile is in the air. The effective range for the M109 howitzer is 17,500 meters, and for the M198 howitzer, 17,740 meters.

**Figure 3-3. Emplacement of ADAMs and RAAMs**

The M692 (long-duration) and the M731 (short-duration) ADAM projectiles deliver AP mines with different SD times. Each ADAM round contains 36 mines. The M731/M731A1 round contains M72 AP mines with 4-hour SD times; the M692/M692A1 round contains M67 AP mines with 48-hour SD times. SD times are preset during the manufacturing process and cannot be changed.

The wedge-shaped ADAM is a bounding-fragmentation mine that deploys up to seven tension-activated trip wires 6 meters away from the mine. After ground impact, trip wires are released and the mine is fully armed. The ADAM contains a metal-jacketed sphere that is filled with 21 grams of composition A5 as its main charge. A liquid-explosive propelling charge positions itself at the bottom of the sphere after impact with the ground. When the mine is jarred or tilted, or when one of its trip wires receives a tension of at least 405 grams, the sphere propels upward 0.6 to 2.4 meters and detonates. The lethal casualty radius is between 6 and 10 meters.

The M741 (short-duration) and the M718 (long-duration) RAAMs are artillery-delivered AT mines. Each RAAM round contains nine mines. The M741/M741A1 round contains M70 AT mines with 4-hour SD times; the M718/M718A1 round contains M73 AT mines with 48-hour SD times. The SD times are preset during the manufacturing process and cannot be changed. The RAAM mine utilizes an SFF warhead, has a magnetic-influence fuse, weighs 1.7 kilograms, and has a small (12 centimeters in diameter by 6 centimeters in height) cylindrical shape.

The new model ADAM and RAAM mines (designated by an A1 suffix) have a 45-second arming time; the older models have a 2-minute arming time. The new model RAAM has a built-in feature that defeats magnetic, signature-duplicating breaching devices.

**Employment**

The ADAM and RAAM systems were designed to provide a flexible, rapid-response mining capability. These systems provide the maneuver commander with the capability to emplace mines directly on top of, in front of, or behind enemy forces. This is one of their greatest advantages. Their responsiveness allows the mission to be executed quickly and allows the commander to effectively influence a rapidly changing battlefield. They also allow the commander to emplace minefields while maintaining maximum standoff from the target. In short, their emplacement does not require committing any force (ground or air) forward. ADAM and RAAM systems may be used for the following purposes:

- Defense.
  - Develop targets for long-range AT weapons.
  - Close gaps and lanes in other obstacles.
  - Delay or disrupt attacking forces.
  - Deny the enemy unrestricted use of selected areas.
  - Disrupt movement and commitment of second-echelon forces.
  - Disrupt and harass enemy $C^2$, logistics (excluding medical), and staging areas.
  - Reinforce existing obstacles.
  - Disrupt or delay river crossings.
- Offense.
  - Supplement flank reconnaissance and security forces to protect flanks along AAs.
  - Suppress and disrupt enemy security elements once contact has been made.
  - Hinder the withdrawal of enemy forces.

— Hinder the ability of the enemy to reinforce the objective area.

The time and the number of rounds required to install effective ADAMs and RAAMs limit their use. Their range is limited to 17,500 or 17,740 meters, depending on which howitzer (M109 or M198, respectively) is used. Many of the deep-interdiction missions that support force-projection doctrine require a greater distance. Due to the large footprint created when the minefield is fired, many mines will scatter outside the planned minefield area. It is therefore necessary to plot the safety zone in order to prevent fratricide. The fire-support element (FSE) is responsible for plotting the safety zone, and the staff engineer should be familiar with the process and the expected results. The staff engineer ensures that the safety zone is plotted on the tactical command post (TCP)/TOC operation overlay.

### Emplacement

ADAM and RAAM mining missions are requested through normal artillery-support channels. Although the actual numbers vary based on the unit and the mission, a representative basic load for an artillery battalion consists of approximately 32 ADAM and 24 RAAM (short SD time) rounds per artillery piece. **NOTE: The rounds with long SD times are normally used for preplanned targets and are issued from an ammunition supply point (ASP) on a mission-by-mission basis.**

Once the proper authorization has been received to employ the mines, requests for ADAMs and RAAMs are processed in the same way as other requests for fire support, including targets of opportunity. Allocate enough time for processing the request and completing firing procedures. This ensures that the enemy has not moved out of the target area before execution. (FM 90-7 contains more information on this process.) The use of ADAMs and RAAMs for preplanned fires requires close coordination among the Assistant Chief of Staff, G3 (Operations and Plans) (G3)/Operations and Training Officer (US Army) (S3), the staff engineer, and FSE sections. Coordination should also be made with the S2 and the S3 during the development of the decision support template (DST) to identify the proper named areas of interest (NAIs), target areas of interest (TAIs), trigger points, and decision points.

There are two critical aspects when emplacing ADAM and RAAM minefields:

- Designing the minefield to achieve the required effect.
- Ensuring the technical correctness of resourcing and delivering the minefield.

The following discussion provides general guidance for designing the minefield to achieve the desired effect and for determining the safety zone to assess the impact on maneuver. Appendix H of FM 6-20-40 serves as the primary source for technically resourcing and delivering artillery-delivered minefields.

ADAM and RAAM minefields can be emplaced to achieve disrupt, fix, turn, and block effects based on the principles outlined in Chapter 2. The engineer is responsible for deciding the required location, the density, the size, the composition, and the duration of the minefield based on the tactical-obstacle plan and the obstacle restrictions of the higher unit. The engineer provides this information to the FSE. Table 3-6 provides guidance on the minefield density and size necessary to achieve the desired obstacle effect.

**Table 3-6. RAAM and ADAM minefield density and size**

| Obstacle Effect | RAAM | | ADAM | | Width (meters) | Depth (meters) |
|---|---|---|---|---|---|---|
| | Area[1] | Linear[2] | Area[1] | Linear[2] | | |
| Disrupt | 0.001 | 0.2 | 0.0005 | 0.1 | 200 | 200 |
| Turn | 0.002 | 0.8 | 0.001 | 0.4 | 400 | 400 |
| Fix | 0.002 | 0.4 | 0.0005 | 0.1 | 200 | 200 |
| Block | 0.004 | 0.6 | 0.002 | 0.8 | 400 | 400 |

[1]Area density = mines per square meter

[2]Linear density = mines per meter

The FSE determines all the technical aspects for delivering the minefield, such as the number of rounds required per aim point, the number of aim points required, the size of the safety zone, and the time required to emplace mines. There is a wide variety of factors involved in determining the number of rounds, the size of the safety zone, and the emplacement time. These factors are the range-to-target time, the battery-to-minefield angle, the high- or low-angle trajectory, and the method of firing (observer adjust or meteorological data plus velocity error [Met+VE] transfer). The FSE must tell the engineer whether the minefield mission is feasible. Feasibility is based on the number of rounds available, the scheme of indirect fires, and the availability of artillery tubes.

The engineer is primarily concerned with two technical aspects of delivery provided by the FSE—the safety zone and the emplacement time. The engineer uses the safety zone and the minefield duration to assess the impact of the minefield on the mobility requirements of the scheme of maneuver. The engineer depicts the safety zone on the obstacle overlay. He also uses the safety zone to identify requirements for minefield marking if the unit leaves or turns over the area before the SD time. The engineer and the FSE use the emplacement time to synchronize the delivery of the minefield with the tactical plan.

## GATOR

The Gator (Figure 3-4) has a longer range than any other SCATMINE system. It provides a means to rapidly emplace minefields anywhere that can be reached by tactical aircraft. The Gator is produced in two versions—the United States Air Force (USAF) CBU-89/B system that contains 94 mines (72 AT and 22 AP) per dispenser and the United States Navy (USN) CBU-78/B system that contains 60 mines (45 AT and 15 AP) per dispenser.

The mines used with the Gator are the BLU-91/B AT mine and the BLU-92/B AP mine. They are similar to the mines used with the Volcano system. The mines are capable of three field-selectable SD times (4 hours, 48 hours, and 15 days). Both types of mines are encased in a plastic, square-shaped protective

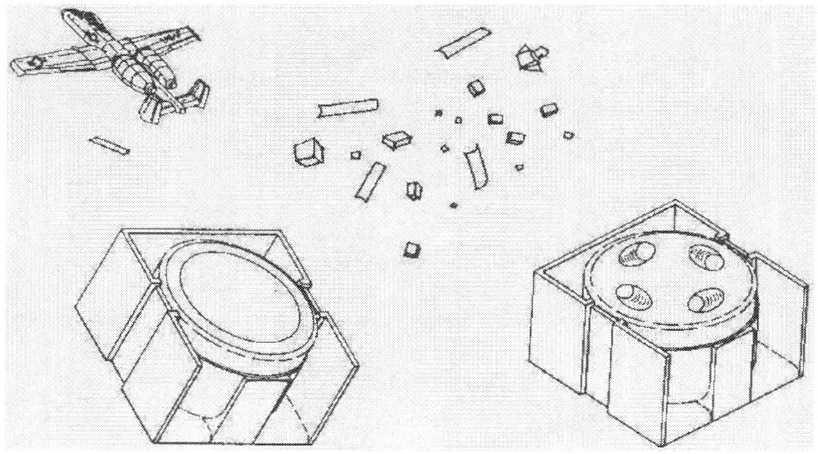

**Figure 3-4. Gator SCATMINE system**

casing that is designed to aid dispersion and lessen ground impact upon delivery.

The mines are contained inside tactical munition dispensers (TMDs) that are attached under the wings of high-performance, fixed-wing aircraft. The TMD is a USAF dispenser that was designed for common use with cluster munitions. The Gator is compatible with the USAF A-10, F-4, F-15, F-16, B-1, and B-52 aircraft and with the USN A-6, A-7, F-4, FA-18, and AV-8B aircraft.

The TMD is released in the air and allowed to fall free. Four linear charges along the edge of the TMD cut the outer casing, and the mines are aerodynamically dispersed. The maximum delivery speed is 800 knots at altitudes of 75 to 1,500 meters. The area of minefield coverage depends on the number of munitions carried, the aircraft speed and altitude, and the altitude where the fuse functions and opens the dispenser. The average area covered is approximately 200 by 650 meters.

## Employment

Gator missions are primarily used at long range to disrupt, fix, turn, or block enemy troop movement beyond the fire-support coordination line (FSCL). For use in interdiction missions beyond the FSCL, submit requests for Gator missions as early as possible to nominate targets for the theater air-tasking order. Gator munitions are well-suited for placing minefields on specific concentrations of forces (artillery, logistic, and $C^2$) that are out of range of conventional artillery.

While the Gator can provide close combat support, deep-interdiction mining is expected to be its primary mission. Gator minefields are normally employed in conjunction with other deep indirect-fire attacks, such as area of interest (AI), battlefield air interdiction (BAI), or joint air-attack team (JAAT). However, a Gator minefield may be employed in conjunction with close air support (CAS) and covered by close indirect- and direct-fire systems. Typical mining missions include—

- Isolating objectives.
- Countering ADA/artillery fires.
- Denying terrain.
- Disrupting and disorganizing support activities.
- Inflicting personnel and equipment losses.

The extended range of the Gator system, together with its speed and responsiveness, makes it one of the most influential weapon systems on the deep battlefield. The primary limitations of the Gator are the availability of high-performance aircraft to emplace the mines and the system's relative ineffectiveness on units in column. During any conflict, aircraft will be in high demand and will not always be immediately available for a Gator mission when required. Communications may also pose a problem because mission execution is a joint US Army-USAF operation.

The Gator is well suited to support contingency operations and amphibious landing operations in an immature theater when there is no danger to friendly forces or host-nation assets. Gator minefields are one of the light-force commander's few durable, long-range antiarmor weapons.

**Emplacement**

As an aircraft-delivered munition, the Gator is a corps asset. The Gator is a BAI mission and is controlled by the tactical air control center (TACC). Missions should be requested as early as possible (no later than 36 hours in advance) through fire-support channels to the corps FSE. As a mine system, Gator missions must be approved by corps. The corps FSE passes the mission to the theater or army air headquarters to be included on the theater air-tasking order for execution. In support of BAI or CAS, Gator sorties may be allocated down to battalion level, with final control exercised by the battalion ALO. Immediate Gator missions can also be requested directly from the maneuver unit's TACC. The same records and reports applicable to other SCATMINE systems are used with the Gator mine system. Close cooperation and coordination among the G3/S3, the staff engineer, and the ALO are required for planning and executing Gator missions.

As with artillery-delivered minefields, the engineer is primarily responsible for identifying the minefield location, size, duration, and density. Minefield density is varied by changing the orientation of the minefield with respect to the target AA. Figure 3-5 illustrates how minefield orientation is changed to achieve a fix or block effect. Normally, Gator is employed as a fix obstacle with a front of 650 meters. Emplacing a fix-obstacle group along a battalion AA (1,500 meters) requires two Gator sorties, each delivering one minefield. Each Gator minefield would have a front of 650 meters and a depth of 200 meters.

The minefields would be delivered at different locations so that the group covers the entire AA and affects the entire enemy battalion.

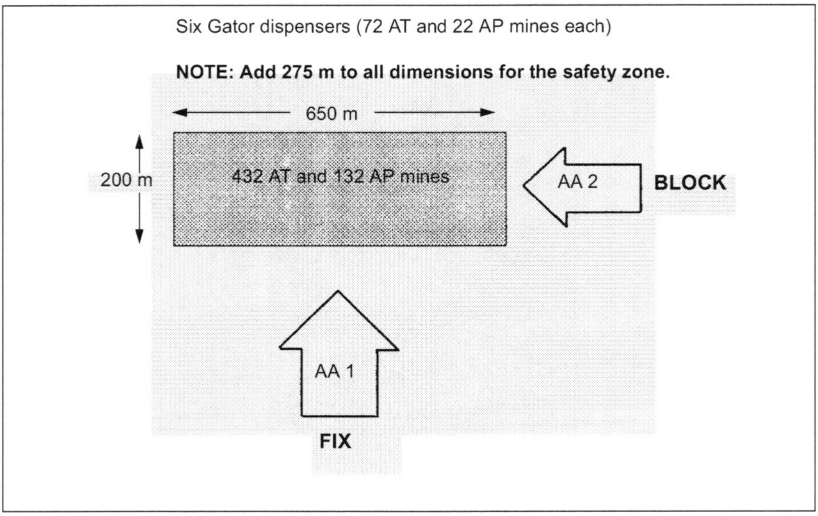


**Figure 3-5. Gator minefield**

## VOLCANO

The Volcano multiple-delivery mine system (Figure 3-6, page 3-18) can be dispensed from the air or on the ground. It can be mounted on any 5-ton truck, an M548 tracked cargo carrier, a heavy expanded mobility tactical truck (HEMTT), a palletized load system (PLS) flat rack, or a UH-60A Blackhawk helicopter. The Volcano uses modified Gator mines and consists of four components (Figure 3-7, page 3-18)—the mine canister, the dispenser, the dispenser control unit (DCU), and the mounting hardware (aircraft also require a jettison kit). The Volcano uses M87 and M87A1 mine canisters. The M87 mine canister is prepackaged with five AT mines, one AP mine, and a propulsion device inside a tube housing. The M87A1 mine canister is prepackaged with six AT mines and a propulsion device. The mixture of mines is fixed and cannot be altered. Mines are electrically connected with a web that functions as a lateral dispersion device as the mines exit the canister. Spring fingers mounted on each mine prevent it from coming to rest on its edge. All canisters are capable of dispensing mines with 4-hour, 48-hour, and 15-day SD times. The SD times are field-selectable prior to dispensing and do not require a change or modification to the mine canister. The arming time is 2 minutes 15 seconds for AT and AP mines. The reload time (not including movement time to the reload site) for an experienced four-man crew is approximately 20 minutes.

Figure 3-6. Volcano mine system

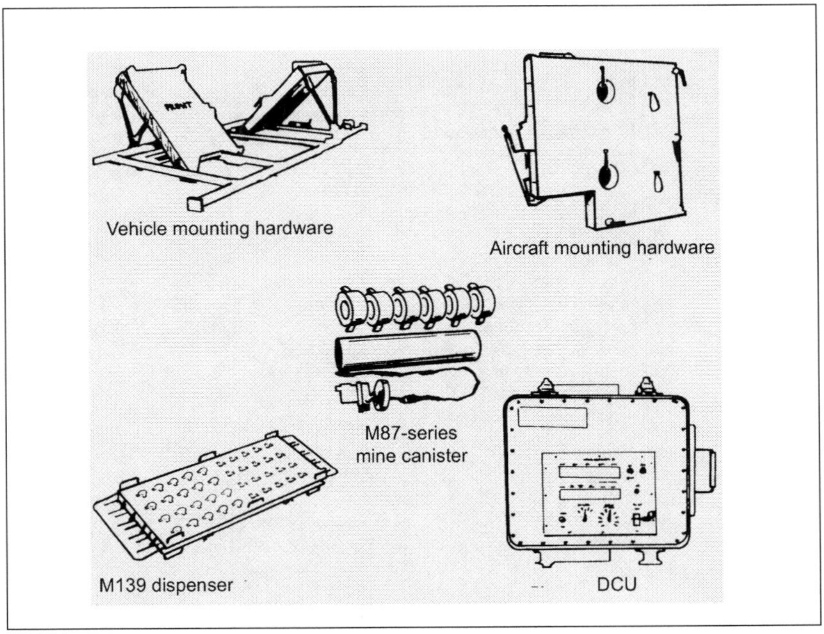


Figure 3-7. Volcano components

The dispenser consists of an electronic DCU and four launcher racks. Four racks can be mounted on a vehicle, and each rack can hold 40 M87-series mine canisters. The racks provide the structural strength and the mechanical support required for launch and provide the electrical interface between the mine canisters and the DCU. Mounting hardware secures the racks to the vehicle or the aircraft. Mounting hardware for the Blackhawk includes a jettison subassembly to propel the Volcano racks and canisters away from the aircraft in the event of an emergency.

The operator uses the DCU to control the dispensing operation electrically from within the carrier vehicle. The DCU provides controls for the arming sequence and the delivery speed and sets mine SD times. The DCU allows the operator to start and stop mine dispensing at anytime. A counter on the DCU indicates the number of remaining loaded canisters on each side of the carrier.

Mines are dispensed from their canisters by an explosive propelling charge. For ground vehicles, the mines are dispensed 25 to 60 meters from the vehicle at ground speeds of 8 to 90 kph. The average time to emplace one ground Volcano load (160 canisters) is 10 minutes.

### Employment

The primary mission of the Volcano is to provide US forces with the capability to emplace large minefields rapidly under varied conditions. The Volcano can be rapidly attached to air or ground vehicles. It is used to emplace tactical minefields; reinforce existing obstacles; close lanes, gaps, and defiles; protect flanks; and deny probable enemy air-defense sites. Volcano minefields are ideal for providing flank protection of advancing forces and for operating in concert with air and ground cavalry units on flank guard or screen missions.

The air Volcano is the fastest method for emplacing large tactical minefields. When employed by combat aviation elements in support of maneuver units, close coordination between aviation and ground units assures that Volcano-dispensed mines are emplaced accurately and quickly. Although mine placement is not as precise as it is with ground systems, air Volcano minefields can be placed accurately enough to avoid the danger inherent in minefields delivered by artillery or jet aircraft. Air Volcano minefields can be emplaced in friendly and enemy territory. They should not be planned in areas of enemy observation and fire because the helicopter is extremely vulnerable while flying at the steady altitude, the speed, and the path required to emplace the minefield. The air Volcano is the best form of an obstacle reserve because a minefield can be emplaced in minutes.

The ground Volcano is designed to emplace large minefields in depth. It is normally employed by combat engineer units. These mounted dispensers are primarily used to emplace tactical minefields oriented on enemy forces in support of maneuver operations and friendly AT fires. The system is vulnerable to direct and indirect fires, so it must be protected when close to the FLOT. It is ideal for use as an obstacle reserve, employed when the enemy reaches a decision point that indicates future movement. Obstacles can then be emplaced in depth on the avenues the enemy is using, leaving other avenues open for friendly movement.

**Emplacement**

The principles and procedures of Volcano emplacement are significantly different for air- and ground-delivery systems. This section outlines the use of the ground Volcano system to emplace disrupt, fix, turn, and block minefields. The air Volcano system is discussed in detail in Appendix D. Both air and ground Volcano systems are capable of emplacing nonstandard minefields. However, the emplacement norms below streamline identifying resource requirements and conducting emplacement drills.

Air and ground Volcano systems emplace a minefield with an average AT linear density of 0.72 mine per meter and an AP linear density of 0.14 mine per meter. These densities may vary slightly since some mines will fail the arming sequence and self-destruct 2 to 4 minutes after dispensing. Additionally, some mines may not orient correctly, will not deliver their full mine effect, and will not produce a K-Kill. The probability of failing the arming sequence and misorienting is relatively small and does not appreciably degrade the minefield's lethality. For tracked vehicles, the AT density yields more than 80 percent probability of encounter. Volcano AT mines do not have AHDs but are highly sensitive to any movement once they are armed. Any attempt to remove the mines will likely result in detonation.

The basic site layout is extremely important, and it is the same for air and ground Volcano minefields. The limits of Volcano minefields are marked before emplacement when the situation (planned targets within the main battle area [MBA] of a defensive operation) allows it. The minefield is not premarked when the situation (offensive operations or situational obstacles) does not allow it. If the mines have not self-destructed, the minefield is marked before the unit leaves the area or turns it over to an adjacent unit. Minefield marking must include the safety zone, which is 40 meters from the start and end points and 80 meters to the left and right of the centerline. The start and end points of the strip centerline are marked based on the minefield front and the number of strips. For a ground Volcano minefield, guide markers are emplaced along the path of the centerline but are offset left to allow the host vehicle to remain on the centerline. When using a ground-delivery system, minefield marking must leave a gap along each centerline for vehicle entrance and exit. The number of guide markers used depends on the terrain and the visibility. Guide markers are not required for an air Volcano minefield because the pilot will use the start and end points of the centerline as reference points.

Figure 3-8 illustrates the emplacement pattern for standard disrupt and fix minefields using the ground or air Volcano. Disrupt and fix minefields use only one centerline to give a minefield depth of 120 meters (ground) or 140 meters (air), not including the safety zone. The strip centerline is 277 meters (ground) or 278 meters (air) long. The host vehicle moves toward the start point, achieving and maintaining the ground or air speed selected on the DCU. The operator depresses the launch switch on the DCU when the vehicle passes the start marker, and he stops dispensing mines when the vehicle passes the end marker. The operator dispenses 40 canisters (20 on each side) along the centerline. One full load of ground or air Volcano emplaces four disrupt or fix minefields. For ground emplacement, the vehicle moves out of the minefield, marks the exit, and waits a minimum of 4 minutes before approaching the minefield. This delay allows faulty mines to self-destruct.

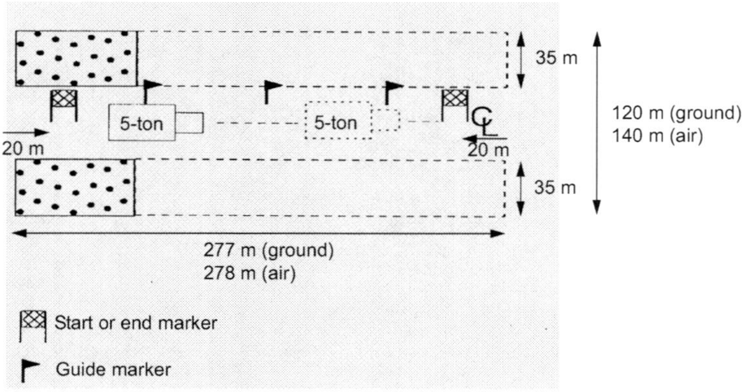


**Figure 3-8. Volcano disrupt and fix minefields**

Turn and block minefields (Figure 3-9, page 3-22) are emplaced using the same basic procedures as those used for disrupt and fix minefields. However, turn and block minefields use two strip centerlines along a front of 555 meters (ground) or 557 meters (air). During site layout, centerlines are separated by at least 320 meters for both ground and air delivery. This gives a total minefield depth of 440 meters (ground) or 460 meters (air). The operator dispenses 80 canisters along each centerline (40 on each side); therefore, turn and block minefields require a total Volcano load of 160 canisters. One full load of ground or air Volcano emplaces one turn or block minefield. Wherever possible, two ground Volcanoes are employed simultaneously on turn and block minefields. When only one ground delivery system is used, the crew must wait 4 minutes after dispensing the first strip before dispensing the second strip. This allows mines that fail the arming sequence to self-destruct. For air delivery, two sorties are also optimal; but demands for sorties elsewhere in the division may preclude the simultaneous employment of two Blackhawks.

## Modular Pack Mine System

The MOPMS (Figure 3-10, page 3-22) is a man-portable, 162-pound, box-shaped mine dispenser that can be emplaced anytime before dispensing mines. The dispenser contains 21 mines (17 AT and 4 AP). The mines have leaf springs along their outer circumference that are designed to push the mines into proper orientation if they land on their side.

Each dispenser contains seven tubes; three mines are located in each tube. When dispensed, an explosive propelling charge at the bottom of each tube expels mines through the container roof. Mines are propelled 35 meters from the container in a 180-degree semicircle (Figure 3-11, page 3-23). The resulting density is 0.01 mine per square meter. The safety zone around one container is 55 meters to the front and sides and 20 meters to the rear.

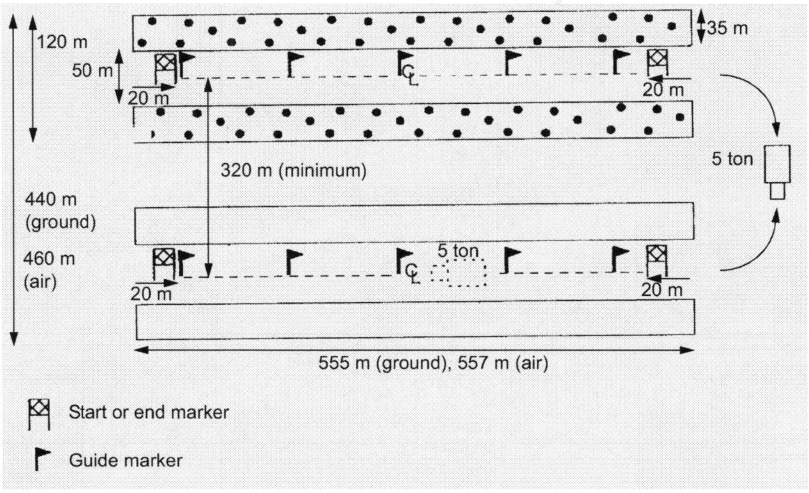


Figure 3-9. Volcano turn and block minefields

Figure 3-10. MOPMS

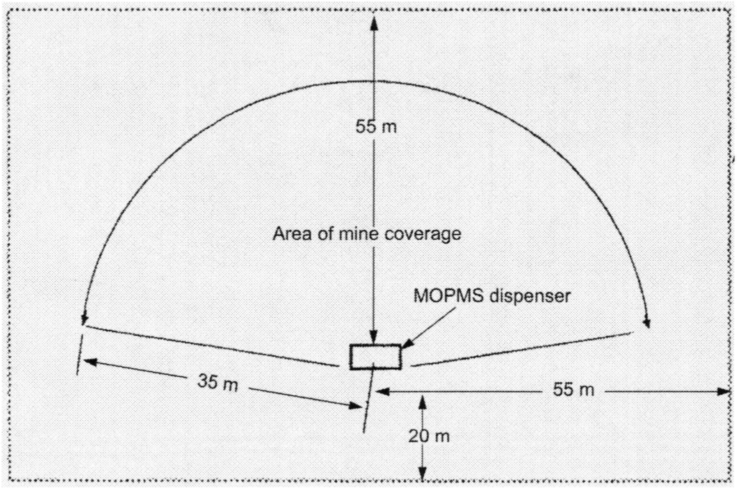


**Figure 3-11. MOPMS emplacement and safety zone**

Mines are dispensed on command using an M71 remote-control unit (RCU) or an electronic initiating device. Once mines are dispensed, they cannot be recovered or reused. If mines are not dispensed, the container may be disarmed and recovered for later use.

The RCU can recycle the 4-hour SD time of the mines three times, for a total duration of approximately 13 hours. Mines with a 4-hour SD time will begin to self-destruct at 3 hours and 12 minutes. All active mines must be recycled within 3 hours of the initial launch or last recycle. This feature makes it possible to keep the minefield emplaced for longer periods if necessary. The RCU can also self-destruct mines on command, allowing a unit to counterattack or withdraw through the minefield, as necessary, rather than waiting until the SD time has expired. The RCU can control up to 15 MOPMS containers or groups of MOPMS containers from a distance of 300 to 1,000 meters via separate pulse-coded frequencies. Coded frequencies defeat threat electronic countermeasures directed against the system.

If the M71 RCU is unavailable, a direct wire link is used in conjunction with an M32, M34, or M57 blasting machine. By using the M32 10-cap blasting machine, one MOPMS dispenser can be detonated at a maximum range of 1,000 meters. The M34 50-cap blasting machine can detonate one MOPMS at a maximum range of 3,000 meters. (Due to internal resistance, the maximum range is decreased by 400 meters for each additional MOPMS connected in series.) The M57 claymore-type FD can fire only one MOPMS at a maximum range of 100 meters. When controlled by direct wire, MOPMS dispensers cannot be command-detonated, and the SD time cannot be recycled.

**WARNING**
**The MOPMS dispenser has seven launch tubes. If all seven tubes are not visible after deployment, mines are jammed in the tube(s). In this event, clear the area and notify EOD. The dispenser is considered to be UXO; do not attempt to recover the dispenser.**

## Employment

The MOPMS provides a self-contained, on-call minefield emplacement capability for all forces. It can be command-detonated, reused (if mines are not dispensed), and directly emplaced to provide complete and certain coverage of small or critical targets. The ability to command-detonate mines or extend their SD time provides an added flexibility not currently available with other SCATMINE systems. With its unique characteristics, the MOPMS is ideally suited for the following minefield missions:

- Emplacing hasty protective minefields.
- Emplacing deliberate protective minefields (cases emplaced, but mines not dispensed).
- Emplacing nuisance minefields (trails, crossing sites, landing zones [LZs], drop zones [DZs], and road junctions).
- Emplacing tactical disrupt and fix minefields.
- Closing gaps and lanes in existing minefields.
- Temporarily closing counterattack routes.
- Supporting ambushes.
- Supporting military operations in built-up areas (MOBA) operations.

When the MOPMS is used to close lanes, the container is positioned and dispensed by personnel in an overwatch position from a safe standoff. The MOPMS is ideally suited for creating a small disrupt obstacle in support of engineers executing a reserved demolition target. Engineers prepare the reserved target for demolition and emplace several MOPMS units on the enemy side, just out of target range. When the last forward element passes through the target, the firing party detonates the charges. If something goes wrong or the firing party needs more time, MOPMS mines can be dispensed to disrupt the enemy before it reaches the target.

The MOPMS provides light and special forces with a versatile, compact system for emplacing nuisance minefields. It can be used in low-, mid-, and high-intensity conflicts and in a variety of environments. The MOPMS cannot be transported long distances by hand because of its weight, so its use is limited.

## Emplacement

MOPMS dispensers are issued as standard Class V munitions and are drawn from an ASP on a mission-by-mission basis. RCUs are organizational issues of equipment and are assigned to engineer and combat arms units. Due to the weight of the system, it will normally be transported by vehicle, as close as possible to the emplacement site, where it can easily be hand-emplaced by four soldiers using the four foldout carrying handles.

To ensure that the minefield will be dispensed in the proper location, the container should be carefully sited by the noncommissioned officer in charge (NCOIC). Several containers can be used together to provide a greater area of coverage or a higher mine density. If mines are not dispensed immediately, containers should be camouflaged and, if possible, buried. When placed in sand or snow, brace the containers to prevent them from moving during mine dispensing. Designate a firing point that gives the operator clear observation of the area to be mined. Firing systems must be inspected according to MOPMS operating instructions. If mines are dispensed immediately, remove empty containers to avoid revealing the minefield location.

The MOPMS can be employed to emplace disrupt and fix tactical minefields. Emplacement procedures are the same as for protective minefields above. However, MOPMS containers are arranged in a specific pattern to achieve the necessary depth, front, and density. Once the minefield is marked (to include the safety zone), MOPMS containers are arranged as shown in Figure 3-12 for a disrupt minefield. The safety zone is 55 meters from the front and sides and 20 meters from the rear of the container. The disrupt minefield uses four MOPMS containers that are spaced 70 meters apart to give a minefield front of 280 meters. Other MOPMS containers are offset from the baseline by 35 meters to give the minefield a depth of 70 meters. All containers are fired using the same RCU or FD.

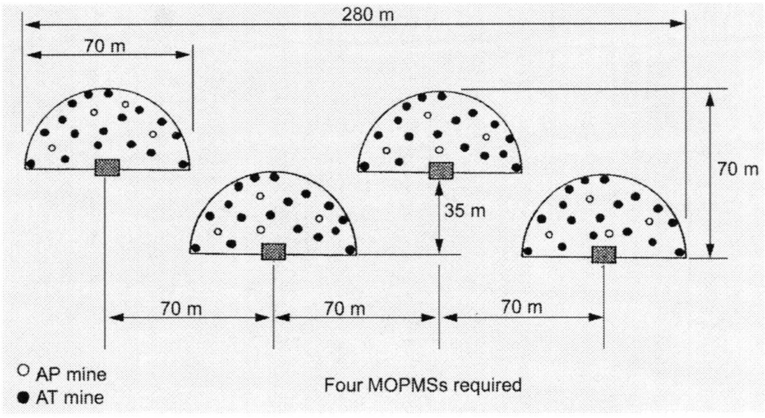


**Figure 3-12. MOPMS in a disrupt minefield**

Figure 3-13 illustrates the arrangement of MOPMS containers for a fix minefield. The basic layout is the same as the disrupt minefield; however, the fix minefield has one additional MOPMS that is placed 70 meters forward of the baseline to act as an IOE. This gives the same 280-meter minefield front but increases the minefield depth to 115 meters.

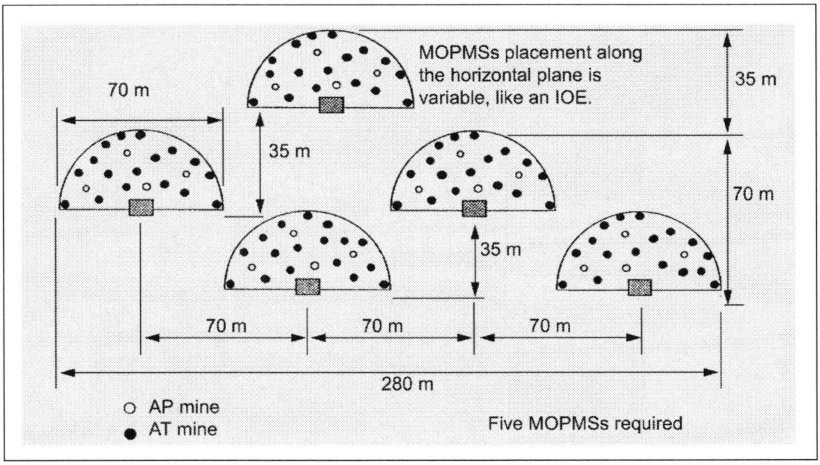


**Figure 3-13. MOPMS in a fix minefield**

MOPMS can be used to construct turn and block tactical minefields using the principles outlined in Chapter 2; however, turn and block minefields require more containers than are normally available to a unit.

## MARKING

The maneuver unit that is responsible for the area of ground in which the minefield is emplaced is also responsible for marking the minefield. This normally requires direct coordination between elements of the maneuver command (usually the engineer) and the delivering/emplacing unit. However, it is unrealistic to expect units to mark artillery-delivered ADAM and RAAM, air-delivered Volcano, or Gator minefields. For this reason, units operating in the vicinity of these minefields must know calculated safety zones and use extreme caution. Scatterable minefields are marked to protect friendly troops as shown in Table 3-7. Ground Volcano minefields are marked according to the guidelines below.

**Table 3-7. Marking scatterable minefields**

| Minefield Location | Marking |
|---|---|
| Enemy forward area | Unmarked |
| Friendly forward area | Sides and rear marked |
| Friendly rear area | All sides marked |

## SAFETY ZONES

A safety zone is an area where a stray or outlying mine has a chance of landing and laying to rest. The commander must prevent friendly forces from maneuvering into the safety zone during the minefield's life cycle. Depending on its specific location on the battlefield, the safety zone may be marked with a fence.

The safety zone around a Volcano minefield is shown in Figure 3-14.

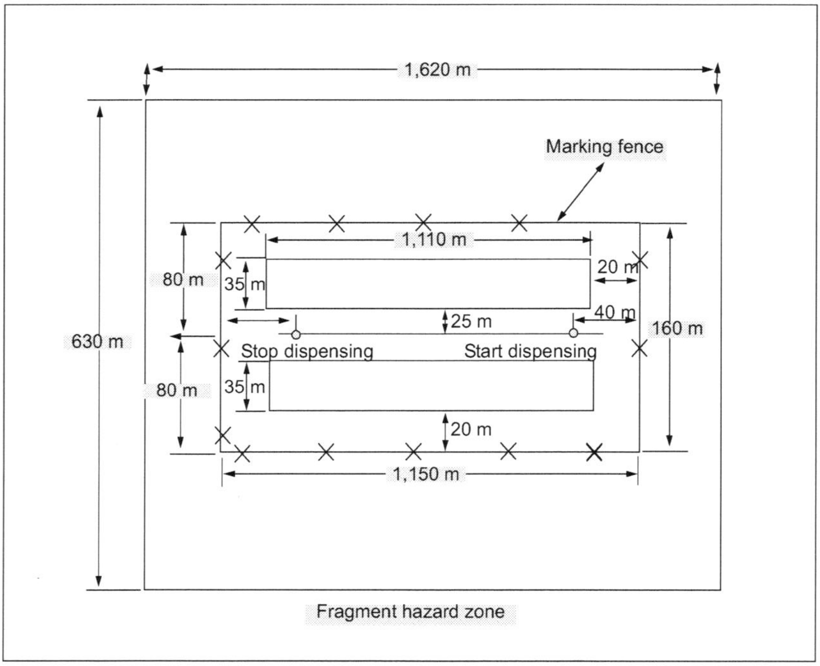


**Figure 3-14. Ground Volcano minefield**

## FRAGMENT HAZARD ZONES

If an AT mine that is oriented on its side self-destructs, the EFP can theoretically travel 640 meters. This is the maximum fragment hazard zone; however, the chances of being struck are negligible at this distance. Tests indicate that the acceptable risk distance is 235 meters from the outer edges of the minefield's safety zone. This fragment hazard zone is also associated with the Gator and MOPMS AT mines. When the MOPMS is used for protective minefield missions, commanders must be made aware of the fragment hazard zone.

Use Table 3-8 to determine safety zones and fragment hazard zones.

**Table 3-8. Safety and fragment hazard zones**

| System | Safety Zone | Fragment Hazard Zone |
|---|---|---|
| ADAM/RAAM | 500 to 1,500 meters from aim point(s) (depends on delivery factors) | 235 meters from the outside dimensions of the safety zone |
| Gator | 925 x 475 meters from aim point(s) | 1,395 x 945 meters from aim point(s) |
| Ground Volcano | 1,150 x 160 meters | 235 meters from start and stop points and the centerline |
| Air Volcano | 1,315 x 200 meters | 235 meters from start and stop points and the centerline |
| MOPMS | See page 3-28 for specific placement. | 235 meters from the outside dimensions of the safety zone |

## FENCING

Fencing for ground Volcano minefields (Figure 3-14, page 3-27) is emplaced 80 meters beyond the centerline of the minefield and 40 meters from the start and stop points. Fencing should be no closer than 20 meters from the nearest mine.

Air Volcano minefields are not normally marked by fencing. However, if air Volcano minefields are emplaced in friendly areas, they are marked with fencing to protect friendly personnel. Fencing is installed before delivering an air Volcano, and it is located 100 meters from the centerline of the minefield and 100 meters from the start and end points. Appendix D contains detailed information pertaining to air Volcano minefields.

Chapter 4

# Special-Purpose Munitions

Special-purpose munitions are hand-emplaced and used to create an expedient obstacle, enhance existing ones, and attack specific types of targets. The commander can employ these munitions to support his scheme of maneuver, to mass firepower, and to disrupt or destroy enemy forces in depth. Special considerations must be made in the planning process to effectively employ special-purpose munitions.

## M18A1 CLAYMORE

The M18A1 claymore munition (Figure 4-2) is a fragmentation munition that contains 700 steel balls and 682 grams of composition C4 explosive. It weighs 1.6 kilograms and can be detonated by command **(Korea Only: or trip wire)**. It is activated by electric or nonelectric blasting caps that are inserted into the detonator well. The claymore projects a fan-shaped pattern of steel balls in a 60-degree horizontal arc, at a maximum height of 2 meters, and covers a casualty radius of 100 meters. The forward danger radius for friendly forces is 250 meters. The backblast area is unsafe in unprotected areas 16 meters to the rear and sides of the munition. Friendly personnel within 100 meters to the rear and sides of the munition should be in a covered position to be safe from secondary missiles. If the M18A1 is employed in a minefield for 72 hours or more, the minefield must be fenced on all sides.

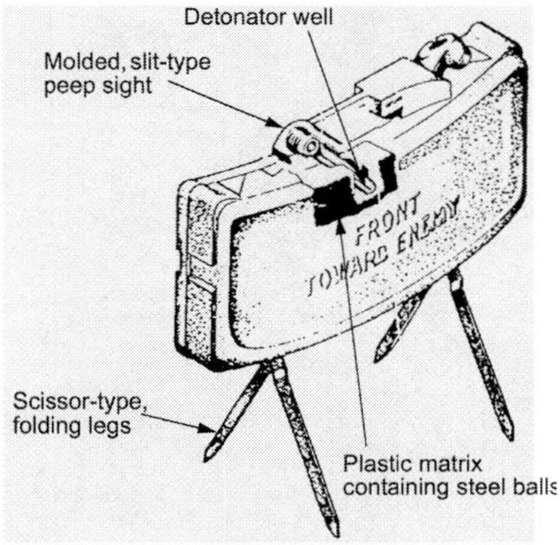


**Figure 4-2. M18A1 claymore**

When employing the M18A1 claymore with other munitions or mines, separate the munitions by the following minimum distances:

- 50 meters in front of or behind other M18A1s.
- 3 meters between M18A1s that are placed side by side.
- 10 meters from AT or fragmentation AP munitions.
- 2 meters from blast AP munitions.

# SELECTABLE LIGHTWEIGHT ATTACK MUNITION

The selectable lightweight attack munition (SLAM) (Figure 4-3) is a multipurpose munition with an antitamper feature. The SLAM is compact and weighs only 1 kilogram, so it is easily portable. The SLAM is intended for use against APCs, parked aircraft, wheeled or tracked vehicles, stationary targets (such as electrical transformers), small fuel-storage tanks (less than 10,000-gallon), and ammunition storage facilities. The EFP warhead can penetrate 40 millimeters of homogeneous steel.

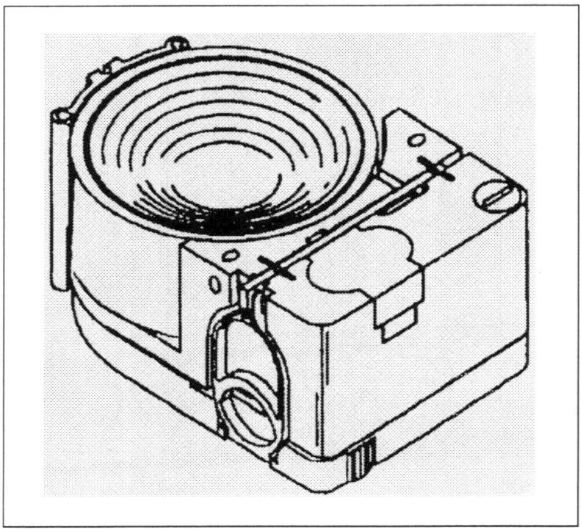

**Figure 4-3. SLAM**

The SLAM has two models—one is self-neutralizing (M2) and the other is self-destructing (M4):

- The M2 is solid green and has no labels, brands, or other distinguishing marks. This device is used by SOF and is not available to other units.
- The M4 is green with a black warhead (EFP) face. This device is normally used by units designated as light, airborne, air assault, crisis response, and rapid deployment.

See Appendix B for a description of major SLAM components.

## OPERATING MODES

The SLAM has four possible employment methods—bottom attack, side attack, timed demolition, and command detonation.

### Bottom Attack

The SLAM has a built-in magnetic sensor, so it can be used as a magnetic-influenced munition against trucks and light armored vehicles (Figure 4-4, page 4-4). It can be concealed along trails and roads where target vehicles operate and can be camouflaged with dry leaves, grass, and so forth without affecting EFP performance. Mud, gravel, water, and other debris that fill the EFP cup have minimal impact on EFP formation and effectiveness as long as the debris does not extend beyond the depth of the EFP cup. The magnetic sensor is designed to trigger detonation when it senses a vehicle's overpass. For the EFP to form properly, it needs a minimum of 13 centimeters from the point of emplacement to the target. The bottom-attack mode is active when the selector switch is set to *4, 10, or 24 HOURS* and the passive infrared sensor (PIRS) cover is in place. The SLAM will self-destruct (M4) or self-neutralize (M2) if the selected time expires before the SLAM is detonated by a vehicle.

**Figure 4-4. SLAM in bottom-attack mode**

### Side Attack

The SLAM is equipped with a PIRS that was specifically developed for the side-attack mode (Figure 4-5). The PIRS detects trucks and light armored vehicles by sensing the change in background temperature when vehicles cross in front of the PIRS port. The PIRS is directional and aligned with the EFP when the device is aimed. The side-attack mode is active when the SLAM selector switch is set to *4, 10, or 24 HOURS* and the PIRS cover is removed to expose the PIRS. The SLAM will self-destruct (M4) or self-neutralize (M2) if the selected time expires before it is detonated by a vehicle.

### Timed Demolition

The SLAM's built-in timer will trigger detonation at the end of a selected time (Figure 4-6). The timed-demolition mode is active when the SLAM selector switch is set to *15, 30, 45, or 60 MINUTES*. In this mode, the magnetic sensor and the PIRS are inoperable, and the SLAM will detonate after the selected time has expired.

### Command Detonation

This mode provides manual warhead initiation using standard military blasting caps and a priming adapter (Figure 4-7). The command-detonation capability bypasses the SLAM's fuse and safing and arming (S&A) assembly.

Figure 4-5. SLAM in side-attack mode

Figure 4-6. SLAM in timed-demolition mode

Figure 4-7. SLAM in command-detonation mode

### Antitamper Feature

The SLAM has an antitamper feature that is only active in the bottom- and side-attack modes. The SLAM will detonate when an attempt is made to change the selector switch's position after arming.

## M93 HORNET

The M93 Hornet (Figure 4-8) is an AT/antivehicular off-route munition made of lightweight material (35 pounds) that one person can carry and employ. The Hornet is a nonrecoverable munition that is capable of destroying vehicles by using sound and motion detection methods. It will automatically search, detect, recognize, and engage moving targets by using top attack at a standoff distance up to 100 meters from the munition. It is employed by combat engineers, rangers, and SOF.

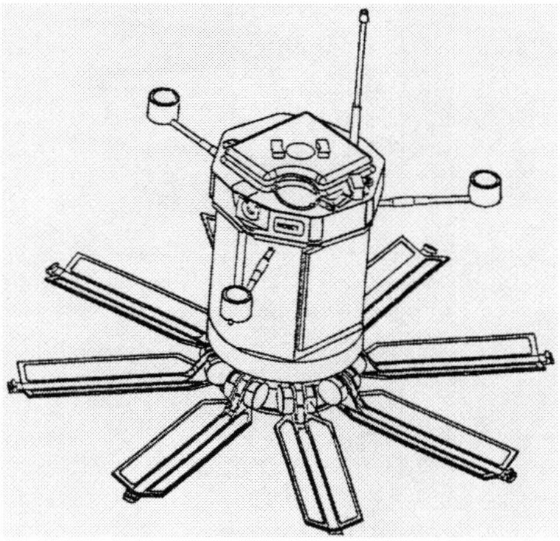

**Figure 4-8. M93 Hornet**

The RCU is a hand-held encoding unit that interfaces with the Hornet when the remote mode is selected at the time of employment. After encoding, the RCU can be used to arm the Hornet, reset its SD times, or destroy it. The maximum operating distance for the RCU is 2 kilometers.

High winds, heavy rain, snow, ice, extreme cold, and extreme heat reduce the Hornet's ability to detect targets at maximum range. Radio-frequency (RF) jamming devices (such as the hand-emplaced, expandable jammer [HEXJAM]), limit the Hornet's communication capabilities if they are placed in the munition field, but they will not affect the Hornet's ability to engage targets and will not damage the system. RF jamming devices affect the remote arming of current Hornet systems using the MOPMS RCU, and they will affect future Hornet's two-way communications capability with the Centurion remote control device.

See Appendix B for a description of Hornet components.

## EMPLOYMENT CONSIDERATIONS

The Hornet's active battery pack is inserted during prearming and has an estimated life of 4 hours. The active battery pack powers the munition from the time it is inserted until the end of the safe-separation time, when the built-in reserve battery is activated. To prevent munitions from becoming duds, do not prearm them too early. Allow adequate time for travelling to the obstacle site, emplacing mines, throwing arming switches, and expiration of safe-separation times.

Once the Hornet is armed and the self-test is performed, the munition will remain active until its SD time expires or until it is encountered. The SD time (4 hours, 48 hours, 5 days, 15 days, or 30 days) is determined by the mission and the commander's intent. The munition will self-detonate after the SD time has expired.

Hornet munitions have an employed life of 60 days in the prearmed mode (remote arming) and 30 days in the armed mode. If the temperature exceeds 100°F, the employed life drops to 15 days in the prearmed mode and 30 days in the armed mode.

## EMPLOYMENT ROLES

Combat engineers or maneuver forces under engineer supervision emplace Hornets in close operations; SOF or rangers emplace Hornets in deep operations. Hornets will be employed throughout the entire depth of the battle space to support Army operations.

### Close Operations

In close operations, the Hornet can be—

- Used to fix the enemy and weaken it along its AA.
- Emplaced as an offensive-support weapon system because of its quick emplacement time and wide attack area.
- Employed rapidly along exposed flanks during a maneuver as a situational obstacle to disrupt the enemy's counterattacks.

- Used as a stand-alone tactical obstacle or as a reinforcement to conventional obstacles.
- Used to disrupt and delay the enemy, allowing long-range weapons to engage more effectively.

### Deep Operations

In deep operations, the Hornet can be—

- Emplaced along key routes in gauntlet obstacles to disrupt and delay threat second-echelon forces, resupply operations, and key lines of communication (LOC).
- Used at $C^2$ and logistics sites to disrupt enemy operations.

### Rear Operations

In rear operations, the Hornet can be emplaced (unarmed) along key routes in preparation for possible retrograde operations.

### Early-Entry Operations

In early-entry operations, the Hornet can be—

- Used as an additional antiarmor weapon to supplement light forces.
- Used along high speed AAs in gauntlet obstacles to buy time and space.

## Tactical Emplacement

There are four basic emplacement scenarios for the Hornet.

### Conventional Minefield Reinforcement

The Hornet can be used to reinforce a conventional turn, block, or fix minefield (Figure 4-9).

Platoon engineers emplace the conventional minefield first, and then they traverse the safe lane that is perpendicular to the minefield. The Hornets are employed in two staggered rows, spaced 100 meters apart, 50 to 100 meters from the front edge (on the enemy side) of the conventional minefield. It is also recommended that a row of Hornets be placed 50 meters behind the minefield to reduce the enemy's breaching capability. (This row will be emplaced after the safe lane is closed.) The emplacing vehicles work toward the safe lane.

Two squads employ Hornets in two rows of ten each. One or more soldiers provide security. Under the supervision of a noncommissioned officer (NCO), four soldiers in each squad vehicle start prearming the Hornets, if necessary. They—

- Rotate the handle.
- Remove the cover.
- Insert the active battery pack and verify functionality via a solid status light.
- Reinstall the active battery-pack cover.

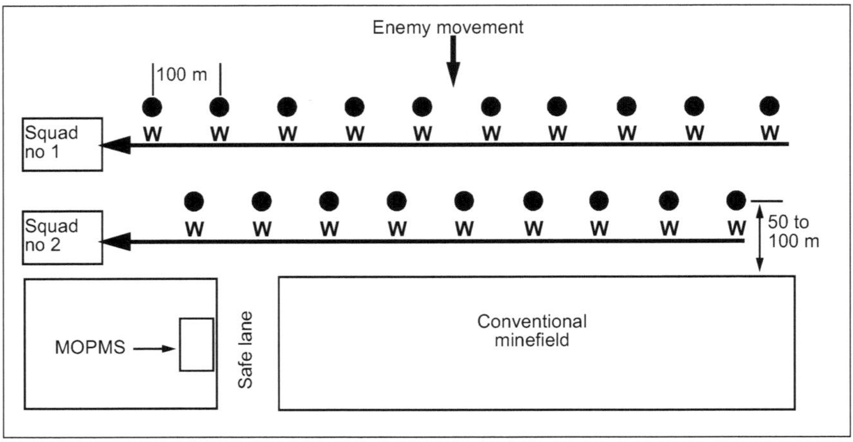


**Figure 4-9. Hornet reinforcing a conventional minefield**

- Select the SD time.
- Encode the Hornet with the M71 RCU and verify functionality via a flashing status light.
- Reinstall the cover.

Each emplacement vehicle moves to the first Hornet emplacement site in each row. The emplacing soldier and the arming soldier dismount. The emplacing soldier is handed a Hornet from the vehicle. He emplaces the Hornet at the designated site and returns to the vehicle.

The arming soldier rotates the handle on the Hornet, removes the cover and the safety and handling (S&H) band, rotates the SD switch to *U*, and pushes the arm switch to *ARM*. He then returns to the vehicle, taking the cover and the S&H band with him. The vehicle travels to the next Hornet emplacement site.

After all the Hornets in the two leading rows have been emplaced and armed, the emplacing vehicles exit through the safe lane and usually secure it with a MOPMS. The emplacement vehicles must be at least 475 meters (safe standoff distance) from the nearest Hornet within 30 minutes. Before sending a remote arming signal, vehicles must wait at least 36 minutes after the arming switch is thrown on the last Hornet emplaced. If a rear row is required, it is emplaced at this time. The Hornets are then remotely armed with the M71 RCU, when required. They are now capable of covering the minefield by fire and engaging threat tracked vehicles.

## Scatterable Minefield Reinforcement

The Hornet can be used to reinforce a Volcano or MOPMS turn, block, or fix minefield (Figure 4-10). Hornet munitions are emplaced, using the same procedures as above, **before** the Volcano or MOPMS minefield is emplaced.

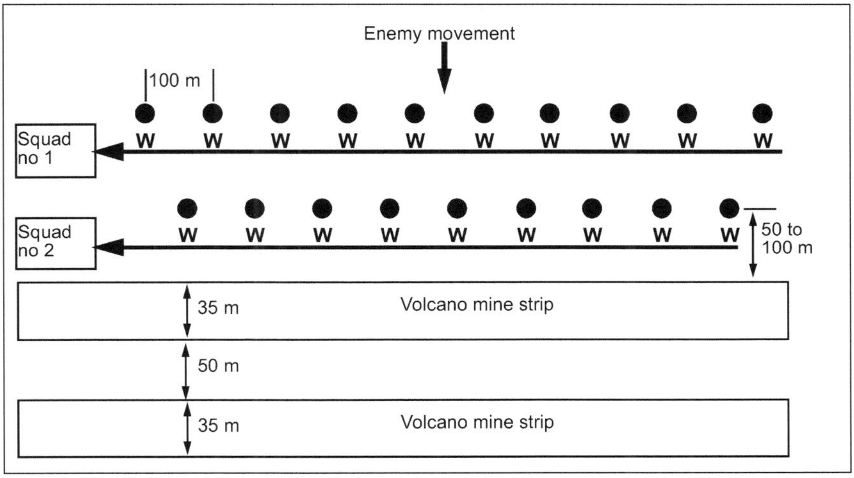


**Figure 4-10. Hornet reinforcing a Volcano minefield**

To ensure that the Volcano dispensing vehicle has sufficient time to reach the safe standoff distance (475 meters), Volcano dispensing should start no later than 30 minutes (minus the Volcano dispensing time) after the first Hornet is armed. This allows Hornet emplacing squads to be finished or nearly finished before the Volcano dispenser begins emplacing the minefield.

## Area-Disruption Obstacle

When the X-pattern is employed, the Hornet is very effective as a disrupting obstacle (Figure 4-11). An area-disruption obstacle is employed to disrupt the enemy's approach prior to the start of the direct-fire battle. It causes disruption and attrition of the advancing threat force and encourages follow-on forces to seek an alternate route. Therefore, multiple area-disruption obstacles will typically be employed to adequately cover the cross-country AA. This requires coordinated action among multiple squads.

An engineer platoon emplaces a Hornet area-disruption obstacle. The obstacle typically consists of 20 Hornets (five clusters of four Hornets each) employed in an X-pattern over a 1- by 1-kilometer area. Individual Hornets are emplaced 100 meters apart. Emplacing this obstacle must be done as a dispense-and-roll operation to ensure that the emplacing vehicles can reach the safe standoff distance (475 meters) from any armed Hornets.

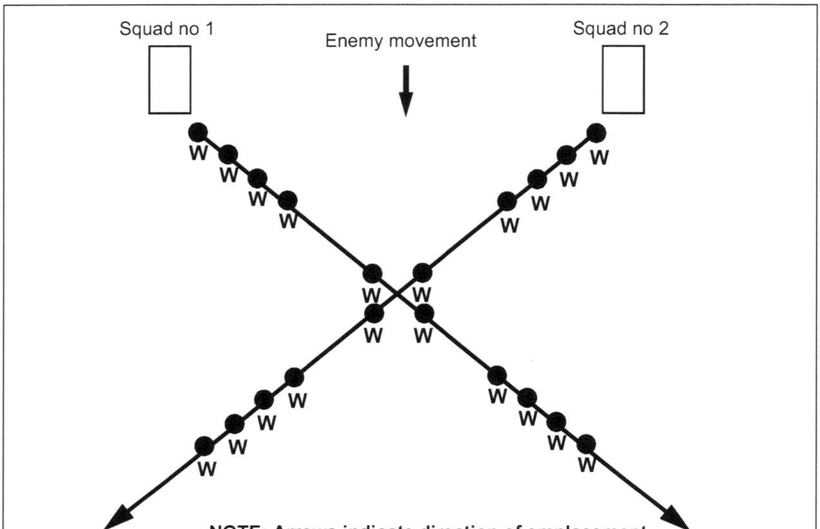


**Figure 4-11. Hornet area-disruption obstacle**

Area-disruption obstacles are normally armed by remote, but they can be manually armed under the following conditions:

- METT-TC requires rapid emplacement and arming.
- Terrain reconnaissance determines that there are no major impediments (rough terrain, vegetation) to maneuver.
- Emplacement is done during daylight hours (mission-oriented protective posture [MOPP] level 0 only).

Hornets are prearmed the same as above. Two squads lay the Hornets in unison, starting with the two emplacement sites closest to the enemy. Each squad drives in a straight line, crossing paths at the middle of the X, and emplaces ten Hornets.

A soldier in the back of each emplacing vehicle throws the arming switch and sets the Hornet down or drops it off (base down) the back of the vehicle. After all the Hornet clusters are emplaced, squad vehicles quickly travel to the 475-meter safe standoff distance (no further than 2 kilometers) to prepare for remote arming. Hornets can be remotely armed 36 minutes after the arming switch is thrown on the last Hornet emplaced. If manual arming is used, Hornets automatically arm at the end of their safe-separation time (5 to 6 minutes after the arming switch is thrown).

### Gauntlet Obstacle

Hornet gauntlet obstacles (Figure 4-12) are emplaced by an engineer platoon and are very effective in constricted terrain along the enemy's AA and at choke points. A Hornet gauntlet typically consists of 40 to 50 Hornets employed in a series of clusters (Figure 4-13). Each cluster contains 3 to 6 Hornets. The Hornets in each cluster are emplaced at 50-meter intervals, perpendicular to the road centerline, on alternating sides of the road/AA, and 25 to 50 meters (depending on the terrain and the vegetation) off the side of the road/AA. The distance between clusters varies from 750 to 2,000 meters so that the advancing threat force is kept guessing about when they will encounter the next cluster.

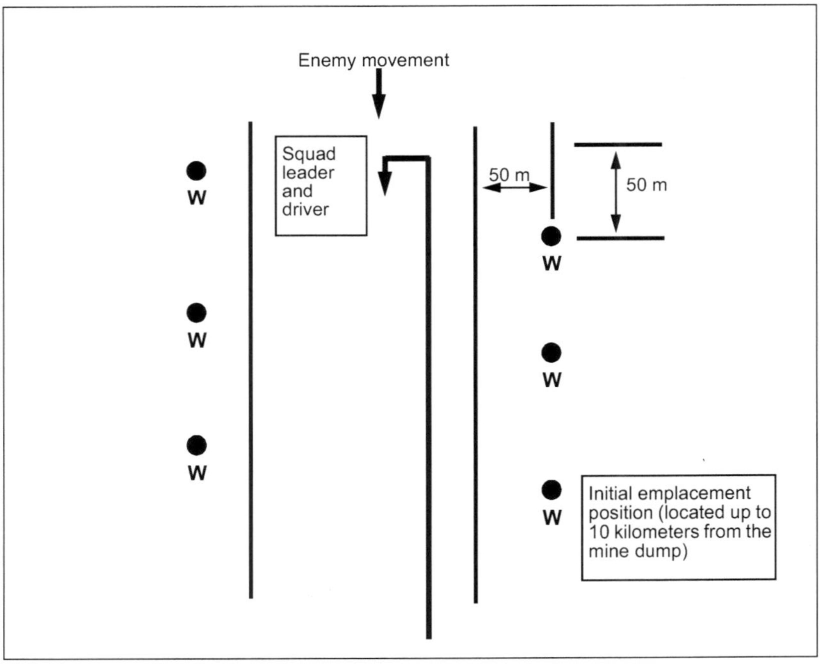


**Figure 4-12. Hornet gauntlet obstacle (one cluster)**

Before laying any Hornets, the munitions are prearmed as above. Soldiers also set the target switch to *HVY* for clusters closest to the enemy, so that the Hornets will only engage heavy tracked vehicles. The intent is to make threat forces commit to a route they perceive to be clear.

Hornets are emplaced beginning on the friendly side of the cluster. The first engineer squad emplaces Hornet munitions beginning with the cluster closest to the enemy. The emplacement vehicle drives even with the first Hornet

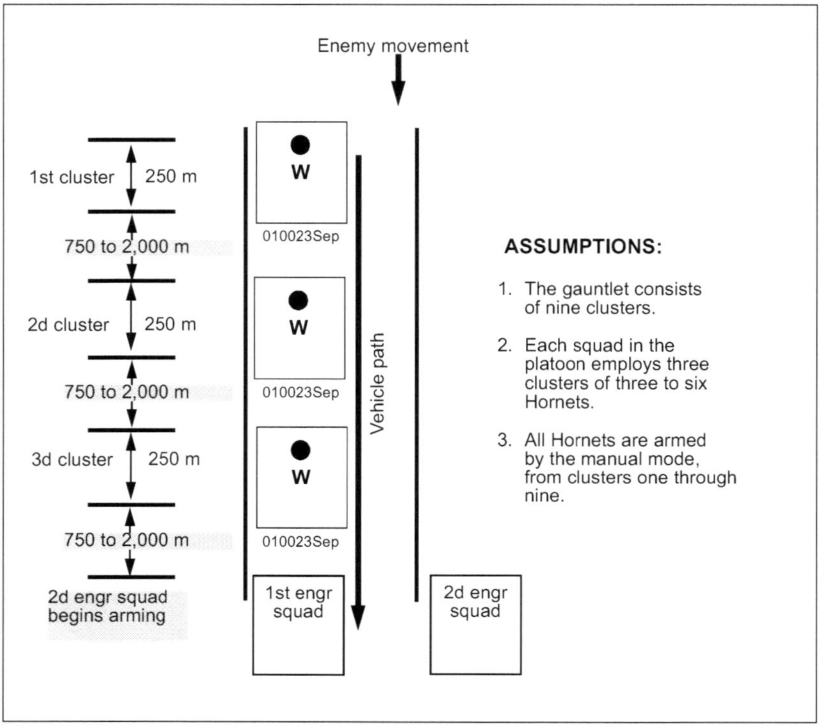


**Figure 4-13. Hornet gauntlet obstacle (platoon)**

emplacement site. The emplacing soldier dismounts, and a soldier in the vehicle hands him a Hornet. The emplacing soldier then proceeds to the Hornet emplacement site. The vehicle travels to a point even with each subsequent emplacement site. A soldier deploys at each emplacing site to lay one of the remaining Hornet munitions in the cluster. The vehicle then turns around and stops even with the last Hornet (on the enemy side) in the cluster.

Upon reaching the Hornet employment location, each emplacing soldier removes the cover and the S&H band, rotates the SD switch to *U*, and on the command (audible or visual signal) of the NCOIC, pushes the arm switch to *ARM*. Once the Hornets' arming switches are thrown, soldiers return to the road, taking the covers and the S&H bands with them, and wait to be picked up by the emplacement vehicle. After all the soldiers are in the emplacing vehicle, the driver quickly travels to the safe standoff distance (475 meters). The Hornet munitions in the first cluster will arm at the end of the safe-separation time (5 to 6 minutes).

The squad repeats the emplacement process for the next Hornet cluster in the gauntlet, taking care not to emplace any Hornets or drive within 475 meters of the previous cluster. Each squad in the platoon typically emplaces three clusters in the Hornet gauntlet, or 9 to 18 total Hornets.

Figure 4-14 shows Hornet emplacement in the battle space. The example used is in support of a defensive position where Volcano mines are used as tactical obstacles.

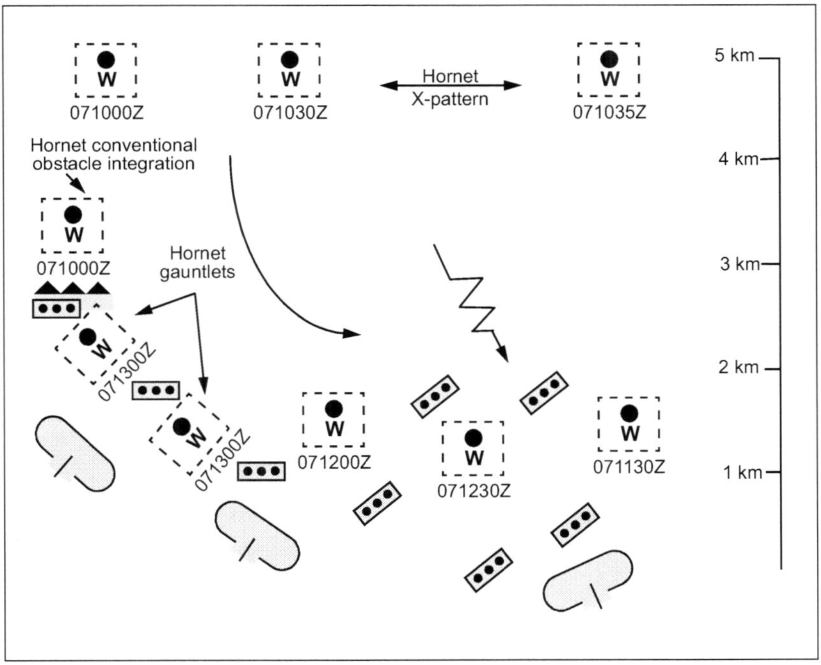


Figure 4-14. Hornet-enhanced turn-and fix-obstacle groups

## Deep-Battle Interdiction Weapon

SOF or ranger units emplace Hornet munitions in the deep battle area as interdiction weapons. A typical mission requires a unit ranging in size from a six-man team to an entire company. The number of Hornets carried by the unit depends on the mission and the mode of insertion (vehicle, aircraft, or dismounted troops); a man can normally carry only one Hornet. Hornets are typically used to support a raid against an enemy position or complex and at bridges or choke points along high-speed AAs used by advancing second-echelon forces or for resupply. In these roles, the Hornets are employed similar to the clusters in a gauntlet obstacle.

## Camouflage and Concealment

The best camouflage and concealment for the Hornet is tall grass and brush. The Hornet can be partially buried if the terrain or the vegetation does not provide effective natural camouflage and concealment. Placing the Hornet in a hole degrades its performance, so it should only be done when Hornets cannot be covered by fires or protected from tampering by dismounted enemy. The following conditions must be met:

- The depth of the hole must not exceed 4 inches, because the acoustic sensors must be above ground level.
- The hole must not restrict the Hornet's ability to rotate and tilt its body and to fire the sublet. To meet this requirement, the hole must be at least 36 inches wide and flat enough to support the munition. Although the Hornet should be placed on a flat surface if possible, it can operate on slopes up to 15 degrees.

Munitions placed at ground level should be no closer to obstructions than the distances shown in Table 4-1.

**Table 4-1. Hornet minimum emplacement distances**

| Maximum Obstruction Height | Minimum Employment Distance from Obstruction |
|---|---|
| 1 m | 3 m |
| 2.4 m | 5 m |
| 6.5 m | 15 m |
| 25 m | 25 m |

When the Hornet is emplaced and concealed, remove all indicators of excess soil and camouflage material before performing the arming sequence.

## Recording and Marking

When the Hornet munition field is completed, the OIC will identify an NCO to be the recorder. The NCO will collect data from the NCOICs of the emplacing squads and complete DA Form 1355 as outlined in Chapter 8. The OIC will ensure that the DA Form 1355 is completed timely and accurately.

Marking the Hornet munition field will be completed as prescribed in Chapter 2. The fence will be no closer than 150 meters from the nearest Hornet munition. Marking must be completed before emplacing the munitions.

This chapter implements STANAG 2990.

Chapter 5

# Conventional Mines

Conventional mines are hand-emplaced mines that require manual arming. This type of mine laying is labor-, resource-, and transport-intensive. Soldiers emplace conventional mines within a defined, marked boundary and lay them individually or in clusters. They record each mine location so that the mines can be recovered. Soldiers can surface lay or bury conventional mines and may place AHDs on AT mines.

## ANTITANK MINES

The M15, M19, and M21 AT mines are used by US forces. They are shown in Figure 5-1, and their characteristics are listed in Table 5-1, page 5-2.

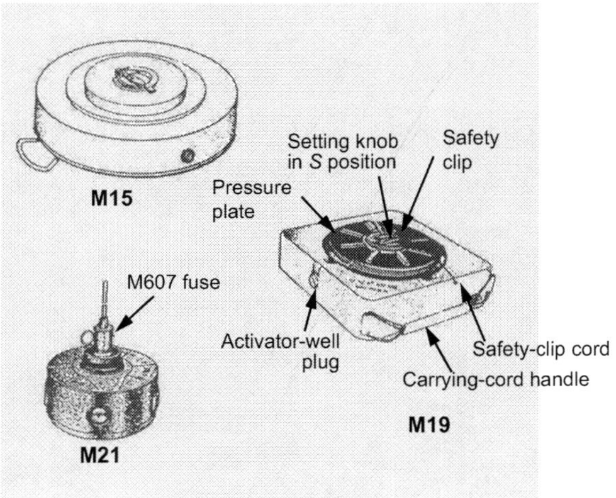


**Figure 5-1. AT mines**

### M15

The M15 AT mine is 337 millimeters in diameter and 125 millimeters high. It weighs 13.5 kilograms and contains 9.9 kilograms of Composition B explosive. The primary fuse well is on the top center of the mine; secondary fuse wells are on the side and bottom. The M15 can contain the following fuses:

**Table 5-1. Characteristics of AT mines**

| Mine | DODIC | Fuse | Warhead | AHD | Explosive Weight | Mine Weight | Mines per Container |
|---|---|---|---|---|---|---|---|
| M15 w/ M603 fuse | K180 | Pressure | Blast | Yes | 9.9 kg | 13.5 kg | 1 |
| M15 w/ M624 fuse | K180 (mine) K068 (fuse) | Tilt rod | Blast | Yes | 9.9 kg | 13.5 kg | 1 |
| M19 | K250 | Pressure | Blast | Yes | 9.45 kg | 12.6 kg | 4 |
| M21 | K181 | Tilt rod or pressure | SFF | Yes* | 4.95 kg | 7.6 kg | 4 |
| *Conventional AHDs will not couple with this mine; however, the M142 multipurpose FD can be emplaced under this mine. | | | | | | | |

- M603 fuse. When the M603 fuse is employed on the primary fuse well, the M15 is a track-width mine that is activated by 158 to 338 kilograms on the pressure plate. This produces an M-Kill.
- M624 fuse. When the M624 fuse (with tilt rod) is employed on the primary fuse well, the M15 is a full-width mine that is activated by a deflection of 20 degrees or 1.7 kilograms of pressure to the tilt rod. Depending on the armor, this produces an M-Kill or a K-Kill.

## M19

The M19 AT mine is a low-metallic, square-shaped mine that measures 332 by 332 millimeters and is 94 millimeters high. It weighs 12.6 kilograms and contains 9.45 kilograms of Composition B explosive, a tetryl booster pellet, and an M606 integral fuse. When the setting knob on the pressure plate is in the *S* (safe) position, the mine cannot function by action of the main fuse. After the safety clip has been removed and the setting knob turned to the *A* (armed) position, a force of 157.5 to 225 kilograms on the pressure plate depresses the Belleville spring and begins the firing chain. A standard FD may be used with the M2 activator in any of the secondary fuse wells on the sides or the bottom of the mine. When the M19 is employed, it is difficult to detect because of its plastic construction. It produces an M-Kill with a blast effect.

## M21

The M21 AT mine is 230 millimeters in diameter and 206 millimeters high. It weighs 7.6 kilograms and has 4.95 kilograms of Composition H6 explosive. The mine is activated by 1.7 kilograms of pressure against a 61-centimeter-long rod on the end of the M607 fuse. It uses an M-S plate to produce a K-Kill. The M21 with tilt rod must be buried or staked (use three stakes at the 12, 4, and 8 o'clock positions) so that enemy vehicles will not tip the mine over. Without the tilt rod, the mine is activated by 130.5 kilograms of pressure on the M607 fuse and produces an M-Kill by blast effect.

## ANTIPERSONNEL MINES

**The M14 and M16 AP mines are used by US forces on the Korean peninsula. They are also used by many other countries. The M16 AP mine is likely to be seen in a modified form. These mines are shown in Figure 5-2, and their characteristics are listed in Table 5-2.**

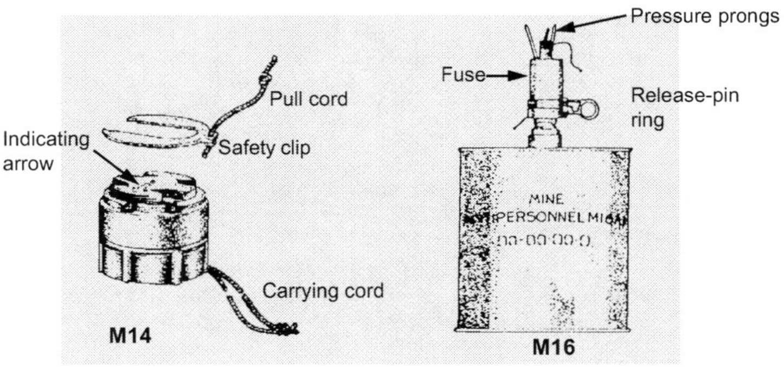


**Figure 5-2. AP mines**

**Table 5-2. Characteristics of AP mines**

| Mine | DODIC | Fuse | Warhead | AHD | Explosive Weight | Mine Weight | Mines per Container |
|---|---|---|---|---|---|---|---|
| M14 | K121 | Pressure | Blast | No | 28.4 g | 99.4 g | 90 |
| M16-series | K092 | Pressure or trip wire | Bounding frag | No | 450 g | 3.5 kg | 4 |

### M14

**The M14 AP mine is a low-metallic blast mine consisting of a main charge (28.4 grams of tetryl) and a plastic fuse with a steel firing pin. It is cylindrical in shape (56 millimeters in diameter and 40 millimeters high) and weighs 99.4 grams. The pressure plate has an indented, yellow arrow that points to the *A* or *S* position on top of the fuse body. A force of 11.5 to 13.5 kilograms depresses the pressure plate and causes the Belleville spring to drive the firing pin into the detonator. The M14 is not designed to kill, but to incapacitate. The M14 AP mine has been modified by gluing a metal washer to the bottom of the mine. The modification was directed to improve the detectability of the mine. Unmodified mines are not authorized for use by US forces.**

### M16

**The M16 AP mine is a bounding fragmentation mine that consists of a mine fuse (M605), trinitrotoluene (TNT) explosive, a propelling charge, and a projectile that are contained in a sheet-steel case. The mine is 103 millimeters in diameter, 199 millimeters high (including the fuse), and weighs 3.5 kilograms. The principal difference between the M16, M16A1, and M16A2 versions are in the construction of the detonators and boosters. The casualty radius is 27 meters for the M16 and M16A1 and 30 meters for the M16A2. A pressure of 3.6 to 9 kilograms applied on one or more of the three prongs of the M605 fuse or a pull of 1.4 to 4.5 kilograms on the trip wire will activate the mine.**

## EMPLACING MINES

The method used to lay and conceal each type of mine depends on the method of mine operations, the type of ground in which the mine is to be laid, and the type of ground cover available for camouflage.

Standard-pattern mine laying is laborious and time-consuming, but it is more effective and flexible than row mine laying and allows better mine concealment. Standard-pattern mine laying is well suited for protective minefields, and it can be used in terrain where the nature of the ground makes row mine laying impractical.

To achieve the maximum effect, mines must be laid where they cannot be seen and where a vehicle or a person exerts enough pressure to detonate them. The following rules should be applied to achieve the maximum effects of mines:

### Mines With Prongs

**Korea Only: If the mine is activated by its prongs, it should be buried flush with the ground so that only the tips of the mechanism are exposed (Figure 5-3). A mine buried in this manner is held firmly upright. The target exerts a direct, downward pressure rather than a sideways thrust. The mine is protected from damage and is difficult to see. If it is buried more deeply, it becomes unreliable because the layer of spoil may prevent the mine mechanism from operating.**

**If the mine is activated by a trip wire, it should be buried so that the trip wire is at least 2 to 3 centimeters above the ground (Figure 5-4).**

### Mines With Pressure Plates

Mines with pressure plates will function when completely buried as long as the cushion of earth above them is not too thick. AT mines are normally buried with the top of the mine approximately 5 centimeters below ground level.

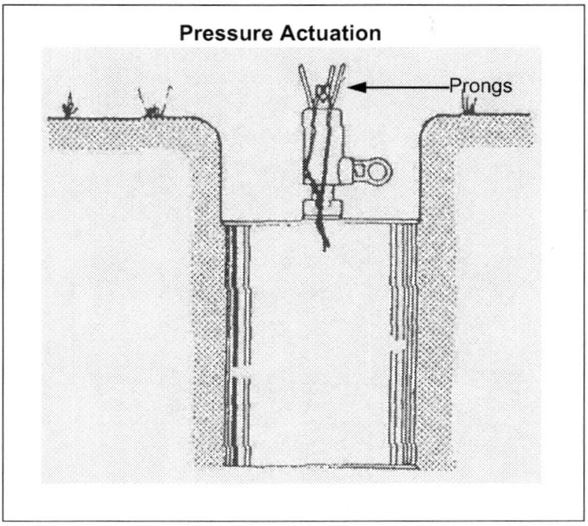


Figure 5-3. Prong-activated AP mine

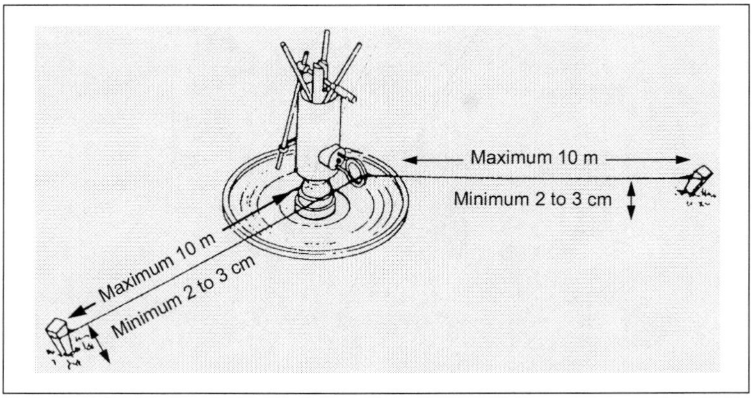


Figure 5-4. Trip-wire-activated AP mine

**Korea Only: AP mines are usually placed in a hole and covered with camouflage material. If the hole is only slightly larger than the mine, the weight of the target may be supported by the shoulder of the hole, and the mine will fail to activate. Such bridging action can be avoided if the hole is dug much wider than the mine (Figure 5-5).**

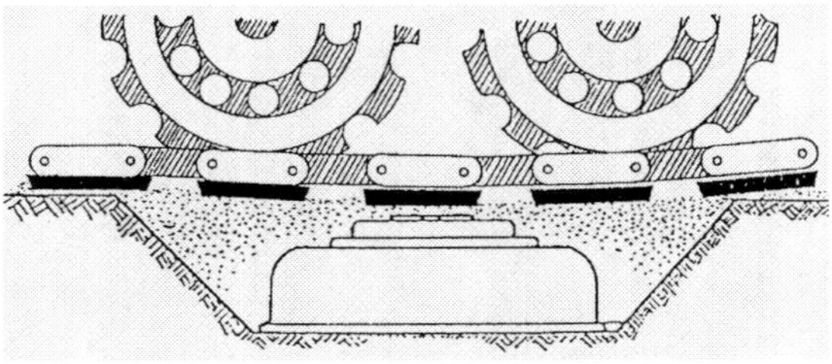

**Figure 5-5. Buried mine with pressure plate**

### Mines With Tilt Rods

Tilt-rod fuses normally require the body of the mine to be buried and the tilt-rod assembly to be clear of the ground (Figure 5-6). A tilt-rod fuse is preferred in areas where vegetation is sufficient to conceal the extension rod. Camouflage materials are carefully used to prevent premature detonation or interference with the normal functioning of the fuse. Extension rods are camouflaged before the mine is armed. If tilt rod mines are surface-laid, they must be staked.

### Bearing Boards

High pressure is required to activate AT mines. When burying a mine in soil that has a low bearing pressure (such as soft sand or clayey soil), it may be necessary to place a board or another bearing plate under the mine. Otherwise, the mine may not detonate when it is forced down.

### Concealment

After digging the hole for a mine, place the spoil in a sandbag to reduce the evidence of mining. If a sandbag is not available, heap the spoil. Camouflage all traces of digging after the mine is laid. If the ground cover is turf or other matted, root material, remove spoil that cannot be hidden. Cut the sod in an X, I, or U shape in the area where the mine is to be placed; lay the mine; and then roll the sod back in place to camouflage the mine. Loose earth over a mine will eventually consolidate, so the mine location should have a small mound immediately after laying (Figure 5-7). Ensure that the mound is inconspicuous and that it blends with the surrounding area. It is very important that you make a final check after concealing each mine so that you can correct faults progressively, because they cannot be corrected later.

AT mines in standard-pattern minefields should be buried. However if conditions dictate, mines with a single-impulse fuse may be laid on the surface. Mines with double-impulse fuses should always be buried, because if

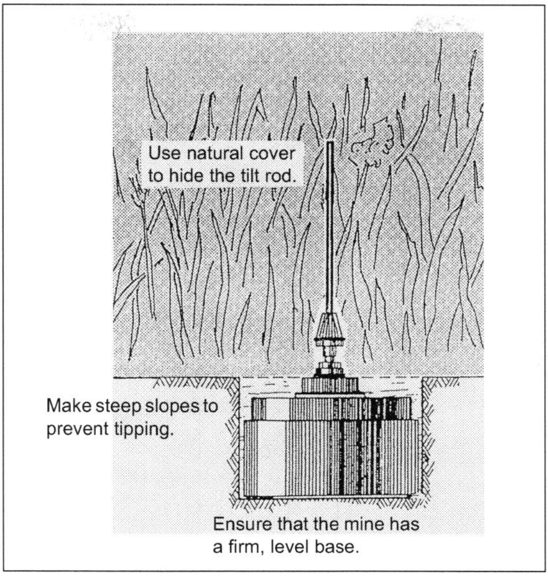


**Figure 5-6. Buried mine with tilt rod**

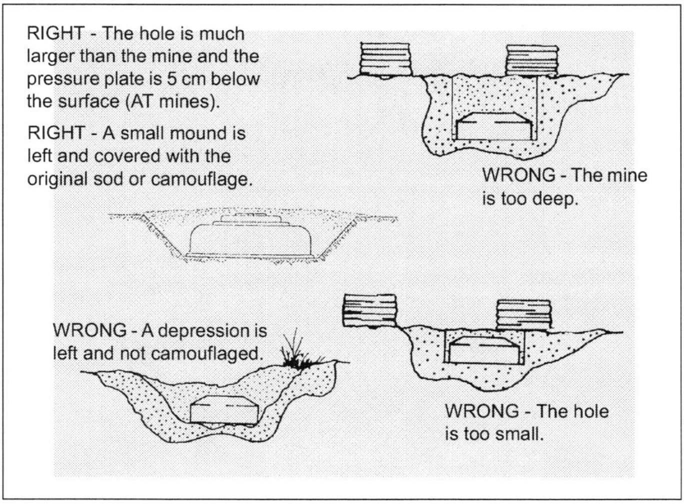


**Figure 5-7. Buried and concealed mines**

they are surface-laid, they may be physically damaged when pressure is exerted by a tracked vehicle. Buried mines also have some resistance to countermeasures, but surface-laid mines have none. Consideration must also be given to sympathetic detonation of AT mines (Table 5-3). US conventional mines do not have integral AHDs, so allow extra time to lay mines with AHDs.

**Table 5-3. Sympathetic detonation chart**

| Type | M16 | M15 | M19 |
|---|---|---|---|
| Surface-laid | NA | 4.0 m | 4.0 m |
| Buried flush | 1.5 m | 2.4 m | 5.5 m |
| Buried 5 cm | NA | 1.5 m | 4.8 m |

The difficulty of burying mines in very rocky ground and the necessity for surface laying will have a bearing on which mines are suitable. For example, small, blast-type AP mines are hard to detect and easy to camouflage. They are much easier to camouflage than larger fragmentation mines. The type of AT mine used will make little difference, because the mine's size will always make camouflage very difficult.

### MANEUVER ASSISTANCE

During large mine-laying operations, engineers seldom have sufficient manpower to carry out all minefield tasks. Other combat arms units must often provide work parties. Engineers must be capable of organizing, controlling, and supervising combined arms work parties. They must also instruct them in new equipment and techniques. Work parties may be integrated with engineers or given certain tasks that are within their capabilities.

When laying a standard-pattern minefield, consider supplementing work parties with other combat arms soldiers to perform the following:

- Executing Class IV/V supply point or mine dump missions. Soldiers uncrate and prepare mines and remove empty boxes and residue.
- Laying. Soldiers position mines within strips and dig holes.
- Marking. Soldiers construct the perimeter fence and emplace mine signs.

Unpacking, preparing, and loading mines are the most time-consuming tasks when laying a row minefield; and they are ideal tasks for other combat arms soldiers.

This chapter implements STANAG 2036.

# Chapter 6

# Row Mining

Row mining is a means of emplacement for tactical minefields. For example, a typical tactical minefield could contain several rows of regularly spaced mines.

## USE

Row mining is not a new idea. It has been used since the beginning of modern mine warfare and is very effective. It is especially effective in support of maneuver-oriented doctrine. Row mining is faster than standard-pattern mining and improves the maneuver commander's flexibility by providing him an obstacle that requires less manpower effort.

Mines may be surface-laid or buried, and they are often laid directly from a slow-moving vehicle. This reduces the time and the personnel required to emplace a minefield. Row mining can be used as a tactical or situational obstacle. Minefields are usually emplaced at or near the FLOT, along flank AAs, to support security operations. Speed and efficiency make row mining a desirable option, and row mining supports current doctrine.

## RULES

Rules governing authority, reporting, recording, and marking are generally the same for row minefields as they are for other minefields. Row mining is simply a method of laying mines.

The most important factor in row mining is the requirement for strict $C^2$. Row mining is potentially the most hazardous form of mine laying. It entails vehicles and personnel moving in and around mines without the safety of a centerline strip. Leaders must place extreme emphasis on safety because the laying procedure is very rapid.

Most of the rules governing row mining are defined in STANAG 2036. A summary of those rules and some additional rules that apply are shown below.

- Rows.
  - There are two types of mine rows—regular and short. Short rows are described under IOE rules below.
  - Regular rows are marked and recorded. They are designated by letters (A, B, and so forth), with Row A being closest to the enemy.
  - The minimum distance between rows of AT mines is 8 meters.
  - **Korea Only: The minimum distance between any row and a row containing AP mines is 15 meters.**

- The distance between the start row marker and the first mine in a row is the mine spacing for that row.
- Start and end row markers are permanent markers and must be made of detectable material.

- Clusters.
  - Clusters are placed on the row centerline and directed toward the enemy side.
  - A cluster in row mining usually consists of one AT mine **(Korea Only: but it may also contain AP mines)**.
  - Cluster composition must remain the same throughout the row.
  - **Korea Only: Different types of AP mines may be used in a cluster.**
  - **Korea Only: The total number of mines in one cluster will not exceed five; no more than one will be an AT mine.**
  - The type of AT mine may vary from one cluster to another.
  - **Korea Only: A cluster of AP mines can be laid in a 2-meter semicircle on the enemy side of the baseline.**
  - When a cluster contains a mine that is equipped with an AHD, the mine is armed before the AHD is armed. The cluster is not armed until all personnel are at least 60 meters away.
  - Omitted clusters do not contain mines. They are recorded on DA Form 1355 (see Chapter 8).
  - Clusters are omitted within lanes, within gaps, in areas less than 2 meters from boundaries and lanes, and in areas where the terrain (trees, rocks) prohibits emplacement.
- IOE.
  - The IOE is located on the enemy side of the minefield.
  - The IOE baseline must be at least 15 meters from Row A.
  - IOE mines are buried.
  - IOE short rows are labeled at start (I1) and end (I1E) points.
  - IOE short rows must be at least 15 meters apart.
- **Korea Only: Trip wires.**
  - **Trip wires can be used in regular rows, but only one mine per cluster can be actuated by a trip wire.**
  - **No more than two trip wires can be used on a mine.**
  - **Trip wires are not considered AHDs.**
  - **Trip wires must be at least 2 meters from a minefield lane, a cluster, another trip wire, an IOE short row, or a minefield perimeter fence.**

— **Trip wires can only be used with AP fragmentation mines.**

- Lanes.
  - — Lanes are sited before laying begins. Lane locations should not be obvious.
  - — Clusters must be at least 2 meters from lane edges.
  - — The number of lanes must be sufficient to ensure that no one lane is overused and turned into an obvious track.
  - — Sufficient mines must be stockpiled so that the responsible unit can seal lanes suspected of being located by the enemy.
- General.
  - — The spacing between mines or clusters can vary from 4 to 10 meters but must remain constant within the row.
  - — Mines and clusters must be at least 15 meters from the perimeter fence.
  - — If the distance between a mine or a cluster and any turning point is less than the spacing for that row, omit that mine or cluster. The mine immediately following a turning point is always located at the mine spacing for that row.
  - — The minefield has two landmarks located to the rear, never to the extreme side or front.
  - — Global-positioning systems (GPSs) can only be used to determine the coordinates for minefield landmarks and reference points (RPs).

**WARNING**
**Do not use GPSs to chart or record minefield perimeter coordinates or to determine safe routes through or around existing minefields.**

  - — If landmarks are more than 200 meters away from the last row or are out of the direct line of sight, intermediate row markers or landmarks are placed at least 75 meters from the last end row marker.
  - — Landmarks can be used for more than one minefield. This must be recorded in the remarks block of DA Form 1355.
  - — Back azimuths are not used to record the minefield.
  - — Measurements are in meters.

## LOGISTICS

### CALCULATIONS

To simplify the calculation process, a minefield requirements computation work sheet (Figure 6-1, pages 6-5 through 6-8) has been developed. This work sheet is provided to the platoon leader or sergeant as a step-by-step guide to the mathematics involved in the logistical computation process. Properly completed, the work sheet provides the number of mines to order (by type), the number of regular strips to be emplaced, cluster composition, the estimated man-hours required to install the minefield, the amount of fencing and marking material required, the number of truckloads required to carry the mines, and the number of rolls of engineer tape required.

Step-by-step procedures for completing the work sheet are shown in Figure 6-2, pages 6-9 through 6-14. Each step is explained in the example to facilitate the understanding of the logic behind the calculations.

Use the following steps to determine the number of AT mines required for a row minefield when not using the standard row minefields discussed later in this chapter. Round the resulting numbers up to the nearest whole number.

**Step 1.** Determine the number of mines required.

$$density \times front = number\ of\ mines$$

**Step 2.** Determine the number of mines per row.

$$front \div mine\ Spacing = number\ of\ mines\ per\ row$$

**Step 3.** Determine the number of rows.

$$number\ of\ mines \div number\ of\ mines\ per\ row = number\ of\ rows$$

**Step 4.** Determine the actual number of mines.

$$number\ of\ mines\ per\ row \times number\ of\ rows = number\ of\ mines$$

**Step 5.** Determine the number of mines to request (includes a 10 percent resource factor).

$$number\ of\ mines \times 1.1 = number\ of\ mines\ to\ request$$

**Step 6.** Determine the number of vehicle loads by using Table 2-8, page 2-45.

**Step 7.** Determine the fencing and marking material required.

**Sample Problem:** Your platoon has been tasked to emplace a 400-meter row minefield with a density of 0.5-0-0 (AT-AP fragmentation-AP blast). You have decided to space the mines 6 meters apart. Determine the number of M15 mines to order and the number of 5-ton dump trucks (with sideboards) required to deliver the mines.

- Step 1. 0.5 × 400 meters = 200 mines
- Step 2. 400 ÷ 6 = 66.6 = 67 mines per row
- Step 3. 200 ÷ 67 = 2.98 = 3 rows
- Step 4. 67 × 3 = 201 mines
- Step 5. 201 × 1.1 = 221.1 = 222 mines
- Step 6. 222 ÷ 204 = 1.08 = 2 5-ton trucks
- Step 7.
  - Concertina: ([400 × 2] + [200 × 2] + 160) × 1.4 = 1,904 meters of concertina
  - Pickets: 1,904 ÷ 15 = 126.9 = 127 pickets
  - Signs: 127 pickets = 127 signs

| MINEFIELD REQUIREMENTS COMPUTATION WORK SHEET | | | |
|---|---|---|---|
| GIVEN | | | |
| Desired density | AT ______ | APF ______ | APB _____ |
| IOE representative cluster | AT ______ | APF ______ | APB _____ |
| Front | ______ meters | | |
| Depth | ______ meters | | |
| Percentage of AHDs | ______% | | |
| Type of mines | AT ______ | APF ______ | APB _____ |
| Type of truck/trailer | ___________ | | |
| Lanes/gaps/traffic tapes | ______ | | |
| Trip-wire safety tapes | ______ | | |
| PART 1. NUMBER OF MINES | | | |
| A. IOE live clusters = front ÷ 9 | ______ ÷ 9 = _____ (round up) | | |
| | AT | APF | APB |
| B. IOE representative cluster ×<br>Number of IOE clusters =<br>Number of mines in IOE | ______<br>× ____<br>______ | ______<br>× ____<br>______ | ______<br>× ____<br>______ |
| C. Desired density ×<br>Minefield front =<br>Mines in regular strips | ______<br>× ____<br>______ | ______<br>× ____<br>______ | ______<br>× ____<br>______ |
| D. Subtotal of mines<br>(line B + line C) | ______ | ______ | ______ |
| E. 10% excess factor = | × 1.10 | × 1.10 | × 1.10 |
| Total number of mines to order<br>(round up for each) | ______ | ______ | ______ |
| PART 2. NUMBER OF REGULAR STRIPS | | | |
| A. Add desired density | AT _____ | + APF _____ | + APB ____= _____ |
| B. 0.6 × line A above | 0.6 × _____ = _____ (round up) | | |
| C. 3 × AT desired density | 3 × _____ = _____ | | |

**Figure 6-1. Minefield requirements computation work sheet**

D. Number of regular strips required = highest number of line B or C _____

PART 3. NUMBER OF AHDs

% AHDs × total number of AT mines _____

PART 4. STRIP CLUSTER COMPOSITION

A. Desired density

AT: 3 × _____ = _____ APF: 3 × _____ = _____ APB: 3 × _____ = _____

B. Cluster composition table

| STRIP | AT | APF | APB | STRIP TOTAL (cannot exceed 5) |
|---|---|---|---|---|
| A | _____ | _____ | _____ | _____ |
| B | _____ | _____ | _____ | _____ |
| C | _____ | _____ | _____ | _____ |
| D | _____ | _____ | _____ | _____ |
| E | _____ | _____ | _____ | _____ |
| F | _____ | _____ | _____ | _____ |
| G | _____ | _____ | _____ | _____ |
| H | _____ | _____ | _____ | _____ |
| I | _____ | _____ | _____ | _____ |
| TOTAL (Cannot exceed desired density × 3 as computed in A above) | _____ | _____ | _____ | _____ |

PART 5. NUMBER OF MAN-HOURS FOR INSTALLATION

Number of mines ÷ emplacement rate = mines per man-hour

Number of AT mines:______ ÷ 4 = _____ (round up)

Number of APF mines:______ ÷ 8 = _____ (round up)

Number of APB mines:______ ÷ 16 = _____ (round up)

_____ + _____ + _____ × 1.2 = _____ man-hours (round up)

**Figure 6-1. Minefield requirements computation work sheet (continued)**

PART 6. AMOUNT OF FENCING AND MARKING MATERIAL

A. Concertina wire or single-strand barbwire

([front × 2] + [depth × 2] + 160) × 1.4 = meters of concertina or single-strand barbwire required

([_____ × 2] + [_____ × 2] + 160) × 1.4 = _____ (round up)

Number of pickets = amount of concertina or single-strand barbwire ÷ 15

_____ ÷ 15 = _____ (round up)

\- OR -

B. Double-strand barbwire

([front × 2] + [depth × 2] + 160) × 2.8 = meters of double-strand barbwire required

([_____ × 2] = [_____ × 2] + 160) × 2.8 = _____ (round up)

Number of pickets = amount of double-strand barbwire ÷ 30

_____ ÷ 30 = _____ (round up)

C. Number of signs = number of pickets = _____

PART 7. NUMBER OF TRUCKLOADS

AT mines

_____ cases per truck × _____ mines per case = _____ mines per truck

_____ mines required ÷ _____ mines per truck = _____ truckloads of AT mines

APF mines

_____ cases per truck × _____ mines per case = _____ mines per truck

_____ mines required ÷ _____ mines per truck = _____ truckloads of APF mines

APB mines

_____ cases per truck × _____ mines per case = _____ mines per truck

_____ mines required ÷ _____ mines per truck = _____ truckloads of APB mines

Total truckloads

_____ AT truckloads + _____ APF truckloads + _____ APB truckloads =

_____ total truckloads required (round up)

**Figure 6-1. Minefield requirements computation work sheet (continued)**

PART 8. AMOUNT OF ENGINEER TAPE

A. Minefield boundaries depth × 2 = _____ × 2 = _____

B. Regular strips front × number of regular strips = _____ × _____ = _____

C. IOE front + (number of IOE clusters × 3) = _____ + (____ × 3) = _____

D. Lanes and gaps depth × 2 × number of lanes and gaps = _____ × 2 × _____ = _____

E. Traffic tapes depth × number of traffic tapes _____ × _____ = _____

F. Trip-wire safety tape front × number of regular strips with trip wire _____ × _____ = ______

G. Subtotal (lines A + B + C + D + E + F)
_____ + _____ + _____ + ______ + _____ + _____ = _____ meters (round up)

H. Number of rolls to order (line G × 1.2)
______ × 1.2 = _____ meters

_____ meters ÷ 170 meters per roll = _____ rolls of engineer tape (round up)

PART 9. SANDBAGS

A. Number of clusters in IOE (from 1A) = _____

B. Number of clusters in minefield = number of clusters in IOE × 3 × number of regular strips (from 2D) = _____

C. Total number of clusters (line A + line B) = _____

D. Number of sandbags = number of clusters × 3 sandbags per cluster (line C × 3) = _____

**Figure 6-1. Minefield requirements computation work sheet (continued)**

Basic information pertaining to the minefield is normally determined by the engineer company commander or the staff engineer. It is provided to the OIC or NCOIC of the emplacing unit during the mission briefing. In this example, the following guidance is given to the emplacing unit:

| | | | |
|---|---|---|---|
| Desired density | AT 1 | APF 4 | APB 8 |
| IOE representative cluster | AT 1 | APF 2 | APB 2 |
| Front | 200 meters | | |
| Depth | 300 meters | | |
| Percentage of AHDs | 10% | | |
| Type of mines | AT M15 | APF M16A2 | APB M14 |
| Type of truck/trailer | 5-ton dump (with sideboards) | | |
| Lanes/gaps/traffic tapes | 1 lane, 1 traffic tape (foot troops) | | |
| Trip-wire safety tapes | 3 | | |

The rest of this work sheet is completed by using the above information.

The regular strip has a cluster density of one cluster every 3 meters. The IOE has a cluster density of one-third that of a regular strip, or one cluster every 9 meters. Therefore, to obtain the number of clusters in the IOE, the length of the strip is divided by 9. Decimals are rounded up to the next higher whole number.

PART 1. NUMBER OF MINES

Step 1.

IOE live clusters 200 ÷ 9 = 23 (rounded up)

The representative cluster composition for the IOE clusters is established and provided by the commander based on METT-TC factors. The number of clusters in the IOE is multiplied by the cluster composition to determine the number of mines, by type, in the entire IOE.

Step 2.

| | AT | APF | APB |
|---|---|---|---|
| IOE representative cluster × | 1 | 2 | 2 |
| Number of IOE clusters = | 23 | 23 | 23 |
| Number of mines in IOE | 23 | 46 | 46 |

The minefield front multiplied by the desired density determines the number of mines in the minefield.

**NOTE: The desired density pertains only to the regular strips and does not take into account the number of mines in the IOE which were calculated in Step 2.**

**Figure 6-2. Step-by-step procedures for completing the minefield requirements computation work sheet**

Step 3.

| | | | |
|---|---|---|---|
| Desired density × | 1 | 4 | 8 |
| Minefield front = | 200 | 200 | 200 |
| Mines in regular strips | 200 | 800 | 1,600 |

The number of mines required for the IOE (Step 2) is added to the number of mines in the regular strips (Step 3).

Step 4. Subtotal of mines

| | | | |
|---|---|---|---|
| (Step 2 + 3) | 223 | 846 | 1,646 |

Ten percent is added to the total number of mines required to allow for damaged items and irregularities in terrain and strip length. This is accomplished by multiplying the total number of mines (Step 4) by 1.1. Decimals are rounded up to the next higher whole number.

Step 5.

| | | | |
|---|---|---|---|
| 10% excess factor = | 1.1 | 1.1 | 1.1 |
| Total number of mines to order | 246 | 931 | 1,811 |

These figures represent the total number of mines, by type, required for the entire minefield. When ordering by the case rather than by individual mines, the total should be divided by the number of mines per case and rounded up to the next whole case. (See Table 2-8, page 2-45.)

PART 2. NUMBER OF REGULAR STRIPS

Step 1.

| | | | |
|---|---|---|---|
| Add desired density | AT 1 | + APF 4 | + APB 8 = 13 |

Each regular mine strip has a cluster every 3 meters; therefore, its density is one-third cluster per meter of front. A total density of 13 mines per meter of front in the previous example would equal 3 × 13 or 39 mines per 3 meters of front. Clusters may contain a maximum of five mines, so the resulting figure must be divided by 5. In short, to determine the minimum number of regular strips required, the total density must be multiplied by three-fifths (3 meters between clusters and five mines per cluster). For ease of calculation, three-fifths is converted to the decimal 0.6. Decimals are rounded up to the next highest whole number.

Step 2.

| | |
|---|---|
| 0.6 × Step 1 | 0.6 × 13 = 8 (rounded up) |

The calculations to determine the minimum number of regular strips previously described are not suitable when the ratio of AT to AP mines is greater than 1:4. For example, if the desired density is 1-1-1, the total density is 3. The minimum number of strips would be 3 × 3/5 = 1.8, rounded up to 2 strips. However, because of the restriction on the number of AT mines per cluster, it is impossible to obtain a density of 1 AT mine per meter of front with only 2 strips. A minimum of 3 regular strips is required. The alternative means of determining the number of regular strips is founded by multiplying the AT desired density by 3.

**Figure 6-2. Step-by-step procedures for completing the minefield requirements computation work sheet (continued)**

Step 3.

3 × AT desired density    3 × 1 = 3

The number of regular strips calculated by the first method and the alternative method are compared, and the higher figure is used as the minimum number of regular strips. The 8 determined by the 3/5 rule is larger than the 3 determined by the alternative method. Therefore, the minimum number of regular strips in the example is 8.

Step 4.

Number of regular strips required = highest number of Step 2 or 3 = 8

PART 3. NUMBER OF AHDs

0.1 × 223 = 23 (rounded up)

PART 4. STRIP CLUSTER COMPOSITION

The cluster composition table is prepared by the OIC of the laying unit to control the allocation of mines to a regular strip. The cluster composition remains constant within a particular strip, but it may vary among different strips. As the mines are allocated by strip, no more than 1 AT mine can be placed in each representative cluster, and each cluster can have a maximum of 5 mines.

A tabular format is prepared (see Figure 6-1, page 6-6) to facilitate the distribution of mines by emplacement personnel. Note that each component of the desired density is multiplied by 3. The number 3 is always used regardless of the minimum number of regular strips because it is the number of mine strips required to give a minefield density of 1 mine per meter of front when a cluster contains only 1 mine of each type. Each mine strip has a cluster every 3 meters; therefore, it has a density of one-third mine per meter when a cluster contains 1 of each type of mine.

Step 1. Desired density

AT: 3 × 1 = 3    APF: 3 × 4 = 12    APB: 3 × 8 = 24

The total of each column in the table cannot exceed the number of mines above. For example, with an APF desired density of 4, 3 × 4 = 12, so the total APF mines in the representative cluster composition for each of the regular strips cannot exceed 12.

PART 5. NUMBER OF MAN-HOURS FOR INSTALLATION

Remember, the total number of mines includes the mines in regular strips and the mines in IOE short strips. The laying rates are—

AT mines: 4 per man-hour.

APF mines: 8 per man-hour.

APB mines: 16 per man-hour.

**Figure 6-2. Step-by-step procedures for completing the minefield requirements computation work sheet (continued)**

The number of man-hours required for each mine type is computed and rounded up. These amounts are totalled and a 20 percent excess factor is included by multiplying the total by 1.2. The resulting figure is the total number of man-hours required for emplacement and represents straight work time only. It does not take into account the time for transportation to and from the emplacement site, meals, and breaks; limited visibility; or NBC conditions. The commander should use his judgment and past experience to determine the time required for transportation, meals, and breaks. When working under limited visibility or NBC conditions, the total man-hours (after the excess factor has been included) should be multiplied by 1.5.

In this example, a total of 357 man-hours is required as determined below. Note that each decimal is rounded to the next higher whole number.

Number of mines ÷ emplacement rate = mines per man-hour

| | |
|---|---|
| Number of AT mines | 246 ÷ 4 = 62 (rounded up) |
| Number of APF mines | 961 ÷ 8 = 121 (rounded up) |
| Number of APB mines | 1,811 ÷ 16 = 114 (rounded up) |
| Total | 62 + 121 + 114 = 297 × 1.2 = 357 (rounded up) |

PART 6. AMOUNT OF FENCING AND MARKING MATERIAL

Standard-pattern minefields must be marked and fenced. The amount of fencing required depends on whether barbwire (single- or double-strand) or concertina is used.

The amount of wire for a single-strand barbwire or a single-strand concertina fence is calculated with the following formula:

([front × 2] + [depth × 2] + 160) × 1.4

The amount of wire for a double-strand barbwire fence is calculated with the following formula:

([front × 2] + [depth × 2] + 160) × 2.8

Step 1.

Concertina Wire or Single-Strand Barbwire

([200 × 2] + [300 × 2] + 160) × 1.4 = 1,624

The number of pickets required is determined by dividing the total amount of fence by 15.

Number of pickets = amount of fence ÷ 15     1,624 ÷ 15 = 109 (rounded up)

Step 2.

Double-Strand Barbwire

([200 × 2] + [300 × 2] + 160) × 2.8 = 3,248 (rounded up)

The number of pickets required is determined by dividing the total amount of fence by 30.

Number of pickets = amount of fence ÷ 30     3,248 ÷ 30 = 109 (rounded up)

**Figure 6-2. Step-by-step procedures for completing the minefield requirements computation work sheet (continued)**

The number of minefield marking signs is equal to the number of pickets.

**NOTE: These calculations determine the marking and fencing materials required for the minefield perimeter only. Additional materials may be required for lanes and gaps.**

PART 7. NUMBER OF TRUCKLOADS

The number of vehicles required depends on the type and amount of mines as well as the type of vehicles available. The total mines, by type, required is divided by the haul capacity of available vehicles to determine the number of truckloads required to transport the mines.

In this example, crated M15 AT mines, M16A2 APF mines, and M14 APB mines are hauled in M930 5-ton dump trucks (with sideboards). (See Table 2-8, page 2-45).

AT mines:

246 mines required ÷ 204 mines per truck = 1.2 truckloads of AT mines

APF mines:

931 mines required ÷ 888 mines per truck = 1.05 truckloads of APF mines

APB mines:

1,811 mines required ÷ 13,770 mines per truck = 0.13 truckloads of APB mines

Total truckloads:

1.2 AT truckloads + 1.05 APF truckloads + 0.13 APB truckloads = 2.38 truckloads = 3 truckloads (rounded up) required

PART 8. AMOUNT OF ENGINEER TAPE

An extensive amount of engineer tape is used to mark the initial layout of a standard-pattern minefield. Engineer tape comes in 170-meter rolls and is used to mark several portions of the minefield.

**NOTE: In this example, only one lane and one roll of traffic tape is required.**

Step 1.

Minefield boundaries depth × 2 300 × 2 = 600

Step 2.

| | | |
|---|---|---|
| Regular strips | front × number of regular strips | 200 × 8 = 1,600 |

Step 3.

| | | |
|---|---|---|
| IOE | front + (number of IOE clusters × 3) | 200 + (23 × 3) = 269 |

Step 4.

| | | |
|---|---|---|
| Lanes and gaps | depth × 2 × number of lanes and gaps | 300 × 2 × 1 = 600 |

Step 5.

| | | |
|---|---|---|
| Traffic tape | depth × number of traffic tapes | 300 × 1 = 300 |

**Figure 6-2. Step-by-step procedures for completing the minefield requirements computation work sheet (continued)**

| | | |
|---|---|---|
| Step 6. | | |
| Trip-wire safety tape | front × number of regular strips with trip wire | 200 × 3 = 600 |
| Step 7. | | |
| Total of Steps 1 through 6 = 600 + 1,600 + 269 + 600 + 300 + 600 = 3,969 meters | | |
| Step 8. | | |
| Add 20% excess | total amount of engineer tape required for the minefield × 1.2 = 3,969 × 1.2 = 4,762.8 = 4,763 (rounded up) | |
| Step 9. | | |
| Total number of rolls | total amount of engineer tape, in meters, from Step 8 ÷ 170 meters per roll = 4,763 meters ÷ 170 = 28.02 = 29 rolls | |
| PART 9. SANDBAGS | | |
| To determine the number of sandbags for the removal of spoil. | | |
| Step 1. | | |
| Number of clusters in IOE (Part 1, Step 1) = 23 | | |
| Step 2. | | |
| Number of clusters in minefield = number of clusters in IOE × 3 × number of regular strips (Part 1, Step 4) | | |
| 23 × 3 × 8 = 552 | | |
| Step 3. | | |
| Total number of clusters (add Steps 1 and 2) = 575 | | |
| Step 4. | | |
| Number of sandbags = number of clusters × 3 sandbags per cluster | | |
| 575 x 3 = 1,725 | | |

**Figure 6-2. Step-by-step procedures for completing the minefield requirements computation work sheet (continued)**

## TASK ORGANIZATION

To maximize the efficiency of the row-mining process, the platoon leader must task-organize his platoon. The organization of the task, as a whole, is intricate and places great demands on the leader. Leave nothing to chance when planning and executing a row minefield, because each situation is different. Make allowances for transporting, handling, and controlling the mines. The officer in charge (OIC) and the squad leaders must be able to exercise control throughout the task under all conditions. Always observe safety.

Organize the platoon into four parties—siting and recording, marking, mine dump, and laying.

### Siting-and-Recording Party

The platoon leader directs the party and is responsible for siting, recording, and reporting the minefield. This party consists of one or two soldiers and a vehicle to carry material. (If a vehicle is not available, increase the party to three soldiers.) Because siting is usually done in daylight, the party must take appropriate physical-security measures. The party starts well ahead of the actual laying, sets out control markers, and avoids using sharp turns. The party marks the vehicle traffic routes to and from the minefield rows.

When siting is complete, the OIC identifies one member of the party to be the recorder and assigns the remaining soldiers to other tasks. The recorder collects data from the laying party NCOIC and completes DA Form 1355 as outlined in Chapter 8. The OIC ensures that the DA Form 1355 is completed timely and accurately.

### Marking Party

This party is composed of an NCOIC and personnel who are not working as members of other teams. After the minefield is sited, the marking party emplaces fence posts, wire, and marking signs.

### Mine-Dump Party

This party is controlled by the platoon sergeant (PSG) and is composed of personnel who are not working as members of other teams. The mine-dump party accounts for all strip packages that arrive from other sources, sets up vehicle mine sets at the mine dump, and hauls supplies as required. The PSG places row packages in a location that maximizes speed and provides concealment for minefield emplacement, and he also ensures that the mine dump is prepared for night operations. The party marks the mine dump's entrance and exit and the routes to them. The PSG verifies the strip feeder reports with the squad leaders upon the completion of each row and passes the reports to the recording party. The PSG is not required to stay at the mine dump continuously; he has the flexibility to move around the area to perform other activities.

The mine-dump party creates vehicle sets by setting aside the number of mines and fuses that are required by each laying vehicle. The party loosens and then hand-tightens arming and shipping plugs, helps load the mines onto laying vehicles, and disposes of residue. Soldiers may also assist the marking party and provide local security. For initial vehicle loads, the mine-dump party may be assisted by the laying party.

### Laying Party

This party consists of an NCOIC, four soldiers, and a vehicle to carry the mines. The NCOIC controls the movement of each laying vehicle. He directs each vehicle to start and stop laying and controls immediate-action drills. The NCOIC initiates a strip feeder report with the PSG or the mine-dump NCOIC, receives azimuths from the siting party, and directs his element to the correct row. He is responsible for replacing the temporary row markers with permanent markers and for ensuring that mines are laid according to the azimuths, mines are spaced correctly, and the strip feeder report is accurate.

The NCOIC ensures that the end row marker is emplaced at the completion of each row, and he closes the strip feeder report with the PSG.

**NOTE: Using tilt-rod fuses requires additional soldiers to stake mines, insert fuses, and arm mines.**

When laying three rows at once, each laying party consists of an APC or an organic squad vehicle and a carrier, sapper, and digging team.

#### *Carrier Team*

This team is comprised of the APC driver and the track commander (TC). They ensure that the APC maintains the proper speed and stays on the proper course.

#### *Sapper Team*

This team is composed of the squad leader and the remaining squad members. It provides personnel to lay and arm mines. Each soldier carries wrenches and fuses. The squad leader supervises laying and tasks personnel who are not needed for laying to other parties.

#### *Digging Team*

The digging team buries mines. It consists of an NCOIC and several soldiers (may be soldiers from supported maneuver units) who are equipped with suitable digging tools. Increase the arming party by two to speed up the laying process or task personnel who are not needed to other parties. **NOTE: If mines are surface-laid, there is no digging team.**

## SITE LAYOUT

Once the platoon leader has coordinated the location of the minefield(s) with the maneuver commander, siting in the minefield can begin. Siting is the first step in the actual laying process and is done for safety and control. Although the minefield may be emplaced at night or during limited visibility, the siting party should site the minefield under favorable conditions, preferably during daylight. Siting consists of identifying landmarks; establishing routes; and emplacing start, end, and intermediate row markers. Actual control measures (stakes or pickets) should not stand out to such an extent that they give away the minefield orientation, but they must be easily discernible to the laying party.

Certain features, like thick woods and deep, wide streams, are natural obstacles. Mine rows should be laid to reinforce terrain and increase the effectiveness of the minefield.

### Mine Rows

Mine rows are labeled with a letter and should be laid in order. Row A is nearest the enemy, followed by rows B, C, D, E, and so forth. When laying tactical minefields, each row has permanent start and end row markers. Intermediate markers may be required, depending on the row length and the terrain. Platoon leaders determine the number of laying vehicles to be employed. The preferred technique is to use three vehicles so that three rows can be laid simultaneously. Using more than three vehicles is beyond the platoon's $C^2$ capabilities and is not considered. The distance between rows is determined by the following factors:

- Depth and density of the minefield.
- Terrain.
- Suitability of the ground.
- Desired obstacle intent.

**NOTE: Rows are spaced 50 meters apart in standardized row minefields (discussed later in this chapter).**

### Mine Spacing

The minefield OIC determines the mine spacing. The desired density, the availability of laying vehicles, the number of rows, and the possibility of sympathetic detonation (Table 5-3, page 5-8) affect the distance between mines. **NOTE: Mines are spaced 6 meters apart in standardized row minefields (discussed later in this chapter).**

### Control Measures

Control measures are temporary markers that are used to guide vehicles and troops during row-mining operations. Markers are constructed of different materials for different uses. For example, use VS17 panels on poles for start and end row markers, and use M133 hand-emplaced minefield marking set (HEMMS) poles with flags for intermediate markers. Use the following temporary markers:

- Start row (does not replace the mandatory permanent marker).
- Start laying (first intermediate marker after the start row marker).
- Intermediate (used between the last row marker and the next visible point, not more than 100 meters away).
- Change of direction or turning point (actually consists of three markers—warning, turning point, and new direction).
- Stop laying.
- End row (does not replace the mandatory permanent marker).

The following materials may be used to construct temporary markers:

- U-shaped pickets.
- HEMMS poles.
- Wooden posts.
- Steel rods.
- Engineer tape.
- VS17 panels.

Control measures for laying mines at night require lights or infrared (IR) equipment as follows:

- Chem-lights placed in U-shaped pickets or hand-held.

- Directional flashlights taped in U-shaped pickets or hand-held.
- HEMMS lights used with U-shaped pickets or poles.
- Lights from a minefield marking set number 2.
- IR reflectors.

**NOTE: The use of control measures should be incorporated into unit standard operating procedures (SOPs).**

### Procedures

The minefield OIC arrives on the site with the siting-and-recording party. He selects Landmark 1 and then sites the left (or right) boundary fence and start row markers (all start and end row markers are permanent markers). The siting-and-recording party takes distances and azimuths to be used in preparing the recording form. If the tactical situation permits and the marking party is ready, emplacement of the fence should begin.

If the minefield is to have an IOE row, the siting-and-recording party proceeds across the IOE and establishes I1, I1E, I2, I2E, and so on until it reaches the end. Personnel proceed down the right (or left) boundary and emplace start row marker A1. Proceeding from A1 to A2, they place intermediate markers as required. Personnel use different colored markers to identify each row (for example: Row A, red light; Row B, green light; Row C, blue light). For IR markings, they use multiple horizontal IR light sources that are spaced at least 6 inches apart (for example: Row A, one light; Row B, two lights; Row C, three lights). When they reach A2, they emplace an end row marker and repeat the procedure from B1 to B2, C1 to C2, and so on until they emplace all the required control measures (Figure 6-3). The siting-and-recording party establishes Landmark 2 and the left (or right) rear fence location. Personnel also assist the PSG in siting mine dumps near the minefield.

## Mine-Laying Vehicles

Soldiers normally lay row minefields from a tactical vehicle. Consider vulnerability, capacity, and trafficability when selecting a vehicle. Before emplacing the minefield and preparing the vehicle for mine laying, drive it in a random pattern across the minefield site. The random pattern deceives the enemy by masking the actual laying pattern. Load enough mines so that each vehicle can complete at least one entire row before reloading, but do not stack fused mines more than two-high.

## Laying a Row Minefield

The following drills demonstrate how to lay a minefield:

### Drill 1

Squad vehicles arrive on the site and proceed down the left (or right) boundary of the minefield to their assigned row. (A separate party must be detailed to install the IOE.) At the start row marker, the squad vehicle moves into position and prepares to lay mines. The squad leader for Row A directs Vehicle 1 to move out.

Mines are laid on the ground at the required spacing, along the temporary markers positioned by the siting-and-recording party.

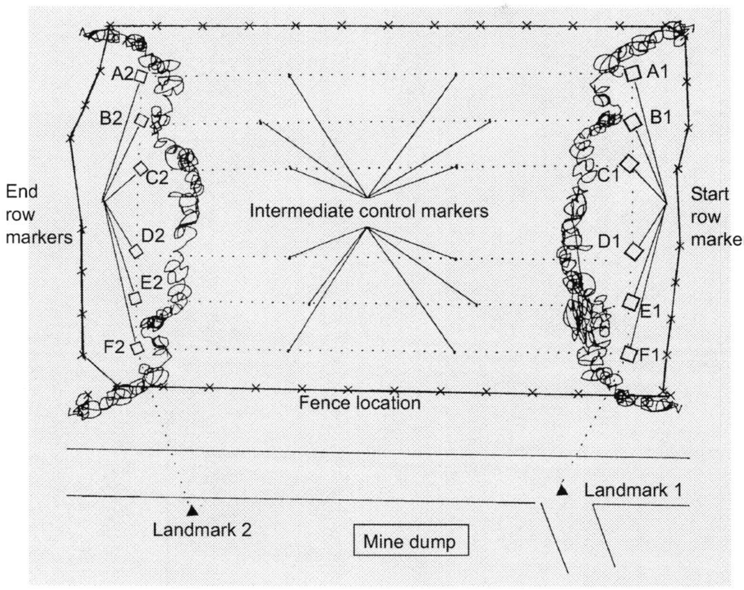


**Figure 6-3. Site layout**

As mines are laid, the arming party moves behind the vehicle and arms the mines. Personnel remove temporary markers installed by the siting-and-recording party and replace the end row markers with permanent markers.

When Vehicle 1 moves a safe distance (approximately 25 meters) along Row A, Vehicle 2 begins to lay mines on Row B. When Vehicle 2 moves a safe distance along Row B, Vehicle 3 begins to lay mines on Row C.

The marking party continues to emplace the left and right boundary fences (Figure 6-4a, page 6-20).

The IOE party exits the minefield outside the left (or right) boundary after it completes the IOE.

When Vehicles 1 and 2 finish their assigned rows, they move past the end row marker and execute a left (or right) turn and wait for Vehicle 3 to complete its row. All vehicles move in column down the left (or right) boundary to the mine dump, load the next row's mines, and then move to their next assigned row. The process of laying and arming mines is repeated (Figure 6-4b, page 6-21).

After the minefield is laid, all the vehicles exit down the left (or right) boundary and out the rear. The marking party completes the rear boundary fence, and the recording party completes DA Form 1355. The OIC or PSG ensures that the DA Form 1355 is complete and accurate and signs it.

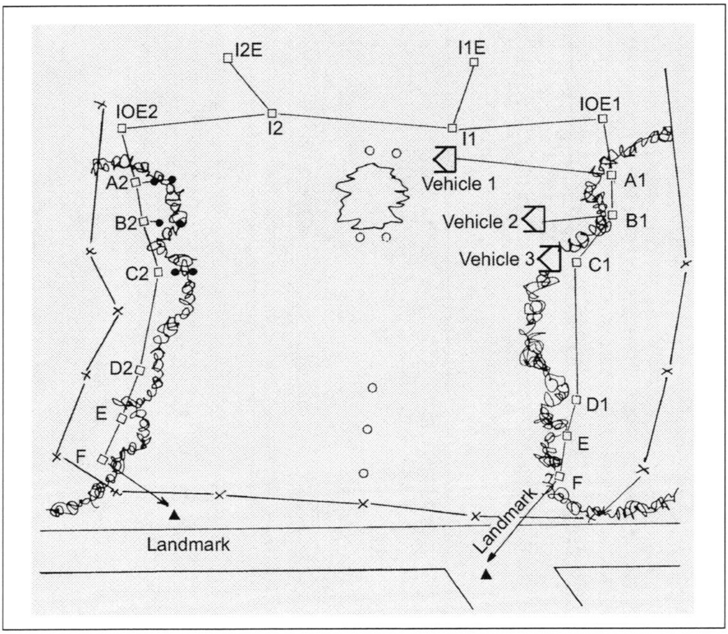


Figure 6-4a. Laying a minefield

## Drill 2

This drill may be used to speed up mine laying; however, strict $C^2$ is vital to ensure security and safety. This method is difficult to use when the terrain is rugged or when weather or visibility is subject to change.

The drill is conducted by three squad vehicles, each laying one row. Row B has turning points and Rows A and C do not. If the minefield has six rows, Row E has turning points and Rows D and F do not. The squad leader (laying leader) in Row B (and Row E, if required) is in charge of the overall laying.

Squad vehicles arrive on the site and proceed down the left (or right) boundary of the minefield to their assigned row. Squad vehicles move into position at start row markers and prepare to lay mines.

The laying leader directs Vehicle 1 to move out on Row A. The sapper team lays mines on the ground at the required spacing. If an IOE is required, the Row A team emplaces the IOE concurrently with Row A and at the same spacing. When Vehicle 1 reaches the IOE short-row start marker, the laying party lays mines along an azimuth designated by the laying leader (Figure 6-5). After the IOE short row is laid, Vehicle 1 returns to Row A and continues laying mines.

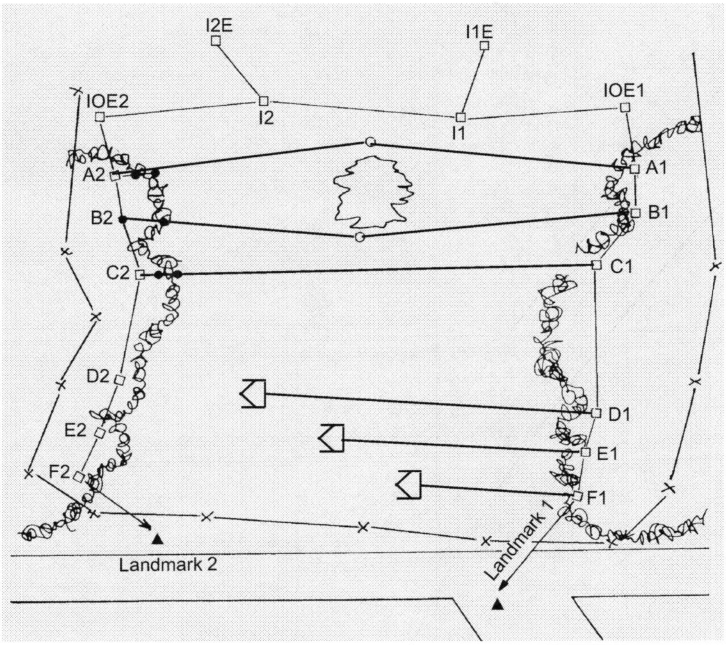


**Figure 6-4b. Laying a minefield (continued)**

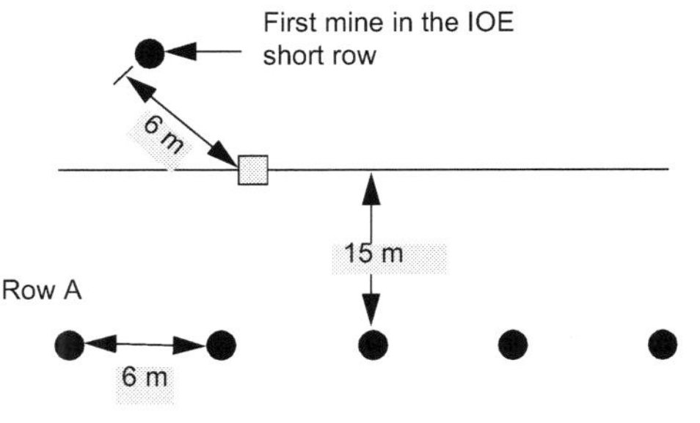


**Figure 6-5. Laying an IOE short row**

After all the mines are laid, the arming party moves behind the vehicle and arms the mines. Personnel remove temporary markers and replace start and end row markers with permanent markers. The arming party must be distinguishable from everyone else. The last member of the arming party should wear a colored vest or carry a specific colored chem-light. No one is allowed behind the last member of the arming party.

The NCOIC completes a strip feeder report (Figure 6-6) and gives it to the recording party. The strip feeder report includes the number of mines laid, the type of mines laid, azimuths of IOE strips and turning points, AHDs emplaced (by cluster number), and any other information (such as omitted mines) the platoon leader requires.

| STRIP FEEDER REPORT FOR STRIP/ROW ________ | | | | | |
|---|---|---|---|---|---|
| Type of Mine | Number of Mines | AHDs by Cluster | IOE-Strip Azimuth | Turning-Point Azimuth | Remarks |
| | | | | | |
| | | | | | |
| | | | | | |
| | | | | | |
| | | | | | |
| | | | | | |
| | | | | | |
| | | | | | |
| | | | | | |
| | | | | | |

**Figure 6-6. Sample strip feeder report**

Vehicle 1 moves down Row A and lays mines until the laying leader directs it to stop. (The laying leader chooses vehicle stops to coincide with the locations of turning points.) The laying leader then directs Vehicle 3 to begin laying mines along Row C. Vehicle 3 lays mines on Row C until the laying leader directs it to stop (somewhere well past Vehicle 1).

Vehicle 2 moves toward Vehicle 1 and begins to lay mines on Row B. He lays mines to within 15 meters of Vehicle 1. Vehicle 2 then turns toward Vehicle 3 and lays mines on Row B to within 15 meters of Vehicle 3. Vehicle 2 then turns back toward Vehicle 1 and continues to lay mines in this pattern until Row B is laid.

Figure 6-7 shows vehicle positions when using the above method to lay a row minefield.

**NOTES:**

**1. At night or during low visibility, Vehicle 1 has two red flashlights and Vehicle 3 has two green flashlights. The flashlights are held side by side, and pointed toward Vehicle 2. The driver of Vehicle 2 moves forward until he is within 15 meters of the lights or until the light holder turns the lights off.**

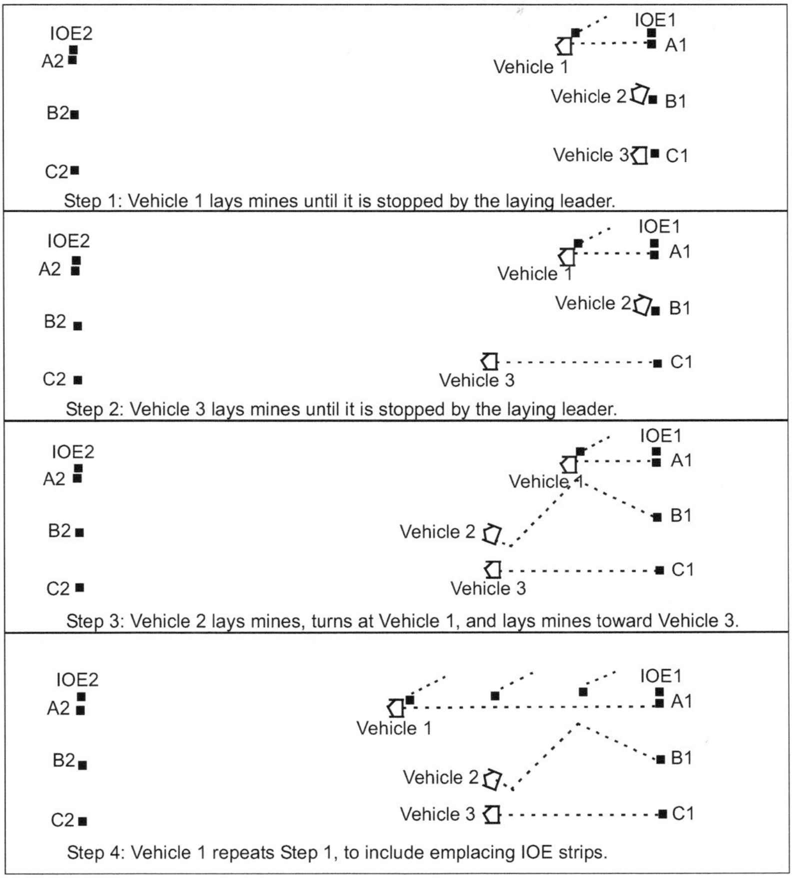


**Figure 6-7. Laying a row minefield**

**2. If the platoon leader feels that low visibility or other reasons preclude the use of vehicle positions as turning points, he may have the siting party emplace turning-point markers (three intermediate markers) for Vehicle 2 to use as a guide. In this event, the three vehicles emplace mines simultaneously.**

After Vehicles 1 and 2 finish their assigned rows, they move past the end row marker, execute a left (or right) turn, and wait for Vehicle 3 to complete its row. All the vehicles move in column down the left (or right) boundary to their next assigned row, if there is one, and continue to lay and arm mines. This process is repeated until the entire minefield is laid. All the vehicles then exit the minefield down the left (or right) boundary and out the rear. The marking party completes the rear boundary fence, and the recording party completes DA Form 1355. The OIC or PSG ensures that the DA Form 1355 is complete and accurate and signs it.

### IMMEDIATE-ACTION DRILL

If the enemy attacks the platoon during minefield emplacement, the laying party should execute the following actions:

- Sapper teams enter vehicles and recover spacing sandbags.
- Vehicle 1 exits the minefield by making a wide turn around the front of the other two vehicles.
- Vehicle 2 follows by making a wide turn around the front of Vehicle 3.
- Vehicle 3 exits the minefield.
- The three squads conduct the immediate-action drill as ordered by the platoon leader.

### SQUAD DRILL

During row mining, the squad in each laying vehicle performs the following actions:

- Squad leader.
  - Directs the squad to start laying mines.
  - Supervises mine arming and placing.
  - Allocates a vehicle, if possible, to help remove spoil from the site.
- Carrier team.
  - Moves the APC to the row start point.
  - Lowers the APC ramp until it is horizontal or opens the rear door. (If using the APC ramp to distribute mines, chains the ramp open to support the weight.)
  - Moves the APC at a low speed (3 to 5 kph) in a straight line toward the row end point.
- Sapper team.
  - Soldier 1 ties the rope to the end of the lowered ramp or the tow pintle.
  - Soldier 2 ties the partially filled sandbag on the other end of the rope. (The rope length from the end of the ramp door to the sandbag is the correct spacing between mines [Figure 6-8]).
  - Soldier 3 (squad leader) positions the team members. Soldier 1 is at the rear of the compartment, Soldier 2 sits on the edge of the APC ramp or open door, and Soldier 4 walks behind the APC.
  - Soldier 1 fuses a mine and passes it to Soldier 2. **(Korea Only: If AP mines are also laid, they are given out simultaneously.)**
  - Soldier 2 records all the mines issued.
  - Soldier 2 places the fused mine on the ground when the sandbag tied to the rope is even with the previously laid mine.
  - Soldier 3 (squad leader) walks behind the vehicle and supervises mine laying.

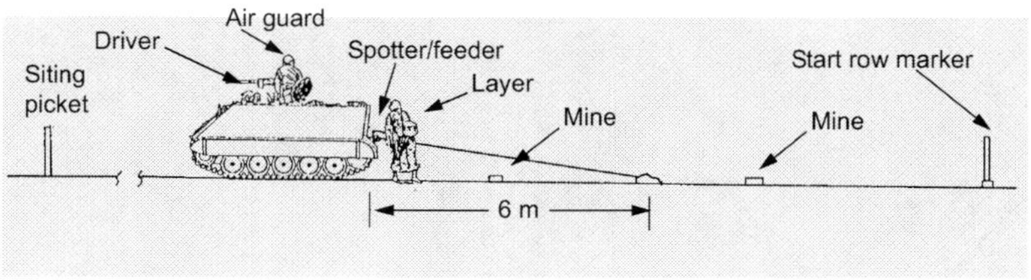


**Figure 6-8. Measuring distances between mines with sandbags**

- Soldier 4 walks behind the vehicle and arms mines.
- After the mine row is armed and camouflaged, Soldier 4 buries pins, clips, and shipping plugs 30 centimeters to the rear of the start row marker.
- The sapper team repeats the above steps until the end of the row is reached.

- Digging team, if needed. (The NCOIC selects the mine to be buried by each soldier and supervises the operation.)
  - Follows the laying party along the friendly side of the row.
  - Digs in mines but leaves them exposed until arming is complete.
  - **Korea Only: Arms AP mines in a cluster before arming AT mines.**

### MARKING, RECORDING, AND REPORTING ROW MINEFIELDS

Marking procedures for row minefields are the same as those for other minefields (see Chapter 2).

Row minefields are recorded on DA Form 1355 (Figures 6-9a and 6-9b, pages 6-26 and 6-27). Reporting procedures for intent, initiation, status, and completion reports are detailed in Chapter 8.

## STANDARDIZED TACTICAL ROW MINEFIELDS

The specific composition of a tactical row minefield depends on METT-TC factors and available resources. To aid in standardization of platoon techniques, four compositions have been developed to match desired obstacle effects. Using standardized minefields facilitates planning the obstacle type, size, and logistical requirements. It is imperative that the design and the effect of these minefields are well understood. They are an integral part of combined arms obstacle doctrine and form the cornerstone of engineer obstacle operations.

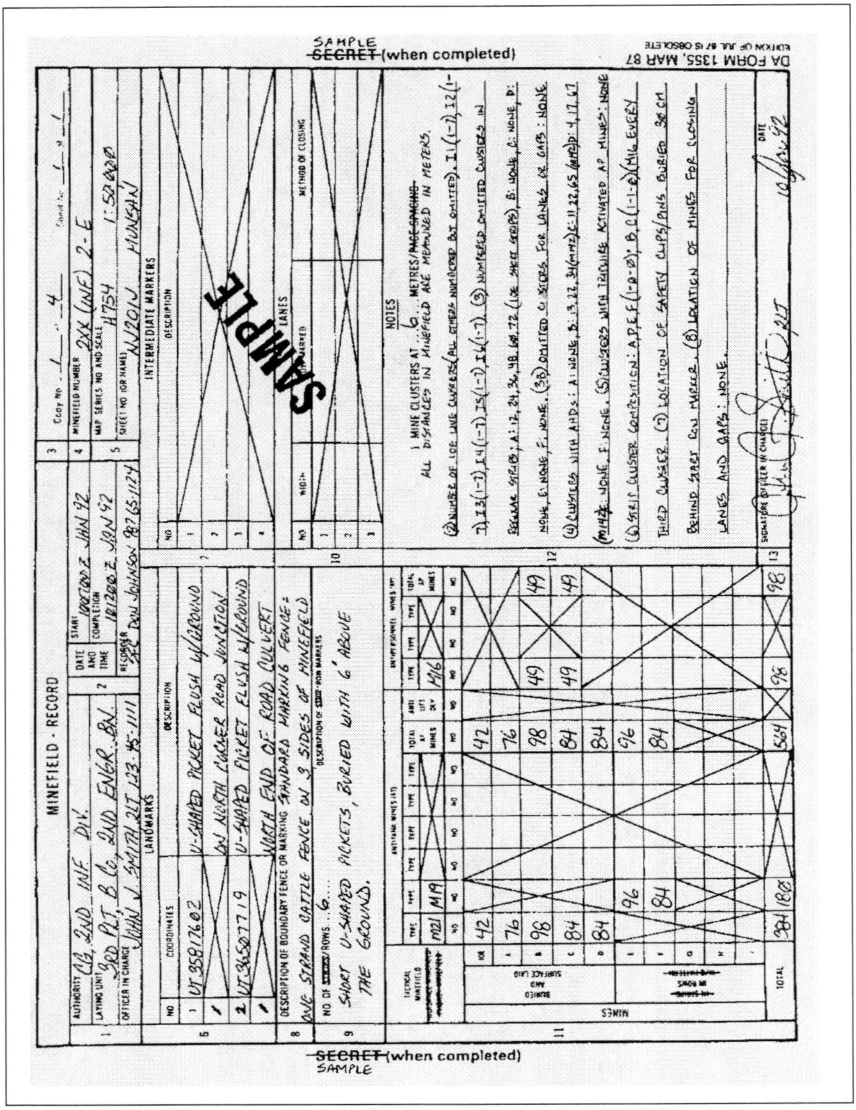

SAMPLE
~~SECRET~~ (when completed)

MINEFIELD - RECORD

| | | | |
|---|---|---|---|
| 1 | AUTHORITY: CG, 2ND INF. DIV. / LAYING UNIT: 3RD PLT, B Co, 2ND ENGR. BN. / OFFICER IN CHARGE: JOHN J. SMITH 2LT 123-45-1111 | 3 | Copy No. 1 of 4; Sheet No. 1 of 1 |
| 2 | DATE AND TIME: START 100700Z JAN 92 / COMPLETION 101200Z JAN 92 / RECORDER SFC DON JOHNSON 987-65-1124 | 4 | MINEFIELD NUMBER: 2XX (INF) 2-E |
| | | 5 | MAP SERIES NO AND SCALE: H754 1:50,000 / SHEET NO (OR NAME): NJ201J MUNSAN |

6 LANDMARKS

| NO | COORDINATES | DESCRIPTION |
|---|---|---|
| 1 | UT 35817602 | U-SHAPED PICKET FLUSH W/GROUND |
| ✓ | | ON NORTH CORNER ROAD JUNCTION |
| 2 | UT 36507719 | U-SHAPED PICKET FLUSH W/GROUND |
| ✓ | | NORTH END OF ROAD CULVERT |

7 INTERMEDIATE MARKERS

| NO | DESCRIPTION |
|---|---|
| 1 | |
| 2 | |
| 3 | |
| 4 | |

8 DESCRIPTION OF BOUNDARY FENCE OR MARKING: STANDARD MARKING FENCE: ONE STRAND CATTLE FENCE ON 3 SIDES OF MINEFIELD

9 NO. OF ~~STRIPS~~/ROWS...6.... DESCRIPTION OF ~~STRIP~~/ROW MARKERS: SHORT U-SHAPED PICKETS, BURIED WITH 6" ABOVE THE GROUND.

10 LANES

| NO | WIDTH | HOW MARKED | METHOD OF CLOSING |
|---|---|---|---|
| 1 | | | |
| 2 | | | |
| 3 | | | |

11 MINES (BURIED AND SURFACE LAID)

| | AT TYPE M21 NO | AT TYPE M19 NO | AT TYPE NO | AT TYPE NO | AT TYPE NO | AT TYPE NO | TOTAL AT MINES NO | ANTI LIFT DEV NO | AP TYPE M16 NO | AP TYPE NO | AP TYPE NO | TOTAL AP MINES NO |
|---|---|---|---|---|---|---|---|---|---|---|---|---|
| IOE | 42 | | | | | | 42 | | | | | |
| A | 76 | | | | | | 76 | | | | | |
| B | 98 | | | | | | 98 | | 49 | | | 49 |
| C | 84 | | | | | | 84 | | 49 | | | 49 |
| D | 84 | | | | | | 84 | | | | | |
| E | | 96 | | | | | 96 | | | | | |
| F | | 84 | | | | | 84 | | | | | |
| G | | | | | | | | | | | | |
| H | | | | | | | | | | | | |
| I | | | | | | | | | | | | |
| TOTAL | 384 | 180 | | | | | 564 | | 98 | | | 98 |

12 NOTES

1 MINE CLUSTERS AT ...6... METRES/~~PACE SPACING~~
ALL DISTANCES IN MINEFIELD ARE MEASURED IN METERS.
(2) NUMBER OF IOE LINE CLUSTERS (ALL OTHERS NUMBERED BUT OMITTED). I1(1-7), I2(1-7), I3(1-7), I4(1-7), I5(1-7), I6(1-7). (3) NUMBERED OMITTED CLUSTERS IN REGULAR STRIPS; A: 12, 24, 36, 48, 60, 72 (IOE SHORT STRIPS), B: NONE, C: NONE, D: NONE, E: NONE, F: NONE. (3B) OMITTED CLUSTERS FOR LANES OR GAPS: NONE. (4) CLUSTERS WITH AHDS: A: NONE, B: 13, 22, 34(M142)C: 11, 22, 65 (M142)D: 4, 17, 67 (M142)E: NONE, F: NONE. (5) CLUSTERS WITH TRIPWIRE ACTIVATED AP MINES: NONE (6) STRIP CLUSTER COMPOSITION: A,D,E,F(1-0-0); B,C(1-1-0)(M16 EVERY THIRD CLUSTER. (7) LOCATION OF SAFETY CLIPS/PINS BURIED 30 CM BEHIND START ROW MARKER. (8) LOCATION OF MINES FOR CLOSING LANES AND GAPS: NONE.

13 SIGNATURE OFFICER IN CHARGE: John J. Smith, 2LT DATE: 10 Jan 92

DA FORM 1355, MAR 87 EDITION OF JUL 87 IS OBSOLETE

~~SECRET~~ (when completed)
SAMPLE

Figure 6-9a. Sample DA Form 1355 for a row minefield (front)

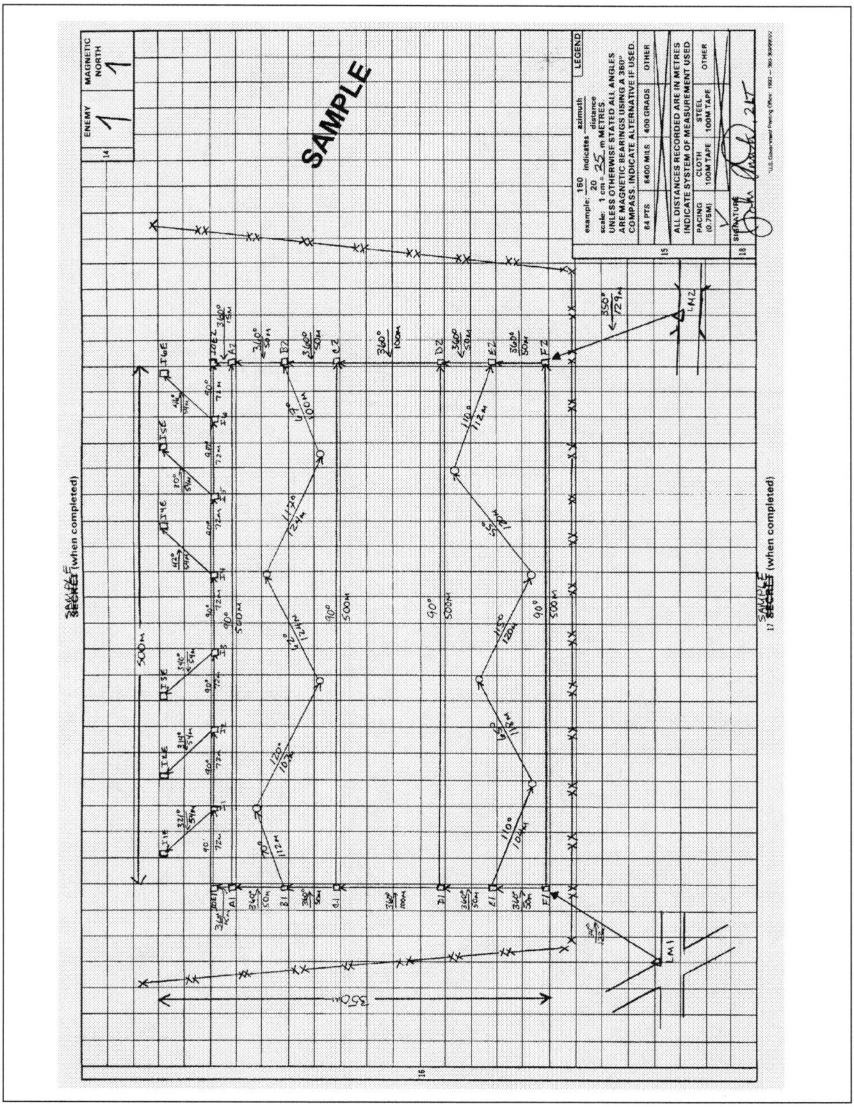


Figure 6-9b. Sample DA Form 1355 for a row minefield (back)

## Disrupt and Fix

Disrupt and fix row minefields (Figure 6-10) are similarly constructed, but a fix minefield has an IOE and does not have AHDs.

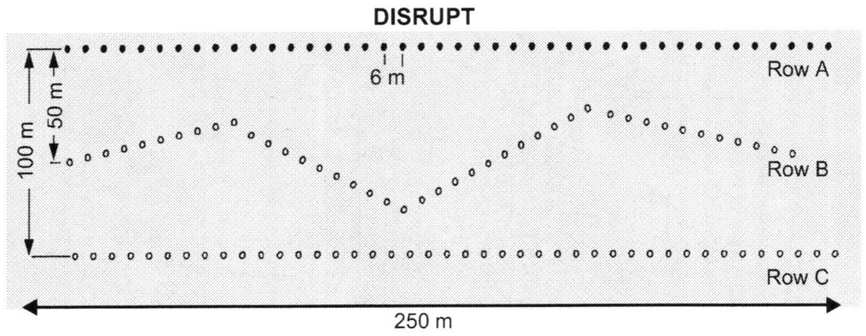


| | Disrupt | Fix |
|---|---|---|
| Front | 250 m | 250 m |
| Depth | 100 m | 120 m |
| Rows of AT full-width mines | 1 | 1 |
| Rows of AT track-width mines | 2 | 2 |
| IOE | No | Yes |
| AHD | No | No |
| Platoon hours required | 1.5 | 1.5 |
| Full-width mines | 42 | 63 |
| Track-width mines | 84 | 84 |
| Density | 0.5 | 0.6 |

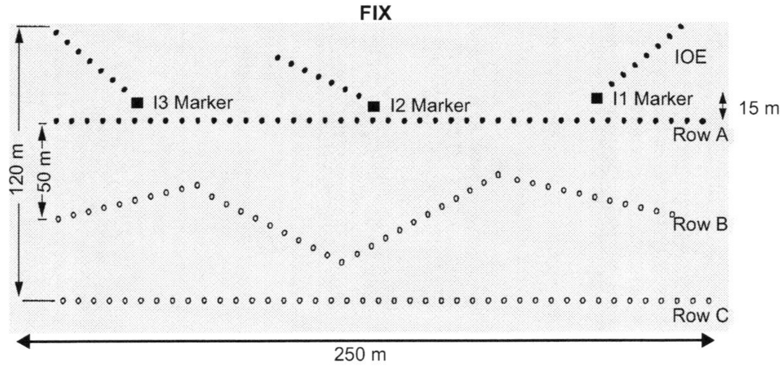


Figure 6-10. Standardized disrupt and fix row minefields

Disrupt and fix row minefields are emplaced as follows:

- Row A.
  - 42 full-width AT mines (tilt-rod) are placed 6 meters apart.
  - No turning points.
  - Mines are surface-laid (staked) or buried.
- Row B.
  - Start and end row markers are emplaced 50 meters behind Row A.
  - 42 track-width AT mines are placed 6 meters apart.
  - Preferably no more than 3 turning points.
  - Mines are surface-laid or buried.
- Row C.
  - Emplaced 100 meters behind Row A.
  - 42 track-width AT mines are placed 6 meters apart.
  - No turning points.
  - Mines are surface-laid or buried.
- IOE (fix minefield only).
  - 3 IOE short rows.
  - A separate IOE baseline.
  - The IOE baseline is on the enemy side, 15 meters from Row A.
  - 7 full-width AT mines are placed 6 meters apart on each IOE short row.
  - Mines are buried.
  - The first IOE short row is placed 48 meters from the IOE end marker, the next short row is 84 meters from the first short row, and the third short row is 84 meters from the second short row.

## TURN

A turn minefield (Figure 6-11, page 6-30) consists of four rows of full-width mines and two rows of track-width mines.

A turn row minefield is emplaced as follows:

- Row A.
  - 84 full-width AT mines (tilt-rod) are placed 6 meters apart.
  - No turning points.
  - Mines are surface-laid (staked) or buried.
- Row B.
  - Start and end row markers are emplaced 50 meters behind Row A.
  - 84 full-width AT mines (tilt-rod) are placed 6 meters apart.

**NOTE: The center of the minefield shown in this figure is omitted for clarity.**

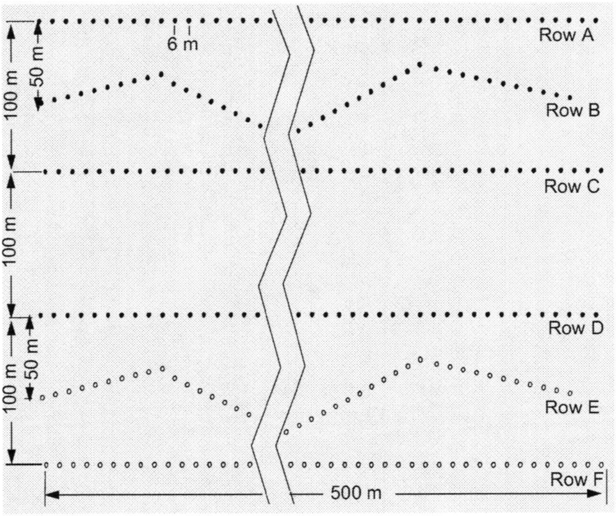


| | |
|---|---|
| Front | 500 m |
| Depth | 300 m |
| Rows of AT full-width mines | 4 |
| Rows of AT track-width mines | 2 |
| IOE | No |
| AHD | No |
| Platoon hours required | 3.5 |
| Full-width mines | 336 |
| Track-width mines | 168 |
| Density | 1.0 |

**Figure 6-11. Standardized turn row minefield**

— Preferably no more than 5 turning points.

— Mines are surface-laid (staked) or buried.

- Row C.

— Emplaced 100 meters behind Row A.

— 84 full-width AT mines (tilt-rod) are placed 6 meters apart.

— No turning points.

— Mines are surface-laid (staked) or buried.

- Row D.
  - — Emplaced 100 meters behind Row C.
  - — 84 full-width AT mines (tilt-rod) are placed 6 meters apart.
  - — No turning points.
  - — Mines are surface-laid (staked) or buried.
- Row E.
  - — Start and end row markers are emplaced 50 meters behind Row D.
  - — 84 track-width AT mines are placed 6 meters apart.
  - — Preferably no more than 5 turning points.
  - — Mines are surface-laid or buried.
- Row F.
  - — Emplaced 100 meters behind Row D.
  - — 84 track-width AT mines are placed 6 meters apart.
  - — No turning points.
  - — Mines are surface-laid or buried.

## BLOCK

A block minefield (Figure 6-12, page 6-32) has an IOE **(Korea Only: and AP mines)** and has 20 percent AHDs in two of its rows of full-width mines. AHDs are placed in Rows B and C for the best effect. **(Korea Only: A block minefield also requires 84 M16 or M14 AP mines and has a density of 0.17 M16/M14 AP mine per linear meter.)**

A block row minefield is emplaced as follows:

- Row A.
  - — 84 full-width AT mines (tilt-rod) are placed 6 meters apart.
  - — No turning points.
  - — Mines are surface-laid (staked) or buried.
- Row B.
  - — Start and end row markers are emplaced 50 meters behind Row A.
  - — 84 full-width AT mines (tilt-rod) are placed 6 meters apart.
  - — Preferably no more than 5 turning points.
  - — Mines are surface-laid (staked) or buried.
- Row C.
  - — Emplaced 100 meters behind Row A.
  - — 84 full-width AT mines (tilt-rod) are placed 6 meters apart.
  - — No turning points.

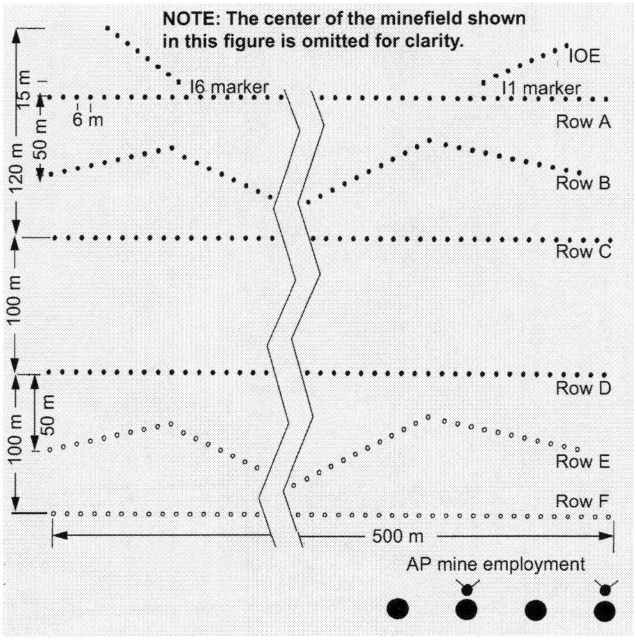


| | |
|---|---|
| Front | 500 m |
| Depth | 320 m |
| Rows of AT full-width mines | 4 |
| Rows of AT track-width mines | 2 |
| IOE | Yes |
| AHD | Yes |
| Platoon hours required | 5 |
| Full-width mines | 378 |
| Track-width mines | 168 |
| AT mine density | 1.1 |
| **Korea Only:** **AP frag mines** | **84** |
| **Korea Only:** **AP mine density** | **0.17** |

**Figure 6-12. Standardized block row minefield**

- — Mines are surface-laid (staked) or buried.
- Row D.
  - — Emplaced 100 meters behind Row C.
  - — 84 full-width AT mines (tilt-rod) are placed 6 meters apart.
  - — No turning points.
  - — Mines are surface-laid (staked) or buried.
- Row E.
  - — Start and end row markers are emplaced 50 meters behind Row D.
  - — 84 track-width AT mines are placed 6 meters apart.
  - — Preferably no more than 5 turning points.
  - — Mines are surface-laid or buried.
- Row F.
  - — Emplaced 100 meters behind Row D.
  - — 84 track-width AT mines are placed 6 meters apart.
  - — No turning points.
  - — Mines are surface-laid or buried.
- IOE.
  - — 6 IOE short rows.
  - — A separate IOE baseline.
  - — 7 full-width AT mines are placed 6 meters apart on each IOE short row.
  - — Mines are buried.
  - — The first IOE short row is placed 72 meters from the IOE end marker, and the five subsequent IOE short rows are placed at 72-meter intervals on the IOE baseline.
- **Korea Only: AP mines.**
  - **— AP mines are placed on two rows of full-width AT mines.**
  - **— AP mines are placed in a cluster around AT mines.**
  - **— One AP mine is placed in front of every other AT mine.**

## HASTY PROTECTIVE ROW MINEFIELDS

Hasty protective row minefields are temporary in nature and are used as part of a unit's defensive perimeter. Units usually use mines from their basic load. If time permits, mines should be buried to increase their effectiveness, but they can be surface-laid. The minefield can be easily recovered because AHDs, nonmetallic mines, or low-metallic mines are not used. Mines are employed outside the hand-grenade range but within the range of small-caliber weapons. All mines are picked up by the emplacing unit upon leaving the area, unless enemy pressure prevents mine retrieval or the minefield is being transferred to a relieving commander.

### RULES

The brigade commander has the initial authority to employ hasty protective row minefields. This authority may be delegated to the battalion or company commander on a mission-by-mission basis.

Most of the rules governing hasty protective row mining are defined in STANAG 2036. A summary of those rules and some additional rules that apply are shown below.

- Rows.
  - Rows are marked and recorded. They are designated by letters (A, B, and so forth) with Row A being closest to the enemy.
  - The minimum distance between rows of AT mines is 8 meters.
  - **Korea Only: The minimum distance between any row and a row containing AP mines is 15 meters.**
  - The distance between the start row marker and the first mine in a row is the mine spacing for that row.
  - An IOE row is not used.
  - Start and end row markers are permanent markers and must be made of detectable material.
- General.
  - The spacing between mines is at least 4 meters to prevent sympathetic detonation. There is no maximum distance between mines within the row.
  - Mines are at least 15 meters from the perimeter fence.
  - The minefield must be fenced on all sides if M18A1 AP mines are employed and the minefield will be in place for more than 72 hours.
  - The minefield has at least one landmark that is located to the rear, never to the extreme side or front.
  - The minefield has an easily identifiable RP (tree, stump, stake).
  - Back azimuths are not used to record the minefield.
  - Measurements are in meters.
  - AHDs, nonmetallic mines, or low-metallic mines are not used.
  - Minefields are recorded on DA Form 1355-1-R (see Chapter 8).

### SITE LAYOUT

Requesting and receiving permission to lay mines is the first step when emplacing a hasty protective row minefield. The next step is conducting a thorough leader's reconnaissance of the proposed minefield area. Identify mine locations that cover likely AAs, enhance key weapon systems, and cover dead space. Establish an easily identifiable RP between the minefield and the unit position. From the RP, visualize mines as running in rows parallel to the unit position (Figure 6-13). After the RP is established and the minefield is visualized, complete recording and emplace the mines, but do not arm them. This procedure simplifies recording and makes retrieval quicker and safer.

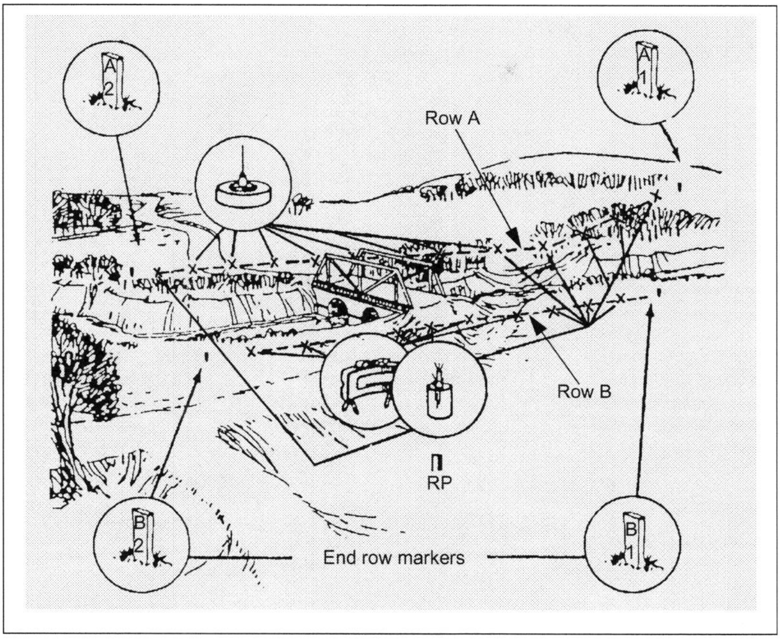


Figure 6-13. Site layout

## Mine Rows

The row closest to the enemy is designated as Row A; succeeding rows are designated B, C, D, and so on. The ends of rows are shown by two end row markers. They are labeled with the letter of the row and the number 1 for the right end of the row or the number 2 for the left end of the row. The rows are numbered from right to left, facing the enemy. The marker should be an easily identifiable object, such as a wooden stake with a nail or a steel picket so that it can be found with an AN/PSS-12 mine detector.

## Laying Procedures

From the RP, the leader measures the magnetic azimuth, in degrees, to a selected point on the right side (facing the enemy) of the tentative minefield. He paces off the distance and records it in meters. This point (B1) marks the beginning of the second row. The leader places a marker at B1 and records the azimuth and the distance on DA Form 1355-1-R.

From B1, the leader measures the azimuth and distance to a second point on the right side of the minefield (facing the enemy). He places a marker at this point (A1), and records the information.

The leader measures the distance and the azimuth from A1 to the location of the first mine in that row. The distance (or spacing) from the end row marker to the first mine is the mine spacing for that row. After the leader records the location, the mine is emplaced, but it is not armed.

The distance and azimuth are measured from the first mine to the second mine, and so on, until all the mines are emplaced and the locations are recorded. This procedure is repeated for the second row. As each mine is recorded, it is assigned a number to identify it in the minefield record.

When the last mine location is recorded for a row, the distance and the azimuth are measured from that point to another arbitrary point, A2 or B2. A marker is placed here in the same manner as A1 and B1. Next, the distance and azimuth from the RP to B2 and from B2 to A2 are measured and recorded.

When all the mines have been placed and recorded, the leader measures the distance and the azimuth between the RP and a permanent landmark that can be found on the map. He records the information on DA Form 1355-1-R. The landmark is used to assist others in locating the minefield if it is transferred or unexpectedly abandoned.

Mines can be armed after recording is complete. Mines nearest the enemy are armed first, allowing soldiers to safely work their way back to the unit position. Pins and clips are buried 30 centimeters behind row markers, the RP, or any easily identifiable, accessible location. Record the location of the pins and clips in the remarks section of DA Form 1355-1-R. The leader then reports the completion of the minefield to higher headquarters.

If the minefield is transferred to another unit, the transferring unit leader briefs the gaining unit leader. The gaining unit leader signs and dates the mines-transferred block on the DA Form 1355-1-R. The form is destroyed when the minefield is removed. If the minefield is abandoned unexpectedly, the DA Form 1355-1-R is forwarded to higher headquarters.

### Mine Removal

When removing mines from a hasty protective row minefield, the leader first determines the best method to use:

- If the minefield has been under constant observation from the time it was laid and has not been tampered with, the squad leader directs the personnel who laid the mines to pick up the same mines. He uses DA Form 1355-1-R preceded by a mine detector to determine the types of mines to be removed and where they are located.
- If the minefield has not been under constant observation, may have been tampered with, or the personnel who laid the mines are not available or do not remember the location of the mines, the squad leader uses DA Form 1355-1-R and a clearance team as outlined in Chapter 11 to locate and remove mines.

The leader retrieves safety devices, shipping plugs, and other items that accompanied the emplaced mines. Using the azimuths and distances provided on the DA Form 1355-1-R, the removal team starts at the RP and moves to B1. They then move from B1 to the mine and remove the mine. If B1 is destroyed, the team moves from the RP to B2. The team then shoots a back azimuth (subtract 180 degrees) from the recorded azimuth from B2 to the first mine and removes the mine. Personnel continue this process until all the mines have been removed. The stakes at Al, B1, A2, and B2 are necessary because it is safer to find a stake than to find an armed mine.

The removal team observes basic safety precautions by maintaining 30 meters between personnel, not running, and moving only in cleared areas. The team starts with the row closest to the defender and works toward the enemy. Personnel—

- Check the sides and bottoms of the mines for AHDs, and disarm or mark the mines as they are found.
- Replace all pins, clips, and other safety devices before the mines are removed from the ground.
- Turn arming dials to *SAFE* or *UNARMED;* or if mines have screw-type fuses, remove the fuses and take them away from the mines.
- Lift the mines from the holes after they have been rendered safe.
  - If a mine was put in place and kept in sight by the individual who removes it, he lifts it directly from the hole after rendering it safe.
  - If a mine has not been kept in sight, the individual attaches a 60-meter rope or wire to the mine, takes cover, and pulls the mine from the hole.
- As each mine is removed, place a tick mark beside it on the DA Form 1355-1-R.
- Assemble all the mines in one location for accountability.

**NOTE: AHDs are not used in hasty protective row minefields. However, as a safety precaution, consider all mines to be equipped with AHDs until proven otherwise.**

The leader confirms the removal of the mines and accounts for the number of mines, by type, as recorded on the DA Form 1355-1-R. The leader may find it necessary to confirm an exploded mine to account for all the mines.

To confirm a mine explosion that was not witnessed, identify the crater or traces of burnt soil made by the detonated mine and place a tick mark beside the mine number on the DA Form 1355-1-R. Ensure that the crater found in the vicinity of the mine was caused by a land mine and not by artillery. A mine crater is normally circular, and it shows traces of burnt soil. The impact and the soil dispersion of artillery is normally elongated.

The squad leader confirms that each mine is disarmed and safe. The removal team cleans and repacks serviceable mines for future use and destroys the others; they repack serviceable mines in their original containers and store them according to the unit SOP and local regulations. The removal team removes and stores the row markers. The leader submits a report to his higher headquarters stating that the minefield has been removed and that the area is clear.

This chapter implements STANAG 2036.

## Chapter 7

# Standard-Pattern Minefields

Emplacing standard-pattern minefields is laborious and time-consuming, but it allows better mine concealment than row mining. Standard-pattern laying is well suited for protective and nuisance minefields. It can be used in terrain where the nature of the ground makes row mining impracticable.

To achieve their maximum effect, mines must be laid so that they cannot be seen and so that a vehicle's wheel or track or a person's foot exerts enough pressure to detonate them.

The method used to lay mines depends on the mine operation, the type of ground in which the mine is to be laid, and the type of ground cover that is available for camouflage.

## COMPONENTS

### MINE STRIPS

The mine strip is the foundation of a standard-pattern minefield. If a mine strip was laid in one straight line, the enemy could easily locate the mines; therefore, mine strips are laid in several segments as shown in Figure 7-1, page 7-2.

When siting, laying, and recording mine strips, all measurements are expressed in meters. Directions are recorded as magnetic azimuths, in degrees.

### MINE CLUSTERS

The cluster is the basic unit of a minefield. It consists of one to five mines that are laid within a 2-meter-radius semicircle (Figure 7-2, page 7-3). When clusters are placed in a mine strip, they are numbered progressively and may consist of—

- One AT mine.
- **Korea Only: One AT mine and one to four AP mines.**
- **Korea Only: One to five AP mines.**

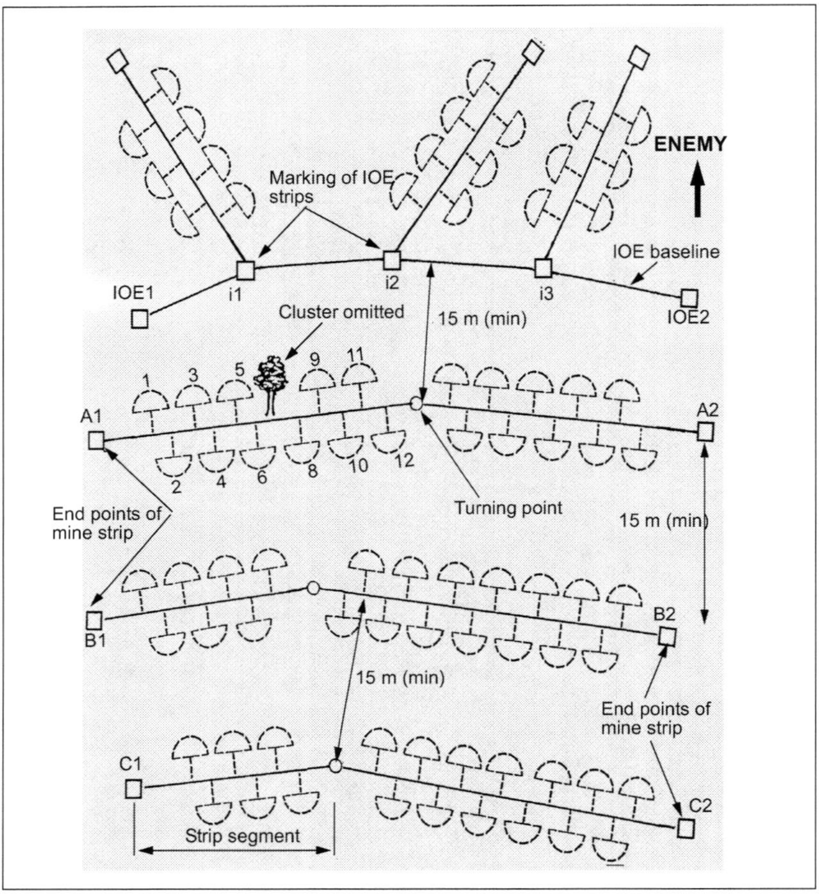


**Figure 7-1. Minefield layout**

Clusters are placed at 6-meter intervals, center to center, to form rows. Two parallel rows, 6 meters apart, form a mine strip. The arrangement of clusters in a mine strip is shown in Figure 7-3.

## RULES FOR POSITIONING CLUSTERS WITHIN A STRIP

The first cluster is placed on the enemy side of the strip centerline, 6 meters from the beginning-of-strip marker. The following clusters are numbered consecutively. Odd-numbered clusters are always on the enemy side of the

| | |
|---|---|
| Individual laying of AT mine (to be at base of cluster) |  |
| **Korea Only: Cluster with one AT mine plus several AP mines within or on a 2-meter semicircle of the AT mine (the AT mine must be the base mine)** | 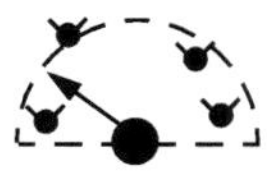 |
| **Korea Only: Individual laying of AP mine (to be at base of cluster)** |  |
| **Korea Only: Cluster with several AP mines within or on a 2-meter semicircle of the center AP mine (the most easily detected mine is the base mine)** | 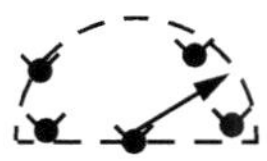 |

**Figure 7-2. Cluster compositions**

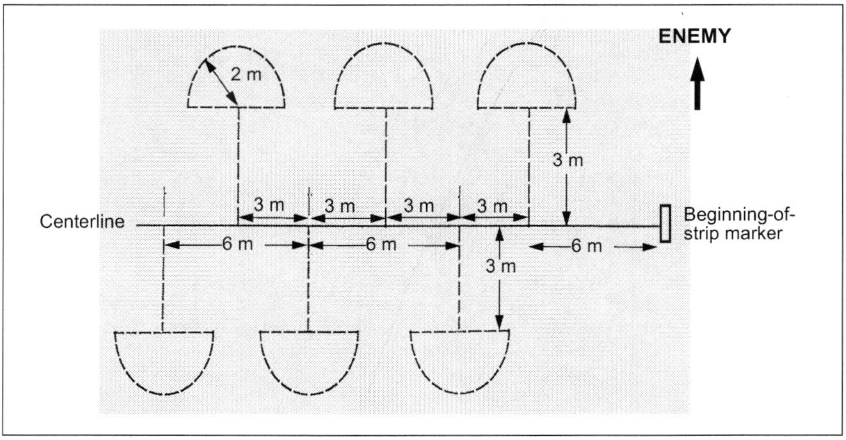


**Figure 7-3. Arrangement of clusters in a mine strip**

strip centerline. The direction of laying follows the numbering (A1 to A2, B1 to B2, and so on).

The IOE is normally the first part of the minefield encountered by the enemy. It consists of a baseline from which short strips are extended (Figure 7-4, page 7-4). Short strips along the IOE deceive the enemy on the minefield's pattern, spacing, and size. IOE placement and composition are largely dictated by the time allowed for laying the minefield, terrain conditions at the laying site, and the tactical situation.

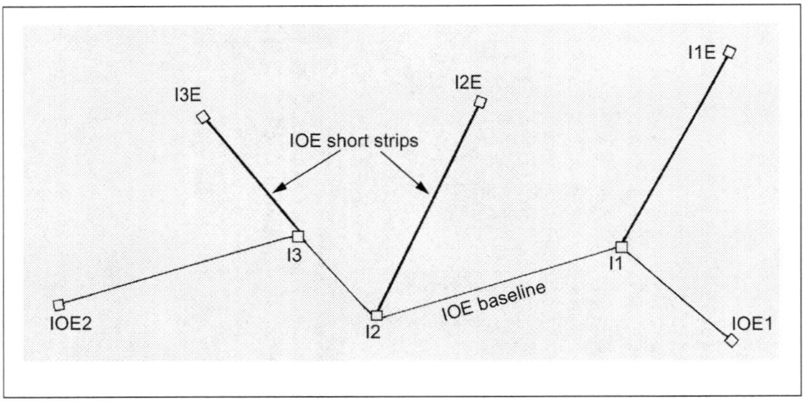


**Figure 7-4. IOE baseline with short strips**

The IOE baseline extends from one end point (IOE1) to another end point (IOE2). The laying direction is indicated by end-of-strip markers. Laying always begins at IOE1. Intermediate or turning points are marked in consecutive order beginning with I1. On the enemy side of the IOE baseline, short strips are extended from turning points at irregular angles. They are identified by turning-point markers.

Turning points should be no more than 45 degrees from the last azimuth. The length of short strips is not standard. A marker is emplaced at the end of each short strip. Markers are numbered in consecutive order beginning with I1E for recording purposes. No trip wires are used in the IOE, but AHDs may be employed.

**Korea Only: AP mines actuated by trip wires are placed on the enemy side of each regular strip. No more than one mine per cluster uses trip wires, and no more than two trip wires extend from the mine. Trip wires are angled toward the enemy and should be at least 2 meters from the cluster, the lane border, and the minefield boundary. Trip wires are only used with AP fragmentation mines; they are not considered to be AHDs.**

## STANDARD-PATTERN MINEFIELD RULES

The following rules apply to tactical and protective standard-pattern minefields. They do not apply to nuisance minefields.

### Clusters

- A cluster is a 2-meter-radius semicircle located 3 meters off the strip centerline.
- There are two types of clusters—live and omitted.

- A live cluster contains one AT mine. **(Korea Only: A live cluster contains as many as five mines, only one can be an AT mine).**
- Omitted clusters do not contain mines, but they are numbered and recorded on DA Form 1355 (see Chapter 8).

- Clusters are omitted within lanes and gaps; in areas less than 2 meters from boundaries, lanes, or another cluster (including the IOE); and in areas where the terrain (trees, rocks) prohibits emplacement.
- The base mine in a live cluster is the first mine laid. It is 3 meters from the strip centerline.
- When a live cluster contains an AT mine, the mine is always used as the base mine. **(Korea Only: If an AT mine is not present, the largest metallic AP mine is the base mine.)**
- The first cluster in a mine strip is located on the enemy side, 6 meters from the beginning-of-strip marker.
- The minimum distance between a cluster and a lane, a gap, a boundary, or another cluster is 2 meters (measured from the edge of the cluster).
- Cluster composition is the number of mines, by type, in any cluster in a specific group.
- Cluster composition remains the same through the entire mine strip and is recorded on DA Form 1355.
- **Korea Only: The types of AP mines may vary within a cluster.**
- The cluster boundary must at least 15 meters from the minefield perimeter fence.
- Clusters are numbered, beginning with the first cluster on the enemy side. Odd-numbered clusters are always on the enemy side of the strip, and even-numbered clusters are always on the friendly side.

**Regular Strips**

- A regular strip (sometimes referred to as a *lettered strip*) consists of a strip centerline and two rows of clusters (Row 1, enemy side; Row 2, friendly side).
- Regular strips are marked and recorded. They are designated by letters (A, B, and so forth), with Strip A being closest to the enemy.
- A standard-pattern minefield contains at least three regular strips.
- The minimum distance between strip centerlines is 15 meters; there is no maximum distance.
- Safety tapes are used to ensure that personnel installing trip wires do not move forward into armed clusters. A safety tape is used behind each regular strip. Safety tapes are 8 meters from the strip centerline (3 meters from the outer edge of the cluster).

- The marking of end points indicates the direction of laying (for example, A1 to A2).

### IOE

- The IOE consists of a baseline from which short strips are extended.
- The IOE is located on the enemy side of the minefield.
- The number of clusters in an IOE is approximately one-third the number used in a regular strip.
- The first cluster along a short strip is placed on the enemy side and must be at least 6 meters from the IOE baseline; the cluster boundary must be at least 2 meters from the IOE baseline (Figure 7-5). If the short strip is exactly parallel to the enemy direction of travel, the NCOIC designates the enemy side of the strip.
- The IOE baseline is labeled at the beginning (IOE1) and end (IOE2) according to the direction mines are emplaced.
- Short strips are labeled at turning points (I1) and at the end (I1E).
- The IOE contains a safety tape that is 2 meters behind the IOE baseline and runs parallel with it.
- The IOE baseline is at least 15 meters from any point of the strip centerline of a regular strip; there is no maximum distance.
- An IOE short strip is at least 15 meters from another IOE short strip; there is no maximum distance.
- AHDs may be employed.

### IOE Short Strips

- Short strips originate from turning points along the IOE baseline.
- The number and length of short strips depend on the tactical situation and the resources available.

### Trip Wires (Korea Only)

- **Trip wires are not used in an IOE.**
- **Trip wires may be used in regular strips, but only one mine per cluster may be actuated by a trip wire.**
- **Trip wires are employed no closer than every third cluster.**
- **No more than two trip wires can be used on one mine.**
- **Trip wires are used only on the enemy side of the strip.**
- **Trip wires are not considered AHDs.**
- **Trip wires are located at least 2 meters from a lane, a safety tape, a cluster, another trip wire, the IOE baseline, and the minefield perimeter fence.**
- **Trip wires can only be used with AP fragmentation mines.**

### Turning Points

- Clusters must be at least 3 meters from turning points.

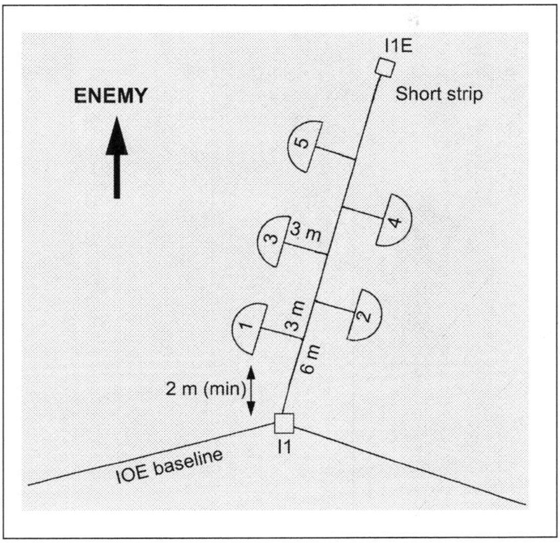


**Figure 7-5. Clusters on an IOE short strip**

- The first cluster after a turning point is laid on the opposite side of the strip centerline from the last cluster before the turning point.
- The angle of any given turning point cannot exceed 45 degrees from the last azimuth. (This ensures a minimum distance of 2 meters between cluster boundaries in the same row.)

**Lanes**

- Minefield lanes (Figure 7-6, page 7-8) are used by dismounted patrols and vehicles.
- Lanes are sited before laying begins.
- Lane locations should not be obvious.
- Clusters are not laid within 2 meters of lane edges.
- Lanes are zigzagged, not straight.
- Lanes cross the strip centerline at approximately right angles.
- Direction changes will not exceed 45 degrees. (This ensures that long vehicles will be able to negotiate turns.)
- The number of lanes must be sufficient to ensure that no one lane is overused and turned into an obvious track.
- Sufficient mines are stockpiled so that the responsible unit can close lanes suspected of being located by the enemy.
- Recommended minefield lane widths are—
  - — Footpath: 1 meter.

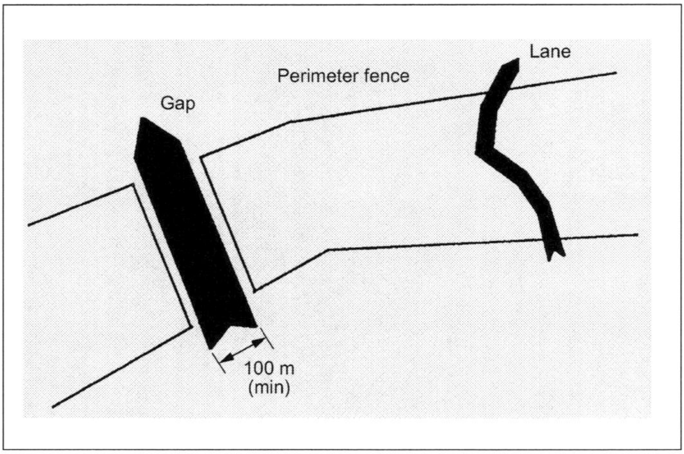


**Figure 7-6. Minefield lanes and gaps**

— One-way vehicle lane: 8 meters.

— Two-way vehicle lane: 16 meters.

**Gaps**

- Minefield gaps (Figure 7-6) are left so that friendly forces can pass through the minefield in tactical formation.
- Gaps must be at least 100 meters wide.
- Gaps are sited before laying begins.
- Gaps are located along recognizable features (fences, tracks, creeks).
- Gaps should run straight through a minefield and not contain bends.
- Sufficient mines must be stockpiled so that the responsible unit can close gaps when necessary.
- Gaps should closely resemble the rest of the minefield so that they will not be discovered by the enemy.
- The ground within a gap should be disturbed with tracks to represent the passage of a mine-carrying vehicle.
- Signs of mine laying (digging, scattered spoil, crates) should be visible in a gap.

**General**

- The farthest extremities of a regular strip determine the minefield front.
- Minefield depth is measured from the first to the last strip and includes the IOE (if applicable).
- Strips can be laid left to right or right to left.
- Back azimuths are not used to record the minefield.

- The minefield will have two landmarks located to the rear, never to the extreme side or front.
- GPSs can be used to determine the coordinates for minefield landmarks and reference points.

**WARNING**
**Do not use GPSs to chart or record minefield perimeter coordinates or to determine safe routes through or around existing minefields.**

- If landmarks are more than 200 meters away from the last regular strip or are out of the direct line of sight, intermediate markers are placed at least 75 meters from the last end-of-strip marker.
- Landmarks can be used for more than one minefield. They are recorded on DA Form 1355.

## LOGISTICAL CALCULATIONS

Emplacing standard-pattern minefields allows the unit to calculate the number of mines required for a minefield accurately.

To simplify the calculation process, a minefield requirements computation work sheet (Figure 6-1, pages 6-5 through 6-8) has been developed. Step-by-step procedures for completing the work sheet are shown in Figure 6-2, pages 6-9 through 6-14.

### CLUSTER CALCULATION

Accuracy is essential when emplacing a standard-pattern minefield. The following check system has been designed so that the minefield OIC can accurately record the number of mines laid. Compare the strip feeder report with the—

- Number of pins and clips returned by the laying party.
- Mine tally sheet.
- Cluster computation.

**EXAMPLE:** You receive a strip feeder report for Strip A. There are 26 clusters in the strip. The strip has three segments (30, 42, and 21 meters) and two turning points.

**Step 1.** Add the total length of the strip as determined by the recording party.

$30 + 42 + 21 = 93$ meters

**Step 2.** The first and last cluster in the strip are located 6 meters in from the end-of-strip markers. Subtract 12 from Step 1.

$93 - 12 = 81$ meters

**Step 3.** Clusters are not located on turning points. Multiply the number of turning points by 3 and subtract it from Step 2.

$81 - (3 \times 2) = 75$ meters

**Step 4.** Divide Step 3 by the cluster spacing.

$75 \div 3 = 25$ clusters

**Step 5.** Add one cluster, because when a line is divided, there is one more interval than spacing.

$25 + 1 = 26$ clusters

From these computations, the minefield OIC is able to cross-check the information on the strip feeder report. In this example, Strip A should have a total of 26 clusters, less any omitted clusters.

## PLATOON ORGANIZATION

The platoon is the basic unit used to install a standard-pattern minefield. Orders to the laying unit OIC specify the proposed location, the length, and the mine type and density. The platoon organization and equipment are shown in Table 7-1.

**Table 7-1. Platoon organization and equipment**

| Personnel | Officer | NCO | EM | Equipment |
|---|---|---|---|---|
| Supervisory | 1 | 1 | 0 | Officer: Map, lensatic compass, notebook, and minefield record forms<br>NCO: Map, lensatic compass, and notebook |
| Siting party | 0 | 1 | 3 | Stakes or pickets, sledgehammers, engineer tape on reels, and nails to peg tape |
| Marking party | 0 | 1 | 2 | Rolls of barbwire or concertina, marking signs, lane signs, wire cutters, gloves, sledgehammers, pickets, and picket pounder |
| Recording party | 0 | 1 | 2 | Sketching equipment, lensatic compass, minefield record forms, map, and metric tape |
| First laying party | 0 | 1 | 6-8 | Notebook for squad leader, picks, shovels, and sandbags |
| Second laying party | 0 | 1 | 6-8 | Same as first laying party |
| Third laying party | 0 | 1 | 6-8 | Same as first laying party |
| Totals | 1 | 7 | 25-31 | |

The OIC performs a map study, and if the situation permits, conducts a ground reconnaissance of the site. He determines locations for mine strips, landmarks, fences, mine dumps, and approaches. Using the minefield requirements computation work sheet (Figure 6-1, pages 6-5 through 6-8), the OIC determines the required number of mines and other materials. He arranges for mines to be delivered and organizes the platoon into siting, laying, recording, marking, and mine-dump parties.

### Siting Party

The siting party places boundary stakes or pickets as strip markers at the beginning and end of each mine strip and at the points where strips change direction. It lays tape on the centerlines of each strip, lane, and traffic path. After siting is complete, the siting party augments other parties.

### Laying Parties

One laying party is responsible for installing, arming, and camouflaging all the mines on a strip or a portion of a strip. Each laying party is then assigned additional strips.

### Recording Party

The recording party obtains the necessary reference data, prepares DA Form 1355 (see Chapter 8), and when needed, installs intermediate markers.

### Marking Party

The marking party erects fences and signs to mark minefield boundaries and lanes. After marking is complete, the marking party augments other parties.

### Mine-Dump Party

The mine-dump party, controlled by the PSG, is composed of personnel who are not working as members of other teams. It accounts for all Class IV and V supplies that arrive from other sources, organizes mine sets at the mine dump, and hauls additional supplies as required. The PSG is not required to stay at the mine dump continuously; he has the flexibility to move around the area to perform other activities.

## MINE-EMPLACEMENT PROCEDURES

Mine-emplacement procedures are shown in Figure 7-7. In the illustration, the minefield is laid from right to left.

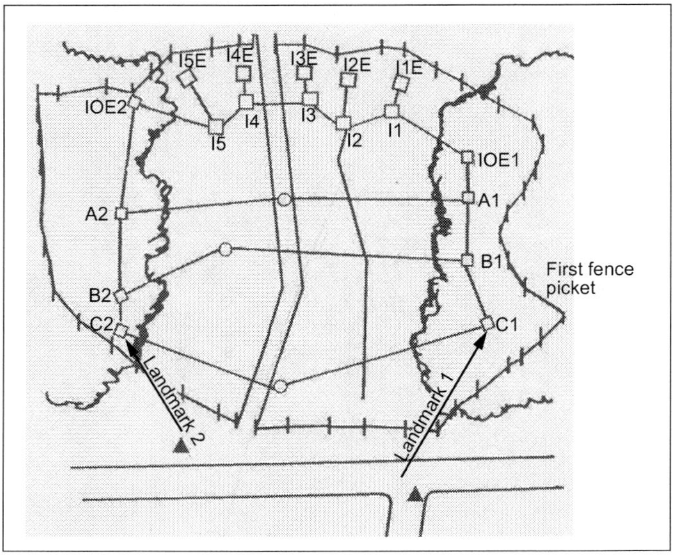


**Figure 7-7. Mine-emplacement procedures**

The OIC arrives at the site with the siting and marking parties. He goes to the right or left (depending on the laying direction) rear boundary of the minefield. This part of the minefield is the farthest from the enemy. The OIC indicates the starting point of the rear strip (this is Strip C in a three-strip minefield), and the siting party emplaces a beginning-of-strip marker.

The OIC designates a starting point for the marking party that is at least 15 meters to the right of the boundary stake. He indicates where the minefield marking fence should be placed. The marking party immediately begins to install fence pickets, working in a counterclockwise direction.

When all the pickets are installed, the marking party encircles the field with a single strand of barbwire (at waist height) and emplaces mine signs. Personnel emplace a second strand of barbwire if required.

From the boundary stake of Strip C, the OIC moves in the direction of the enemy and establishes the starting point of Strip B. Strip centerlines should not be parallel nor less than 15 meters apart. Two members of the siting party emplace a beginning-of-strip marker at the starting point of Strip B. The remaining two members lay tape between the two strips and fasten the tape to the ground at frequent intervals to prevent movement. This procedure is followed until the beginning-of-strip markers of the three regular strips (C, B, and A) and the IOE on the right-hand side of the minefield have been installed.

At the IOE beginning-of-strip marker (IOE1), the OIC gives the siting party a sketch of the minefield and instructions on siting the IOE baseline and strip centerlines. The NCOIC and one other member of the siting party immediately begin setting stakes to indicate the IOE baseline. The centerline laying team lays tape on the IOE baseline, leaving tape reels where tapes run out. Short strips extending from the IOE baseline are established at the same time. Each short strip starts with a marker that is designated as I1, I2, and so forth and ends with a marker that is designated as I1E, I2E, and so forth. Turning points are not used on short strips.

Upon reaching the IOE end-of-strip marker (IOE2), the NCO moves away from the enemy side, establishes the left boundary stake of Strip A, stakes out Strip A, and repeats the procedure until all strip centerlines are taped. All stakes are driven flush with the ground.

While the IOE is being taped, the recording party obtains reference data for the DA Form 1355. Working behind the siting party, the recording party starts from Landmark 1 (designated by the OIC) and proceeds to C1. After C2 has been sited, the recording party proceeds from Landmark 2 to C2 to establish the distance and the azimuth. Finally, the recording party ties C1 and C2 to both landmarks in case one of the landmarks is removed or destroyed. The amount of detail obtained by the recording party depends on the tactical classification of the minefield and any special orders. Aerial photographs taken of the minefield before the tracing tape is removed become valuable supplements to the DA Form 1355.

As soon as laying parties arrive at the site with mines, they establish mine dumps a minimum of 150 meters apart and 50 meters behind the field. They uncrate and stack AT mines and place fuses and detonators in separate boxes.

Fuse types are not mixed. **(Korea Only: AP mines are left in their crates, and the crate lids are removed.)**

When the siting party completes the centerline staking, it installs lane tapes and traffic tapes, respectively. Lane tapes are used by tactical vehicles and patrols. Traffic tapes are used by laying personnel to assist in camouflage and to reduce the amount of traffic on strip centerlines. Traffic tapes are laid perpendicular to the minefield trace at 100-meter intervals.

### MINE EMPLACEMENT

The laying party must know the cluster composition of the strip, the location of any omitted cluster, and future lane locations. When the centerline tape for a regular strip has been installed, the NCOIC designates all but two members of the laying party to emplace mines in the ground. The remaining two soldiers, usually the most experienced, are designated as fusers and are responsible for arming mines. Layers carry the maximum load of mines to be used as base mines in the clusters. Fusers carry the fuses and the detonators.

The NCOIC moves to the right or left (depending on the laying direction) beginning-of-strip marker of the strip and organizes the layers into one column to his rear, directly on the centerline. He measures 6 meters along the centerline for the first cluster and, pointing perpendicular from the centerline and in the direction of the enemy, indicates the placement of the base mine. The first layer on the enemy side places a mine on the ground, 3 meters from the centerline.

The NCO measures 3 more meters and indicates the placement of the second base mine on the opposite (friendly) side of the strip. The first layer on that side places a base mine on the ground. As the initial load of mines is laid, each layer returns to the nearest mine dump for another load. Fusers follow behind layers and insert mine fuses, but they do not arm the mines. This procedure is followed until the end-of-strip marker on the far side of the minefield is reached.

**Korea Only: The NCO tells layers the number and type of mines to be placed next to the base mine in each cluster. As AP mines are being placed, the NCO proceeds along the strip and ensures that the proper number of AP mines is placed in each cluster. The NCO places a spool of trip wire next to the mines that are to be activated by trip wire.**

When all the mines are positioned in clusters, one layer is assigned to dig the holes for all the mines in a cluster. He places the spoil from the holes in sandbags and leaves the sandbags beside the base mine in each cluster. The layer checks the positioning of the mines in the holes, removes the mines from the holes, and places the mines beside the holes. **(Korea Only: The layers anchor trip wires with nails or stakes and wrap the loose ends of trip wires around the fuses.)**

When digging has progressed at least 25 meters from the first mine laid, the arming procedure begins. Fusers arm all the mines in a cluster, beginning with the mine farthest from the centerline and work backward. They place all the mines in the holes **(Korea Only: attach trip wires)** and arm and camouflage the mines. They place filled sandbags on the centerline of the strip, opposite the base mine. Fusers keep their back toward the centerline. Other personnel must remain at least 25 meters from the fusers (Figure 7-8).

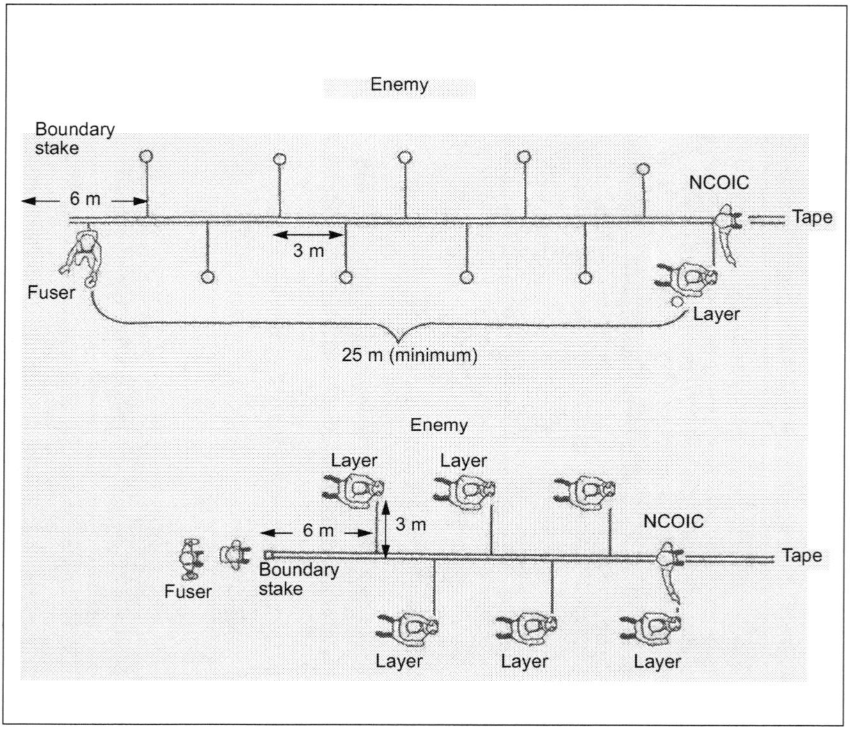


**Figure 7-8. Laying and fusing mines**

Mines located in lanes are not initially buried. They are placed aside to prevent confusion when counting clusters. The mines can be buried after the lane is closed. Upon completing the arming operation, fusers give the safety clips to the NCO, who verifies that all the mines have been armed and camouflaged. The NCO checks the strip and ensures that sandbags, tape, and debris have been picked up. The NCO gives the safety clips to the PSG, who buries them 30 centimeters to the rear of the beginning-of-strip marker.

All mines and other explosive items are recorded upon issue. They are summarized on a mines tally sheet (see Table 7-2). If more than one mine dump is established, a mines tally sheet is kept at each dump and the information is later transferred to a master tally sheet.

**Table 7-2. Sample mines tally sheet**

| Strip/Laying Party | Movement of Mines | M15 | M16 | Trip Wire | AHD Type |
|---|---|---|---|---|---|
| IOE strip/ Party 1 | No forecasted | 23 | 46 | — | — |
| | No issued | 23 | 46 | — | — |
| | No returned | — | — | — | — |
| | No used | 23 | 46 | — | — |
| Strip A/ Party 2 | No forecasted | 81 | 154 | 35 | — |
| | No issued | 81 | 154 | 35 | — |
| | No returned | 7 | 6 | 9 | — |
| | No used | 74 | 148 | 26 | — |
| Strips B, C, D, and so forth | No forecasted | | | | |
| | No issued | | | | |
| | No returned | | | | |
| | No used | | | | |
| Total | No forecasted | | | | |
| | No issued | | | | |
| | No returned | | | | |
| | No used | | | | |
| Date _______ Rank _____ Name ____________________ Signature ____________________ | | | | | |

The PSG ensures that the number of mines used per tally sheet is entered on the DA Form 1355.

When a lane is no longer required through a minefield, it is closed by a lane-closure team that consists of one NCO and two other soldiers. Before closure starts, the NCO checks the minefield record to ascertain the—

- Width of the lane.
- Cluster composition of each strip.
- Total number of mines required.
- Number of strips that intersect the lane.
- Azimuth of each strip.
- Distance between strips, along the lane centerline.
- Location of the mine dump.

The procedures for lane closure is as follows (see Figure 7-9, page 7-16):

- The lane-closure team moves along the safe-lane centerline until it arrives at the strip safe-lane marking picket.
- The NCO lays out a strip centerline tape and a tape along both sides of the safe lane to mark its boundaries.
- The team lays clusters.
- The team recovers the tapes.

- The above steps are repeated at successive strips.
- The NCO amends the DA Form 1355.

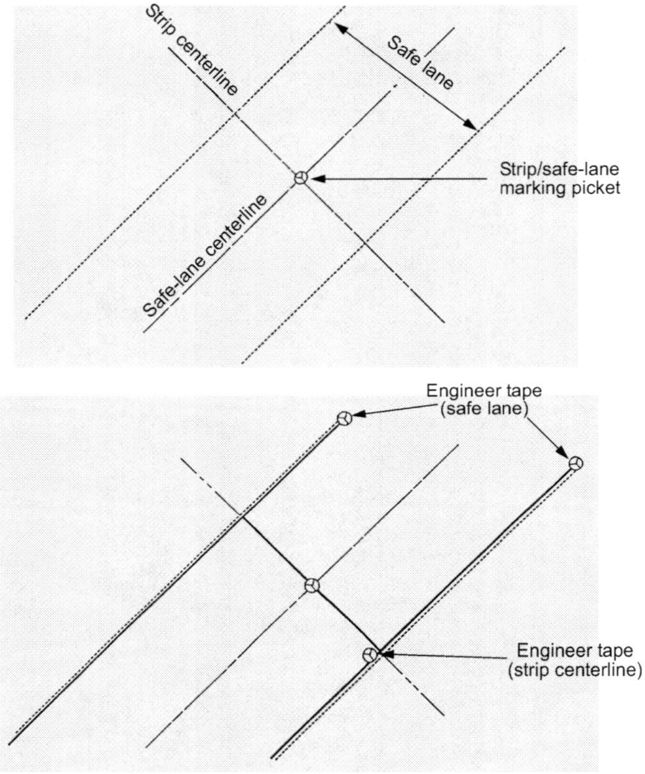


**Figure 7-9. Lane closure**

**NOTE: When gaps must be closed, fences are temporarily erected along the side boundaries. They are removed later to avoid the indication of a passage through the minefield.**

NCOICs do not act as working members of their parties; they ensure that—

- No one moves back into a mined area.
- Any irregularity, such as an omitted cluster, is recorded.

- All safety devices are recovered and checked against the minefield record.

Safety tapes are used to create a network of safe routes through a minefield. They may be removed progressively but are normally left in place until the minefield is complete.

**Korea Only: When trip wires are used, safety tapes are laid between strips where the trip wire will be positioned (including Strip A and the IOE baseline). If trip wires are not used, safety tapes are recommended but are not mandatory.**

## NUISANCE MINEFIELDS

### SITING

Consider the following factors when siting a nuisance minefield:

- The effort needed by the enemy to bypass a mined area.
- The importance of an area or a route to the enemy.
- Achievement of the goal (use the minimum amount of effort needed to maximize casualties and the effect on enemy morale).

Remember, the more ingenuous the methods of concealment, the longer it will take to lay mines.

Observation and covering fires are not essential and are seldom feasible for nuisance minefields. Their value depends on effective siting and concealment to cause surprise.

The minefield OIC is responsible for detailed siting and the design of a nuisance minefield. He must consider the minefield from the enemy's point of view and assess the courses open to the enemy when he encounters it. Such considerations may expose weaknesses in the initial plan and bring about a change to the proposed minefield layout or may lead to a decision to site the minefield elsewhere.

### LOCATION

In wooded or hilly terrain, the enemy's logistics transport will normally be confined to existing routes. Nuisance mines at selected sites along roads can impose considerable delay on the enemy and have a cumulative effect on his resources and morale. The best sites for axial mining are—

- Natural defiles and constructed areas that are difficult to bypass (cuts, embankments, causeways, fords, forest tracks, built-up areas).
- In the vicinity of road craters, AT ditches, and any obstacles that have to be cleared.
- Around culverts.
- Demolished bridges, particularly the home bank, including likely adjacent crossing places and alternative building sites.
- Likely assembly areas.
- Covered approaches or dead space.

- In the vicinity of enemy-needed fuel, supplies, and engineering materials that cannot be destroyed or removed.
- Railroads (in or near culverts, bridges, sharp turns, tunnels, steep grades). Mines should be laid where enemy trains cannot bypass the mined area on branches or spurs.

### LAYING

There is no requirement for recording the precise location of individual nuisance mines. Recording mine positions that are laid to a pattern is easy and quick. Pattern laying should be used when it can be done without prejudicing concealment.

When the number of mines to be laid on the site makes it impracticable or undesirable to lay mines in a pattern, they may be scatter-laid unless otherwise directed. Scatter-laying by hand is useful in road blocks, bridge abutments, and craters when it would be difficult and wasteful to lay mines in a pattern. Again, scatter-laying along routes to be denied to the enemy will add considerably to the delay imposed. All available types of AT mines **(Korea Only: and AP mines)** are used to make nuisance minefields complex and difficult to remove. Combinations of mine types should be varied constantly so that each minefield presents a clearance problem. Deeply buried mines can be included; however, they take much longer to lay. These mines may be worthwhile around craters, where the enemy is likely to need earthmoving equipment.

If a nuisance minefield is laid in a standard pattern, standard procedures are followed. If mines are selectively positioned, procedures must be tailored to suit the situation. In all occasions, however, the following rules should be observed:

- The intended position for each mine is clearly marked on the ground before laying begins.
- Laying parties work in pairs, and each pair is detailed to lay specific mines.

### INSPECTION AND MAINTENANCE

Mines that are left in the ground for an extended period of time may deteriorate and malfunction for one or more of the following reasons:

- Moisture may have entered the igniter or body of the mine and either neutralized the explosive or corroded the metal parts. Such actions may be aggravated by local factors (soil acidity, wide temperature swings).
- Frost or heat may have subjected the mine to mechanical strain and caused distortion.
- Insects or vegetation may have caused obstructions.
- Animals may have turned mines over or detonated them.

Technical inspections should only be made by experienced engineers or EOD personnel. When a minefield deteriorates below the operating level, additional mine strips and/or rows must be added to restore its effectiveness. They are sited to the front or rear of the existing minefield to increase its depth. New mine strips and/or rows are treated as a separate, additional minefield.

Technical inspections of minefields are normally done at three-month intervals, and personnel work in pairs. They are dcne more frequently during extreme weather conditions.

## HANDOVER

Minefield handover is an extremely important task. The following items need to be addressed between emplacing and overwatching units:

- Discussing minefield compositions.
- Discussing the minefield configuration and walking and/or riding the minefield trace.
- Discussing lane closure, if applicable.
- Training the unit on how to the close lane, if applicable.
- Discussing obstacle protection against enemy dismounted patrols.
- Signing over the written report (Figure 2-26, page 2-54).
- Discussing indirect fires.
- Reporting completion of handover to higher headquarters.
- Forwarding a copy of the written report and the DA Form 1355 to higher headquarters.

This chapter implements STANAG 2036.

Chapter 8

# Reporting and Recording

Reporting and recording mine and minefield/munition field information is critical to the success of a unit's mission. It not only provides tactical data for the commander, but it also provides force-protection information to subordinate and adjacent units.

## MINEFIELD/MUNITION FIELD REPORTS

A minefield/munition field report is an oral, electronic, or written communication that concerns friendly or enemy mining activities. The report format is specified by the local command. It is submitted by the emplacing unit commander through operational channels to the G3/S3 of the authorized headquarters. The headquarters integrates the report with terrain intelligence and disseminates it along with tactical intelligence. The report is sent by the fastest, most secure means available. Figure 8-1, page 8-2, summarizes the minefield/munition field report flow at the division level and below.

### REPORT OF INTENTION

When planning to emplace a minefield/munition field, the unit must submit a report of intention to notify their higher headquarters. The report doubles as a request when it is initiated at levels below emplacement authority. The report includes—

- Tactical purpose of the minefield/munition field.
- Estimated number and type of mines to be emplaced.
- Location.
- Proposed start and completion times.
- Type of minefield/munition field.
- Whether mines are surface-laid or buried.
- Whether AHDs are used.
- Location and width of lanes and gaps.

Conventional minefields/munition fields that are part of an OPLAN or a general defense plan (GDP) approved by the authorizing commander do not require a report of intention because inclusion in an OPLAN or a GDP implies an intention to lay.

### REPORT OF INITIATION

A report of initiation is mandatory. It informs higher headquarters that emplacement has begun and that the area is no longer safe for friendly movement and maneuver. The report specifies the time emplacement began

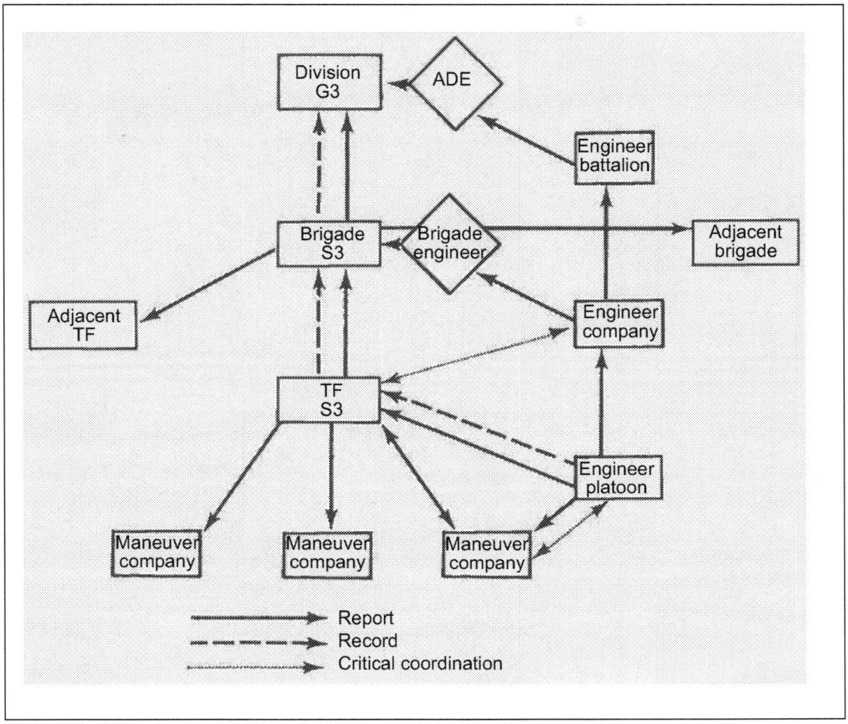


**Figure 8-1. Conventional minefield/munition field reporting chain**

and identifies the location and the target number of minefields/munition fields.

### REPORT OF COMPLETION

A report of completion is usually an oral report to the authorizing commander. It indicates that the minefield/munition field is complete and functional. A report of completion is immediately followed by a completed DA Form 1355 or 1355-1-R.

### REPORT OF TRANSFER

Minefield/munition field responsibility is transferred from one commander to another in a report of transfer (Figure 2-26, page 2-54). This report is signed by the transferring and receiving commanders. It includes a certificate stating that the receiving commander was shown or otherwise informed of all the mines within the transferring commander's zone of responsibility. The report states that the receiving commander assumes responsibility for those mines.

The report of transfer is sent to the next higher commander who has authority over the transferring and receiving commanders.

### REPORT OF CHANGE

A report of change is made immediately upon any change or alteration to a previously reported minefield/munition field. It is sent to the next higher commander and through channels to the headquarters that keeps the written minefield/munition field record. A report of change is made by the commander responsible for surveillance and maintenance of the minefield/munition field.

### PROGRESS REPORTS

During the emplacing process, the commander may require periodic reports on the amount of work completed.

## MINEFIELD/MUNITION FIELD RECORDS

Most conventional minefields/munition fields are recorded on DA Form 1355; hasty protective row minefields/munition fields are recorded on DA Form 1355-1-R.

The laying unit prepares the minefield/munition field record. The OIC signs the form and forwards it to the next higher command as soon as possible. Once the information is entered on the form, the form is classified SECRET, NATO SECRET, or SECRET-ROKUS. (When used for training, the record is marked SAMPLE.) The number of copies prepared depends on the type of minefield/munition field emplaced and the local procedures. Unit SOPs should provide advanced guidance on how minefield/munition field information is to be passed to higher, lower, and adjacent commands. Minefield/munition field records are circulated on a need-to-know basis. Large minefields/munition fields are recorded on two or more DA Forms 1355.

Once the minefield/munition field is removed, the date of removal, the time removal was complete, and the name of the unit that removed the minefield/munition field will be placed in the remarks block of the form. The form will then be forwarded to the unit's central control cell for mine clearance information. The central control cell will retain the form for two years and then destroy it.

When changes are made to an existing minefield/munition field, a new record must be prepared. This record is marked REVISED. It shows the minefield/munition field as it is after changes. The original minefield/munition field number remains unchanged. Some changes which require a new record are—

- Relocation of mines in safe lanes.
- Relocation of safe lanes.
- Changed lane or minefield/munition field marking.
- Inclusion of the minefield/munition field into a larger minefield/munition field system.
- Removal or detonation of mines.
- Addition of mines to the minefield/munition field.

Conventional minefield/munition field records are forwarded through operational channels to theater army headquarters (TAHQ) where they will be maintained on file by the theater engineer. If a TAHQ has not been established, minefield/munition field records are maintained on file with the assistant corps engineer of the AO where the minefield/munition field is located.

## MINEFIELD RECORD

DA Form 1355 consists of a single sheet that is printed on both sides and folded in half. The front side contains blocks of tabular data, the inside is a graph of 1-centimeter squares for a scaled sketch of the field, and the back side includes a computation formula and instructions. The scale for plotting minefields/munition fields depends on the size of the field. To avoid using two sheets for the sketch, adjust the scale so that one form will support the sketch. For very large minefields/munition fields, two sheets may be required. The system of measurement and the scale size must be indicated in the legend block. A second form may be used to support any additional information in the notes block. Any blocks or lines not used on the form must be crossed out to avoid unauthorized entries on the form.

An example of a completed DA Form 1355 is shown in Figures 8-2a through 8-2c, pages 8-5 through 8-7.

The following step-by-step instructions are provided for completing DA Form 1355:

**Block 1.** Enter complete data on the laying authority and the laying unit. The OIC block should include rank, name, and social security number (SSN).

| | |
|---|---|
| 1 | AUTHORITY: CG 2nd INF DIV |
| | LAYING UNIT: B Co 2nd ENGR BN 1st PLT |
| | OFFICER IN CHARGE: 1LT R. YOUNG 762-01-1352 |

**Block 2.** Enter the date-time group (DTG) for start and completion times. The recorder block includes rank, name, and SSN.

| | | |
|---|---|---|
| 2 | DATE AND TIME | START 090630Z JAN 90 |
| | | COMPLETION 091800Z JAN 90 |
| | RECORDER SFC F. LING 550-72-1332 | |

**Block 3.** Enter the copy and the sheet number. The number of copies prepared depends on the unit SOP and the minefield/munition field classification.

| | |
|---|---|
| 3 | Copy No. 1 of 4    Sheet No 1 of 1 |

**NOTE: The minefield/munition field record is forwarded by the laying unit. One copy is retained by the overwatching unit, one copy by the next higher command, one copy by the unit central control cell (for mine clearance information), and one copy by the proper national territorial authority.**

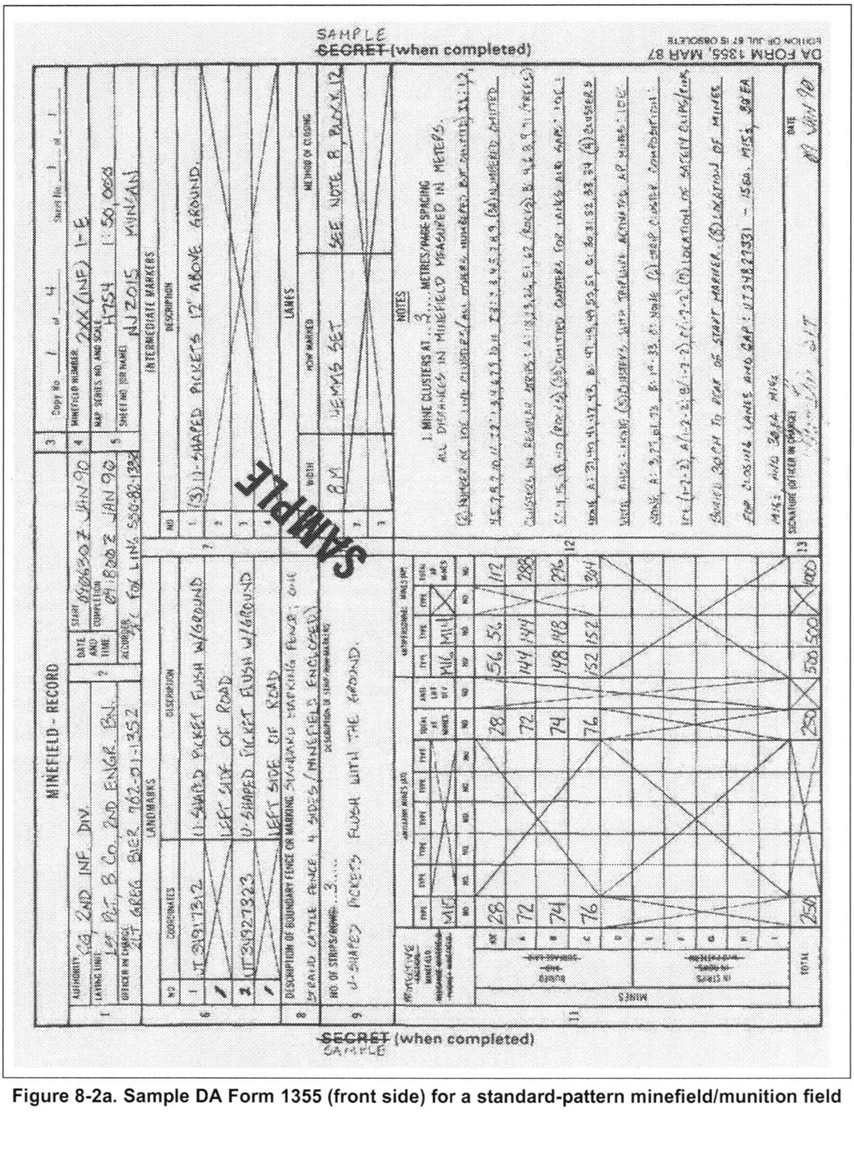
SAMPLE
~~SECRET~~ (when completed)

DA FORM 1355, MAR 87

EDITION OF JUL 87 IS OBSOLETE

MINEFIELD - RECORD

SAMPLE

~~SECRET~~ (when completed)
SAMPLE

Figure 8-2a. Sample DA Form 1355 (front side) for a standard-pattern minefield/munition field

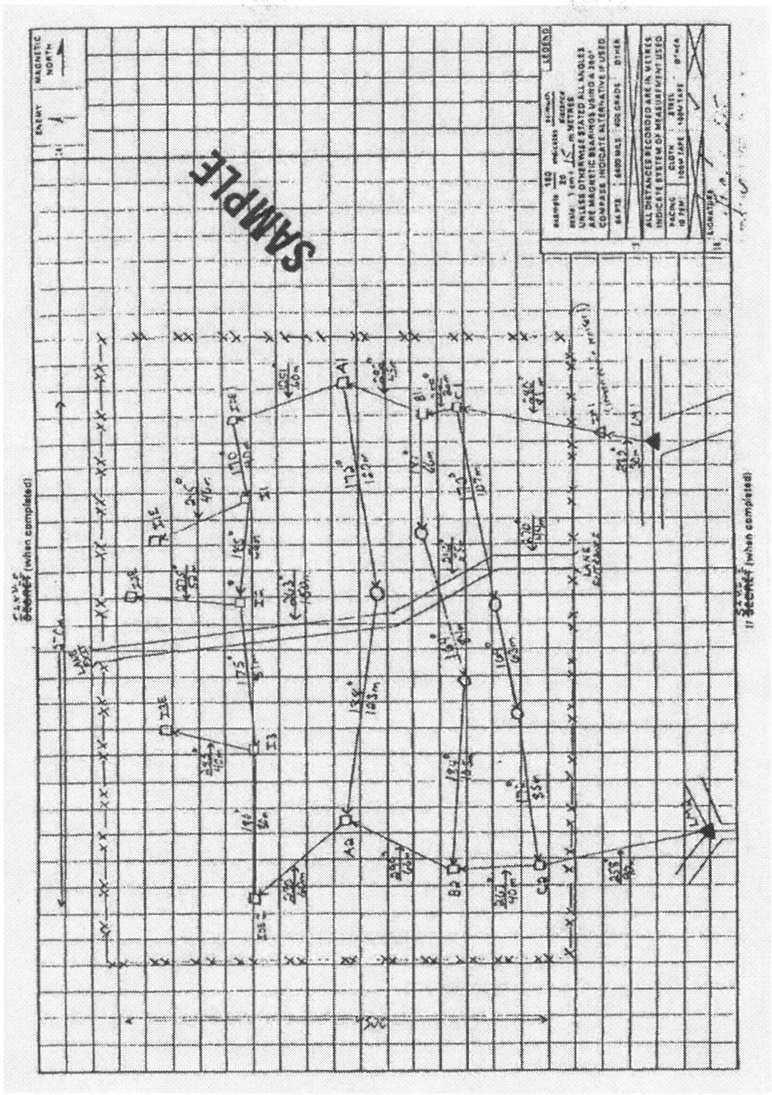


Figure 8-2b. Sample DA Form 1355 (inside) for a standard-pattern minefield/munition field

SAMPLE
~~SECRET~~ (when completed)

**MINEFIELD REQUIREMENTS COMPUTATION FORMULA**

| | | AT | APF | APB |
|---|---|---|---|---|
| Desired Density | | a____ | b____ | c____ |
| IOE Representative Cluster | | a____ | b____ | c____ |
| Front ____ meters | | | | |
| Depth ____ meters | | | | |
| AHD ____ % | | | | |
| *1. No. of IOE Clusters = Front ÷ 9 | ____ | | | |
| 2. No. of IOE clusters x IOE Representative Cluster | | a____ | b____ | c____ |
| 3. Mines in Minefield = Front x Desired Density | | a____ | b____ | c____ |
| 4. Subtotal of Mines Required = Line 2 + Line 3 | | a____ | b____ | c____ |
| *5. Mine Rejections, Strip Length Variances = Line 4 x .1 | | a____ | b____ | c____ |
| 6. Total AT Mines Required = Line 4 + Line 5 | | a____ | b____ | c____ |
| 7. Add a + b + c of "Desired Density" = | ____ | | | |
| *8. Line 7 x .6 = | ____ | | | |
| 9. AT Mine "Desired Density" x 3 = | ____ | | | |
| 10. No. of Regular Strips = Highest No. of Line 8 or 9 | ____ | | | |
| *11. No. of AHD = % AHD x Total AT Mines (Line 6) | ____ | | | |
| 12. Strip cluster Composition = Desired Density x 3 | | a____ | b____ | c____ |

* NOTE: Round up to the next whole number.

| | STRIP | AT | APF | APB | ROW TOTAL (Cannot exceed 5) |
|---|---|---|---|---|---|
| "CLUSTER | A | a____ | b____ | c____ | ____ |
| | B | a____ | b____ | c____ | ____ |
| | C | a____ | b____ | c____ | ____ |
| | D | a____ | b____ | c____ | ____ |
| COMPOSITION | E | a____ | b____ | c____ | ____ |
| | F | a____ | b____ | c____ | ____ |
| | G | a____ | b____ | c____ | ____ |
| | H | a____ | b____ | c____ | ____ |
| TABLE" | I | a____ | b____ | c____ | ____ |
| | J | a____ | b____ | c____ | ____ |
| COLUMN TOTAL (Totals cannot exceed Line 12) | | a____ | b____ | c____ | ____ |

**TABULAR DATA** (Numbers correspond to numbered blocks on front of form).
1. Enter complete data on authority of laying and on the laying unit. OIC blanks will include name, rank, and SSN.
2. Enter date-time groups for starting and completion times. Recorder blanks will include name, rank, and SSN.
3. Enter copy and sheet numbers. Number of copies will depend upon unit SOP and the classification of the minefield. The number of sheets will depend upon the length and the depth of the minefield versus scale.

4. Enter minefield number as follows:

Designation of unit authorizing installation
Number of obstacle
Status of obstacle
(E=Executed, P=Proposed, U=Under Construction). 3/147-Inf-2-E

5. Enter map data as stated on map(s) used.
6. Enter complete data on at least two landmarks with 8 digit grid coordinates. Cross out unused blocks.
7. Enter description(s) of any intermediate markers used. When a landmark is more than 200 meters from the minefield or the strip/row reference stake cannot be seen from the landmark [illegible] intermediate marker must be used. If possible, the intermediate marker should not be closer than 75 meters to the [illegible] reference stake. Cross out unused blocks.
8. Enter the word "Standard" when the standard marking fence [illegible] the boundary marking if other than the standard marking fence is used.
9. Enter the number of strips/rows laid other than IOE. Describe the [illegible] (Line out words not applicable).
10. Enter the width, marking, and closing provisions for each lane; when [illegible] give the type and number of mines for closing. The location of these mines is described in the "Notes" (Line 12). [illegible] are 1 meter wide, one-way vehicular lanes are 8 meters and two-way vehicular lanes are 16 meters. Cross [illegible] blocks.
11. Enter type of minefield by crossing out lines not needed. Indicate method of laying by marking out incorrect descriptions. Enter types of mines as AT, APF, APB. (Enter chemical mines under AT mines). For each type of mine, enter number of mines and antihandling devices installed in the IOE and in each Strip or Row. Strips or Rows will be lettered serially, starting with the first one laid. Enter totals. Cross out unused blocks.
12. Enter under Notes information which would be useful to personnel clearing the minefield. Appropriate items include location of chemical mines, location of AT mines with antihandling devices, location AP mines with tripwires, clusters in IOE which contain mines, where safety devices are buried, strip cluster composition and numbered omitted clusters in regular strips.
13. OIC enters signature, rank and date.
14. Enter arrows for the direction of the enemy and magnetic north. The enemy arrow should always point within the top 180 degrees of the paper; the north arrow should follow one of the lines of the graph.
15. Enter scale of sketch for standard pattern minefields the sketch should be drawn to a scale of about 1cm=10 meters.
16. Sketch in the following, as applicable.
a. Show directional arrows as follows:
(1) Landmarks (or intermediate markers) to strip markers at starting and finishing points of the last strip laid or to the nearest or farthest mine in a group.
(2) From landmarks (or intermediate markers) to fence or boundary markers.
(3) From landmarks to intermediate markers, if used.
(4) For each straight line section of a lane centerline.
(5) Between markers of starting points of adjacent strips, including IOE, and between finishing points of adjacent strips, including (IOE).
(6) For each segment of a strip or of the IOE, label all directional arrows with magnetic azimuth in degrees and distance in meters. Express as a fraction (247 degrees/90 meters). Recorded from friendly to enemy side and from right to left/or left to right.
b. Show approximate location of protective fence or boundary markers.
c. Show length and depth of minefield in meters. These dimensions indicate the extremities of the minefield.
d. Show a grid intersection and give grid coordinates.
e. Show trace of shoreline and direction and approximate rate in meters per second of water current, for mines laid underwater.
17. Enter security classification of the form. (If the form is used for training, enter the word SAMPLE.)
18. OIC enters signature and rank.

SAMPLE

Sample
~~SECRET~~ (when completed)

Figure 8-2c. Sample DA Form 1355 (back side) for a standard-pattern minefield/munition field

**Block 4.** The minefield/munition field obstacle numbering system (Table 8-1) consists of 11 characters and 1 obstacle status symbol. It shows the type of obstacle, the belt and the zone in which it is located, and the headquarters that established the zone. Abbreviations for obstacle types are shown in Table 8-2.

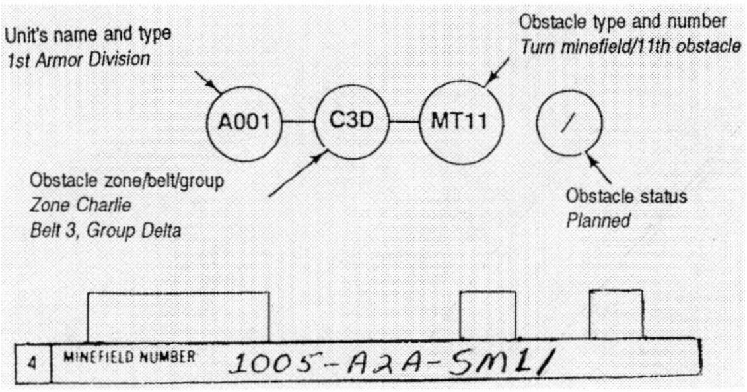


**Example:** Obstacle number I005-A2A-SM21/ indicates that the 5th Infantry Division planned the obstacle in Zone A. It is the 21st obstacle in Group A, Belt 2, and has not been executed. The obstacle is a MOPMS.

**Table 8-1. Minefield/munition field obstacle numbering system**

| Character | Description |
|---|---|
| 1 through 4 | Alphanumeric description of the headquarters type and numerical designation that established the obstacle zone; character 1 designates the unit type with a letter:<br>• A = armor division or brigade<br>• I = infantry division or brigade<br>• C = cavalry division<br>• R = cavalry regiment<br>• Z = corps |
| 5 | Letter indicating the obstacle zone |
| 6 | Number indicating the belt number in the obstacle zone |
| 7 | Letter indicating the group in the obstacle belt |
| 8 and 9 | Letters indicating the obstacle type (see Table 8-2) |
| 10 and 11 | Two numbers indicating the obstacle number in the group |
| 12 | One of four characters indicating the obstacle status—<br>• / = planned obstacle<br>• — = obstacle being prepared<br>• + = prepared obstacle (This sign is for reserve demolition targets and may indicate a readiness state of safe or armed.)<br>• X = completed obstacle |

**Table 8-2. Abbreviations for obstacle types**

| B—Bridge Demolition | | W—Wire Obstacle | |
|---|---|---|---|
| BA | Abutment | WA | Double apron |
| BS | Span | WB | Booby-trapped |
| BC | Abutment and span | WF | Tanglefoot |
| **M—Minefield/Munition Field** | | WG | General-purpose, barbed tape |
| MD | Disrupt | WN | Nonstandard |
| MT | Turn | WR | Road block |
| MF | Fix | WT | Triple standard |
| MB | Block | **R—Road Crater** | |
| MN | Nonstandard | RH | Hasty |
| MP | Protective | RD | Deliberate |
| MQ | Nuisance | RM | Mined |
| MS | Standard pattern | **M—Miscellaneous** | |
| **S—Scatterable Minefield/Munition Field** | | AD | AT ditch |
| SA | ADAM | AR | Rubble by CEV gun |
| SB | Gator | AB | Rubble by blade |
| SR | RAAM | AT | Abatis |
| SF | ADAM and RAAM | AE | Rubble by explosives |
| SM | MOPMS | AM | Movable MOBA obstacle (car, bus) |
| SV | Volcano | AN | Expedient nonstandard |
| **H—Hand-Emplaced Munitions** | | AL | Log crib, log obstacle |
| HH | Hornet | AP | Post obstacle (hedgehog, tetrahedron) |
| HS | SLAM | AH | Log hurdle |

**Block 5.** Enter the map data as stated on the map(s) used.

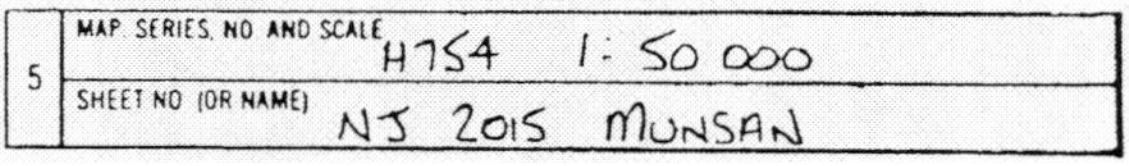
5
MAP SERIES, NO AND SCALE H754 1:50 000
SHEET NO (OR NAME) NJ 2015 MUNSAN

**Block 6.** Enter the grid coordinates and a description of at least two landmarks. If the landmarks are roads, trails, or routes, enter their name or number. This makes identification easier when removing the minefield/munition field. When recording minefields/munition fields, GPSs can only be used to determine the coordinates for minefield/munition field landmarks and RPs.

> **WARNING**
> **Do not use GPSs to chart or record minefield/munition field perimeter coordinates or to determine safe routes through or around existing minefields/munition fields.**

| 6 | LANDMARKS | | |
|---|---|---|---|
| | NO | COORDINATES | DESCRIPTION |
| | 1 | UT 34917312 | U-SHAPED PICKET FLUSH WITH GROUND |
| | 2 | | NEXT TO ROAD |
| | 2 | UT 34927323 | U-SHAPED PICKET FLUSH WITH GROUND |
| | 4 | | NEXT TO ROAD |

**Block 7.** Enter the description of intermediate markers, if applicable. When a landmark is more than 200 meters from the minefield/munition field, or a strip or row reference stake cannot be seen from the landmark, an intermediate marker must be used. If possible, the intermediate marker is at least 75 meters from the strip or row reference stake.

| 7 | INTERMEDIATE MARKERS | |
|---|---|---|
| | NO | DESCRIPTION |
| | 1 | 3 U-SHAPED PICKETS 12" ABOVE GROUND |
| | 2 | |
| | 3 | |
| | 4 | |

**Block 8.** Enter the word STANDARD when a standard marking fence is used. Describe the boundary marking if a standard marking fence is not used. (Use two sides and the rear for a tactical minefield/munition field; use four sides for a protective minefield/munition field.)

| | |
|---|---|
| 8 | DESCRIPTION OF BOUNDARY FENCE OR MARKING STANDARD<br>4 SIDES MINEFIELD FULLY ENCLOSED |

**Block 9.** Enter the number of strips or rows laid. (Do not include the IOE.) Describe the strip or row markers. Cross out words that do not apply.

| | |
|---|---|
| 9 | NO. OF STRIPS/~~ROWS~~: ..3..... DESCRIPTION OF STRIP/~~ROW~~ MARKERS<br>U-SHAPED PICKETS FLUSH WITH GROUND |

**Block 10.** Enter the width, the marking, and the provisions for each lane. When appropriate, give the types of mines and the number of each type of mine for closing. (The location of these mines is described in Block 12.)

| 10 | LANES | | | |
|---|---|---|---|---|
| | NO | WIDTH | HOW MARKED | METHOD OF CLOSING |
| | 1 | 8m | HEMMS SET | 15×M15, 30×M16, 30×M14 |
| | 2 | | | |
| | 3 | | | |

**Block 11.** Enter the type of minefield/munition field by crossing out the lines that are not needed. Indicate the method of laying by crossing out incorrect descriptions. Enter the types of mines and the number of each type of mine. Also enter the number of AHDs installed in the IOE and in each row. Letter the strips or rows sequentially, starting with the first one laid. Enter totals.

| 11 | Protective ~~TACTICAL~~ MINEFIELD ~~NUISANCE MINEFIELD~~ ~~PHONEY MINEFIELD~~ | | ANTITANK MINES (AT) | | | | | | | | ANTIPERSONNEL MINES (AP) | | | |
|---|---|---|---|---|---|---|---|---|---|---|---|---|---|---|
| | | | TYPE M15 | TYPE | TYPE | TYPE | TYPE | TYPE | TOTAL AT MINES | ANTI-LIFT DEV | TYPE M16 | TYPE M14 | TYPE | TOTAL AP MINES |
| | | | NO | NO | NO | NO | NO | NO | NO | NO | NO | NO | NO | NO |
| | MINES | BURIED ~~AND SURFACE LAID~~ IOE | 28 | | | | | | 28 | | 56 | 56 | | 112 |
| | | A | 72 | | | | | | 72 | | 144 | 144 | | 288 |
| | | B | 74 | | | | | | 74 | | 148 | 148 | | 296 |
| | | C | 76 | | | | | | 76 | | 152 | 152 | | 304 |
| | | D | | | | | | | | | | | | |
| | | IN STRIPS ~~IN ROWS~~ ~~W/O PATTERN~~ E | | | | | | | | | | | | |
| | | F | | | | | | | | | | | | |
| | | G | | | | | | | | | | | | |
| | | H | | | | | | | | | | | | |
| | | J | | | | | | | | | | | | |
| | TOTAL | | 250 | | | | | | 250 | | 500 | 500 | | 1000 |

If the type of munition field is Hornet, enter "Hornet" above the word "tactical" and line out "nuisance minefield" and "phoney minefield." (See Figures 8-3a and 8-3b, pages 8-12 and 8-13.)

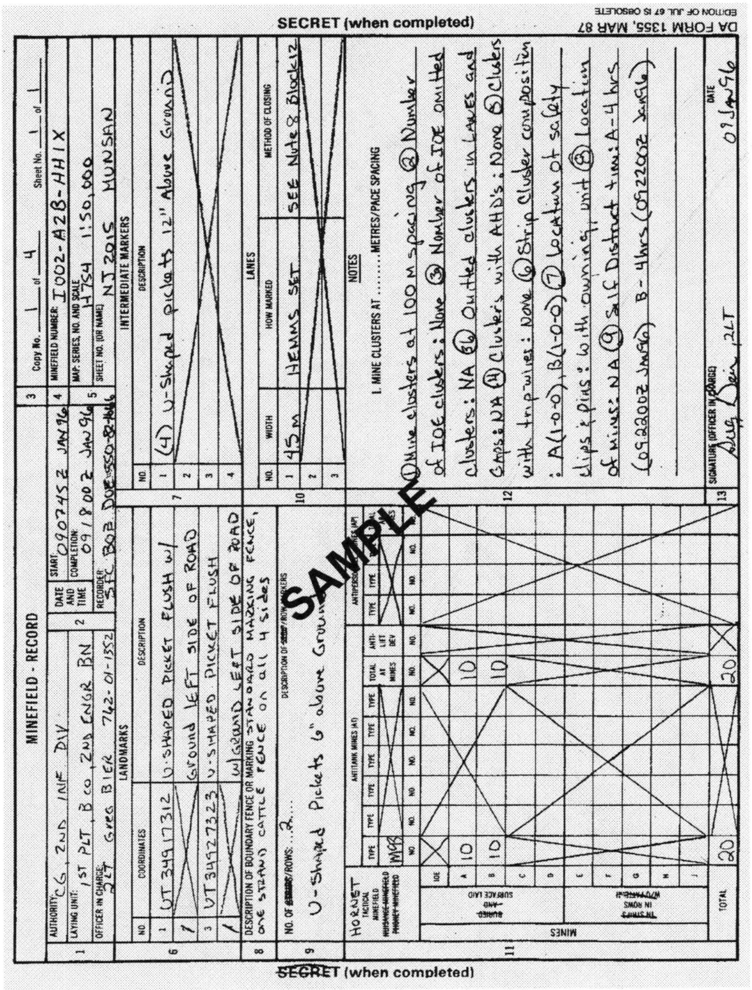

SECRET (when completed)

MINEFIELD - RECORD

1 AUTHORITY: CG, 2ND INF DIV
LAYING UNIT: 1ST PLT, B CO, 2ND ENGR BN
OFFICER IN CHARGE: 2LT Greg Bier 762-01-1552

2 DATE AND TIME — START: 090745Z JAN96 — COMPLETION: 091800Z JAN96 — RECORDER: SFC BOB DOE 550-82-

3 Copy No. 1 of 4 Sheet No. 1 of 1

4 MINEFIELD NUMBER: I002-A2B-HH1X
MAP: SERIES, NO. AND SCALE: H754 1:50,000

5 SHEET NO. (OR NAME): NJ 2015 MUNSAN

6 LANDMARKS

| NO | COORDINATES | DESCRIPTION |
|---|---|---|
| 1 | UT 34917312 | U-SHAPED PICKET FLUSH w/ Ground LEFT SIDE OF ROAD |
| 2 | X | X |
| 3 | UT 34927323 | U-SHAPED PICKET FLUSH w/GROUND LEFT SIDE OF ROAD |
| 4 | X | X |

7 INTERMEDIATE MARKERS

| NO | DESCRIPTION |
|---|---|
| 1 | (4) U-Shaped pickets 12" above Ground |
| 2 | X |
| 3 | X |
| 4 | X |

8 DESCRIPTION OF BOUNDARY FENCE OR MARKING: ONE STRAND CATTLE FENCE, STANDARD MARKING FENCE on all 4 sides

9 NO. OF ROWS: 2 — DESCRIPTION OF ROW MARKERS: U-Shaped Pickets 6" above Ground

10 LANES

| NO. | WIDTH | HOW MARKED | METHOD OF CLOSING |
|---|---|---|---|
| 1 | 45 m | HEMMS SET | SEE Note 8 Block 12 |
| 2 | X | X | X |
| 3 | X | X | X |

11 MINES — HORNET TACTICAL MINEFIELD
ANTITANK MINES (AT): TYPE M93 — A: 10; B: 10; TOTAL: 20
TOTAL AT MINES — A: 10; B: 10; TOTAL: 20

12 NOTES
1. MINE CLUSTERS AT ........METRES/PACE SPACING
(1) Mine clusters at 100 M spacing (2) Number of IOE clusters: None (3) Number of IOE omitted clusters: NA (3b) Omitted clusters in Lanes and Gaps: NA (4) Clusters with AHD's: None (5) Clusters with tripwires: None (6) Strip cluster composition: A(1-0-0), B(1-0-0) (7) Location of safety clips & pins: with owning unit (8) Location of mines: NA (9) Self Destruct time: A-4 hrs (092200Z Jan96) B-4hrs (092200Z Jan96)

13 SIGNATURE (OFFICER IN CHARGE): Greg Bier, 2LT — DATE: 09Jan96

SAMPLE

DA FORM 1355, MAR 87 — EDITION OF JUL 67 IS OBSOLETE

SECRET (when completed)

Figure 8-3a. Sample DA Form 1355 (front side) for a Hornet minefield/munition field

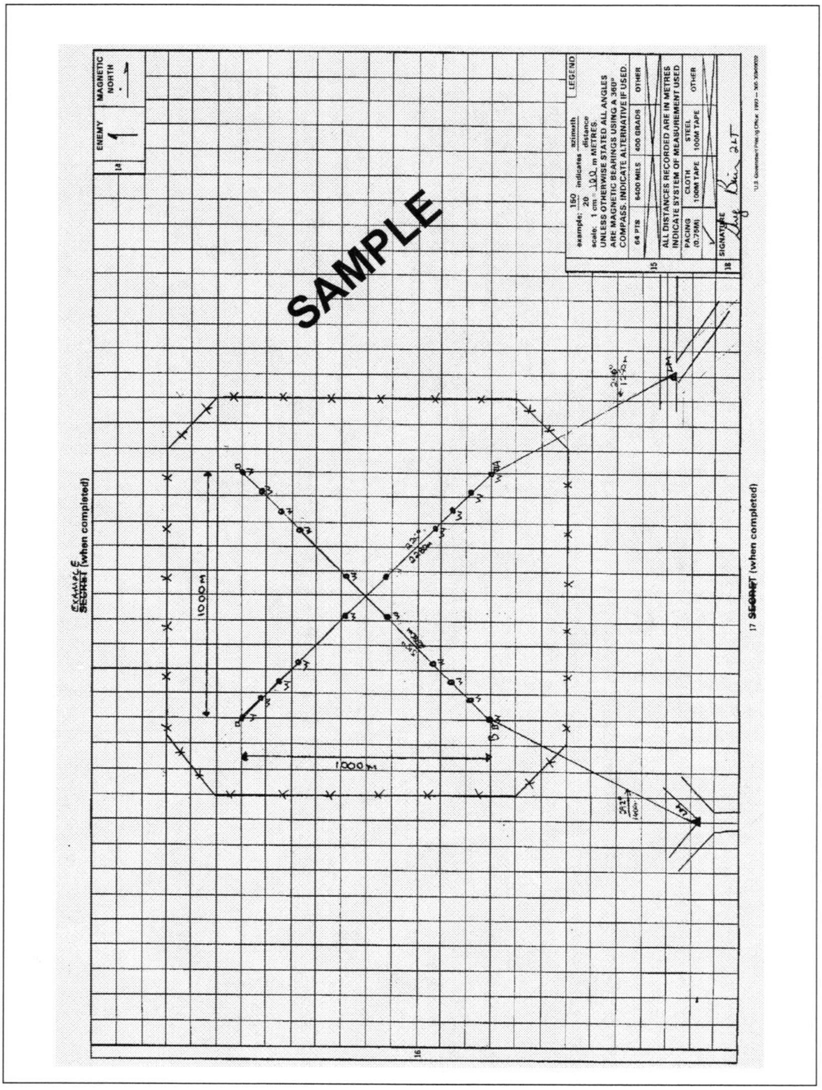


Figure 8-3b. Sample DA Form 1355 (back side) for a Hornet minefield/munition field

**Block 12.** As a minimum, enter the following items in Block 12. Also enter any additional information that would be useful when removing the minefield/ munition field.

1. Mine clusters at ____ meters spacing.

2. Number of IOE live clusters (all others numbered but omitted).

3a. Number of omitted clusters in the IOE and in regular strips and why they were omitted.

3b. Omitted clusters in lanes and gaps.

4. Clusters with AHDs (identify the mine and the type of device used).

5. Clusters with trip-wire-actuated AP mines.

6. Strip cluster composition.

7. Location of safety clips and pins (buried 30 centimeters to the rear of each start strip or row marker).

8. Location of mines for closing lanes and gaps.

9. SD time for special munitions.

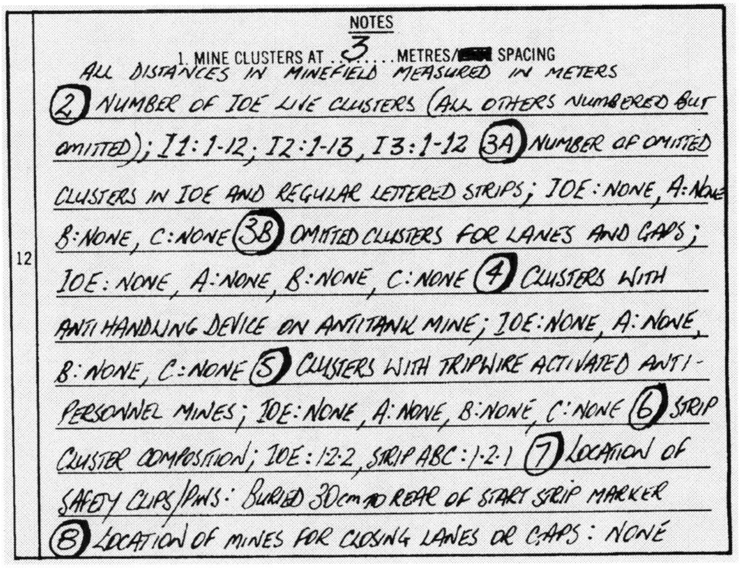

NOTES

12

1. MINE CLUSTERS AT ..3.... METRES/[redacted] SPACING

ALL DISTANCES IN MINEFIELD MEASURED IN METERS

(2) NUMBER OF IOE LIVE CLUSTERS (ALL OTHERS NUMBERED BUT OMITTED); I1: 1-12; I2: 1-13, I3: 1-12 (3A) NUMBER OF OMITTED CLUSTERS IN IOE AND REGULAR LETTERED STRIPS; IOE: NONE, A: NONE B: NONE, C: NONE (3B) OMITTED CLUSTERS FOR LANES AND GAPS; IOE: NONE, A: NONE, B: NONE, C: NONE (4) CLUSTERS WITH ANTI HANDLING DEVICE ON ANTITANK MINE; IOE: NONE, A: NONE, B: NONE, C: NONE (5) CLUSTERS WITH TRIPWIRE ACTIVATED ANTI-PERSONNEL MINES; IOE: NONE, A: NONE, B: NONE, C: NONE (6) STRIP CLUSTER COMPOSITION; IOE: 1-2-2, STRIP ABC: 1-2-1 (7) LOCATION OF SAFETY CLIPS/PINS: BURIED 30cm TO REAR OF START STRIP MARKER (8) LOCATION OF MINES FOR CLOSING LANES OR GAPS: NONE

**Block 13.** The emplacing unit OIC signs and dates the form.

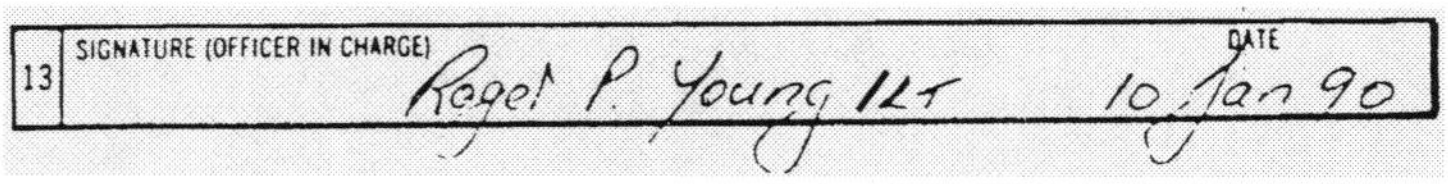
13 SIGNATURE (OFFICER IN CHARGE) Roger P. Young 1LT DATE 10 Jan 90

The front side of DA Form 1355 is now complete. Continue on the inside as follows:

**Block 14.** When completing the sketch, enter arrows for the direction of the enemy and the magnetic north. The enemy arrow will always point within the top 180 degrees of the form; the north arrow will follow one of the graph lines.

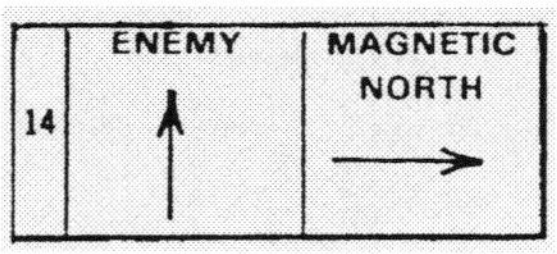
14 ENEMY | MAGNETIC NORTH

**Block 15.** If a compass was not available, enter what was used in the information block. Indicate the system of measurement and the scale used.

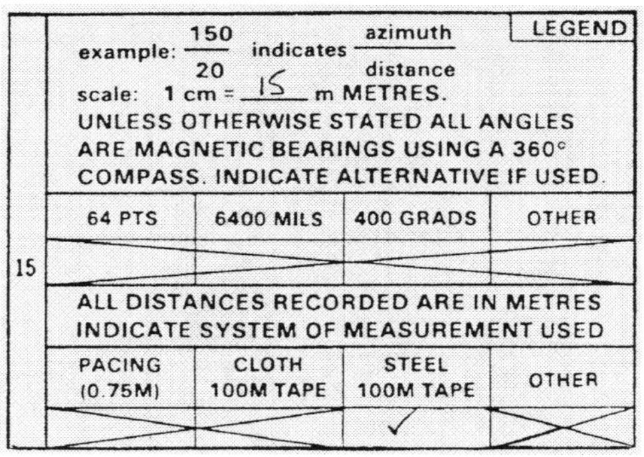
15

LEGEND

example: 150/20 indicates azimuth/distance

scale: 1 cm = 15 m METRES.

UNLESS OTHERWISE STATED ALL ANGLES ARE MAGNETIC BEARINGS USING A 360° COMPASS. INDICATE ALTERNATIVE IF USED.

| 64 PTS | 6400 MILS | 400 GRADS | OTHER |
|---|---|---|---|
| X | X | X | X |

ALL DISTANCES RECORDED ARE IN METRES INDICATE SYSTEM OF MEASUREMENT USED

| PACING (0.75M) | CLOTH 100M TAPE | STEEL 100M TAPE | OTHER |
|---|---|---|---|
| X | X | ✓ | X |

**Block 16.** Enter the information shown below on the sketch:

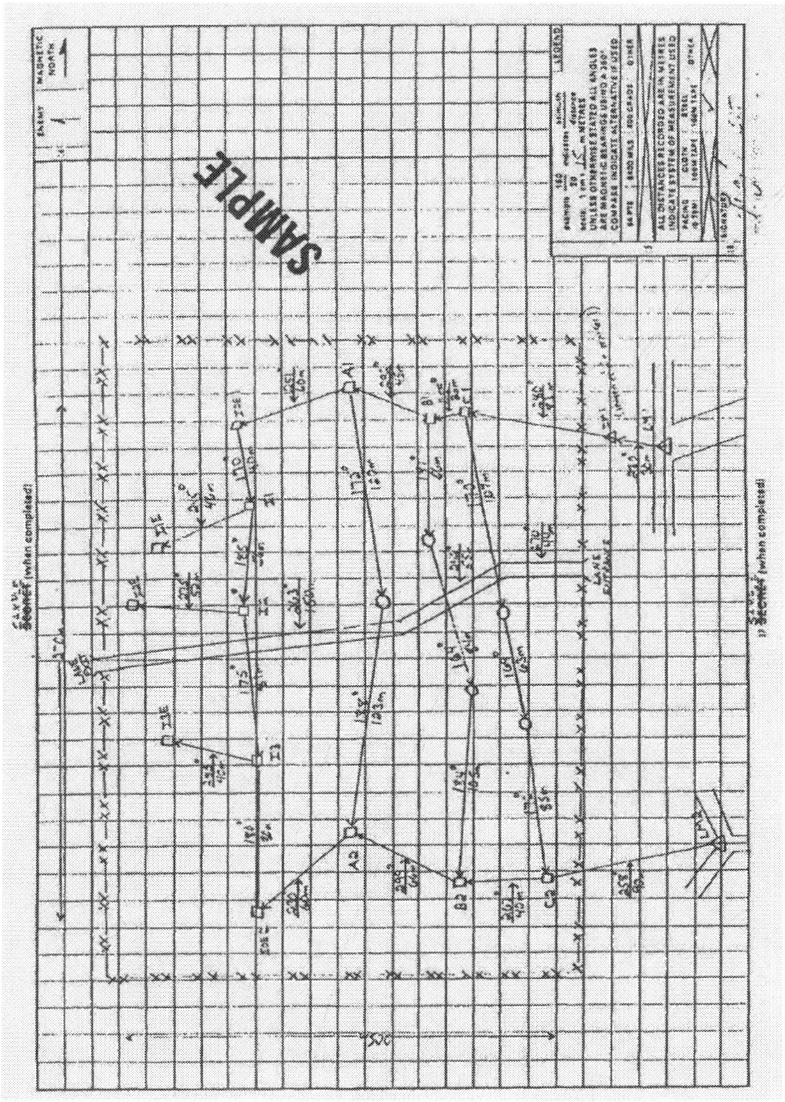

**Block 17.** Enter the security classification of the form. If the form was used for training, enter the word SAMPLE.

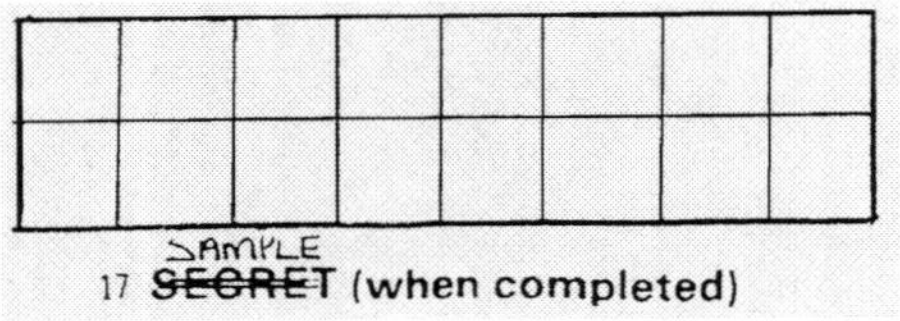


**Block 18.** The emplacing unit OIC signs the signature block.

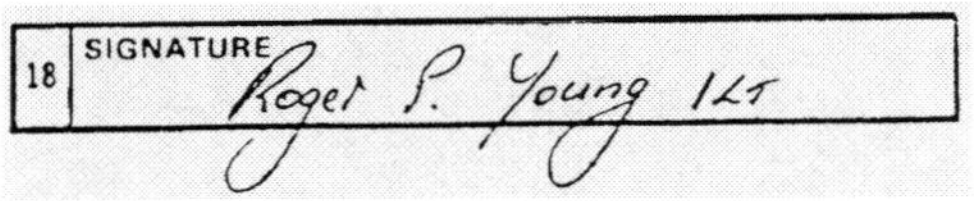


## Hasty Protective Row Minefield Record

Hasty protective row minefields/munition fields are recorded on DA Form 1355-1-R (Figure 8-4, page 8-18). DA Form 1355-1-R will be printed on 8½- by 11-inch paper. A copy for local reproduction is available in the back of this FM. DA Form 1355-1-R is also available on the Army Electronic Library (AEL) CD-ROM (EM 0001) and the USAPA web site (www.usapa.army.mil).

Use the following formula to determine the scale used on DA Form 1355-1-R:

*Distance from RP to the farthest point in the minefield + 10 meters ÷ 4 = scale*

**Example:** 90 meters + 10 meters = 100 meters ÷ 4 = 25 meters

The number 4 is a constant and represents the four concentric rings on DA Form 1355-1-R. Ten is added as a safety margin to ensure that the minefield/munition field sketch is entirely contained within the largest ring. The distance between rings is 2 centimeters; therefore, the scale used in this example is 2 centimeters = 25 meters.

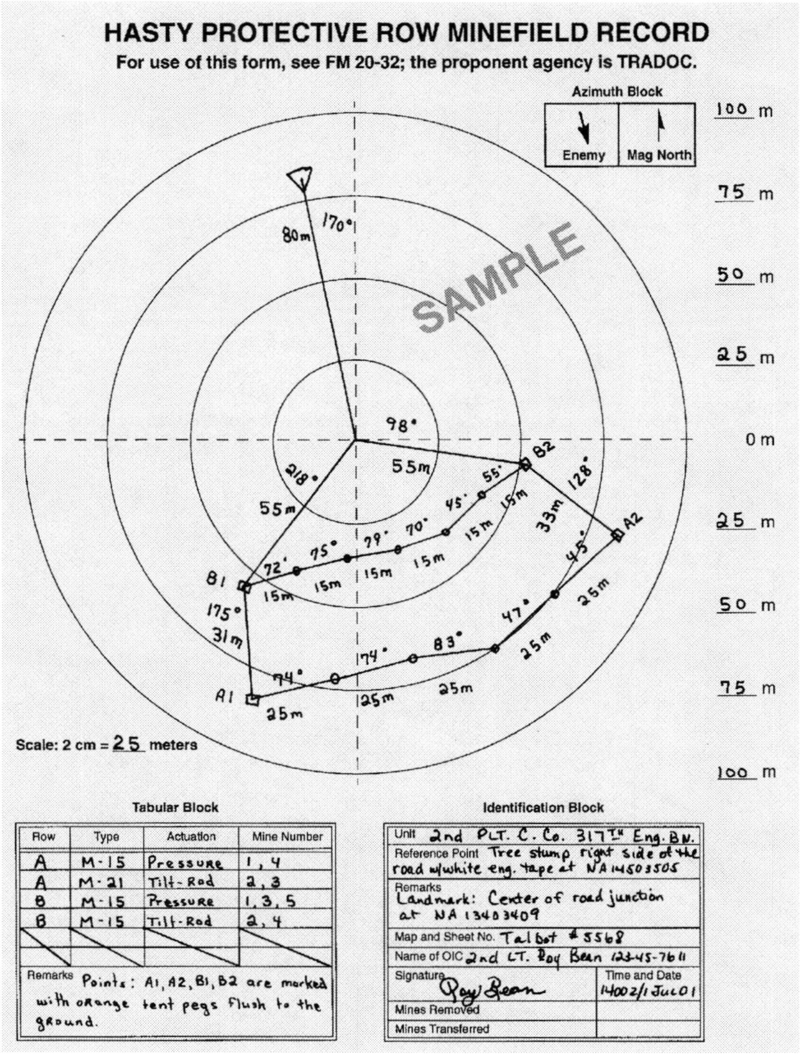

## HASTY PROTECTIVE ROW MINEFIELD RECORD

**For use of this form, see FM 20-32; the proponent agency is TRADOC.**



**Tabular Block**

| Row | Type | Actuation | Mine Number |
|---|---|---|---|
| A | M-15 | Pressure | 1, 4 |
| A | M-21 | Tilt-Rod | 2, 3 |
| B | M-15 | Pressure | 1, 3, 5 |
| B | M-15 | Tilt-Rod | 2, 4 |
| | | | |

Remarks Points: A1, A2, B1, B2 are marked with orange tent pegs flush to the ground.

**Identification Block**

Unit 2nd PLT. C. Co. 317TH Eng. BN.

Reference Point Tree stump right side of the road w/white eng. tape at NA 14503505

Remarks Landmark: Center of road junction at NA 13403409

Map and Sheet No. Talbot #5568

Name of OIC 2nd LT. Roy Bean 123-45-7611

| Signature | Time and Date |
|---|---|
| Roy Bean | 14002/1 Jul 01 |
| Mines Removed | |
| Mines Transferred | |

Figure 8-4. Sample DA Form 1355-1-R

The following step-by-step instructions are provided for completing DA Form 1355-1-R (see Figure 8-5):

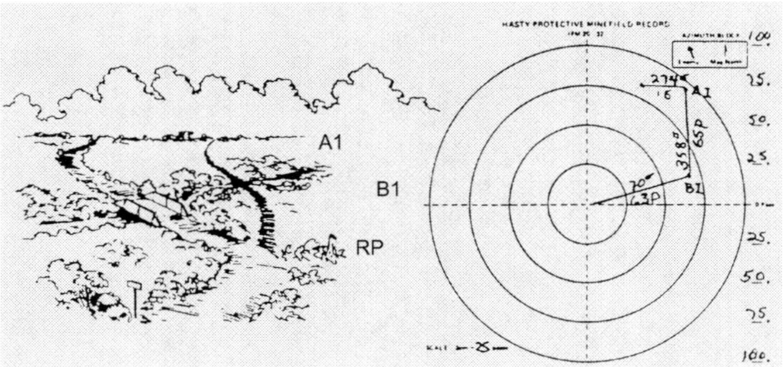


**Figure 8-5. Hasty protective row minefield/munition field record**

From the RP, the leader measures the magnetic azimuth, in degrees, to a specific point on the right side (facing the enemy) of the tentative minefield/munition field. He determines the distance and records it in meters. He identifies the method of measurement in the remarks block of DA Form 1355-1-R (for example, distances were paced out and multiplied by 0.75 to convert the measurement to meters). This point, called B1 (if there are two rows), marks the beginning of the second row. A marker is placed at B1, and the leader records the azimuth and the distance.

The leader measures the azimuth and the distance from B1 to a second point on the right side of the minefield/munition field (facing the enemy). A marker is placed at this point and recorded as A1.

The leader measures the distance and the azimuth from A1 to the location of the first mine in that row and records the information.

The distance (or spacing) from the end row marker to the first mine is the spacing between all the mines in that row. After the location is recorded, the mine is emplaced, but it is not armed.

The leader measures the distance and the azimuth from the first mine to the second, and so on, until all the mines are emplaced and their locations are recorded. This procedure is repeated for the second row. As each mine is recorded, it is assigned a number to identify it in the minefield/munition field record. When the last mine location is recorded for a row, the distance and the azimuth are measured from that point to another arbitrary point, A2 or B2. A marker is placed here in the same manner as A1 and B1. Next, the distance and the azimuth from the RP to B2 and from B2 to A2 are measured and recorded.

When all the mines have been emplaced and recorded, the leader measures the distance and the azimuth between the RP and a permanent landmark that can be found on the map. He records the information on DA Form 1355-1-R. The landmark is used to help locate the minefield/munition field if it is transferred or unexpectedly abandoned. Finally, the tabular and identification blocks are completed.

Mines can be armed after recording is complete. The mines nearest the enemy are armed first, allowing soldiers to safely work their way back to the unit's position. Pins and clips can be buried 30 centimeters behind row markers, the RP, or any easily identifiable, accessible location. Note the location in the remarks section of DA Form 1355-1-R. The leader then reports the completion of the minefield/munition field to higher headquarters.

If the minefield/munition field is transferred to another unit, the receiving unit leader is briefed by the transferring unit leader. The receiving unit leader signs and dates the mines transferred block on the DA Form 1355-1-R. When the minefield/munition field is removed, the form is destroyed. If the minefield/munition field is abandoned unexpectedly, the DA Form 1355-1-R must be forwarded to higher headquarters.

### NUISANCE MINEFIELD

It is unnecessary to record the precise location of individual nuisance mines that are laid in a pattern. Recording the position of a minefield/munition field that is laid in a pattern (either a standard pattern or one adopted for the situation) is easy and quick. However, the number of mines to be laid on site may make it impracticable or undesirable to lay mines in a pattern. In this case, they may be laid individually and their positions accurately recorded.

Figures 8-6a and 8-6b, pages 8-21 and 8-22 provide an example of a completed nuisance minefield/munition field record.

## SCATTERABLE MINEFIELD/MUNITION FIELD REPORTING AND RECORDING

The speed and responsiveness of SCATMINE employment require accurate, uniform, and timely reports. All known information on scatterable minefields/munition fields is simply and rapidly reported to all affected units. Although SCATMINES have SD capability, they are still recorded; and the information is disseminated to prevent casualties to friendly forces.

Since the locations of individual SCATMINEs are unknown, they cannot be recorded in detail. For most systems, a safety zone is calculated from one or more aim points. For example, a RAAM minefield/munition field is recorded based on the target location (the grid coordinates given to the firing battery). The size of the minefield/munition field depends on the number of rounds fired, the number of aim points, and the angle of fire. An artillery-delivered minefield/munition field is recorded by plotting it on a map, based on the aim point and the safety zone specified in the scatterable minefield/munition field report and record that was prepared by the emplacing unit. A ground-Volcano minefield/munition field can be recorded more accurately by plotting each of the minefield/munition field corner points rather than an aim point.

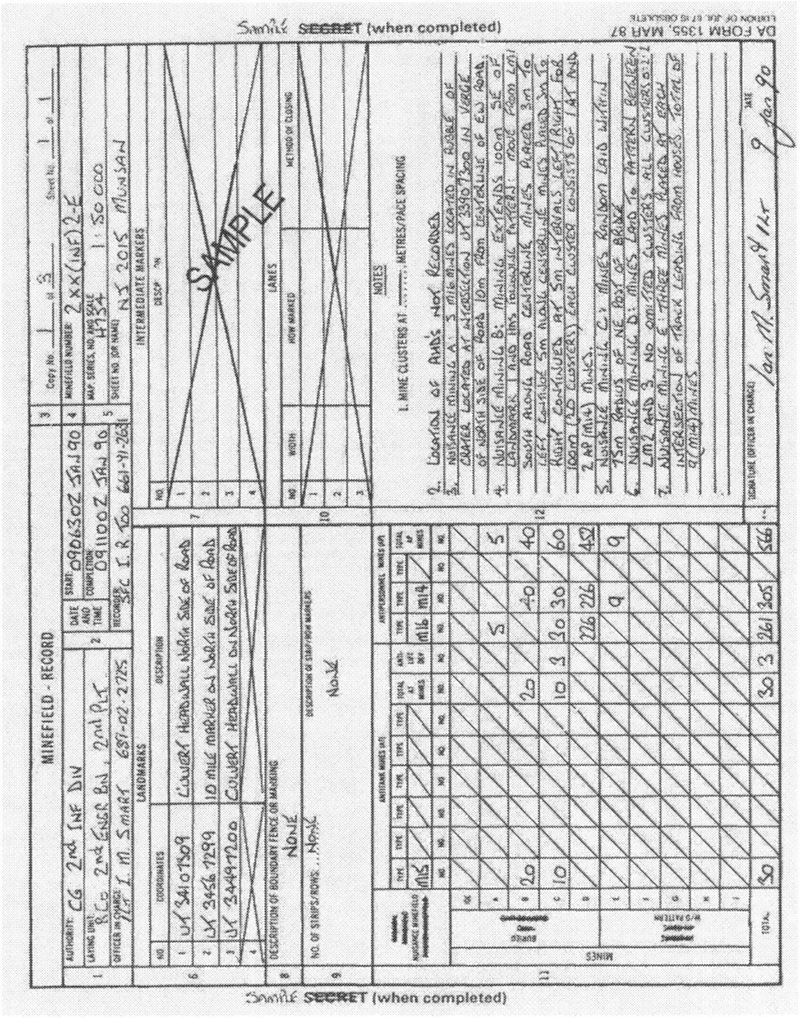

Sample SECRET (when completed)

MINEFIELD - RECORD

| | |
|---|---|
| 1 AUTHORITY | CG 2nd INF DIV |
| LAYING UNIT | B Co 2nd ENGR BN, 2nd PLT |
| OFFICER IN CHARGE | 2LT I. M. SMART 651-02-2725 |
| 2 DATE AND TIME START | 090630Z JAN 90 |
| COMPLETION | 091100Z JAN 90 |
| RECORDER | SPC I. R. Too 661-71-2631 |
| 3 Copy No. | 1 of 3 |
| Sheet No. | 1 of 1 |
| 4 MINEFIELD NUMBER | 2 XX (INF) 2-E |
| MAP SERIES, NO. AND SCALE | M754 1:50 000 |
| 5 SHEET NO. (OR NAME) | NY 2015 MUNSAN |

6 LANDMARKS

| NO | COORDINATES | DESCRIPTION |
|---|---|---|
| 1 | UT 3410 7309 | CULVERT HEADWALL NORTH SIDE OF ROAD |
| 2 | UT 3456 7299 | 10 MILE MARKER ON NORTH SIDE OF ROAD |
| 3 | UT 3449 7200 | CULVERT HEADWALL ON NORTH SIDE OF ROAD |
| 4 | | |

7 INTERMEDIATE MARKERS

| NO | DESCP | GR |
|---|---|---|
| 1 | | |
| 2 | | |
| 3 | | |
| 4 | | |

8 DESCRIPTION OF BOUNDARY FENCE OR MARKING: NONE

9 NO. OF STRIPS/ROWS: NONE — DESCRIPTION OF STRIP/ROW MARKERS: NONE

10 LANES

| NO | WIDTH | HOW MARKED | METHOD OF CLOSING |
|---|---|---|---|
| 1 | | | |
| 2 | | | |
| 3 | | | |

11 MINES

| | AT TYPE | AT NO | AT TYPE | AT NO | AT TYPE | AT NO | TOTAL AT MINES | ANTI-LIFT DEV NO | AP TYPE | AP NO | AP TYPE | AP NO | TOTAL AP MINES |
|---|---|---|---|---|---|---|---|---|---|---|---|---|---|
| | M15 | | | | | | | | M16 | | M14 | | |
| A | | | | | | | | | | 5 | | | 5 |
| B | | 20 | | | | | 20 | | | | | 40 | 40 |
| C | | 10 | | | | | 10 | 3 | | 30 | | 30 | 60 |
| D | | | | | | | | | | 226 | | 226 | 452 |
| E | | | | | | | | | | | | 9 | 9 |
| F | | | | | | | | | | | | | |
| G | | | | | | | | | | | | | |
| H | | | | | | | | | | | | | |
| I | | | | | | | | | | | | | |
| TOTAL | | 30 | | | | | 30 | 3 | | 261 | | 305 | 566 |

12 NOTES

1. MINE CLUSTERS AT ........ METRES/PACE SPACING
2. LOCATION OF AHD'S NOT RECORDED
3. NUISANCE MINING A: 5 M16 MINES LOCATED IN RUBBLE OF CRATER LOCATED AT INTERSECTION UT 3390 7300 IN VERGE OF NORTH SIDE OF ROAD 10m FROM CENTERLINE OF E-W ROAD
4. NUISANCE MINING B: MINING EXTENDS 100m SE OF LANDMARK 1 AND HAS FOLLOWING PATTERN: MOVE FROM LM1 SOUTH ALONG ROAD CENTERLINE MINES PLACED 3m TO LEFT, CONTINUE 5m ALONG CENTERLINE MINES PLACED 3m TO RIGHT CONTINUED AT 5m INTERVALS LEFT/RIGHT FOR 100m (20 CLUSTERS) EACH CLUSTER CONSISTS OF 1 AT AND 2 AP (M14) MINES.
5. NUISANCE MINING C: MINES RANDOM LAID WITHIN 75m RADIUS OF NE POST OF BRIDGE
6. NUISANCE MINING D: MINES LAID IN PATTERN BETWEEN LM2 AND 3, NO OMITTED CLUSTERS ALL CLUSTERS 0:1:1
7. NUISANCE MINING E: THREE MINES PLACED AT EACH INTERSECTION OF TRACK LEADING FROM HOUSES. TOTAL OF 9 (M14) MINES.

SIGNATURE (OFFICER IN CHARGE): Ian M. Smart 1Lt — DATE: 9 Jan 90

DA FORM 1355, MAR 87 — EDITION OF JUL 67 IS OBSOLETE

Sample SECRET (when completed)

Figure 8-6a. Sample DA Form 1355 (front side) for a nuisance minefield/munition field

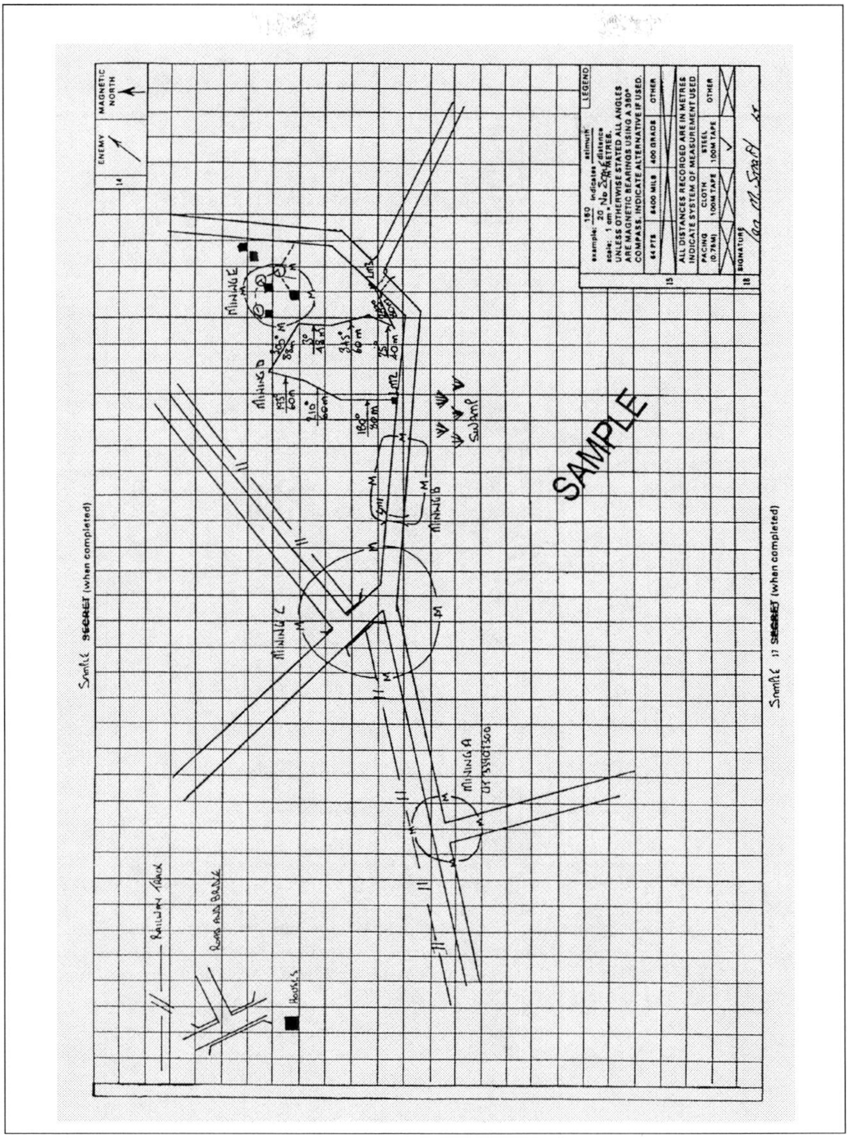


**Figure 8-6b. Sample DA Form 1355 (inside) for a nuisance minefield/munition field**

To facilitate reporting and recording of scatterable minefields/munition fields, a simple, uniform procedure is used. This procedure combines the report and the record into one document (Figure 8-7) that is applicable for all delivery systems.

| Line No | Information Required | Data |
|---|---|---|
| 1 | Approving authority | Enter the approving authority, such as CDR 3AD. |
| 2 | Target/obstacle number | If the minefield/munition field is part of an obstacle plan, enter the obstacle number, such as 2XXX0157. This number represents II Corps, target number 157. If the minefield/munition field is not a part of an obstacle plan or does not have a number, then leave this line blank or enter NA. |
| 3 | Type of emplacing system | Enter the type system that emplaced the minefield/munition field, such as artillery or Volcano. |
| 4 | Type of mines | Enter AP for antipersonnel mines and AT for antitank mines. If both types of mines are used, enter AP/AT. |
| 5 | Life cycle | Enter the DTG the minefield/munition field was emplaced and the DTG the last mine SDs. |
| 6-14 | Aim point/corner points of minefield/munition field | If the system used to emplace the minefield/munition field uses a single aim point to deliver the mines, enter that aim point, such as MB 10102935. If the system has distinct corner points (Volcano), enter those corner points, such as MB 17954790, MB 18604860, MB 18504890, and MB 18054895. |
| 15 | Size of safety zone from aim point | If an aim point is given in Line 6, enter the size of the safety zone from that aim point. **Example:** Artillery emplaces a minefield/ munition field from aim point MB 10102935, and the safety zone is 1,000 x 1,000 m. Enter 500 m so that personnel plotting or receiving the information can plot the coordinates and go 500 m in each direction from the aim point to plot the safety zone. |
| 16 | Unit emplacing mines/ report number | Enter the unit emplacing mines and the report number, such as BCO 23 ENGR BN 4. Reports should be numbered consecutively. This would be the fourth minefield/munition field that B Company has emplaced. |
| 17 | Person completing report | Enter the person's name completing the report, such as SFC Jones. |
| 18 | DTG of report | Enter the DTG of the report, such as 160735ZOCT90. |
| 19 | Remarks | Include any other items the reporting unit may feel are important. |

**Figure 8-7. Scatterable minefield/munition field report and record work sheet**

In addition to the scatterable minefield/munition field report and record, the SCATMINWARN (a sample is shown in Figure 8-8, page 8-24) notifies effected units that SCATMINEs will be emplaced. These two reports are the only reports used with scatterable mines.

A completed scatterable minefield/munition field report and record for an ADAM/RAAM artillery mission is shown in Figure 8-9, page 8-24. Note that on line 6, only one grid coordinate is given. It is the aim point used when the mission was fired. Also note that the 500-meter distance from the aim point (line 15) designates a safety zone that is 1,000 by 1,000 meters.

| Line | Message |
|---|---|
| Alpha | Emplacing system |
| Bravo | AT (Yes or No) |
| Charlie | AP (Yes or No) |
| Delta | 4 aim or corner points |
| Echo | Grid coordinates of aim points/corner points and size of the safety zone |
| Foxtrot | DTG of the life cycle |

**Figure 8-8. Sample SCATMINWARN**

| Line No | Information Required | Data |
|---|---|---|
| 1 | Approving authority | CDR 3AD |
| 2 | Target/obstacle number | 2XXX0157 |
| 3 | Type of emplacing system | Artillery |
| 4 | Type of mines | AT/AP |
| 5 | Life cycle | 0816102-082020OCT90 |
| 6 | Aim point/corner points of minefield/munition field | MB 10102935 |
| 7 | | |
| 8 | | |
| 9 | | |
| 10 | | |
| 11 | | |
| 12 | | |
| 13 | | |
| 14 | | |
| 15 | Size safety zone from aim point | 500 m |
| 16 | Unit emplacing mines/report number | 2/48FA/2 |
| 17 | Person completing report | SFC Hollins |
| 18 | DTG of report | 061645ZOCT90 |
| 19 | Remarks | NA |

**Figure 8-9. Scatterable minefield/munition field report and record for an ADAM/RAAM artillery mission**

The SCATMINWARN provides affected units with the necessary warning to plan and execute their operations. The information is kept to a minimum to ensure rapid dissemination. The report may be sent orally, digitally, or hard copy. It is sent before or immediately after the mines have been emplaced. A completed SCATMINWARN for an artillery mission is shown in Figure 8-10.

| Line | Message |
|---|---|
| Alpha | Artillery |
| Bravo | Yes |
| Charlie | Yes |
| Delta | One |
| Echo | MB 10102935 500 m |
| Foxtrot | 081610Z-081920ZOCT90 |

**Figure 8-10. Sample SCATMINWARN for an artillery mission**

## MINEFIELD/MUNITION FIELD OVERLAY SYMBOLS

The symbols contained in Figure 8-11, pages 8-26 through 8-30, are extracted from FM 101-5-1 and are provided for posting mine data on maps and overlays.

| Description | Symbol |
|---|---|
| **Minefield/Munition Fields** | |
| **Korea Only: AP mine** |  |
| AT mine |  |
| AT mine with AHD |  |
| Directional mine (arrow points in direction of main effect) |  |
| Mine cluster | 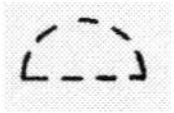 |
| Mine, type unspecified |  |
| Trip wire | 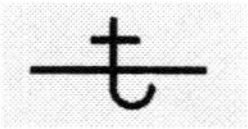 |
| **Control Measures** | |
| Zone | 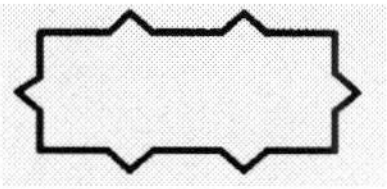 |
| Belt | 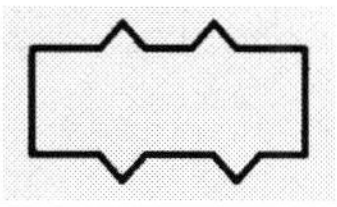 |
| Restrictions | 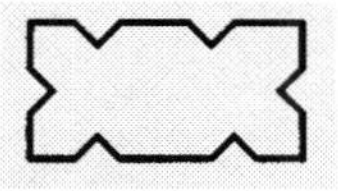 |

**Figure 8-11. Minefield/munition field overlay symbols**

| Description | Symbol |
|---|---|
| Block effect |  |
| Turn effect |  |
| Disrupt effect | 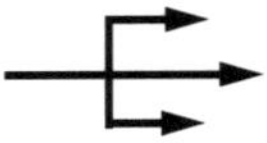 |
| Fix effect |  |
| **Conventional** | |
| A planned minefield/munition field consisting of unspecified mines | 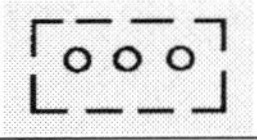 |
| A completed minefield/munition field consisting of unspecified mines | 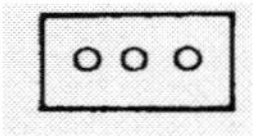 |
| Scatterable minefield/munition field (DTGs used for SD times) | 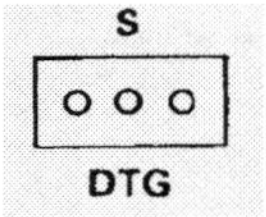 |
| Conventional AP minefield/ munition field reinforced with SCATMINEs | 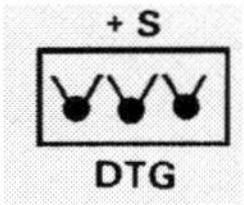 |
| Tactical AP row minefield/ munition field (outline drawn to scale) | 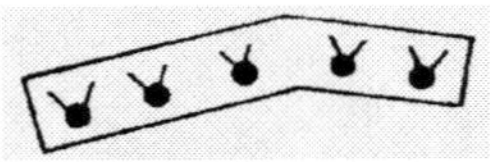 |

**Figure 8-11. Minefield/munition field overlay symbols (continued)**

| Description | Symbol |
|---|---|
| Tactical minefield/munition field of scatterable AT mines, effective until 101200Z |   |
| Completed AT minefield/munition field (drawn away from the location and connected by a vector) | 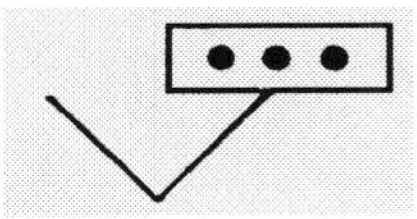 |
| Executed Volcano minefield/munition field (DTG used for SD time) | 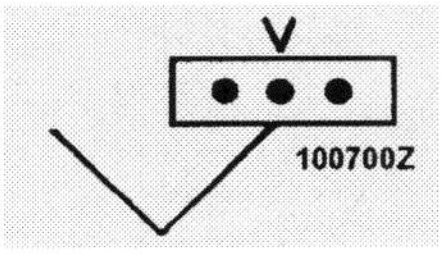  |
| Lane in conventionally laid AT minefield/munition field | 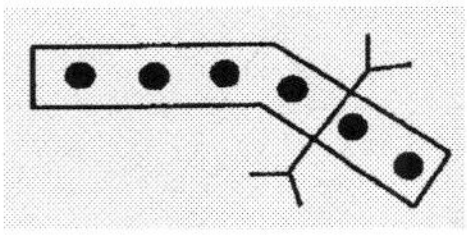 |
| Gap in conventionally laid AT minefield/munition field (DTG opened to DTG closed) | 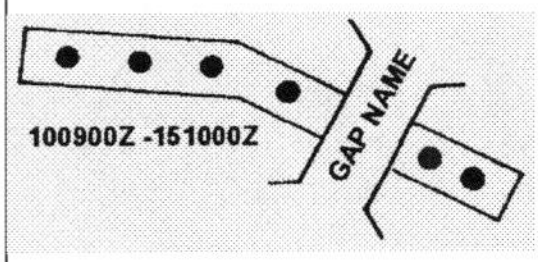  |

**Figure 8-11. Minefield/munition field overlay symbols (continued)**

| Description | Symbol |
|---|---|
| AT ditch reinforced with AT mines | 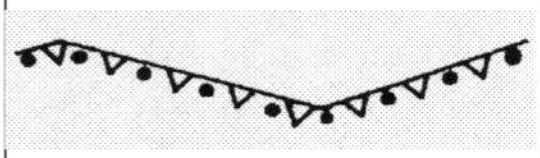 |
| **UXO** | |
| UXO area | 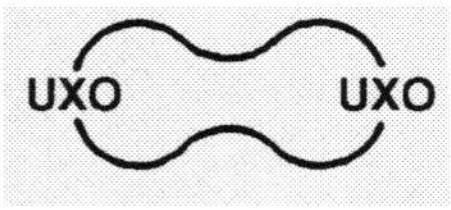  |
| **Nuisance** | |
| Nuisance minefield/munition field | 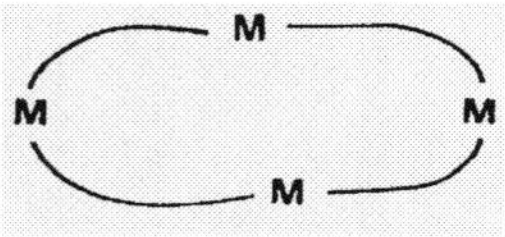  |
| Demolished crossroads with nuisance mines | 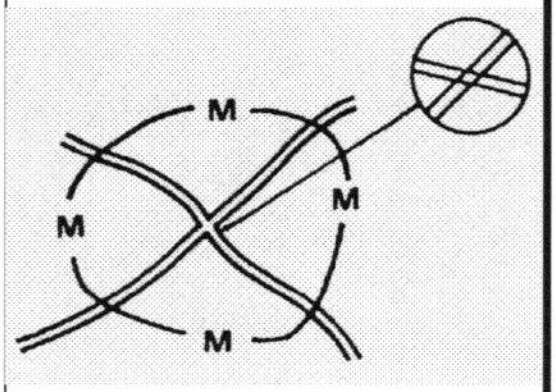  |
| **Phony** | |
| Phony minefield/munition field | 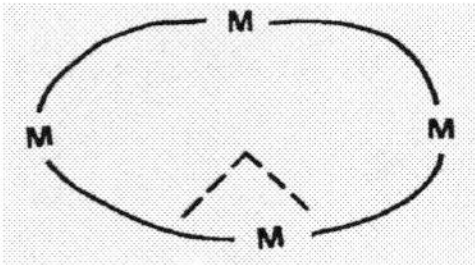  |

**Figure 8-11. Minefield/munition field overlay symbols (continued)**

| Description | Symbol |
|---|---|
| Phony minefield/munition field, fenced | 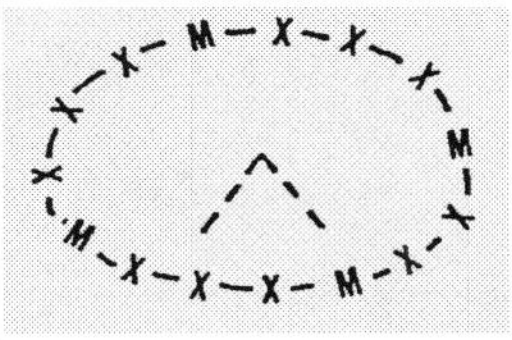 |
| **Hornet Symbology** | |
| Planned |  |
| Unarmed |  |
| Armed |  |
| Expended |  |
| Armed munition field (DTG used for SD time) | 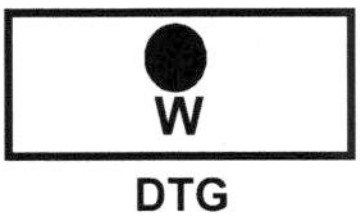 |

**Figure 8-11. Minefield/munition field overlay symbols (continued)**

# PART TWO

# Counteroperations

This part of the manual provides overall guidance for conducting counteroperations by US forces. The types of breaching and clearing operations conducted, the tasks performed, and the equipment required are described in detail. Responsibilities and planning considerations are outlined for each operation.

## Chapter 9

## Countermine Operations

Countermine operations are undertaken to breach or clear a minefield. All the tasks fall under breaching or clearing operations and include detecting, reporting, reducing, proofing, and marking.

### DEFINITIONS

#### OBSTACLE

The term *obstacle* is used often in this chapter because the same breaching and clearing operations are used for minefields and other obstacles. For the purpose of this manual, breaching and clearing tactics, techniques, and procedures (TTP) focus solely on minefields.

#### REDUCTION

Reduction is the act or actions taken against an obstacle that diminishes its original effect. For example, creating a lane in a minefield would yield a reduction of the minefield obstacle.

#### BREACHING

Breaching is the employment of TTP to project combat power to the far side of an obstacle. It is a synchronized combined arms operation that is under the control of the maneuver commander.

#### AREA CLEARANCE

Area clearance is the total elimination or neutralization of an obstacle or portions of an obstacle. Clearing operations are not conducted under fire. They are usually performed by follow-on engineer forces after a breaching operation or anytime in a friendly AO where an obstacle is a hazard or hinders movement.

#### ROUTE CLEARANCE

Route clearance is the removal of mines along preexisting roads and trails.

#### MINE NEUTRALIZATION

Mine neutralization occurs when a mine is rendered incapable of firing on passage of a target. The mine may still be dangerous to handle.

### PROOFING

Proofing is done by passing a mine roller or other mine-resistant vehicle through a lane as the lead vehicle. It verifies that a lane is free of mines.

### DEMINING

Demining is the complete removal of all mines and UXO within a geopolitical boundary after hostilities cease.

## BREACHING OPERATIONS

Breaching is a synchronized combined arms operation that is under the control of the maneuver commander. FM 3-34.2 provides combined arms commanders and staffs with doctrine TTP that are needed to successfully overcome obstacles. Breaching operations make maneuver possible in the face of enemy obstacle efforts. Since obstacles may be encountered anywhere, maneuver forces integrate breaching operations into all movement plans.

When possible, enemy minefields are bypassed to maintain the momentum and conserve critical countermobility assets. However, when making the decision to bypass rather than breach, consider the likelihood of friendly units being channelized into kill zones. Bypassing is done by maneuvering around a minefield or, if aviation assets are available, moving over the minefield. When maneuvering around an obstacle, attempt to locate a portion of the force in overwatch positions to cover the bypass of the main element. Even when the decision is made to conduct a breach, scouts should continue to reconnoiter for bypass routes.

The first step in understanding breaching operations is to know the obstacle breaching theory. Knowing the theory behind breaching operations equips the engineer and the maneuver commander with fundamentals that are needed to integrate breach into the tactical planning, preparation, and execution of an operation.

Successful breaching operations are characterized by the application of the following tenets of breaching:

### INTELLIGENCE

In any operation where enemy obstacles interfere with friendly maneuver, obstacle intelligence (OBSTINTEL) becomes a priority intelligence requirement (PIR). Finding enemy obstacles or seeing enemy obstacle activity validates and refines the S2's picture of the battlefield. OBSTINTEL helps determine enemy intentions, plans, and strength. The force engineer is the unit's expert on enemy countermobility, and he assists the S2 in templating enemy obstacles and analyzing OBSTINTEL.

When collecting OBSTINTEL, reconnaissance is a combined arms activity that includes engineers. An engineer squad moves with scouts or the patrol and conducts dismounted reconnaissance of templated or discovered obstacles. Additional information on reconnaissance can be found in FM 5-170. Reconnaissance teams gather the following OBSTINTEL information from the reconnaissance:

- Minefield location. Plot the perimeter location on a large-scale map and refer to recognizable landmarks.

- Perimeter description. Describe how the perimeter is fenced. If it is unfenced, describe how it is marked. If it is unmarked, show how it was recognized.
- Nuisance mines. If you discover a nuisance mine forward of the minefield's outer edge, remember, there may be others. Assembly areas might also be mined.
- Types of mines. Indicate whether mines are AT or AP or have unknown fuses (self-neutralized or SD). If possible, recover specimens of unknown or new mines and note the details.
- Details of any other devices. Describe booby traps, trip wires, and flares.
- Laying method. Indicate whether mines are buried or surface-laid.
- Density and pattern. Include the mine spacing and the number of mine rows; estimate the mine density based on this information.
- Minefield depth. Provide the distance between strips or rows and describe markers.
- Safe lanes and gaps. Plot the location of suspected safe lanes and gaps and describe their markings.
- Ground conditions. Include information on general ground conditions.
- Other obstacles. Plot the location and the construction of other obstacles.
- Enemy defenses. Describe the enemy's location and size. Include the location of enemy direct-fire weapons.

**NOTE: Engineers engaged in reconnaissance for OBSTINTEL should rarely, if ever, be used to reduce obstacles during the reconnaissance.**

Each reconnaissance team commander submits a detailed OBSTINTEL report to the next higher headquarters when the reconnaissance is complete. A sample OBSTINTEL report is shown in Figure 9-1.

| Letter Designation | Explanation |
|---|---|
| A | Map sheet(s). |
| B | Date and time the information was collected. |
| C | Type of minefield (AT, AP, or mixed). |
| D | Grid references of minefield extremities, if known. |
| E | Depth of minefield. |
| F | Estimated time required to clear the minefield. |
| G | Estimated material and equipment required to clear the minefield. |
| H | Routes for bypassing the minefield, if any. |
| I-Y | Grid references of lanes (entry and exit) and width of lanes, in meters. |
| Z | Additional information such as types of mines used, unknown mines, or types of booby traps. |

**Figure 9-1. Sample OBSTINTEL report**

## FUNDAMENTALS

Suppress, obscure, secure, and reduce (SOSR) are the breaching fundamentals that must be applied to ensure success when breaching against a defending enemy. This TTP will always apply but may vary based on specific situations.

### Suppress

Suppression is the focus of all available fires on enemy personnel, weapons, and equipment to prevent effective fires on friendly forces. The purpose of suppression is to protect forces who are reducing the obstacle and maneuvering through it and to soften the enemy's initial foothold.

### Obscure

Obscuration hampers enemy observation and target acquisition, and it conceals friendly activities and movement. Obscuration smoke deployed on or near the enemy position minimizes the enemy's vision.

### Secure

The force secures the breaching site to prevent the enemy from interfering with reduction and to prevent the enemy's passage through lanes created during reduction. In general, enemy tactical obstacles are secured by fires, and protective obstacles are secured by force.

### Reduce

Reduction means creating lanes through or over the obstacle to allow passage of the attacking force. The lanes must be sufficient to allow the force to cross and accomplish the mission. (Table 9-1 provides information on lane widths.) The unit reducing the minefield marks the minefield and lane locations and reports their conditions to higher headquarters.

**Table 9-1. Lane widths**

| Lane | Width | Remarks |
|---|---|---|
| Assault footpath | 1 meter | Allows passage of dismounted troops so that they may continue an attack or secure the far side of a minefield while vehicle lanes are being breached |
| Initial lane | 4.5 meters | Allows passage of breaching and assault forces (initial lanes are widened and marked as soon as the tactical situation allows) |
| Intermediate lane | 4.5 meters | Upgrades sign posting and further defines exit markers |
| Full lane | 10 meters | Allows passage of two-way traffic |

## ORGANIZATION

The commander organizes the force to accomplish SOSR breaching fundamentals quickly and effectively. This requires him to organize support, breach, and assault forces with the necessary assets to accomplish their missions.

### Support Force

The support force is primarily responsible for eliminating the enemy's ability to interfere with the breaching operation. It must isolate the battlefield with fires and suppress enemy fires that are covering the breach location. This involves massive direct and indirect fire to destroy enemy vehicles and personnel who are able to bring fire on the breach force.

### Breach Force

The breach force creates lanes that enable the assault force to pass through the obstacle and continue the attack. It is also responsible for marking lanes (length and entry points) to speed passage of assault and follow-on forces. The breach force is a combined arms force. It includes engineers, breaching assets, and enough maneuver force to provide local security. The breach force must be able to deploy and begin reducing the obstacle as soon as enemy fires have been suppressed. It must be capable of creating a minimum of one lane for each assaulting company or two lanes for an assaulting TF. At a minimum, the lanes must be marked and their locations and conditions reported to higher headquarters and follow-on units as prescribed in the unit's SOP. The commander should expect a 50 percent loss of mobility assets in close combat.

### Assault Force

The assault force destroys or dislodges the enemy on the far side of the obstacle. It secures the far side by physical occupation in most breaching operations. The assault force may be tasked to assist the support force with suppression while the breach force reduces the obstacle. The assault force must be sufficient in size to seize objectives that eliminate fires on the breaching site.

## MASS

Breaching is conducted by rapidly applying concentrated force at a designated point to crack the obstacle and rupture the defense. Massed combat power is directed against an enemy weakness. Smoke and terrain help isolate the enemy force that is under attack. The commander also masses engineers and breaching equipment to reduce the obstacle. The breach force is organized and equipped to use several different reduction techniques in case the primary technique fails (a key vehicle is destroyed or casualties render dismounted engineers ineffective). Additional reduction assets are present to handle the unexpected (50 percent over the requirement are normally positioned with the breach force).

## SYNCHRONIZATION

Breaching operations require precise synchronization of the SOSR breaching fundamentals by support, breach, and assault forces. Failure to synchronize effective suppression and obscuration with reduction and assault can result in rapid, devastating losses of friendly troops in the obstacle or the enemy's fire sack. The commander cannot adequately synchronize his force's application of combat power in the short time available to him when he encounters an obstacle. The number of decisions that he must make while under fire in an unclear situation will rapidly overwhelm him. Even with a force trained to execute a combined arms breach, synchronizing all the necessary tasks remains a complex, difficult process. The commander uses the following principles to ensure synchronization through proper planning and force preparation:

- Detailed reverse planning.
- Clear subunit instructions.
- Effective $C^2$.
- A well-rehearsed force.

## CLEARING OPERATIONS

Clearing is the total elimination or neutralization of mines from an area. It is not usually conducted under enemy fire, but it can be conducted by engineers during war or after hostilities as part of nation assistance.

A limited clearing operation can be conducted by follow-on engineers after the breaching force has reduced the minefield and secured the area. Engineers initially improve existing breach lanes by widening and marking them, and they also clear and mark new lanes through the minefield. The clearing operation supports the continued passage of forces.

A clearing operation is also conducted to eliminate all the mines in a minefield (previously identified, reported, and marked in a friendly AO) that hinders mobility or is a hazard to friendly forces.

### UPGRADE OF BREACH LANES

Lane clearance is more deliberate than lane breaching and normally takes longer. Follow-on engineers upgrade breach lanes to improve existing lanes through minefields and to create new lanes. This clearing operation is intended to further reduce the minefield so that follow-on units can pass through it quickly.

The breach force that initially reduced the obstacle and marked the lanes turns over the lanes to follow-on engineers. Follow-on engineers can expect lane widths of 4.5 meters. The total number of lanes depends on the size of the lead assault force. Two to four assault lanes are normal if the lead unit was brigade-size.

If forces continue to pass through existing lanes while further reduction and clearance is conducted, follow-on engineers first begin reducing new lanes. At a minimum, two lanes are required for an assaulting TF and one lane is required for an assaulting company/team.

A limited amount of mechanical breaching assets is available for clearing operations. Follow-on engineers will probably not have tank-mounted mine-clearing blades (MCBs) or mine-clearing rollers (MCRs). The main mechanical clearing asset is an armored dozer with a mine rake. Mine-clearing line charges (MICLICs) are used if available. Engineers conducting clearing operations—

- Ensure that lanes are a minimum of 100 meters apart.
- Reduce additional lanes by using the equipment and techniques outlined in Chapter 10.
- Widen lanes to 10 meters to allow two-way traffic.
- Mark breach lanes by using the original marking system or the division SOP. (Marking procedures are outlined in Chapter 10.)
- Emplace entrance, exit, and left and right lane markers to provide day and night capability.

Traffic control is critical during lane reduction and when shifting lanes to improve existing lanes. Engineers conducting reduction and clearance may also provide guides at the lanes. Control procedures are outlined in FM 90-13-1.

To eliminate the danger of forces entering the minefield adjacent to lanes, the minefield is marked with fencing (barbwire or concertina) and mine markers.

Marking is emplaced across the front, on both sides, between lanes, and to the left and right of the crossing site as far out as practical.

Engineers may also help remove damaged vehicles from minefield lanes. Recovery vehicles should be available near lanes for this purpose.

### AREA CLEARANCE

Clearing operations are done when engineers receive a mission to clear an area of mines or to clear a specific minefield in a friendly AO. The minefield was reported and may already be marked on all sides. The worst case would be if the minefield was reported but not marked and its limits were unknown. The engineer unit receiving the mission bases plans on available information and prepares equipment based on the estimate. Detailed techniques and procedures for area and route clearance operations are outlined in Chapter 11.

Actions at the minefield begin with a thorough reconnaissance to identify the minefield limits and the types of mines. This is a time-consuming process that is hazardous to shortcut. Identified limits are marked with an expedient system of single-strand barbwire or concertina. In this situation, since all mines must be destroyed, the unit takes a systematic approach to clearing mines. The procedure depends on the types of mines and whether the mines are buried or surface-laid.

If mines are magnetic- or seismic-fused, mechanical assets are used. Pressure mines can be destroyed by using hand-emplaced explosives. When a manual procedure is used, eliminate trip wires on AP mines with grapnel hooks before moving forward to detect mines.

Using the manual procedure, engineers visually detect mines or detect them with mine detectors and probes. They also mark mines for destruction by explosives. Chapter 11 contains information on minesweeping procedures.

After the mines are destroyed, engineers proof used lanes and routes to ensure that all the mines were eliminated. This is done by using a mine roller or another blast-resistant device. Proofing is discussed further in Chapter 10.

### DEMINING

Demining is the complete removal of all mines and UXO to safeguard the civilian population within a geopolitical boundary after hostilities cease. It is an extremely manpower- and time-intensive operation and is sometimes contracted. Although not a formal Army mission or function, SOFs may provide special expertise in training demining organizations, acting as advisors, and taking the lead in providing clearance equipment or techniques that can be useful in demining operations. Demining TTP are outlined in TC 31-34.

Chapter 10

# Minefield Reduction

Reduction is the physical creation of a lane through a minefield. It is a fundamental of breaching operations as discussed in Chapter 9 and in FM 3-34.2. A number of tasks (detecting, reporting, reducing, proofing, and marking) directly support or are included in minefield reduction.

## DETECTING

Detection is the actual confirmation and location of mines. It may be accomplished through reconnaissance, or it may be unintentional (such as a vehicle running into a mine). Mine detection is used in conjunction with intelligence-gathering operations, minefield bypass reconnaissance, and breaching and clearing operations. There are four types of detection methods—visual, physical (probing), electronic, and mechanical.

### VISUAL

Visual detection is part of all combat operations. Personnel visually inspect the terrain for the following minefield indicators:

- Trip wires.
- Signs of road repair (such as new fill or paving, road patches, ditching, culvert work).
- Signs placed on trees, posts, or stakes. Threat forces mark their minefields to protect their own forces.
- Dead animals.
- Damaged vehicles.
- Disturbances in previous tire tracks or tracks that stop unexplainably.
- Wires leading away from the side of the road. They may be firing wires that are partially buried.
- Odd features in the ground or patterns that are not present in nature. Plant growth may wilt or change color, rain may wash away some of the cover, the cover may sink or crack around the edges, or the material covering the mines may look like mounds of dirt.
- Civilians. They may know where mines or booby traps are located in the residential area. Civilians staying away from certain places or out of certain buildings are good indications of the presence of mines or booby traps. Question civilians to determine the exact locations.
- Pieces of wood or other debris on a road. They may be indicative of pressure or pressure-release FDs. These devices may be on the surface or partially buried.

- Patterns of objects that could be used as a sighting line. The enemy can use mines that are fired by command, so road shoulders and areas close to the objects should be searched.

### PHYSICAL

Physical detection (probing) is very time-consuming and is used primarily for clearing operations, self-extraction, and covert breaching operations. Detection of mines by visual or electronic methods should be confirmed by probing. Use the following procedures and techniques when probing for mines:

- Roll up your sleeves and remove your jewelry to increase sensitivity. Wear a Kevlar helmet, with the chin strap buckled, and a protective fragmentation vest.
- Stay close to the ground and move in a prone position to reduce the effects of an accidental blast. When moving into a prone position—
  - Squat down without touching your knees to the ground.
  - Scan forward up to 2 meters and to the sides up to 3 meters for mine indicators.
  - Probe the area around your feet and as far forward as possible.
  - Kneel on the ground after the area is found to be clear, and continue probing forward until you are in a prone position.
- Use sight and touch to detect trip wires, fuses, and pressure prongs.
- Use a slender, nonmetallic object as a probe.
- Probe every 5 centimeters across a 1-meter front.
- Gently push the probe into the ground at an angle that is less than 45 degrees.

**DANGER**
**Use extreme caution when probing. If the probe is pushed straight down, its tip may detonate a pressure fuse.**

- Apply just enough pressure on the probe to sink it slowly into the ground.
- If the probe encounters resistance and does not go into the ground freely, carefully pick the soil away with the tip of the probe and remove the loose dirt by hand. Care must be taken to prevent functioning the mine.
- When you touch a solid object, stop probing and use two fingers from each hand to carefully remove the surrounding soil and identify the object.
- If the object is a mine, remove enough soil to show the mine type and mark its location. Do not attempt to remove or disarm the mine. Use explosives to destroy detected mines in place, or use a grappling hook and rope to cause mines to self-detonate. Do not use metal grappling hooks on magnetic-fused mines.

Probing is extremely stressful and tedious. The senior leader must set a limit to the time a prober can actually probe in the minefield. To determine a reasonable time, the leader must consider METT-TC factors, weather conditions, the threat level, the unit's stress level, and the prober's fatigue level and state of mind. As a rule, 20 to 30 minutes is the maximum amount of time that an individual can probe effectively.

### ELECTRONIC

Electronic detection is effective for locating mines, but this method is time-consuming and exposes personnel to enemy fire. In addition, the suspected mines must be confirmed by probing.

#### AN/PSS-12 Mine Detector

The AN/PSS-12 mine detector (Figure 10-1) can only detect metal, but most mines have metal components in their design. The detector can locate and identify plastic or wooden mines by a slight metallic signature. Employment and operation procedures for the AN/PSS-12 are discussed in Appendix F, and technical data is available in TM 5-6665-298-10. The detector is hand-held and identifies suspected mines by an audio signal in the headphones.

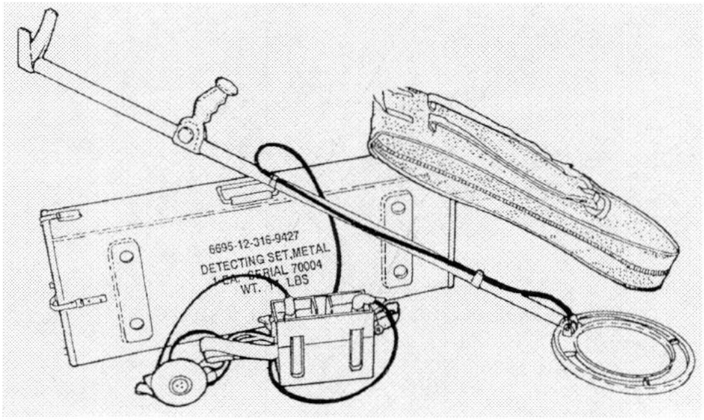

**Figure 10-1. AN/PSS-12 mine detector**

As in probing, consideration must be taken for the maximum amount time an individual can operate the detector. The leader considers METT-TC factors, weather conditions, the threat level, the unit's stress level, and the individual's fatigue level and state of mind. As a rule, 20 to 30 minutes is the maximum amount of time an individual can use the detector effectively.

#### Airborne Standoff Minefield Detection System

The Airborne Standoff Minefield Detection System (ASTAMIDS) (Figure 10-2, page 10-4) provides US forces with the capability to detect minefields rapidly. Environmental conditions must be favorable for aircraft and ASTAMIDS operations. ASTAMIDS can be mounted on a UH-60 Blackhawk helicopter, an unmanned aerial vehicle (UAV), or a fixed-wing aircraft. The system detects and classifies thermal and other anomalies as suspected minefields along routes or in areas of interest. ASTAMIDS can be used to protect advancing forces and can operate in concert with air and ground units in reconnaissance missions.

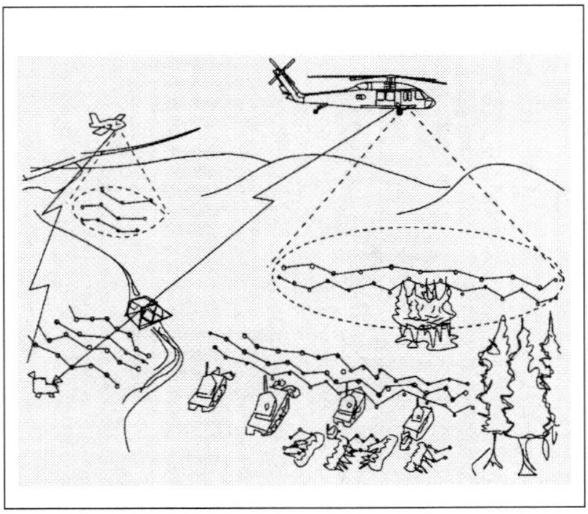

Figure 10-2. ASTAMIDS

***System Components***

ASTAMIDS hardware and software components consist of a sensor with associated electronics and the minefield-detection algorithm and processor (MIDAP). Surrogate equipment includes an air-data package (GPS, radar altimeter, inertial measurement unit [IMU]), a power supply, a work station(s), a digital data recorder, mounting racks, and a modified floor for the specific aircraft.

Operators view the data displayed on the monitors, communicate with the aircrew, and perform other functions (such as changing data tapes and producing reports). The aircrew must maintain an altitude of 300 feet and an airspeed of approximately 70 knots for the system to detect mines accurately within the sensor's ground swath (approximately 215 feet wide). The system has a 2-hour operational capability, based on standard flight time for the mission profile.

***Employment Concept***

ASTAMIDS is a fast method for detecting tactical minefields. When it is employed by aviation elements in support of maneuver units, close coordination between aviation and ground units assures that minefield detection is reported accurately and quickly. ASTAMIDS is not as precise as ground detection systems, but it is accurate enough to help mitigate the dangers inherent with minefields. It can be used in both friendly and enemy territories. The use of a Blackhawk ASTAMIDS in areas of threat observation and fire must be planned and coordinated very carefully, because a helicopter is extremely vulnerable while flying the mission profile required for detection (steady altitude, speed, and path).

Once airborne and at its start point, the ASTAMIDS system is placed in the correct detection mode, based on the intended mission (route or area reconnaissance). When the system indicates an initial detection, the operator communicates it to the pilot. The pilot then flies a verification pass over the indicated area. If the system again indicates a detection, the pilot resumes the mission (route reconnaissance) or continues the survey pattern to determine the minefield borders (area reconnaissance). If no detection is indicated on the verification pass, the operator instructs the pilot to resume the flight plan.

## Interim Vehicle-Mounted Mine Detector

The interim vehicle-mounted mine detector (IVMMD) is used in all levels of conflict and OOTW. The IVMMD is mounted on a blast- and fragmentation-survivable vehicle; it is designed to detect and mark buried and surface-laid, metallic AT mines. The primary mission of the IVMMD is to detect mines during route clearance. The system should not be used when operating in an environment where the enemy employs mines that are not pressure-fused.

### *System Components*

A complete IVMMD (Figure 10-3) consists of one mine-detection vehicle (MDV), one towing/mine-detection vehicle (T/MDV), three mine detonation trailers, a spare-wheel module for the MDV, a spare-wheel module for the T/MDV, and a container of spare parts.

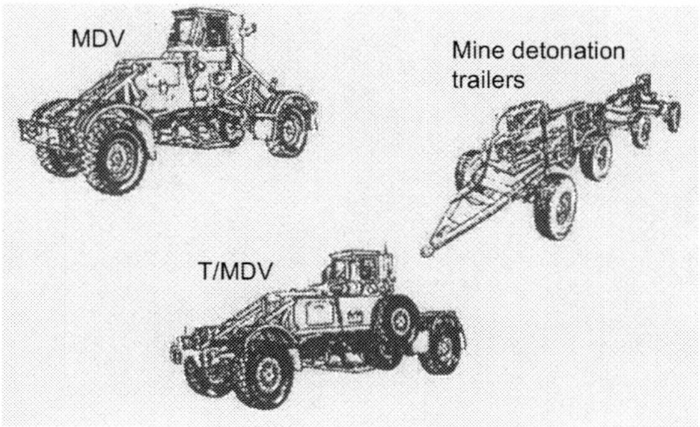


**Figure 10-3. IVMMD components**

The MDV's only mission is to detect mines. It can negotiate vertical slopes up to a 20 percent grade. The MDV employs a 4-cylinder engine and a manual transmission to propel the 4.8-ton vehicle with a 3-meter-wide detection array. The detection array consists of two separate induction coils (one for the left side and one for the right side) that detect magnetic fields below the vehicle.

The detection array is suspended between the two axles of the vehicle. When the detector encounters a metallic object, the operator is notified by an audible signal in the earphone. A visual signal appears on the dashboard that denotes which side of the array detected the object. The operator then stops the vehicle, backs it up, and reencounters the metallic object. (The MDV has two detection modes—the locate mode is used to identify the metal object, and the pinpoint mode is used to find the center of the object.) When the operator encounters the strongest signal, he activates the marking system (a nozzle mounted on the rear frame and centered on each detection array) that deploys a water-based ink onto the roadway.

The T/MDV has a 6-cylinder engine and the same detection and marking system as the MDV. The T/MDV tows three detonation trailers. The recommended maximum operating speed while towing the trailers is 20 kph. The T/MDV (with trailers) can negotiate vertical slopes up to a 20 percent grade; however, going down such slopes is difficult. The T/MDV must be in first gear, and the trailer brakes must be deployed to decrease the speed of the vehicle when going down a slope.

The mine detonation trailers are very heavy, and they are specifically designed to apply heavy ground pressure that initiates pressure-activated mines. Each trailer has two axles of different lengths so that the three trailers provide a full 3-meter-wide proofing capability behind the T/MDV. If a mine detonates underneath the trailers, the wheel bolts are designed to sheer so that repair is limited to replacing a single wheel.

The detection array is suspended between the two axles of the MDV. Although the vehicle is designed to produce very little ground pressure, it will detonate most pressure-fused mines, depending on the sensitivity of their fuses.

- The MDV produces 27.9 pounds per square inch (psi) of ground pressure when the tires are inflated to 14.5 psi and 21.8 psi of ground pressure when the tires are inflated to 8.7 psi.
- The T/MDV produces 49.8 psi of ground pressure when the tires are inflated to 29 psi and 23.4 psi of ground pressure when the tires are inflated to 8.7 psi.

### *Employment Concept*

The IVMMD is used to support route-clearance operations. Clearance operations ensure that LOC are safe for the passage of personnel and equipment. The IVMMD should not be used during hours of limited visiblity, because it hampers the operator's ability to see surface-laid mines and visual signatures that indicate mining activities.

## MECHANICAL

The track-width mine roller is a mechanical minefield-detection system. It is most effectively deployed to lead columns on route movement, but it can be used to precede tactical formations. In column movement, unit vehicles travel a narrow path, and one or two mine rollers can effectively detect mines in the path. Mine rollers can also be used to detect minefields in front of deployed tactical formations; however, more than one roller is required for a good probability of detection.

## REPORTING

Intelligence concerning enemy minefields is reported by the fastest means available. Spot reports (SPOTREPs) are the tactical commander's most common source of minefield intelligence. They originate from patrols that have been sent on specific minefield reconnaissance missions or from units that have discovered minefield information in the course of their normal operations. The information is transmitted to higher headquarters.

## REDUCING

Minefield reduction and clearing equipment is broken down into explosive, mechanical, electronic, and manual. Combat engineers and the operators of breach assets practice and become proficient in these reduction means. They integrate them into the breach drills of the units they support. The team applies different TTP to breach drills and prepares and rehearses them as part of the TF plan.

### EXPLOSIVE

#### M58A4 Mine-Clearing Line Charge

The MICLIC (Figure 10-4, page 10-8) is a rocket-propelled, explosive line charge. It is used to reduce minefields that contain single-impulse, pressure-activated AT mines and mechanically activated AP mines. It clears a 14- by 100-meter path. The MICLIC has a 62-meter standoff distance from the launcher to the detonation point. The MICLIC's effectiveness is limited against prong AP mines, magnetically activated mines (including some SCATMINEs), top-attack mines, side-attack mines, and mines containing multiple-impulse or delay-time fuses. It also has little effect on other obstacles, such as log and concrete barriers, antivehicular ditches, and walls. The shock effect and the psychological impact of the detonation make the MICLIC a useful weapon in a close fight or in MOBA.

The MICLIC is mounted on a rubber-tired trailer, or two MICLICs can be mounted on an armored vehicle-launched bridge (AVLB), with the bridge downloaded, using a fabricated I-beam frame (procedures for mounting the MICLIC on the AVLB are outlined in TM 9-1375-215-14&P). This is called an armored vehicle-launched MICLIC (AVLM) (Figure 10-5, page 10-8), and it is the preferred system because no trailer is involved to hinder the mobility of the vehicle.

Towing vehicles for the trailer-mounted MICLIC are a combat engineer vehicle (CEV), an M113 APC, M2 and M3 Bradleys, an M9 armored combat earthmover (ACE), a 5-ton wheeled vehicle, and a 2½-ton wheeled vehicle. The trailer limits the MICLIC's mobility in rough terrain and degrades the maneuverability of the towing vehicle, thereby increasing vulnerability. Since the MICLIC is critical to the breach, it is kept under the protection of the force and is moved to the breach site along easily trafficable, covered, and concealed routes. This effectively prevents the towing vehicle from performing any other task (firing or maneuvering) or serving as an engineer squad vehicle unless MICLIC employment is the squad's only mission. This is an important consideration when selecting the towing vehicle because this vehicle must be solely dedicated to the mission.

Figure 10-4. MICLIC

Figure 10-5. AVLM

The MICLIC can be fired from within an armored vehicle without exposing soldiers to fires; however, the prefiring preparations must be done in advance at a covered and concealed location near the breach site. The initiating cable is brought into the vehicle through the hatch, which must be left ajar, or through the portal of the periscope, which has been removed. Therefore, the crew is not afforded nuclear, biological, chemical (NBC) protection. When the MICLIC is fired from a wheeled vehicle, however, the crew must move to a covered position outside the backblast area. The special-purpose cable on the firing control switch is long enough to allow adequate standoff.

The vehicle operator must be proficient in all aspects of preparing and deploying the MICLIC, including the critical aspect of selecting the optimum breach site. Although the operator will be directed to the breach site by the engineer platoon leader or the breach force commander, ensuring that he can independently accomplish the task will simplify the operation and greatly enhance its likelihood of success. The towing vehicle and the operator must be selected well in advance and be dedicated solely to the task. The operator must be included in all rehearsals and planning sessions and, if possible, during leaders' reconnaissances.

Each MICLIC trailer transports and fires one charge, and then it must be reloaded. The AVLM can fire both MICLICs before reloading. The loaded charge container weighs 1,283 kilograms, so a lifting device such as a 5-ton wrecker or a HEMTT is needed. Reloading, which can be done by an experienced crew in about 20 minutes, entails loading a rocket on the rail and lifting a new charge container onto the launcher. The reloading operation must be done in a covered and concealed location.

The exact limits and depth of an enemy minefield are seldom known before the breach. This is particularly true when the situation is unclear, and the minefield is encountered simultaneously with enemy contact. The first and only indication that a unit is in a minefield may be when a vehicle encounters a mine. The leading edge of the minefield still may be an uncertainty, because the vehicle could have hit a mine in an interior row. The number of MICLICs needed to clear a single lane through a minefield depends on the minefield depth:

- Clearing a lane through a minefield less than 100 meters deep requires one MICLIC (Figure 10-6, page 10-10). The leading edge of the minefield is identified and, if possible, confirmed by reconnaissance. The MICLIC is deployed from a minimum standoff distance of 62 meters from the leading edge of the minefield.
- Clearing a lane through a minefield more than 100 meters deep or of uncertain depth requires two or more MICLICs (Figure 10-7, page 10-10). If the leading edge cannot be identified, the MICLIC is deployed 100 meters from the possible edge or stricken vehicle. When the first MICLIC is detonated, a second MICLIC moves 25 meters into the first MICLIC's path and fires its charge. This extends the lane an additional 87 meters. Additional MICLICs are used for minefields of extreme depth, and each one moves down the lane 25 meters into the path created by the previous charge.

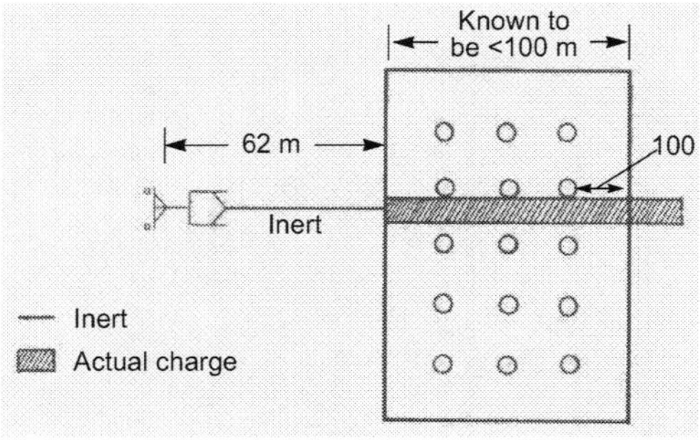


**Figure 10-6. MICLIC employment in a minefield less than 100 meters deep**

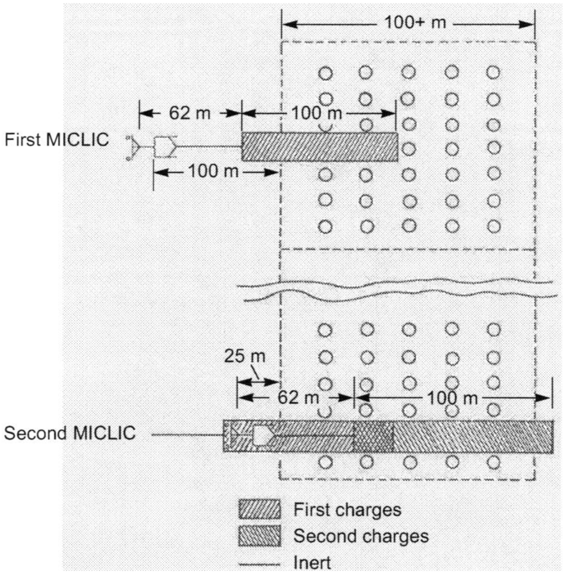


**Figure 10-7. MICLIC employment in a minefield of uncertain depth or greater than 100 meters**

The neutralization of mines by blast depends on the peak pressure and the impulse. For the MICLIC, the impulse is at a maximum of 3 meters from the line charge (on both sides) and decreases the closer it gets toward the line charge, to a minimum of 1 meter from the line charge. This decrease on impulse causes a *skip zone* (Figure 10-8). This does not mean that neutralization is equal to zero percent; it means that it is not equal to 100 percent. Mines that are buried deeper than 10 centimeters and located 1 to 2 meters from the line charge have a high probability of not being neutralized.

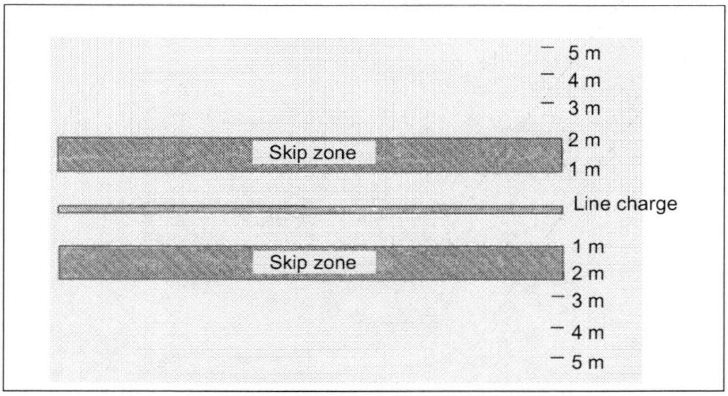


**Figure 10-8. Skip zone**

## Antipersonnel Obstacle Breaching System

The Antipersonnel Obstacle Breaching System (APOBS) (Figure 10-10) is a man-portable device that is capable of quickly creating a footpath through AP mines and wire entanglements. The APOBS is normally employed by combat engineers, infantry soldiers, or dismounted armored cavalry personnel. The APOBS provides a lightweight, self-contained, two-man, portable line charge that is rocket-propelled over AP obstacles from a standoff position away from the edge of the obstacle.

For dismounted operations, the APOBS is carried in 25-kilogram backpacks by no more than two soldiers for a maximum of 2 kilometers. One backpack assembly consists of the rocket-motor launch mechanism, containing a 25-

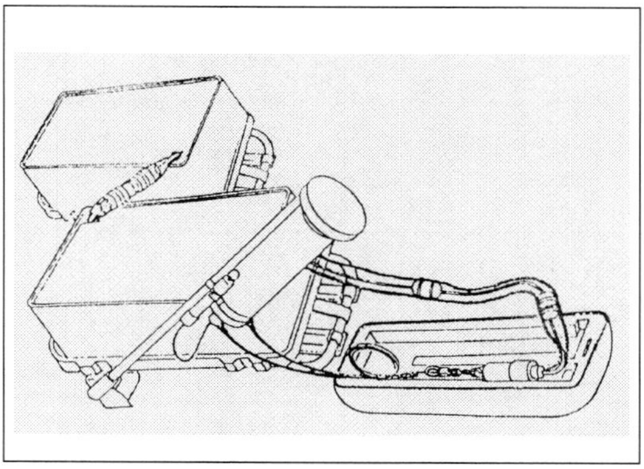

Figure 10-9. APOBS

meter line-charge segment and 60 attached grenades. The other backpack assembly contains a 20-meter line-charge segment and 48 attached grenades. The total weight of the APOBS is approximately 54 kilograms. It is capable of breaching a footpath that is approximately 0.6 by 45 meters and is fired from a 25-meter standoff.

### M1A1/M1A2 Bangalore Torpedo

The bangalore torpedo (Figure 10-11) is a manually emplaced, explosive-filled pipe that was designed as a wire breaching device, but it is also effective against simple pressure-activated AP mines. It is issued as a demolition kit and consists of ten 1.5-meter tubes. Each tube contains 4 kilograms of high explosives and weighs 6 kilograms. The kit clears a 1- by 15-meter lane.

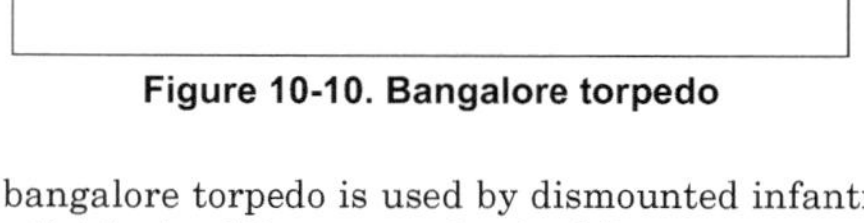

Figure 10-10. Bangalore torpedo

The bangalore torpedo is used by dismounted infantry and engineer troops. An individual soldier or a pair of soldiers connects the number of sections needed and pushes the torpedo through the AP minefield before priming it. A detailed reconnaissance is conducted before employing the bangalore torpedo to ensure that trip wires have not been used.

The bangalore torpedo generates one short impulse and is not effective against pronged, double-impulse, or pressure-resistant AP and AT mines.

**WARNING**
**Do not modify the bangalore torpedo. Cutting the bangalore in half or performing any other modification could cause the device to explode.**

### MECHANICAL

MCBs and MCRs are fielded as armor battalion sets that contain 12 MCBs and 4 MCRs. Blades clear lanes through minefields, while rollers are used to detect minefields and proof lanes created by other means. Rollers are not a good primary system for lane reduction because multiple mine detonations destroy the roller system and the vehicle pushing it. (The roller is designed to resist two conventional-mine or three scatterable-mine strikes, depending on the mine type.)

The CEV, the ACE, and dozer blades were not designed for breaching minefields and should be employed only as a last resort, because using them is extremely hazardous to the crew and equipment. However, CEVs or ACEs can effectively clear a lane through AP scatterable minefields because they sustain little or no damage and offer protection to the crew. When using a dozer to clear a path through minefields, the operator is exposed to mine effects. Before clearing begins, the operator's cabin should be sandbagged or *up-armored* and the lane should be cleared of trip wires. When using an engineer blade to clear a path through a scatterable minefield, use the skim technique (Figure 10-12). Start skimming 100 meters from the suspected minefield leading edge.

### Mine-Clearing Blade

The MCB (Figure 10-13) is used to remove land mines from the minefield. It consists of a blade arrangement with scarifying teeth to extract mines, a moldboard to cast mines aside, and leveling skids to control the depth of the blade.

The MCB lifts and pushes mines, which are surface-laid or buried up to 31 centimeters deep, to the side of the track-width lanes. The blade has three depth settings—21, 25, and 31 centimeters. The blade creates a 1.5-meter cleared path in front of each track. Figure 10-14, page 10-16, shows inside clearance distances between tracks of common track vehicles in relationship to the uncleared area left by the MCB.

Mines armed with AHDs, magnetic fuses, or seismic fuses may be activated when they are lifted by the blade; and they may disable the blade. Mines lifted by the blade are left in the spoil on each side of the furrowed path and remain a hazard until they are removed. Double-impulse mines that are lifted into the spoil on the side have a probability of functioning into the hull of the plowing vehicle. The skid shoe for each blade exerts adequate pressure to activate

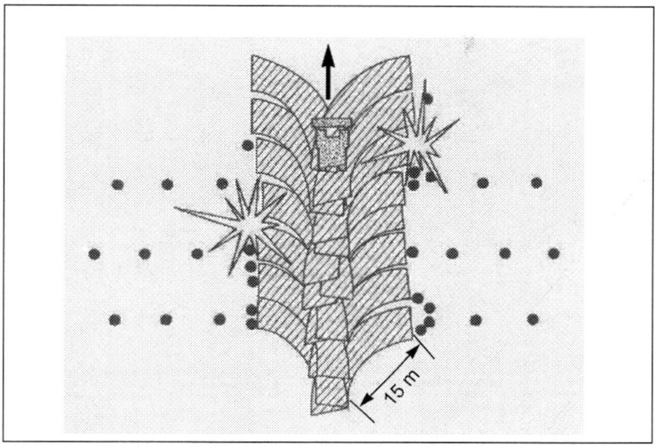


Figure 10-11. Skim technique

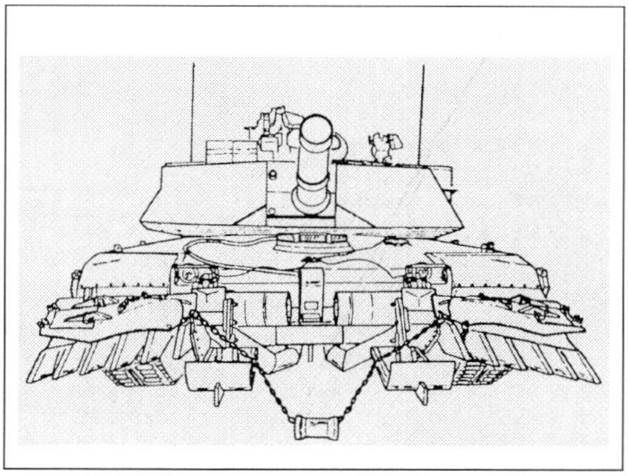

Figure 10-12. MCB

most single-impulse mines, which effectively clears a section of the centerline by explosive detonation. This action may disable the blade. Multiple-impulse pressure fuses encountered by the skid shoe are not defeated. A dog-bone assembly between the blades defeats tilt-rod mines. The improved dog-bone assembly (IDA) projects a magnetic signature and defeats tilt-rod and magnetic mines.

The MCB weighs approximately 3,150 kilograms and can be mounted on an M1 tank without special preparation or modification. Mounting requires lift capability and takes up to an hour, so it must be mounted well in advance of the mission. It is not easy to mount or transfer the MCB to another tank under battlefield conditions.

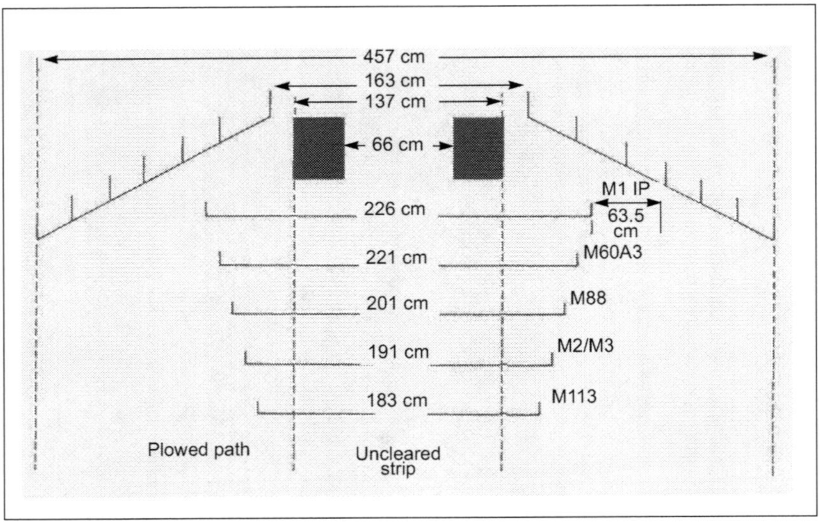


**Figure 10-13. Mine-blade width compared to track-vehicle widths**

Once mounted, an electric motor raises and lowers the blade. When it is in the raised position, it minimally effects the M1's maneuverability and speed. This will not greatly impact the employment of the weapon system except when the blade is in operation. The MCB is also equipped with an emergency, quick-disconnect feature.

The M1 should perform plowing operations from 8 to 10 kph, depending on soil conditions. It cannot maneuver but must continue in a straight path through the minefield to avoid damaging the blade. The main gun must be traversed to the side during plowing because mine detonation under the blade may cause the gun to be thrown violently into the air, damaging the tube. The area selected for the lane must be relatively flat and free of rocks or other obstacles.

The operator begins plowing approximately 100 meters from the estimated minefield leading edge. He creates a lane extending another 100 meters beyond the estimated minefield far edge to ensure that the lane extends through the entire minefield. Multiple vehicles crossing the breach will deepen the cut made by the MCB, and pressure-fused mines left in the uncleared strip will be dangerous. The uncleared strip should be cleared as soon as possible.

### Mine-Clearing Roller

The MCR (Figure 10-15) consists of a roller assembly, a mounting kit, and a hand winch kit. The roller assembly weighs approximately 9,072 kilograms and consists of two push beams mounted to the front of the tank. The rollers are designed to defeat most single-pulse, pressure-activated AT and AP mines. The roller creates a 1.1-meter-wide cleared path in front of each track.

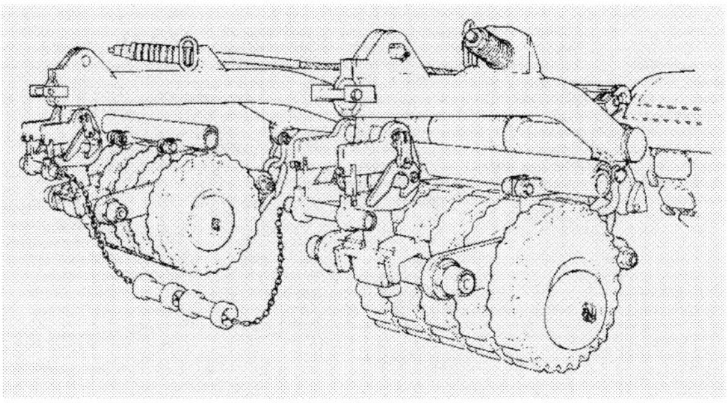

Figure 10-14. MCR

Figure 10-16 shows inside clearance distances between tracks of common track vehicles in relationship to the uncleared area left by the MCR. A dog-bone-and-chain assembly between the rollers defeats tilt-rod mines. The IDA can be fitted to the roller. The roller is designed to withstand multiple mine explosions before damage; however, this depends on the size of the mines. Large blasts may destroy the roller or the vehicle or injure the crew.

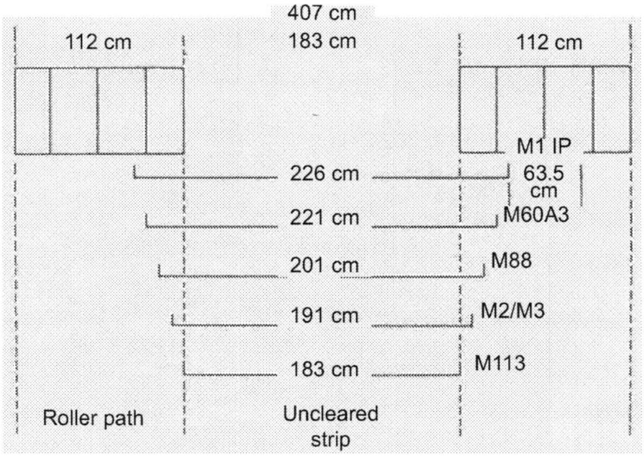


Figure 10-15. Mine-roller width compared to track-vehicle widths

The roller can be mounted on an M1 or M60 tank that is modified with a permanently attached mine-roller mounting kit. Mounting the roller to a tank is a cumbersome, time-consuming operation because it is very difficult under battlefield conditions and requires lift capability. The roller tank is limited to a speed of 5 to 15 kph. When employed in a suspected minefield, the MCR must travel in a relatively straight path, because tight turns may cause the roller to deviate from the path of the track and leave the tank vulnerable to mines. Ground fluctuations, bumps, and berms may cause the roller to lift from the ground and miss mines.

The MCR is not designed to negotiate gaps on its own; however, it can be used on AVLB caution crossings. In this situation, the curbing from the bridge is removed. To prevent damage to the bridge's hydraulic line, the tank driver uses a strap to lift the dog bone and chain when crossing the bridge. The main gun must be traversed to the side when a mine encounter is possible or imminent, because a mine blast can throw the roller or parts of the roller violently into the air and damage the tube. The main gun should only be fired from a temporary halt.

When the situation and the mission permit, MCRs may be employed as lead vehicles to detect minefields. This is most viable when the supported element is traveling in a column. The roller may also be used to lead a supported element traveling in a tactical formation other than a column, but it is less effective than other methods because—

- Vehicles not directly behind the roller may encounter mines passed by the roller.
- The roller may travel well into or completely through a widely spaced minefield without encountering a mine, thus giving the formation a false sense of security.
- A mine encountered by the roller may not be on the leading edge of the minefield.
- The roller vehicle is extremely vulnerable because it can only use its weapon system from a temporary halt.

Rollers are best used to proof lanes in obstacles that are breached by other means, such as a MICLIC or an MCB. A roller pulling a trailer-mounted MICLIC can proof a lane created by a MICLIC that was launched by another vehicle. The roller then fires the second MICLIC and proofs its own lane.

If rollers participate in a deliberate breach operation or if the force incorporates rollers into a hasty breach plan, rollers must be mounted before rehearsals. Unmounted rollers that not being used for the mission are carried in the TF formation on M916 tractor trailers. Rollers require lift capability (such as an M88), a secure location, and 30 to 60 minutes to mount on a tank that is fitted with a mounting kit.

## Panther

The M60 Panther (Figure 10-17) is one of several developmental countermine systems used by US forces during operations *Joint Endeavor* and *Joint Task Force Eagle*. The Panther is a remotely controlled vehicle with mine rollers, and it is used to proof lanes and assembly areas. The system consists of a turretless M60 tank, Israeli mine rollers, an antimagnetic actuating device, and an RCU that is mounted in a separate vehicle. Additionally, a remote video camera allows the operator to see the road ahead.

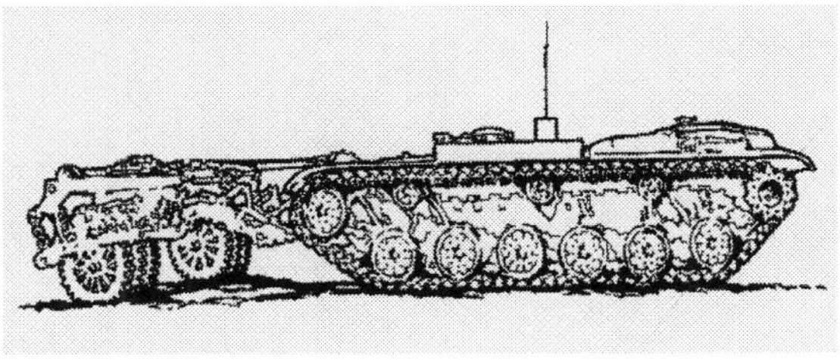

**Figure 10-16. Panther**

During route clearance or proofing operations, the Panther is the lead vehicle on the route. It is followed closely by an armored control vehicle, usually an M113. The control vehicle contains the Panther operator, the RCU, and the monitor. The monitor displays the route being proofed or cleared through a camera mounted on the Panther. The Panther is controlled from the commander copula or troop hatch of the control vehicle. The control vehicle should be approximately 200 to 300 meters behind the Panther, and its hatches should be secured open. Crew members in the control vehicle should be wearing Improved Body Armor System, Individual Countermine (IBASIC) protective garments.

Mine rollers can be raised for limited travel while mounted on the Panther. If the distance is excessive, the rollers must be transported on a cargo carrier. Rollers must be adjusted before every mission to ensure that they have contact with the ground and that their weight is uniformly distributed. To ensure proper coverage and overlap of rollers, at least three passes should be conducted. Passes should have a minimum of 30 centimeters overlap. Inside roller distances are the same as the MCR.

## MiniFlail

The MiniFlail (Figure 10-18. page 10-20) is a remotely operated, line-of-sight, AP-mine and UXO neutralization system that was developed for use by US light forces. It can clear at a rate of 1,200 square meters per hour. The MiniFlail detonates or disables AP mines from a safe operating distance. The MiniFlail neutralizes by striking objects with a rotating chain assembly, called a *flail*, and clears a foot path approximately 1.1 meters wide. The system neutralizes AP mines and UXO by detonation, mechanical destruction, or displacement from the cleared lane. The MiniFlail is approximately 1.3 meters wide, 1.3 meters high, 3 meters long, and weighs 1,100 kilograms. The system is operated by a hand-held controller that has a maximum range of 300 meters. It is fully armored with a material similar to Kevlar, and the tires are filled with foam. The flail is a self-articulating, hydraulically powered shaft with 84 chains; each chain is 0.5 meter long.

Figure 10-17. MiniFlail

## M1 Grizzly

The Grizzly (Figure 10-19) provides a hasty capability for breaching complex obstacles of mines, wire, posts, rubble, and tank ditches to create a lane for other vehicles to follow. The Grizzly's primary features are a full-width, 4.2-meter MCB and a power arm. The power arm has a reach of 9 meters and a bucket capacity of 1.2 cubic meters. Its primary missions are to reduce berms and fill AT ditches.

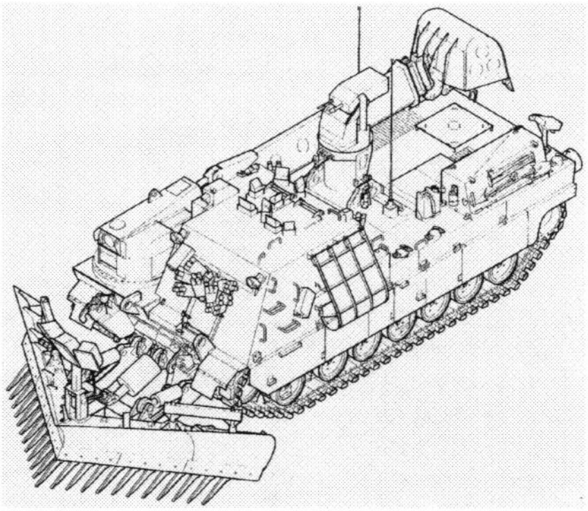

Figure 10-18. Grizzly

The system is armed with an M240 7.62-millimeter machine gun and an MK19 40-millimeter grenade launcher. The Grizzly has an M1-series tank chassis.

The Grizzly lifts and pushes mines, which are surface-laid or buried up to 31 centimeters deep, to the side of full-width lanes. The blade has multiple depth settings, depending on the mission, and it creates a 4.2-meter-wide cleared path. When plowing, the Grizzly is restricted to less than 10 kph, depending on soil conditions. The operator begins plowing approximately 100 meters from the estimated minefield leading edge. He creates a lane extending another 100 meters beyond the estimated minefield far edge to ensure that the lane extends through the entire minefield.

The Grizzly has integrated digital features to enhance battlefield awareness. Some of the digital features are thermal and video cameras, ground-speed sensors, terrain-mapping sensors, and an integrated commander's control station.

## Combat Engineer Vehicle with Full-Width Mine Rake

**NOTE: The CEV with full-width mine rake will be used in Korea only.**

This system consists of a wedge-shaped rake that is mounted to a CEV blade (Figure 10-20).

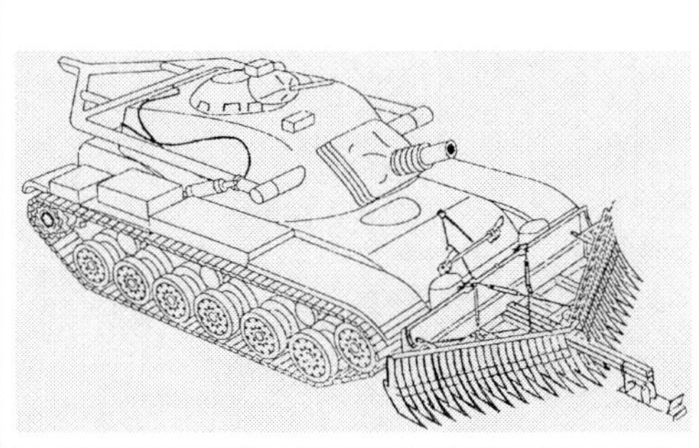

**Figure 10-19. CEV with mine rake**

The rake weighs 2,025 kilograms and is lifted off its transport vehicle with a HEMTT, a wrecker, an M88, or a CEV boom. The CEV crew uses basic-issue items to install the rake, and installation takes approximately 30 minutes. The rake has a skid shoe to maintain a raking depth of 31 centimeters. The CEV with mine rake provides a vehicle-width clearance (4.5 meters) at 5 to 10 kph. The rake has a quick-disconnect feature. It lifts surface-laid and buried mines (up to 31 centimeters deep) and pushes them off to both sides.

The CEV with mine rake is used to clear lanes during minefield breaching. While it can be employed as the first breaching asset into a minefield, a MICLIC should be used first to eliminate as many mines as possible. The rake is then used to proof the lane. The system can pull a MICLIC and fire it before proofing. Raking begins 100 meters before the minefield and continues 100 meters beyond the suspected limit. The CEV maintains a straight course through the minefield. If the skid shoe is damaged, the operator reduces speed and manually controls the blade depth. This is very difficult and risky.

### Mine-Clearing/Armor-Protection Kit

The mine-clearing/armor-protection kit (MCAP) consists of two parts—the mine-clearing rake and the armor protection.

The rake uses a tine that is mounted on a diagonal beam. The rake assembly is designed to sift through the soil, lift out mines, and windrow buried and surface-laid mines to the right of the vehicle. The system clears a 30-centimeter-deep path through a minefield. The rake has a skid shoe that acts as a depth control guide for the operator.

The armor protects the crew against mine blast, small-arms fire, and artillery fire. Protection is also provided for the engine, the fuel tank, and exposed hydraulic cylinders and lines. Ballistic glass blocks are provided at each vision port to permit unrestricted view and operation of the vehicle and the equipment.

The MCAP is mounted on a D7 dozer to perform minefield breaching and lane widening. Proofing the lane must be conducted after the dozer has cleared the lane. Some AP mines may still be left in the lane.

#### ELECTRONIC

The Field-Expedient Countermine System (FECS) is a series of copper coils that fit over the front of tracked and wheeled vehicles. Power is supplied by the vehicle's battery. The coil emits a large magnetic signature that detonates magnetically fused mines located 2 to 5 meters in front of the vehicle. The FECS is designed to defeat magnetically influenced mines only and must be used with other countermine systems.

#### MANUAL

When stealth is required or advanced mechanical equipment is unavailable, manual breaching procedures can be used. Engineers use hand-emplaced explosives, grapnel hooks attached to ropes, probes, mine detectors, and hand-emplaced marking equipment to manually breach obstacles. This is the only method that works in all situations and under all conditions because certain types of terrain, weather, and sophisticated fuses can severely degrade the effectiveness of rollers, plows, and line charges.

### Surface-Laid Minefields

The enemy possesses a significant mechanical, mine-burying capability. It has the capacity and the propensity for the labor-intensive effort required to bury mines by hand; however, the enemy often lays mines on the surface. Buried mines are usually found in a highly prepared defense that requires a deliberate breach operation. Training and execution of surface and buried minefield breaches should always assume the presence of AHDs and trip wires until proven otherwise.

From covered positions, the engineers first use grapnel hooks to check for trip wires in the lane. The limited range of the tossed hook requires the procedure to be repeated through the estimated width of the obstacle. A demolition team then moves through the lane. The team places a *line main* down the center of the lane, ties the line from the explosives into the line main, and places blocks of explosives next to surface-laid mines. After the mines are detonated, the team makes a visual check to ensure that all of the mines were cleared before directing a proofing roller and other traffic through the lane.

Manual procedures must be well-practiced. Members of the demolition team are assigned special tasks, such as grappler, detonating-cord man, and demolitions man. All of the members should be cross-trained on all the procedures. Demolitions are prepared for use before arriving at the breach site. An engineer platoon uses squads in series through the minefield to clear a lane for a company team. The platoon must rehearse reduction procedures until execution is flawless, quick, and technically safe. During the breach, the engineer platoon will be exposed in the lane for 5 minutes or more depending on the mission, the minefield depth, and the platoon's level of training.

## Buried Minefields

Manually reducing a buried minefield is extremely difficult to perform as part of a breaching operation. It is usually part of a clearance operation. If the mine burrows are not easily seen, mine detectors and probes must be used to locate the mines. The mines are then destroyed by hand-emplaced charges. As an alternative, the mines can be removed by using a grappling hook and, if necessary, a tripod (Figure 10-21). Using a tripod provides a vertical lift on the mine, making it easier to pull the mine out of the hole.

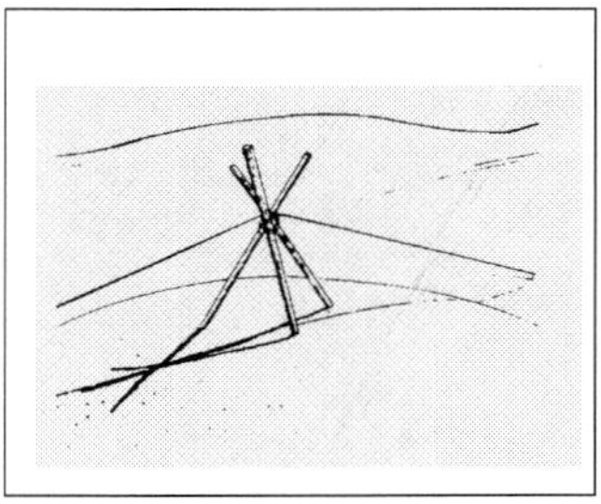

**Figure 10-20. Tripod**

The platoon leader organizes soldiers into teams with distinct, rehearsed missions including grappling, detecting, marking, probing, and emplacing demolitions and detonating cord. The platoon is exposed in the obstacle for long periods of time.

## Grappling Hook

The grappling hook (grapnel) is a multipurpose tool that is used for manual obstacle reduction. Soldiers use it to detonate mines from a standoff position by activating trip wires and AHDs. After the grapnel is used to clear the trip wires in a lane, dismounted engineers can move through the minefield, visually locate surface-laid mines, and prepare the mines for demolition. In buried minefields, soldiers grapple, then enter the minefield with mine detectors and probes.

There two types of grapnel hooks—hand-thrown and weapon-launched.

### *Hand-Thrown*

A 60+-meter light rope is attached to the grapnel for hand-throwing. The throwing range is usually no more than 25 meters. The excess rope is used for the standoff distance when the thrower begins grappling. The thrower tosses the grapnel and seeks cover before the grapnel and rope touch the ground in case their impact detonates a mine. He then moves backward, reaches the end of the excess rope, takes cover, and begins grappling. Once the grapnel is recovered, the thrower moves forward to the original position, tosses the grapnel, and repeats the procedure at least twice. He then moves to the end of the grappled area and repeats this sequence through the depth of the minefield.

### *Weapon-Launched*

A 150-meter light rope is attached to a lightweight grapnel that is designed to be fired from an M16A1 or M16A2 rifle using an M855 cartridge. The grapnel is pushed onto the rifle muzzle, with the opening of the retrieval-rope bag oriented toward the minefield. The firer is located 25 meters from the leading edge of the minefield, and he aims the rifle muzzle at a 30- to 40-degree angle for maximum range. Once fired, the grapnel will travel 75 to 100 meters from the firer's position. After the weapon-launched grapnel hook (WLGH) has been fired, the firer secures the rope, moves 60 meters from the minefield, moves into a prone position, and begins to grapnel. The WLGH can be used only once, but it can be reused up to 20 times for training (blanks are used to fire the grapnel for training).

Multiple grapplers can clear a lane of trip wires quickly and thoroughly, but they must time their efforts and follow procedures simultaneously, if possible. A hit on a trip wire or a pressure fuse can destroy the hook and the cord, so engineers should carry extras.

## PROOFING

Proofing is done by passing a mine roller or another mine-resistant vehicle through the minefield as the lead vehicle to verify that a lane is free of mines. An MCB, a Panther, a MiniFlail, or an MCR can be used to proof lanes. If the risk of live mines remaining in the lane does not exceed the risk of loss to enemy fires while waiting, proofing may not be practical. Some mines are resistant to some breaching techniques (for example, magnetically fused mines may be resistant to the MICLIC blast), so proofing should be done when the time available, the threat, and the mission allow.

During a limited clearing operation, proof upgraded breach lanes following a breach. After the minefield is completely cleared, proof the routes used through the area.

## MARKING

This section implements STANAGs 2036 and 2889.

This section provides commanders with a standard system for marking breach lanes and bypasses. It centers around a systematic, phased upgrade of lane marking. Each upgrade conforms to the tactical requirements for that phase of the attack, from initial reduction of the obstacle to the passage of larger follow-on forces, as well as the return traffic necessary to sustain the force. Additional guidelines are discussed in FM 3-34.2.

Marking breach lanes and bypasses is critical to obstacle reduction. Effective lane marking allows the commander to project forces through the obstacle quickly, with combat power and $C^2$. It gives the assaulting force confidence in the safety of the lane and helps prevent unnecessary minefield casualties.

There are two critical components of the lane-marking system:

- Lane-marking pattern (location of markers indicating the entrance, the lane, and the exit).
- Marking device (type of hardware emplaced to mark the entrance, the lane, and the exit).

The lane-marking system outlined in this section centers around standardized marking patterns rather than the marking device. Standardizing the marking pattern is critical to offensive operations. A common lane pattern—

- Enables cross attachments and adjacent units to recognize breach lanes easily with minimal knowledge of a particular unit's tactical SOP.
- Gives all forces a standardized set of visual cues that are needed to pass through a lane safely while maintaining their momentum.
- Facilitates quick conversion to the lane-marking requirements of STANAGs 2889 and 2036 (discussed later in this chapter).

The standard lane-marking hardware is decided by unit commanders. This gives units greater flexibility and allows them to adopt marking devices that are tailor-made for their type of unit and operational focus (such as an armored or light force, a mounted or dismounted attack, limited visibility, thermal capability). However, regardless of the type of device used, it must support the standard lane-marking pattern outlined in the following paragraphs. Therefore, commanders should consider these guidelines and examples before developing or adopting their own marking system.

## LANE-MARKING TERMS

The definitions in the following paragraphs provide a common basis for discussing lane marking.

### Entrance Markers

Entrance markers indicate the start of a reduced lane through an obstacle. They signify the friendly-side limit of the obstacle and the point at which movement is restricted by the lane width and path. Entrance markers are placed to the left and the right of the entrance point and spaced the width of the reduced lane. They must be visually different from handrail markers to help the force distinguish this critical point in the lane.

### Handrail Markers

Handrail markers define the lane path through the obstacle and indicate the limits of the lane width. As a minimum, mounted and dismounted lanes will have a left handrail marker. Mounted and dismounted forces moving through the lane should keep the left handrail marker immediately to their left. As the operation progresses, lane marking may be upgraded to include left and right handrail markers.

### Exit Markers

Exit markers indicate the far side of the reduced lane through an obstacle. Like entrance markers, exit markers must be distinguishably different from handrail markers; however, the exit may be marked the same as the entrance. Exit markers are placed to the left and the right of the exit point and spaced the width of the reduced lane. This visual reference is critical when only the left handrail is marked. The combination of entrance markers, left handrail markers, and exit markers provide the driver and the tank commander with visual cues so that they can safely pass through a reduced lane.

### Entrance Funnel Markers

Entrance funnel markers augment entrance marking. The V formed by a funnel marker forces the platoon into a column and helps drivers and tank commanders make last-minute adjustments before entering a lane.

### Final-Approach Markers

Final-approach markers are highly visible, robust markers that augment the visual signature of entrance funnel markers. They are critical when initial assault forces must maneuver to the breaching site. Normally, the initial assault force can observe the breaching area but cannot clearly distinguish entrance funnel markers. Final-approach markers provide the assault force commander with a highly visible RP toward which to maneuver his formation. They also signal company team commanders to begin changing from combat column to column formation, with platoons in combat column.

### Far Recognition Markers

Far recognition markers are highly visible markers that are located between the final-approach marker and the friendly unit. They are primarily used when passing forces are denied direct observation of the final-approach marker due to distance, visibility, or terrain. When possible, far recognition markers should be different from the final-approach marker. Far recognition markers indicate the point at which forces begin changing their formation to posture for the passage. A single far recognition marker may serve up to two initial breach lanes. Once lanes are upgraded to two-way traffic, far recognition markers are required for each two-way lane. When a far recognition marker serves more than one lane, a guide or a traffic-control post (TCP) is collocated with the far recognition marker that is nearest to the breach.

### Guides and Traffic-Control Posts

A TCP or a guide consists of a two-man team with communications means. The team assists the commander in controlling the movement of forces. When possible, military police (MP) should man TCPs. However, the commander may initially use other personnel as guides to man critical far recognition markers until the MP establish full TCPs. TCPs and guides provide the commander with a man on the ground who controls traffic flow to the appropriate lanes. When there are multiple lanes branching off a single far recognition marker, the TCP can assist in breaking parts of the formation off into various lanes. The TCP can also help modify the traffic flow when lanes have been closed for maintenance, for lane expansion, or by enemy SCATMINEs. The guide or TCP must give the assault force commander the azimuth and distance to the final-approach marker, identify the device used for the final-approach marker, and provide the level of the lane-marking pattern. For light forces, guides may physically escort passing units from the far recognition marker to the lane entrance.

## LEVELS OF LANE MARKING AND PATTERNS

The three standard levels of marking for breach lanes and bypasses are initial, intermediate, and full.

Each lane-marking level provides an increase in lane signature and capability. Lane requirements change as a breaching operation matures from an initial breach to the forward passage of large combat forces.

Initial lane-marking requirements are driven by the nature of the fight through the obstacle. Marking must be rapid, providing only the bare minimum signature needed to pass small units who make up the initial assault force. This contrasts with the lane requirements of later phases of an offense where larger units are passed to subsequent objectives. Here, the lane signature must be more extensive and more visible, because it must guide larger forces over a greater distance to the lane's entrance without interruption. Two-way traffic becomes a priority for the simultaneous forward passage of combat units as well as the return traffic (such as ambulances and empty supply vehicles) that is necessary to sustain the force. Lane-marking limits must be absolutely clear to the most inexperienced driver or crewman. A fully developed lane must support two-way traffic and be completely marked.

Bypasses are not marked the same as lanes. They are marked with directional panels indicating the direction of the bypass. The limits of the mine threat must be marked to prevent friendly forces from entering the minefield. Marking the direction of the bypass and the minefield limits will enable the maneuvering element to bypass the minefield without having to unnecessarily defile through a marked lane. Further information on bypass marking can be found in FM 3-34.2.

Commanders must be aware of how the needs of the force change with the operation so that they can anticipate lane-marking and lane-capability requirements. Integrating the levels of lane marking into the overall breaching plan ensures that the unit's needs are satisfied. Forces necessary to mark, maintain, and upgrade lanes must be allocated and tasked with the mission. The phases of the scheme of maneuver and the service-support plan are the basis for analyzing lane requirements. The following paragraphs describe lane-marking patterns in detail and provide guidelines on when the commander should upgrade lane marking and lane capability.

### Initial Lane Marking

Initial lane marking (Figure 10-22, page 10-28) is emplaced by the breach force immediately after the lane is reduced and proofed. It provides a signal to the assault force commander that the lane is ready for traffic. Initial lane marking is kept to a minimum, centering on markings needed to pass immediate assault forces through the lane to seize the initial foothold on the objective. Normally, the assault force can observe the breach and does not need the more visual signature of a mature lane marking. The initial lane-marking pattern has the following markers:

- Entrance.
- Exit.
- Left handrail.
- Entrance funnel.
- Final-approach.

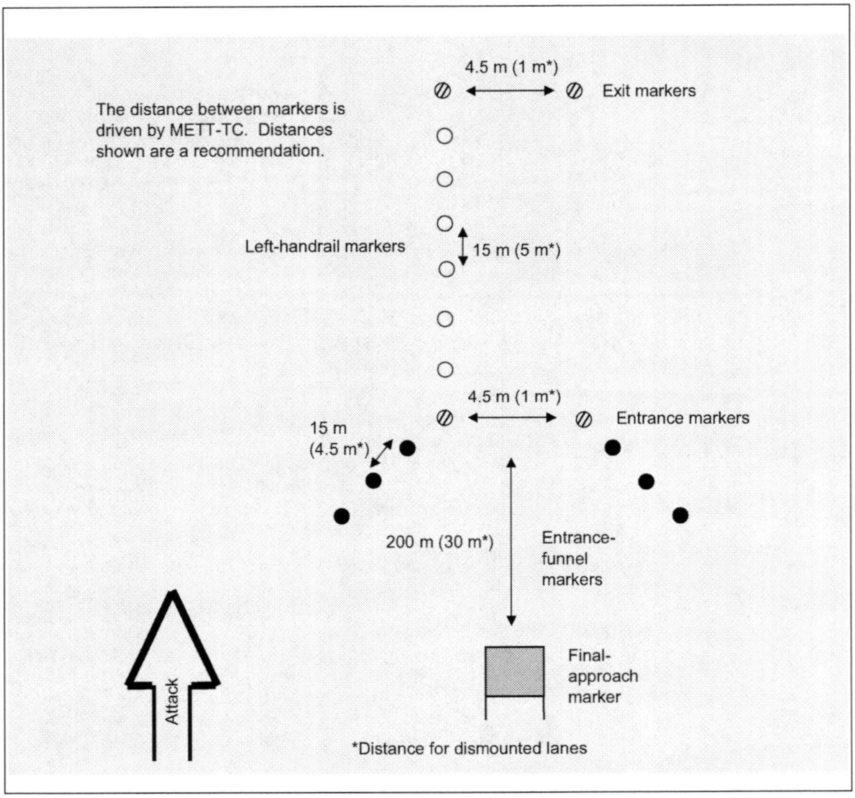


**Figure 10-21. Initial lane marking**

The entrance, left handrail, and exit markers are the first markers emplaced by the breach force because they define the location and the limits of the reduced lane.

- Entrance markers are placed to the left and the right of the reduced lane's entrance point, and they are spaced the width of the lane (4.5 meters for mounted lanes, 1 meter for dismounted lanes).
- Left handrail markers are placed at the left limit of the lane, along the entire path. Handrail markers are placed at 15-meter intervals for mounted forces and at 5-meter intervals for dismounted forces. Commanders may have to modify the intervals based on the terrain, the visibility, the lane length, and the lane path.
- Exit markers are placed to the left and the right of the reduced lane's exit point, and they are spaced the width of the lane (4.5 meters for mounted lanes, 1 meter for dismounted lanes).

Once the entrance, left handrail, and exit markers are emplaced, the breach force emplaces the entrance funnel markers and the final-approach marker.

- Entrance funnel markers are placed at 15-meter intervals for mounted forces and at 5-meter intervals for dismounted forces. They are placed diagonal to the lane entrance and form a 45-degree V (Figure 10-22).
- The final-approach marker is centered on the lane and placed at least 200 meters from the lane entrance for mounted forces. For dismounted forces, the nature of the attack may initially preclude using a final-approach marker; however, as soon as the mission allows, a final-approach marker is placed 30 meters from the entrance. Final-approach markers for mounted and dismounted forces must be placed on high ground to ensure that they are clearly visible. The commander may modify the recommended distance for the final-approach marker, based on the terrain and the visibility.

**Intermediate Lane Marking**

Upgrading initial lane marking to intermediate lane marking (Figure 10-23, page 10-30) is triggered by one of two key events—the commitment of larger combat forces who are unable to directly observe the breach or the rearward passage of sustainment traffic (casualty evacuation and vehicle recovery). Intermediate lane marking has two goals:

- Increasing the lane signature to help the passage of larger, more distant combat forces.
- Providing sufficient marking for two-way, single-lane traffic.

Intermediate lane marking builds on initial lane marking by adding right handrail markers, exit funnel markers, far recognition markers, and a farside final-approach marker.

The commander sets the priority of marker emplacement based on the situation. If the scheme of maneuver requires the immediate passage of larger combat forces, the right handrail markers and the far recognition marker may be the priority. On the other hand, if it is necessary to ground evacuate casualties or to recover vehicles, emplacing right handrail markers, exit funnel markers, and a farside final-approach marker may be required first.

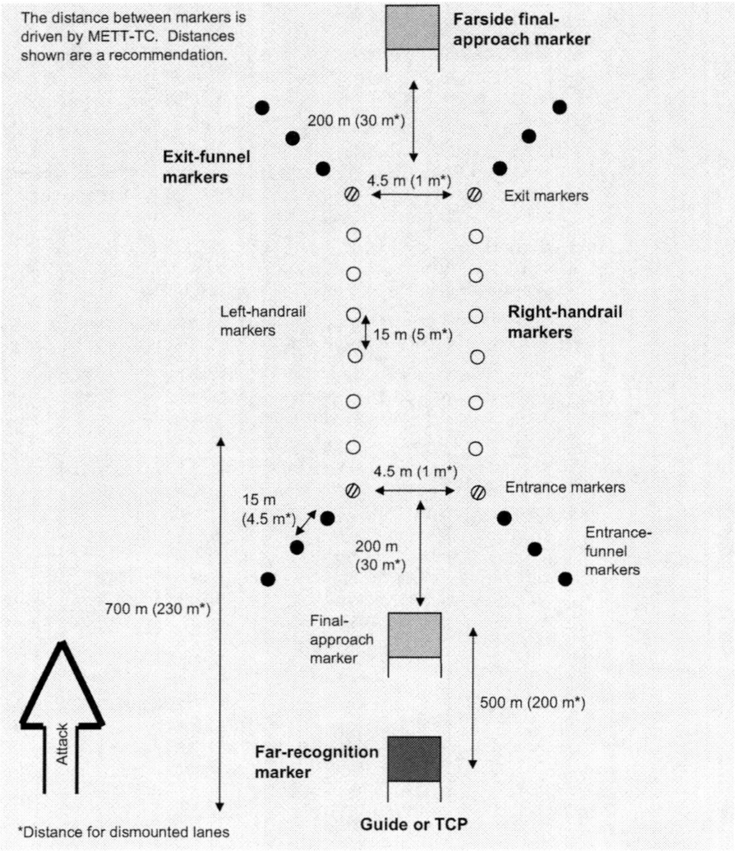


**Figure 10-22. Intermediate lane marking**

When upgrading to intermediate marking, the first step is to emplace the right handrail markers. Right handrail markers define the rightmost limit of the lane. They are placed the width of the lane as defined by the entrance and exit markers. The right handrail follows a path parallel to the left handrail through the obstacle. Right handrail markers are placed at the same interval as left handrail markers.

Exit funnel markers and a farside final-approach marker are emplaced to mirror the entrance markers. Exit funnel markers prevent the premature deployment of the passing force into combat formation before it is safely outside the obstacle. They also become the entrance funnel markers for rearward passing traffic, giving these forces the visual cues needed to line themselves up on the lane. The exit funnel markers are augmented by a farside final-approach marker to help rearward passing forces clearly identify the lane from their side. The farside final-approach marker is centered on the lane and placed 200 meters (mounted forces) or 30 meters (dismounted forces) from the exit.

A far recognition marker completes intermediate lane marking. It provides commanders with a visual signature or a series of signatures for guiding their movement toward the lane. For mounted forces, the far recognition marker nearest to the breach lane is placed 500 meters from the lane entrance or on the nearest terrain feature. Dismounted forces may require a system of guides instead of far recognition markers for passing combat forces; however, far recognition markers must be emplaced as soon as possible to reduce guide requirements for passing mounted sustainment traffic. This gives the assault force commander the space needed to transition his formation to companies in combat column. Far recognition markers may be emplaced before or concurrent with exit markers, based on the mission and the situation.

The commander collocates guides or TCPs at the far recognition marker when he feels the situation requires more positive control over traffic flow. Commanders should plan for the use of full-time guides once they have upgraded to intermediate marking. TCPs become mission-critical during limited visibility or in restrictive terrain. They should also be used when a single far recognition marker feeds more than one breach lane. TCPs must be manned with a minimum of two soldiers and must have FM communications with the controlling headquarters. It is essential that soldiers acting as guides or TCPs know the—

- Azimuth and distance to the breach lane and the 8-digit grid coordinate of the lane.
- Level of lane marking.
- Type of final-approach marker used.
- Traffic-control plan and march order.
- Up-to-date status of lane marking, maintenance, and so forth.

## Full Lane Marking

Expanding breach lanes to full (two-way) lane marking (Figure 10-24, page 10-32) is resource-intensive and is not normally a part of an initial breach operation. A fully matured lane is one that will support uninterrupted, two-way traffic. Expanding a breach lane to a full lane involves expanding the width of the lane to accommodate two-way traffic and modifying the marking pattern to give forward and rearward passing forces the same visual signature. Upgrading to a full lane is normally assigned to follow-on engineer forces, since it is usually beyond the immediate capability of engineers with forward units.

Upgrading intermediate lane marking to full lane marking begins by temporarily closing the lane, rerouting traffic, and expanding the lane width. The initial reduced and proofed lane is always expanded to the left, in relation to the direction of the attack. Engineers reduce and proof the obstacle beginning at the left handrail to give a total lane width of 10 meters (5 meters

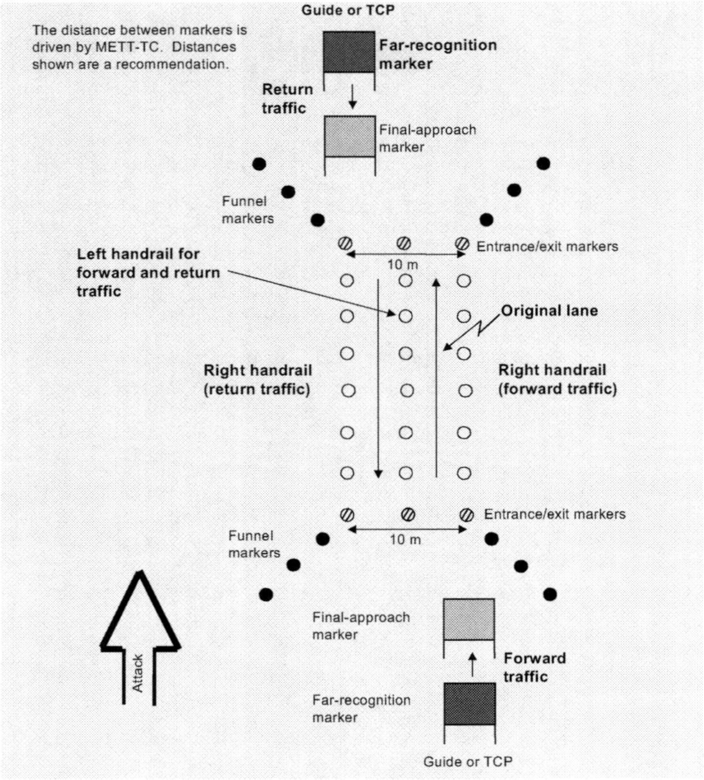


**Figure 10-23. Full lane marking**

each way). The expansion width requirement is the same for armored and light forces, because both forces must be able to pass mounted sustainment and combat forces during this phase.

Once the engineers expand the lane width to 10 meters, they ensure that entrance, exit, handrail, funnel, and final-approach markers are replaced on the return lane. All markings are the same as described in previous paragraphs.

The full lane-marking pattern has three entrance and three exit markers. They are placed the width of forward and return lanes and are visually different from other markers. Units must be trained to recognize that three entrance markers indicate a two-way traffic lane and that they should always use the rightmost lane.

Entrance and exit funnel markers are placed slightly different from previous marking patterns. They extend out from the entrance and exit markers on the right side only.

Final-approach markers are placed 200 meters from, and centered on, entrances of forward and return lanes. This helps forces clearly identify the entrance points from either direction.

Far recognition markers are placed a maximum of 500 meters from the lane entrance or on the nearest terrain feature from forward and return final-approach markers.

### COMMANDER'S GUIDANCE FOR LANE MARKING

Table 10-1 provides a summary of lane-marking levels, guidelines on unit responsibilities, and events that trigger lane upgrade. In the table, *who* refers to the unit responsible for lane upgrade marking and *when* describes events that trigger the need to upgrade.

**Table 10-1. Lane-marking levels, unit responsibilities, and trigger events**

| | Initial | Intermediate | Full (Two Way) |
|---|---|---|---|
| Who | TF breach force | TF breach force | Brigade |
| When | Obstacle is reduced | Passing battalion- or company-size forces | Passing brigade- or battalion-size forces |
| | Passing platoon- or company-size assault forces | Passing force which cannot see the lane | Situation requires uninterrupted sustainment traffic |
| | | Passing TF combat trains | |
| Markers | Entrance | Add right handrail | Expand lane width to 10 meters |
| | Exit | Add exit funnel | Adjust entrance/exit |
| | Left handrail | Add farside final approach | Adjust left/right handrails to new width |
| | Entrance funnel | Add far recognition | Add far recognition |
| | Final approach | Add guides or TCPs | Add farside guides or TCPs |

### LANE-MARKING DEVICES

The majority of lane marking in the field is done by using nonstandard marking devices. When adopting a nonstandard marking device, commanders should consider the guidelines summarized in Table 10-2.

**Table 10-2. Guidelines for lane-marking devices**

| Marker | Mounted Forces | Dismounted Forces |
|---|---|---|
| Handrail and funnel markers | Visible by TC and driver (buttoned up) from 50 meters | Visible by a dismounted soldier in a prone position from 15 meters |
| | Quick and easy to emplace, minimizing the need to expose soldiers outside the carrier | Lightweight, quick, and easy to emplace (a dismounted soldier should be able to carry enough markers for the lane and still be able to fire and maneuver) |
| Entrance and exit markers | Visible by TC buttoned up from 100 meters | Visible by a dismounted soldier from 50 meters |
| | Visually different from handrail and funnel markers | Visually different from handrail and funnel markers |
| | Quick and easy to emplace (may require soldiers to dismount to emplace) | Lightweight, quick, and easy to emplace |
| | Easily man-portable | |
| Final-approach and far recognition markers | Visible by TC (not buttoned up) from 500 meters | Visible by a dismounted soldier on the march from 100 meters |
| | Visually different from each other | Visually different from each other |
| | Visually alterable to facilitate traffic control through multiple lanes | Visually alterable to facilitate traffic control through multiple lanes |

Figure 10-25 shows some of the devices that can be utilized for lane marking, and they are easily procured or fabricated. This is not an inclusive listing but is intended to show commanders some of the options.

Some general requirements for lane marking are—

- Markers must be able to withstand the rigors of the terrain, the weather, and the battlefield.
- Markers should be easy to modify, using minimal manpower and equipment, when visibility is limited.
- Lane-marking panels should have thermal and IR reflective marking so that they can be easily identified during limited visibility.
- Enhancements for limited visibility should be a constant source rather than a pulsating strobe. Strobes do not make the marking pattern readily apparent, particularly when approaching from an angle.

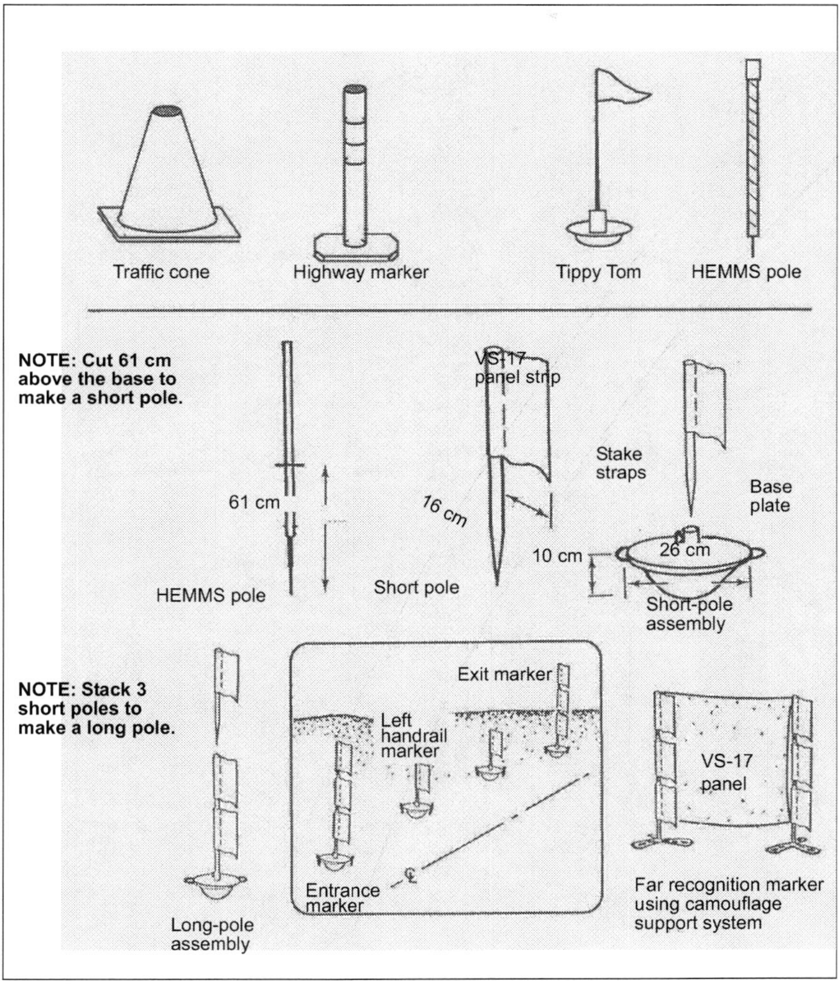


**Figure 10-24. Marking devices**

The following standard marking sets are available through normal supply channels:

- Minefield marking set number 2, line item number (LIN): M49096, NSN: 9905-00-375-9180.
- HEMMS, LIN: M49483, NSN: 9905-01-019-0140.

## Marking Requirements of the North Atlantic Treaty Organization

The following paragraphs paraphrase the lane-marking requirements outlined in STANAGs 2889 and 2036. They also establish the procedures used by US forces to modify intermediate and full lane marking to STANAG standard. With the combined nature of warfare, commanders need to be aware of their responsibilities for marking hazardous areas, particularly breach lanes.

STANAGs 2889 and 2036 state that the type of marking device, pattern, and lighting used to mark breach lanes in forward areas is at the discretion of national authorities or the authorized commander. This gives commanders who are participating in a combined operation the flexibility to mark lanes consistent with their respective Army's standard. It also outlines minimum requirements for the lane-marking pattern before it is used by troops of other nations; however, commanders must plan for converting a lane to NATO standard as early as possible. When converting to NATO standard, the STANAG directs commanders to use lane-marking devices as stated below. Within an offensive operation, marking a lane to NATO standard will not normally occur until after the lane is matured to a full lane.

### Marking Pattern and Device

The intermediate lane marking discussed earlier satisfies the minimum lane-marking pattern that must be used before forces from another country are able to pass through a lane. STANAGs 2889 and 2036 state that regardless of the marking device used, the entrance point, exit point, and left and right handrails are the minimum required lane signature. Therefore, once the lane is marked to the intermediate level, allied forces can use the lane without any additional marking.

STANAG 2889 requires that commanders convert marking devices to NATO standard as early as possible. Figure 10-26 shows a NATO standard marker. The marker is placed at right angles to the direction of travel, so that the white portion of the arrow points inward to the lane, indicating the safe side of the lane. The red portion is outward, indicating the lane limit or dangerous side of the lane. STANAG 2889 also requires that markers be large enough to be visible from 50 meters under most daylight conditions and have a field life of 60 days.

### Conversion to NATO Standard Marking

To convert intermediate and full lane marking to NATO standard, affix NATO markers to long pickets and replace the existing entrance, exit, funnel, and handrail markers one for one (Figure 10-27).

Two NATO markers are used for entrance and exit markers to make them distinctly different. One NATO marker is affixed to each funnel marker and to

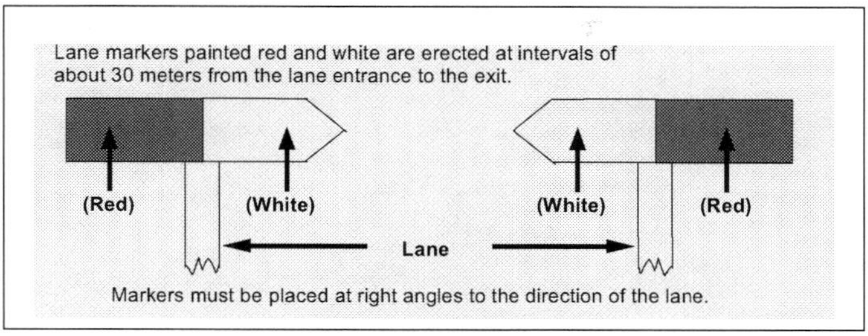


**Figure 10-25. NATO standard marker**

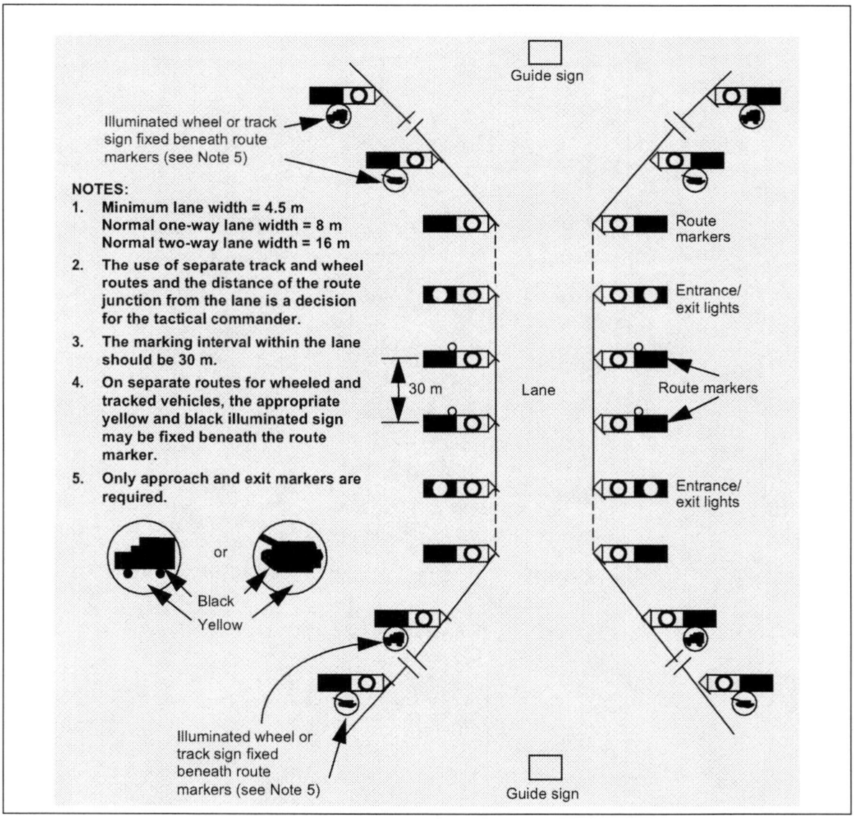


**Figure 10-26. NATO lane-marking conversion**

each left and right handrail marker. When converting full lane marking, the center handrail is marked with a modified NATO marker. The combination of a modified center handrail marker and directional arrows at each lane entrance provides allied forces with the signature necessary to distinguish two separate lanes. In addition, a barbwire or concertina fence (one strand minimum) is laid 1 meter above the ground to connect funnel markers, entrance markers, handrail markers, and exit pickets.

NATO uses white or green lights to illuminate markers at night (Figure 10-28). Entrance and exit markers are marked with two green or white lights placed horizontally, so that the safe and dangerous markings on them are clearly visible. One white or green light is used on funnel and handrail markers. The commander decides whether the light is placed on top of the NATO marker or placed so that it illuminates the markers. Lights must be visible from a minimum of 50 meters under most conditions and have a continuous life of 12 hours.

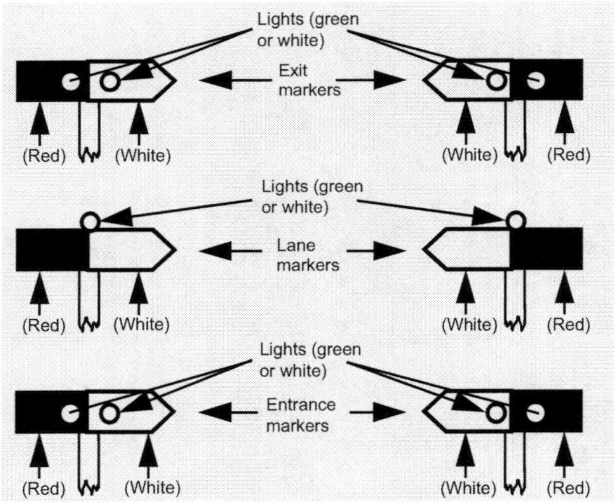


**Figure 10-27. NATO standard marking for limited visibility**

The mission to convert intermediate or full lane marking to NATO standard is normally assigned to corps-level engineer battalions working in the division rear area. In special cases, divisional engineer battalions may be tasked with NATO marking.

Chapter 11

# Route and Area Clearance

The ability to move forces and material to any point in an AO is basic to combat power and often decides the outcome of combat operations. Maneuver relies on the availability of LOC within an AO; and during OOTW, clear LOC is essential to the movement of forces. Units must conduct route and area clearance to ensure that LOC enables safe passage of combat, combat support (CS), and CSS organizations. Clearance operations are best-suited for rear-area and stability support operations.

## ROUTE CLEARANCE

Route clearance is a combined arms operation. Units must clear LOC of obstacles and enemy activity that disrupt battlefield circulation.

### PLANNING

The principles of breaching operations (Chapter 9) apply to the development and execution of the route-clearance mission. The breaching tenets (intelligence, fundamentals, organization, mass, and synchronization) should be the basis for planning.

#### Intelligence

Incorporating the IPB and METT-TC factors into route-clearance operations will enable units to predict what the enemy will do and where it will do it. The IPB and the EBA offer ideal methods for establishing a SITEMP. After the S2 and the engineer identify the most probable threat sites, the S2 designates them as NAIs. These NAIs are the focus of the reconnaissance effort. Engineers work in concert with other reconnaissance assets to confirm the presence or absence of ambushes, UXO, and minefields. The information gathered from the IPB and the reconnaissance effort determines the method and the type of route clearance necessary. It also helps the commander determine any outside resources (EOD, SOF) that he may need.

#### Fundamentals

SOSR may not be executed, but it is planned as it is in breaching operations. Units must be prepared to execute SOSR fundamentals as necessary.

#### Organization

Task organization for a route clearance is similar to the task organization for a deliberate breach. The clearance company team is organized into breach, support, and assault forces. The breach force conducts clearing operations, the support force isolates the area being cleared, and the assault force performs security functions beyond the clearance site (traffic control points) and assists the breach force in disengagement, as required. Table 11-1 shows a sample task organization for a route clearance.

**Table 11-1. Sample task organization for a route clearance**

| Team | Support Force | Assault Force | Breach Force |
|---|---|---|---|
| Heavy | • Mechanized infantry platoon with dismount capability<br>• Armor platoon | • Mechanized infantry platoon<br>• Engineer squad<br>• Mortar section<br>• Medical team (two ambulances)<br>• PSYOP team<br>• FIST<br>• MP element | • Engineer platoon with organic vehicles<br>• Armor platoon with plows and rollers |
| Light/Heavy | Two infantry platoons (light) | • Bradley platoon with dismount capability<br>• Engineer squad<br>• 60-mm mortar section<br>• Medical team (two ambulances)<br>• PSYOP team<br>• Forward observer<br>• MP element | • Engineer platoon with organic vehicles<br>• Armor platoon with plows and rollers |
| Light | Two infantry platoons (light) | • AT/MP section with M60/MK19 mix<br>• 60-mm mortar section<br>• Medical team (two ambulances)<br>• PSYOP team<br>• Forward observer<br>• MP element | • Engineer squad (+)<br>• Infantry platoon (light)<br>• AT/MP section with M60/MK19 mix |

## Mass

Sufficient maneuver and engineer assets must be allocated to the clearance company team. The length and the width of the route and the type of clearance to be conducted determine the size of the sweep team. Clearing a Class A military road with the deliberate sweep technique requires at least two engineer squads due to the total lane width to be cleared and the requirement for the rotation of mine-detector operators. Depending on the type of sweep operations, the commander can expect a 50 percent loss of sweep assets. Normally, as in breaching, a 50 percent redundancy of engineer assets should be allocated to the sweep team.

## Synchronization

All aspects of synchronization should be implemented when planning route clearance. It is especially important that rehearsals be conducted at the combined arms level. Rehearsals should include—

- Reaction to enemy contact.
- Reaction to an ambush.
- Communications exercise.
- Fire support (obscuration smoke, immediate suppression fires, critical friendly zones for counterfire radar, and no-fire area around the clearance site).

- CSS (maneuver, casualty evacuation, marking materials, and demolitions resupply).

## PLANNING CONSIDERATIONS

The purpose of breaching is to project combat power to the farside of an obstacle, and breaching usually occurs under enemy fire. Route-clearance operations focus on opening and maintaining LOC to ensure the safe passage of combat, CS, and CSS organizations. Like breaching, route-clearance operations require extensive BOS coordination. The following planning considerations should be used by brigade and battalion TFs when planning route-clearance operations:

### Intelligence

- Identify choke points, bridges, tunnels, critical road junctions, and other built-up areas. These are the most suspect areas for obstacle emplacement. However, depending on the enemy's overall mission, it may not always emplace obstacles at these locations. This is especially true if the enemy's goal is to psychologically disrupt our convoys.
- Maintain a situation map with a graphics overlay that reflects the most current intelligence information.
- Maintain an incident map with a graphics overlay to facilitate a pattern analysis.
- Maintain a threat order-of-battle database, such as how the enemy will disrupt unit LOC.
- Develop a detailed R&S plan that incorporates modern battlefield techniques and systems, such as ground sensors, forward-looking airborne radar, and satellite images. As a minimum—
  - Coordinate for UAV support, if available.
  - Develop infiltration routes to support recon and security at likely enemy ambush sites.
  - Develop an estimate of impact to civilians on the battlefield (COBs). COBs include local nationals, nongovernment organizations (NGOs), and private volunteer organizations (PVOs).
  - Conduct a daily flight over the area to provide up-to-the-minute intelligence. When available, coordinate ASTAMIDS coverage.
  - Coordinate with the USAF to periodically check the route (for example, using an AC-130 Specter gunship).
- Provide an intelligence update to company team leaders before departure. This should be in the form of a 1:50,000 enemy SITEMP overlay (confirmed and suspected/templated).
- Establish liaison with the host nation, NGOs, and SOF.

### Maneuver

- Provide personnel for TCPs.

- Clear and secure flanks (at least 100 meters) and the farside of suspected and known obstacle locations.
- Close the route to US-controlled traffic during route-clearance operations to minimize the target presented to enemy forces.
- Identify and clear potential sniper positions before beginning obstacle reduction or clearance.
- Provide security for the cleared route.
- Give operational control (OPCON) of aviation assets to the route-clearance commander for clearance-support missions.
- Plan the building of static security points along the cleared route to reduce the probability of reseeding.

**Fire Support**

- Plan smoke for templated locations.
- Position mortars to ensure continuous coverage of the operation (move and set up with the support force).
- Prepare fires within the tactical rules of engagement.
- Ensure that the route-clearance team has a FIST coordinator. The clearance commander should locate the FIST element well forward in the order of march.
- Designate obstacle clearance sites as critical friendly zones for counterfire radar and no-fire areas.

NOTES:

**1. Priority targets shift in conjunction with company team movement on the main supply route (MSR).**

**2. Clearance of fires is the responsibility of the maneuver commander of the sector where the target is located.**

**3. Adequate Q-36 radar coverage is necessary for deliberate sweep operations.**

**Mobility/Survivability**

- Provide detailed OBSTINTEL on minefields. It must include the—
  - — Description of mines or explosive devices most likely encountered.
  - — Composition and pattern of obstacle(s).
  - — Enemy actions or techniques used during obstacle emplacement.
- Conduct deliberate sweep operations 100 meters past the obstacle or suspected threat.
- Report, clear, and mark mines, obstacles, and explosive devices to facilitate unimpeded movement.
- Ensure that lane marking meets the standards outlined in Chapter 10 and that materials and techniques are standard throughout the route.

- Consider including road repair equipment and material as part of the sweep element (for example, a 5-ton dump truck filled with soil and an ACE to spread the soil).
- Keep all radios, electronic equipment, and aviation assets at a safe distance during reduction operations.
- Block uncleared roads and trails that branch from the route being cleared. This protects units from inadvertently traveling an uncleared route.
- Debrief the chain of command and the TF S2 on the location, the composition, and the orientation of all obstacles cleared and encountered. This assists the S2 and the engineer in IPB/EBA pattern analysis.

**Air-Defense Artillery**

- Consider the possibility of an air attack.
- Use the following passive air-defense measures:
  - Eliminate glare by using mud, tape, cardboard, or camouflage nets to cover headlights, mirrors, and portions of windshields.
  - Reduce dust clouds by reducing speed.
  - Plan routes that offer natural concealment.
  - Use air guards.
- Increase the distance between vehicles.
- Incorporate Stinger missile teams into the support force.

**Combat Service Support**

- Ensure that clearance operations are supported by a logistical/combat health support (CHS) package from the brigade support area.
- Plan for air and ground evacuation of casualties. The preferred evacuation method is by air; the routine method is by ground.
  - Conduct an air-mission brief with air ambulance assets, to include pickup zones and markers. Rehearse procedures for evacuation requests.
  - Ensure that the medical team consists of one or two ambulances. Locate the medical team with the support force.
  - Identify the ambulance exchange point along the route to be cleared.
- Ensure that all personnel wear flak vests or IBASIC (Figure 11-1, page 11-6).
- Ensure that all vehicles have tow cables in the front and the rear for extraction purposes.
- Ensure that all vehicles carrying troops have hardening (sandbags on floors and sides).

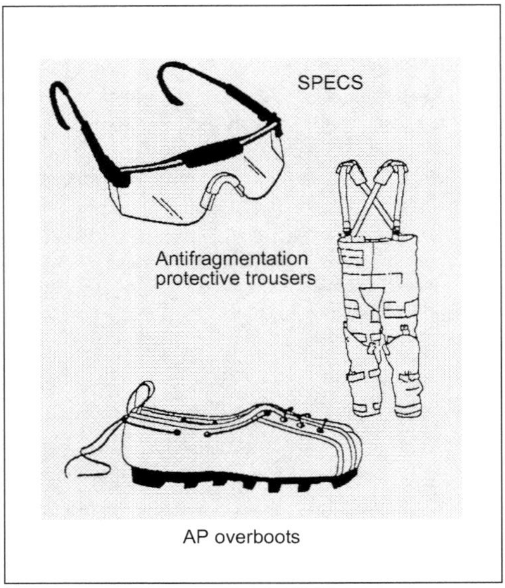


**Figure 11-1. IBASIC**

- Provide MP and explosive-sniffing dogs to help in clearance and provide security for convoys during and after clearing operations.

**Command and Control**

**NOTE: The company team commander is required to operate on three separate frequencies—battalion command network, company team command network, and fire-support network.**

- Designate, recognize, and include minefield indicators (Chapter 10) as part of company team rehearsals.
- Designate a reserve force (at least platoon-size) that is mechanized or air-assault capable.
- Ensure that proper rehearsals are planned and conducted according to FM 3-34.2. As a minimum, the clearance force should rehearse actions on the obstacle, actions on enemy contact, casualty evacuation, and the control of COBs.
- Ensure that the tasked unit has a clear understanding of the mission, intent, and end state. For example, the clearing unit commander should understand that his unit must clear the road width, including the shoulders, and secure the route.
- Assign clearance responsibilities to brigade and battalion assets.
- Ensure that the maneuver commander/TF S3—

- Controls the movement of all personnel and equipment along the route (travel authorization is coordinated through the S4).
- Prepares a mine risk assessment of the mission before issuing the OPORD. (An example of a mine risk assessment is shown in Appendix F.)
- Tracks the status of routes (red, amber, green) in the TF sector, based on the amount of time since the route was cleared and the intelligence and enemy situations.
- Tracks the progress of the clearance operation and integrates it into the maneuver and CSS plans.
- Determines the route length, using clearly definable start and end points.
- Sets priorities for the route-clearance element.
- Coordinates with adjacent units, the host nation, NGOs, PVOs, and SOF.

#### Special Operations

- Ensure that psychological operations (PSYOP)/civil affairs (CA) support the counterintelligence effort by conducting civilian interviews.
- Direct civilians along the MSR to the displaced-personnel holding areas or along the routes that the brigade has indicated for use.
- Employ PSYOP/CA teams forward to disperse civilians and provide traffic management to isolate the route during clearance operations.

### TASK ORGANIZATION

A brigade or battalion TF normally conducts clearance-in-zone operations. To clear a route, the battalion TF focuses a company team as the main effort on the proposed MSR. Table 11-1, page 11-2, shows a sample task organization for a route clearance.

#### Support Force

This force is comprised of two maneuver platoons and the maneuver company team XO. The support force provides flank security, rear security, and protection to the breach force. It neutralizes hostile forces encountered by the company team. In rugged terrain or highly mined areas, moving the assault force on the flanks would be too risky. Aviation assets can provide flank security while ground forces provide rear security. The assault force also searches for suspected off-route mines.

#### Assault Force

This force is comprised of a maneuver platoon, an engineer squad, a mortar section, a medical team, a PSYOP team, an EOD team (or one that is on call), and a forward observer. The assault force's mission is the same as in a breaching operation (Chapter 9).

## Breach Force

This force is comprised of a maneuver platoon (including the commander) and an engineer platoon (minus). The breach force sweeps the route and reduces mine and explosive threats. It is further task-organized into sweep teams.

A sweep team is a trained detection team that searches for mines and explosive devices. The organization of the sweep team depends on the type of sweep mission and the length, the width, and the surface composition (pavement, gravel, dirt) of the area to be swept. A platoon-size element can normally clear a 4.5-meter-wide path, and a squad-size element can normally clear a 1.5-meter-wide path. If the route is wider or time does not permit multiple passes of the route, additional engineer assets are required. Table 11-2 outlines personnel and equipment requirements for a sweep team.

**Table 11-2. Personnel and equipment requirements for a sweep team**

| Personnel | Support Personnel | Equipment |
|---|---|---|
| • NCOIC<br>• Mine-detector operators<br>• Probers/markers<br>• Radio operator<br>• Demolition teams | • Medics<br>• Vehicle operator | • One panel marker<br>• Operational map with required maneuver graphics<br>• Four smoke grenades (minimum)<br>• Six mine detectors (includes three backups) and extra batteries<br>• Two grappling hooks with 60 meters of rope each<br>• One demolition kit or bag for each demolition man<br>• Six probes<br>• Mine marking material<br>• Lane marking tape<br>• Ten stakes |

### *Platoon-Size Sweep Team*

The normal configuration for a platoon-size sweep team is twelve soldiers in a modified column (Figure 11-2). The platoon leader supervises the entire operation. This configuration is best suited for sweeping routes in friendly territory that is not under constant surveillance.

When sweeping in areas where 100 percent coverage is required, the team should establish a clear lane from which to operate. This is conducted by the grapple hook being thrown one additional time into the suspected minefield. The cord on the grapple hook is pulled taunt, with the free end secured to a stake.

This establishes a left or right physical boundary marker for the mine-detector operator to use as a guide. The first mine-detector operator will sweep a 1.5-meter lane left or right of this cord, up to the grapple hook. At the hook, a second stake will be emplaced to hold the boundary line and the grapple hook will be thrown again and pulled for trip wires as before. A fourth throw will serve as the boundary marker through the suspected minefield. This procedure will continue until the first mine-detector operator is out of the minefield on the farside.

The boundary marker will then be secured to stakes, using the lane marking tape. A second boundary maker will be placed 1 meter to the left or right of the original boundary marker, depending on which side was swept. This will

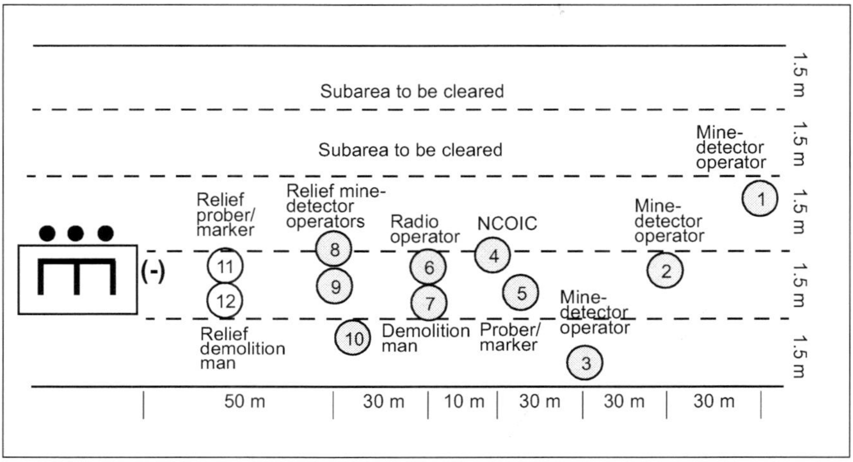


**Figure 11-2. Platoon-size sweep team**

establish a fixed 1-meter lane. The 0.5-meter area that was not included in the original lane is to ensure that no gaps in coverage were missed on the edge of the first lane.

The unit will then stretch a piece of marking tape out from across the clear footpath, moving it left or right 1.5 meters and staking it down on both ends. This will be repeated as many times as necessary.

Always mark lanes that have not been cleared with a piece of marking tape across the entry to that lane to prevent personnel from using the lane before it has been swept. After multiple lanes are physically marked, the unit can perform a platoon-size sweep. Lanes should be swept and cleared by working out from the original 1-meter footpath. As the cleared area becomes bigger, the marking tape is removed to allow the passage of personnel and equipment though the cleared lane.

As subsequent lanes are being swept, the marking tape will need to be secured on the outside edge of the lane. Always conduct a thorough sweep of the spot where the stake will be placed. Stake intervals will depend on terrain and atmospheric conditions.

Soldiers 1, 2, and 3 (mine-detector operators) lead the sweep team in echelon. Each sweep team covers 1.5 meters of front, and sweep teams are spaced 30 meters apart to prevent fatalities from accidental detonation by other mine-detector operators. If required, a fourth mine-detector operator can be added to the detection column.

Soldiers 4 (NCOIC) and 5 (prober/marker) follow 30 meters behind the last mine-detector operator (Soldier 3) and are centered in the cleared lane. The prober/marker is responsible for marking the cleared lane on both sides.

Soldiers 6 (radio operator) and 7 (demolition man) follow 10 meters behind Soldiers 4 and 5 and are centered in the cleared lane.

Soldiers 8, 9, 10 (relief mine-detector operators), 11 (relief prober/marker), and 12 (reserve demolition man) follow 30 meters behind Soldiers 6 and 7. If a fourth mine-detector operator is added to the column, an additional relief mine-detector operator must also be added.

The remaining platoon members help the support force or act as a reserve force, as required. They should first be integrated into the sweep team as a relief element and then moved forward as needed.

### *Squad-Size Sweep Team*

The normal configuration for a squad-size sweep team is seven soldiers in a modified column (Figure 11-3). The squad leader supervises the entire sweep operation. This configuration is designed for sweeping routes in friendly territory that is not under constant surveillance.

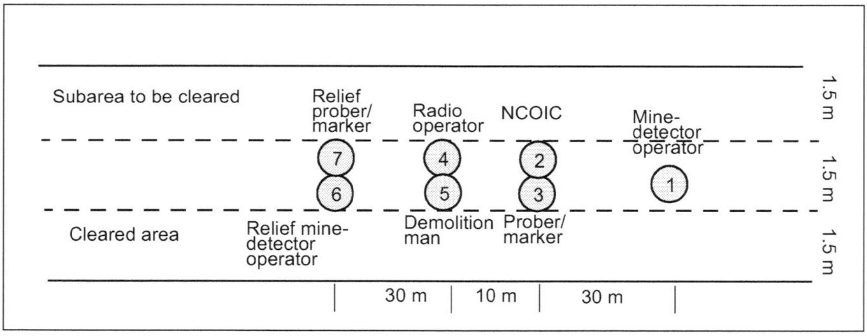


**Figure 11-3. Squad-size sweep team**

Soldier 1 (mine-detector operator) leads the sweep team and covers a 1.5-meter-wide path.

Soldiers 2 (NCOIC) and 3 (prober/marker) follow 30 meters behind Soldier 1 and are centered in the cleared lane. The prober/marker is responsible for marking the cleared lane on both sides.

Soldiers 4 (radio operator) and 5 (demolition man) follow 10 meters behind Soldiers 2 and 3 and are centered in the cleared lane.

Soldiers 6 (relief mine-detector operator) and 7 (relief prober/marker) follow 30 meters behind Soldiers 4 and 5. If the squad cannot use seven team members, the relief prober/marker position can be eliminated from the formation.

The engineer platoon can configure the platoon into squad-size sweep teams and place them in echelon (Figure 11-4).

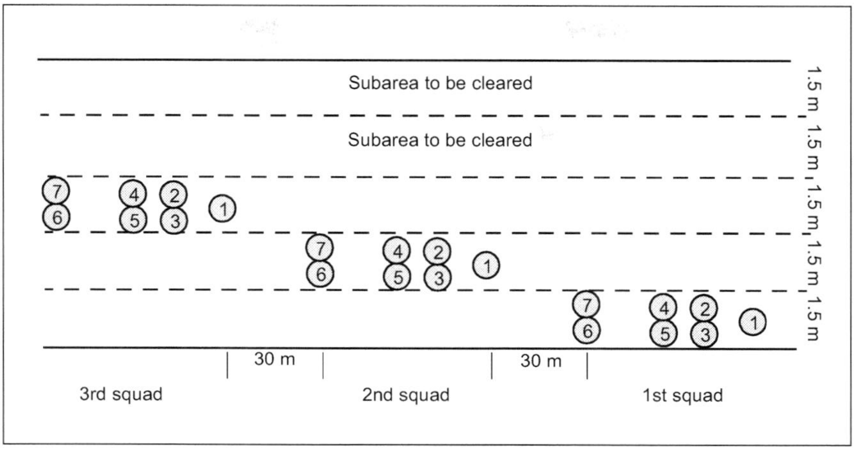


**Figure 11-4. Sweep teams in echelon**

## METHODS AND TYPES

The information gathered from the IPB and the reconnaissance effort determines the method and the type of route clearance to conduct. The determination is based on the situation, the time available, the threat level, and available assets.

During OOTW, it is recommended that former warring faction (FWF) engineer-equivalent clearance teams precede US forces clearance teams within the FWF's AO. Do not assume that FWF clearance teams will be thorough in their clearance operation. Treat the route as unsafe until US or allied force clearance teams have proofed the route to confirm that it is cleared.

### Methods

There are three methods of route clearance—linear, combat, and combined. The method employed depends on the situation, the time available, and the clearance assets available. The maneuver force should always establish static security positions at critical locations following the completion of route clearance.

### *Linear Clearance*

In linear clearance (Figure 11-5), sweep and security teams begin route clearance at Point A and complete it at Point B. This method provides the best assurance of route coverage. Although this is an effective method, it is not the most secure method in a high-threat environment. It is also time-intensive and constrains the maneuver commander's flexibility.

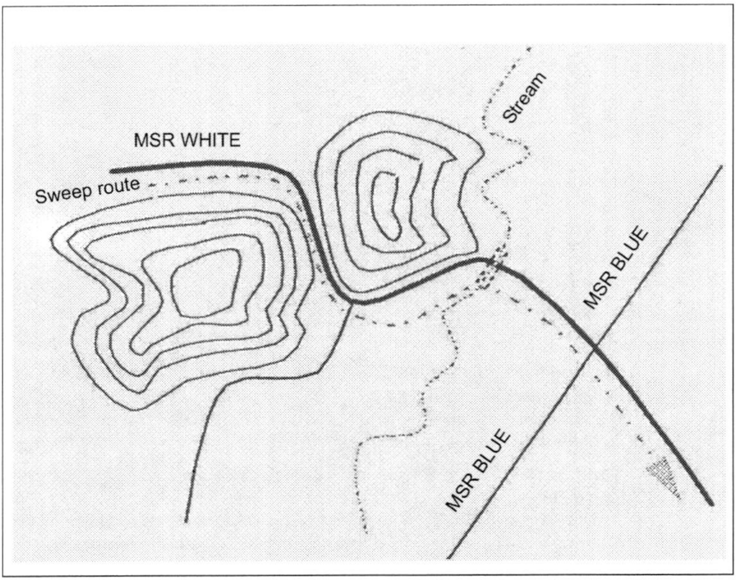


**Figure 11-5. Linear clearance method**

### *Combat Clearance*

Whereas linear clearance focuses on a specific route, combat clearance (Figure 11-6) focuses on specific points along a route. As mentioned previously, IPB and EBA can identify areas for likely mine and ambush locations. These areas become NAIs or objectives for combat clearance missions. The combat clearance method divides a route into sections according to the number of suspected high-threat areas. Once the sweep element (maneuver and engineer forces) secures and sweeps these areas, the route is considered clear. Combat forces can patrol the route from these objectives to ensure that the route is secure, and if necessary, the sweep element can sweep the surrounding area if a minefield is found. Following the seizure of these objectives, the commander must assume a moderate risk that the S2 and the force engineer have identified all high-threat areas and that the route is clear of mines. Combat clearance is ideal for dismounted (light) forces since it provides the maximum use of surprise and concealment.

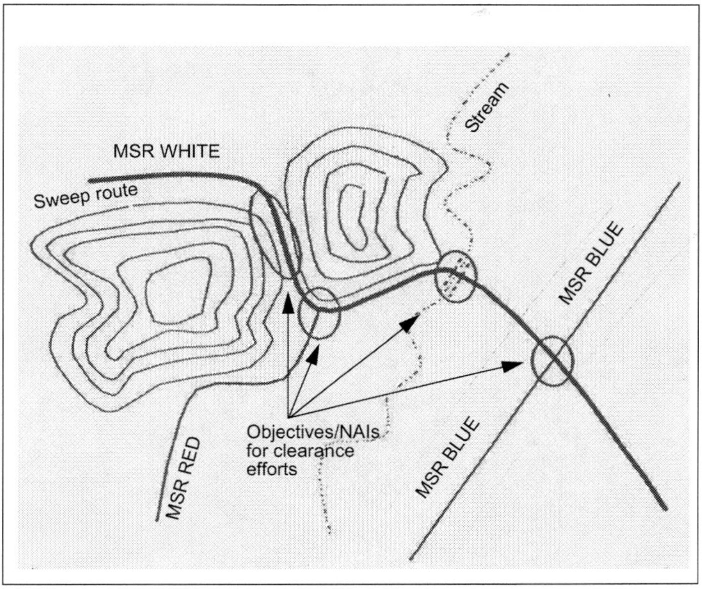


**Figure 11-6. Combat clearance method**

### *Combined Clearance*

This method combines the complete clearance capabilities of the linear clearance method with the security and surprise elements of the combat clearance method. Combined clearance is a two-phase, force-intensive operation and may require a battalion-size effort, depending on the length of the route. First, identify high-threat areas through IPB/EBA and target them as NAIs and/or objectives to secure. Then, clear obstacles and enemy forces before the movement of sweep elements. The sweep team moves down the road and clears any obstacles missed or not identified during the planning process. The main advantage of this method is that the TF commander has immediately secured MSRs, allowing him to *push out* (expand forces out past the secured area and secure additional areas) and find the enemy with a degree of confidence that follow-on forces will be much safer.

### Types

There are two types of sweep operations—deliberate and hasty. Deliberate and hasty clearance operations can be modified to meet the time and equipment limitations of the TF, but the commander assumes greater risk when the clearance type is modified.

***Deliberate***

A deliberate sweep (Figure 11-7) is very thorough and includes a complete sweep of the entire road (shoulders, culverts, ditches, and bridges). It is the most time-consuming sweep operation and relies on electronic (primary) and visual (secondary) detection systems.

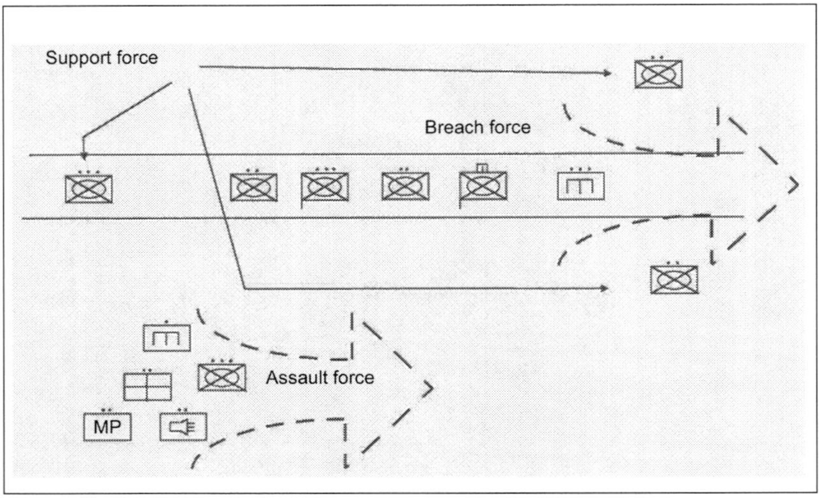


**Figure 11-7. Deliberate route clearance**

The platoon sweep team (Figure 11-2, page 11-9) is dismounted to focus its attention on the entire length of the route. The support force (company-size) secures at least 100 meters on the flanks and 100 meters forward to clear possible enemy direct-fire systems and overwatching elements in front of the breach force. This not only allows the breach force to focus solely on the route but also clears the area of off-route and command-detonated mines.

If enemy contact is made, the support force fixes the threat while the assault force reacts. The sweep teams withdraw to a location that provides concealment and/or security. Mechanical detection provides a third means of detection and is the method used to proof the route after the sweep team has passed through the area. The deliberate sweep includes a route reconnaissance and looks at all areas of a route, including bypasses. The deliberate sweep focuses on thoroughness rather than speed. This method is very slow and tedious and should only be used when time is not a factor; 80 to 100 meters can be covered per hour.

***Hasty***

A hasty sweep (Figure 11-8) consists of visual inspection, physical search or probing, and the use of mine detectors. It is the fastest, most risky method and is suited for an armored or mechanized team. It relies primarily upon visual detection (thermal sights or the naked eye) for minefield identification. The breach force looks for mines, wire, and other minefield indicators. The road surface, culverts, ditches, and bridges are inspected and searched. Visual detection is accompanied by a mechanical proofing system. Electronic mine detectors are used by sweep teams to check all suspected areas.

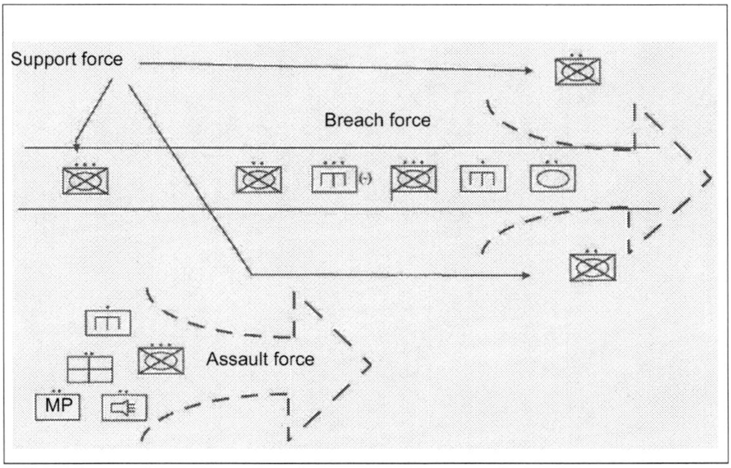


**Figure 11-8. Hasty route clearance**

The support force includes a maneuver platoon that provides overwatching fire and/or security. Actions upon enemy contact are the same as in a deliberate sweep. The primary objective of this technique is speed, moving approximately 3 to 5 kph. This method is extremely similar to the instride breach method when encountering minefields.

The sweep team focuses on identifying immediate risks to traffic, neutralizing those risks, and continuing on with the mission. A hasty sweep is used during the combat clearance method to validate the areas that were not deliberately cleared by the sweep team. It is also used if the METT-TC analysis does not permit a deliberate sweep or if the need for a road to be opened is urgent. Time and distance factors may be imposed. A light force may not have an MCR system but can conduct the same sweep method with an improvised roller system, or the force can use a sandbagged, 5-ton truck moving backwards as a last-resort method. Using MCRs or their equivalent is absolutely imperative due to the high risk of encountering a minefield. The mine rake or plow is not a satisfactory substitute because it destroys road surfaces.

## AREA CLEARANCE

Area clearance is also a combined arms mission. Clearing operations occur when engineers receive a mission to clear an area of mines or to clear a specific minefield in a friendly AO. In most cases, the minefield has been reported and may already be marked on all sides. The engineer unit receiving the mission must base its plans on available information and prepare equipment based on the estimate.

## PLANNING

Planning area-clearance operations is very similar to planning breach operations. Commanders and staffs plan and coordinate all the breaching tenets.

### Intelligence

Intelligence is particularly important for discovering the types of mines and mine fuses the enemy employs. The engineer uses this information to determine which clearance and neutralization techniques offer the best chance for success and minimize the risk to the sweep teams. Intelligence also helps the commander determine the need for outside resources, such as EOD and SOF elements.

The information needed for area-clearance operations includes—

- Minefield location.
- Minefield orientation.
- Presence of wire as an obstacle.
- Location of gaps and bypasses.
- Minefield composition (buried or surface-laid mines, AT or AP mines, AHDs, trip wires, and minefield depth).
- Types of mines, employment techniques, fuses, and booby-trap configurations.
- Information on whether or not the minefield is marked and, if it is marked, the material used.
- Possibility of hostile forces in the area.

### Fundamentals

You must plan to apply the four fundamentals of breaching operations (SOSR), but you may not execute all of them due to the lack of enemy presence.

### Organization

Task organization is similar to that used for route-clearance operations. The breach force is the clearance element, and the support force is responsible for all security and maneuver responsibilities. There is no assault force in area-clearance operations. Table 11-3 shows a sample task organization for an area clearance. The size of the force can be tailored, based on the probability of contact.

### Mass

Sufficient maneuver and engineer assets must be allocated to the clearance company team. The length and width of the route and the amount of time

**Table 11-3. Sample task organization for an area clearance**

| Team | Support Force | Breach Force |
|---|---|---|
| Heavy | • Mechanized infantry platoon<br>• Mortar section<br>• FIST<br>• Armor platoon | • Engineer platoon with organic vehicles<br>• Medical team (two ambulances)<br>• EOD team |
| Light/heavy | • Bradley platoon with dismount capability<br>• 60-mm mortar section<br>• Forward observer<br>• One infantry platoon (light) | • Engineer platoon with organic vehicles<br>• Medical team (two ambulances)<br>• EOD team |
| Light | • AT/MP section with M60/MK19 mix<br>• 60-mm mortar section<br>• Forward observer<br>• Two infantry platoons (light) | • Engineer squad (-)<br>• Medical team (two ambulances)<br>• EOD team |

available determines the size of the sweep team. A platoon is normally used to clear a 200- by 300-meter minefield; additional assets are required to clear larger and multiple minefields.

**Synchronization**

All aspects of synchronization should be implemented when planning area clearance. It is especially important that rehearsals be conducted at the combined arms level. Rehearsals should include—

- Reaction to enemy contact.
- Reaction to an ambush.
- Communications exercise.
- Fire support (obscuration smoke, immediate suppression fires, critical friendly zones for counterfire radar, and no-fire area around the clearance site).
- CSS (maneuver, casualty evacuation, marking materials, and demolitions resupply).

## PLANNING CONSIDERATIONS

Area-clearance BOS planning considerations are parallel to route-clearance planning considerations. Brigade and battalion TF staffs should use the planning considerations outlined on pages 11-3 through 11-7 plus the ones outlined below:

- Intelligence. Focus on the most probable enemy attack method and AAs.
- Maneuver.
  - — Clear and secure flanks (at least 500 meters) and the farside of the area to be cleared.
  - — Provide security for the cleared area.

- Fire support. Ensure that the area-clearance team has a FIST coordinator. The FIST should be collocated with the support force OIC.
- Mobility/survivability. Establish minefield control points along the area to be cleared.
- CSS.
  - Ensure that the medical team consists of one or two ambulances and that it is located with the breach force.
  - Ensure that all personnel wear flak vests or IBASIC (Figure 11-1, page 11-6).
- C2.
  - Determine the area length, using clearly definable perimeter points.
  - Coordinate with adjacent units, the host nation, NGOs, PVOs, and SOF.

## TASK ORGANIZATION

The battalion TF will focus a company team (minus) as the main effort to conduct area clearance.

### Support Force

This force is comprised of two maneuver platoons and an OIC. The support force provides flank security, forward security, and protection for the breach force. It neutralizes hostile forces that are encountered by the company team. The support force secures the area 500 meters beyond the area to be cleared. METT-TC factors will affect the actual distance based on the threat and the weapon systems. The support force OIC establishes static security positions around the area until the clearance operation is complete. He also has control of fires and the responsibility to neutralize any hostile force.

### Breach Force

The breach force is comprised of an engineer platoon that is organized into sweep teams, a medical team, and an EOD team (or one that is on call). The sweep team (squad-size) is organized as shown in Figure 11-3, page 11-10. The breach force's mission is to sweep and clear the area of mine and explosive threats.

## METHODS AND TYPES

The breach force OIC determines the perimeter of the area to be cleared and ensures that it is marked. The OIC divides the area into sections to be cleared (Figure 11-9). The sections should be no larger than 40 meters wide and 100 meters long. This is an optimal-sized area for a sweep team to clear at one time. The OIC assigns squad-size sweep teams to each section.

The squads clear their assigned sections using the sweeping techniques discussed earlier in this chapter. As the sections are cleared, they are marked for safety and control purposes. This process is continued until the entire area is cleared. Progress is reported to the company team commander as required.

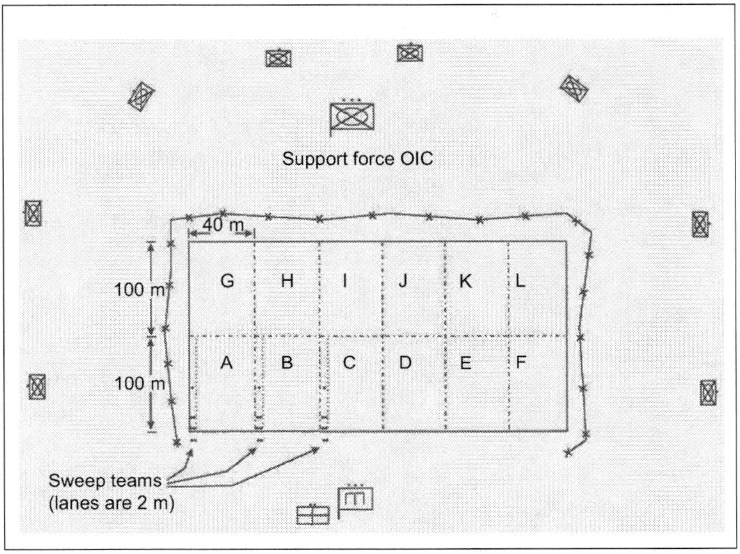


**Figure 11-9. Area clearance site layout**

# IMPROVISED MINE THREAT

Mines are not always employed conventionally by military forces organic to the host nation or its enemies. In many cases, they are also employed by terrorists against allied forces or the host-nation populace. In these cases, the threat increases because of the improvised methods in which the mines were emplaced. In conventional emplacement of mines, a pattern emerges from the emplacing force's doctrine, and the threat can easily be reduced by using this knowledge. There is less pattern in the case of improvised mining methods, and this makes detection and removal very difficult.

Improvised mining has many different employment techniques. In most of the techniques shown below, a UXO can easily be employed in place of a mine:

- Coupling mines. Coupling is done by linking one mine to another, usually with detonating cord. When the initial mine is detonated, it detonates the linked mine. This technique is done to defeat countermine equipment.
- Boosting mines. Buried mines are stacked atop one another, and the farthest mine from the surface is fused. This reduces the probability of detection and increases the force of the blast.
- Sensitizing AT mines. On some nonmetallic AT mines, the pressure plate can be cracked and the spring removed or the mine's explosive can be cut into smaller blocks and employed as powerful AP mines.

The pressure plate can be removed from metallic AT mines and employed in the same manner. Alternatively, a pressure-fused AP mine can be placed on the top of an AT mine.

- Mixing training mines with live mines. Hostile forces can employ training mines at the start of a minefield and emplace live mines toward the end. The sweep element falsely believes that the minefield is phony and becomes complacent in its reduction activities. When this technique is used, live mines are painted to resemble training mines.
- Daisy-chaining mines. Command-detonated AP mines are commonly used in daisy chaining. Hostile forces link the mines with trip wires or detonating cord. When the initial mine is detonated, the other mines will detonate.

## MINE LOCATIONS

Hostile forces normally place more than one mine in each mined area. Do not focus the detection effort solely on a horizontal mine threat, such as on the ground or in culverts. The mine threat is also vertical, such as in trees or attached to an overpass. Clearance efforts must accommodate the three-dimensional battlefield. Mines may be placed in—

- Frequently used roadways.
- Brush and other traffic obstructions placed on roadways.
- Bridge bypasses and fording sites.
- Road junctions.
- Obvious turnarounds, bypasses, culverts, ditches, and shoulders.
- Key logistic points (water, fuel, food).
- Debris along a route.

## DISPOSITION OF MINES

The following actions should be taken when a suspected mine is found:

- Mark the suspected mine location; do not leave any mine unmarked.
- Search for electric wires and trip wires in the immediate area. Trace the wires in both directions to determine if items are attached to them. If there is nothing attached and the IPB does not state otherwise, cut loose trip wires and electric wires.

**DANGER**

**Never cut taut trip wires. Alert the security element to search for an enemy that may be manning a command-detonated mine. Keep troops away from the mine until all the wires are traced and cut. Be alert for booby traps and ambush. If booby traps are found, use the clearance procedures outlined in Chapter 13.**

- Probe the suspected mine location and uncover enough of the object to identify it. Other personnel should stay at least 30 meters away.

- If the object is a mine, the prober withdraws and notifies the OIC. The OIC decides to bypass the mine, destroy it in place, remove it with a grapnel, or notify EOD for hand neutralization.
- If the object is debris, get in a protected position and carefully remove the debris with a grapnel hook. Be alert for booby traps or AHDs wired to the debris.

### MINE-REMOVAL TECHNIQUES

A mine can be bypassed, detonated in place, pulled out by a rope or a wire, or neutralized and removed by hand. The method used depends on the location of the mine, the type of the mine and the fuse, and the tactical situation. Methods of removal and actions on finding a mine should be addressed in the OPORD and rehearsed prior to executing the mission.

Trip-wire and tilt-rod mines can be detonated by throwing a grapnel, with a rope attached, past the trip wire or tilt rod and pulling the grapnel back to actuate the mine. Grapnels may be improvised from any available material, such as a bent drift pin or scrap material.

A hand-emplaced charge is the standard demolition material used to destroy a mine in place (see FM 5-250). A 1-pound block of explosive placed next to a mine is sufficient to detonate most mines. A charge can be placed next to each mine in a group, then the charges can be connected and fired simultaneously.

Rope or wire can be used to pull a mine out of its installed position. This is a safe method and only detonates mines that are equipped with AHDs. It also reduces noise and cratering. A tripod (Figure 10-21, page 10-23) makes it easier to pull a mine out of a hole on the first attempt. Use the following procedures to remove mines:

- Uncover only enough of the mine to expose a handle or a projection. Attach a 60-meter length of rope or wire to the mine or engage a grapnel. If there is no projection, engage a grapnel on the bottom side of the mine, opposite the direction of pull.

**DANGER**
**Do not move the mine while uncovering it or attaching the rope because movement might detonate an AHD.**

- Ensure that the covered area is not mined. Take cover and lie in a prone position at least 50 meters from the mine. Pull the rope to remove the mine from the hole.
- Wait at least 30 seconds before leaving cover and approaching the mine if the mine type is unknown. This guards against the possibility of a delay firing mechanism.
- Dispose of the mine according to the unit directive or SOP.

### HAND NEUTRALIZATION

Appendix A discusses procedures for hand neutralization of US mines. Foreign mines and booby traps should only be neutralized by EOD personnel.

Mines are neutralized by hand, when—

- Units are conducting a covert breach.
- The mine is located on a bridge, building, or other facility required for use by friendly forces.
- Neutralization by other means is not possible.
- The mine can be positively neutralized by hand and is required for reuse.
- The mine type is unknown and recovery must be attempted for intelligence purposes.
- Chemical mines are located in areas where contamination would restrict the use of the area by friendly troops.

## SAFETY

The following safety procedures should be observed during route and area clearance:

- Personnel should wear helmets and flak jackets to protect them against fragmentation. Sweep team members should wear IBASIC, if available.
- Vehicle floorboards should be sandbagged.
- Vehicles should be dispersed at 50-meter intervals. This ensures that a mine detonated by one vehicle will not cause casualties in other vehicles.
- One person at a time should be allowed at a suspected mine location.
- Personnel should assume that mines and explosive devices are equipped with AHDs until proven otherwise.
- Personnel should not run and should move only in previously cleared areas.
- Armored vehicles should have their hatches open to vent the pressure pulse from a mine detonation.
- Soldiers should wear ballistic and laser protective spectacles (BLPS) or lightly tinted, protective eyewear to reduce eye fatigue and improve their ability to recognize mine indicators.

## REPORTS

Dissemination of information is the key to battlefield management. Units encountering minefields or explosive devices should follow a five-step process—stop, secure, mark, report, and avoid. Units must provide adequate information to their higher headquarters to ensure that follow-on elements are well informed. Information must include the known or suspected minefield location, types of mines (if known), the marking method, the time the minefield was encountered, and any additional information that may be of use to the clearing unit.

Division and maneuver brigades must establish a central control cell for mine clearance information. This cell receives and gathers all mine and explosive threat data within the unit's AO. Mine-contact reports are reported through maneuver command channels with a priority of *flash* or *immediate*. The information is then jointly controlled in the operations cell and the central mine control cell by the engineer staff officer, the G3/S3, and the Assistant Chief of Staff, G2 (Intelligence) (G2)/S2. The mine contact database is jointly maintained by the engineer and the G2/S2, who subsequently conduct pattern analysis and integrate it into intelligence and operational updates.

The central mine control cell performs the following actions:

- Maintains a current situation map and overlay that depict friendly and enemy mines and obstacles.
- Maintains and updates enemy obstacle (Figure 11-10, page 11-24) and route status (Figure 11-11, page 11-25) information.
- Receives and maintains minefield recording forms (US and foreign) within the unit's operational area (this includes host-nation minefield data, if available).
- Maintains a mine-contact database. (This could be a clearinghouse for future operations.)
- Processes, analyzes, and updates information; disseminates the information to subordinate commanders and staff.

### SITUATION REPORT

Clearing units submit a situation report to higher headquarters if enemy activity is encountered or if an explosive device is discovered. This information should be tracked in the TOC and the CTCP. Information must be disseminated to subordinate units, especially CSS elements.

### PROGRESS REPORT

The clearing unit submits progress and completion reports until the clearance operation is complete. Progress reports must be timely and accurate. Report format and frequency are established in the OPORD before the clearance mission is executed.

**ENEMY OBSTACLE REPORT**
As of: ________

| Obstacle No | Obstacle Location | Reporting Unit | Info DTG | Obstacle Description (Type of Mines/Wire) | Depth | DTG Marked | DTG Reduced | DTG Cleared | Lane/ Bypass Grids | Remarks |
|---|---|---|---|---|---|---|---|---|---|---|
| | | | | | | | | | | |
| | | | | | | | | | | |
| | | | | | | | | | | |
| | | | | | | | | | | |
| | | | | | | | | | | |
| | | | | | | | | | | |
| | | | | | | | | | | |
| | | | | | | | | | | |
| | | | | | | | | | | |
| | | | | | | | | | | |
| | | | | | | | | | | |
| | | | | | | | | | | |
| | | | | | | | | | | |

Figure 11-10. Sample enemy obstacle report

**ROUTE STATUS REPORT**

As of: ______

| Route Name | Responsible Unit | Status (Red, Amber, Green) | As Of | Start Point | End Point | Remarks |
|---|---|---|---|---|---|---|
| | | | | | | |
| | | | | | | |
| | | | | | | |
| | | | | | | |
| | | | | | | |
| | | | | | | |
| | | | | | | |
| | | | | | | |
| | | | | | | |
| | | | | | | |
| | | | | | | |
| | | | | | | |
| | | | | | | |

Figure 11-11. Sample route status report

## MINE INCIDENT REPORT

A mine incident includes any unplanned activity involving a mine, UXO, or a booby trap. It also includes near misses that could have resulted in damage or injury. The mine incident report (Figure 11-12) is a technical report that follows a serious incident report (SIR), and it should be submitted as soon as possible (local SOP will indicate time requirements).

**MINE INCIDENT REPORT**

DATE:
FROM: THRU: TO:

REFERENCE SIR #__________

| | | |
|---|---|---|
| A. Incident DTG | A1. dd/time/zone/mm/yy | |
| B. Incident location | B1. Map sheet/UTM/grid reference (8 digit) | Include a site sketch as an attachment. |
| | B2. Location (road, field, building) | |
| | B3. Emplacement (buried, surface-laid, off-route) | |
| C. Effects (to complement information already in the SIR) | C1. Casualties (rank, name, date of awareness training, time in the mission area, protection equipment used) | |
| | C2. Vehicle damage (number, type, extent of damage) | Include a photo if possible. |
| | C3. Collateral damage | |
| D. Device suspected | D1. Type of mine (AT, AP, make, model) | |
| | D2. Type of booby trap (pull, release, pressure) | |
| | D3. Type of UXO (dropped, thrown, projected) | |
| | D4. Unknown (detail, color, shape, size) | |
| E. Circumstances | E1. Activity at the time of the incident | |
| | E2. Degree of previous use of the route, area, location | |
| | E3. Date of previous clearance and proofing by engineers | |
| | E4. Where the route, area, or location is monitored | |
| F. Recommendations | F1. Recommendations to prevent reoccurrence | |
| G. Miscellaneous | G1. Any other pertinent data | |

**Figure 11-12. Sample mine incident report**

# PART THREE

# Special Mining Operations

Part three provides tactical and technical information on special-mining operations, such as using booby traps and expedient devices. It also discusses mining in rivers, urban terrain, and unique environments. Restrictions and responsibilities are outlined in detail for the employment and the clearance of special mines and devices.

## Chapter 12

## Mining Operations in Special Environments

Mines are emplaced and encountered in all environments, some of which need special consideration to understand effective employment, detection, and/or removal.

### STREAMBED AND RIVER MINING

#### EMPLOYMENT

Conventional AT mines are much more effective in water than on land because water transmits the shock effect better than air. Vehicle support members, tracks, and wheels are damaged by a mine blast. Small vehicles are overturned and almost completely destroyed. Because water amplifies and transmits shock waves, mines equipped with pressure-actuated fuses are subject to sympathetic detonation at greater distances in water than on land.

M15 and M19 AT mines can be used for streambed and river mining. The M21 AT mine should not be used because it is very difficult to arm and disarm underwater, and it can be easily functioned by drifting debris. To avoid sympathetic detonation, AT mines must be at least 14 meters apart in water that is less than 61 centimeters deep, and at least 25 meters apart in water that is deeper than 61 centimeters. The mined areas are chosen to take advantage of stream and adjacent area characteristics. Water depth within the minefield should not exceed 1 meter because it is difficult to work in deeper water, and pressure-actuated fuses are usually ineffective against waterborne vehicles.

Current velocity must be considered when emplacing mines in a streambed or a river. If the mines are placed deeper than 45 centimeters, they must be recovered by engineer divers:

- A lightweight diver has diving restrictions based on current velocity.
- A scuba diver is restricted to a maximum current velocity of 0.5 meter per second.

- A surface-supplied diver is restricted to a maximum current velocity of 1.3 meters per second.

Seasonal current velocity should also be considered if the minefield is to be in place for an extended period of time. Additional information on diving restrictions can be found in FMs 20-11 and 5-490.

Since sand in inland waters continuously moves downstream, it may be difficult to locate and remove mines planted on sandbars or downstream from sandbars. If the site has a muddy bottom, the mud should not be deeper than 46 centimeters and there must be a hard base underneath it. The enemy is unlikely to choose a fording point where vehicles mire easily. If underwater obstacles (gravel, rock, stumps) are bigger than the mine, the area cannot be easily mined. If such areas must be used, place the mines so that they are exposed to vehicle wheels or tracks. Armored vehicles usually enter and exit streams at points where the incline is less than 45 percent. After entering a stream, vehicles often travel upstream or downstream before exiting. Carefully examine riverbank formations and underwater obstacles to predict the trail a vehicle will use to ford the stream.

### EMPLACEMENT

When emplacing mines in streams and rivers, always work in pairs. Prepare the mine on land near the emplacement site. Coat fuse threads and wells with silicone grease (a waterproof lubricant) or a heavy grease to minimize the chances of water leaking into the mine. Waterproof joints between the pressure plate and the mine case with silicone grease. As a rule of thumb, waterproofed mines are reliable up to 3 months when immersed without waterproof coverings. Secure the mine with outriggers to prevent drifting:

- Construct field-improvised outriggers with—
  - Two green limbs that are about 3 centimeters in diameter and 1 meter long. Green limbs are recommended because they are stronger and less likely to float than those which are dried out and dead. (Steel pickets, sign posts, fence rails, or similar items having the proper dimensions may also be used.)
  - Two pieces of clothesline, manila line, or similar material that are about 1 meter long.
- Fasten the limbs to the underside of the mine and secure them with the line (Figure 12-1).
- Approach the emplacement position from the downstream side. To prevent dragging the outrigger or contacting objects in the stream, carry the mine by grasping its sides, not by its carrying handle.
- Place the mine and the outrigger on the stream bottom. Stake down outriggers after they are emplaced to prevent drifting. If staking is impossible, place sandbags or large rocks on the outriggers for better anchorage.
- Arm the fuse.

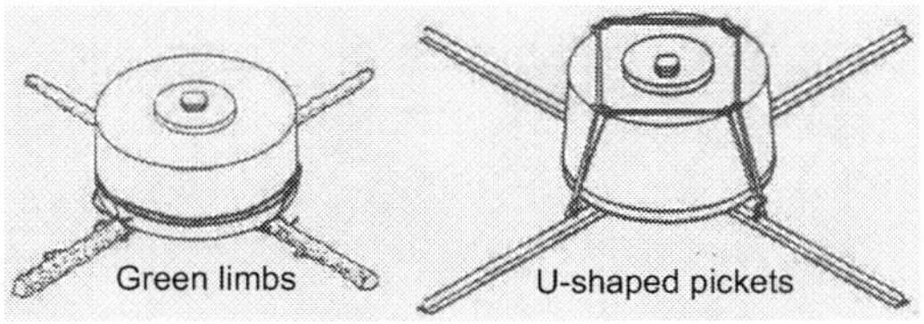


**Figure 12-1. Outrigger techniques**

## Recovery

> **WARNING**
> **Mines may have drifted downstream and/or been tampered with by enemy forces. Removal by any method other than explosive breaching (see TM 9-1375-213-12) is extremely hazardous and is not recommended. If the situation demands recovery by hand, proceed with utmost caution.**

A two-person recovery team—

- Slowly proceeds 2 meters downstream from where the mine was emplaced and then carefully probes for the mine. If the mine was placed deeper than 45 centimeters, it must be recovered by engineer divers.
- Removes any foreign material from the top of the mine and disarms it.
- Carries the mine ashore and removes the fuse and the detonator.

Engineer divers normally emplace new mines or mines that have not been submersed in water; but if the mine and the fuse show no evidence of damage or deterioration, the mine can be resused. If the mine is reused, mark it to indicate that it has been immersed in water; for example, place the letter *W* on the pressure plate.

## Recording

The minefield is recorded on DA Form 1355 (Figures 12-2a and 12-2b, pages 12-4 and 12-5).

## Safety

In addition to normal safety measures, underwater mining requires evaluation of the tactical situation and application of special safety techniques. The turbidity, the velocity, and the depth of the water and the condition of the bottom require that laying-party personnel be able to swim well. Prolonged immersion of personnel, especially in cold temperatures, must be avoided. Sudden drop-offs, rocks, and other objects that are likely to cause personnel to lose their footing must be considered. Other safety measures include the following:

- Work in pairs.

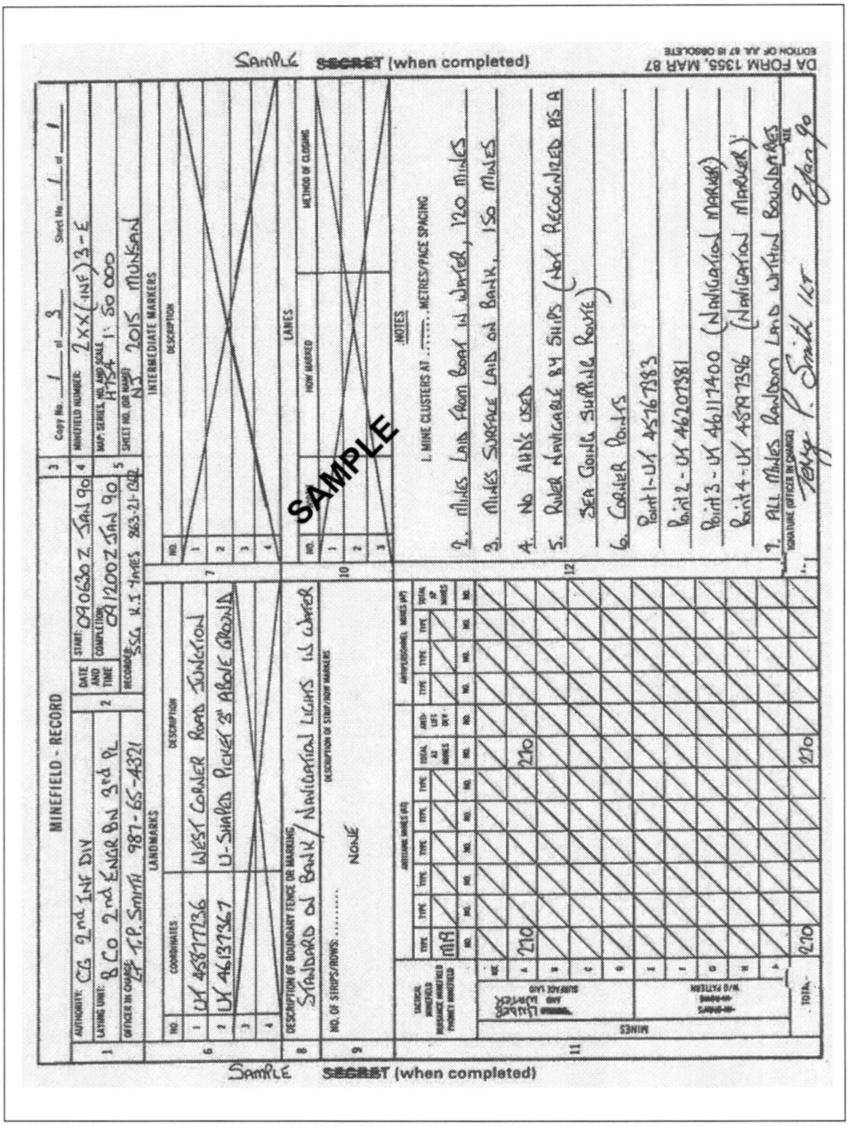


Figure 12-2a. Sample DA Form 1355 (front side) for river mining

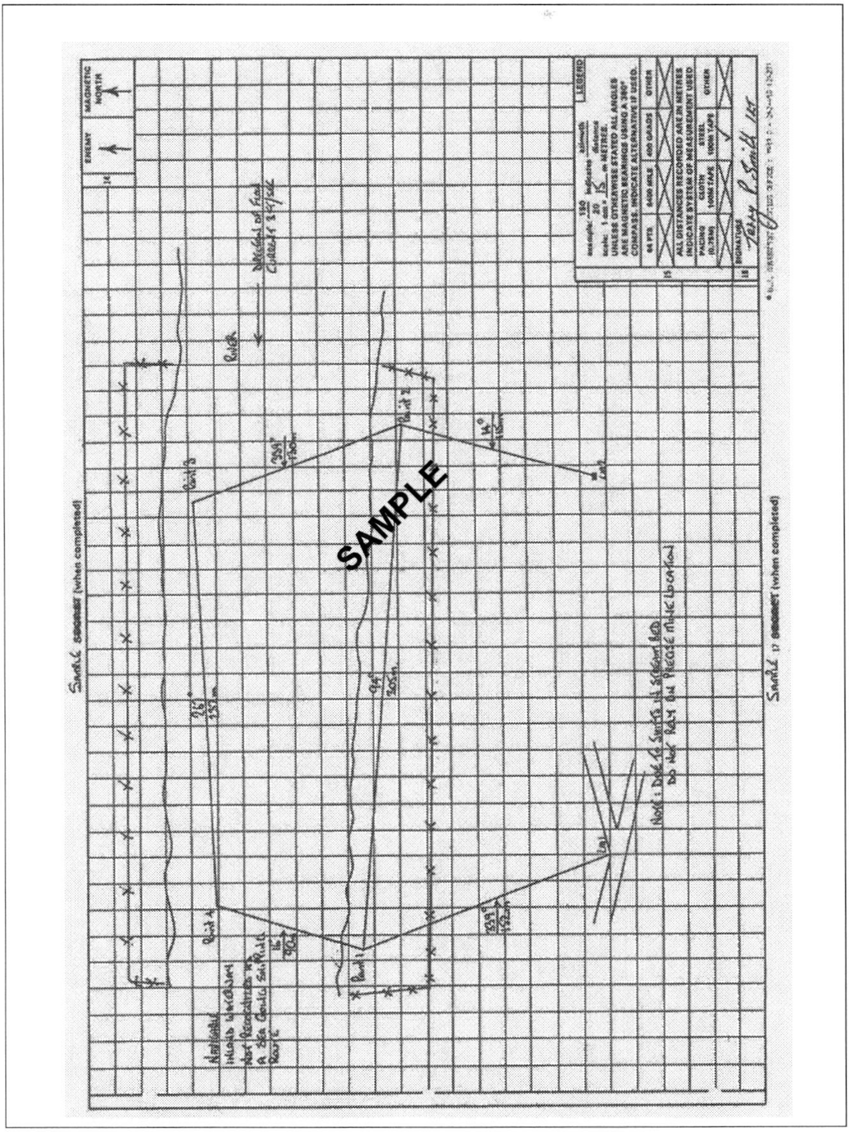


Figure 12-2b. Sample DA Form 1355 (inside) for river mining

- Emplace mines from upstream to downstream to prevent personnel and equipment from being swept into the mined area.
- Stay on the downstream side of the mine when arming the fuse.
- Place the mine as flat as possible on the bottom to prevent drifting. Use green saplings or other nonbuoyant material to construct outriggers.
- Do not arm the mine before it is laid.
- Carry the mine horizontally or edgewise to the current to reduce water resistance on the mine's pressure plate.

## URBAN-TERRAIN MINING

Characteristics of urban areas (such as a high proportion of hard-surfaced roads) prohibit a simple transition from open- to urban-area mine employment techniques and doctrine. The advantages of abundant cover and concealment, maneuver restrictions, and observation already possessed by the defender of an urban area can be significantly enhanced by the proper use of mines. Terrain modified through the process of urbanization provides a unique battle environment.

The major characteristics of urban terrain which are likely to impact on mine warfare include the following:

- Multistoried buildings add a vertical dimension to the battle. Basements and floors become part of the battle scene. The vulnerability of combat vehicles increases because attack from above or below is likely.
- Fighting is done at close range, often face-to-face, and seldom exceeds 50 meters. Some weapons, particularly large-caliber weapons, are unsuitable at a short range.
- Sewers, subways, and tunnels provide covered and concealed passageways for the movement of troops on both sides. Detailed knowledge of the location and the status of these tunnels is needed to successfully wage an urban battle.
- Streets and parking lots are modified to withstand continuous use by vehicles. Major routes and lots are paved. A high density and complex pattern of streets provide numerous avenues of advance. Burying mines is extremely difficult. Most mines are surface-laid and camouflaged with rubble and debris to avoid detection.
- Movement by vehicle is difficult. Streets are littered by rubble and cratered if the city has been bombed or subjected to artillery attack. Bridges and overpasses are likely to be destroyed or blocked. Traffic flow is highly channelized.
- Extensive map and chart data are needed by the commander. For example, the commander should know the locations of telephone, electric, gas, water, and sewer connections; substations; and generating and pumping stations.

## Antipersonnel Mines

US policy prohibits the use of non-self-destructing AP mines for all US forces except those on the Korean Peninsula. However, US forces can expect to encounter AP mines that are emplaced by other countries in support of MOBA. They are employed to block infantry approaches through or over underground passageways; open spaces; street, roof, and building obstacles; and dead spaces.

When AP mines are encountered **(Korea Only: or used)** in MOBA, mine locations are recorded on DA Form 1355 as shown in Figure 12-3)

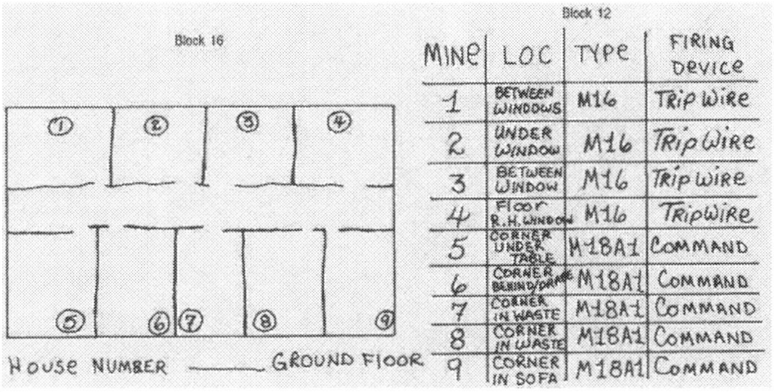


**Figure 12-3. Building sketch and mine plan (DA Form 1355)**

### Underground Passageways

Subways, sewers, cellars, and utility tunnels provide protected movement routes for troops. In large cities where underground systems are numerous and complex, limited manpower may dictate that forces employ obstacles to block key passageways with wire and AP mines. (See Figure 12-4, page 12-8.)

### Open Spaces

Open spaces include gaps between buildings, courtyards, residential yards, gardens, parks, and parking lots. In some cases, mines can be concealed in rubble or buried. However, the characteristics of most terrain surfaces, coupled with limited time and resources, dictate that mines be surface-laid. (See Figure 12-5, page 12-8.)

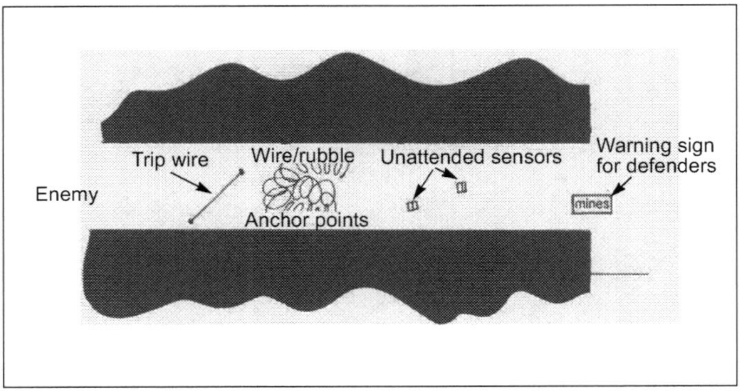


Figure 12-4. Underground passageway

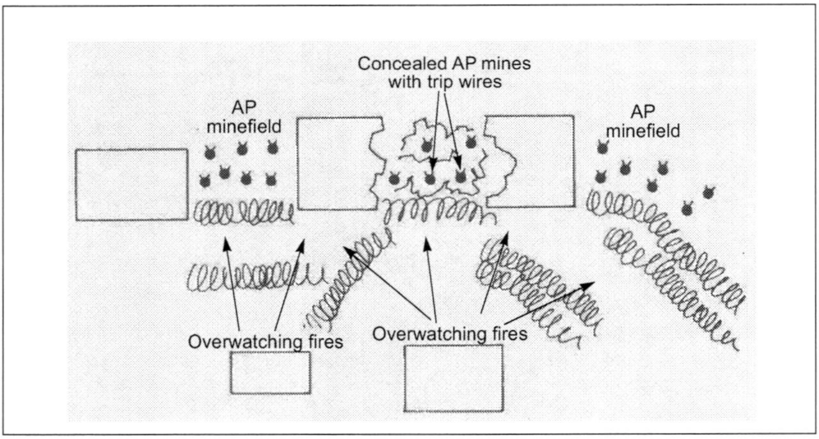


Figure 12-5. Open spaces

## Street Obstacles

Hand-emplaced AP mines can be emplaced on street surfaces, on railroad lines, and in areas along shallow waterways. (See Figure 12-6.)

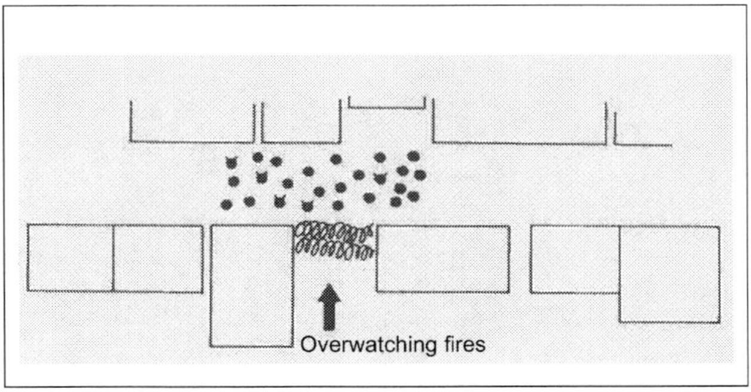


Figure 12-6. Street obstacles

## Roof Obstacles

Mines and booby traps supplement wire obstacles to deny operations that require air assault onto rooftops. They also prevent occupation on roofs that afford good observation points and fields of fire. (See Figure 12-7.)

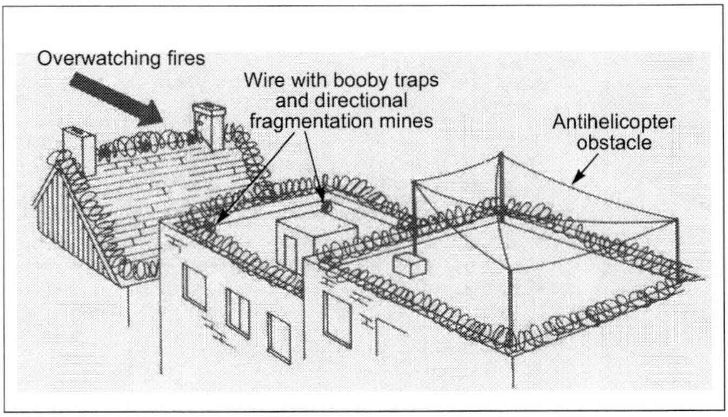


Figure 12-7. Roof obstacles

## Building Obstacles

Building obstacles include areas within and adjacent to buildings. Forces can lay mines in conjunction with wire obstacles to deny infantry access to covered routes and weapon positions (Figure 12-8).

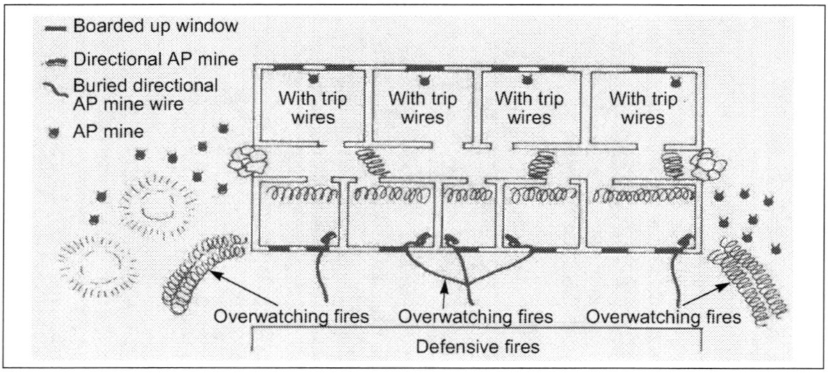


**Figure 12-8. Building obstacles**

## Dead Spaces

Obstacles and mines can be emplaced to restrict infantry movement in areas that cannot be observed and in areas that are protected from direct fire.

## Employment

The following AP mines are effective in urban terrain:

- **M14 (used by US forces in Korea only). Its small size makes it ideal for obscure places, such as stairs and cellars. It can be used in conjunction with metallic AP and AT mines to confuse and hinder breaching attempts. (See Figure 12-9.)**
- **M16 (used by US forces in Korea only). With trip-wire actuation, its lethal radius covers large areas such as rooftops, backyards, and cellars. An added advantage can be gained by attaching twine or wire to the release-pin ring to expediently rig the mine for command detonation. (See Figure 12-10.)**
- M18A1 (claymore). Numerous innovative applications of claymore munition deployment can be found for defensive warfare in urban areas (Figure 12-11, page 12-12). With remote firing, a series of claymore mines along a street establishes a highly effective ambush zone. Mines can also be employed on the sides of buildings, in abandoned vehicles, or in any other sturdy structure. Numerous opportunities exist for effectively sited, well-concealed mine employment above the terrain surface. Claymore munitions can be used to fill the dead space in the FPF of automatic weapons. They present a hazard when used in confined, built-up areas. Exercise caution when using them close to friendly forces because there is a danger of backblast.

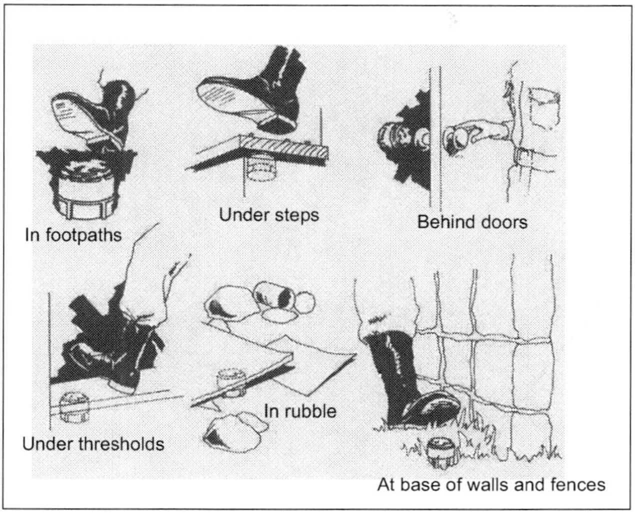


**Figure 12-9. Probable M14 AP mine emplacement**

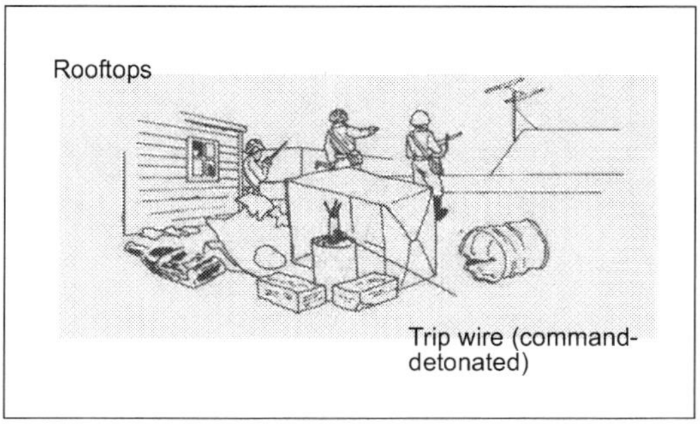


**Figure 12-10. Probable M16 AP mine emplacement**

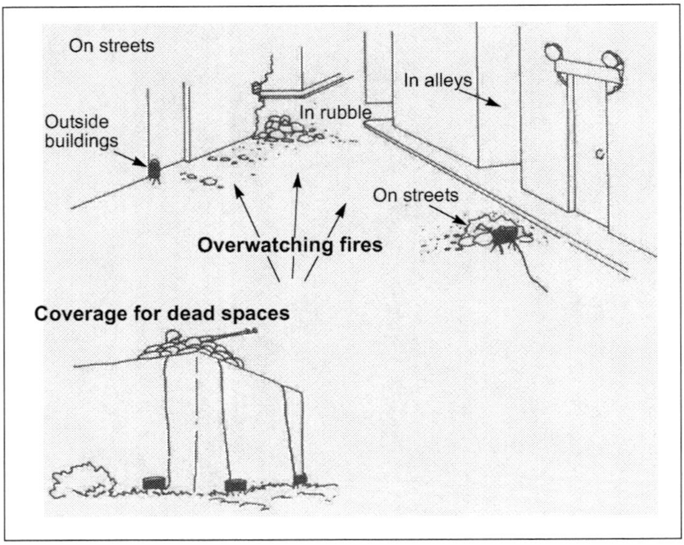


Figure 12-11. Probable M18A1 munition emplacement

### Conventional Antitank Mines

Enemy tanks, infantry fighting vehicles (IFVs), and direct-fire support weapons are restricted to streets, railroad lines, and, in some instances, waterways. (See Figure 12-12.) M15, M19, and M21 mines are used primarily in tactical and nuisance minefields; but they are occasionally used in protective minefields. They should be employed with other obstacles and covered by fire. Conventional AT mines emplaced in streets or alleys block routes of advance in narrow defiles. Concealment of large AT mines is accomplished by placing them in and around rubble and other obstacles. Extensive labor requirements generally prohibit burying mines in difficult terrain types.

In dispersed residential areas, obstacles are required to reduce the enemy's infantry mobility through and between houses and in open areas. They also prevent armored vehicles from moving between houses and along streets. AT minefield patterns should extend outward from the streets, incorporating open areas between buildings and streets to prevent easy bypass of street obstacles.

Significant labor and mine materials are required to deploy conventional mines between widely spaced buildings, in high-rise construction, and in industrial and transportation areas. Therefore, SCATMINEs should be seriously considered as viable alternatives. Some situations, such as the one shown in Figure 12-13, provide opportunities for the effective employment of mines in tactical and nuisance minefields.

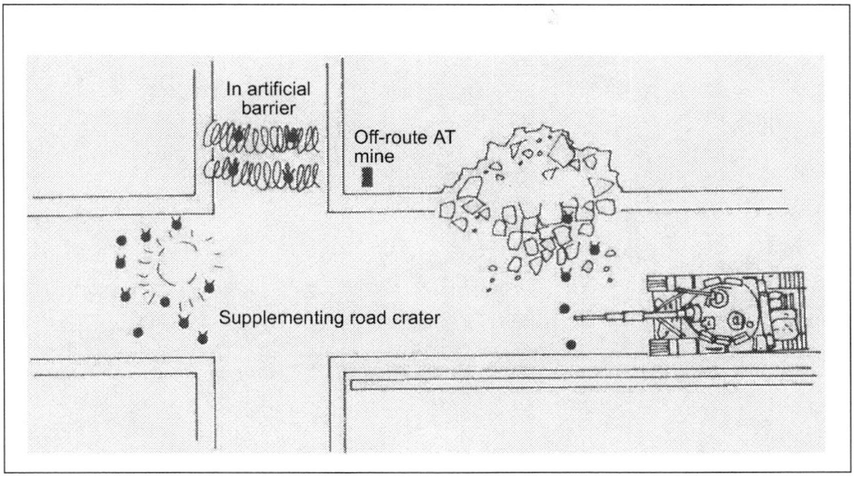


Figure 12-12. AT mine emplacement in urban areas

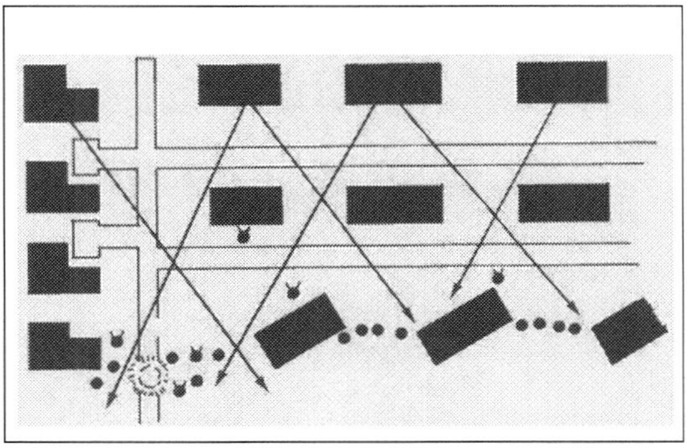

Figure 12-13. AT mine emplacement in industrial and transportation areas

SCATTERABLE MINES

### Area-Denial Artillery Munitions and Remote Antiarmor Mines

In addition to the advantages (such as reducing required resources and emplacement time) applicable to all SCATMINE systems, ADAMs and RAAMs have two specific advantages. They are the most rapidly deployed SCATMINE systems, and preplanning artillery-delivered minefields increases the rate at which nuisance minefields can be emplaced. Secondly, these mines can be delivered under enemy fire. Employment of ADAMs and RAAMs is most effective when the enemy's intentions are known and their forces are committed to an avenue of advance. (See Figure 12-14, page 12-14.)

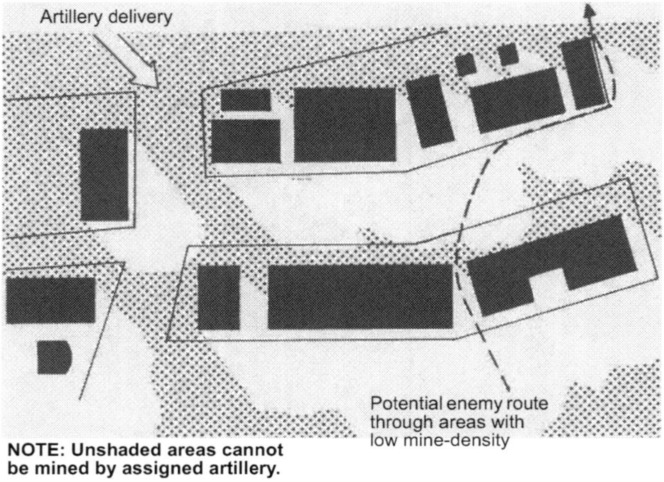


**Figure 12-14. ADAM/RAAM employment**

Using ADAMs/RAAMs in urban terrain involves five specific problem areas:

- Difficulty in precise minefield siting. Accurate siting is extremely critical due to the typically restrictive avenues of advance and may be futile due to the difficulty in adjusting artillery rounds in an environment that obscures observation. Further, buildings tend to create unmined shadow zones.
- Uncertainty of ADAM/RAAM survivability upon impact with a building or ground surfaces that are characteristic in urban areas.
- Likely availability of artillery firing units. ADAM/RAAM emplacement may not be a priority of the maneuver commander, because his supporting units may not have enough ADAM/RAAM munitions on hand. Assuming the availability of artillery assets for an ADAM/RAAM mission could prove disastrous for defending forces.
- High detectability of these mines on bare and lightly covered surfaces. This permits the enemy to seek out unmined passageways or pick through lightly seeded areas. If you use the doctrinal guidelines for emplacing artillery-delivered mines on top of the advancing enemy or immediately in front of them, the desired obstacle intent (disrupt, turn, fix, block) and enhanced fires are achieved.
- Difficulty in achieving a good, random pattern. Hard-surfaced areas cause mines to bounce and roll. Some mines (especially AT mines) will land on top of buildings and are ineffective.

### Air Volcano

The primary advantage of the air Volcano system is its capability to site and emplace minefields accurately. This depends on the helicopter's maneuverability over the selected minefield terrain and the proper coordination between ground forces and aviation support. Disadvantages include vulnerability and the high replacement cost of the helicopter. However, in view of the system's operational concept, employment in urban terrain (which provides little exposure of the helicopter) actually increases the practicality of employing this system in urban areas. Mine survival rate on impact with a hard surface is another potential problem.

### Ground Volcano

Three aspects of the ground Volcano distinguish it from other SCATMINE systems:

- The dispenser is organic to supporting combat engineers, making it readily available to support the maneuver commander's defensive plan.
- Delivery siting is accurately pinpointed to the ground.
- Better opportunities exist to record the presence of a minefield. In contrast to artillery-delivered and air Volcano systems, the ground Volcano is delivered by engineers who are normally located with and report directly to the maneuver commander.

Some primary factors may degrade ground Volcano deployment in urban terrain. The requirement to emplace minefields before an actual attack in order to reduce system vulnerability is the most significant factor. This makes the minefield detectable and provides more reaction time for the enemy to alter their scheme of maneuver. The delivery of mines depends on terrain trafficability. The prime mover and the launch vehicle must negotiate the terrain over which mines are to be dispensed.

### Modular Pack Mine System

The MOPMS is ideally suited for employment in urban terrain (Figure 12-15, page 12-16). The module can be hidden from enemy view, and the mines can be dispensed after attackers are committed to a route of advance. Additionally, mines can be emplaced rapidly under enemy fire. In contrast to other SCATMINE systems, the commander controls when and where mines are dispensed and how they are detonated, regardless of the enemy situation.

### Gator

When considered for employment in urban terrain, Gators encompass the same problems as artillery-delivered and air Volcano mine systems.

## DECEPTION MEASURES

Phony minefields can be established rapidly with negligible effort and cost. They have the distinct advantage of blocking the enemy but not friendly forces. Although it is difficult to fake a surface-laid minefield, expedients such as soup pans, seat cushions, and cardboard boxes have historically proven effective in delaying and channelizing attacking forces. These objects, as well

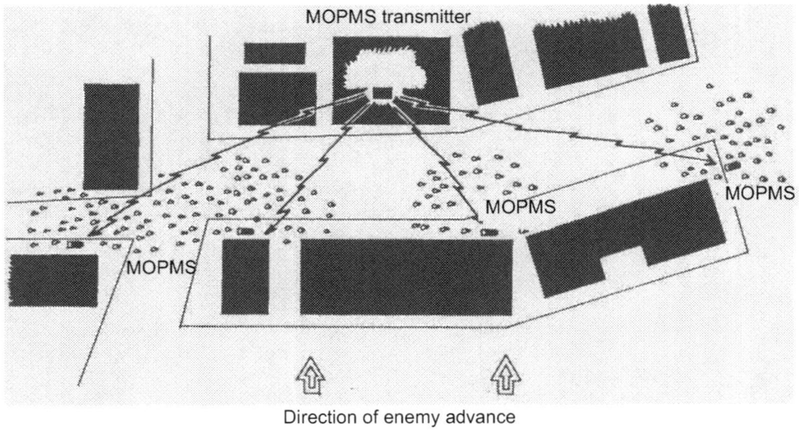


**Figure 12-15. MOPMS employment**

as other ones readily available in urban areas, can be used as phony minefields or used to cover real mines. A more realistic phony minefield could be created with inert or training mines.

Inadequate minefield camouflage in urban terrain is viewed as a critical constraint in deploying conventional mines and SCATMINEs. Smoke can be deployed from various dispensers, but it must be dense and accurately employed and released.

## SPECIAL ENVIRONMENTS

### COLD REGIONS

Mine employment in cold regions poses special problems—the principal one being emplacement. Mine burial is extremely difficult in frozen ground. The freezing water in soil causes it to have high strength and penetration resistance, so digging times are greatly increased if not impractical. However, there are several means to overcome this problem. In some cases, the minefield can be laid out before the soil freezes. To do this, dig holes for each individual mine and insert a plug into the hole to protect its shape and prevent it from being filled in. A wide variety of material can be used for plugs. Ideally, the plug should be economical, easy to remove, and rigid enough to maintain the depth and shape of the hole. Sandbags, plastic bags filled with sand or sawdust, or logs make excellent plugs. If the minefield cannot be prechambered, mechanical means can be used to dig holes. When available, civilian construction equipment (particularly large earth augers) can be used to drill holes for mine emplacement.

To assure detonation of buried, pressure-actuated mines, they should be placed in a hole that is shallow enough for the pressure plate to be above ground. Covering spoil should be a maximum of 1 centimeter deep.

When burial is impossible, mines are placed on the surface. Heavy snow cover may reduce the effectiveness of both buried and surface-laid mines by causing them to be bridged. Mines laid in deep snow should be placed as close as possible to the surface and supported by boards or compacted snow. Waterproof mines before emplacement in cold regions. Mines can also be placed in plastic bags before burial. In some cases, a layer of ice may form over the top of the pressure plate. Although the load required to break the ice is slightly higher than that required to activate the fuse, thin layers of external ice will have little effect on mine functioning. When possible, tilt-rod mines should be used in cold regions because they are less susceptible to ice and snow. Magnetic mines are not significantly affected by snow, although cold weather decreases battery life.

Camouflaging a minefield in a cold region is difficult. Mines should be painted white when snow is expected to remain on the ground for extended periods of time. Minefield signature tracks should be swept away, or deliberate tracks should be made to give the impression of a safe area.

**Korea Only: When trip-wire mines are employed in snow, the wire should be about 10 meters long, with a slight amount of slack left in the wire. The trip wires should be supported approximately 46 centimeters above the ground to avoid degradation by snowfall.**

## JUNGLES

Fuses and explosive components deteriorate very rapidly in jungle climates. As a result, mines and mine material require more frequent and extensive maintenance and inspection. Waterproof mines employed in humid climates. The rapid growth of vegetation hinders maintenance recovery and removal. Dense vegetation may cause mines to become inoperable or windblown foliage can detonate them. FM 90-5 provides detailed information on jungle operations.

## DESERTS

In desert climates, fuses and explosive components deteriorate slowly. The terrain and the situation determine how mines will be emplaced. Mine boards will normally be required to provide support in soft, shifting sand. Mines emplaced in the desert have a tendency to shift position, and the spacing between mines and rows should be increased to prevent sympathetic detonation. Blowing sand may expose buried mines or cover surface-laid mines. Sand may also cause mines to malfunction. It is important to realize the difficulty of accurately recording minefield locations in vast, open, desert areas void of recognizable terrain features. More mines are required for desert operations. Typically, desert minefields are much larger and have a lower density than those used in Europe or Korea. FM 90-3 provides detailed information on desert operations.

Chapter 13

# Booby Traps and Expedient Devices

During war and OOTW, booby traps can be found anywhere at anytime. They can kill or incapacitate their unsuspecting victims. This chapter provides information on booby-trap employment concepts, detection techniques, marking and recording procedures, and removal guidelines.

This chapter also provides an overview of expedient devices and their employment considerations.

## SECTION I. SETTING BOOBY TRAPS

US policy restricts the use of booby traps by US personnel. This does not preclude their use by other countries, so US forces may encounter them during operations.

The use of booby traps is limited only by the imagination of the force employing them. They—

- Are usually explosive in nature.
- Are actuated when an unsuspecting person disturbs an apparently harmless object or performs a presumably safe act.
- Are designed to kill or incapacitate.
- Cause unexpected, random casualties and damage.
- Create an attitude of uncertainty and suspicion in the enemy's mind, thereby, lowering his morale and inducing a degree of caution that restricts or slows his movement.

Many booby traps are constructed using military equipment and ammunition. Improvised traps are used during counterinsurgency missions in low-intensity conflicts.

The corps commander is the employment authority for booby traps. He can delegate this authority to the division commander. If authority is given to set booby traps, US personnel will adhere to the rules for international law applicable to armed conflict. There are several uses of booby traps that are prohibited. Remember, these restrictions are not observed by all countries; US personnel must still be cautious when approaching objects in areas where booby traps are supposedly prohibited.

International law prohibits the use of booby traps as follows:

- Booby traps and other devices are prohibited if they are attached to or associated with—

- Internationally recognized protective emblems, signs, or signals.
- Sick, wounded, or dead personnel.
- Burial or cremation sites or graves.
- Medical facilities, equipment, or supplies.
- Children's toys or other portable objects or products that are designed for their feeding, health, hygiene, clothing, or education.
- Food or drink.
- Kitchen utensils or appliances except in military establishments, military locations, or supply depots.
- Objects that are clearly religious in nature.
- Historic monuments, works of art, or places of worship.
- Animals or their carcasses.

- Booby traps are prohibited in cities, villages, and other areas that contain civilians if combat between ground forces is not taking place or does not appear to be imminent, unless—
  - Booby traps are placed on or in the close vicinity of a military objective.
  - Measures (guards, warning, or fences) are taken to protect civilians from booby-trap effects.

## TACTICS

Booby traps are psychological weapons. They make the enemy cautious and slow it down. These actions, in turn, cause enemy casualties. Do not waste time attempting to set elaborate traps that are undetectable or impossible to disarm. Also, do not waste time developing difficult sites, because simple traps usually have the same chance of catching the enemy. Even if booby traps are detected and cleared, their aim is achieved.

The principles governing the use of booby traps and nuisance mines are identical, so consider using them in conjunction with one another. They have characteristics that make them suitable for use in different situations:

- Nuisance mines are quicker to lay and safer to use than booby traps, and they are normally used in outside locations where they can be buried.
- Booby traps are normally used in urban areas, structures, and places where mines are easily detected.

Booby traps and nuisance mines are particularly suited for defensive operations. They are used to—

- Slow the enemy's advance.
- Deny the enemy use of facilities and material.
- Warn of enemy approach.

- Deter the enemy from using ground not covered by direct fire.
- Plan defensive operations.

In offensive operations, booby traps and nuisance mines are employed on an opportunity basis during raids and patrols. Formal instruction is not usually issued by the staff.

Exercise caution when using bobby traps in offensive operations because they may hinder the operation. In advance and pursuit operations, they are primarily used by patrols and raiding parties. They slow down enemy follow-up actions and hinder the enemy's repair and maintenance teams after raids.

The following considerations pertain to defensive operations but may be relevant to offensive operations and must be considered when briefing troops:

- Booby trapping is rarely given a high priority and is usually peripheral to other engineer tasks.
- Nuisance mines are more cost-effective than booby traps, unless booby traps are used in situations that allow their full potential to be exploited. If it is easier, use nuisance mines instead of booby traps.

To maximize the effect of booby traps and nuisance mines, the staff provides engineer commanders with the following information:

- Purpose. Booby traps are time-consuming and dangerous to set. Do not waste time and effort setting traps that are unlikely to be actuated or that are not specifically designed to achieve the required aim. For example, if booby traps are being used against troops, small, simple traps designed to incapacitate will achieve this result just as well as complicated ones with large charges. If the aim is to destroy vehicles, use mines.
- Location. The precise location for booby traps and nuisance mines can only be determined by the setting unit. Areas must be delineated and recorded so that there is no threat to friendly forces in the event of reoccupation.
- Time setting starts and time available for setting. The time setting starts affects other engineer tasks, and the length of time available for setting governs the number of men required.
- Number of safe routes required. Safe routes are important during general withdrawals where authority has been given to booby-trap positions as they are evacuated. They also provide safe areas for the covering force to launch counterattacks.
- Likelihood of reoccupation. Even if the enemy has not detonated booby traps, they might have interfered with them. Therefore, do not set booby traps when areas are to be vacated to meet short-term tactical requirements or when reoccupation is expected soon.

Intelligence personnel provide information to assist the setting unit in maximizing the effect of booby traps. The nature and the type of traps required depend on the enemy unit. For example, while paying particular attention to dead space and defilade positions, use mines or widely dispersed traps (with large charges) against a mechanized enemy. Conversely, use small traps and AP mines (in places that afford cover) against an infantry enemy.

## SITING

If the first obstacle or installation the enemy strikes is booby-trapped or nuisance-mined, he is delayed while he clears it. The enemy is further delayed by an increased degree of caution. His troops know that additional traps and mines can be encountered. Booby traps and nuisance mines are generally located—

- In and around buildings, installations, and field defenses.
- In and around road craters or any obstacle that must be cleared.
- In natural, covered resting places along routes.
- In likely assembly areas.
- In the vicinity of stocks of fuel, supplies, or materials.
- At focal points and bottlenecks in the road or rail systems (particularly the ones that cannot be bypassed).

The setting-party commander is responsible for the detailed siting and design of booby traps. Consider all the information about the enemy soldier and his operating procedures when selecting places and objects to trap. Also, consider the traps from the enemy's point of view and assess the courses open to the enemy when he encounters them. This can expose weaknesses in your initial plan and bring about changes to the proposed layout, or it can result in a different location being selected. In addition, determine the effort required by the enemy to bypass the traps. This shows whether the imposed delay justifies the effort required to set the booby traps in the selected location.

## TYPES OF TRAPS

Booby traps are designed to—

- Be actuated by persons carrying out their normal duties.
- Take advantage of human nature.

The following booby traps can often be detected because they are designed to make the person do something:

- Bait. Usually consists of objects that arouse someone's interest, such as attractive or interesting items that have apparently been left behind or discarded during a rapid evacuation.
- Decoy. The most common decoy consists of two traps—one designed to be detected, the other designed to actuate when personnel deal with the first one. The first trap can be a dummy. A classic form of a decoy is to place booby traps or nuisance mines in locations from which the decoy mine can be removed.
- Bluff. A bluff is a hoax and usually consists of a dummy trap.

- Double bluff. A double bluff only appears to be a bluff. Personnel believe the trap is safe or can be disarmed. For example, a number of traps can be set that are disarmed when the detonating cord is removed from the charge. The double bluff is achieved by setting another trap that appears to be the same, but it actually explodes when the detonating cord is removed from the charge. Double bluffs rely on a reduced awareness and alertness caused by repetition.

## COMPONENTS AND PRINCIPLES

There are two initiation methods for explosive booby traps—electric and nonelectric. Both methods can be constructed using many different types of FDs. FDs can be secured to the charge (direct connection) or located away from it (remote connection). They are actuated by one or more methods. It is impossible to describe every type of trap that may be encountered; however, most are constructed and operated by using components similar to those listed below:

- FD.
- Power source (battery, for example).
- Connection (usually detonating cord or electric wires).
- Blasting cap.
- Main charge.

Figure 13-1, page 13-6, shows how typical electric and nonelectric traps can be made.

## ACTUATION METHODS

Many sophisticated booby-trap devices are now being manufactured that operate on vibration, sound, temperature change, and other methods. Current intelligence on the booby trap being used in the AO should be gathered so that countermeasures can be developed and practiced. Most FDs found in the combat zone are simple mechanisms designed to be actuated by pull, pressure, pressure release, or tension release (Figure 13-2, page 13-7).

## METHODS OF CONNECTION

Procedures can be varied when it is safe to do so. For example, instead of connecting the FD to a charge already in position, preconnect trap components and then position the trap.

Small charges (up to 1 kilogram) are sufficient for AP traps, but larger quantities can be used to increase their effect. Shrapnel can be produced by packing stones, scrap metal, nails, or other material around the charge. AT traps require large charges (up to 6.75 kilograms for wheeled vehicles and 11.25 kilograms or more for tracked vehicles).

### REMOTE

Follow the procedures listed below when assembling a remotely connected trap using an M142 FD (similar to the illustration in Figure 13-3, page 13-7):

- Design the trap and collect necessary materials.

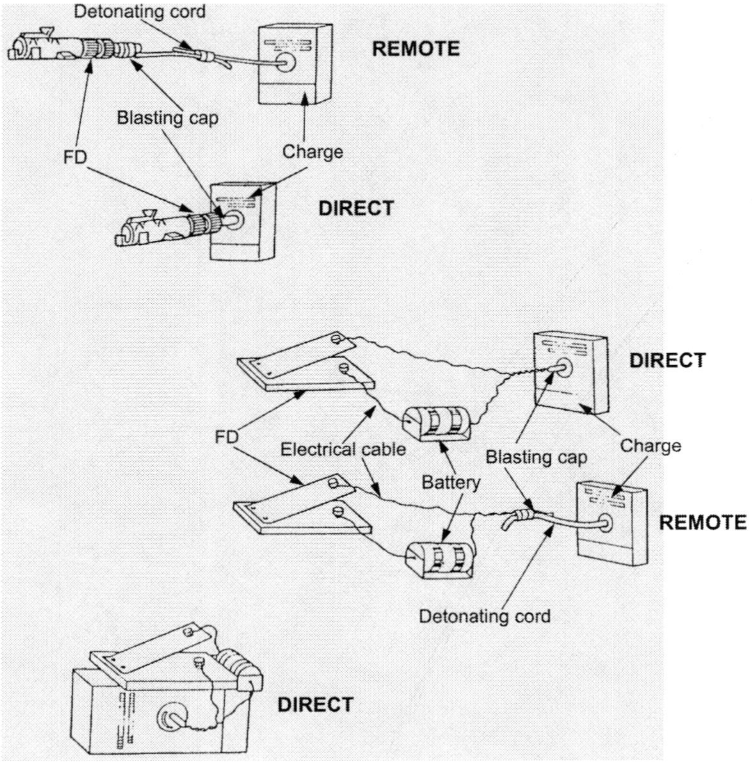


**Figure 13-1. Typical electric and nonelectric booby traps**

- Test the M142 FD.
- Lay the detonating cord from the charge location to the FD location.
- Position the charge.
- Connect the detonating cord to the charge.
- Prepare the coupler.
- Tape a length (46 centimeters, minimum) of detonating cord to the coupler's blasting end.
- Prepare and position the M142, set it to operate in the desired manner, and remove the round- or square-headed pin.

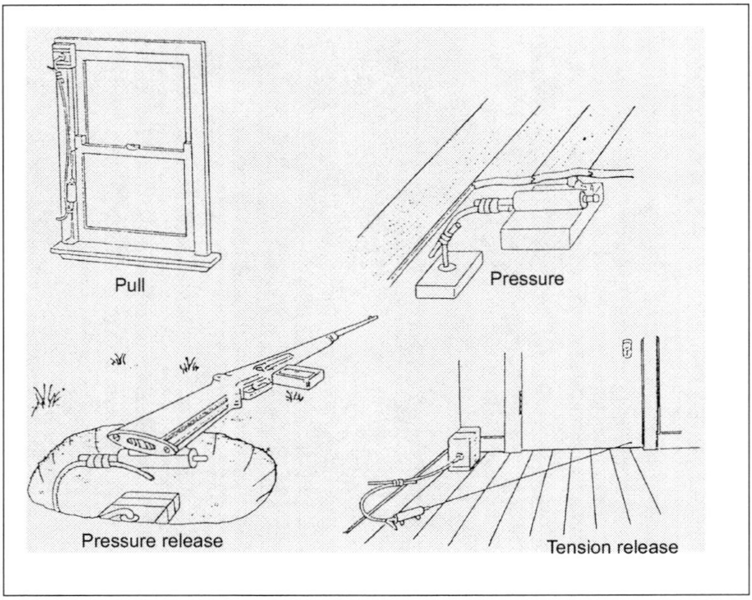


Figure 13-2. Methods of actuation

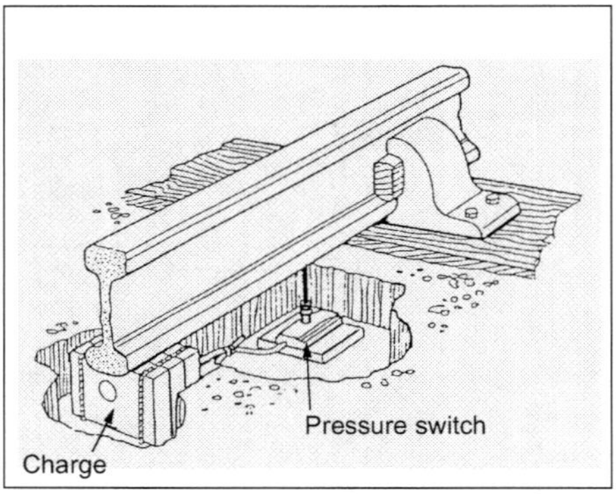


Figure 13-3. Remotely connected traps

- Ensure that the two detonating cords are not touching but can be conveniently connected when necessary.
- Remove any materials and other signs of laying, and fully camouflage the area.

- Arm the FD by removing the positive safety pin.
- Arm the trap by connecting the two detonating cords.

### DIRECT

Follow the procedures listed below when assembling a nonelectric, directly connected trap using an M142 FD (similar to the illustration in Figure 13-1, page 13-6):

- Design the trap and collect necessary materials.
- Test the M142 FD.
- Prepare and position the charge for coupling to the FD, and prepare the explosive to receive a blasting cap.
- Prepare the coupler and insert it into the charge.
- Prepare the M142, set it to operate in the desired manner, and remove the round- or square-headed pin.
- Remove any materials and other signs of laying, and fully camouflage the area.
- Arm the trap by removing the positive safety pin.

**DANGER**
**Do not attempt to further camouflage the area after the trap is armed.**

## PLANNING, SETTING, AND RECORDING

### TIMELINESS

The setting-party commander must be given all the information and sufficient time to carry out a reconnaissance. This allows his plan and the soldier's briefing to be as complete as possible.

### ORDERS AND BRIEFING

Once the commander has finished the reconnaissance, he makes a detailed plan, prepares the order, and briefs soldiers. The following points are covered:

- Enemy intelligence.
- Number and types of traps to be set.
- Location and design of traps (precise or general).
- Tasking and allocation of areas.
- Recording, marking, and arming procedures.
- Control measures.
- Time requirements.
- Material and equipment availability.
- Rehearsal and equipment preparation.

### REHEARSAL

Whenever possible, rehearse laying procedures. Thoroughly exercise control and safety measures so that any flaws in the system will be discovered. Lay practice traps and carry out arming and recording procedures. If you cannot rehearse at the trap location, hold the rehearsal in a similar area.

### ORGANIZATION AND PROCEDURE

Setting booby traps is a dangerous task and it must be carefully controlled. Exact drills cannot be prescribed, but the following procedures should be followed as closely as possible:

- Control point. The commander must establish a control point upon arrival at the trap area. The control point forms a headquarters and a material holding area. Also, safe routes start at the control point. If the rehearsal is conducted at another area, assemble all the troops at the control point before setting starts. At this time, relate the ground to the plan, confirm control measures, and check FDs and equipment for serviceability and adequacy.
- Control of parties. Each setting party consists of one or two men. Clearly define the area where each setting party will work (use tape, if necessary). Position traps so that accidental detonation will not cause friendly casualties. Strictly control the entry to areas where traps are being set. In most situations, only the commander, setting parties, and a recorder enter areas. Troops who are not assigned tasks should remain at the control point unless otherwise authorized by the commander.
- Marking. As with nuisance minefields, mark booby-trapped areas until they are evacuated. When possible, use standard booby-trap signs (Figure 13-4). When signs are not available, use temporary markings (any barrier or fence) to clearly indicate a booby-trapped area. Remove the temporary marking after the area is evacuated and before the enemy reconnoiters the area.

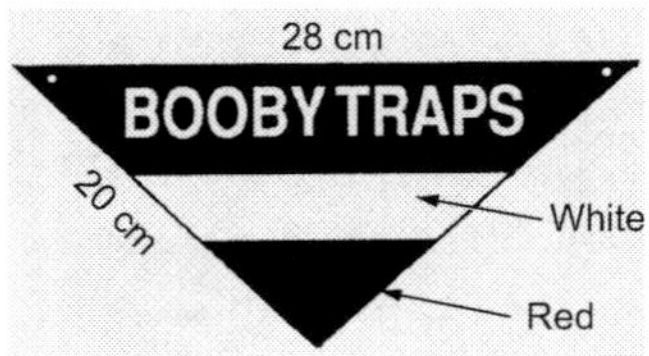


**Figure 13-4. Standard booby-trap sign**

- **Setting.** Setting parties transport required explosives and material to the site. They do not arm FDs or traps until ordered to do so. If the setting party consists of one man, he remains with the trap until it is armed or until the commander directs him to leave. In a two-man party, one man returns to the control point and reports the completion of the task, and the other man remains with the trap. A setting party enters and leaves its area by the same route. If the route is not clearly apparent, personnel lay tapes to guide the commander and the recorder to the area.

- Inspection and arming. The setting-party commander inspects traps for safety and camouflage before giving the order to arm them. He devises an arming plan that enables the setting parties to withdraw to the control point in a safe and logical manner. For example, in open areas, start arming at the furthest mine from the control point; in buildings, arm from the top down.

## REPORTING AND RECORDING

Booby traps do not distinguish between friend and foe. Correct reporting and recording procedures must be observed to avoid friendly casualties. The procedures for recording booby traps are the same as for nuisance mines (see Chapter 8). When traps are set in areas occupied by the enemy, they are not recorded. In this case, it is sufficient to note their location and type in after-action reports.

### Tactical Reports

Information about booby-trapped areas must be provided in stages from the time they are established until they are cleared. This procedure allows up-to-date information to be annotated on operational maps at all concerned units and headquarters (see Table 13-1).

### Booby-Trap Records

The setting-party commander compiles records for all booby-trapped areas on DA Form 1355 (Figures 13-5a and 13-5b, pages 13-12 and 13-13). Strictly observe the instructions printed on the back of the form. Booby-trap records provide detailed information about the composition of an area, and they allow the area to be cleared quickly, with minimum casualty risk. They contain complete information on the number and types of traps set, as well as accurate and precise details on the location and design of individual traps. If a single record is not large enough to record all the traps set in a definable area, clearly note this fact and reference serial numbers of adjacent records. Submit the completed DA Form 1355, in triplicate, to the next higher headquarters.

**Table 13-1. Tactical reports**

| Report | By Whom and When Initiated | To Whom Forwarded | Details to be Included |
|---|---|---|---|
| Intention to set (required) | Division commander, on deciding to set booby traps | • Corps HQ<br>• Controlling HQ | • Purpose<br>• General area to be booby-trapped (by grid coordinates)<br>• Estimated number and types of traps<br>• Estimated starting date and time<br>• Estimated completion date and time |
| Siting (optional) | • Setting unit, on completion of reconnaissance<br>• Division ordering setting, only if materially different from intention-to-set report | • Division ordering setting<br>• Next higher HQ | • Boundaries of booby-trapped area (by grid coordinates)<br>• Number and approximate positions of safe routes, if any<br>• Estimated number and types of traps |
| Start of setting (required) | • Setting unit, when setting starts<br>• Division ordering setting | • Division HQ ordering setting<br>• Next higher HQ | • Time setting started<br>• Estimated time of completion<br>• Number and types of traps being set<br>• Any change to the detail given in the siting report, and whether or not it was a result of orders |
| Completion (required) | • Setting unit, when setting is completed<br>• Division HQ ordering setting | • Division HQ ordering setting<br>• Next higher HQ | • A trace on the largest scale map available, showing the boundaries of the trapped area<br>• Full details of the area and buildings trapped (with sketch maps)<br>• Full details of traps set, including design and location<br>• Full details of all safe routes<br>• Total number and types of traps<br>• Time and date of completion |
| Changes (required) | Division HQ ordering change | • Next higher HQ<br>• Any other unit holding information on existing traps | Full details of change |

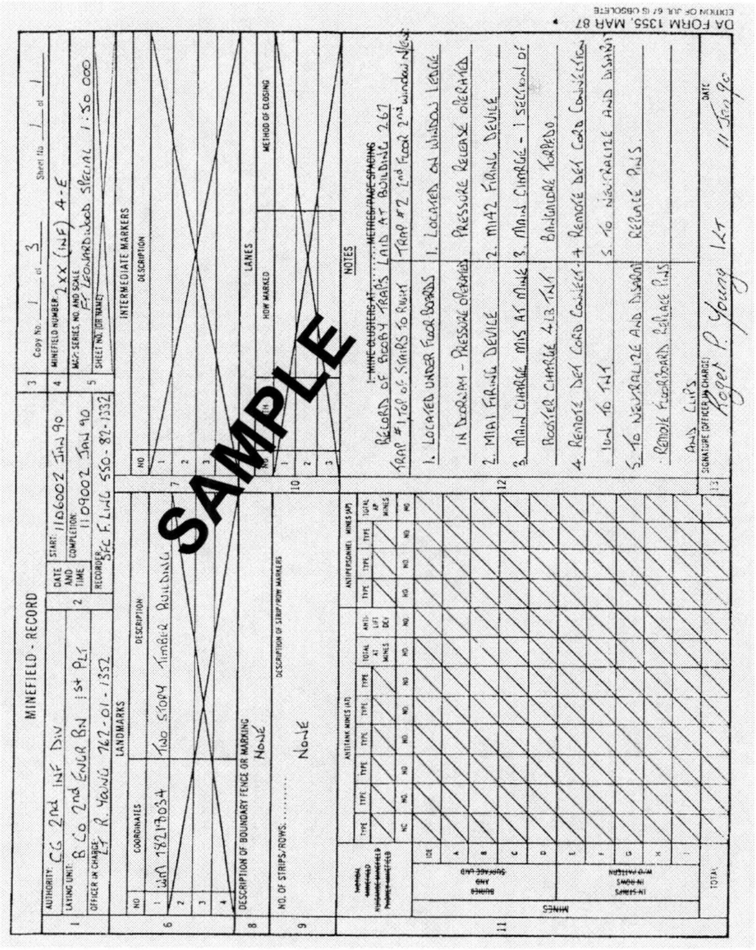

MINEFIELD - RECORD

1 AUTHORITY: CG 2nd INF DIV
LAYING UNIT: B Co 2nd ENGR BN 1st PLT
OFFICER IN CHARGE: LT R. YOUNG 762-01-1352

2 DATE AND TIME START: 1106002 JAN 90
COMPLETION: 1109002 JAN 90
RECORDER: SFC F. LING 550-82-1332

3 Copy No 1 of 3 Sheet No 1 of 1

4 MINEFIELD NUMBER: 2XX (INF) 4-E

5 MAP SERIES, NO. AND SCALE: FT LEONARDWOOD SPECIAL 1:50 000
SHEET NO. (OR NAME)

6 LANDMARKS

| NO | COORDINATES | DESCRIPTION |
|---|---|---|
| 1 | WN 18217034 | TWO STORY TIMBER BUILDING |
| 2 | | |
| 3 | | |
| 4 | | |

7 INTERMEDIATE MARKERS — NO / DESCRIPTION

8 DESCRIPTION OF BOUNDARY FENCE OR MARKING: NONE

9 NO. OF STRIPS/ROWS ........ DESCRIPTION OF STRIP/ROW MARKERS: NONE

10 LANES — NO / WIDTH / HOW MARKED / METHOD OF CLOSING

11 MINES — ANTITANK MINES (AT): TYPE, TOTAL AT MINES, ANTI-LIFT DEV; ANTIPERSONNEL MINES (AP): TYPE, TOTAL AP MINES
~~NORMAL MINEFIELD~~ / ~~NUISANCE MINEFIELD~~ / ~~PHONEY MINEFIELD~~; ~~BURIED AND SURFACE LAID~~; ~~IN STRIPS / IN ROWS / W/O PATTERN~~; STRIP A–K; TOTAL

12 NOTES

~~MINE CLUSTERS AT ........ METRES/PACE SPACING~~

RECORD OF BOOBY TRAPS LAID AT BUILDING 267

TRAP #1 TOP OF STAIRS TO RIGHT
1. LOCATED UNDER FLOOR BOARDS IN DOORWAY - PRESSURE OPERATED
2. M1A1 FIRING DEVICE
3. MAIN CHARGE M15 AT MINE BOOSTER CHARGE 4 LB TNT
4. REMOVE DET CORD CONNECTION TO TNT
5. TO NEUTRALIZE AND DISARM REMOVE FLOORBOARDS REPLACE PINS AND CLIPS

TRAP #2 2nd FLOOR 2nd WINDOW NEXT
1. LOCATED ON WINDOW LEDGE PRESSURE RELEASE OPERATED
2. M142 FIRING DEVICE
3. MAIN CHARGE - 1 SECTION OF BANGALORE TORPEDO
4. REMOVE DET CORD CONNECTION
5. TO NEUTRALIZE AND DISARM REPLACE PINS.

13 SIGNATURE (OFFICER IN CHARGE): Roger P. Young 1Lt DATE: 11 Jan 90

SAMPLE

DA FORM 1355, MAR 87 EDITION OF JUL 67 IS OBSOLETE

Figure 13-5a. Sample DA Form 1355 (front side) for a booby-trapped area

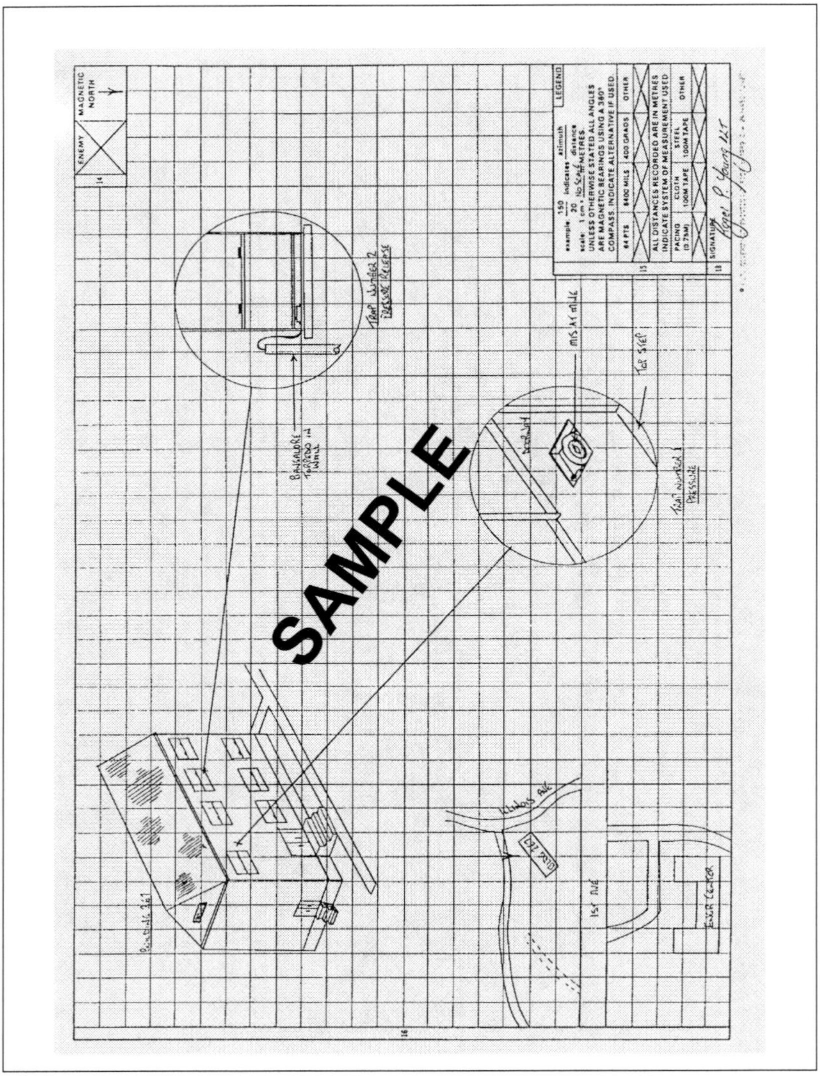


Figure 13-5b. Sample DA Form 1355 (inside) for a booby-trapped area

## SECTION II. CLEARING BOOBY TRAPS

Soldiers must be aware of the threat that is presented by booby traps. They must also receive sufficient training to recognize locations and items that lend themselves to booby-trapping. Avoid overemphasis and strike a balance between what is possible and what is probable. For example, in medium-level operations, overemphasis of the booby-trap threat can be counterproductive by slowing momentum. This causes casualties that might otherwise have been avoided. In low-intensity conflicts where the use of booby traps and improvised explosive devices are probably widespread, training must be given high priority and emphasized at all levels.

### SITES

Although many booby-trapped sites are similar, the items selected, the reason for their use, and the scale of the threat are quite different. In medium-level conflicts, booby traps are mainly used in recently contested areas. The items selected and the reasons for using traps are the same as those taught to our troops. Therefore, by anticipating the presence of traps, it is possible to isolate and bypass trapped areas. If this is not practicable, you can plan countermeasures, such as avoiding convenient and covered resting places along routes where mines can be located. At bridge or ferry sites that cannot be avoided, you must ensure that they are free of traps.

In low-intensity conflicts where booby traps are used to cause casualties, delays, or disruptions, no items or areas can be considered safe. Quality collective training in booby-trap awareness is necessary for all units. Rapidly disseminate booby-trap incident reports to all levels. This allows personnel to develop an understanding of the enemy's method of operation and a feel for what might or might not be targets.

### SAFETY

The following rules and safety procedures can save your life and the lives of others—learn and remember them:

- Suspect any object that appears to be out of place or artificial in its surroundings. Remember, what you see may well be what the enemy wants you to see. If you did not put it there, do not pick it up.
- Examine mines and booby traps from all angles, and check for alternative means of firing before approaching them.
- Ensure that only one man works on a booby trap.
- Do not use force. Stop if force becomes necessary.
- Do not touch a trip wire until both ends have been investigated and all devices are disarmed or neutralized.
- Trace trip wires and check for additional traps located along and beneath them.
- Treat all parts of a trap with suspicion, because each part may be set to actuate the trap.

- Wait at least 30 seconds after pulling a booby trap or a mine. There might be a delay fuse.
- Mark all traps until they are cleared.
- Expect constant change in enemy techniques.
- Never attempt to clear booby traps by hand if pulling them or destroying them in place is possible and acceptable.

## INDICATIONS

Successful detection depends on two things—being aware of what might be trapped and why, and being able to recognize the evidence of setting. The first requirement demands a well-developed sense of intuition; the second, a keen eye. Intuition, like mine sense, is gained through experience and an understanding of the enemy's techniques and habits. A keen eye is the result of training and practice in the recognition of things that might indicate the presence of a trap. The presence of booby traps or nuisance mines is indicated by—

- Disturbance of ground surface or scattered, loose soil.
- Wrappers, seals, loose shell caps, safety pins, nails, and pieces of wire or cord.
- Improvised methods of marking traps, such as piles of stones or marks on walls or trees.
- Evidence of camouflage, such as withered vegetation or signs of cutting.
- Breaks in the continuity of dust, paint, or vegetation.
- Trampled earth or vegetation; foot marks.
- Lumps or bulges under carpet or in furniture.

## DETECTION

Detection methods depend on the nature of the environment. In open areas, methods used to detect mines can usually detect booby traps. Look for trip wires and other signs suggesting the presence of an actuating mechanism. In urban areas, mine detectors are probably of little use. You have to rely on manual search techniques and, if available, special equipment.

## CLEARING METHODS

The method used to neutralize or disarm a trap depends on many things, including time constraints, personnel assets, and the type of trap. Remember, a trap cannot be considered safe until the blasting cap or the detonating cord has been removed from the charge. This is your first objective and is particularly important for electric traps, which may contain a collapsing circuit.

Use the safest method to neutralize a trap. For example, if the FD and the detonating cord are accessible, it is usually safer to cut the detonating cord. This method does not actuate the trap, but inserting pins in the FD might.

## COMBAT CLEARANCE

Clearing booby traps and nuisance mines in AOs is done primarily by engineers. Therefore, engineer advice is important during the planning stages of any operation where booby traps are likely to be encountered. Intelligence regarding the possible presence and types of traps must be provided to engineer units as soon as possible. This allows the unit to take necessary action and provide relevant training. Clearance of booby traps cannot be undertaken as a secondary task, because engineer clearing teams might require protection that necessitates combined arms training. Before engineer planning can start, the staff provides commanders with the latest intelligence information and, if possible, the following information:

- Amount of clearance required.
- Acceptable damage.
- Time requirements.
- Availability of special equipment.
- Security requirements.

Intelligence information regarding the nature, type, and location of traps has a direct bearing on the number of clearing parties necessary and the degree of protection required. For example, in built-up areas where traps have to be cleared in buildings that offer protection from enemy fire, direct protection is usually provided by the normal combat situation. On the other hand, in open areas where clearing parties may be required to clear traps covered by direct enemy fire, protection arrangements must be more specific.

Engineer commanders must be aware of the time needed to clear various types of traps in differing terrain situations. Remember, increasing the number of clearance parties may not necessarily reduce the time required to clear traps. This is particularly true when traps are set close together or set deep along a narrow front that is the only available route.

Initially, clear areas of immediate tactical importance and traps that present a specific threat. For example, clear only the portion of a building required for observation and those traps presenting an immediate hazard. This enables clearing parties to concentrate on other areas of tactical importance.

Clearing traps by hand is the only way that damage can be avoided and security guaranteed. When it is vital to avoid equipment or structure damage, consider using available EOD assets. It is often necessary to balance the requirement to remain silent and avoid damage with the requirement to maintain momentum.

When traps are being cleared in direct-support combat situations, they are normally dealt with by using unit resources and locally manufactured or acquired aids. Specified equipment is rarely available. Equipment varies with the situation but usually consists of selected items from Table 13-2. In areas with a high incidence of booby traps, assemble and reserve special clearing kits.

Table 13-2. Clearing equipment

| Item | Remarks |
| --- | --- |
| **Unit Equipment** | |
| Mine detector | |
| Mine probe/knife/bayonet | |
| Eye protection | |
| Body armor | Flak jacket, Kevlar helmet |
| Booby-trap signs | |
| Mine marking tape | |
| Mine markers | Chalk, felt-tipped pens |
| Electrician's adhesive tape | |
| Flashlight | |
| Tape measure | |
| Mine marking cones | |
| Trademan's tools | Saws, pliers, hammers, screwdrivers |
| Self-developing camera and film | |
| **Locally Manufactured/Acquired** | |
| Trip-wire feeler | |
| Grapnel and rope | |
| Hand mirrors | |
| Pins, wire, nails | For use as safety pins |
| Meat hooks | For use as block and tackle |

## CLEARANCE IN SECURE AREAS

### POLICY AND PLANNING

Formal clearing procedures must be followed in secure areas. This type of clearance can be done by engineers or EOD personnel. It is subject to time constraints when traps or suspected traps are located in urgently required installations or facilities, such as supply dumps or telecommunication centers.

Commanders reconnoiter their areas of responsibility, make detailed plans, prepare orders, and brief their men. The following points are covered:

- Intelligence.
- How the operation is to be conducted (include acceptable damage and methods of clearing).
- Action on finding traps.
- Marking.
- Disarming procedures.
- Tasking and allocation of areas.
- Equipment available.
- Control measures.

- Time requirements.
- Rehearsal and equipment preparation.

## Control Point

The operation commander establishes a control point upon arrival at the area to be cleared. The control point functions as the headquarters and the material holding area. It is also the point from which all clearance starts. Its suitability might not have escaped the enemy's notice, so it must be cleared before it can be used.

## Control and Size of Parties

The size of clearing parties varies depending on the location being cleared. The following rules apply:

- Each party is controlled by an NCO.
- Only one party works in a particular subarea.
- The distance between parties is sufficient so that detonation in one area does not endanger persons in other areas.
- The NCOIC of each party is in visual, radio, or voice contact with every person in his party.
- Only one person works on a booby trap.

## Marking

Booby traps and nuisance mines can be placed in diverse locations. It is impossible to standardize a method for marking areas, individual traps, or safe routes. Any form of prominent, permanent marking can be used.

### Uncleared Areas

The perimeter marking of uncleared areas can take any clearly recognized form. Standard minefield or booby-trap signs suspended from a single-strand fence are recommended. The spacing of the signs is the same as standard minefield marking (see Chapter 2). As the area is cleared, the fence should be progressively moved.

### Individual Traps

Because booby traps can have more than one means of actuation, do not attempt to place a cone or other marker over any part of a trap. Use sufficient signs to ensure that the trap can be detected and accurately located. In buildings, clearly mark rooms containing traps and, where possible, indicate the exact location of traps.

### Internal Marking

The internal marking system depends on the area being cleared. One good system is to divide the total area into subareas, clear and mark safe lanes between the subareas, and then use the lanes as safe routes.

## Clearing of Open Areas

Roads, road shoulders, bridges, obstacles, and other structures must be cleared in open areas. The main threat comes from nuisance mines, so regard each potential site as a nuisance minefield and use established minefield clearing procedures. After the site is cleared, adapt the drill to suit the situation.

### Clearing of Buildings

Buildings are excellent locations for booby traps. Booby traps are concentrated inside the building, so carefully organize search and clearing procedures. In most cases, buildings are required for use, and excessive damage must be avoided.

A two-man clearing party can clear an average-size structure. No one else should enter the building until it is cleared. In very large buildings, two or more parties can be employed, if they work as far apart as possible and have clearly defined boundaries. As a general rule, two walls or two floors is the minimum distance between parties.

### Exterior Reconnaissance and Entry

Before approaching a building, check the surrounding area for booby traps and nuisance mines. Carry out a reconnaissance to determine the point of entry and clear the way to it. When selecting the point of entry, consider the following:

- Doorways. Never consider doorways to be safe unless the door is fully open and the entrance is clear. If a house is built on a concrete slab, it is not likely to have a pressure FD in the floor.
- Windows. Windows are excellent locations for booby traps. Pay particular attention to the ground outside and the floor inside because they are classic sites for pressure FDs. Use the following procedures if access must be gained through a window:
  - Pull the window if it is unlocked and can be moved. If it is locked, use a small charge or a heavy object to break the glass.
  - Select a stationary window that cannot be opened versus a window that can be opened when both types are available, because the window that can be opened is more likely to contain a booby trap.
  - Deal with blinds and curtains in a manner similar to procedures used for windows.
  - Use a mirror to examine inside the window frame before entering the building.
- Mouseholes. If you decide not to enter the structure through a door or a window, use explosives to make a mousehole in the wall, roof, or floor. This offers a remote, safe method of creating an access point, but it can also detonate nearby traps. Exercise caution if entering through the roof of a two-story building, because it is far more difficult to clear booby traps when going down steps than it is when going up them.

### Search Techniques

It is impossible to establish a search drill that is suitable for use in all buildings. It is essential, however, for each working party to develop its own drill and follow it. Domestic dwellings should be searched in the following order:

1. Floors and furniture.
2. Walls (including doors, windows, fireplaces, and cupboards).
3. Fittings (including light switches and pictures).
4. Ceilings.

Mark each area or item *SAFE* as it is cleared. This can be conveniently done by using chalk or a felt-tipped pen.

Use the following techniques and precautions when searching buildings. They can save your life—learn and remember them!

- Check both sides of a door before opening it. Do this by drilling a hole through the door and using a mirror to check the other side. You can also check or open doors by pulling or blowing the lock and hinges with a small charge.
- Examine floor coverings for signs of disturbance. The presence of FDs is often indicated by loose floorboards, bulges or tears in carpets, or loose tiles.
- Use a pulling cable to move furniture and to open cupboard doors and drawers.
- Check upholstered furniture and beds by remotely dropping a heavy object onto them.
- Treat every switch with suspicion, since electrical wiring provides a ready-made circuit for booby traps. To explode all the traps connected to the normal power supply, disconnect the power at the fuse board, turn all the switches on, and then reconnect the power. Repeat the procedure with the switch turned off in case the switch has been reversed. Remember, this procedure will not disclose traps that use a battery. Exercise caution when using switches, even if the power is disconnected.
- Leave doors, windows, cupboards, and drawers open after they have been cleared.
- Mark all routes, areas, and items that have been cleared.
- Check plumbing by remotely turning on all water taps and allowing the water to run for at least one minute.
- Check toilet tanks before flushing.
- Beware of light-sensitive devices in dark places (closets, cabinets, basements, attics, chimneys).
- Allow at least 30 seconds for an explosion after pulling anything, because there might be a delay fuse.

### Clearing Installations and Facilities

Clearing by hand is necessary in installations and facilities (fuel dumps, ammunition dumps, electric substations) where an explosion could result in the loss of resources. In other situations, the item's importance or the resulting damage might not be obvious. For example, a small charge placed against the control valves of a dam or against the main cable entering a telephone exchange results in unforeseen damage that can take days to repair. Therefore, you should seek a specialist's advice, if possible, when clearing booby traps in industrial areas and unfamiliar locations.

### Clearing Obstacles

If an enemy has time to create obstacles, he also has time to set booby traps and lay nuisance mines. The obstacle itself is usually clear of traps to encourage a false sense of security and lead troops into more dangerous areas. Therefore, regard all obstacles as booby-trapped until proven otherwise. The simplest, safest way to deal with movable obstacles is to pull them. Before an obstacle can be pulled, you must first clear the area from which the pull will be made.

### Clearing Secure Areas

When clearing secure areas and time is not a major factor, use specialized clearance equipment as much as possible. The following equipment might be available for use:

- Cameras. Cameras have a wide range of applications. They can be used with different types of film, such as infrared and ultraviolet, to disclose evidence that is indiscernible to the naked eye. For example, infrared photography reveals differences in the heat emitted by objects and can often disclose recent digging and buried or concealed objects.
- Explosive detector dogs (EDDs). Although EDDs can detect minute quantities of explosives and the presence of trip wires, they are trained to detect the charge and not the FD. This extremely limits their usefulness in detecting booby traps. They also tend to become confused if the area contains explosive odors other than those emitting from booby traps.
- Electronic countermeasures. Electronic countermeasures can be used to explode electric booby traps and to prevent remotely controlled, improvised explosive devices from being detonated by radio.
- Robots. In their simplest form, robots can be used to detonate or neutralize booby traps. More sophisticated models can be remotely controlled to carry out simple tasks, such as videotaping or cutting wires.
- Body armor.
- Electric meters.
- X-ray equipment.

## CLEARANCE METHODS

- Pulling. This method uses a grapnel and a rope to pull the trap. It is used when the resulting damage is acceptable. It is the safest method and is particularly applicable to traps set in open areas. Do not disturb any part of a booby trap when placing the grapnel and pulling the cable. Carefully select the site from where the pull is to be made because it might be mined or trapped. When a booby trap is pulled and does not explode, wait at least 30 seconds before approaching it in case delay devices have been used. Disposal of unexploded traps depends on their condition when inspected. The procedure for pulling booby traps is similar to that for pulling mines (see Chapter 11).
- Trip wires. Check the area for AP devices before proceeding. Place a grapnel hook as close as possible to the trip wire. Do not touch the trip wire until the pulling party is in a covered area.
- Pull and release. Pull away objects that conceal and operate pull and release mechanisms.
- Pressure mechanism. Pull pressure mechanisms from under objects that conceal and operate them. If this is impossible, blow them in place. In many cases, it might be easier to pull the charge rather than the FD. Take extreme care when attempting this, because additional mechanisms are often concealed in or under the main charge.
- Destroying in place. When destroying booby traps in place, explode a small charge near the booby trap's charge. Again, use this method only if damage from the explosion is acceptable. When it is impossible to place the explosive close enough to ensure detonation of the main charge, carefully place it alongside the mechanism. Do not assume the main charge is safe to handle just because the mechanism has been destroyed. Actuate pressure mechanisms by suspending one-half pound of explosive above the pressure plate.
- Clearing by hand. This method involves neutralizing, disarming, removing, and disposing of traps without causing damage. It is extremely hazardous and should only be used when pulling or destructing traps in place is impossible or unacceptable. Clearance should only be conducted by EOD personnel or experienced engineers. Carefully examine all aspects of the trap before deciding how to clear it.
- Explosive line charge. Using this device produces quick results when only a narrow path is required through a booby-trapped area. It gives clearance for the same distance to either side, only where it is in contact with the ground.
- Armor. This method is used where traps with small charges (designed as AP devices) are located in open areas. Armored vehicles track back and forth over the area. This shortens the clearing time with little risk of casualties.

- Fire. If traps are set in grass or dense vegetation, fire can be used to burn away camouflage material and expose traps. In most cases, sufficient heat is generated to burn or explode the traps. Unexploded traps are considered extremely sensitive and must not be cleared by hand.

## IMPROVISED TRAPS

In low-intensity conflicts, there is a high incidence of improvised mines and booby traps. It is impossible to provide a complete list of FDs and improvised demolition material that can be used. However, most improvised traps are variations of those described below.

An electric FD requires a current to be passed between two contacts. The ways in which this can be achieved are limited only by the imagination. Examples of simple electric FDs that can be manufactured using household items or appliances are shown in Figure 13-6, page 13-24. Examples of improvised mechanical FDs are shown in Figures 13-7 and 13-8, page 13-25.

The simplest and most accurate method of incorporating delay is to use some form of clock or timer. However, any mechanism or chemical reaction that takes a measurable time to complete its function can be used. Examples of simple improvised delay devices are shown in Figures 13-9 and 13-10, page 13-26.

## NONEXPLOSIVE TRAPS

Nonexplosive traps are typically used in tropical or rain-forest regions. Ideal construction materials abound, and concealment in surrounding vegetation is relatively easy. No prescribed procedures exist for clearing nonexplosive traps. Each trap must be cleared according to its nature.

### PUNJI

The punji (Figure 13-11, page 13-27) is one of the simplest traps. It is normally made from locally available material, such as sharpened stakes, nails, or heavy-gauge wire. It is placed in concealed pits or in places that give cover from fire. The enemy normally smears the spikes or cutting edges with excrement or poison. After clearing a punji, sterilize cuts and abrasions immediately and see a medical officer as soon as possible.

### CLOSING TRAP

The side-closing trap (Figure 13-12, page 13-27) consists of two wooden slats that are studded with spikes. The slats slide along a pair of guide rods controlled by heavy bands. When the prop holding the slats apart is dislodged, the slats spring together and implant the spikes into the portion of the body passing between them.

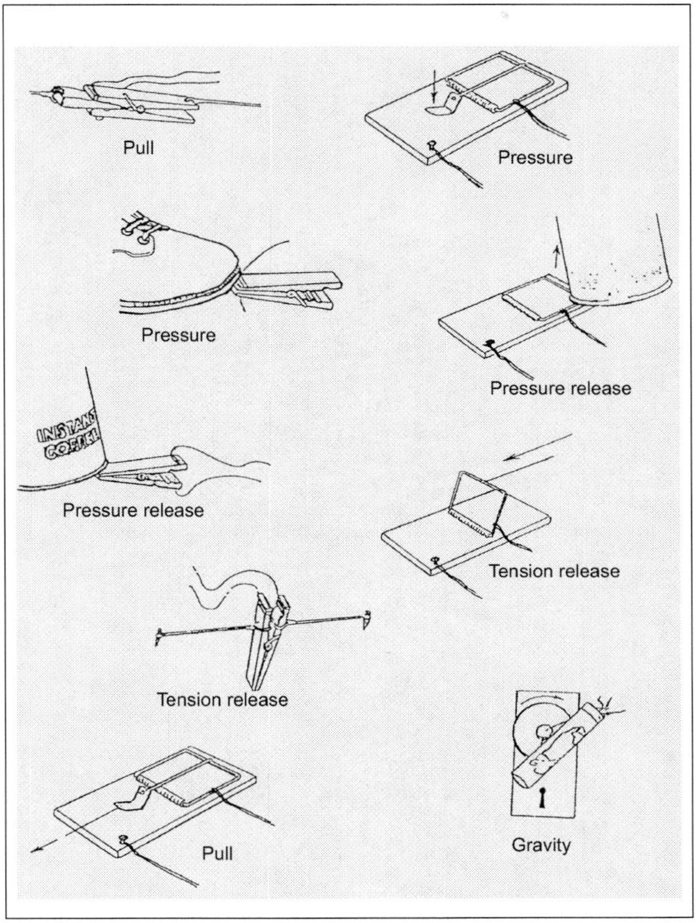


**Figure 13-6. Improvised electrical FDs**

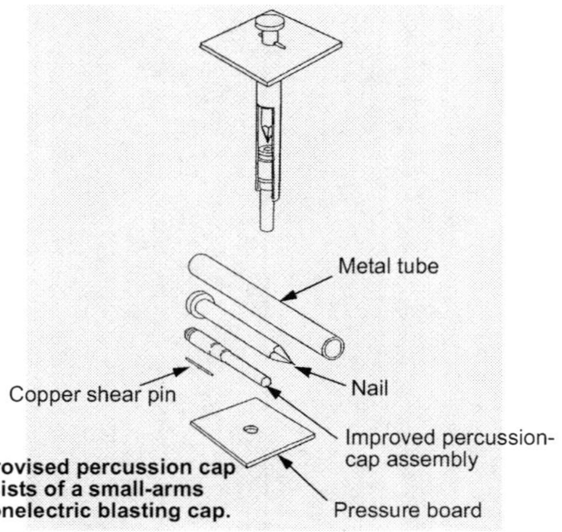


**NOTE: An improvised percussion cap assembly consists of a small-arms round and a nonelectric blasting cap.**

**Figure 13-7. Improvised nonelectric FDs (shear-pin operated)**

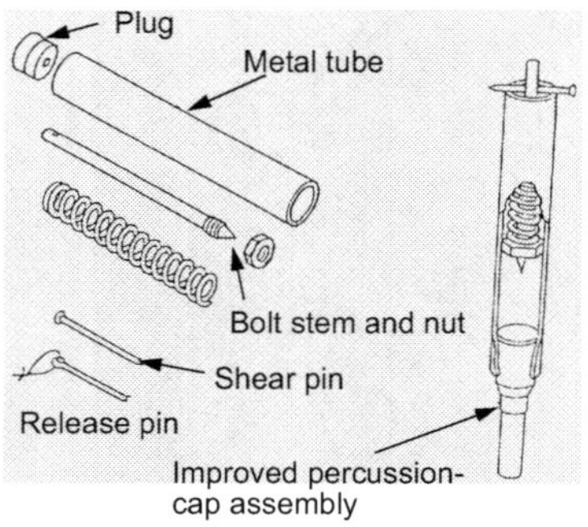


**Figure 13-8. Improvised nonelectric FDs (spring-operated)**

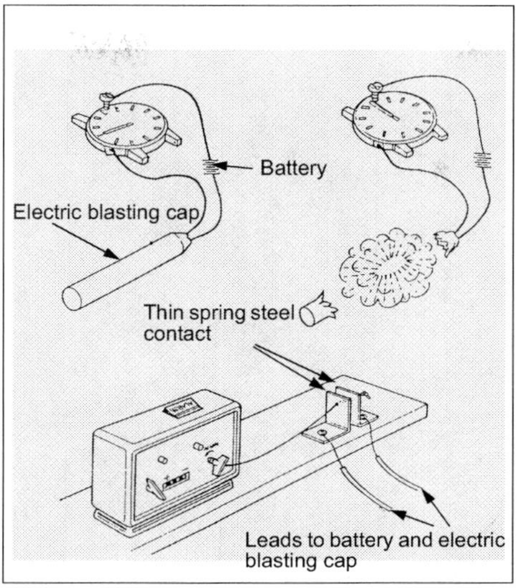


Figure 13-9. Improvised, electric delay devices

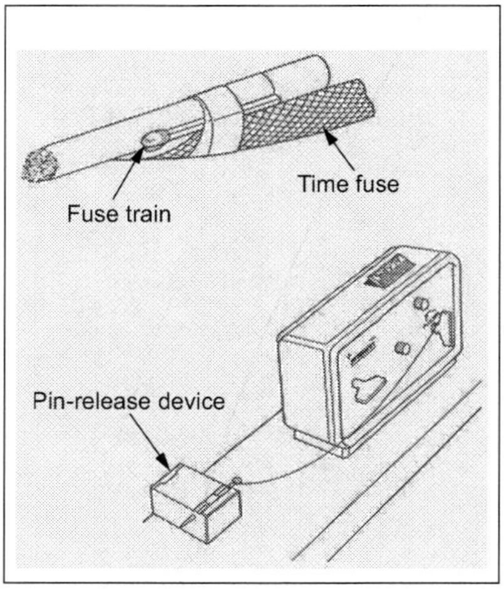


Figure 13-10. Improvised, nonelectric delay devices

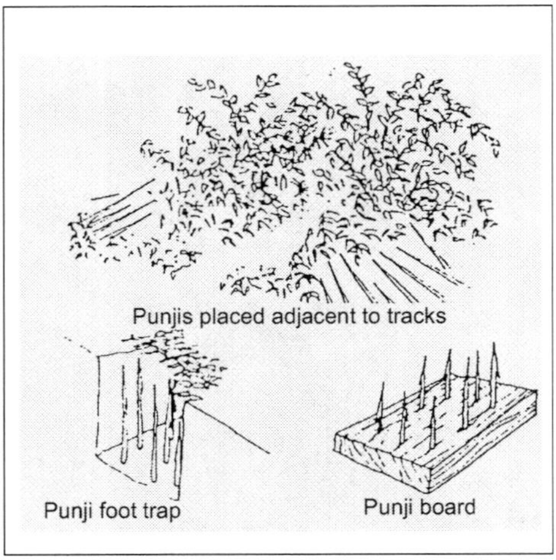


Figure 13-11. Typical punjis

Figure 13-12. Side-closing trap

### Spike Board

The spike board (Figure 13-13) is used in a pit and consists of a treadle board with one end spiked. When a man steps on the board, the spiked end flies up and strikes him in the face or chest.

Figure 13-13. Spike board

### Venus Flytrap

The Venus flytrap consists of a rectangular framework with overlapping barbs emplaced in a pit. The one shown in Figure 13-14 is made from a metal container that is sunk into the ground until the top is flush. It is then covered with grass and/or leaves. The barbs inflict injury, especially when the victim attempts to withdraw his leg from the trap. These traps are typically located on tracks and along road edges.

Figure 13-14. Venus fly trap

## SECTION III. EXPEDIENT DEVICES

Expedient devices are constructed in the field with locally available material. They are employed against vehicles or personnel in the same manner as other mine systems. Expedient devices—

- Supplement a unit's low supply of conventional mines.
- Hinder reconnaissance, clearance, and neutralization of minefields.
- Create enemy attitudes of uncertainty and suspicion to lower morale and slow movement).

### AUTHORIZATION

Because expedient devices have nonstandard design and functioning, take special precautions to protect friendly forces. Consider neutralization, disarming requirements, and adequate marking procedures. The use of expedient devices is restricted under the Convention of Conventional Warfare. Expedient devices have the same international restrictions as booby traps. The corps commander is the employment authority for expedient devices. He can delegate this authority to the division commander. If authority is given to use expedient devices, US personnel will adhere to the rules for international law that are applicable to armed conflict.

### EMPLOYMENT AND CONSTRUCTION TECHNIQUES

If issued mines are not readily available on the battlefield, expedient devices can be manufactured in the field. Construction varies based on available materials and the ingenuity of the personnel who are fabricating the devices. Expedient devices pose a potential safety hazard to friendly forces—those who are constructing them and those who may later encounter them. Construction should be performed by personnel who are familiar with the materials being used. Innovative designs should be checked and tested before arming and emplacing the devices.

As a minimum, test the fusing mechanism separately to ensure that it functions as designed. Improper fuse operation is the most common cause of malfunction. Also, test the fuse and the firing chain (base charge, blasting cap, and detonating cord) without the main charge to ensure proper operation. Emplace the device after satisfactory performance of the firing mechanism. First, emplace heavy items (such as artillery shells) that are used as the main charge, and then add the firing mechanism. Take care when moving or emplacing expedient devices because their nonstandard manufacture and potentially faulty construction make them highly sensitive to jars and shocks. Construct devices at the emplacement site whenever possible.

Expedient devices are prepared in the field using standard US FDs, detonators, and demolition materials. All devices discussed in this chapter can be made to function electrically or nonelectrically using modernized demolition initiators (MDIs). AP devices must be command-detonated.

### HIGH-EXPLOSIVE, ARTILLERY-SHELL DEVICE

The HE, artillery-shell device (Figure 13-15) can be readily adapted to expedient mining. Remove the artillery fuse and replace it with a standard FD and a length of detonating cord or with an MDI blasting cap. If properly assembled, a destructor may also be used. If a destructor is not available, firmly pack the fuse well with composition C4 explosive and insert a length of knotted detonating cord or a blasting cap.

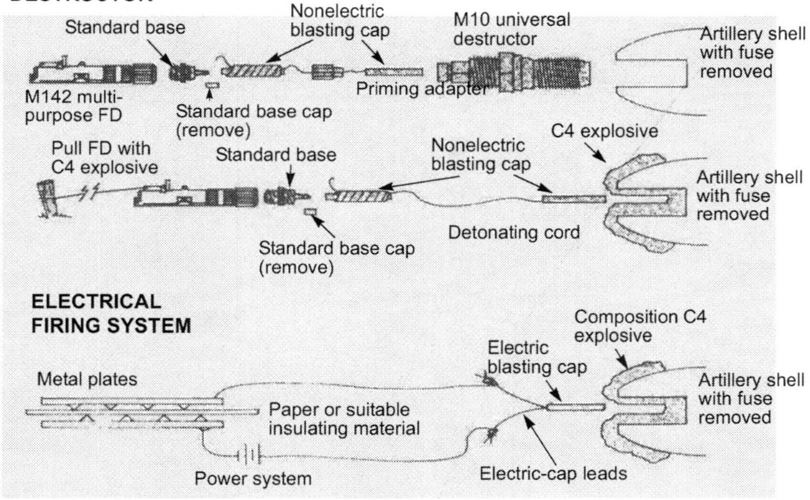


**NOTE: For command-detonation, an M34 blasting machine can replace the metal plates and the battery.**

**Figure 13-15. HE, artillery-shell device**

The device can be activated by a variety of methods depending on the type of FD used. When MDI blasting caps are used, the device is command-detonated. The device can also be adapted to function electrically by adding an electric cap and a power source.

This device can be used as an AT or an AP device. When used as an AP device, it must be command-detonated.

**NOTE: Use only serviceable US ammunition that has remained in the possession of US forces. Never use captured ammunition or UXO found on the battlefield. It may be armed, booby-trapped, or deteriorated.**

**PLATTER CHARGE**

The platter charge (Figure 13-16) consists of a suitable container that is filled with uniformly packed explosive and placed behind a platter. The platter is metal (preferably round, but square is satisfactory) and weighs 1 to 3 kilograms. The explosive required is equal to the weight of the platter. The container may not be necessary if the explosive can be held firmly against the platter (tape can be used). The charge should be primed from the exact rear center, and the blasting cap should be secured with a small amount of C4 to ensure detonation.

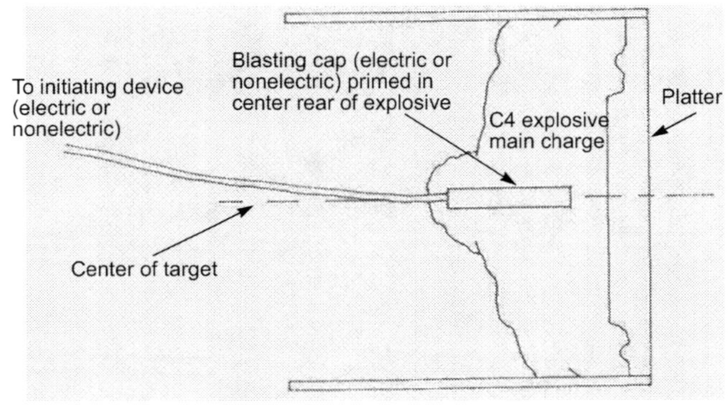


**Figure 13-16. Platter charge**

The charge should be aimed at the direct center of the target. The effective range (primarily a matter of aim) is approximately 35 meters for a small target. With practice, experienced personnel can hit a 55-gallon drum (a relatively small target) at 25 meters with about 90 percent accuracy.

The platter charge can be used as an AT or an AP device. When used as an AP device, it must be command-detonated.

**IMPROVISED CLAYMORE**

For the improvised claymore device (Figure 13-17, page 13-32), a layer of plastic explosive is attached to the convex side of a suitably dense, curved base (such as wood or metal). A hole must be made in the exact rear of the base. A blasting cap is placed in the hole to prime the device. Shrapnel is fixed to the explosive with a suitable retainer (cloth, tape, mesh screen).

The device must be command-detonated. Command detonation is best achieved with electrical priming or an MDI. A blasting device is attached to the electric cap via firing wires laid at least 50 meters from the device. Ensure that personnel have adequate cover when detonating the improvised claymore.

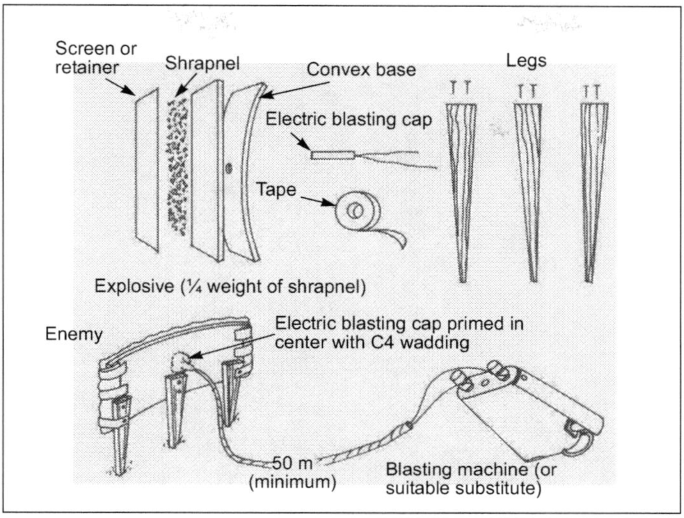


**Figure 13-17. Improvised claymore device**

### GRAPESHOT ANTIPERSONNEL DEVICE

Place shrapnel in the bottom of a cylindrical container to make a grapeshot AP device (Figure 13-18). The shrapnel is tamped and held in place with a suitable separator (wadding). Explosive (approximately one-quarter the weight of the shrapnel) is packed to a uniform density behind the wadding. The device is primed in the center of the explosive with an electric cap or an MDI.

**NOTE: The United Nations Convention of Certain Conventional Weapons mandates that all fragment munitions produce fragments that are visible by X ray (such as metal or rock).**

This device must be command-detonated. The explosive propels the shrapnel outward from the container. The grapeshot is very effective against personnel targets.

### BARBWIRE ANTIPERSONNEL DEVICE

The barbwire AP device (Figure 13-19) can be made directional by placing the wire against an embankment or a fixed object. This causes the force of the explosion to expel the barbwire fragments in the desired direction. One roll of standard barbwire is placed into position, and one block of C4 is placed in the center of the roll and primed. This device must be command-detonated.

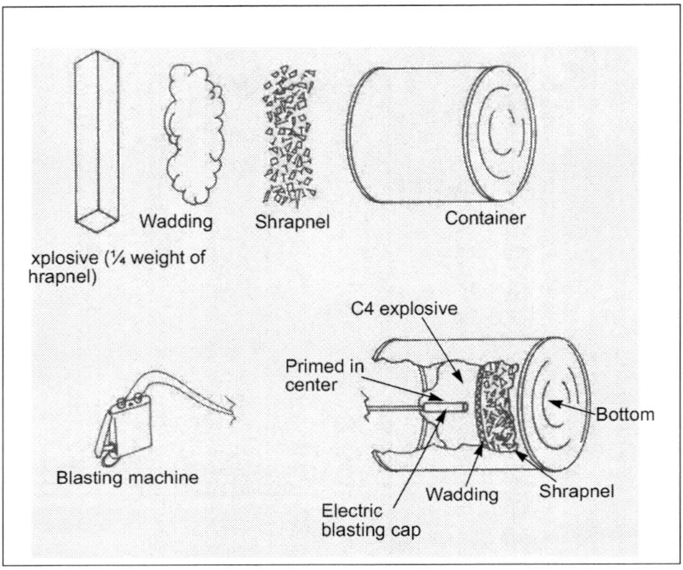


**Figure 13-18. Grapeshot AP device**

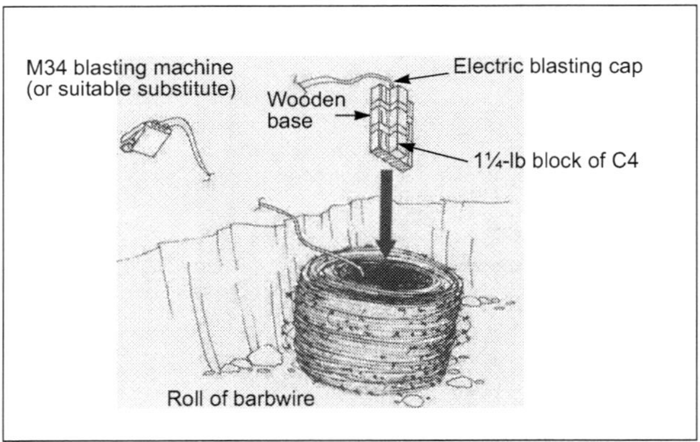


**Figure 13-19. Barbwire AP device**

Appendix A

# Installation and Removal of US Mines and Firing Devices

This appendix provides installation and removal procedures for AT mines, AP mines, and FDs. The scope of this appendix is limited to US hand-emplaced mines that require manual arming.

**WARNINGS**

**1. If there is a problem when performing any installation or removal step, notify the NCOIC.**
**2. If you hear a click when removing the safety clip or if the pressure plate snaps downward so it is level with the body of the mine, notify the NCOIC. DO NOT use the mine.**
**3. If the safety clip cannot be reinserted, notify the NCOIC.**
**4. DO NOT apply pressure to the pressure plate, tilt rod, or fuse at any time.**
**5. Before attempting to disarm and remove the mine, check for AHDs, damage, and malfunctions. If any of these conditions exist, notify the NCOIC. DO NOT attempt to disarm the mine.**
**6. If you feel a jar or hear a metallic click when removing the locking safety pin, stop and notify the NCOIC. The firing pin has gone forward and is resting on the positive safety pin. DO NOT remove the positive safety pin.**
**7. After removing the positive safety pin, proceed with extreme caution. The slider pin can detonate the mine if it is accidentally pushed in.**
**8. When attaching trip wires to the release-pin ring on the fuse, leave a little slack in the wires. This prevents pull on the release-pin ring, which could set off the mine when the safety pins are removed.**
**9. Ensure that the extension rod is vertical and is not tilted in any direction. A 20-degree tilt of the extension rod will detonate the mine.**
**10. Ensure that the safety fork moves freely. If there is pressure on the fork, DO NOT remove it.**
**11. DO NOT apply pressure to the pressure plate of the fuse when inserting it into the fuse well.**
**12. If the setting knob is difficult to turn, DO NOT force it; notify the NCOIC.**
**13. DO NOT adjust the setting knob while the detonator is in the detonator well.**
**14. If any cracks are noted in the plastic collar, slowly and carefully reassemble the stop and safety pin on the fuse. Carefully remove the extension rod and the fuse from the mine. Give the fuse to the NCOIC and replace it with a new fuse.**
**15. Before cutting loose trip wires, look at each end to ensure that there are no electric-producing devices that might initiate another system.**
**16. Arm and disarm all mines while in the prone position.**

# SECTION I. ANTIPERSONNEL MINES

AP mines can kill or incapacitate their victims. They can be fused by pressure, wire, or command detonation and contain a blast, bounding-fragmentation, or direct-fragmentation warhead.

## M14

The M14 mine (Figure A-1) is a low-metallic, blast AP mine. It has a plastic body and an integral plastic fuse with a steel firing pin.

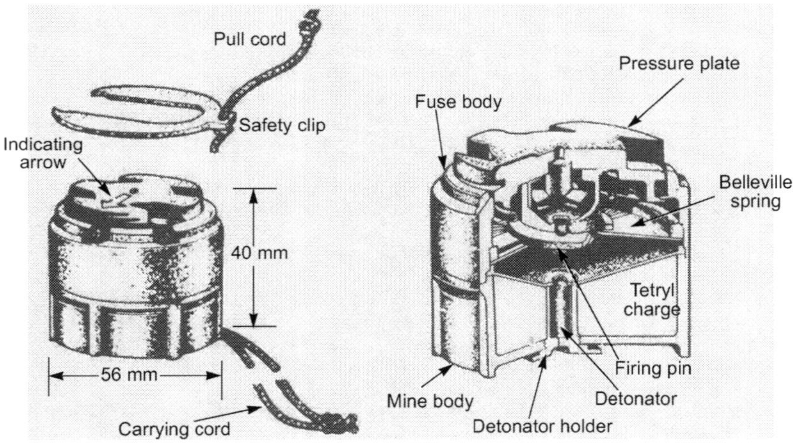


**Figure A-1. M14 AP mine**

### CHARACTERISTICS

| Main Charge | Diameter | Height | Weight | No Mines per Box | Weight per Box |
|---|---|---|---|---|---|
| Tetryl, 28.4 g | 56 mm | 40 mm | 99.4 g (fused) | 90 | 19.8 kg |

- **Korea Only: The M14 is employed in tactical and nuisance minefields.**
- The M14's size allows for employment in large numbers, and rapid concealment is possible.
- The M14 is buried to prevent the target from knocking it over.

- The M14 is not designed to kill but to penetrate a boot or a foot.
- The M14 is difficult to detect because of plastic construction.
- The M14 requires an operating force of 11.5 to 13.5 kilograms to activate.

INSTALLATION

> **WARNING**
> **Emplace and remove the mine while in the prone position.**

- Inspect the mine.
  - Do not use the mine if it is dented, cracked, or damaged.
  - Use the M22 wrench (Figure A-2) to remove the shipping plug from the detonator well.

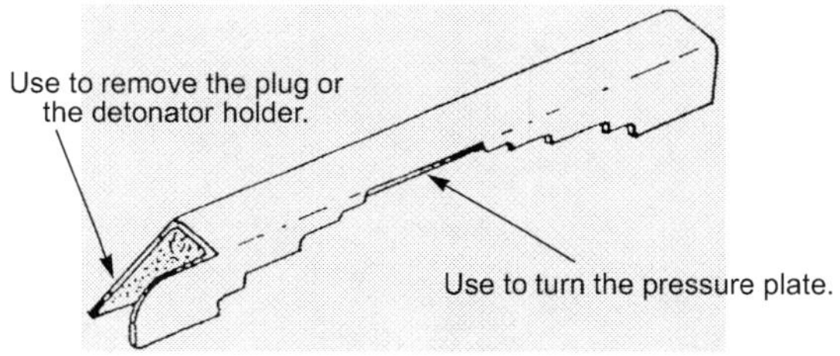


**Figure A-2. M22 wrench**

  - Inspect the position of the firing pin. Do not use the mine if the firing pin extends into the detonator well.
  - Inspect the detonator well for foreign material. If foreign material is present, remove it by carefully tapping the mine against the palm of your hand. If you cannot remove the debris, replace the shipping plug and do not use the mine.
- Test the pressure plate.
  - Ensure that the arrow is in the *SAFE* position.
  - Use the M22 wrench to turn the pressure plate from the *SAFE* position to the *ARMED* position (Figure A-3, page A-4).
  - Grasp the mine in one hand and remove the safety clip with the other hand. Listen for a click, indicating that the firing pin has dropped. Recheck the fuse well (Figure A-4, page A-4).
  - Replace the safety clip.

— Use the M22 wrench to turn the pressure plate back to the *SAFE* position.

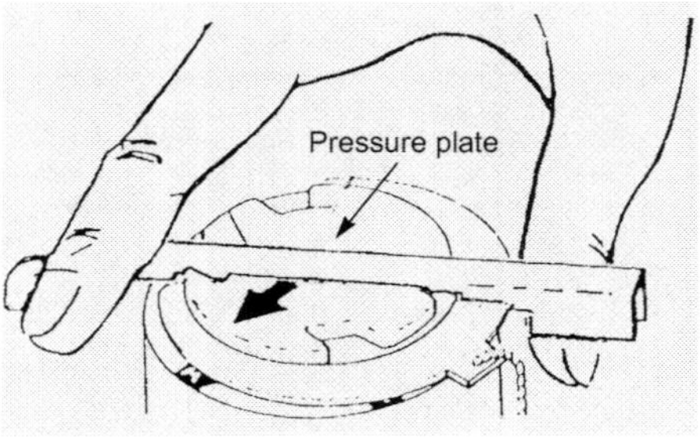


**Figure A-3. M14 mine in *ARMED* position**

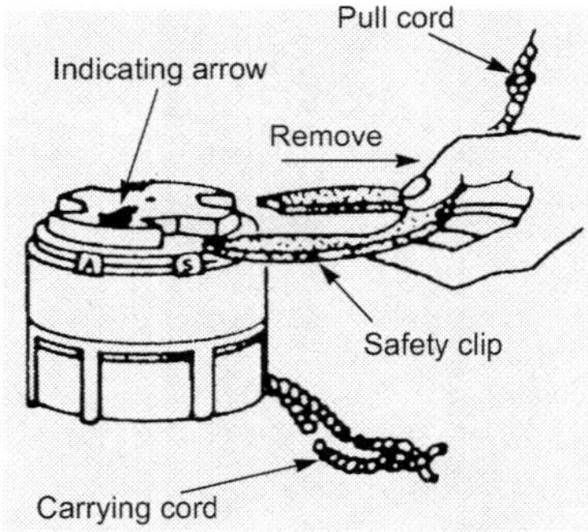


**Figure A-4. Removal of safety clip**

- Ensure that a metallic washer is attached (glued) to the bottom of the mine. If a washer is not attached, notify the NCOIC and do **not** use the mine. Apply silicone on the outside of the washer to prevent water damage, and screw the detonator into the well on the bottom of the mine (Figure A-5). **NOTE: Ensure that the gasket is tightly wedged between the detonator and the washer and that the washer is properly seated between the gasket and the body of the mine so that water cannot enter the mine.**
- Dig a hole to fit the mine.

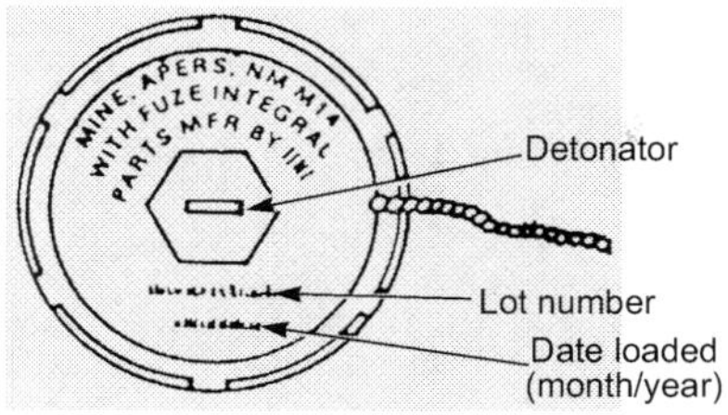


**Figure A-5. Bottom view of M14 mine**

— Dig a hole approximately 10 centimeters in diameter and deep enough (approximately 3.5 centimeters) so that the pressure plate extends above the ground.

— Inspect the ground at the bottom of the hole.

> Ensure that the ground is hard enough to support the mine when pressure is applied to the pressure plate.

> Place a flat object in the bottom of the hole for the mine to rest on if the ground is too soft. Allow additional depth for the objects.

- Arm the mine with an M22 wrench by turning the pressure plate to the *ARMED* position.
- Place the mine in the hole.
- Remove the safety clip carefully, while holding the mine body firmly in the hole.
- Camouflage the mine.

**REMOVAL**

- Disarm the mine.

— Clear the soil away from the mine carefully.

— Grasp the body of the mine firmly with one hand, and insert the safety clip with the other hand.

— Use the M22 wrench to turn the pressure plate to the *SAFE* position.

- Remove the mine from the hole.

— Turn the mine over, and carefully remove the detonator from the detonator well.

— Screw the shipping plug into the detonator well.

— Give the detonator to the NCOIC.

## M16

The M16-series mine (Figures A-6, page A-6) is a bounding-fragmentation AP mine. Once actuated, the mine is propelled out of the ground (to a height of approximately 1.8 meters) and explodes. The mine consists of an M605 fuse, a propelling charge, and a projectile that are contained in a sheet-steel case. The fuse is screwed into the top of the case and extends through the center of the projectile to the bottom of the case, where the propelling charge is located. The remaining space inside the case is occupied by the projectile. Earlier versions of the mine are also available for issue. The principal difference between the old and new versions is the construction of the detonators and boosters.

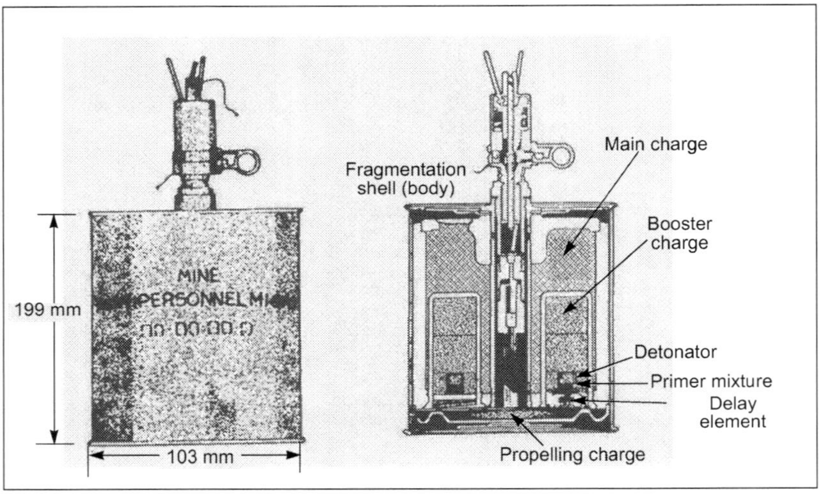


Figure A-6. M16A1 AP mine

### CHARACTERISTICS

| Main Charge | Diameter | Height | Weight | No Mines per Box | Weight per Box |
|---|---|---|---|---|---|
| TNT | 103 mm | 199 mm | 3.5 kg | 4 | 20.25 kg |

- **Korea Only: The M16 is employed in protective, tactical, and nuisance minefields.**
- The M16 is used to defeat dismounted assaults and breaching operations.
- The M16 is pressure-actuated (3.6 to 9 kilograms) or pull-actuated (1.4 to 4.5 kilograms).

- The M16 and M16A1 have a casualty radius of 27 meters; the M16A2 has a casualty radius of 30 meters. The danger radius for friendly forces is 183 meters for all M16 mines.

INSTALLATION

**WARNING**
**Emplace and remove the mine while in the prone position.**

- Inspect the mine.
  - — Do not use the mine if it is dented, cracked, or damaged.
  - — Use the closed end of an M25 wrench to remove the shipping plug from the fuse well (Figure A-7).

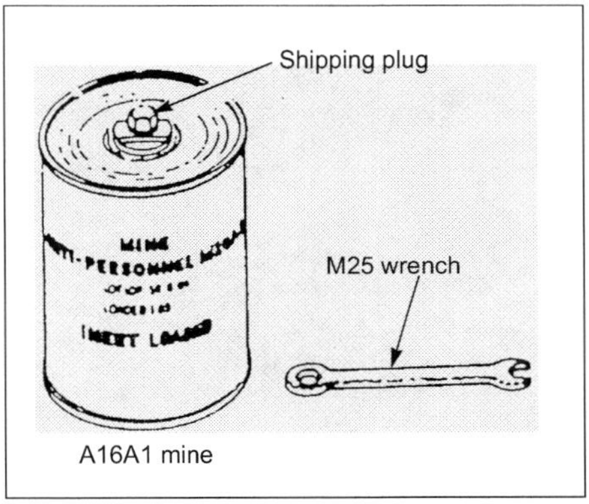


**Figure A-7. M16A1 mine and M25 wrench**

- — Inspect the fuse well for foreign material. If foreign material is present, turn the mine upside down and gently tap the bottom with your hand to dislodge the material. If you cannot remove the debris, replace the shipping plug and do not use the mine.
- — Examine the fuse assembly carefully for evidence of damage or missing safety pins. Ensure that the safety pins move freely in their holes and that the rubber gasket is around the fuse base (Figure A-8, page A-8).

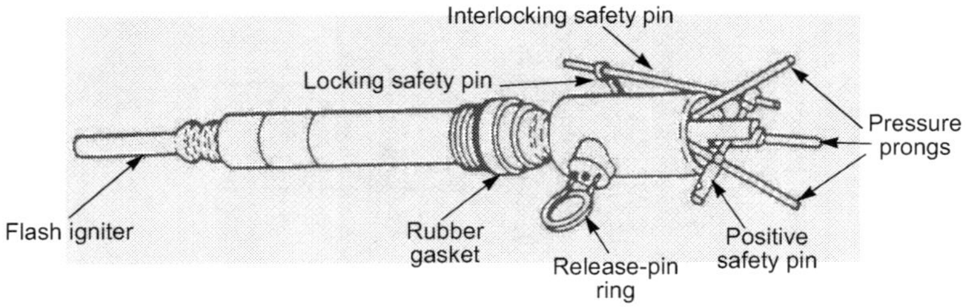


Figure A-8. M605 fuse

> **WARNING**
> **DO NOT use any fuse that was manufactured before 1957.**

- Fuse the mine.
  - Use the open end of the M25 wrench to tighten the bushing adapter on the fuse well.
  - Screw the fuse assembly into the fuse well by hand and tighten the assembly with the M25 wrench. Ensure that the rubber gasket is between the fuse body and the bushing adapter. **NOTE: For long-term use, smear a thin layer of silicone grease or similar lubricant on the fuse and threads.**
- Dig a hole to fit the mine.
  - Pressure installation. Dig the hole so that only the pressure-prong tips are above ground level.
  - Trip-wire installation. Dig the hole so that the release-pin ring is above ground level.
- Emplace the mine.
  - Place the mine in the hole and ensure that the safety pins remain in place.
  - Cover the mine with soil to the bottom of the release-pin ring.
  - Press the soil firmly around the sides of the mine.

**Pressure Installation**

- Remove the locking safety pin (Figure A-9). The interlocking safety pin will come free.

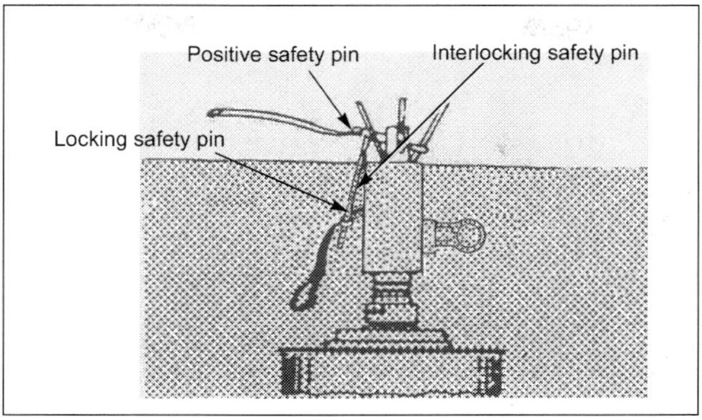


**Figure A-9. Safety pins**

- Arrange the pull cord on the positive safety pin so that it withdraws easily. Remove the metal collar.
- Finish covering the mine with soil until only the pressure prongs are above ground level.
- Camouflage the mine, place excess soil in sandbags, and remove sandbags from the area.
- Arm the mine by removing the positive safety pin (Figure A-9).
- Give the safety pins and the shipping plug to the NCOIC.

**Trip-Wire Installation**

- Install the trip wire (Figure A-10).

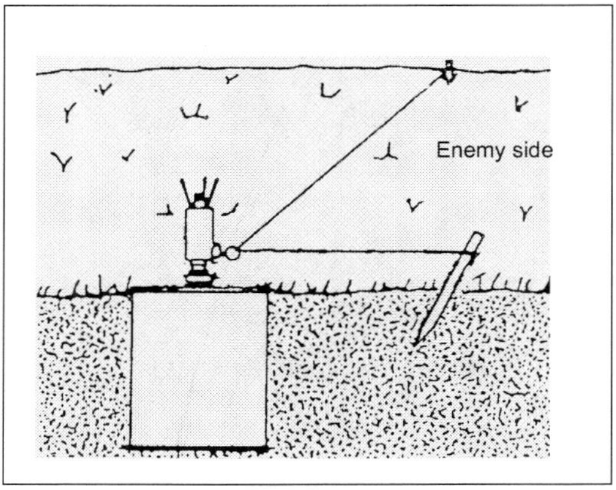


**Figure A-10. Buried mine with a trip wire**

- Cover the mine with soil and press it firmly around the sides of the mine. Leave the release-pin ring and the pressure prongs exposed.
- Tie off trip wires, approximately 10 meters from the mine. The wires should form a wide V, with the opening toward the enemy.
- Attach trip wires to the release-pin ring on the fuse.
- Remove the locking safety pin. The interlocking safety pin will come free.
- Arrange the pull cord on the positive safety pin so that it withdraws easily.
- Camouflage the mine, place excess soil in sandbags, and remove sandbags from the area.
- Arm the mine by removing the positive safety pin.
- Give the safety pins and the shipping plug to the NCOIC.

**REMOVAL**

> **WARNING**
> **Before attempting to disarm and remove the mine, ensure that the metal collar (Figure A-11) over the top of the striker on the M605 fuse is in place.**

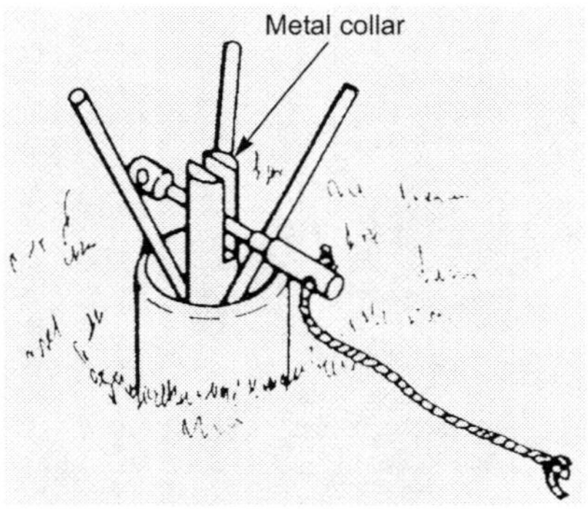


**Figure A-11. Metal collar on an M605 fuse**

- Disarm the mine.
  - Clear the soil carefully from the top of the fuse to the positive safety-pin hole. When using the M605 fuse, clear away all the soil from the fuse area.
  - Insert the positive safety pin through the positive safety-pin hole.
  - Insert the locking safety pin through the locking safety-pin hole.
  - Cut the slack trip wires that are attached to the release-pin ring.
- Check for AHDs.
  - Hold the mine body firmly in place with one hand.
  - Feel for AHDs with the other hand by digging around the sides of and underneath the mine.
- Remove the mine.
  - Remove the mine from the hole. Ensure that the safety pins remain in place.
  - Remove the M605 fuse with the M25 wrench.
  - Replace the shipping plug in the fuse well.

## SECTION II. ANTITANK MINES

AT mines are designed to immobilize or destroy tanks and vehicles and their crews. They perform this function by producing an M-Kill or a K-Kill. An M-Kill is achieved by destroying one or more of the vehicle's vital drive components (usually breaking the track on a tank), causing the target to be immobilized. The weapon system and the crew are not destroyed in an M-Kill; the weapon system is immobile but continues to function. A K-Kill results when the weapon system or the crew is destroyed.

Conventional AT mines are distinguished by their effects and their fusing systems. Blast AT mines, such as the M15 and M19, derive their effectiveness through the blast generated by their detonation. These usually produce an M-Kill, but a K-Kill may result. Mines such as the M21 use a shaped charge or an SFF designed to penetrate the underside of a vehicle's armor. A K-Kill normally results unless the mine detonates under the vehicle's track.

### M15

The M15 (Figure A-12, page A-12) is a blast AT mine that is contained in a round sheet-steel casing. The primary fuse well is located in the top center of the mine. There are two secondary fuse wells—one on the side and one on the bottom. The primary fuse well accepts the M603 pressure-actuated fuse. Standard FDs can be used in the secondary fuse wells with the M1 activator. The M624 tilt-rod-actuated fuse can also be used with this mine.

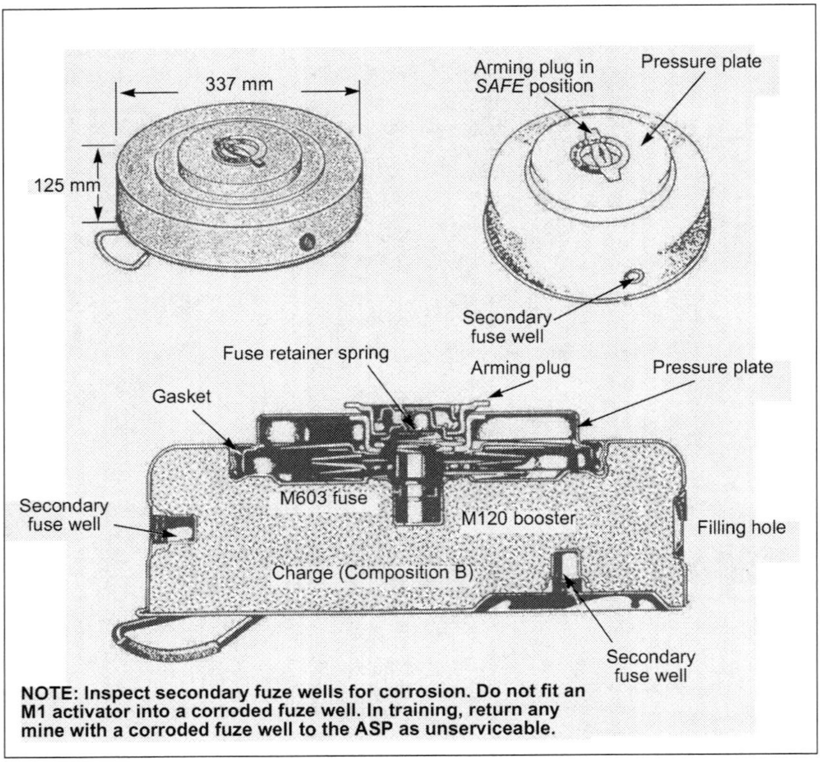


Figure A-12. M15 AT mine

CHARACTERISTICS

| Main Charge | Diameter | Height | Weight | No Mines per Box | Weight per Box |
|---|---|---|---|---|---|
| Comp B, 9.9 kg | 337 mm | 125 mm | 13.5 kg | 1 | 18 kg |

- The M15 is employed in protective, tactical, and nuisance minefields.
- The M15 is surface-laid or buried.
- The M15 requires a force of 158 to 338 kilograms to detonate the M603 fuse and a force of 1.7 kilograms to deflect the tilt rod and detonate the M624 fuse.
- The M15 is designed to defeat heavy tanks.
- The M15 produces an M-Kill upon contact.

## INSTALLATION USING THE M624 FUSE

> **WARNING**
> **Emplace and remove the mine while in the prone position.**

- Inspect the mine.
  - Do not use the mine if it is dented, cracked, or damaged.
  - Use the M20 wrench (Figure A-13) to remove the arming plug.

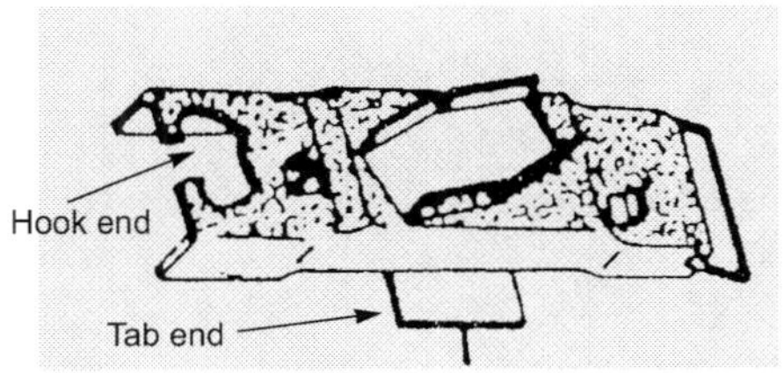


**Figure A-13. M20 wrench**

  - Inspect the fuse well for foreign material. If foreign material is present, turn the mine upside down and gently tap the bottom with your hand to dislodge the material. If you cannot remove the debris, replace the arming plug and do not use the mine.
  - Ensure that the booster retainer ring is seated in the fuse well. If the retainer ring is missing, replace the mine.
- Inspect the fuse.
  - Remove the M624 fuse from the metal shipping container and inspect it for serviceability.
  - Inspect the plastic collar of the fuse by looking down through the top of the pressure ring. If the safety pin is missing or improperly assembled, do not use the fuse (Figure A-14, page A-14).
  - Do not use the fuse if the plastic collar appears to be cracked.
- Fuse the mine.
  - Remove the M624 fuse from its fiber sleeve.
  - Remove the end closure on the M624 fuse. **NOTE: For long-term emplacement, coat the fuse threads and gasket with silicone grease before removing the end closure (Figure A-15, page A-14).**

Figure A-14. Correct safety-pin configuration

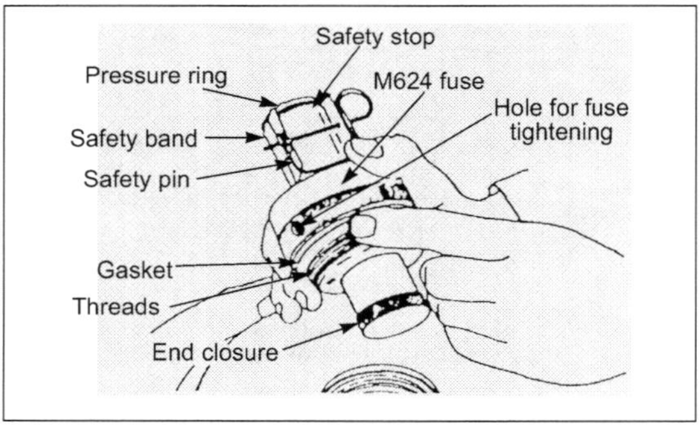


Figure A-15. Greasing the M624 fuse

- Screw the fuse hand-tight into the fuse well.
- Remove the extension rod from its packaging.
- Tighten the fuse by inserting the unthreaded end of one extension rod piece into the hole on the side of the fuse. Turn the fuse a quarter turn (Figure A-16).
- Remove the extension rod for further use after the fuse is secure.

**NOTE: The M15 AT mine (with the M624 fuse) can be buried or surface-laid. If surface-laid, it must be staked in place.**

- Dig a hole to fit the mine.
  - Dig a hole deep enough so that the top of the pressure plate will be at ground level.

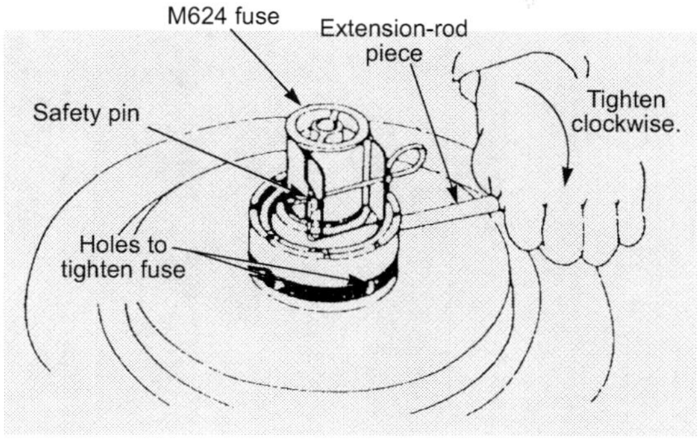


**Figure A-16. Tightening the fuse with the extension rod**

- Dig the sides of the hole at a 45-degree angle to prevent vehicles from bridging the mine.
- Emplace the mine.
  - Place the mine in the hole.
  - Cover the mine with 2 centimeters of soil (Figure A-17).

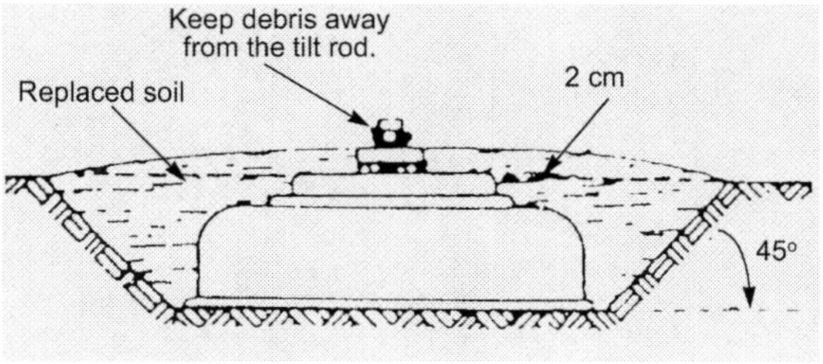


**Figure A-17. M15 mine in the hole**

**NOTE: The M15 AT mine (with the M624 fuse) can be used in the tilt-rod or pressure role. In the tilt-rod role only, assemble all three pieces of the extension rod (Figure A-18, page A-16) and thread the extension rod into the threaded pressure ring of the fuse (Figure A-19, page A-16).**

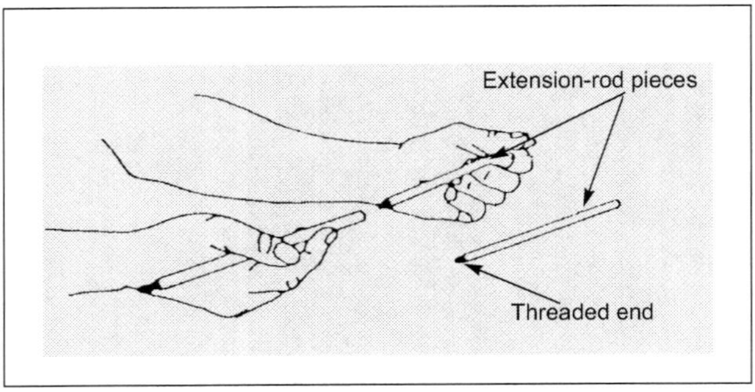


**Figure A-18. Extension-rod assembly**

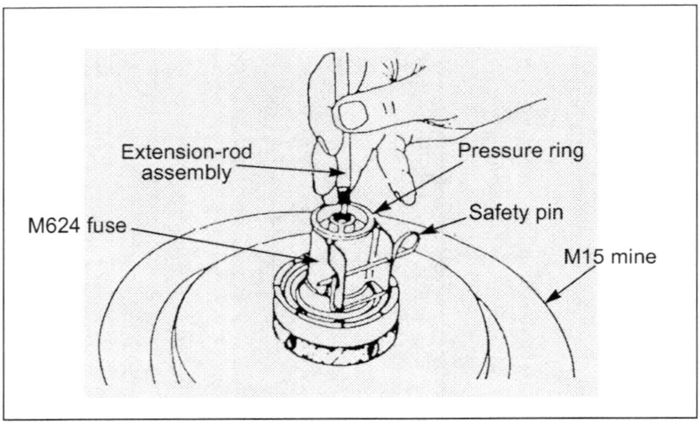


**Figure A-19. Assembly of the extension rod into the fuse ring**

- Arm the mine.
  - Use your right hand to raise the safety pin to the horizontal position, and grasp the safety band and safety stop with your left hand. Note the position of the thumb in Figure A-20.
  - Remove the safety pin with your right index finger, pulling it to the right.
  - Remove the safety stop carefully while holding the safety band in place.
  - Remove the safety band.
- Camouflage the mine.
  - Camouflage the mine with twigs, grass, or other material in the area. Place mines with extension rods in tall grass, if possible. Ensure that no pressure is applied to the tilt rod or the fuse.

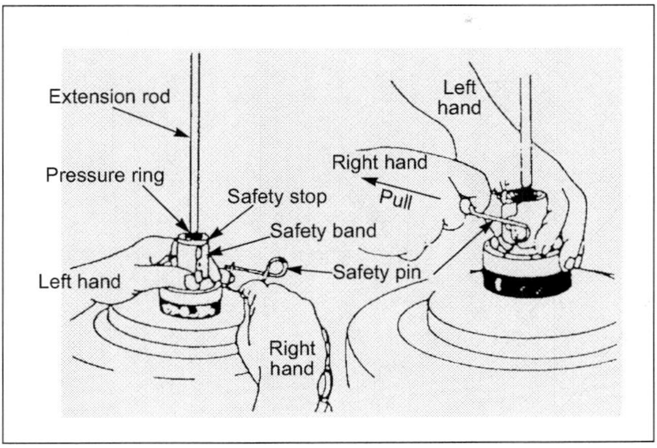


**Figure A-20. Removal of safety pin**

— Place excess soil in sandbags and remove them from the area.

— Give the band, the stop, the pull-ring assembly, the arming plug, and the end closure to the NCOIC.

### REMOVAL USING THE M624 FUSE

- Disarm the mine.
  - — Clear camouflage away from the mine carefully.
  - — Assemble the band, the stop, and the safety-pin assembly on the fuse so that the pressure ring is immobilized.
  - — Remove the extension rod.
- Check for AHDs.
  - — Hold the mine firmly in place with one hand, without putting pressure on the fuse.
  - — Feel for AHDs with the other hand by digging around the sides of and underneath the mine.
- Remove the mine.
  - — Remove the mine from the hole.
  - — Remove the fuse from the mine; use the extension rod if necessary.
  - — Replace the end closure on the fuse.
  - — Install the arming plug into the fuse well.

### INSTALLATION USING THE M603 FUSE

- Inspect the mine.
  - — Do not use the mine if it is dented, cracked, or damaged.

- Use the M20 wrench to remove the arming plug from the mine.
- Inspect the fuse well for foreign material. If foreign material is present, turn the mine upside down and gently tap the bottom with your hand to dislodge the material. If you cannot remove the debris, replace the arming plug and do not use the mine.
- Ensure that the booster retainer ring is seated in the fuse well. If the retainer ring is missing, replace the mine.

- Perform a function check with the arming plug.
  - Turn the setting knob to the *ARMED* position. Ensure that the shutter bar moves across the bottom of the arming plug (Figure A-21)
  - Turn the setting knob to the *SAFE* position. Ensure that the shutter bar moves back across the bottom of the arming plug (Figure A-22). **NOTE: If the shutter bar does not go into the *SAFE* or *ARMED* position, notify the NCOIC.**

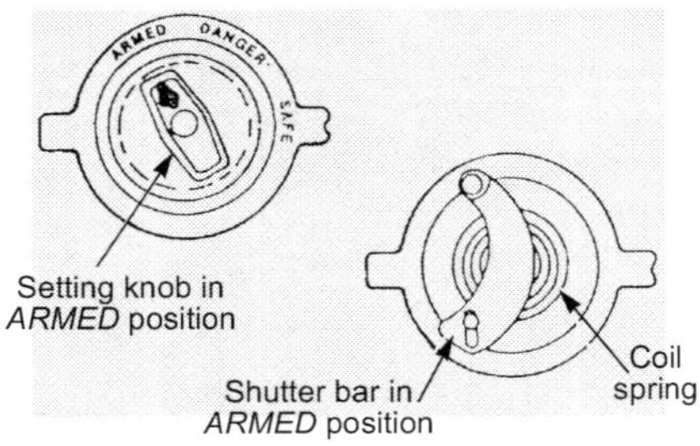


**NOTE: A coil spring may not be present in older models.**

**Figure A-21. *ARMED* position**

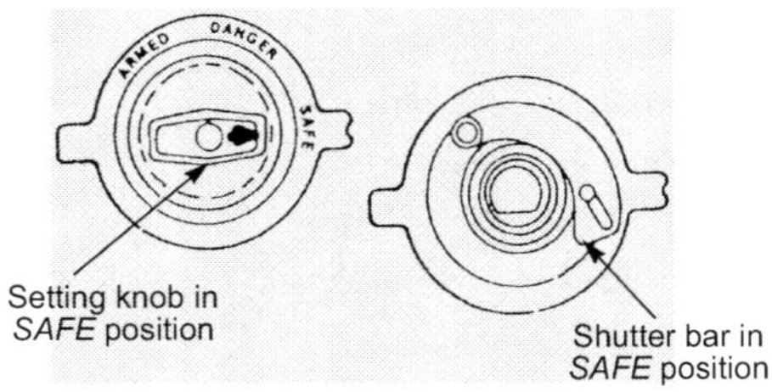


**Figure A-22. *SAFE* position**

- Fuse the mine.
  - — Remove the M603 fuse from the metal shipping container and inspect it for serviceability. The green end of the detonator must show in the bottom of the fuse.
  - — Remove the safety fork; use the hooked end of an M20 wrench if necessary (Figure A-23).

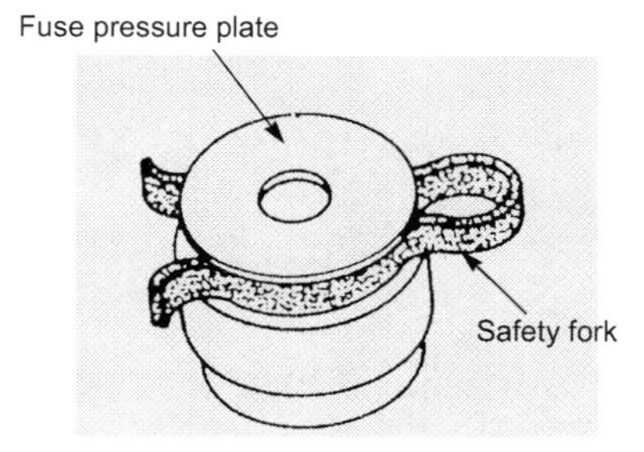


**Figure A-23. Safety fork**

  - — Insert the fuse into the fuse well carefully until it seats securely on top of the booster retaining ring.
  - — Perform a clearance test using the tab end of the M20 wrench (Figure A-24, page A-20).

**WARNING**
**If the fuse pressure plate interferes with the tab end of the M20 wrench, investigate the cause and notify the NCOIC. DO NOT arm the mine.**

**NOTE: For long-term emplacement, smear a thin layer of silicone grease or similar lubricant on the arming plug, the threads, and the gasket.**

  - — Ensure that the setting knob is in the *SAFE* position.
  - — Screw the arming plug into the mine by hand. Ensure a watertight seal by tightening the arming plug with the M20 wrench.
- Dig a hole to fit the mine.
  - — Dig a hole deep enough so that the top of the pressure plate is about 3 centimeters below ground level.
  - — Dig the sides of the hole at a 45-degree angle to prevent vehicles from bridging the mine (Figure A-25, page A-20).
- Emplace the mine.

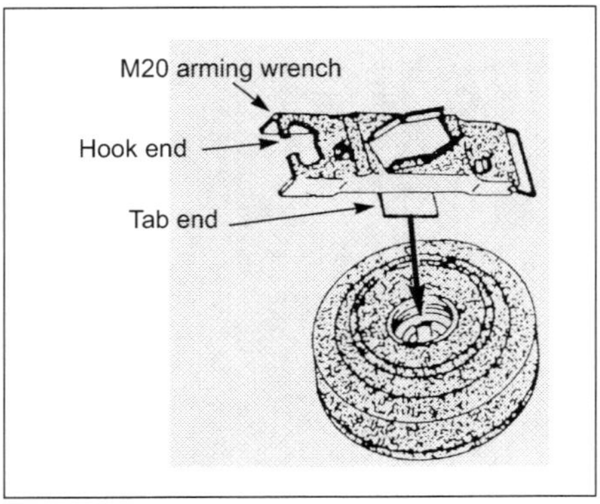


**Figure A-24. Clearance test**

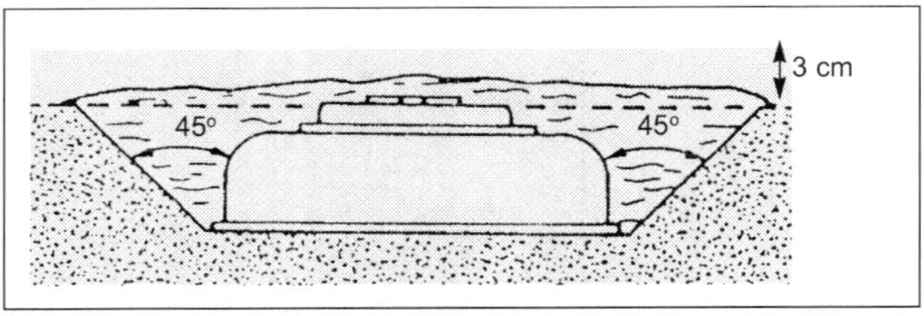


**Figure A-25. M15 mine in the hole**

- — Place the mine in the hole.
- — Cover the mine with soil until it is level with the top of the pressure plate.

- Use the M20 wrench to arm the mine by turning the setting knob from the *SAFE* position to the *ARMED* position.
- Camouflage the mine.
    - — Cover the mine with 3 to 5 centimeters of soil.
    - — Camouflage the mine, place excess soil in sandbags, and remove sandbags from the area.
    - — Give the safety clip to the NCOIC.

### REMOVAL USING THE M603 FUSE

- Disarm the mine.
    - — Clear the soil from the top of the mine carefully.
    - — Hold the mine firmly in place with one hand, without putting pressure on the pressure plate.
    - — Feel for AHDs with the other hand by digging around the sides of and underneath the mine.

   — Use the M20 wrench to turn the setting knob to the *SAFE* position.

- Remove the mine.
   — Remove the mine from the hole.
   — Use the M20 wrench to turn the arming plug counterclockwise, and remove the arming plug.
   — Remove the M603 fuse from the fuse well and replace the safety fork.
   — Install the arming plug.

## M19

The M19 AT mine (Figure A-26) is housed in a square, plastic case and holds 9.45 kilograms of Composition B (HE charge). It consists of an M606 integral pressure fuse and two secondary fuse wells—one in the side and one in the bottom. The fuse body contains a pressure plate, a Belleville spring, a setting knob, a step plate, a firing-pin assembly, and a detonator.

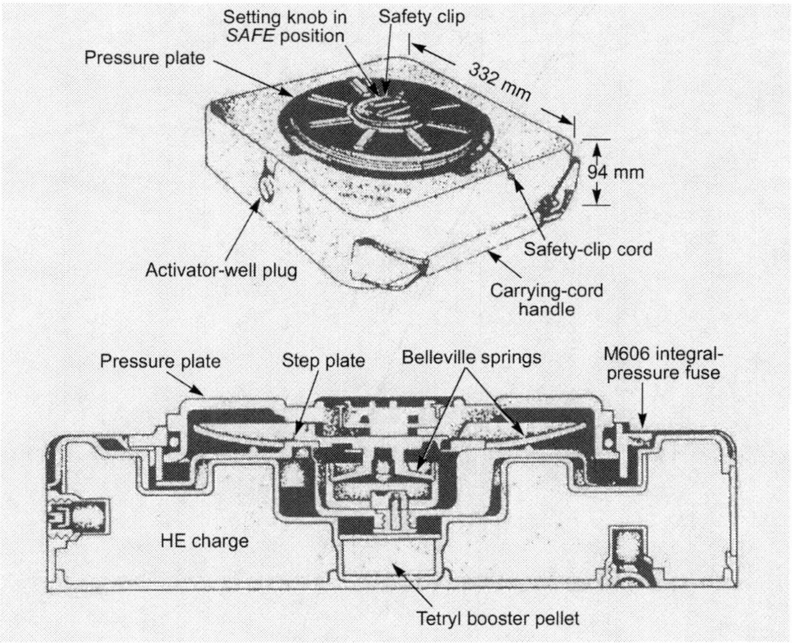


**Figure A-26. M19 AT mine**

## CHARACTERISTICS

| Main Charge | Diameter | Height | Weight | No Mines per Box | Weight per Box |
|---|---|---|---|---|---|
| Comp B, 9.45 kg | 332 mm | 94 mm | 12.6 kg | 2 | 32.5 kg |

- The M19 is employed in tactical and nuisance minefields.
- The M19 is buried or surface-laid.
- The M19 requires a force of 157 to 225 kilograms to detonate.
- Standard FDs may be used with the M2 activator in any of the secondary fuse wells of the M19.

## INSTALLATION

**WARNING**
**Emplace and remove the mine while in the prone position.**

- Inspect the mine.
  - — Do not use the mine if it is dented, cracked, or damaged.
  - — Remove the M606 fuse from the fuse well by turning it counterclockwise a quarter turn (Figure A-27).
  - — Ensure that the rubber gasket is on the M606 fuse.

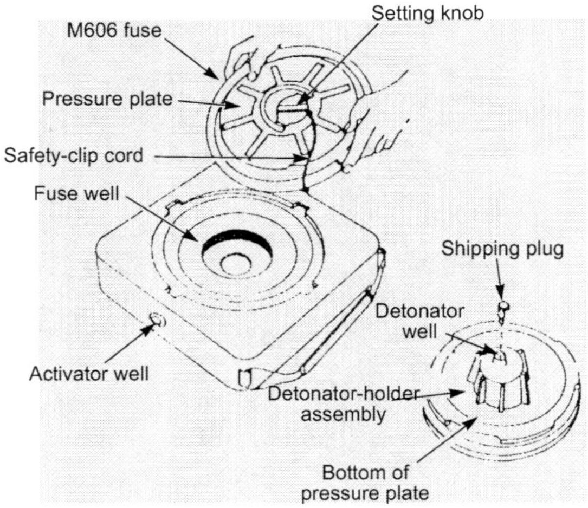


**Figure A-27. Removal of the pressure plate**

— Remove any foreign material from the fuse well.

— Ensure that the setting knob is in the *SAFE* position and that the safety clip is in place.

— Use the M22 wrench to remove the shipping plug from the detonator well.

— Inspect the detonator well for foreign material. If foreign material is present, gently tap the pressure plate with your hand to dislodge the material.

- Test the position of the firing pin (Figure A-28).

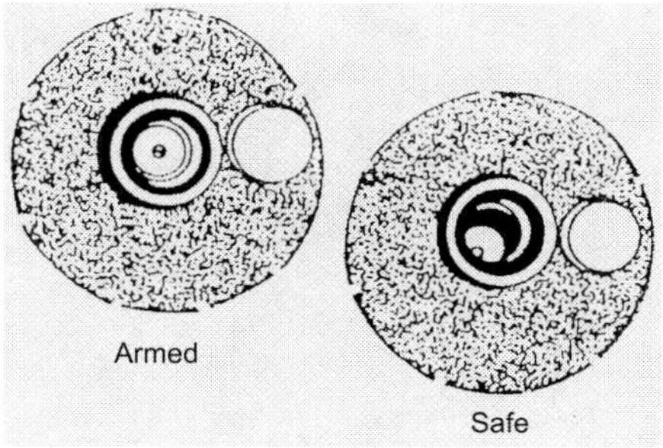


**Figure A-28. Firing pin**

— Ensure that the firing pin is at the edge of the well when the setting knob is in the *SAFE* position.

— Remove the safety clip.

— Use the M22 wrench to turn the setting knob to the *ARMED* position. Ensure that the firing pin is in the center of the well.

— Use the M22 wrench to turn the setting knob back to the *SAFE* position. Ensure that the firing pin moves back to the side of the well. **NOTE: If the firing pin is not in the correct position when the setting knob is in the *ARMED* or *SAFE* position, notify the NCOIC.**

— Replace the safety clip.

- Use the M22 wrench to screw the M50 detonator into the detonator well.
- Use the M22 wrench to tighten the M606 fuse into the fuse well.
- Dig a hole to fit the mine.

— Dig a hole deep enough so that the top of the pressure plate will be even with or slightly below ground level.

- — Dig the sides of the hole at a 45-degree angle to prevent vehicles from bridging the mine.
- Emplace the mine.
  - — Place the mine in the hole.
  - — Cover the mine with soil until it is level with the top of the pressure plate.
- Arm the mine.
  - — Remove the safety clip.
  - — Use the M22 wrench to turn the setting knob from the *SAFE* position to the *ARMED* position.
- Camouflage the mine.
  - — Cover the mine with 3 centimeters of soil.
  - — Camouflage the mine, place excess soil in sandbags, and remove sandbags from the area.
  - — Give the safety clip and the shipping plug to the NCOIC.

**REMOVAL**

- Disarm the mine.
  - — Clear the soil from the top of the mine carefully.
  - — Hold the mine firmly in place with one hand, without putting pressure on the pressure plate.
  - — Feel for AHDs with the other hand by digging around the sides of and underneath the mine.
  - — Use the M22 wrench to turn the setting knob to the *SAFE* position.
  - — Replace the safety clip on the M606 fuse.
- Remove the mine.
  - — Remove the mine from the hole.
  - — Use the M22 wrench to remove the M606 fuse by turning it counterclockwise and lifting it out of the fuse well.
  - — Use the M22 wrench to remove the detonator from the detonator well.
  - — Replace the shipping plug in the detonator well.
  - — Replace the pressure plate in the mine.

## M21

The M21 AT mine (Figure A-29) utilizes a direct-energy warhead that is designed to produce a K-Kill. It is used in conjunction with the M607 fuse. The M21 produces a K-Kill against heavy tanks, unless the mine is activated under the track. It can be buried with a tilt rod, or it can be surface-laid with or without a tilt rod. If the M21 is surface-laid with a tilt rod, it must be staked to prevent it from being knocked over and causing the warhead to be directed away from the target. The M21 is not compatible with any mechanical mine-dispensing system.

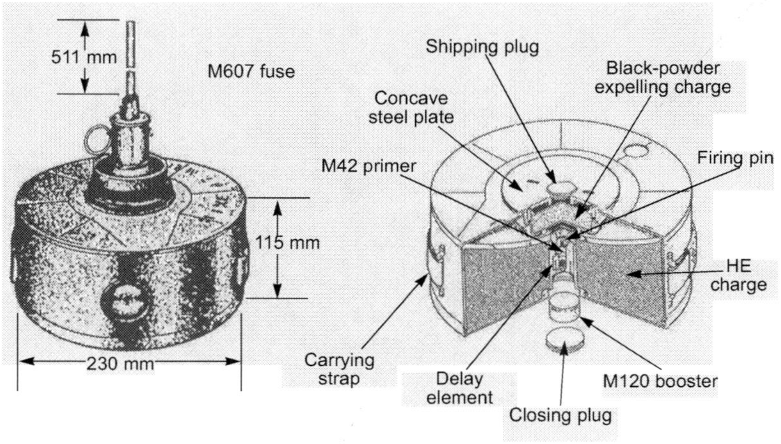


**Figure A-29. M21 AT mine**

### CHARACTERISTICS

| Main Charge | Diameter | Height | Weight | No Mines per Box | Weight per Box |
|---|---|---|---|---|---|
| Comp H6, 4.95 kg | 230 mm | 115 mm | 7.8 kg | 4 | 41 kg |

- The M21 requires a minimum of 130.5 kilograms of pressure to detonate.
- The M21 can be used with a tilt-rod assembly, requiring 1.7 kilograms of pressure on the extension rod to cause a 20-degree deflection.
- The M21 is the only conventional US AT mine with a direct-energy warhead.

INSTALLATION

> **WARNING**
> **Emplace and remove the mine while in the prone position.**

- Inspect the mine.
  - Do not use the mine if it is dented, cracked, or damaged.
  - Ensure that the cotter pins on the fuse pull-ring assembly and the fuse-closure assembly are securely in place (Figure A-30).
  - Inspect the fuse to ensure that the neck portion behind the tile collar is not cracked.

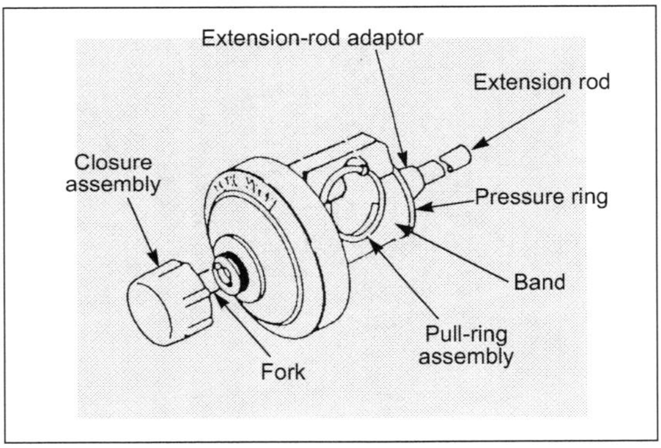


**Figure A-30. M607 fuse**

- Insert the booster.
  - Use the screwdriver end of the M26 wrench (Figure A-31) to remove the closing plug from the bottom of the mine.

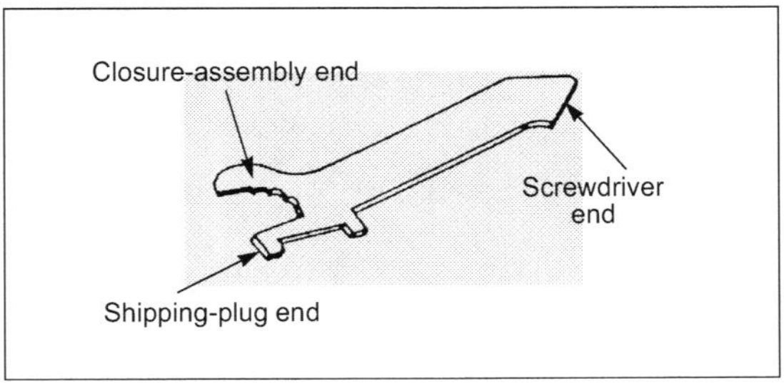


**Figure A-31. M26 wrench**

— Inspect the booster well for foreign material. If foreign material is present, gently tap the top of the mine with your hand to dislodge the material. If you cannot remove the debris, replace the closing plug and do not use the mine.

— Insert the M120 booster (with the washer side toward the fuse) into the booster well.

— Replace the closing plug with the M26 wrench.

- Fuse the mine.

  — Remove the shipping plug from the fuse well with the M26 wrench.

  — Inspect the fuse well for foreign material. If foreign material is present, gently shake the mine to dislodge the material. If black powder falls out of the fuse well or you cannot remove the foreign material, do not use the mine.

  — Remove the closure assembly from the M607 fuse with the M26 wrench. Ensure that the gasket remains in place on the fuse.

  — Screw the fuse hand-tight into the fuse well.

- Dig a hole to fit the mine.

  — Dig a hole deep enough so that the top of the mine will be at ground level (Figure A-32).

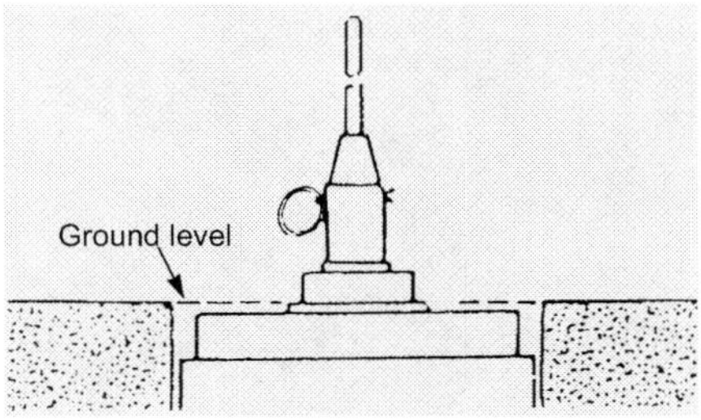


**Figure A-32. Buried M21 mine**

  — Check the bottom of the hole to ensure that the ground is solid enough to support the mine. If necessary, place a flat object under the mine to provide a firm foundation. Allow additional depth for the object.

- Emplace the mine.

  — Place the mine in the hole.

  — Cover the mine with soil until it is level with the top of the mine. Ensure that no soil falls around or under the plastic collar.

— Press the soil firmly around the sides of the mine.

- Assemble the extension rod. **NOTE: For pressure operation, do not use the extension-rod assembly.**

  — Screw the extension rod onto the M607 fuse.

  — Ensure that the extension rod is pointing straight up.

- Arm the mine.

  — Squeeze the end of the cotter pin together on the pull ring.

  — Remove the cotter pin by holding the fuse firmly in one hand and pulling the pull ring with the other hand.

  — Remove the band and the stop slowly and carefully from the neck of the fuse (Figure A-33).

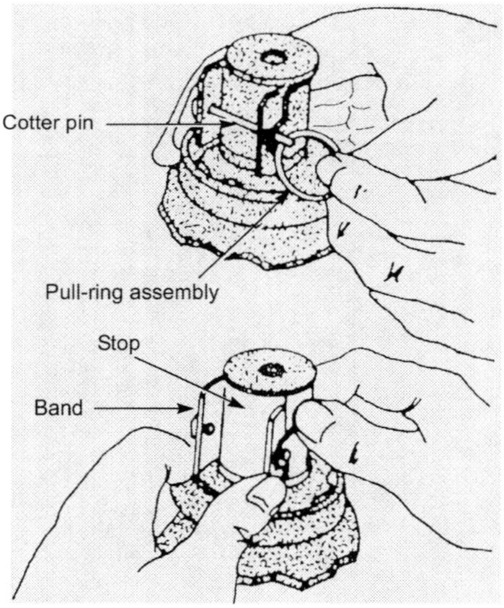


**Figure A-33. Removing the band and the stop**

- Camouflage the mine.

  — Camouflage the mine with twigs, grass, or other material in the area. Place mines with extension rods in tall grass, if possible. Ensure that no pressure is applied to the tilt rod or the fuse.

  — Place the excess soil in sandbags and remove them from the area.

— Give the band, the stop, the pull ring, the shipping plugs, and the closure assembly to the NCOIC.

**REMOVAL**

- Disarm the mine.
  - — Clear the camouflage away from the mine carefully.
  - — Attach the band and the stop to the fuse.
  - — Insert the cotter pin into the band and the stop. Spread the ends of the cotter pin.
  - — Remove the extension rod.
- Check for AHDs.
  - — Hold the mine firmly in place with one hand, without putting pressure on the fuse.
  - — Feel for AHDs with the other hand by digging around the sides of and underneath the mine.
- Remove the mine.
  - — Remove the mine from the hole.
  - — Remove the fuse from the mine.
  - — Install the closure assembly on the fuse.
  - — Install the shipping plug into the fuse well of the mine.
  - — Remove the closing plug from the bottom of the mine.
  - — Remove the booster from the mine.
  - — Install the closing plug into the booster well.

## SECTION III. FIRING DEVICES AND ACTIVATORS

An FD performs the function of a mine fuse by providing an alternative means to detonate the mine. It is normally used in conjunction with a standard fuse so that a mine will have two separate explosive chains. The purpose of the second firing chain is to prevent the enemy from disarming or removing mines after emplacement. When used for this purpose, the FD is called an AHD and it is designed to function by detonating the attached mine or another explosive charge nearby if unauthorized personnel attempt to remove or tamper with the mine. **NOTE: US forces will not employ AHDs on AP mines.** Both the M19 and the M15 have two secondary fuse wells for attaching an FD and an activator.

There are two standard US FDs—M5 pressure release and M142 multipurpose. They utilize a spring-loaded striker and a standard base and are designed to function in one or more of the following modes:

- Pressure.
- Pressure release.
- Tension.
- Tension release.

## M5 PRESSURE-RELEASE FIRING DEVICE (MOUSETRAP)

The M5 FD (Figure A-34) is activated by the release of pressure. Lifting or removing a restraining weight releases the striker to fire the cap.

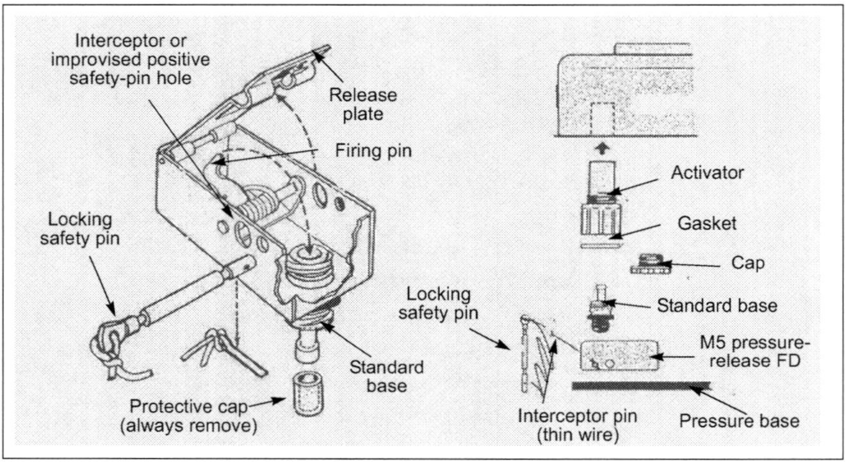


**Figure A-34. M5 FD**

### CHARACTERISTICS

- Case: Metal.
- Color: Olive-drab.
- Length: 445 millimeters.
- Width: 239 millimeters.
- Height: 175 millimeters.
- Internal action: Mechanical with hinged striker release.
- Initiating action: Removal of restraining weight, 2.25 kilograms or more.
- Accessories: Pressure board.
- Safeties: Safety pin and hole for interceptor pin.
- Packaging: Four complete FDs and four plywood pressure boards are packaged in a paper carton, five cartons are packaged in a fiberboard box, and 10 fiberboard boxes are shipped in a wooden box.

### INSTALLATION

- Dig a hole deep enough to bury the mine on a firm foundation. The pressure plate should be slightly above ground level.

> **WARNING**
> **Ensure that the mine and the FD are resting on a firm foundation before removing the pins.**

- Remove the protective cap from the standard base. Assemble the FD.
- Use a coat hanger as a positive safety pin. Place the mine in the hole. Leave enough room to remove the pins.
- Bury and camouflage the mine.
- Arm the mine.
- Remove the locking safety pin (Figure A-35).

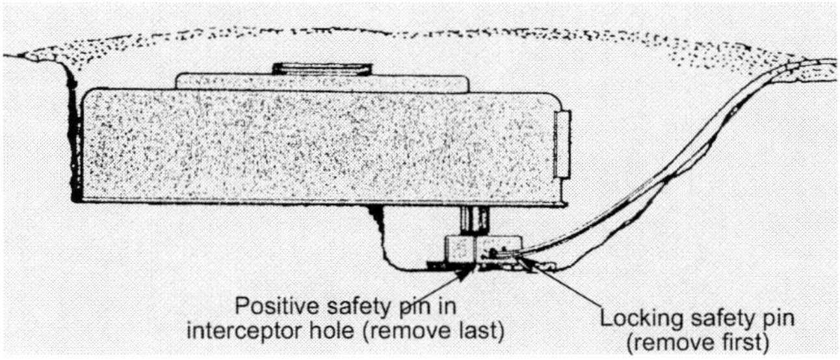


**Figure A-35. Arming the M15**

- Remove the positive safety pin (interceptor pin). **NOTE: If the positive safety pin is difficult to remove or if you hear a click when removing the locking safety pin, carefully remove the mine and replace the FD.**

### REMOVAL

- Uncover the mine carefully and inspect it for tampering. Locate and carefully uncover the FD.

> **WARNING**
> **DO NOT release the pressure being applied to the device.**

- Insert the positive safety pin into the interceptor hole, and then insert the locking safety pin into the safety-pin hole.

- Disarm and remove the mine. Recover the FD.

## M142 MULTIPURPOSE FIRING DEVICE

The M142 FD (Figure A-36) can be designed to function in the following modes:

- Pressure (11 kilograms or more).
- Pressure release (between 2 and 67 kilograms).
- Tension (3 kilograms or more).
- Tension release.

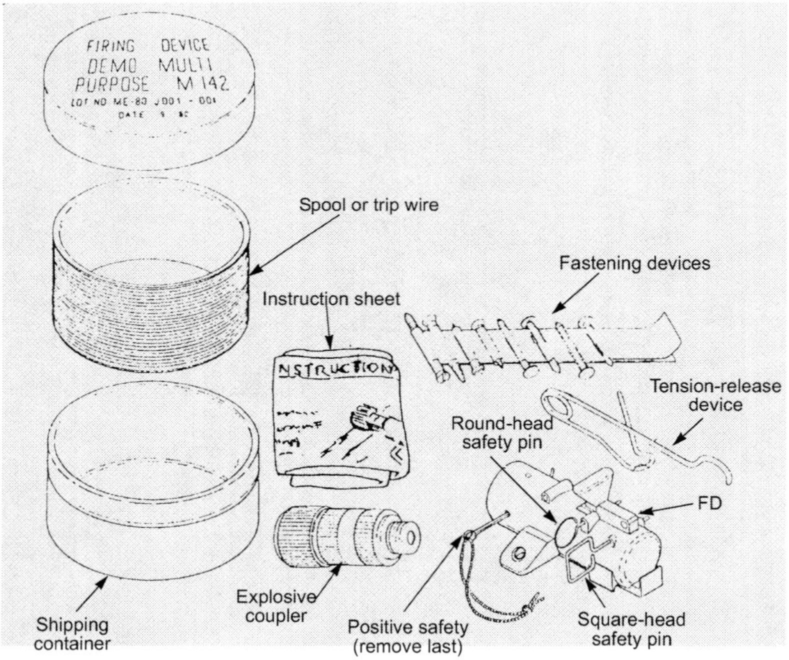


**Figure A-36. M142 FD**

Although primarily intended for booby-trap applications, the M142 can be readily adapted as an AHD. The M142 comes with a coupling device and a primer that accepts a standard nonelectric blasting cap. The initiating action sets off an explosive chain that passes from the FD and the primer to the blasting cap, and then via the detonating cord to the main charge. However, the coupling device with the primer will not initiate the detonating cord alone without a blasting cap attached, and it is not adaptable to any activator or secondary fuse well. When the M142 is used as an AHD, the coupling device is removed and an M1 or M2 standard base is used.

### CHARACTERISTICS

- Case: Plastic.
- Color: Olive-drab.
- Diameter: 190.5 millimeters.
- Length: 571.5 millimeters.
- Internal action: Spring-driver striker.
- Safeties: Positive safety pin, square-head pivot pin, round-head pivot pin, and alternative safety-pin hole.
- Accessories: Nail and screw fasteners, coupling assembly, tension-release attachment, 15-meter spool of trip wire, and vinyl instruction sheet.
- Packaging: Round, metal can containing FD with accessories.

### ARMING AND DISARMING

Arming and disarming procedures vary based on the activation mode. Detailed instructions are printed on a weatherproof, vinyl sheet included in each FD package.

## M1 AND M2 ACTIVATORS

When FDs are employed with M15 and M19 AT mines, they require the use of an M1 or M2 activator.

Activators are essentially detonator boosters that are designed to magnify the explosive force generated by an FD with a standard base and transfer the force to the main charge. Activators may be used with either type of FD to supply an AT mine with a secondary fuse for antihandling purposes. The M1 activator is used with the M15 AT mine, and the M2 activator is used with the M19 AT mine. The activator also performs the function of an adapter for attaching the FD to the mine. One end of the activator is threaded externally for insertion in the secondary well of the mine; the other end is threaded internally to receive the standard base coupling of the FD.

The M1 activator (Figure A-37, page A-34) is 54 millimeters long (with cap), is made of olive-drab plastic, contains a detonator, and has a threaded closing plug and a gasket. It has a cylindrical, unthreaded cap that is cemented to the opposite end of the body and contains a tetryl booster charge. The threaded end, which screws into the mine, is 25 millimeters in diameter.

The M2 activator is similar to the M1 except that it contains an HE pellet, and its overall length, with cap, is 53 millimeters.

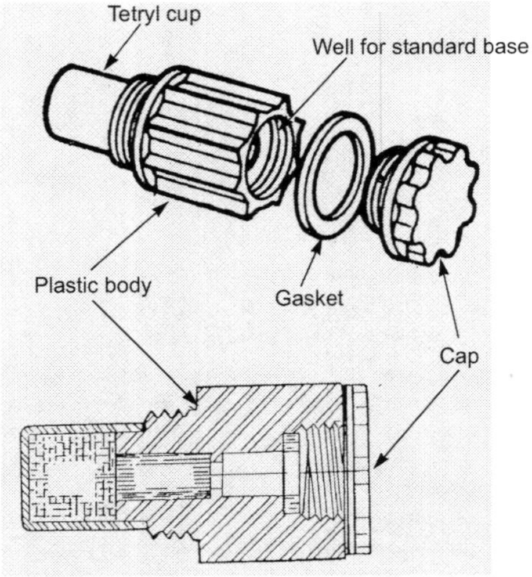


Figure A-37. M1 activator

Appendix B

# Controls and Components of Special-Purpose Munitions

This appendix provides characteristics and detailed descriptions of US special-purpose munitions. The use of these munitions is outlined in Chapter 4.

## SELECTABLE LIGHTWEIGHT ATTACK MUNITION

The SLAM is a multipurpose munition with antidisturbance and antitamper features. There are two models of the SLAM—one is self-neutralizing (M2) and the other is self-destructing (M4). The M2 is solid green and has no labels, brands, or other distinguishing marks. The M4 is green with a black warhead (EFP) face.

Employment methods for the SLAM are outlined in Chapter 4.

Figure B-1 describes and illustrates the major components of the SLAM.

| Component | Description |
|---|---|
| Mounting holes (1) | The mounting holes are used to secure the carrying strap or the mounting wire to the SLAM when attaching the SLAM to trees and so forth. |
| Bore sights (2) | Two bore sights and an omega sight are located on the top of the SLAM and are used to aim the SLAM at targets. |
| Selector switch (3) | The selector switch is used to select operating modes and times. It has eight detent positions. The switch is against a stop (in the shipping position), which is the only switch position that allows the SLAM to fit in the reusable environmental protective pack. Turning clockwise, there are three positions for selecting the operating time (4, 10, and 24 hours). Setting any of these positions will select an internal sensor mode of operation, which is a magnetic sensor for mine mode and a passive infrared sensor for side-attack mode. These three positions will cause the SLAM to self-destruct (M4) or self-neutralize (M2) at the end of the selected operating time. Continuing clockwise, the last four positions select an internal timer, which sets the minutes until demolition. These positions are 15, 30, 45, and 60 minutes. |
| Activation-lever shear pin (4) | There is a shear pin mounted across the SLAM's lever slot. If the shear pin is sheared, thereby breaking the seal, the lever may have been pulled and the SLAM may be an electronic dud. If the shear pin is broken, it should only be used in the command-detonation mode. |
| Safety pin (7) | The safety pin slides from the body and starts the SLAM's timing. It is pried from its latch with the tip of the lever. Once the safety pin is pulled, it cannot be reinserted. |

**Figure B-1. SLAM components**

| Component | Description |
|---|---|
| Passive infrared sensor (8) and cover (9) | The SLAM is equipped with a passive infrared sensor that detects trucks and light armored vehicles by sensing the change in background temperature as vehicles cross in front of the SLAM. The sensor is directional and is aligned with the EFP. The sensor is active when the SLAM is operating with the selector switch set to 4, 10, or 24 hours and the sensor cover is removed to expose the infrared sensor (such as, during the side-attack mode). The SLAM will self-destruct (M4) or self-neutralize (M2) if the selected time expires before it is detonated by vehicle passage. |
| Blasting-cap well and plug (10) | The threaded plug seals the blasting-cap well. It is removed to mount a standard military blasting cap with a priming adapter. |
| Warhead (11) | The warhead is an EFP that is designed to defeat light armored vehicles. The EFP forms within the first 5 inches of flight and has an effective range of 25 feet. |
| Housing assembly (12) | The housing assembly contains the fusing, electronics, and S&A components. It also provides a structural interface for the warhead, the sights, the activation lever, the passive infrared sensor, the selector switch, and the safety pin. |

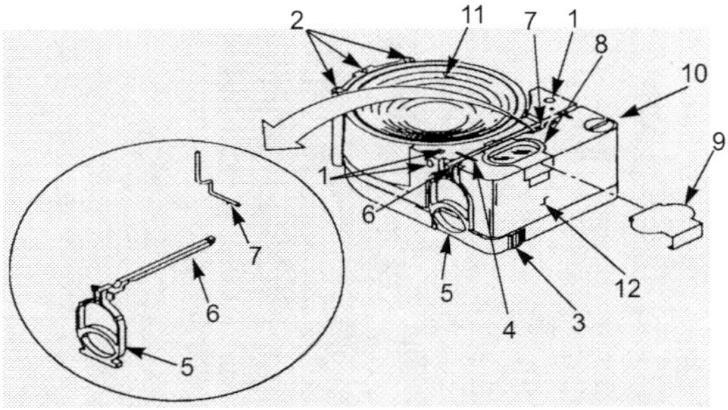


Figure B-1. SLAM components (continued)

# M93 HORNET

The M93 Hornet is a lightweight (35 pounds) AT/antivehicular munition that one person can carry and employ. It is a one-time use, nonrecoverable munition that is capable of destroying vehicles using sound and motion as detection methods. The Hornet will automatically search, detect, recognize, and engage moving targets, using top attack at a maximum standoff distance of 100 meters. It is employed by units equipped with an M71 RCU. The RCU is a hand-held encoding unit that interfaces with the Hornet when the remote mode is selected at the time of employment. After encoding, the RCU can be used to arm the Hornet, reset SD times, and destruct the Hornet. Employment methods of the Hornet are outlined in Chapter 4.

Figure B-2 describes and illustrates the major components of the Hornet. Figure B-3, page B-4, describes and illustrates the controls and indicators of the Hornet.

| Component | Description |
|---|---|
| Support legs (1) | Support legs are used to stabilize the Hornet when it is deployed. |
| Active battery-pack cover (2) | The active battery-pack cover provides a seal to protect and secure the active battery pack. The latch is lifted up to remove the cover, the active battery pack is installed, and the cover is then reinstalled and latched down. A line secures the battery-pack cover to the control panel of the munition. |
| SD switch (3) | The SD switch is a six-position rotary switch that is used to select the SD time and unlock the arm control switch. The SD switch is also used to unlock the arming lever. This is done by rotating the switch to the setting "U." A red lock element is extended 1/8 inch from the side of the munition when the SD switch is in the unlock position. The SD time is preset to Setting 1 when the Hornet is shipped. SD times are as follows:<br>Setting — Time<br>1 — 4 hours<br>2 — 48 hours<br>3 — 5 days<br>4 — 15 days<br>5 — 30 days |
| Arm control switch (4) | The arm control switch consists of an arming lever interlocked with the SD switch and the S&H band assembly to prevent inadvertent actuation. Until the S&H band assembly is removed and the SD switch is placed in the unlock position, the arming lever cannot be moved to the arm position. An internal lock secures the arming lever in the arm position. |
| Microphones (5) | When the geophone seismic sensor detects a potential target, usually at ranges up to 600 meters, it alerts the munition to start listening with the three microphones that extend from the munition body. They track the two loudest noise sources that are heard. |
| Antenna (6) | The antenna provides a means for the Hornet to receive M71 RCU commands. |
| Capture screws (7) | These are four flat-head screws that secure the bottom plate to the munition body. They are removed along with the bottom plate to access the battery compartment. |
| Bottom plate (8) | The bottom plate provides a seal to protect and secure the battery compartment and connect the batteries once they are installed. |
| D-cell batteries (9) | The battery compartment houses four D-cell batteries. A drawing on the inside of each battery tube shows battery orientation. |
| Dowel pin (10) | The dowel pin ensures that the bottom plate is in the correct orientation to properly connect the batteries. |

**Figure B-2. Hornet components**

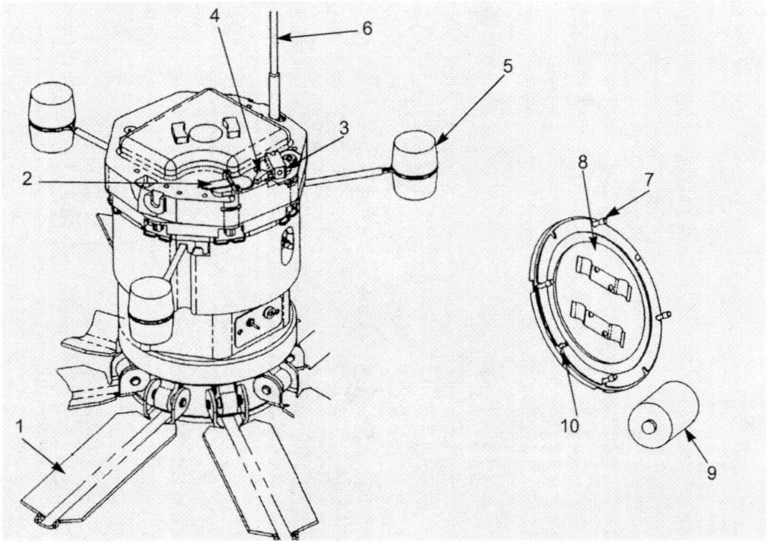


**Figure B-2. Hornet components (continued)**

| Component | Description |
|---|---|
| Magnetic coupling device (MCD) (1) | This device is used as part of the RCU interface. The RCU interface consists of the MCD and keyed tabs. In the remote arming mode, the RCU is placed on top of the MCD and minefield code data is transferred to the munition. Upon successful encoding, the status light begins to flash. |
| Target switch (2) | The target switch is a toggle switch used to select the type of target engagement. This gives the operator the choice between detecting and destroying only heavy armored vehicles or all vehicles. |
| Manual select switch (3) | The manual select switch is a push-button switch, protected by a plastic cover that must be removed to access the switch. Successful activation of the switch will cause the status light to flash. This switch is used to allow the operator to employ the Hornet without the RCU. |
| Status light (4) | The status light is a visual indicator for the operator during the munition setup. It is a green light-emitting diode (LED) that indicates a self-test was successfully performed or an operating-mode selection was successfully selected. |
| SD switch (5) | See Figure B-2, page B-3. |
| Arming lever (6) | See Figure B-2. |
| Active battery-pack cover (7) | See Figure B-2. |

**Figure B-3. Hornet controls and indicators**

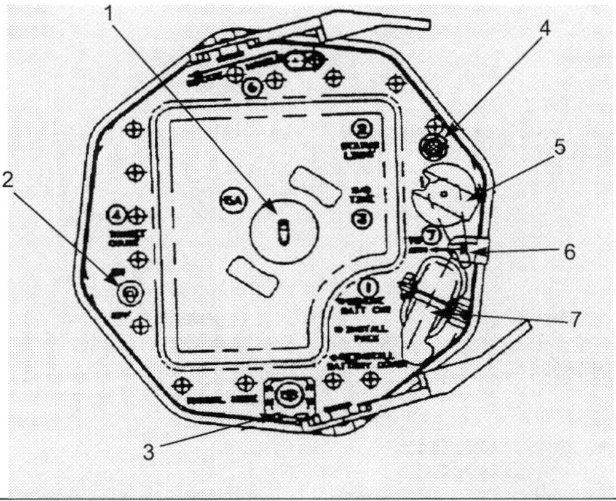


Figure B-3. Hornet controls and indicators (continued)

Appendix C

# Threat Mine/Countermine Operations

This appendix is intended to complement the information presented in other manuals on threat obstacle tactics. It applies to most threat armies and their surrogates. Commanders should use this information to give added realism to unclassified training, although obstacle employment norms can change with METT-TC factors for a given AO. Therefore, preoperational training on templating, intelligence, reconnaissance, and reduction procedures must be based on the best information available before deployment.

Appendix G contains a compilation of countermine data.

## MINE OPERATIONS

Threat formations contain considerable organic minefield emplacement capability. Threat rapid-mining capability presents a serious challenge to friendly maneuver.

To lay mines and place obstacles rapidly during offensive operations, threat armies form a special team from regimental and divisional assets. This team is called a *mobile obstacle detachment* (MOD). The MOD places AT mines on the most likely avenues for armored attacks or counterattacks. MODs are positioned on the flanks of a march formation for rapid deployment and are normally close to AT reserves. During the march, MODs reconnoiter avenues into the flanks and identify the most likely avenues for tank movement. At secured objectives, MODs reinforce existing obstacles and place new obstacles to assist in the defeat of counterattacks.

The combined arms commander orders the organization of MODs and determines their composition based on the combat situation and available troops. Engineer elements in a division MOD come from the divisional engineer battalion and normally consist of three armored tracked mine layers known as *GMZs* (Figure C-1, page C-2). This platoon-sized element has two or three trucks that carry mines for immediate resupply. For the regimental MOD, the regimental engineer company normally provides a platoon-sized unit equipped with two or three GMZs. The platoon travels in BTR-50/60s and has 600 AT mines.

The GMZ dispenses mines at a predetermined spacing of 5.5 meters. Mine-laying helicopters also support the MOD. The HIP and HIND-D helicopters carry two or three dispenser pods of AP or AT mines. Artillery-fired SCATMINEs can also support the MOD. Three GMZs can lay a 1,200-meter, three-row minefield, containing 624 mines, in 26 minutes. Doctrinally, this minefield would be broken into several minefields, each 200 to 300 meters long.

Threat armies use obstacles extensively throughout the depth of their defense, and their tactics are chosen well. Shallow obstacles are reduced quickly and easily. For example, a shallow, one-row minefield is essentially reduced by blowing one or two mines in the row. A threat rapidly emplaced minefield

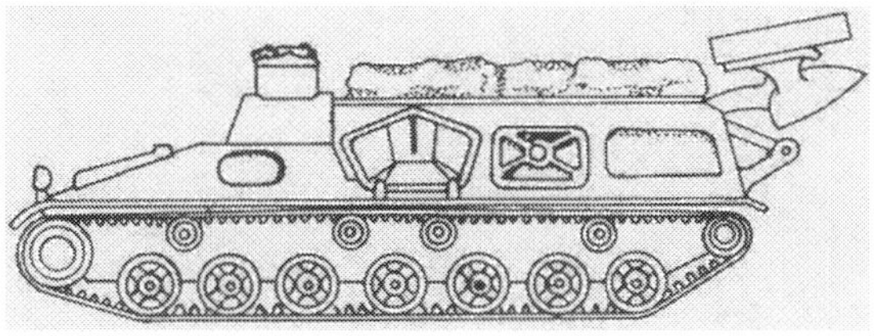

Figure C-1. GMZ armored tracked mine layer

consists of three or four 200- to 300-meter rows, spaced 20 to 40 meters apart, with mines spaced 4 to 6 meters apart. As a rule, the minefield covers the depth of a football field.

Table C-1 provides detailed information on standard threat AT and AP minefields. Terrain and tactical situations dictate the actual dimensions and distances of minefields.

**Table C-1. Normal parameters for threat-style minefields**

| AT Minefields | |
|---|---|
| Front (situation-dependent) | 200 to 300 meters |
| Depth | 40 to 120 meters |
| Number of rows | 3 or 4 |
| Distance between rows | 20 to 40 meters |
| Distance between mines | 4 to 6 meters for antitrack mines; 9 to 12 meters for anithull mines |
| Outlay, normal | 550 to 750 antitrack mines per kilometer; 300 to 400 antihull mines per kilometer |
| Outlay, increased effect | 1,000+ antitrack mines per kilometer; 500+ antihull mines per kilometer |
| Probability of destruction | 57% for antitrack mines (750 per kilometer); 85% for antihull mines (400 per kilometer) |
| **AP Minefields** | |
| Front (situation-dependent) | 30 to 300 meters |
| Depth | 10 to 150 meters |
| Number of rows | 3 or 4 |
| Distance between rows | 5+ meters for blast mines; 25 to 50 meters for fragmentation mines |
| Distance between mines | 1 meter for blast mines; 50 meters (or twice the lethal radius of fragmentation) for fragmentation mines |
| Outlay, normal | 2,000 to 3,000 HE/blast mines per kilometer; 100 to 300 fragmentation mines per kilometer |
| Outlay, increased effect | 2 to 3 times the normal outlay |
| Probability of destruction | 15 to 20% for HE/blast mines (2,000 per kilometer); 10 to 15% for fragmentation mines (100 per kilometer) |

Figure C-2 shows a standard rapidly emplaced minefield. The threat army typically uses such a minefield when they are in a hasty defense (offense is temporarily stalled).

Figures C-3 shows a standard antitrack minefield.

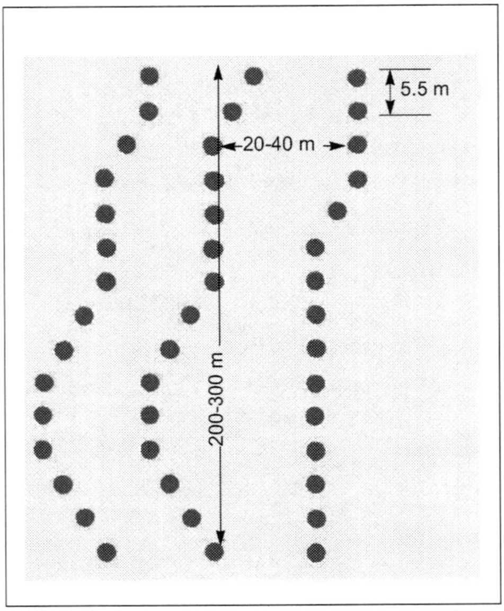


**Figure C-2. Threat-style rapidly emplaced minefield**

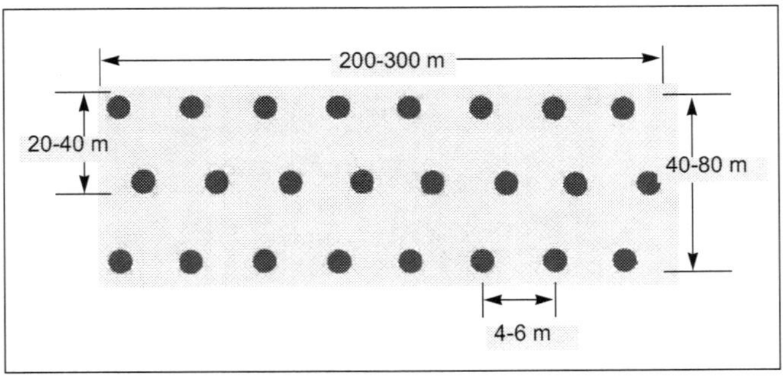


**Figure C-3. Threat-style antitrack minefield**

Figure C-4 shows a standard antihull minefield.

Figure C-5 shows a standard AP minefield.

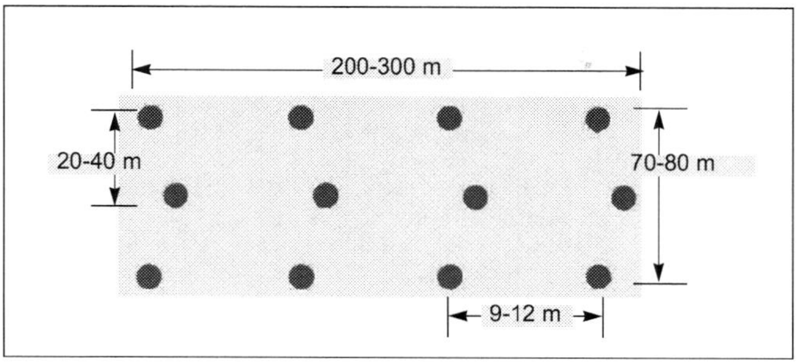


**Figure C-4. Threat-style antihull minefield**

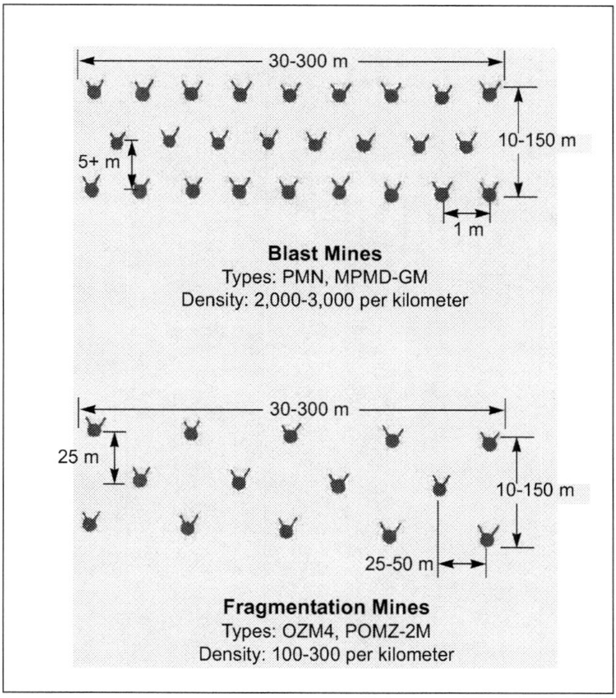


**Figure C-5. Threat-style AP minefield**

Threat armies also emplace mixed minefields. They are not the same as US mixed minefields. Threat armies normally emplace three rows of AT mines, then several rows of AP mines. AT and AP mines **are not** mixed in the same row.

Threat engineers use two fundamental drills to emplace mines:

- When emplacing armed mines, the drill uses a crew of five sappers. The first crew member (the senior man and operator) is in the mine-layer's seat and monitors the operation of the mine layer and the motion of the mines in the guide chute. He also sets the mine spacing and controls the actions of the GMZ. The second and third members take mines out of containers and place them in the intake chute at intervals between the guide tray's drive chain. The GMZ driver steers the vehicle along the indicated route at the established speed.
- When emplacing unarmed mines, two or three additional sappers are assigned to arm the mines. After emplacing the mines, one sapper trails the mine layer, marks emplaced mines with pennants, and partially camouflages the mines. The remaining sapper(s) then arm the mines.

Special precautions are taken when emplacing AP minefields. Threat doctrine only allows PMN mines to be surface-laid from mine layers. POMZ-2M mines are emplaced with the truck-and-tray technique. Extra effort is required to assemble, emplace, and deploy the trip wire and to camouflage the POMZ-2M mine.

Using three GMZs, a threat MOD can emplace 1,200 meters of a three-row AT, surface-laid minefield, containing 624 AT mines, in 26 minutes. This does not include the 12- to 15-minute reload and travel times. Travel and reload times increase during limited visibility.

Threat forces can also have ground-emplaced SCATMINE capability. One such system is the UMZ SCATMINE system (Figure C-6, page C-6). There are three UMZ truck-mounted SCATMINE systems in each combat regiment. The UMZ consists of six firing modules mounted on the back of a Zil-131 truck. Each module has 30 firing tubes, for a total of 180 firing tubes per system. Depending on the type of minefield desired, the UMZ can lay 180 to 11,520 mines without reloading. The UMZ can launch an AT or AP minefield 30 to 60 meters from the vehicle while the truck is driving 10 to 40 kph. It takes two men 1½ to 2 hours to reload the UMZ. One UMZ can lay a three-row minefield, 150 to 1,500 meters long, depending on the type of mine that is used.

UMZ vehicles are usually deployed together as a mobile obstacle/mine-laying detachment. The UMZ is used to lay minefields that protect subunit positions and flanks and the boundaries between subunits. UMZ-laid minefields also cover firing lines and gaps in combat formations. The UMZ can quickly close gaps in existing minefields and increase the density of mines on armor avenues of approach.

For hand-emplaced SCATMINEs, there is a man-portable SCATMINE dispenser. The PKM weighs 2.63 kilograms (without the mine canister) and consists of a single launch tube with a base mount, a blasting machine, and a reel of electric ignition wire. The operator loads a propelling charge and a mine canister into the launch tube and mounts the tube on the edge of a trench or firing parapet. He then aims the tube, connects the ignition wire to

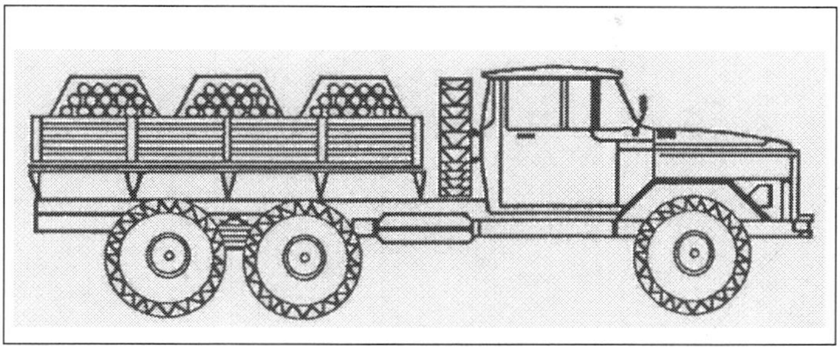

Figure C-6. UMZ SCATMINE system

the tube, and moves to a safe distance. At an initiating point, the operator connects the ignition wire to the blasting machine and initiates the system. The PKM propels the canister 30 to 100 meters, depending on the type of mine. It lays an AP minefield that is 10 by 20 meters (POM-1S mine canister), 10 by 40 meters (POM-2S mine canister), or 20 by 10 meters (PFM-1S mine canister). It takes a trained operator 5 minutes to set up the PKM and create a minefield. The PKM can also be used to launch the PTM-1S and PTM-3 AT mine canisters.

Threat forces use the PKM to lay minefields that protect subunit positions and flanks and the boundaries between subunits. PKM-laid minefields also cover firing lines and gaps in combat formations. The PKM can quickly close breaches in existing minefields and increase the density of mines on armor avenues of approach.

The type and complexity of an obstacle depends on the installing unit. Maneuver and artillery soldiers usually install simple single-system minefields that are protective in nature. Engineer soldiers install complex obstacles that can include AHDs. Engineer obstacle placement is usually equipment-intensive. Threat engineer effort generally concentrates on tactical obstacles unless maneuver soldiers are unable to employ the necessary protective obstacles. Threat units continue to improve the obstacles, supporting their positions by marking the friendly side of the obstacles, burying mines, and adding AHDs.

## CHEMICAL MINES

Chemical land mines are AP mines with command- or target-detonated fuses, and they are filled with a persistent chemical (nerve or blister) agent. US policy prohibits their use by US personnel. However, this does not preclude their use by other countries, and US forces may encounter them during operations. When used, they are normally used in defense and retrograde operations. They are mixed with HE mines to form a HE chemical minefield. Chemical mines are normally encountered in tactical or nuisance minefields, and some countries use them in protective minefields. When an integrated HE chemical minefield is laid, it serves the following purposes:

- Chemical mines discourage the use of explosive, rapid mine-clearing devices because they create a chemical hazard in the area.

- HE mines reduce the speed of enemy forces crossing the minefield. Speed is further reduced by forcing the enemy to use protective clothing and masks.

Chemical mines will usually be added to existing HE minefields by laying additional strips of chemical mines in a random pattern or by adding HE chemical strips or rows to the front or rear of existing fields (Figure C-7).

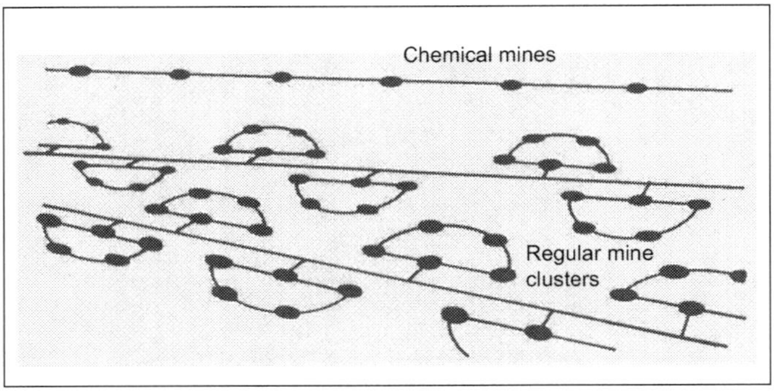


**Figure C-7. Chemical-mine employment**

No particular branch is responsible for clearing chemical mines. Planning chemical countermine operations is a brigade-level responsibility. When reducing chemical mines, consider prevailing and expected wind conditions. Commanders must ensure that friendly troops are protected when chemical agents are released. The release of chemical agents occurs as a result of enemy fire or friendly breaching attempts. Contact-actuated chemical mines are not likely to create a major downwind hazard because only single mines or small groups may be set off at one time.

## COUNTERMINE OPERATIONS

In offensive operations, threat engineers clear lanes through obstacles when they cannot be bypassed. Although clearing obstacles applies to the march and the defense, the most critical performance of this task occurs during the attack. Engineers can be required to clear mines delivered by air, artillery, and rockets well ahead of NATO's forward edge. They must breach obstacles contained within NATO strongpoints. Threat forces must also clear their own minefields when making the transition from defense to offense. In the offense, threat forces breach or bypass remotely delivered minefields in their form-up areas or routes of movement to the attack line. They also breach obstacles along the forward edge of the battle area and deep within NATO defenses.

Although clearing passages through obstacles is a primary task for threat engineers, any maneuver element may encounter mines. Engineers may not be able to respond to every encounter, so maneuver troops are also required to breach through remotely emplaced obstacles.

### ORGANIZATION

A movement support detachment (MSD) supports the movement of maneuver forces. It is task-organized from divisional or regimental engineer assets and can be platoon- to company-size. The MSD is equipped with route- and mine-clearing vehicles and devices. Depending on the mission (which comes directly from the combined arms commander or the chief of engineer services), an MSD is capable of filling craters, clearing minefields, preparing bypasses around major obstructions, and identifying NBC-contaminated areas.

The divisional engineer battalion can form two or three MSDs. During marches, MSDs travel in advance of the main body and clear obstructions reported by division reconnaissance elements. When they are deployed on main routes, they are under the protection of an advance guard or forward security element. When deployed on other routes, the leading regiments provide MSDs from organic engineer assets. An MSD at this level might consist of an engineer platoon, with one or two dozers and up to three tanks fitted with dozer blades. MSDs can be protected by a platoon of infantry or tanks and are usually accompanied by chemical-reconnaissance personnel. They can detect, mark, and breach hasty minefields that are not properly covered by fire. If MSDs encounter properly defended minefields, their clearing capabilities are limited.

Each battalion forms an obstacle-clearing group to create gaps in explosive and nonexplosive obstacles. Normally a part of a battalion-level MSD, the group follows first-echelon companies in APCs and creates gaps for those forces. These units may possess BAT-M vehicles with BTU bulldozer blades (Figure C-8) or KMT-series mine plows (Figure C-9).

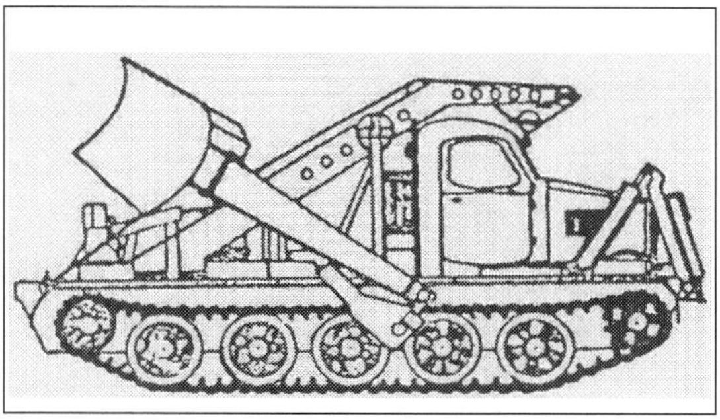

**Figure C-8. BAT-M with BTU bulldozer blade**

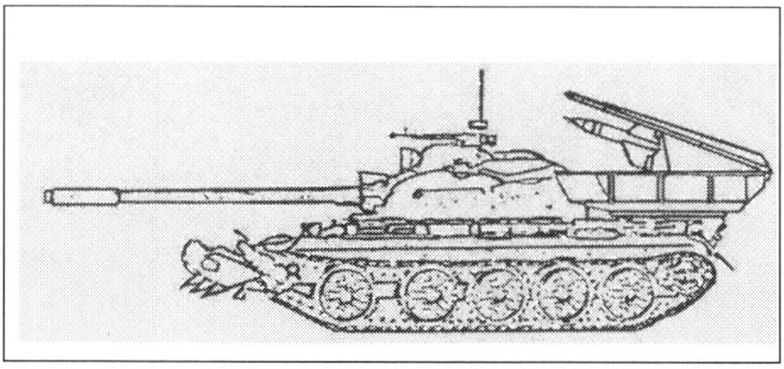

**Figure C-9. KMT-4 plow**

An obstacle-clearing detachment is created when more resources are needed to clear obstacles and debris. This usually occurs in urban environments and under conditions of massive destruction. An obstacle-clearing detachment is similar to an MSD, but its sole mission is to clear debris. Like an MSD, its composition depends on the mission scope, the mission objective, and the tempo of the offensive.

The divisional engineer battalion of the motorized rifle or tank division has a sapper company to clear obstacles. The company commander receives a mission to clear minefields. He then determines the exact location of the obstacle, ascertains the assets to devote to the task, and plans the methodology for success. Teams may be created to manually breach lanes using probes, IMP portable mine detectors (Figure C-10), and shovels. Larger tasks may necessitate the use of vehicle-mounted DIM mine detectors (Figure C-11), armored vehicle mine plows and/or rollers (Figure C-12, page C-10), and explosive line charges. When necessary or more practical, mines are explosively destroyed in place.

Figure C-10. IMP portable mine detector

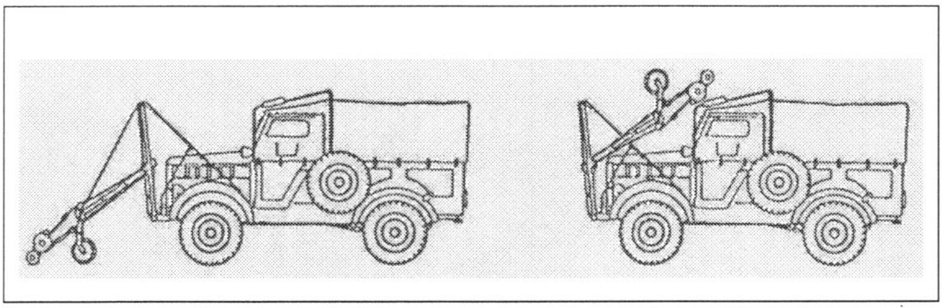

Figure C-11. DIM mine detector

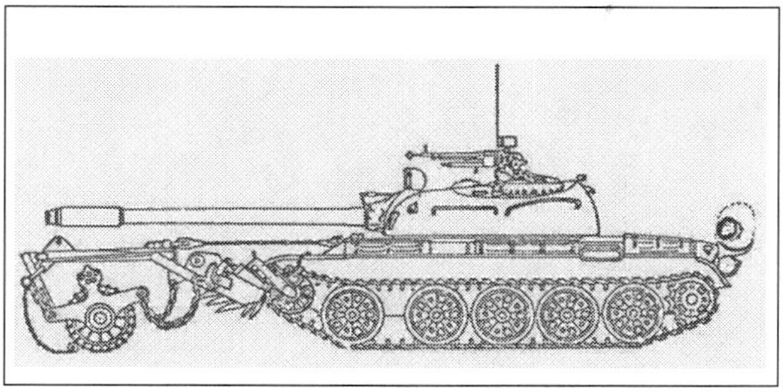

Figure C-12. KMT-5 plow-roller combination

The engineer company of the motorized rifle or tank regiment has breaching equipment such as KMT-series mine plows and rollers and BTU bulldozer blades located in its technical platoon. Because of limited assets in the technical platoon, coupled with the responsibility of forming its own MSD, the regiment can receive a sapper section from the divisional sapper company. An additional IMR armored engineer tractor (Figure C-13), BTR-50/60, and M1979 armored mine clearer (Figure C-14) and manual breaching equipment come with the sapper section.

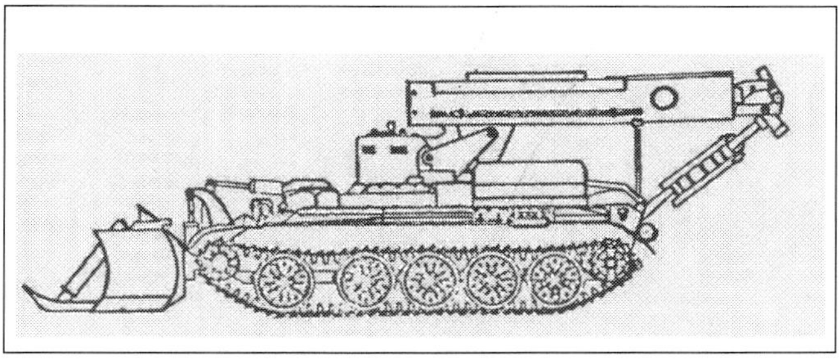

Figure C-13. IMR armored engineer tractor

Maneuver units usually breach remotely emplaced obstacles by using attached, built-in breaching equipment (BTUs and KMTs). In order to carry out this task successfully, all subunit commanders must organize constant reconnaissance, notify subordinates about mined areas in a timely manner, train personnel on the means and methods for handling remotely emplaced mines, and clear terrain in a timely manner. They must also train their own teams for independent actions when removing combat equipment from mined areas. Plows in the threat army are considered maneuver-force assets, and one plow is assigned to each tank platoon. The BMP has recently been equipped with track-width mine plows, but the allocation has not been determined.

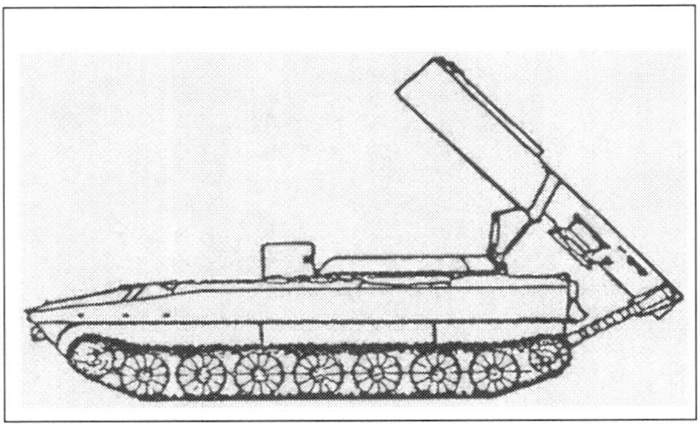

Figure C-14. M1979 armored mine clearer

## EQUIPMENT

Several pieces of equipment are used by threat armies to detect and clear mines.

### BAT-M Dozer

The BAT-M dozer (Figure C-8, page C-8) is a modified artillery tractor with a hydraulically operated bulldozer blade and crane. It is sometimes called a *roader* by Russians. The BAT-M dozer clears obstacles, fills craters, prepares bridge approaches, and performs other heavy pioneer tasks. It can also be configured for snowplowing.

The second generation BAT-M is the BAT-2. The BAT-2 is able to carry an 8-man engineer squad and operate in an NBC environment. It is replacing the BAT-M.

### KMT-Series Plows and Rollers

#### *KMT-4*

The KMT-4 mine-clearing plow (Figure C-9, page C-8) was developed in the 1960s to fit on a T-545/55 tank. It actually consists of two plows (one mounted in front of each track), and each plow has five attached teeth. When the plow is lowered, the teeth dig into the ground and remove mines from the path of the tank. A plow is lighter than a roller and permits tanks to retain their cross-country mobility. The estimated clearing speed is 10 kph, and the depth of clearance is 10 centimeters.

Three plows are issued per tank company (one per platoon). However, these assets are normally held in the engineer company of a tank or MRR.

#### *KMT-5*

The KMT-5 mine-clearing plow-roller combination (Figure C-12, page C-10) consists of two plows and two rollers attached to the front of a tank hull. The plows or the rollers can be used, depending on terrain features, the type of soil, and the mine fuse. Plows and rollers cannot be used simultaneously. The rollers function against pressure-fused mines. The system can survive 5 to 6 kilograms of explosives, five or six times. The KMT-5 also includes a luminous lane-marking device for night operations.

### *KMT-6*

The KMT-6 mine-clearing plow was introduced with the T-64 and T-72 tanks in the early 1970s. It has operating characteristics similar to those of the KMT-4.

### *KMT-10*

The KMT-10 mine-clearing plow is fitted to the BMP-2 infantry combat vehicle.

## IMP Portable Mine Detector

The IMP portable mine detector (Figure C-10, page C-9) weighs 7 kilograms and can detect mines buried to a depth of 45 centimeters. It has a tubular search head (one transmitting and two receiving coils encased in plastic) and a four-section handle. Power is furnished by four flashlight batteries that permit 20 hours of continuous operation. Two tuning controls are mounted on the handle. The coils in the search head compromise an induction bridge and are initially balanced for zero coupling. When the head passes over a metallic object, the induction bridge becomes unbalanced and produces an audible signal in the headset.

## DIM Vehicle-Mounted Mine Detector

The DIM vehicle-mounted mine detector (Figure C-11, page C-9) is primarily used to clear roads during convoys and road marches. It sweeps at a speed of 10 kph with a 2.2-meter width. It can detect metallic mines at a depth of 25 centimeters. The brakes on the DIM automatically engage when a mine is detected. Cross-country use of the DIM is limited.

## IMR Armored Engineer Tractor

The IMR armored engineer tractor (Figure C-13, page C-10) is mounted on a modified T-54/55 chassis. The turret is removed and a hydraulic crane, which can be fitted with either a grab or an excavator bucket, is emplaced. An adjustable, hydraulically operated blade is mounted on the front. The crane operator is provided with an armored cupola. The IMR can operate in an NBC environment.

## M1979 Armored Mine Clearer

The M1979 armored mine clearer (Figure C-14, page C-11) is mounted on the chassis of an amphibious 122-millimeter, 2S1 self-propelled howitzer. It has a turret-like superstructure that contains three rockets on launch ramps. These, along with the upper part of the superstructure, are hydraulically elevated for firing. The rocket range is estimated at 200 to 400 meters. Each rocket is connected to 170 meters of mine-clearing hose via a towing line. The hose is folded and stowed in the uncovered base of the turret and connected to the vehicle with a cable. The cable allows the vehicle crew to reposition the hose after launching.

This chapter implements STANAG 2990.

Appendix D

# Air Volcano

The air Volcano system provides a three-dimensional capability that allows units to emplace minefields in deep, close, and rear operations. It provides US forces with the capability to employ minefields rapidly under varied conditions. The air Volcano can be used to emplace tactical minefields; reinforce existing obstacles; close lanes, gaps, and defiles; protect flanks; and deny the enemy use of potential air-defense sites. Volcano minefields are ideal for flank protection of advancing forces and for operating in concert with air and ground cavalry units on flank-guard or screen missions.

## COMPONENTS

The air Volcano system (Figure D-1) consists of an M87-series mine canister, an M139 dispenser, and vehicle-specific mounting hardware (UH-60 Blackhawks require a jettison kit).

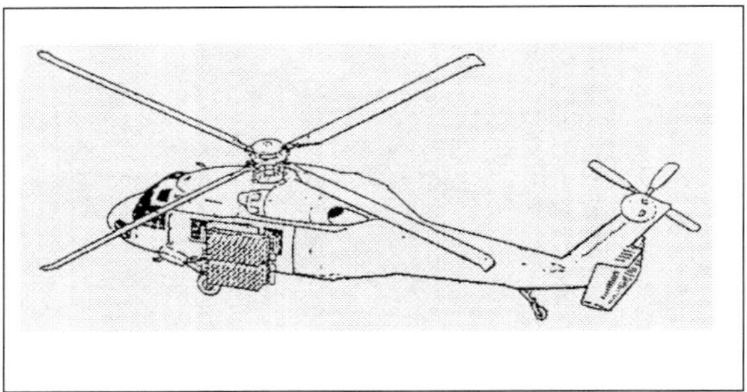

**Figure D-1. Air Volcano system**

### M87-Series Mine Canister

The M87-series mine canister is the same canister used for the ground Volcano system:

- M87. Prepackaged with five AT mines, one AP mine, and a propulsion device inside a tube housing.
- M87A1. Prepackaged with six AT mines and a propulsion device inside a tube housing.

The mixture of mines is fixed and cannot be altered in the field. The mines in each canister are electrically connected with a web that functions as a lateral dispersion device as the mines exit the canister. Spring fingers mounted on each mine prevent the mine from coming to rest on its edge. AT mines have a delay-arm time of 2 minutes 30 seconds; AP mines have a delay-arm time of 4 minutes. All canisters are capable of dispensing mines with 4-hour, 48-hour, or 15-day SD times. SD times are selected prior to dispensing and do not require a change or modification in the base M87-series mine canister.

### M139 Dispenser

The M139 dispenser consists of an electronic DCU and four launcher racks; each rack holds 40 M87-series mine canisters. The racks provide the structural strength and the mechanical support required for launch and provide the electrical interface between the mine canisters and the DCU. Mounting hardware for the UH-60A Blackhawk includes a jettison subassembly to propel the Volcano racks and canisters away from the aircraft in the event of an emergency.

The operator uses the DCU to electrically control the dispensing operation from within the aircraft. The DCU provides controls for the arming sequence and the delivery speed selection, and it sets mine SD times. The DCU allows the operator to start and stop mine dispensing at anytime. A counter on the DCU indicates the number of canisters remaining on each side of the aircraft.

Mines are dispensed from their canisters by an explosive propelling charge, 35 to 70 meters from the aircraft's line of flight. The aircraft flies at a minimum altitude of 1.5 meters, at speeds of 20 to 120 knots. It can deliver up to 960 mines per sortie.

## LIMITATIONS

The total weight of the air Volcano system is 2,886 kilograms. An aircraft will be close to its maximum gross weight when it contains the Volcano system and a full crew. Based on weather and environmental conditions, the aircraft may be required to execute the mission without a full fuel load, thus reducing enroute time.

The flight crew cannot operate the M60D machine gun with the air Volcano system installed, and it takes 3 to 4 hours to install the air Volcano system on a UH-60A Blackhawk.

## EMPLOYMENT

The air Volcano is the fastest method for emplacing large tactical minefields. When employed by combat aviation elements in support of maneuver units, close coordination between aviation and ground units ensures that mines are emplaced accurately and quickly. Although placement is not as precise as it is with ground systems, air Volcano minefields can be placed accurately enough to avoid the dangers inherent in minefields emplaced by artillery or jet aircraft.

Air Volcano minefields can be emplaced in friendly and enemy territories. They should not be planned in areas of enemy observation and fire, because the helicopter is extremely vulnerable while flying at the steady altitude, the speed, and the path required to emplace the minefield. The air Volcano is the best form of a situational obstacle because of its short emplacement time. Its employment varies depending on the type of operation.

## DEEP OPERATIONS

### Employment

The air Volcano is employed in deep operations to—

- Disrupt enemy formations along key AAs and choke points.
- Fix enemy formations in EAs to enhance target acquisition by attack helicopters, CAS, artillery, or a combination of weapon platforms.
- Provide area denial of possible enemy artillery and ADA sites that will affect future friendly schemes of maneuver.

AH-64 security is essential to air Volcano missions because the target area must not be in the enemy formation's direct line of sight. It may require two suppression of enemy air defense (SEAD) missions to get the air Volcano across the FLOT to the target location. This removes the security aircraft from the battle. All air Volcano missions require planning, designation, and control during the execution of air corridors (routes).

### Aviation Configuration

The air Volcano can be used in deep operations, but it has distinct limitations. The aircraft can maintain a speed of 80 to 90 knots (UH-60A) for 90 to 120 kilometers, depending on the wind and the temperature (see Table D-1, page D-4).

The aircraft cannot employ the door guns for self-defense with the Volcano system mounted. The attack aircraft averages 100 to 120 knots while targeting enemy formations that are 150 to 200 kilometers deep. In order for the air Volcano to go deep, the attack aircraft must slow down or special security/escort aircraft must be assigned for protection. The security/escort can be AH-64s or OH-58D Kiowa Warriors (KWs). KWs have many advantages—they are not a primary killing platform, and they are very good at quickly securing an area. The air Volcano requires one or two security aircraft for protection, and there should also be a backup aircraft for the mission.

### Fire-Support Coordination

The division main (light force) or brigade main (mechanized force) FSE is responsible for coordinating and executing fires in support of air Volcano missions. Upon completion of the target meeting, the FSE, the assistant division/brigade engineer, and the G3/S3 air representative will coordinate to ensure that the air-coordination/tasking order will support the mission and the planned SEAD fires.

If the mission is a deep aviation attack, the aviation element FSE is responsible for coordinating through the forward command post to the division/brigade main FSE. If the forward command post has *jumped forward* or has not deployed, the aviation element FSE will coordinate directly with the division/brigade main FSE.

**Table D-1. Air Volcano capabilities and limitations**

| Pressure Altitude (ft) | Temperature (°C) | Maximum Weight (lb)[1] | Cruise Speed (knots/kph) | Fuel Burn Rate (lb per hr) | Endurance (hr + min)[2] | Maximum Range (nm/km)[3] | Planning Combat Radius (nm/km)[4] |
|---|---|---|---|---|---|---|---|
| 0 | +20 | 22,000 | 80/148 | 965 | 1 + 57 | 154/285 | 60/111 |
| | | | 90/166 | 1,040 | 1 + 48 | 160/296 | |
| | | | 100/182 | 1,145 | 1 + 35 | 158/292 | |
| | +30 | 22,000 | 80/148 | 975 | 1 + 56 | 156/289 | 59/109 |
| | | | 90/166 | 1,045 | 1 + 48 | 160/296 | |
| | | | 100/182 | 1,160 | 1 + 34 | 156/289 | |
| | +40 | 21,600 | 80/148 | 995 | 1 + 53 | 150/277 | 58/107 |
| | | | 90/166 | 1,055 | 1 + 45 | 156/289 | |
| | | | 100/182[7] | 1,175 | 1 + 33 | 154/285 | |
| 2,000 | +20 | 22,000 | 80/148 | 960 | 1 + 58 | 157/290 | 60/111 |
| | | | 90/166 | 1,060 | 1 + 45 | 157/290 | |
| | | | 100/182 | 1,155 | 1 + 35 | 158/292 | |
| | +30 | 21,000[5] | 80/148 | 1,000 | 1 + 28 | 117/217 | 45/83 |
| | | | 90/166 | 1,045 | 1 + 23 | 124/229 | |
| | | | 100/182[7] | 1,175 | 1 + 12 | 120/222 | |
| | +40 | 20,000[6] | 80/148 | 1,005 | 1 + 21 | 28/52 | 7/13 |
| | | | 90/166[7] | 1,075 | 1 + 19 | 29/53 | |
| | | | 100/182[7] | 1,150 | 1 + 16 | 27/49 | |

**NOTE: Based on UH-60A with an aircraft torque factor of 97.5.**

[1]Maximum weight for deep operations only; no interdiction capability.

[2]Endurance includes 20-minute fuel reserve.

[3]Does not compensate for winds. This is the maximum range line-of-site distance.

[4]The general planning figures used for time-on-target missions; includes compensation for 10-knot winds, makeup for lost time, and a 5-minute on-station time.

[5]1,800 pounds of fuel at takeoff

[6]700 pounds of fuel at takeoff

[7]Operation at this air speed may be limited to 30 minutes due to target limits.

LIMITATIONS:

- There are no modifications to mount the M60D machine gun on the air Volcano aircraft.
- Crews cannot use the M60D with the system mounted on an aircraft.
- Crews cannot fly the aircraft with full tanks above +30°C and 2,000-ft pressure altitude.
- Aircraft may require a rolling takeoff and landing, depending on ambient weather conditions.

## Close Operations

### Employment

The air Volcano is employed in close operations to fix enemy formations in EAs, turn advancing formations into desirable terrain that supports the friendly scheme of maneuver, disrupt formations enough to slow the enemy down, and block key AAs. Multiple missions may be required, depending on the intended effect. The air Volcano can be used to reseed existing minefields or to close lanes and gaps. The target area must be clear of friendly forces before an air Volcano mission is executed.

Use of the air Volcano in close operations should be a primary planning consideration. It can quickly reach the outer edge of the forward operating base where AAs need a minefield obstacle. The threat level will be lower, and the station time will increase.

**Aviation Configuration**

Two air Volcano aircraft should be used (one primary, one backup). The requirement for security aircraft depends on METT-TC factors, but security should be used whenever possible.

**Fire-Support Coordination**

The forward command post FSE coordinates and executes fires in support of air Volcano missions. The FSE, the engineer liaison officer, and the G3/S3 representative coordinate to ensure that the air coordination/tasking order supports the mission and the planned SEAD fires. The division/brigade main will be available to support the forward command post as necessary.

The brigade/TF FSE is responsible for coordinating through the forward command post to the division/brigade main FSE. If the forward command post has jumped, the brigade/TF FSE coordinates directly with the division/brigade main FSE.

### REAR OPERATIONS

**Employment**

The primary purposes of the air Volcano in rear areas is to protect key terrain from possible airborne/air-assault forces and to fix/disrupt enemy forces long enough to allow the tactical combat force or ready-reserve force time to react and meet the changing enemy situation.

The least preferred employment method is to deliver tactical minefields to brigade and corps support areas. This employment tactic is normally used when all other available assets have been exhausted. The flexibility of the air Volcano system makes it ideal for employment against a mounted Level III threat in the rear. The target area should be out of the direct view/fire of the threat and on a choke point that allows cover for the reacting forces.

**Aviation Configuration**

The air Volcano aircraft could be employed individually or with security/escort aircraft. The use of OH-58D KWs as security aircraft allows units to develop the situation and helps place minefields in the proper location to assist inbound attack aircraft or fires. If the air Volcano aircraft is not provided security aircraft, it is recommended that ground forces provide covering fires.

**Fire-Support Coordination**

The division/brigade rear FSE coordinates and executes fires in support of air Volcano missions. The FSE, the engineer liaison officer, and the G3/S3 representative coordinate to ensure that the air coordination/tasking order supports the mission and the planned SEAD fires. The division/brigade main will be available to support the division/brigade rear as necessary.

The headquarters element that controls the rear area coordinates with the division/brigade rear FSE. The division/brigade rear FSE coordinates with the division/brigade FSE for fire support and air assets.

## MINEFIELD EFFECTS

### Turn

A turn minefield manipulates enemy maneuver in a desired direction. It forces or entices enemy formations to move in a different direction rather than breach the obstacle. This means the bypass must be easily identified. Turn minefields are extremely lethal, with approximately 80 percent probability of mine encounter. The typical width is 557 by 320 meters for air Volcano. Figure D-2 shows two turn minefields combined to create a turn-effect obstacle group. It takes 160 canisters (800 AT/160 AP mines) to emplace one turn minefield. One air Volcano aircraft can lay one turn minefield (see Table D-2).

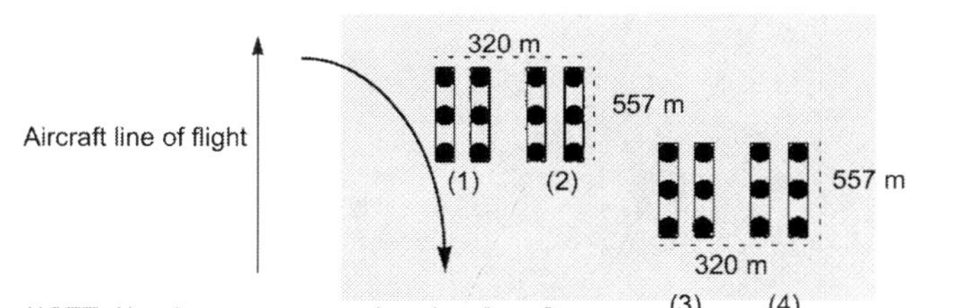


Figure D-2. Turn obstacle

Table D-2. Air Volcano minefield data

| Type of Minefield | Depth (m) | Frontage of Minefield (m) | Number of Strips | Canisters per Strip | Total Canisters | Minefields per Aircraft |
|---|---|---|---|---|---|---|
| Disrupt | 120 | 278 | 1 | 40 (20 each side) | 40 | 4 |
| Fix | 120 | 278 | 1 | 40 (20 each side) | 40 | 4 |
| Turn | 320 | 557 | 1 | 80 (40 each side) | 160 | 1 |
| Block | 320 | 557 | 1 | 80 (40 each side) | 160 | 1 |

### Block

A block minefield (Figure D-3) is designed to stop an enemy advance along a specific AA or allow it to advance at an extremely high cost. Block minefields are obstacles with intensive integrated fires. They should be employed in a complex obstacle scheme with road craters or bridge demolitions enhancing the effectiveness of the minefield. One air Volcano aircraft can lay a 557- by 320-meter block minefield, using all 160 canisters (800 AT/160 AP mines). The probability of mine encounter is more than 80 percent. One Volcano aircraft can lay one block minefield, making two passes side by side (see Table D-2).

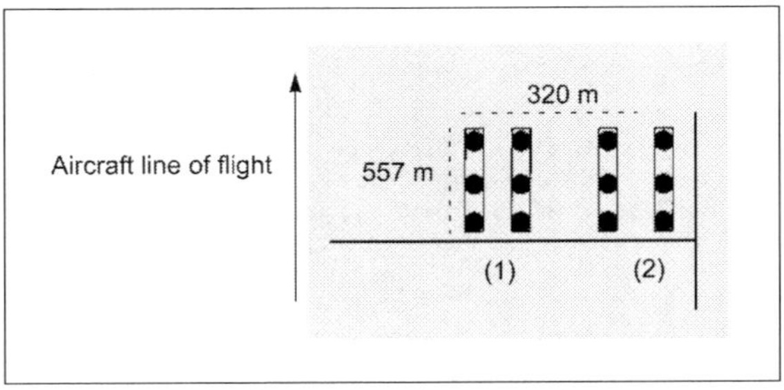


**Figure D-3. Block obstacle**

### Disrupt

A disrupt minefield (Figure D-4) fractures and breaks up enemy formations. It causes premature commitment of reduction assets, interrupts $C^2$, and alters timing. A disrupt minefield is not resource- or time-intensive. The probability of mine encounter is approximately 50 percent, and the typical width is 278 by 120 meters. It takes 40 canisters (200 AT/40 AP mines) to emplace one disrupt minefield. One air Volcano aircraft can lay four disrupt minefields (see Table D-2).

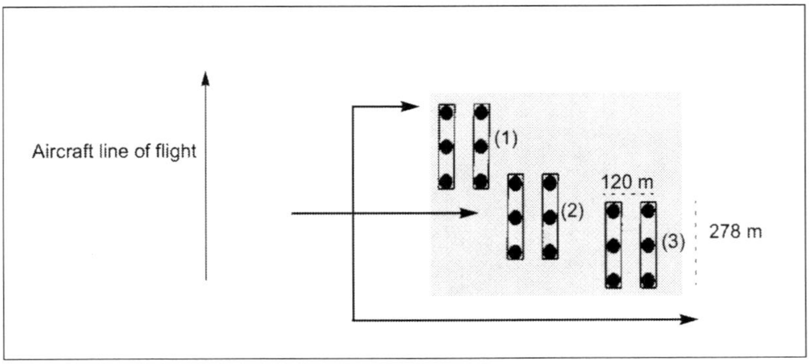


**Figure D-4. Disrupt obstacle**

### Fix

A fix minefield (Figure D-5) slows enemy formations within a specified EA. It gives friendly forces time to acquire, target, and destroy enemy formations. A fix minefield is employed in depth and causes enemy formations to react and breach repeatedly. The air Volcano fix minefield does not look impenetrable to enemy formations. The probability of mine encounter is 50 percent, and the typical size is 278 by 120 meters. It takes 40 canisters (200 AT/40 AP mines) to emplace one fix minefield. One air Volcano aircraft can lay four fix minefields (see Table D-2, page D-6).

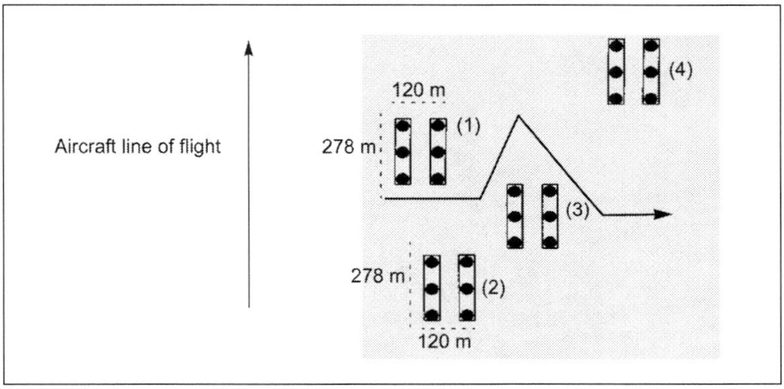


**Figure D-5. Fix obstacle**

## Linear

If the threat situation allows, the aircrew makes a pass to confirm the minefield end points and the suitability of the terrain. In a high-threat situation, the aircrew emplaces the minefield on the first pass. This hasty minefield is linear in configuration and is 1,115 by 75 meters. All 160 canisters are fired.

## PLANNING

## Responsibilities

### *Division Commander*

The division commander approves air Volcano employment and integration into deep, close, and rear operations. He is also the authority for SCATMINE employment.

### *Maneuver Brigade Commander*

When authority is delegated by the division commander, the maneuver brigade commander is responsible for employing air Volcano in close operations and in supporting follow-on missions. He is responsible for approving target nominations to be submitted to the division. The maneuver brigade commander receives the air Volcano aircraft and its crew in OPCON status.

***Aviation Brigade Commander***

The aviation brigade commander is responsible for integrating the air Volcano into deep aviation attacks and for shaping EAs in the division AO. He executes the air Volcano missions in deep, close, and rear operations. The aviation brigade commander submits target nominations to the division targeting cell through the aviation brigade engineer and provides support for transporting and loading Volcano mines.

***Division Engineer***

The division engineer is responsible for target nominations that support division missions or objectives in deep operations. He—

- Submits the nominations to the division targeting cell.
- Determines minefield characteristics.
- Incorporates the Volcano minefield, DTG of SD times, and safety zones into the division obstacle plan.
- Estimates the requirements for mine canisters and Class IV/V supplies.
- Determines the intent of the Volcano minefield as it is integrated into the division obstacle plan.
- Disseminates the SCATMINWARN to adjacent and subordinate units before the minefield is laid and one hour before the SD sequence of the minefield is initiated.

***Brigade Engineer***

The brigade engineer is responsible for target nominations that support objectives within his respective fight. He—

- Submits target nominations to the assistant division engineer.
- Synchronizes air Volcano missions into brigade operations.
- Assists the brigade FSE in planning SEAD, CAS, and Apache escort.
- Provides logistics estimates to the brigade S4 for coordination of ammunition requirements.
- Posts the operations map with the minefield's location, aviation graphics, the safety zone, and the DTG of SD times.
- Disseminates the SCATMINWARN to adjacent and subordinate units before the minefield is laid and one hour before SD sequence of the minefield is initiated.

***Deputy Fire-Support Coordinator***

The deputy FSCOORD recommends commanding general/chief of staff approval for target nominations developed by the division targeting cell. He is responsible for submitting division-approved air Volcano target nominations to the G3 air for inclusion in the air coordination order. He coordinates SEAD, CAS, and intelligence and electronic warfare for division-directed air Volcano missions. The deputy FSCOORD is also responsible for—

- Targeting (intelligence and asset coordination).
- The air tasking order.
- The G3 air coordination order.
- G2 collection and assessment.

### *G3 Air*

The G3 air synchronizes coordination and deconfliction of division air space for air missions, SEAD, and CAS. When necessary, he submits the air coordination order to higher headquarters with the division-approved target list.

### *Emplacing Unit/UH-60 Company*

The emplacing unit or the designated UH-60 company sets up and loads the air Volcano system in conjunction with the forward-area refuel point. The Volcano system is loaded on the UH-60 at the designated point. The emplacing unit/UH-60 company is also responsible for—

- Maintaining the unit basic load at the forward-area refuel point or ATP.
- Preparing the scatterable minefield report and record, forwarding it to the authorizing commander or the aviation brigade engineer, and verifying that the assistant division/brigade engineer receives the entire report.
- Coordinating air routes and corridors.
- Requesting SEAD and security aircraft.
- Planning the air-mission coordination meeting to refine or develop the aviation scheme of maneuver.
- Ensuring that pilots attend the unit's rehearsal.
- Posting aviation graphics on the current operations overlay.

## Process

All target nominations are submitted through the division targeting cell. In the brigade, all target nominations go to the assistant division engineer for submission to the board. Nominations should be submitted 96 hours prior to the execution time (see Table D-3).

- The targeting cell validates the targets based on the SITEMP and recommends approval through the commanding general/chief of staff.
- Upon approval, the deputy FSCOORD turns the target numbers and the air requirements over to the G3 air, who adds the targets to the air coordination order.
- The deputy FSCOORD begins to plan SEAD/CAS requirements for division missions.
- The G3 issues the division WO so that units can be prepared to execute specific air Volcano missions.

**Table D-3. Planning process (H-hour sequence)**

**H-96+00.** Submit target nominations/receive approval. Submit CAS requests for enemy ADA sites along ingress and egress routes.

**H-72+00.** Receive mission. Division WO issued.
- Conduct air-mission coordination meeting.
- Conduct S2 update on enemy situation and pass down to executing unit.
- Submit additions or changes to the CAS request to FSE for enemy ADA sites.
- Post the air Volcano mission in the air coordination order.
- Disseminate the SCATMINWARN.

**H-48+00.** Mount the system. The system takes 3 to 4 hours with a crew of three on a hard surface. The time is doubled in a field environment and requires three or four personnel to assist in loading.

**H-24+00.** Conduct detailed mission analysis and planning. Issue WO to attack assets, including—
- NLT time for mission planning and upload completion.
- Time and location of air-mission brief.
- Units attending.
- Minefield location and type.
- Establishment of $C^2$ relationship.
- Face-to-face coordination with attack assets.
- Liaison officers planning multiple routes to minefield.
- Liaison officers coordinating with engineers, ADA, and FSEs for support. G4/S4 must be notified of reload plans and any other Class III/V requirements.
- Intelligence update.

**H-20+00.** Volcano integrated into deep attack or air-assault matrix by the aviation brigade.

**H-8+00.** Conditions-check matrix approved by the commanding general or designated authority, including—
- Weather.
- Threat suppression along route.
- Attack support.
- FLOT crossing coordination completion.
- FLOT initial crossing point.
- SEAD fire coordination.
- Mines available.
- Integration into execution matrix.
- Crews updated on ADA threat.
- Backup system and aircraft available.
- CAS coordination.
- Rehearsal completion with attack assets.
- Intelligence update.
- Close operation mission, including—
  - — Location of friendly forces.
  - — Type of marking and when it will be in place.

**H-6+00.** Conduct air-mission brief, including—
- S2 intelligence update.
- ADA threat update.
- Rehearsal held after brief.
- Brigade engineer and brigade fire-support officer being present.

**H-4+00.** Latest time for link-up of aviation assets. Fly to forward-area refuel point or tactical assembly area for mine load-up. Remember that—
- Loading canisters takes 1 to 3 hours with a crew of four to eight. Time varies greatly based on whether or not the canisters are in shipping containers.
- The loading time in a field environment is decreased if one pad is dedicated to air Volcano aircraft loading and arming.

**H-0+30.** Cross FLOT (estimated; exact time is based on route), to include—
- SEAD fired.
- CAS sorties.

**H-hour.** Minefield deployed, to include—
- Aircraft reports to the supporting brigade engineer if an infantry brigade mission.
- Aircraft reports to the aviation brigade engineer if a division mission.
- Engineer disseminates the scatterable minefield report and record to appropriate units.

- The division issues a fragmentary order (FRAGO) to order the execution of the air Volcano mission.
- The planning staff gives the aviation brigade a date and a time for the air Volcano system to be uploaded and prepared for employment.
- Air Volcano minefields are integrated into the scheme of maneuver as a directed, situational, or reserve obstacle as stated in FM 90-7. This includes integration into the COA that is synchronized during war gaming and included in the OPORD as part of the rehearsal.

## Logistics

Air Volcano munitions are transported and handled in the same manner as Class V mines and explosives. The only significant difference in handling is whether air Volcano operations are conducted from the tactical assembly area or the forward operating base.

Echelon-above-corps transportation assets will transfer the air Volcano to corps storage areas (ammunition). Based on forecasts by the division ammunition officer, the corps support area will push air Volcanos to ASPs or ATPs in the division area (tactical assembly area). ASPs and ATPs in or near the tactical assembly area are normally operated by a direct-support ammunition company attached to the corps support group (forward). ATPs in brigade support areas are normally operated by forward support battalions. Mines are moved from the corps support area to ASPs/ATPs by transportation units from the corps support group (rear). Further throughput of mines from ASPs to ATPs is supported by transportation assets from the corps support group (forward). Based on the division ammunition officer's forecast and the availability of transportation, the corps support area will attempt to transfer bulk mines to the ASP/ATP that is best positioned to support requirements. ATPs issue mines to using units while ASPs handle bulk ammunition. ASPs can do emergency issue to using units, but this is usually done only for units using large quantities of bulk ammunition and having their own ammunition transportation support available (division or corps artillery).

## Preparation and Coordination

If an air Volcano mission has been approved for a unit, an aviation brigade liaison officer coordinates with the S3 air, the engineer, and the air defense officer to outline air Volcano requirements.

### *Logistical Requirements*

The engineer planner calculates Class IV/V supplies and requests them from the Assistant Chief of Staff, G4 (Logistics)(G4)/S4. The engineer and the assistant division aviation officer coordinate the location (forward-area refuel point, tactical assembly area, forward operating base) of the ATP where the UH-60 will be loaded and fueled. The ALO must provide the amount of air time available and the fuel required, and he must be prepared to discuss emplacement times based on Volcano locations. The air-mission brief will facilitate dissemination of this information.

### *Concept of the Operation*

The scheme of maneuver, fires, and engineer operation must be outlined. The scheme of engineer operations outlines exact grid coordinates, the obstacle intent, and the minefield composition and size. The air Volcano can be emplaced under enemy contact, but additional control measures must be used to protect the aircraft and the crew. If the air Volcano is triggered by enemy action, the DST must be briefed during the air-mission brief. The DST must outline NAIs, including the type of sensors and triggers (long-range surveillance device [LRSD], SOF, UAV), TAIs, decision points, and execution criteria.

### *Control Points and Markers*

The initial point, the approach marker, and minefield markers are designated. Approach and minefield markers must be visible from the air and be distinctly different from one another:

- The initial point is an easily identifiable terrain feature used for coordinating the entry point of the UH-60 into the sector.
- The approach marker allows the aircraft to set the altitude, the speed, and the final orientation to the minefield. The approach marker can also be a terrain feature.
- Minefield markers establish the limits of the desired minefield. Deep area minefields do not require minefield markers.

### *Terrain Analysis*

A combined (aviation, maneuver, fire-support, and engineer planners) terrain analysis should be conducted using Terrabase or a similar product. Terrabase enables planners to analyze the effects of the terrain in a three-dimensional format. AAs, terrain references, TAIs, NAIs, line-of-sight profiles, and minefield locations can be confirmed during this analysis.

### *Air-Mission Brief*

The air-mission brief is the most critical planning and coordination meeting at the execution level, and it must occur no later than H-6 in the planning process. During the air-mission brief—

- $C^2$ is established.
- Updates and changes to the situation are exchanged between the brigade or TF engineer, the fire-support officer, and the air-mission commander.
- The engineer and the air-mission commander use four control techniques (discussed later in this appendix) to ensure mission success. The primary technique for emplacement and the responsibilities for each control technique are outlined.
- Radio frequencies; points of contact; code words; identification, friend or foe, modes; and challenge and passwords are exchanged and disseminated.

- The forward-area refuel point location, the security aircraft, flight routes, and lethal/nonlethal SEAD are identified (if already approved).

### *Mounting*

The system must be mounted no later than H-48 in the planning process.

## Execution

### *Siting*

The key to proper emplacement is the location of the minefield in relation to existing terrain features. If appropriate, the minefield should tie into an existing terrain feature to prevent easy bypass or fording. Using Terrabase enhances siting and emplacement procedures in deep operations. Reconnaissance must be conducted to verify the location.

### *Movement*

**Loading.** The launcher rack functions as the carrier and launcher platform for 40 mine canisters. The rack has 40 keyholes for mine canisters, a green latch that latches the mine canister to the rack, and a red latch that arms the mine canisters. The rack has two electrical receptacles—one for the connector and one for the launcher rack cable from the DCU. While looking at the canister side of the rack, rows are 1 through 4 from bottom to top and columns are 1 through 10 from left to right.

**Arming.**

- Due to the weight of the Volcano mine system, a large, open area that is clear of obstacles must be selected. The site should have a hard surface if possible. If a hard surface is unavailable, inspect the ground to ensure that it is firm enough to support the weight of the aircraft. Perforated steel planking or two pieces of 1-inch plywood (4 by 4 feet) may be used as a field-expedient surface in soft areas.
- Concentrations of nonessential personnel or frequently traveled vehicular routes should not be within 1,000 meters of the site. This distance is based on the total weight of explosives and the safe fragmentation distance found in FM 5-250. When using the M88 training canisters, the minimum distance is 30 meters.
- Two Underwriters Laboratories, Incorporated (UL)-listed 10BC fire extinguishers and a grounding rod (minimum safety equipment) must be available at the arming point. This equipment is provided by personnel who deliver the mines.
- The number of personnel allowed access to the site should be held to a minimum. All personnel involved in the arming will receive a safety brief that includes—
  - Ammunition handling and inspection procedures.
  - Loading procedures.
  - Emergency procedures and rendezvous points.
- Emergency procedures.

— Fire. In the event of a fire away from the mines, attempt to contain or extinguish the fire by any available means. If the fire is near the mines or in them, clear the area to a minimum distance of 1,000 meters and notify fire-fighting personnel immediately. When training with M88 canisters, clear the area to a minimum distance of 30 meters.

— Accidental discharge. Immediately clear the area to a distance of 640 meters and notify EOD. The mines arm approximately 2½ minutes after firing. When training with M88 canisters, terminate arming until the problem can be identified and corrected.

— Failure to fire. Remove the canister from the aircraft and place it in the dud pit. Notify EOD immediately. When training with M88 canisters, remove the canister from the aircraft, separate it from the other canisters, repack it, and return it to the ASP.

- Site layout (Figure D-6).

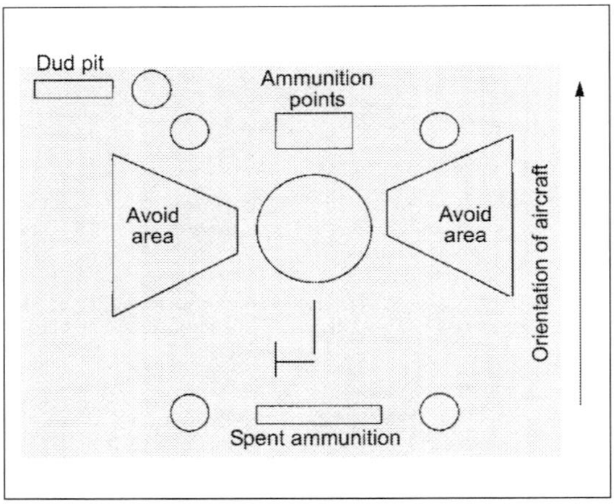


**Figure D-6. Site layout**

— Berming of the site is not required for a tactical arming point.

— The following rules apply when the site is located next to a refuel point:

> A minimum of 1,000 meters must exist between arming points and refuel points when the total quantity of explosives is less than 600 kilograms. For quantities greater than 600 kilograms, refer to FM 5-250.

**NOTE: Each M87 canister contains 3.4 kilograms of explosives; a full load (160 canisters) contains 550 kilograms of explosives.**

    > The refuel point for armed aircraft must be located at least 375 meters from other aircraft refueling points.
    > Parked, armed aircraft must be at least 36 meters from other armed aircraft to prevent the detonation of explosives on adjacent aircraft. This distance will not prevent damage to adjacent aircraft; a 130-meter distance is required to prevent damage by fragments and to ensure that the aircraft remains operational.
  - A dud pit (bermed when possible) for damaged or misfired ammunition should be established beyond the ammunition points.
  - Arming points should be laid out as shown in Figure D-6.

**Dearming.** After the mission is complete, the aircraft returns to the arming point for dearming. Spent canisters should be discarded at least 30 meters from the aircraft, at the 4- and 8-o'clock positions. Live canisters should be returned to ASPs for future use or repackaging. Canisters that misfire should be placed in the dud pit.

**Flight Planning and Preflight.**

- The flight crew analyzes the mission using METT-TC factors and determines the flight profile to be used during mine emplacement. It will select (or have designated) one or more of the following control measures to be used during mine emplacement:
  - Visual identification (start and stop markers on the ground).
  - Time-lapse (tables to determine the minefield length).
  - Number of canisters fired.
  - Doppler/GPS (start and stop coordinates).
- The crew member(s) will ensure that the air Volcano is installed properly, that all installation checks are completed, and that mine canister pallets are loaded as directed by the pilot or the SOP.
- The flight crew conducts ground checks according to the checklist in TM 1-1520-237-10 to confirm proper operation of the air Volcano prior to takeoff.

**Before Arrival at the Target Area.**

- During the equipment check, the crew chief turns on the DCU power-control switch, verifies that no malfunctions were indicated during the initial built-in test, and turns off the DCU power-control switch.
- After completion of *run-up* with the aircraft at flight idle, the crew chief turns on the DCU power-control switch.
- Before arrival at the release point, the pilot will make the following checks (listed on the Volcano card [a sample is shown in Figure D-7]):
  - Verify that the DCU is on.
  - Verify that the mine SD time is properly set.

| VOLCANO CARD | | | |
|---|---|---|---|
| | **Emplacement Date:** | | |
| N | **Unit** | **Engr** | **ASP** |
| | Call Sign | | |
| | Frequency | | |
| | Location | | |
| | Remarks | | |

Not to scale

| | 1st pass | 2nd pass | 3rd pass | 4th pass |
|---|---|---|---|---|
| Initial Point | A | D | G | J |
| SD Time | | | | |
| Ground Speed | knots<br>kph | knots<br>kph | knots<br>kph | knots<br>kph |
| No Tubes | | | | |
| Altitude | | | | |
| Track | | | | |
| Start Time | B | E | H | K |
| Stop Time | C | F | I | L |

**Figure D-7. Sample Volcano card**

— Announce the ground speed in knots or kph, as required. (The pilot and the crew chief will acknowledge.)

— Announce the number of canisters the crew chief will count down. If deployment is 40 canisters per run (20 per side) and there are two runs, the crew chief will count down from 80 to 60 on the first run and 60 to 40 on the second run. (The crew chief will acknowledge.)

— Announce the altitude for employment. (The pilot will acknowledge.)

— Announce the course for the delivery track. (The pilot will acknowledge.)

- Announce the delivery time based on the setting of the ground speed. (The crew chief will acknowledge.)

- Before arrival at the initial point, the crew chief will—
  - Ensure that the DCU fire-circuit switch's safety pin and streamer are removed.
  - Ensure that the DCU fire-circuit switch is enabled.
  - Place the interface control-panel arming switch to the *ARM* position. He will verify that the jettison advisory light indicates *armed* and that no fault codes are displayed on the DCU.

**At the Target Area.**

- The pilot simultaneously announces "mark," presses the go-around switch, and starts timing the run when he is over the minefield start point.
- The pilot maintains a ground speed of 5 kph/3 knots and an altitude of 3 meters during the mine-dispensing pass. The pilot is responsible for flying the aircraft within the prescribed limitations.
- The crew chief announces the mine-canister count, as the canisters are dispensed, by counting down in 10s. He then announces the last three canisters. For example, the pilot announces a canister count of 60, the crew chief calls out "80, 70, 3, 2, 1, mark."
- The pilot terminates mine dispensing when the grid location is reached.

**After Mission Completion.**

The crew chief will—

- Place the interface control-panel arming switch to the *SAFE* position and verify that the armed advisory capsule is extinguished.
- Place the DCU fire-circuit switch to the *OFF* position. Install the safety pin and the streamer.
- Prepare and submit a SCATMINWARN.

## EMPLACEMENT

The air Volcano is fast and flexible, but it is difficult to accurately dispense mines within the confines of the minefield marking. The desired obstacle-effect norms for the air Volcano require extensive planning, preparation, coordination, and positive control during emplacement. The critical aspect of the air Volcano is getting the right amount of mines in the specified location and in the desired density.

The detailed coordination focuses on positive control. Positive control of an air Volcano mission requires a redundancy of control techniques to minimize errors in minefield size and location. These control techniques must compensate for poor visibility, wind speed and direction, and navigational errors.

The following control techniques are used by the engineer and the air-mission commander to ensure that Volcano minefields match specific obstacle-effect norms. Units rely on these techniques to accomplish the mission, and they are part of the Volcano air-mission brief:

- Visual identification. Focuses on the visual identification of minefield emplacement. As part of the preparation for a Volcano minefield, an engineer element erects airfield panel markers to mark start and end points. This provides a visual signal for the engineer and the air-mission commander to start and stop firing Volcano canisters. The pilot depresses the launch switch over the first marker to start firing and depresses it again over the second marker to stop firing. This control technique is good for open terrain with adequate visibility and little canopy coverage.
- Time lapse. Focuses on when to stop firing Volcano canisters. The UH-60's air speed and the type of minefield being laid determines the amount of time it takes to lay a minefield. The air Volcano has six air speed settings—20, 30, 40, 55, 80, and 120 knots. Table D-4 shows the time required to lay minefields and the full load time.

**Table D-4. Air Volcano dispensing times based on air speed**

| Knots | Disrupt and Fix Minefields | Turn and Block Minefields | 160 Canisters per Load |
|---|---|---|---|
| 20 | 27 seconds | 54 seconds | 108 seconds |
| 30 | 18 seconds | 36 seconds | 72 seconds |
| 40[1] | 13 seconds | 27 seconds | 54 seconds |
| 55 | 9 seconds[2] | 18 seconds | 39 seconds |
| 80 | 6 seconds[2] | 13 seconds[2] | 27 seconds |
| 120 | 4 seconds[2] | 9 seconds[2] | 18 seconds |
| Width of minefield (meters) | 278.8 | 557.5 | 1,115 |
| No passes per minefield | 1 | 2[3] | 1 |
| No canisters per pass | 40 | 80 | 160 |

[1]Recommended air speed

[2]Recommended only if absolutely necessary

[3]Blackhawks in pairs can lay turn and block minefields in one pass, firing 80 canisters each.

The following example is provided to show how Table D-4 is used:

**Example:** The mission is to install an air Volcano disrupt minefield. The UH-60 is traveling at 40 knots (this is entered on the DCU), and the pilot initiates (depresses the launch switch) at the identification of the Volcano start marker or the grid location on the ground. The pilot depresses the launch switch a second time after 13 seconds have elapsed.

- Number of canisters fired. Focuses on when to stop firing Volcano canisters. The number of Volcano canisters dispensed also determines when firing is terminated. There is a digital readout on the DCU (for the left and right side) that shows the number of canisters remaining. The pilot stops firing when the required number of canisters have been fired (see Table D-4, page D-19).

Using the disrupt minefield example above, the UH-60 starts the mission with a full load (80 canisters on each side of the aircraft). The pilot initiates (depresses the launch switch) at the identification of the Volcano start marker or the grid location on the ground. The pilot depresses the launch switch a second time after 20 canisters have been expended on each side. The DCU counts down from the total number of canisters. When the DCU reads 60 right/60 left, the pilot depresses the switch to end the firing process. Ideally, the timing of delivery and the number of canisters fired are done simultaneously. As the crew chief counts down the timer, the pilot and the crew chief monitor the number of canisters remaining on the DCU digital readout.

- Doppler/GPS. Focuses on when to start and stop firing Volcano canisters using the UH-60's Doppler/GPS guidance and navigation set. This set provides the present position or destination in latitude and longitude (degrees and minutes) or grid coordinates. As part of the preparation for the Volcano minefield, exact grid coordinates are needed to determine the approach points and the limits of the minefield. These coordinates are provided to the air-mission commander during the air-mission brief. The pilot enters the grid coordinates into the Doppler/GPS on the primary and the backup aircraft. During execution, the air-mission commander monitors the Doppler/GPS and determines the time to target, when to initiate firing, and when to terminate firing.

### OUTSIDE FRIENDLY TERRITORY

Reconnaissance of the proposed site for the air Volcano minefield will be conducted before mines are emplaced. This could include—

- LRSD.
- Apache gun tapes.
- UAV overflights.
- Imagery.

Key terrain or landmarks are used to identify start and end points for aviation assets. It is unlikely that military marking will be employed based on the proximity to enemy forces and the probability of early detection if man-made markers are present. Fencing the minefield is not required until the area has been secured by friendly forces.

### WITHIN FRIENDLY TERRITORY

Within friendly territory, air Volcano minefields should be fenced and marked with NATO identification signs to protect friendly forces.

Fencing is installed before the air Volcano minefield is delivered, and it is located 100 meters from the centerline of the minefield and 100 meters from the start and end points (Figure D-8).

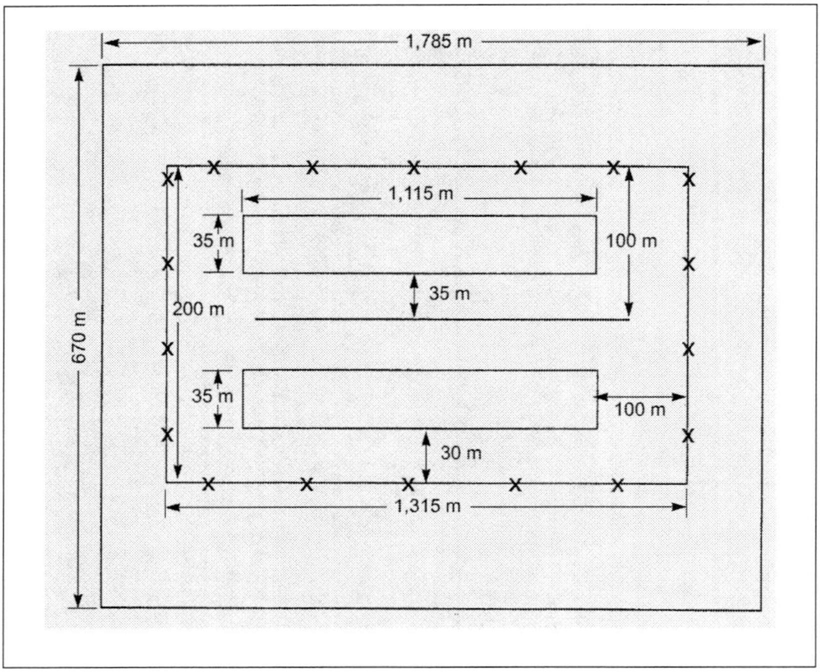


**Figure D-8. Fencing for an air Volcano minefield**

Start and end points should be marked with man-made devices such as VS-17 panels. During limited visibility, start and end points should be marked with infrared or heat-producing sources. Key terrain features and landmarks should still be used to identify start and end points.

Fencing the minefield is not viable when the minefield duration is short or civilians on the battlefield are an issue. In this case, CA, public affairs, and PSYOP personnel should be involved in letting friendly personnel know the minefield location. This could include—

- Leaflet drops.
- CA teams disseminating information.
- Host-nation support.
- PSYOP.

# REPORTING

## SCATTERABLE MINEFIELD WARNING

The emplacing unit is responsible for issuing the SCATMINWARN (Figure 8-7, page 8-23) to adjacent units and higher headquarters. The brigade engineers and the assistant division engineer assist in this process. They disseminate the warning based on whether or not it is a brigade (brigade engineer) or division (aviation brigade engineer) mission. To ensure that all units are informed, the assistant division engineer forwards the SCATMINWARN to the G3 for dissemination through operational channels.

## SCATTERABLE MINEFIELD REPORT AND RECORD

The aircraft emplacing the minefield reports initiation and completion times to the engineer of the emplacing unit. The engineer prepares the scatterable minefield report and record (Figure 8-8, page 8-23) and forwards it through his unit to the assistant division engineer. The assistant division engineer forwards the report to the G3 who provides the information to higher and subordinate units through operational channels.

Appendix E

# Safety and Training

Mine training is inherently dangerous, in part, because several different types of mines and fuse systems are used throughout the world. Detailed safety instructions for each type of mine are provided throughout this manual. This appendix merely points out the safety aspects of live-mine training that are common to all types of mines.

Conduct mine training as if the mines were live. This is the only way soldiers form a habit of correctly and safely handling mines and gain a true appreciation of the requirements and the time it takes to perform an actual mine-warfare mission. Live-mine training gives soldiers the confidence they need to handle mines and their components. Accidents can usually be traced to ignorance, negligence, deliberate mishandling, overconfidence, mechanical failure, or fright. The first four can be overcome by training and proper supervision. Mechanical failure rarely happens; but if it does, it can be controlled by training and proper supervision. The last item, fright, is mastered through well-controlled, live-mine training.

## STORAGE

There are three types of mines used in mine training:

- Inert. Does not contain explosives.
- Practice. Contains an LE charge or a smoke-producing component to simulate detonation.
- HE. Involves actual mines used in combat

Conventional mines are painted to enhance concealment, retard rusting of exposed metal parts, and help identify the type of mine and filler (HE, LE, or chemical agent). Older manufactured mines are painted according to the Five-Element Marking System; newer mines use the Standard Ammunition Color-Coding System (see Table E-1, page E-2).

**NOTE: Mines that are color-coded and marked according to the old system have been on hand for several years. Ensure that all ammunition, whether color-coded according to the old or new system, is properly and fully identified.**

Always handle mines with care. The explosive elements in fuses, primers, detonators, and boosters are particularly sensitive to mechanical shock, friction, static electricity, and high temperatures. Boxes and crates containing mines should not be dropped, dragged, tumbled, walked on, or struck. Do not smoke within 50 meters of a mine or its components.

**Table E-1. Mine color-coding system**

| Type of Ammunition | Five-Element Marking System (Old) | Standard Ammunition Color-Coding System (New)* |
|---|---|---|
| Persistent casualty chemical agent | Gray with green markings and two green bands | Gray with green markings and two 12-mm green bands |
| Nerve agents | Gray with green markings and two or three green bands | Gray with green markings and three 12-mm green bands |
| Incendiary | Gray with violet markings and one violet band | Light red with black markings and one yellow band |
| HE | Olive drab with yellow markings | Olive drab with yellow markings |
| Practice mines | Blue with white markings | Blue with white markings |
| Inert mines | Black with the word *INERT* in white | Blue with the word *INERT* in white |
| *Chemical ammunition containing an HE has one 6-mm yellow band in addition to the other markings. | | |

When it is necessary to leave mines in the open—

- Set them on dunnage at least 5 centimeters above the ground.
- Place a waterproof cover (such as canvas) over them, and leave enough space for air circulation.
- Dig drainage trenches around stacks of mines to prevent water from collecting under them.
- Protect mines and their components against moisture by waterproofing them with grease coatings, tar paper, or tarpaulins.

Additional maintenance procedures are as follows:

- Do not open mine boxes in a magazine, at an ammunition dump, or within 30 meters of an explosive store. Use copper or wooden safety tools, if available, to unpack and repack mines.
- Do not fuse mines within 30 meters of an explosive or ammunition holding area. Mines can be fused at the mine dump.
- Use specifics authorized by the US Army Materiel Command and applicable TMs to disassemble mines and their components.
- Remove safety pins, safety forks (clips), and other safety devices as the last step when arming the mine; and replace them before the mine is moved again. These devices prevent accidental initiation of the mine while it is being handled.
- Place tape over open fuse cavities and secondary fuse wells. Ensure that they are clear of obstruction and free of foreign matter before attempting to install the fuse, the detonator, or the FD.
- Take steps to prevent moisture or water from accumulating around the mine and subsequently freezing if the temperature fluctuates around freezing. Mines usually function satisfactorily at temperatures between 40 and 160°F. Most mines are not appreciably affected by temperature changes, but mines can become neutralized by ice formations (see Chapter 12).

- Observe proper procedures when recovering mines. Ensure that components do not show evidence of damage or deterioration.
- Ensure that practice or inert mines or their components are not present when live mines or their components are being used.
- Do not mix inert mines with live mines.
- Do not display live mines or their components in museums, demonstrations, models, or similar layouts. Only inert equipment can be used for displays.
- Handle explosive materials with appropriate care. The explosive elements in primers, blasting caps, and fuses are particularly sensitive to shock and high temperatures.
- Assemble activators, standard bases, and FDs before installing them. Do not carry them in the pockets of your clothing.
- Do not point FDs at anyone.
- Camouflage the mine before removing the positive safety pin when possible.

**NOTE: Additional storage and safety precautions are outlined in TM 9-1300-206.**

## LIVE-MINE TRAINING

**NOTE: No live-mine training is authorized with M14 mines. Units outside Korea will not use live M16A1 mines in tactical or protective minefield training.**

Live-mine training is conducted by preparing, laying, arming, neutralizing, and disarming live mines (with live fuses and components) in a training environment.

Supervisors must adhere to the following safety considerations when conducting live-mine training:

- Only personnel who are qualified and certified according to the local range SOP are allowed to supervise activities or training in which live mines or their components are used.
- Minimum personnel requirements to conduct live-mine training are—
  - Range officer (OIC).
  - Range safety officer (RSO).
  - One NCO supervisor for each arming bay.
  - Mine-explosive breakdown NCO.
  - One medic per four arming bays.
  - Guards, as required by the range SOP.
- Sound organization is a must before live-mine training can begin. The OIC and supervising NCOs conduct a demonstration/briefing to ensure that the practice runs smoothly.

- The training officer must foresee hazards that can occur through personnel nervousness or material failure. The commander should conduct a risk assessment according to AR 385-10.
- The OIC takes his place at the control point or post. Once he is satisfied that all safety regulations have been observed, he orders the first detail to start training.
- Soldiers are trained on inert and practice mines before arming live mines, according to the guidelines established by the Standards in Training Commission.
- Fuses are not inserted into mines until ordered by the OIC.
- An NCO supervisor must be present when soldiers arm live mines. He ensures that soldiers adhere to the proper procedures and regulations.
- Only one soldier arms a mine at any given time.
- Personnel disarm one mine before arming the next one.
- Personnel never arm an M16 AP mine in the trip-wire mode during live-mine training.
- Personnel never remove the positive safety pin from the M16 AP mine during live-mine training.
- Instructors inspect fuses and mines for serviceability before starting practice.
- Instructors inspect mines and their components for damage and excessive wear after each student has gone through the station. Replace the mine and the fuse if damage or wear is found.
- All personnel wear a helmet (with the serviceable chin strap fastened) and body armor when arming and disarming mines.
- Ear protection is not permitted in the arming bays. The student must be able to hear the supervisor and certain distinct noises (such as a firing pin dropping).
- Instructors post guards at all entrances to the range. The guards communicate with the RSO by radio, wire, voice, or signal. No one enters the range without permission from the RSO.
- Instructors keep mine records and inventory sheets. They maintain accountability of all mines and fuses, before and after each exercise.
- Instructors draw and return supplies; check equipment for issue; and ensure that live mines are safe, serviceable, and unarmed. They ensure that the requirements contained in AR 385-63, range regulations, and SOPs are observed and that no one does anything to prejudice safety.
- Instructors clearly mark the word *LIVE* on all live mines and their components that are used for live-mine training. Live mines are maintained separately from practice and inert mines.

- Live AHDs are not used with live mines during training, but they can be used with practice and inert mines.
- Arming and disarming are conducted in the prone position.
- Waiting personnel are located in a bunker, behind a suitable barricade, or at a safe distance from live-mine training.
- Supervisors ensure that live-mine training is not rushed. There are no shortcuts. Supervisors must allow soldiers ample time to arm and disarm mines. Most soldiers are already in a high state of stress from dealing with live munitions, and rushing them only serves to heighten their stress level.

## LIVE-MINE DEMONSTRATIONS

Live-mine demonstrations show mine characteristics and capabilities using M16 and M18 AP mines and M15, M19, and M21 AT mines. The appropriate authority must authorize the demonstration, and firing personnel must be fully conversant with all safety and technical aspects pertaining to live-mine firing.

An OIC and an RSO are appointed for each activity involving live-mine firing. The amount of explosive contained in the mine cannot exceed the maximum amount allowed for the range, and one mine is fired at a time.

Upon arriving at the range, the instructor and his assistants establish areas according to the following rules (signs are posted for large demonstrations):

- Firing point. Sited outside the danger area and near the OIC to facilitate coordination, commentaries, and firing.
- Spectator area. Sited outside the danger area and within earshot of the commentator. It is large enough to provide a good view of the explosion.
- Supply area. Any suitable area away from spectators.
- Explosive area. Sited away from supplies and spectators.
- Mine area. Mines are set out in full view of the OIC and spectators. Individual mines are at least 25 meters apart.
- Target area. Targets are positioned and inspected by spectators before the blasting cap is inserted into the mine.

### M16 Antipersonnel Mine

- Safety distance. 300 meters.
- Firing procedures.
  - Roll out 300 meters of firing cable and attach it to a stake or picket in the ground (leave at least 1 meter of free end). Test the firing cable for continuity.
  - Place the mine in the ground (dig in level with the surface). Remove the shipping plug.
  - Test a blasting cap (under a sandbag) with the demolition test set.

— Attach the ends of the blasting-cap leads to the ends of the electric cable and insulate the joints with tape. Place the blasting cap into the fuse well (see Figure E-1, page E-6).

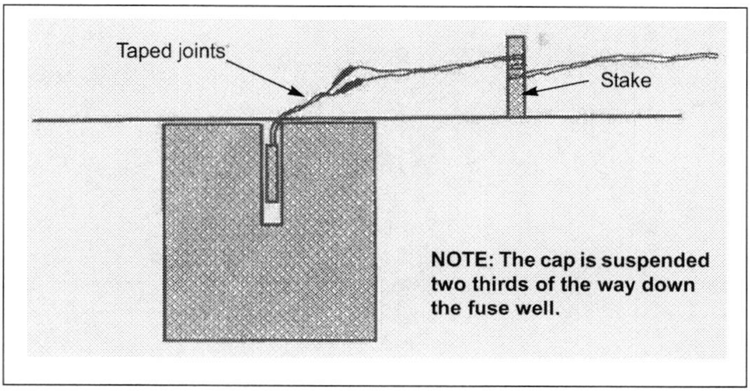


**Figure E-1. M16 AP mine**

- Suggested target. A circle of tar paper, 6 meters in diameter, supported by 1.8-meter pickets. Spectators can later view shrapnel effects.

**NOTE: The procedure detailed here dispenses with the M605 igniter. The mine cannot be detonated by pull or pressure. The expulsion charge and millisecond delay fuses are still operated, and the mine bounds out of its casing (which remains in the ground) before exploding in the air. Although the normal firing delay is removed, it does not detract from the demonstration. The blasting cap is suspended two-thirds of the way down the fuse well to initiate the expelling charge and delay elements.**

- Misfires. In the event of a misfire, the RSO disposes of the mine by placing a block of C4 as close to the mine as possible, without touching it. He destroys the mine by normal nonelectric means.

### M18A1 ANTIPERSONNEL MUNITION

- Safety distance. 300 meters.
- Firing procedures.
  - — Roll out 300 meters of firing cable and attach it to a stake or picket in the ground (leave at least 1 meter of free end). Test the firing cable for continuity.
  - — Place the mine on the ground and ensure that the front of the mine faces away from the firing point. Remove the shipping plug.
  - — Test an electric blasting cap (under a sandbag) with the demolition test set.

— Attach the ends of the blasting cap leads to the ends of the electric cable and insulate the joints with tape. Place the blasting cap into the detonator well (see Figure E-2).

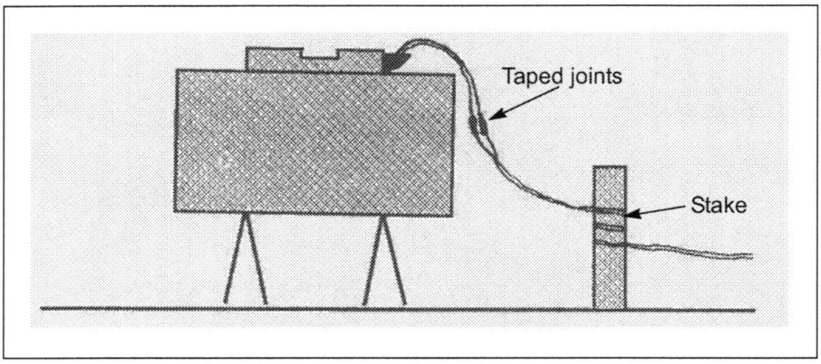


**Figure E-2. M18A1 AP mine**

- Suggested target. Several E-type silhouette targets, 15 to 100 meters from the mine.

**NOTE: The procedure detailed here applies only to demonstration firings. Standard accessories are used on all other occasions. The mine explodes instantaneously and clearly illustrates the sound of an M18A1 explosion.**

- Misfires. In the event of a misfire, the RSO disposes of the mine by placing a block of C4 as close to the mine as possible, without touching it. He destroys the mine by normal nonelectric means.

**M15, M19, AND M21 ANTITANK MINES**

- Safety distance. 1,000 meters.
- Firing procedures.
  - — Roll out 1,000 meters of firing cable and attach it to a stake or picket in the ground (leave at least 1 meter of free end). Test the firing cable for continuity.
  - — Place the mine in the ground and leave the top exposed. A target is used only when the mine can be placed without disturbing the target. A derelict vehicle is a suitable target.
  - — Place a block of C4 on top of M15s and M19s (see Figure E-3, page E-8).
  - — Remove the shipping plug from the booster well of the M21 and pack the well with C4. Insert an electric blasting cap into the C4 (see Figure E-4, page E-8).

**NOTE: Do not remove safety devices. Keep arming dials in the *SAFE* position. The mine explodes instantaneously and clearly demonstrates the blast/shaped-charge effect.**

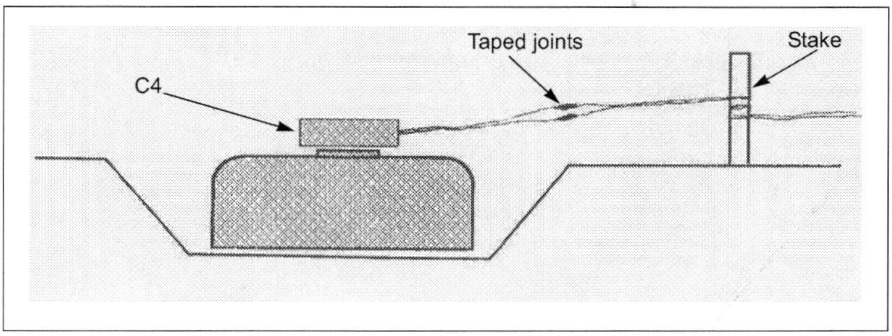


Figure E-3. M15 and M19 AT mines

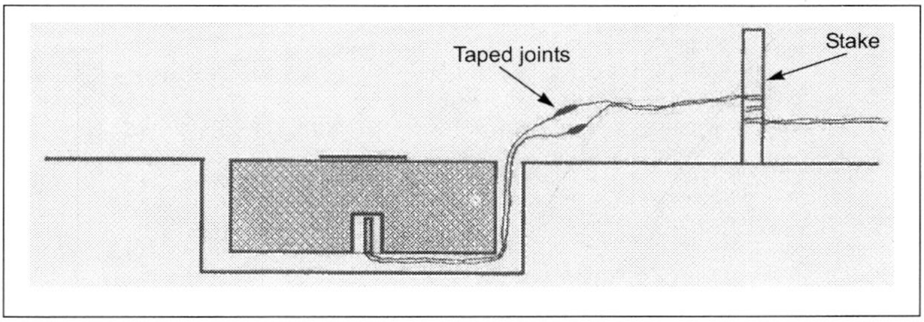


Figure E-4. M21 AT mine

- Misfires. In the event of a misfire, the RSO disposes of the mine by placing a block of C4 as close to the mine as possible, without touching it. He destroys the mine by normal nonelectric means.

## RISK ASSESSMENT FOR LIVE-MINE DEMONSTRATIONS

The following risk assessment is provided as a guideline for live-mine demonstrations using an M16 AP mine. It must be carefully reviewed before conducting a demonstration. Live-mine demonstrations can be conducted in a safe manner. The risk of injury to personnel is significantly minimized if you adhere to established procedures.

During the demonstration, mines are not armed with standard fuses. They are activated by electric blasting caps placed inside the fuse wells.

A demonstration shows the effectiveness of an M16 AP mine. Spectators do not handle the mines or explosives. To show the effectiveness of an M16 mine, a sheet of paper is placed in a semicircle around the mine. Spectators remain in bunkers or at a safe distance while mines are primed with electric blasting caps and detonated. After the mines have been detonated and the RSO has cleared the area, spectators are allowed to view the results of the detonated mine. Misfires are handled by the RSO.

Figure E-5 is a risk assessment prepared by the Department of Transportation.

## QUALITATIVE RISK ASSESSMENT

Qualitative risk-assessment techniques are used to place a value on the level of risks created by hazards in an operation. The principal qualitative technique is the risk-assessment code (RAC) described in MILSTD-882C. This method was established as a common way to set priorities for Department of Defense (DOD)-wide hazard-abatement programs and uses an RAC matrix format to combine the concepts of frequency and severity into a single, numerical code. It is very useful in comparing different risks (such as those from different programs) or even differences (such as health versus safety risks).

RACs are implemented for the Army in Army Regulation (AR) 385-10, which describes the two qualities of hazard severity and hazard probability as follows:

- Category I—CATASTROPHIC. May cause death or loss of a facility. In this case, loss does not mean a period of interrupted service; it means destruction of the facility or operation.
- Category II—CRITICAL. May cause severe injury, severe occupational illness, or major property damage.
- Category III—MARGINAL. May cause minor injury, minor occupational illness, or minor property damage.
- Category IV—NEGLIGIBLE. Probably would not affect personnel safety or health, but is in violation of specific standards.

Mishap probability is the probability that a hazard will result in a mishap, based on an assessment of factors such as location, exposure in terms of cycles or hours of operation, and affected population. The expression combines the idea of the probability of an event and the exposure to the event. These probabilities are expressed as letters conforming to the following system:

- Subcategory A—Likely to occur immediately.
- Subcategory B—Probably will occur in time.
- Subcategory C—May occur in time.
- Subcategory D—Remote.
- Subcategory E—Improbable.

The two qualities are combined to yield an RAC by using the following table:

| Risk-Assessment Code Table | | | | | |
|---|---|---|---|---|---|
| Hazard Severity | Mishap Probability | | | | |
| | A | B | C | D | E |
| I | 1 | 1 | 2 | 3 | 4 |
| II | 1 | 2 | 3 | 4 | 5 |
| III | 2 | 3 | 4 | 5 | 5 |
| IV | — | — | — | — | — |

In using the RAC system, it is important to note that the IA and IIA classifications are termed "imminent danger." Though their RAC codes of 1 are the same as that of the IB entry, their immediacy makes them more critical. The codes are useful in assessing an operation as it begins, but they must be updated as the operation continues, the facility ages, and so forth to account for degrading condition or performance.

For risk managers, there are some important organizational/management considerations to RAC codes. AR 40-10 also contains an RAC system, but due to a difference in definitions, the health RAC code may be a lower number, indicating a higher degree of risk. This is important to managers who are comparing health risks to others risks under AR 385-10; the health issue would always win if no compensation or consideration was factored into the codes. From the managerial standpoint, it must be remembered that RACs are judgmental and not necessarily held to be the same by different managers or evaluators. When differences in perception occur, the differences are likely based on either the understanding of the operation's behavior or the criteria for selecting the probability and severity. It is wise to listen to the basis of others' RAC choices and attempt to develop a common understanding.

**Figure E-5. Excerpt from Risk-Assessment Techniques Manual, prepared by the Department of Transportation's Transportation Safety Institute, August 1986**

Qualitative risk-assessment techniques are used to prepare estimates of risk levels using performance data, when available, to improve the accuracy of risk estimates used in risk-acceptance decision making. These assessments are numeric values representing the safety risk of an Army activity, system operation, or comparable endeavor, based on actuarial or derived numeric data. Though RACs are numerical, they are derived from judgments and are not demonstrable in records of performance. If it is desirable that performance be measured, it is necessary that quantified estimates of risk levels be established, that risk levels be predictive so that future performance has a base of comparison, and that risk levels be assigned numeric values.

**Figure E-5. Excerpt from Risk-Assessment Techniques Manual, prepared by the Department of Transportation's Transportation Safety Institute, August 1986 (continued)**

## RISK ASSESSMENT FOR LIVE-MINE TRAINING

The United States Army Engineer Center, Department of Instruction, obtained information for the following risk assessment from the Collective Training Branch, Department of Training and Doctrine, and from the Engineering Branch for Engineer Officer Basic Course demolitions training. Hazards are identified and analyzed on preliminary hazard-analysis work sheets (see Figures E-6 through E-15, pages E-11 through E-21). Risk-assessment codes are assigned to each hazard based on the severity and the probability of occurrence.

References used in the risk-analysis process include this manual and the following publications:

- DA Pamphlet (Pam) 350-38.
- AR 385-10.
- AR 385-16.
- AR 385-63.
- Soldier Training Publication (STP) 5-12B1-SM.
- Training Circular (TC) 25-8.
- TM 9-1345-203-12.
- TM 43-0001-36.

**Operation Name:** Arming the M15 AT Mine
**Date Prepared:** 11 October 1990
**Sheet Number:** 1 of 1

| Hazards | System Effects | Causal Factors | RAC | Actions | Controlled RAC | Standards/Comments |
|---|---|---|---|---|---|---|
| Damaged mine used | None | Lack of training/improper training<br><br>Improper supervision<br><br>Incomplete or no inspection | IV B (-) | Do not use mine.<br><br>Contact QASAS to investigate.<br><br>Turn mine in for investigation/ disposal. | IV D (-) | Proper training is conducted.<br><br>Soldiers are proficient on inert mines first.<br><br>One instructor is assigned to one soldier.<br><br>All training is conducted in the prone position.<br><br>Mine training is done after identified by a unit's METL and only with select personnel.<br><br>Training is conducted in the proper environment.<br><br>Soldiers are in the proper protective gear.<br><br>No trip wires or AHDs are used with an M15 mine.<br><br>Using the cap to perform the fuse depth check is not authorized. Use the wrench. |
| Too much force on extension rod/ pressure ring (M624 fuse) | Mine detonates | Soldier bends/pushes on rod or fuse | I B (1) | Do not remove the safety stop and band if the extension rod is used.<br><br>**OR**<br><br>Remove the extension rod before removing safety stop and band. | I D (3) | |
| Too much pressure on pressure plate (M603 fuse) | Mine detonates | Soldier applies excessive pressure | I D (3) | See comments. | I D (3) | |
| Fuse improperly seated in the fuse well (M603 fuse) | Mine detonates | Excessive debris/corrosion in well<br><br>Lack of training/improper training<br><br>Improper supervision<br><br>Incomplete or no inspection | I C (2) | Perform depth check with wrench. | I D (3) | |
| Improper handling of fuse (M603 fuse) | Fuse detonates | Pressure applied to fuse with safety fork removed | II D (4) | Handle fuse properly. | II D (4) | |

**Figure E-6. Preliminary hazard-analysis work sheet (arming M15)**

**Operation Name:** Disarming the M15 AT Mine
**Date Prepared:** 11 October 1990
**Sheet Number:** 1 of 1

| Hazards | System Effects | Causal Factors | RAC | Actions | Controlled RAC | Standards/Comments |
|---|---|---|---|---|---|---|
| Camouflage removed | Mine detonates | Too much pressure on pressure plate<br><br>Damage to the mine | I D (3) | Do not apply pressure to pressure plate.<br><br>See comments. | I D (3) | See comments for arming an M15 AT mine (Figure E-8, page E-13).<br><br>See risk assessment for command-detonating a mine in place with explosives (Figure E-16, page E-21). |
| Unable to turn the mine to SAFE or reinstall safety ring, safety stop, or safety pin | Mine detonates | Too much resistance (M603 fuse) is applied to turning indicator.<br><br>Safety band (M624), safety stop, or safety pin has become damaged.<br><br>Foreign material has entered the fuse well. | I C (2) | Perform controlled detonation in place. | None | |
| Fuse removed/ safety fork installed | Fuse detonates | Pressure is placed on the pressure plate of fuse.<br><br>Fuse is not removed. | II D (4) | Follow proper procedures. | II D (4) | |

**Figure E-7. Preliminary hazard-analysis work sheet (disarming M15)**

**Operation Name:** Arming the M16-Series AP Mine
**Date Prepared:** 11 October 1990
**Sheet Number:** 1 of 1

| Hazards | System Effects | Causal Factors | RAC | Actions | Controlled RAC | Standards/Comments |
|---|---|---|---|---|---|---|
| Damaged mine used | None | Lack of training/improper training<br>Improper supervision<br>Incomplete or no inspection | IV C (-) | Do not use mine.<br>Contact QASAS to investigate.<br>Turn in the mine for investigation/ disposal. | None | Proper training is conducted.<br>Soldiers are proficient on inert mines first.<br>One instructor is assigned to one soldier.<br>All training is conducted in the prone position.<br>Mine training is done after the need is identified by a unit's METL and then only with select personnel.<br>Training is conducted in the proper environment.<br>Soldiers are in the proper protective gear.<br>No trip wires or live AHDs are used with an M16 mine.<br>In the event of damage or loss of any safety pins, stop training with this particular mine and destroy it. |
| Faulty fuse used (M605) | Mine detonates | Safety pins missing<br>Prongs bent<br>Fuse head does not turn freely<br>Pins installed incorrectly | I B (1) | Do not use fuse.<br>Contact QASAS to investigate.<br>Turn mine in for investigation/ disposal.<br>Conduct quality control check before training. | I D (3) | |
| Pressure applied to prongs after the positive safety pin is removed | Mine detonates | Pressure is applied inadvertently | I B (1) | Do not remove the positive safety pin at any time. | I D (3) | |
| Fuse improperly armed | Mine detonates | Individual pushes in on release pin while or after positive safety pin is removed<br>Wrong sequence for removing pins<br>Locking safety pin is removed, click is heard, soldier continues | I B (1) | Do not remove the positive safety pin at any time. | I D (3) | |

**Figure E-8. Preliminary hazard-analysis work sheet (arming M16)**

**Operation Name:** Disarming the M16-Series AP Mine
**Date Prepared:** 11 October 1990
**Sheet Number:** 1 of 1

| Hazards | System Effects | Causal Factors | RAC | Actions | Controlled RAC | Standards/Comments |
|---|---|---|---|---|---|---|
| Camouflage removed | Mine detonates | Too much pressure on M605 fuse pressure prongs | I C (2) | Provide close supervision.<br><br>Do not rush the procedure.<br><br>Do not camouflage mine. | I D (3) | Do not fully arm this mine; leave the positive safety pin in the fuse at all times.<br><br>See comments for arming an M16 AP mine (Figure E-10, page E-15).<br><br>See risk assessment for command-detonating a mine in place with explosives (Figure E-16, page E-21). |
| Replace safety pins | Mine detonates | Pins not replaced<br><br>Pull on release-pin ring<br><br>Pressure on fuse pressure prongs<br><br>Pins replaced in wrong order | I C (2) | Provide close supervision.<br><br>Do not rush the procedure. | I D (3) | |
| M605 fuse removed | Mine detonates | Safety pins not properly in place, thus allowing pressure to be applied to the fuse, pressure prongs, or release pin | I C (2) | Provide close supervision.<br><br>Assure that pins are secure and replaced correctly. | I D (3) | |

**Figure E-9. Preliminary hazard-analysis work sheet (disarming M16)**

**Operation Name:** Arming the M19 AT Mine
**Date Prepared:** 11 October 1990
**Sheet Number:** 1 of 1

| Hazards | System Effects | Causal Factors | RAC | Actions | Controlled RAC | Standards/Comments |
|---|---|---|---|---|---|---|
| Damaged mine used | None | Lack of training/improper training<br>Improper supervision<br>Incomplete or no inspection | IV B (-) | Do not use mine.<br>Contact QASAS to investigate. | IV B (-) | Proper training is conducted.<br>Soldiers are proficient on inert mines first.<br>One instructor is assigned to one soldier.<br>All training is conducted in the prone position.<br>Mine training is done after identified by a unit's METL and only with select personnel.<br>Training is conducted in the proper environment.<br>Soldiers are in the proper protective gear.<br>No trip wires or AHDs are used with a live M19 mine. |
| Improper handling of detonator | Detonator detonates | Soldier handles detonator improperly<br>Lack of training/improper training<br>Improper supervision<br>Incomplete or no inspection | I D (4) | Handle detonator properly. | III D (5) | |
| Too much pressure on pressure plate | Mine detonates | Soldier applies too much pressure | I D (3) | See comments. | I D (3) | |
| Weakened Belleville spring | Mine detonates at pressure lower than designed pressure | Belleville spring weakening | I C (2) | Inform soldiers of less pressure required to detonate. | I D (3) | |

**Figure E-10. Preliminary hazard-analysis work sheet (arming M19)**

**Operation Name:** Disarming the M19 AT Mine
**Date Prepared:** 11 October 1990
**Sheet Number:** 1 of 1

| Hazards | System Effects | Causal Factors | RAC | Actions | Controlled RAC | Standards/Comments |
|---|---|---|---|---|---|---|
| Camouflage removed | Mine detonates | Too much pressure on pressure plate<br><br>Damaged or malfunctions | I D (3) | Do not camouflage mine. | I D (3) | See comments for arming an M19 AT mine (Figure E-12, page E-17).<br><br>See risk assessment for command-detonating a mine in place with explosives (Figure E-16, page E-21). |
| Unable to turn the mine to SAFE | Mine detonates | Too much resistance turning the dial indicator<br><br>Foreign material in the fuse well | I C (2) | Perform controlled detonation in place. | I D (3) | |
| Fuse assembly removed and detonator replaced with shipping plug | Detonator detonates | Dropped fuse assembly with detonator<br><br>Dropped the detonator | II C (3) | Handle fuse assembly and detonator properly. | II D (4) | |

**Figure E-11. Preliminary hazard-analysis work sheet (disarming M19)**

**Operation Name:** Arming the M21 AT Mine
**Date Prepared:** 11 October 1990
**Sheet Number:** 1 of 1

| Hazards | System Effects | Causal Factors | RAC | Actions | Controlled RAC | Standards/Comments |
|---|---|---|---|---|---|---|
| Damaged mine used | None | Lack of training/improper training<br><br>Improper supervision<br><br>Incomplete or no inspection | IV B (-) | Do not use mine.<br><br>Contact QASAS to investigate.<br><br>Turn mine in for investigation/disposal. | IV B (-) | Proper training is conducted.<br><br>Soldiers are proficient on inert mines first.<br><br>One instructor is assigned to one soldier.<br><br>All training is conducted in the prone position.<br><br>Mine training is done after identified by a unit's METL and only with select personnel.<br><br>Training is conducted in the proper environment.<br><br>Soldiers are in the proper protective gear.<br><br>No trip wires or AHDs are used with a live M21 mine. |
| Too much force on extension rod/pressure ring (M607 fuse) | Mine detonates | Soldier bends/pushes on rod or fuse | I B (1) | Do not remove the safety stop and band if the exten-sion rod is used.<br><br>**OR**<br><br>Remove the extension rod before removing the safety stop and band. | I D (3) | |
| Fuse improperly seated in the fuse well | Mine detonates | Excessive debris/corrosion in well<br><br>Lack of training/improper training<br><br>Improper supervision<br><br>Incomplete or no inspection | I C (2) | See comments. | I D (3) | |
| Improper handling of fuse | Fuse detonates | Pressure applied to fuse while mine is armed | II D (4) | Handle fuse properly. | None | |

**Figure E-12. Preliminary hazard-analysis work sheet (arming M21)**

**Operation Name:** Disarming the M21 AT Mine
**Date Prepared:** 11 October 1990
**Sheet Number:** 1 of 1

| Hazards | System Effects | Causal Factors | RAC | Actions | Controlled RAC | Standards/Comments |
|---|---|---|---|---|---|---|
| Camouflage removed | Mine detonates | Too much pressure<br><br>Tilted extension rod<br><br>Damaged or malfunctions | I B (1) | Do not use extension rod.<br><br>Ensure close supervision. | I D (3) | See comment for arming an M21 AT mine (Figure E-14, page E-19).<br><br>Stop training with this particular mine and destroy it in the event of damage or loss of stop, band, or cotter pin.<br><br>See risk assessment for command-detonating a mine in place with explosives (Figure E-16, page E-21). |
| Band, stop, and cotter pin replaced | Mine detonates | Too much pressure<br><br>Tilted extension rod<br><br>Damaged stop, band, or cotter pin | I B (1) | Do not use extension rod.<br><br>Ensure close supervision. | I C (2) | |
| Extension rod and/or fuse removed | Mine detonates | Too much pressure<br><br>Tilted extension rod | I C (2) | Do not use extension rod.<br><br>Ensure proper placement of band, stop, and cotter pin. | I D (3) | |
| Booster removed | Booster detonates | Soldier improperly handles booster | II D (4) | Handle booster properly. | None | |

**Figure E-13. Preliminary hazard-analysis work sheet (disarming M21)**

**Operation Name:** Preparing charges and priming explosives to be used for mine demolition (detonated in place)
**Date Prepared:** 11 October 1990
**Sheet Number:** 1 of 1

| Hazards | System Effects | Causal Factors | RAC | Actions | Controlled RAC | Standards/Comments |
|---|---|---|---|---|---|---|
| Defective wire or faulty test of wire | No detonation | Wire will not carry electrical charge to detonate explosive | IV D (-) | Do not use wire (replace it). | None | After testing, ensure that wire ends are twisted together.<br><br>Only one demolitions person and one safety person will be at the mine when the blasting cap is inserted into the demolition.<br><br>RSO is responsible for clearing misfires.<br><br>RSO keeps a *misfire kit* under his control. |
| Electric blasting cap check done incorrectly or not at all | Blasting cap detonates | Cap improperly handled<br><br>Cap not put under sandbag | III C (4) | Handle cap properly.<br><br>Place cap under sandbag. | IV D (-) | |
| Blasting cap attached incorrectly to firing wire | Blasting cap detonates | Cap/firing wire improperly handled<br><br>End of firing wire not shunted | III C (4) | Ensure that firing wire is shunted.<br><br>Ensure proper handling of blasting cap. | IV D (-) | |
| Blasting cap inserted incorrectly/too forcefully into 1-pound demolition charge | Blasting cap and/or demolition charge detonates | Too much pressure on blasting cap<br><br>Cap improperly handled | I D (3) | Ensure proper handling of blasting cap. | I D (3) | |
| Charge does not detonate | No detonation | Faulty blasting cap<br><br>Faulty blasting machine<br><br>Faulty firing wire | IV C (-) | OIC declares a misfire.<br><br>Follow RSO procedures for a misfire. | IV C (-) | |

**Figure E-14. Preliminary hazard-analysis work sheet (command detonation)**

**Operation Name:** Mine arming/disarming peripheral factors (not all-inclusive)
**Date Prepared:** 11 October 1990
**Sheet Number:** 1 of 2

| Hazards | System Effects | Causal Factors | RAC | Actions | Controlled RAC | Standards/Comments |
|---|---|---|---|---|---|---|
| Good weather | No direct | Temperature between 45 and 70°F<br><br>Clear or partly cloudy<br><br>Wind speed less than 5 mph | IV A (-) | None | None | This list is not conclusive and will depend on the particulars of the unit, the training, and the range facilities.<br><br>Additional peripheral factors a unit commander may wish to consider are—<br>• Level of proficiency.<br>• Time of event (day or night).<br>• Availability and extent of emergency-response assets.<br>• Train up (rehearsals and dry runs).<br>• Terrain.<br>• Location of instructor in relation to soldier during the training. |
| Minimal weather | No direct | Temperature between 32 and 44°F or 71 and 80°F<br><br>Drizzle<br><br>Wind speed between 5 and 15 mph | III A (2) | Ensure close supervision.<br><br>Do not rush training to get in shelter.<br><br>Postpone or cancel training. | III C (4) | |
| Bad weather | Wet, miserable soldiers<br><br>Wet munitions | Temperature less than 32°F or greater than 80°F<br><br>Moderate to heavy rain<br><br>Snow or ice<br><br>Wind speed greater than 15 mph | I C (2) | Postpone or cancel training. | I D (3) | |
| Lack of references | The sequential arming or disarming not done correctly, causing the fuse and/or mine to detonate | Current doctrine not available<br><br>Questions or confirmation of techniques not quantified | I B (1) | Stop training if unsure of a task.<br><br>Have proper references available. | I D (3) | |

**Figure E-15. Preliminary hazard-analysis work sheet (peripheral factors)**

**Operation Name:** Mine arming/disarming peripheral factors (not all-inclusive)
**Date Prepared:** 11 October 1990
**Sheet Number:** 2 of 2

| Hazards | System Effects | Causal Factors | RAC | Actions | Controlled RAC | Standards/Comments |
|---|---|---|---|---|---|---|
| Lack of support materials or components | Improper procedures or practices introduced, causing the fuse and/or the mine to detonate | Not enough or incorrect fuses, wrenches, and/or washers used | I C (2) | Ensure that proper support material is available.<br><br>Stop training if the proper material is not available. | I D (3) | |
| Condition of soldiers | Improper procedures or practices introduced, causing the fuse and/or the mine to detonate | Long amount of training with little or no sleep<br><br>Strenuous training conducted before or during live-mine training | I D (3) | Ensure that soldiers have adequate sleep.<br><br>Allow for breaks during training.<br><br>Stop training if soldiers appear to be heavily fatigued. | None | |
| Protective clothing | Improper or lacking protective clothing, increasing the severity of an accident if it does occur | Improper or lacking protective clothing | I C (2) | All soldiers participating in mine training will have Kevlar helmets, flak vests, and boots. | None | |

**Figure E-15. Preliminary hazard-analysis work sheet (peripheral factors) (continued)**

Appendix F

# Mine Awareness

Mine awareness should actually be entitled mine/UXO awareness. If only mines are emphasized, ordnance (bomblets, submunitions) may be overlooked, and it has equal if not greater killing potential. The main objective of mine awareness is to save lives, so it is important to ensure that soldiers are well-informed and thoroughly trained. This appendix outlines the tasks needed for soldiers and units to survive in a mined/UXO environment.

Mine awareness should be emphasized at all levels of command, and it involves soldier and leader skills. Soldier skills are a mix of individual and collective tasks that are required for an element to maintain its combat effectiveness in and around a mined environment. It is important to note that a soldier's basic mine-awareness skills are critical to his and the unit's survival. Leader skills involve planning missions, assessing situations, and tracking/disseminating mine information. A unit must be proficient in all mine-awareness skills to effectively operate in a mined environment.

## SOLDIER

Soldier skills involve individual and collective tasks that are required for basic survival in a mined/UXO environment. They include minefield indicators, probing techniques, mine-detector operation, extraction drills, survival rules, casualty treatment, and evacuation drills.

### VISUAL INDICATORS

Mine/UXO indicators are part of all combat operations. Understanding and recognizing mine indicators could determine whether or not a soldier becomes a casualty. The following may indicate the presence of mines/UXO:

- Trip wires.
- Signs of road repair (such as new fill or paving, road patches, ditching, culvert work).
- Signs placed on trees, posts, or stakes. Threat forces mark their minefields to protect their own forces.
- Dead animals.
- Damaged vehicles.
- Disturbances in previous tire tracks or tracks that stop unexplainably.
- Wires leading away from the side of the road. They may be firing wires that are partially buried.

- Odd features in the ground or patterns that are not present in nature. Plant growth may wilt or change color, rain may wash away some of the cover, the cover may sink or crack around the edges, or the material covering the mines may look like mounds of dirt.
- Civilians. They may know where mines or booby traps are located in the residential area. Civilians staying away from certain places or out of certain buildings are good indications of the presence of mines or booby traps. Question civilians to determine the exact locations.
- Pieces of wood or other debris on a road. They may be indicative of pressure or pressure-release FDs. These devices may be on the surface or partially buried.
- Patterns of objects that could be used as a sighting line. The enemy can use mines that are fired by command, so road shoulders and areas close to the objects should be searched.

## PROBING

Probing is very time-consuming and is used primarily for clearing operations, self-extraction, and covert breaching operations. Detection of mines by visual or electronic methods should be confirmed by probing. Use the following procedures and techniques when probing for mines:

- Roll up your sleeves and remove your jewelry to increase sensitivity. Wear a Kevlar helmet, with the chin strap buckled, and a protective fragmentation vest.
- Stay close to the ground and move in a prone position to reduce the effects of an accidental blast. When moving into a prone position, the prober should—
  - — Squat down without touching his knees to the ground.
  - — Scan forward up to 2 meters and to the sides up to 3 meters for mine indicators.
  - — Probe the area around his feet and as far forward as possible.
  - — Kneel on the ground after the area is found to be clear and continue probing forward until he is in a prone position.
- Use sight and touch to detect trip wires, fuses, and pressure prongs.
- Use a slender, nonmetallic object as a probe.
- Probe every 5 centimeters across a 1-meter front.
- Gently push the probe into the ground at an angle that is less than 45 degrees.

**DANGER**
**Use extreme caution when probing. If the probe is pushed straight down, its tip may detonate a pressure fuse.**

- Apply just enough pressure on the probe to sink it slowly into the ground.

- If the probe encounters resistance and does not go into the ground freely, carefully pick the soil away with the tip of the probe and remove the loose dirt by hand. Care must be taken to prevent functioning the mine.
- When a solid object is touched, stop probing and use two fingers from each hand to carefully remove the surrounding soil and identify the object.
- If the object is a mine, remove enough soil to show the mine type and mark its location. Do not attempt to remove or disarm the mine. Use explosives to destroy detected mines in place or use a grappling hook and rope to cause mines to self-detonate. Metal grappling hooks should not be used on magnetic-fused mines.

Probing is extremely stressful and tedious. The senior leader must set a limit to the time a prober is actually probing in the minefield. To determine a reasonable time, the leader must consider METT-TC factors, weather conditions, the threat level, the unit's stress level, and the prober's fatigue level and state of mind. As a rule, 20 to 30 minutes is the maximum amount of time that an individual can probe effectively.

## AN/PSS-12 Metallic Mine Detector

The AN/PSS-12 mine detector (Figure F-1, page F-4) is a man-portable metallic mine-detection system that is used to detect AT and AP land mines. Its search head contains two concentric coils—the transmitting (outer) coil and the receiving (inner) coil. During operation, the transmitting coil emits a magnetic field symmetrically above and below the search head. When this field encounters metal objects, it induces currents in these objects. When the induced currents reach the receiving coil, an auditory signal is provided to the operator. The detector is able to detect large metal mines from a distance of several feet, but can detect low-metal mines from only a few inches.

> **WARNING**
> **It is important to understand that the mine detector is only effective when there is a sufficient amount of metal in the mine.**

### Unpacking

The system is stored and transported in a single carrying case.

- Open the pressure-relief valve in the carrying case.
- Release the latches on the carrying case and open the top.
- Remove the bag that contains system components.
- Unzip the bag and ensure that all components are present (Figure F-2, page F-4).
- Remove the following items from the bag carefully:
  - Telescopic pole and search-head assembly with cable and plug.
  - Electronic unit.

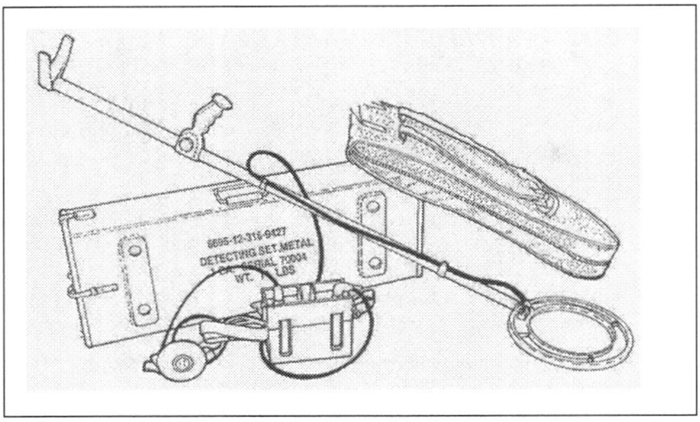


Figure F-1. AN/PSS-12 metallic mine detector

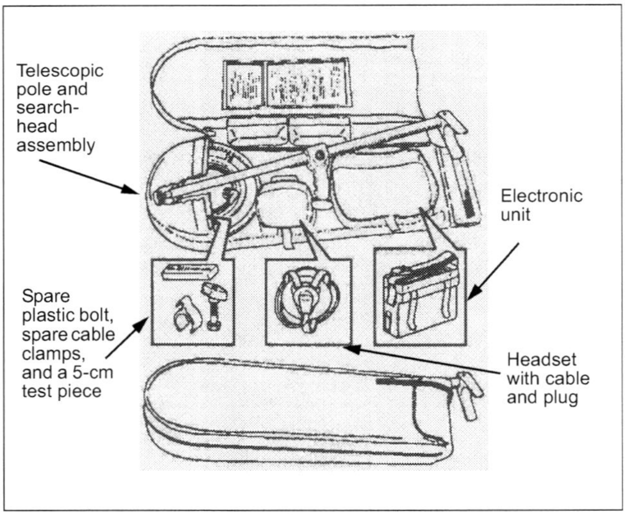


Figure F-2. AN/PSS-12 packed components

— Headset with cable and plug.

- Ensure that the bag contains the following spare parts and test items:
  — Spare plastic bolt.
  — Spare cable clamps.
  — 5-centimeter test piece.
- Inspect the search head for cracks or damage.

- Inspect cable connectors for damage and check for bent pins.
- Inspect hook and pile material on headphones for serviceability.
- Inspect cables to ensure that they are not cut, broken, or frayed.
- Inspect the electronic unit for cracks, damage, and completeness. This includes ensuring that all switches and knobs are present and functional.
- Ensure that telescopic pole is not bent, dented, or damaged and can be extended and retracted.

**Electronic-Unit Setup**

- Ensure that the power switch on the electronic unit is in the *OFF* position (Figure F-3).

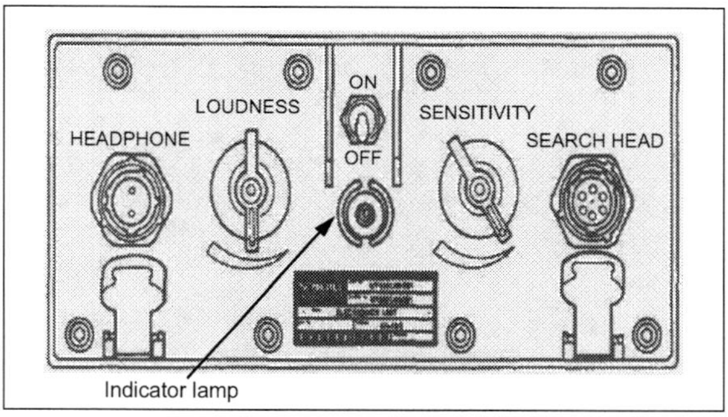


**Figure F-3. Electronic unit**

- Release the latches on the battery-compartment cover and remove the cover (Figure F-4).

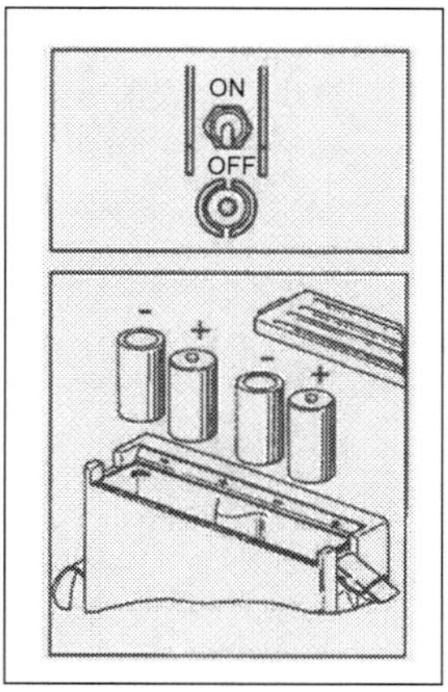


**Figure F-4. Battery installation**

- Insert batteries according to markings.

> **WARNING**
> **Ensure that the battery cover is completely closed and the latches are in the proper position. This prevents the inadvertent opening of the battery compartment during operation. Failure to do this could result in injury, damage to the equipment, and/or improper operation.**

- Reinstall and latch the battery cover. Ensure that the latches are in the proper closed position.
- Press in the catch (located just below the arm support) to unlock the telescopic pole from the transport position. Turn the outer tube until the catch snaps into the guide groove. Pull the telescopic pole out to a suitable length. Ensure that the telescopic pole is locked in one of three fixed positions.
- Connect the headphones and the detector-head cables to the electronic unit. Fit the protective caps of the cable connectors to the corresponding rubber caps on the electronic unit.
- Put on the headphones. Place the straps of the headphones over the forehead and on top of the head. The hook and pile material goes behind the head at the nape of the neck.
- Connect the detector-head cable to the lower (plastic) portion of the telescopic pole.
  - — Attach the lower cable clamp 5 centimeters above the wing nut.
  - — Attach the second cable clamp 5 centimeters down from the junction of the lower and upper (aluminum) portion of the telescopic pole.
  - — Attach the third cable clamp between the other two cable clamps. Do not attach the cable to this middle clamp. This clamp will become the sensitivity marker clamp.

> **WARNING**
> **The cable should not be attached to the aluminum portion of the pole, because it causes the sensitivity of the mine detector to fluctuate.**

- Adjust the position of the handle by loosening the knurl nut. Retighten the nut when the handle is properly positioned.
- Adjust the position of the search head so that it will be parallel to the ground.
- Attach the electronic unit to the operator's load-bearing equipment (LBE) belt using the belt clips. Right-handed soldiers should put the electronic unit on the left hip. Left-handed soldiers should put the electronic unit on the right hip. Put the safety strap over the shoulder.

## Electronic-Unit Controls and Indicators

The AN/PSS-12 electronic unit has the following controls and indicators (Figure F-3, page F-5):

- On/off switch. This switch is the power control for the AN/PSS-12. It must be in the ON position when detection is being conducted.
- Loudness control. This is the volume control for the audible alarm. It should be adjusted to a comfortable listening level during use.
- Sensitivity control. The sensitivity control is used to adjust the detection characteristics of the mine-detector head. Details on how to set the sensitivity are provided below.
- Indicator lamp. This lamp indicates low battery voltage or a system malfunction. It provides a short flash when the unit is turned on and flashes continuously when the batteries are low or there is a malfunction.
- Audible signals.
  - The system provides an audible signal through the headset when the search head is over or very near a metal object. The signal is a continuous tone that at first resembles a growl. The tone increases to a higher pitch as the search head is moved closer to a metal source (such as a mine) and decreases in pitch when moved away.
  - The AN/PSS-12 has a check tone that is provided every 1 to 2 seconds. The tone resembles a clicking sound, and its purpose is to continuously inform the operator that the system is functioning satisfactorily. If the check tone disappears or its frequency decreases, discontinue searching and determine the malfunction.
  - A flashing indicator lamp and a continuous tone (the same tone heard with a target return, except it does not abate when moved away) on the headphone indicates low battery voltage. If the unit's indicator lamp flashes, change the batteries and readjust the unit.

## Operation

Once the system is assembled, it is ready for adjustments prior to operation. Remove rings, watches, and other jewelry before adjusting or using the system.

### *Power Up/Sensitivity Adjustments*

- Turn the sensitivity and loudness knobs completely counterclockwise.
- Look at the indicator lamp on the electronic unit and turn the on/off switch to the ON position. The lamp should give a short flash. If it does not, ensure that the batteries are correctly installed or insert new batteries. If the lamp flashes continuously, the battery voltage may be low.
- Hold the search head approximately 0.6 meter above the ground, and turn the sensitivity knob clockwise until a continuous tone is heard.

While this is being done, adjust the loudness control to a comfortable listening level.

- Turn the sensitivity knob slightly counterclockwise until the tone ceases. The check tone (a clicking sound) should still be heard every 1 to 2 seconds. Readjust the loudness control if necessary. If the check tone disappears or its frequency decreases, discontinue searching and determine the malfunction.
- Set the sensitivity in either of the following two ways:
  - The preferred method is to use the most difficult to detect type of mine (made safe without changing the metal content) that is expected to be encountered.
  - In the absence an actual mine, use the 5-centimeter test piece that comes with the AN/PSS-12 packed components.
- Bury the mine (if you have a mine) at the deepest depth that threat mines are expected to be found. If you do not have a mine, bury the 5-centimeter test piece vertically so that the metal portion is at the deepest depth which threat mines are expected to be found. In the absence of more specific information, bury the test piece at a depth of 5 centimeters). This places the top of the test piece flush with the ground's surface.
- Place the detector head lightly on the ground directly above the mine/ test piece and adjust the sensitivity knob so that the mine is detected and you have an easily heard signal from the headset. Now, move the detector away from the mine/test piece until the signal ceases to be heard.
- Float the detector head lightly on the ground at a 0.3-meter-per-second movement rate across the mine/test piece. With the detector head in motion, check to see if the signal can again be easily heard. If not, increase the sensitivity until the signal can be easily heard.

**NOTE: The higher the sensitivity is set (between just being able to detect the mine/test piece and where the signal from the soil first becomes audible) the less the chance of missing a mine, but the more the chance of a false alarm rate.**

**WARNING**

**In some conditions, you may not be able to detect a target mine/test piece at the deepest depth the threat mine is expected to be encountered. Do not use the AN/PSS-12 to detect mines in those conditions.**

**NOTE: Adjusting the sensitivity with the mine/test piece buried in soil that is similar to the soil where the detector will be used is very important. Otherwise, the sensitivity setting is simply a guess, because different soil types and moisture content influence the sensitivity of the detector.**

***Sensitivity Level***

The AN/PSS-12's sensitivity setting drifts over time. During use, soldiers should check the detector's sensitivity every 1 to 2 meters of forward advance in the mine lane.

- With the sensitivity properly set using a buried mine/test piece, the detector is moved to an adjacent place on the ground over a clear area.
- To check the detector, a metal object (bayonet or other metal object) is slid down the lower (plastic) portion of the telescopic pole until the operator hears the same auditory signal that was emitted when the search head was placed over the buried mine/test piece.
- At this point on the lower plastic shaft, the middle plastic cable clamp (the middle clamp previously attached without the cable) is positioned. This clamp now becomes the sensitivity marker clamp.
- At approximately every 1 to 2 meters of forward advance in the mine lane (or more often if desired), the same metal object (bayonet) is moved down the shaft to the sensitivity marker clamp and the operator listens to determine if the same auditory signal is emitted. When the same auditory signal is not heard, the sensitivity knob is adjusted up or down until the same auditory signal is replicated and maintained.

**NOTE: It is important that the operator maintain a consistent body posture and angle between the telescopic pole and the search head each time a sensitivity check is made.**

**DANGER**
**Low batteries may reduce detector performance well before the indicator light comes on. If you need to frequently adjust the sensitivity knob to maintain a constant sensitivity setting or if the indicator light comes on, discard all batteries and replace them with new ones. Failure to do this could result in injury or death.**

**DANGER**
**Discontinue searching and readjust the sensitivity if the check tone disappears or its frequency decreases. Failure to do this could result in injury or death.**

***Search Methods***

- Sweep across the lane, keeping the entire search head in light contact with the ground. Sweeping speed should be approximately 0.3 meter per second. Each sweep must be a minimum of 1.5 meters and must overlap the edge of the lane by at least one-half the width of the search head. Each sweep should advance forward at no more than 6 inches per sweep. The sweep and advance must be maintained as described, otherwise gaps will be left between sweep paths or at the edge of the lane and a low-metal mine can be missed.

- Perform a sensitivity check at approximately every 1 to 2 meters of forward advance in the mine lane (or more often if desired) and make adjustments as necessary. It is important that the operator maintain a consistent body posture and angle between the telescopic pole and the search head each time a sensitivity check is made.
- Pat the search head lightly on the ground (each pat advancing no more than one-half the width of the search head) in low vegetation or rocky or uneven ground where smooth sweeping cannot be performed. Ensure that the wing nut is tight, so that the position between the head and telescopic pole is fixed.

**NOTE: Many conditions of vegetation may not allow getting the search head on the ground. Do not push through vegetation to get the search head on the ground. Different environments will require slightly different techniques. If you cannot get the search head on or very near the surface of the ground, the detector will not detect low-metal mines.**

**NOTE: The AN-PSS/12 detector's performance is reduced when the cable between the electronic unit and the top cable clamp is permitted to hang unrestrained during sweeps. The soldier can control this problem by holding the cable in his free hand.**

**WARNING**

**Always be aware of the potential for booby traps. The search head should never be swept where its path cannot be visually cleared. If trip wires are a threat, other techniques to detect trip wires should be used before sweeping.**

### *Alarm Investigation/Mine Identification*

When an auditory signal is heard, the soldier must investigate further to determine if the signal is a reliable indication of a mine.

- End the sweep procedure and begin the investigation procedure at the first auditory indication from the detector of metal in the ground. The purpose of the investigation is to determine if the signal is repeatable and, therefore, a likely indication of a mine. If so, the investigation continues with the purpose of gaining more information concerning the size, type (high-metal or low-metal mine), and specific location of the signal source.

**NOTE: In some cases, the detector may provide a weak signal that cannot be repeated during the initial investigation process. Do not give up on the reliability of the signal until the sensitivity of the detector has been checked to ensure that it has not drifted to the low side.**

**NOTE: A footprint is defined as the entire area on the ground where a mine or metallic source causes the detector to generate an auditory signal. Small footprints, often as small as 10 to 15 centimeters in diameter, will indicate low-metal mines. Footprints as large as 60 to 122 centimeters in diameter may indicate high-metal mines.**

- Note the location on the ground (at the center of the search head) where the signal is first heard. Then attempt to develop a set of points on the ground that identifies the source footprint.
  - — Develop this footprint by first moving the search head away from the signal source until no signal is heard and then sliding the search head toward the signal from several clock positions.
  - — Note the specific location of the center of the search head at each location where the auditory signal begins. Once the signal is detected, the detector is no longer slid toward the potential source to avoid coming closer than necessary to a potential mine.
  - — Repeat this process and mark or note the locations on the ground until the size and shape of the footprint can be understood. Typically, the pattern will resemble a semicircle with the 6 o'clock position nearest the operator.

**WARNING**
**The large footprints of high-metal mines may mask signals from low-metal mines within the footprint. Always assume that there are low-metal mines within the large footprint area.**

- Identify the center of very large footprints (if important). The airborne technique may be quicker than the method previously discussed to identify the perimeter.
  - — The search head must be adjusted so that it can be held above the ground with the head parallel to the surface.
  - — The detector can then be manipulated above the source until the signal can be heard at only a single point. Move the search head in a crossing pattern, which should produce a smaller and smaller signal area as the pattern is repeated at gradually increasing heights. Near the surface, the signal can be heard over a broad lateral area, but as the search head is raised higher off the ground, this area becomes progressively smaller. As the search head is raised higher off the ground, the area where the signal can still be heard is reduced to a point.
  - — The center of the mine or metal source should be directly below this point. Do not ignore the potential for low-metal mines to be hidden in this large footprint on the ground.

**WARNING**
**Always be aware of the potential for booby traps and UXO. The search head should never be swept where its path cannot be visually cleared. If trip wires are a threat, other techniques to detect trip wires should be used before sweeping.**

***Other Operational Actions***

- Keep mine detectors at least 2 meters apart during setting, adjustment, and operational phases to prevent interference.
- Change the batteries immediately if the indicator lamp flashes, a constant audible tone is heard, or the check tone (clicking sound) stops and readjust the sensitivity. The search sensitivity is affected long before the lamp begins flashing, and the unit will be unusable until fresh batteries are installed.
- Ensure that only the lower part of the telescopic pole is used when the equipment is operated by a soldier in the prone position.
- Turn the unit off after completing the search operations.

**Disassembly and Packing**

- Ensure that the on/off switch on the electronic unit is in the *OFF* position.
- Disconnect the headphones from the electronic unit.
- Detach the cable connection on the electronic unit for the magnetic search head, and replace the protective caps on the plug and socket.
- Release the electronic unit's battery-cover latches, and remove the battery cover.
- Remove the batteries, and ensure that none of the battery cases have ruptured; if they have, notify your supervisor. Reinstall the battery cover and latch it.
- Remove the cable clips from the telescopic pole.
- Collapse the telescopic pole to its travel length, and turn its outer tube until it is locked by the catch. Loosen the plastic restraining bolt, and fold in the magnetic search head.
- Pack the components in the carry bag as shown in Figure F-2, page F-4. For long-term storage, do not put batteries in the carry bag. Close and zip the carry bag.
- Place the carry bag in the metal transport case, and latch the case. Close the pressure-relief valve.

As in probing, the senior leader must set a limit to the time an individual can use the mine detector. The time limit is determined by METT-TC factors, weather conditions, the threat level, the unit's stress level, and the individual's fatigue level and state of mind. As a rule, 20 to 30 minutes is the maximum amount of time an individual can use the detector effectively.

## EVACUATION DRILLS

A well-developed, well-rehearsed evacuation drill is necessary to extract an individual or a unit from a mined area. Units must develop evacuation drills for dismounted and mounted operations. Each type of operation should include two drills—one using a mine detector (mounted extraction) and one without using a mine detector (dismounted extraction).

### Mounted Extraction

- The convoy commander halts the convoy and reports to higher headquarters.
- No vehicles move and no troops dismount unless directed to do so.
- Elements provide 360-degree security from vehicles.
- Troops thrown from vehicles should not move. Personnel are extracted by using dismounted evacuation procedures if electronic detectors are not available.
- The senior leader, if engineers are not available, assesses the situation and directs vehicles to back up along the entry-route tracks. If an immediate threat exists, occupants of damaged vehicles evacuate out the rear of the vehicle and along the vehicle-entry tracks. If no immediate threat exists, occupants of damaged vehicles remain in the vehicle until it is extracted.
- Engineers, if available, sweep the area and provide a cleared path for movement. They—
  - Clear a lane that is wide enough for the towing vehicle.
  - Use all available tow cables to increase the distance before towing if an M88 is unavailable. Remember, an M88 has a wider track base than other tracked vehicles.
  - Ensure that all towing-shackle sets are complete and mounted.
  - Ensure that the towing vehicle has tow cables on the front and the rear if possible.
  - Ensure that rear cables are attached to the lower mounts. This allows the crew to recover the vehicle without touching the ground.
  - Pull the vehicle out at least two-vehicle lengths before switching to a tow bar. When towing a vehicle after a mine strike, the chance of fire is greater because of possible damage to the vehicle.
  - Provide first aid and conduct casualty evacuation or have medics provide treatment and medical evacuation.
  - Guide vehicles through the safe area.
  - Mark, record, and report the threat.
- Continue the mission.

## Dismounted Extraction

- All personnel freeze and crouch into a low-silhouette position. Be cautious when making this movement to ensure that you do not detonate another mine. If a protective mask is worn on your hip, do not allow it to come in contact with the ground because contact may detonate a mine. Do not help casualties because you could also become a casualty.
- The leader designates a security element and a soldier to assist in casualty evacuation.
- Soldiers extract along the path they entered. If possible, they step in the same places as before; if impossible, they probe their way out.
- The security element, consisting of individuals who are not in the minefield, sets up security for the unit.
- The soldier extracting the casualty—
  - Probes a 1-meter-wide path to the casualty and marks the cleared path with foot powder or marking tape as it is probed.
  - Probes around the casualty to clear the area.
  - Performs first aid.
  - Carries the casualty out of the minefield along the cleared path. (Stretcher parties do not enter the area unless a 2-meter-wide path has been cleared to the casualty.)
- The unit marks the threat and assembles back at the rally point.
- The unit reports the incident when it is 50 to 100 meters away from the minefield. If soldiers are in the minefield and radio transmission is required, move the transmitter at least 300 meters from the minefield. This prevents accidental mine detonation from the radio signal.
- The unit provides first aid and conducts casualty evacuation or has medics provide treatment and medical evacuation.
- The unit continues the mission.

## Extraction from Scatterable Mines

- The individual who discovers the mine initiates the alarm according to the unit SOP.
- Unit personnel at the command post receive the alarm and alert others.
- The unit TOC requests counterbattery fire if the mines are artillery-delivered.
- Vehicle commanders check the immediate area. They do not dismount. Inspect the vehicle for mines and/or trip wires. Note and record the location of all mines found on or around the vehicle.
- Personnel leave any vehicle that is touching or is blocked by AT mines in place until the remainder of the unit is out of the minefield.

- Unit leaders identify unmovable vehicles and designate one or more lanes for remaining personnel and vehicles to use when exiting the minefield.
- Leaders identify a clearance team to extract remaining vehicles and personnel. The clearance team—
  - Uses visual means to locate mines and marks vehicle lanes at least 4 meters wide. Personnel mark lanes according to the tactical situation and the threat in the area; however, they mark them so that team personnel can reenter the minefield and recover equipment and vehicles.
  - Destroys or removes all mines in lanes, using a grapnel hook or other means as necessary, in the sequence directed by the team leader. Personnel detonate unmovable mines to reduce personnel injury and equipment damage.
- Vehicle commanders direct personnel to ground-guide vehicles out of the minefield. Ground guides—
  - Ensure that individual elements move only when directed by the chain of command.
  - Place equipment that is not in contact with a mine or a trip wire onto vehicles.
  - Direct vehicles to the designated exit lane or, if safer, allow vehicles to exit the minefield on their own.
- Clearance team personnel, aided by unit personnel, remove equipment and vehicles remaining after initial extraction from the minefield. They—
  - Reenter the minefield using the same exit routes.
  - Detonate the minimum number of mines necessary to remove vehicles and equipment from the minefield.
  - Avoid touching mines. Personnel should take all possible precautions to ensure that mines are not jarred.
  - Place sandbags near mines, if possible, to minimize damage to vehicles and equipment.
  - Remove mines from equipment by using a line or other remote means. Ensure that the entire team is far enough from the mine to avoid casualties if it explodes.
  - Place explosive charges to minimize vehicle damage when detonating mines on the ground.
- Clearance team personnel—
  - Clear sufficient mines to allow for mission accomplishment if the position cannot be evacuated.
  - Clear and mark communication lanes between positions.
- Continue the mission.

## Survival Rules

Many of our allies have extensive experience in mine operations. Among them is Canada. They have produced several manuals and videotapes on mine awareness and have developed the following *survival rules*. They are very practical and can be applied to our soldiers as well.

- *If you did not drop it, do not pick it up.*
- *All terrain and structures are potentially mined or booby-trapped.*
- *Beware of areas associated with basic human needs. They could be mined or booby-trapped.*
- *Immediately report all confirmed or suspected mines.*
- *Leave mine disposal to the EOD personnel and combat engineers.*
- *Avoid touching or removing foreign objects, no matter how attractive. They could be mined or booby-trapped.*
- *Avoid verges because they could be mined. Stay on the traveled road.*
- *Mark and avoid UXOs if possible. Consider them unstable.*
- *Develop and rehearse effective evacuation drills.*

A convoy provides better protection against mine and UXO threat than a single vehicle. In convoy movement, some rules of thumb should be followed:

- The lead vehicle should be one of the heaviest (2½-ton, 5-ton) vehicles in the unit and be hardened against a mine threat.
- A high-mobility, multipurpose, wheeled vehicle (HMMWV) or a HEMMT should not lead a convoy unless absolutely necessary. These vehicles are extremely vulnerable to mine and UXO threat and are difficult to harden without commercial products.

## Casualty Treatment and Evacuation

Casualty treatment and evacuation should be a part of every mission order. Incorporate the following elements into the unit SOP, and ensure that soldiers understand them. Emphasize that rushing to help a mine victim can lead to the rescuer becoming a casualty.

- Reassure the casualty.
- Do not panic and create another casualty.
- Call higher headquarters at the earliest opportunity, and request a specialist engineer and medical help as soon as possible.
- Extract yourself and mark the path as you go.
- Reenter along the marked path.
- Clear a path to the casualty.
- Administer first aid.
- Recover the casualty.

- Mark the area after it is clear, record it on the map, and report to higher headquarters.

Medical personnel and combat lifesavers should anticipate and train for the following injuries:

- Blast injuries with fragments embedded.
- Burns.
- Fractures.
- Traumatic amputations.
- Blunt trauma.
- Psychological anxiety reaction.

# LEADER

Leader skills in effective planning, risk management, and mine-data tracking/ dissemination can greatly enhance force protection and reduce casualties in a mined environment. Unit leaders must train themselves and their subordinates on mine-awareness tasks.

## RISK MANAGEMENT

Risk management reduces the frequency of mine and UXO strikes and diminishes the physical effects when they do occur. It is used to identify hazards, define risks, identify methods for control, and identify responsibilities for implementation. The risk-management process enables commanders and staffs to define acceptable risk levels and implement controls until risks are commensurate with the mission. Risk management is a simple, five-step process:

- Identify hazards.
- Assess the risk of each hazard.
- Make risk decisions and develop controls.
- Implement controls.
- Supervise.

### Identify Hazards

This is often the most difficult part of risk management. A mine hazard is the condition that results from the interactions of a mine, a catalyst (such as activation from a soldier or vehicle), and a common spatial relationship. These hazards are defined in terms of mine types (AT, AP, fuses, metal content, AHD), how a soldier might encounter mines (dismounted or mounted and the type of vehicle), and the locations where encounters would be most likely. Mines are seldom employed in isolation, so complete hazard definition includes the complex obstacle and the covering fires.

### Assess the Risk of Each Hazard

This requires determining the probability of a mine strike and the effects of the strike. An effective risk assessment is critical for evaluating the combat effectiveness of a unit in a mined environment. Risk-assessment criteria is developed by using Table F-1. A sample risk assessment is shown in Table F-2, page F-19.

**Table F-1. Risk-assessment criteria**

| | | | Hazard Probability | | | | |
|---|---|---|---|---|---|---|---|
| | | | Frequent (A) | Probable (B) | Occasional (C) | Remote (D) | Improbable (E) |
| | Single Item | | Likely to occur frequently | Will occur several times during life of item | Likely to occur sometime during life of item | Unlikely, but possible to occur during life of item | So unlikely, can assume occurrence will not take place |
| | Fleet/Inventory | | Occurs continuously | Will occur frequently | Will occur several times | Unlikely but reasonably expected to occur | Unlikely to occur, but possible |
| Hazard Severity | CATASTROPHIC (death or system loss) | I | High | High | High | High | Medium |
| Hazard Severity | CRITICAL (severe injury or occupational illness or major damage) | II | High | High | High | Medium | Low |
| Hazard Severity | MARGINAL (minor injury or occupational illness or minor damage) | III | High | Medium | Medium | Low | Low |
| Hazard Severity | NEGLIGIBLE (less than minor injury, illness, or damage) | IV | Medium | Low | Low | Low | Low |

### Make Risk Decisions and Develop Controls

This step requires decision makers to identify actions that can reduce the probability and/or severity to acceptable levels. This may be accomplished by taking actions to reduce the probability of a mine strike or by providing more protection to the soldier or materiel to reduce severity of a mine strike. Often, it is a combination of the two.

### Implement Controls

Leaders must apply the identified controls to reduce the probability and severity of a mine strike.

**Table F-2. Sample risk assessment**

**DIRECTIONS: Circle the appropriate number in each section and total them. Appropriate directions are at the end of the assessment block.**

TRAINING (Circle One)
1. Experience working in mined areas
2. No experience working in mined areas, but trained
3. No experience or training working in or around mined areas

TYPE OF AREA (Circle One)
1. Area known by friendly forces to be clear of mines
2. Old confrontation line or suspected mined area
3. Area known to be mined

LIGHT AND WEATHER (Circle One)
1. Daylight with clear weather
2. Daylight with poor weather
3. Darkness

MISSION (Circle One)
1. One that your soldiers have done before
2. One that your subordinate leaders have done before
3. An unfamiliar type of mission

ON OR OFF ROAD (Circle One)
1. Approved division/corps route
2. Paved road
3. Unpaved road
4. Cross-country

SLEEP (Circle One)
1. 6 hours of sleep in last 24
2. 4 hours of sleep in last 24
3. 2 hours of sleep in last 24
4. Awake for more than 24 hours.

TYPE OF VEHICLES IN CONVOY (Circle One)
1. Armored
2. Mixed, armored vehicles leading
3. Wheeled
4. On foot

AVAILABILITY OF MINE INFORMATION (Circle One)
1. Updated mined-area graphics in each vehicle
2. Updated mined-area graphics in the lead vehicle or a reliable, knowledgeable guide is available
3. No upgraded mined-area graphics are available

GROUND COVER (Circle One)
1. Dry, hard ground with short grass
2. Dry, hard ground with long grass or vegetation
3. Wet, muddy ground or snow less than 10 centimeters deep
4. Snow more than 10 centimeters deep

ROAD USE (Circle One)
1. Heavy tracked vehicles or trucks recently used the road
2. Light wheeled vehicles recently used the road
3. No traffic observed on the road; some tire marks
4. No traffic observed on the road; no tire marks

**10-16** Continue the mission. Keep following training and common-sense rules that apply to working around mines.
**17-24** Continue the mission. Remember to stress safety and mine awareness when you brief soldiers. Ensure that leaders maintain positive control of personnel.
**25-35** Consider postponing the mission until better conditions are attained. If you must continue the mission, constantly stress mine awareness and safety. Ask higher headquarters for engineer support to accomplish the mission. Conduct mine-awareness training.

### Supervise

This step ensures that controls are implemented and that a measure of quality control exists to ensure a quantified level of clearance.

The key to using risk management successfully is to employ it at each echelon—from the commander, through the tactical planner, to the soldiers executing the mission. Each level identifies hazards, eliminates or reduces hazards as feasible, and communicates the residual hazards to the next lower echelon. As such, each echelon works as a filter to control unacceptable risks.

- Training provides soldiers with an understanding of equipment limitations and plays a critical role in the risk-management process. Capabilities and limitations of Army systems are taken into consideration during the development of doctrine and TTP.
- Risk management at the tactical planning level requires a thorough knowledge and awareness of the hazards and potential controls that can be employed. The planning process requires a methodical and disciplined technique to identify the hazards and develop appropriate controls for operating in a mined environment. The controls for countermine operations, discussed in Chapters 9 and 11 and in FM 3-34.2, provide a framework for risk-managing hazards associated with mines.
- The execution level is the culminating point of risk management. It is where soldiers and leaders employ the systems provided to accomplish the mission. The amount of residual hazards remaining after the filtering process from echelons above may well determine success. The individual soldier is the last element to control any residual hazards.

Optimizing the components of risk management at the tactical planning level is more challenging as emerging technologically dependent systems bring more variables into the mission. While tactical intelligence is the key element in identifying mine-related hazards, technical knowledge is the key element in assessing the risks associated with mine hazards. This knowledge assimilates the tactical intelligence with the capabilities of the unit's equipment, the performance of threat mines, and the protection provided to our soldiers by their vehicles or personal protective equipment. The staff engineer, using his engineer $C^2$ system to risk-manage each COA, provides the maneuver commander and his staff with information on risks and potential controls early on in the planning process. Each subsequent commander must perform the same analysis and incorporate the mine threat into risk management.

## RECORDING AND MINE-DATA TRACKING

Obtaining and disseminating information are the keys to battlefield management. Units encountering minefields or explosive devices should follow a five-step process—stop, secure, mark, report, and avoid. Units must provide adequate information to their higher headquarters to ensure that follow-on elements are well informed. Information must include known or suspected minefield locations, types of mines (if known), the marking method, the time encountered, and any additional information that may be of use to the clearing unit.

Division and maneuver brigade engineer planning cells must establish a central control cell for mine-clearance information. The central control cell—

- Maintains a current situation map and overlay that depicts friendly and enemy mines and obstacles.
- Maintains and updates information on minefield tracking and route status.
- Receives and maintains minefield recording forms within the unit's AO (includes host-nation minefield data if available).
- Maintains a database of mine information.
- Processes, analyzes, updates, and disseminates the information to subordinate commanders and staff.

### MINE-INCIDENT REPORT

A mine incident includes any unplanned activity involving a mine, UXO, or booby trap. It also includes near misses that could have resulted in potential damage or injury. The mine-incident report is a technical report that follows a serious-incident report. The report should be submitted as soon as possible (the local SOP will indicate time requirements). A sample mine-incident report is shown in Figure 11-12, page 11-26.

## TRAINING

Modern combat is complex, lethal, and demanding. Soldiers must be capable of performing their missions in any type of battlefield environment. Current doctrine and TTP provide soldiers with guidelines to accomplish their tasks and quality equipment provides the means. The common thread that connects doctrine, tactics, and equipment is quality training. To fight and win, units must train their soldiers to execute all wartime missions successfully. They must use every training opportunity to improve soldier, leader, and unit task performance. Without quality training, no amount of world-class equipment can make the soldier effective or make him survive in a wartime environment. Even the best doctrine in the world is worthless unless soldiers receive effective training. This is especially true with mine-awareness training.

Soldiers must be trained to think mine awareness as well as perform mine-related actions. Decisions, actions, and reactions must become automatic to every soldier. This requires that all soldiers receive mine-awareness training early in their careers. It must begin at early entry training with basic individual tasks and continue through advanced unit training with collective tasks. Soldiers who survive mine threats can survive mine warfare, but it requires continual training.

### INDIVIDUAL TRAINING

#### Basic Mine-Awareness Training

This is the most important phase for preparing soldiers to survive in a mine environment. Individual soldiers must be trained to a level that meets the environment they will face on the modern battlefield. Basic mine-awareness tasks (mine detection, survival rules, minefield indicators, and procedures for self-extraction from a minefield) must begin early in the soldier's career.

With an estimated 80 to 100 million mines deployed worldwide, countermine considerations must become second nature to mission planners. From CSS operations in the rear area to close combat operations in the main battle area, there is no place on the battlefield that is safe from mines. This is especially true with the threat of SCATMINEs that are deployed from aircraft and indirect-fire sources. The need for initial countermine training will get more critical as the modern battlefield becomes more lethal.

### Annual Mine-Awareness Training

Units exercising military skills at the National Training Center, Joint Reserve Training Center, or the Combat Maneuver Training Center can reinforce and evaluate their awareness training. Soldiers should be fully trained on the individual and collective tasks that are required to support their unit's mine-related missions before deploying to the combat training centers. An effective method to meet this challenge is a well-planned annual training program that trains soldiers on individual and collective mine-awareness tasks. The plan should address the requirements for initial collective-task training and sustainment training.

### Predeployment Mine-Awareness Training

With an ever-changing global environment, the probability of projecting US forces to various geographical locations is very real. It is extremely critical that units be situationally aware. Each flash point in the world contains its own type of threat. Many of these threats can be neutralized or reduced through quality predeployment training. Quality predeployment mine-awareness training enhances the ability of the unit to perform its mission and increases the confidence of soldiers.

Units must train intensively when alerted for a deployment. Predeployment training is the time to polish and refine training and to focus on theater-specific operations. Predeployment training is intended to augment individual and collective training, not replace it.

### In-Theater Mine-Awareness Training

Training does not end with predeployment training. Units must train in the country where they deploy. They need to plan and conduct routine sustainment training on individual and collective mine-awareness tasks. This is an essential ingredient of force-protection operations. In-theater mine-awareness training reinforces the soldier's existing skills and places mine survival in the forefront of each soldier's mind.

## LEADER TRAINING

Officers and NCOs must receive mine-awareness training in basic and career courses. Commissioned and noncommissioned leaders must be trained on the use and the employment of the equipment that they will use in the field. Officer basic courses and officer and NCO advanced courses must include access to modeling/simulation training. Using modeling/simulation enables the leader to practice decision-making, employment, and sustainment operations in a simulated mine environment. This develops the situational awareness required for timely, accurate decisions.

## UNIT TRAINING

The Army Training and Evaluation Program remains the most effective measure of individual and collective training effectiveness short of actual combat. It provides leaders with the opportunity to verify the effectiveness of virtual and actual training without endangering soldiers with mine effects. Units must include mine-threat scenarios in their home-station training exercises. Basic missions include minefield detection, reduction, marking, proofing, and recording. Commanders are required to assess their unit's state of proficiency in mine-awareness tasks on a routine basis.

Mine awareness is a critical, perishable skill. It must be trained effectively and sustained on a continuous basis. If a unit properly trains its soldiers on mine awareness, it will maintain its force, boost its soldiers' confidence in their abilities, and accomplish its mission more effectively. Mine awareness is not a liability; it is an investment in a unit's future.

Appendix G

# Countermine Data

This appendix provides a compilation of countermine information to assist commanders in the decision-making process for countermine operations.

## BREACHING ASSETS VERSUS THREAT OBSTACLES

Tables G-1 and G-2, pages G-2 through G-8, provide general guidance on the effect of breaching assets versus threat obstacles.

## FOREIGN MINE DATA

Tables G-3 through G-7, pages G-9 through G-12, provide foreign mine characteristics.

## FOREIGN MINEFIELD EMPLACEMENT DATA

Table G-8, page G-13, provides foreign minefield emplacement data. Information is presented in the probability of occurrence based on historical data.

## FOREIGN MINE DELIVERY SYSTEMS

Table G-9, pages G-14 and G-15, contains characteristics of foreign mine delivery systems.

**NOTE: Blank areas in the above tables indicate that the information is not applicable or that it was unavailable when this manual was published.**

**Table G-1. Mounted breaching assets versus threat obstacles**

| Asset | Track-Width AT Mine | Full-Width AT Mine | Side-Atk AT Mine | Cmd-Det Blast Mine | AT Ditch and Other Gaps | Wire Obstacles | Remarks |
|---|---|---|---|---|---|---|---|
| MCB with dog-bone assembly | Is more than 90 percent effective to 12 inches deep at speeds less than 10 kph (soil dependent) | Do not use against magnetic-influence mines; is more than 99 percent effective against tilt-rod mines | Do not use | Do not use | May be effective at filling in some shallow gaps or trenches | Do not use | The MCB provides almost 100 percent clearance of simple pressure-fused AT mines when used with the MICLIC.<br>The MCB has minimal impact on the M1's maneuverability and speed when not plowing.<br>The main gun should be traversed to the side to prevent damage in the event of a mine detonation when plowing.<br>The reduction site should be relatively flat and free of large rocks and other obstructions.<br>Plowing operations should begin about 100 m short of the suspected minefield and continue 100 m past its suspected end (typically 2 to 2 1/2 minutes to cover 300 to 400 m overall).<br>Plowing operations should be performed in a straight line to keep the vehicle out of uncleared ground.<br>Fifteen percent of pressure-fused AT-mine encounters across the MCB's width can be expected to detonate under the skid shoe, probably disabling the blade.<br>The MCB is vulnerable to coupled mines and may be stopped by log-hurdle obstacles laid within a minefield as part of a complex obstacle.<br>A second plow should not be used to clear the center of the lane because this will simply push mines into the area cleared by the first plow. |

**Table G-1. Mounted breaching assets versus threat obstacles (continued)**

| Asset | Track-Width AT Mine | Full-Width AT Mine | Side-Atk AT Mine | Cmd-Det Blast Mine | AT Ditch and Other Gaps | Wire Obstacles | Remarks |
|---|---|---|---|---|---|---|---|
| MCB with IDA | See above | Is effective against magnetic-influence mines; is more than 99 percent effective against tilt-rod mines | Do not use | Do not use | See above | Do not use | These AT mines are fitted with magnetic-influence fuses and AHDs may disable the blade.<br>Dismounts should expect to encounter some pressure-fused AP mines in the breached lane that may be missed by the blade.<br>The MCB clears a 180 inch-wide lane but leaves a 26 inch-wide area uncleared of simple pressure-fused mines and a 54 inch-wide area uncleared of double-impulse mines (under the skid shoes) in the middle of the breached lane.<br>The berms and furrows left by the MCB aid in immediately marking the breached lane.<br>Live mines remain in both berms and may roll back into the cleared lane. Over time, these may vibrate free and roll back into the breached lane. |
| MCR (including Panther) with dog-bone assembly | Is more than 90 percent effective to 12 inches deep at speeds less than 5 to 15 kph (soil dependent)<br>Has reduced effectiveness against some (mostly Italian) blast-hardened mines<br>Do not use against double-impulse mines | Do not use against magnetic-influence mines; is more than 99 percent effective against tilt-rod mines | Do not use | Do not use | Do not use | Do not use | Normally used to detect the leading edge of a conventionally laid tactical minefield and to proof lanes/routes cleared by other means.<br>Use caution when crossing on an AVLB.<br>May become stuck or ineffective in muddy conditions.<br>The MCR is vulnerable to coupled mines.<br>The MCR is designed to withstand up to two detonations per side.<br>The MCR clears a 160 inch-wide lane but leaves a 72 inch-wide area uncleared in the middle of the breached lane. **NOTE: The M113 also has 72 inches between tracks and must use extreme caution when following the MCR.** |

**Table G-1. Mounted breaching assets versus threat obstacles (continued)**

| Asset | Track-Width AT Mine | Full-Width AT Mine | Side-Atk AT Mine | Cmd-Det Blast Mine | AT Ditch and Other Gaps | Wire Obstacles | Remarks |
|---|---|---|---|---|---|---|---|
| MCR with IDA | See above | Is effective against magnetic-influence mines; is more than 99 percent effective against tilt-rod mines | Do not use | Do not use | Do not use | Do not use | Multiple rollers are required to provide the overlap necessary for proofing the route's entire width when the MCR is used for proofing during route-clearance operations.<br>Vehicle speed is limited to 5 to 15 kph.<br>The main gun should be traversed to the side to prevent damage in the event of a mine detonation when rolling a suspected mined area.<br>The IDA may be damaged by detonated mines. |

**Table G-2. Dismounted breaching assets versus threat obstacles**

| Asset | Pressure-Fused AP Mine | Trip-Wire/Break-Wire-Fused AP Mine | Influence-Fused AP Mine | Cmd-Det Blast Mine | Wire Obstacles | Remarks |
|---|---|---|---|---|---|---|
| APOBS | Effectively clears 95 percent, including blast-resistant types | Effectively cuts 99 percent of trip wires in its path | Is ineffective | Is ineffective | Clears a footpath | Provides a 0.6- x 45-m footpath from a 25-m standoff<br>Can be deployed by two men in 2 minutes<br>Consists of two manpack units weighing approximately 55 lbs each |
| Bangalore torpedo | Effectively clears 85 percent of simple pressure-fused mines; is less effective against blast-hardened mines | Effectively cuts 99 percent of trip wires in its path | Is ineffective | Is ineffective | Clears a footpath | Provides a 0.6- x 15-m footpath<br>Can be deployed by a squad in 5 minutes<br>Consists of ten 1.5-m-long sections weighing 13 lbs each<br>Is ineffective against AT mines |
| Grapnel hook and WLGH | Is ineffective | Effectively clears 99 percent of trip wires/break wires in a single pass | Is ineffective; however, it may snag the wires between the control unit and its mines | Is ineffective; however, it may snag the wires between the observer and his mines | Is somewhat effective at clearing trip-wire-fused mines integrated with wire obstacles | Has a range of 80 to 100 m when launched<br>Can be thrown 15 to 20 m when hand thrown<br>Must consider the lethal range/radius (up to 90 m in some cases) of fragmenting AP mines that may be encountered, and select a covered position or move back an adequate distance before pulling the grapnel<br>Helps increase confidence by grappling an area several times<br>Is hampered by thick vegetation/overhead obstructions which may reduce the range/ effectiveness of the WLGH |

Table G-2. Dismounted breaching assets versus threat obstacles (continued)

| Asset | Pressure-Fused AP Mine | Trip-Wire/Break-Wire-Fused AP Mine | Influence-Fused AP Mine | Cmd-Det Blast Mine | Wire Obstacles | Remarks |
|---|---|---|---|---|---|---|
| Probe | Is effective at locating all buried mines; however, extreme caution should be used when probing for small mines (less than 3-inch dia) and sensitive mines (more than 10 lbs actuating force) | Is ineffective; the soldier must rely on clearance by other means and visual detection | Do not use | Ensure that observer is neutralized before approaching | Is ineffective | |
| Wire cutter | Is ineffective | Do not cut a slack wire until both ends have been checked if break-wire-fused mines may be present; do not cut taught trip wires until both ends have been checked for tension-release FDs | Do not use | Ensure that observer is neutralized before approaching | Is effective; sound can be minimized by wrapping the wire with a cloth before cutting it | |
| Gunfire | Is effective against surface-laid munitions | Is effective against surface-laid munitions | Do not use | Is effective against surface-laid munitions | Is ineffective | Is used as a last resort<br>Is a relatively slow method for clearing visually detected ordnance<br>Ensures protection of weapons operator<br>Works better with larger caliber weapons (up to 25 mm), incendiary ammo works best<br>May be the only option open to artillery, MP, and transportation units for dealing with SCATMINEs in the absence of engineers and reduction assets |

**Table G-2. Dismounted breaching assets versus threat obstacles (continued)**

| Asset | Pressure-Fused AP Mine | Trip-Wire/Break-Wire-Fused AP Mine | Influence-Fused AP Mine | Cmd-Det Blast Mine | Wire Obstacles | Remarks |
|---|---|---|---|---|---|---|
| MiniFlail | Is effective against simple pressure-fused AP mines<br>Is less effective against blast-hardened AP mines | Is effective | Is ineffective | Is armored against fragmenting threats<br>Is vulnerable to command-detonated blast mines | Do not use | Do not use if the presence of AT mines is suspected, because AT mines can destroy the MiniFlail.<br>Can clear at a rate of about 1,500 square meters per hour |
| AN/PSS-12 | Is very effective at finding metallic mines<br>Is less effective against low-metal mines | Cannot detect monofilament fishing line used for trip wires; it is undetectable by the AN/PSS-12. Visual detection and other neutralization means should be used first in any area suspected of having trip wires. | Do not use | Ensure that observer is neutralized before approaching | Is ineffective | Is less effective against low-metal mines and in areas contaminated with metal fragments (artillery, mortar, small arms) or areas that have heavily mineralized soils |

**Table G-2. Dismounted breaching assets versus threat obstacles (continued)**

| Asset | Pressure-Fused AP Mine | Trip-Wire/Break Wire-Fused AP Mine | Influence-Fused AP Mine | Cmd-Det Blast Mine | Wire Obstacles | Remarks |
|---|---|---|---|---|---|---|
| Demolitions | Place at least a 1-pound charge within 6 inches of simple pressure-fused mines<br>Ensure that the charge is placed within an inch of blast-hardened mines | Place at least a 1-lb charge within 6 inches of the mine after the mine at the end of a trip wire has been located<br>Can use elevated charges if necessary against the claymore and stake-type mines | Do not use | Ensure that observer is neutralized before approaching<br>Can use elevated charges if necessary against claymore-type mines | Is effective when used to make an improvised bangalore torpedo | |
| Blast overshoes | May be effective against very small (up to 1 oz of explosive) mines<br>Is ineffective against mines with more than 1-oz explosive weight | Can reduce casualty severity against fragmenting mines | Can reduce casualty severity against fragmenting mines | Ensure that observer is neutralized before approaching<br>Can reduce casualty severity | Is ineffective | |
| IBASIC | Will protect personnel near the individual who detonates a pressure-fused AP mine | Can reduce casualty severity against fragmenting mines | Can reduce casualty severity against fragmenting mines | Ensure that observer is neutralized before approaching<br>Can reduce casualty severity | Is ineffective | |

**Table G-3. Foreign track-width AT mines**

| Origin | Mechanically Planted | Scatterable | Metallic | Nonmetallic | Simple Pressure | Double Impulse | Blast Hardened | AHD |
|---|---|---|---|---|---|---|---|---|
| Russia | TM-46/57/62 | PTM-1 | TM-46/57/62M | TM-62B/D/P | TM-46/57/62 | | TM-46/57 | TMN-46 |
| China | Type 72 | Type 84B | Type 72 | Type 72 | Type 84B | Type 72 | Type 72 | |
| Yugo | TMM-1 | | TMM-1 | TMA-1/2/3/4/5 | TMA-1/2/5 | | TMA-3/4 | TMM-1, TMA-1/2/3/4/5 |
| Czech | PT-MI-K, PT-MI-BA III | | PT-MI-K, PT-MI-D | PT-MI-BA II/III | PT-MI-BA III, PT-MI-D | | PT-MI-K, PT-MI-BA II | PT-MI-K, PT-MI-D |
| Italy | VS-1.6/2.2, TC-.6/6, SB-81 | VS-1.6, SB-81 | | VS-1.6/2.2, TC-.6/6, SB-81 | | | VS-1.6/2.2, TC-3.6/6, SB-81 | VS-1.6/2.2 |
| France | M-51, M-52 | | | M-51, M-52 | M-51, M-52 | | | M-51 |
| UK | L9 Barmine, MK-7 | | MK-7 | L9 Barmine | L9 Barmine, MK-7 | L9 Barmine, MK-7 | | L9 Barmine |
| Ger | PM-60, DM-11 | | DM-21 | PM-60, DM-11 | PM-60, DM-11, DM-21 | | | PM-60, DM-11, DM-21 |

**Table G-4. Foreign full-width AT mines**

| Origin | Mechanically Planted | Scatterable | Tilt Rod | Magnetic Influence | Seismic Influence | Coupled | AHD |
|---|---|---|---|---|---|---|---|
| Russia | TM-62/72/89 | PTM-3 | TMK-2 | TM-62/72/89 | TM-62/72/89 | | TM-62/72/89 |
| China | | | Type 84C | Type 84A | | | Type 84A |
| Yugo | TMRP-6 | KB-2 | TMRP-6 | TMRP-6, KB-2 | | | TMRP-6, KB-2 |
| Czech | | | PT-MI-P | | | | |
| France | HPD series | Lance | M-51, M-52, Model 1954 | HPD series | | Model 1953 | HPD series, M-51, Lance |
| Hungary | UKA-63 | | UKA-63 | | | | UKA-63 |
| Italy | VS-HCT/HCT-2, SB-MV | VS-SATM1, VS-HCT | | VS-HCT/HCT-2, SB-MV, VS-SATM1 | | Fuse option | VS-HCT/HCT-2, SB-MV, VS-SATM1 |
| Sweden | FFV-028 | | | FFV-028 | | Fuse 15 | FFV-028 |
| UK | L9 Barmine | HB-876 | L9 Barmine, MK-7 | L9 Barmine, HB-876 | | | L9 Barmine, MK-7, HB-876 |
| Germany | DM-31 | AT-2, MIFF | | DM-31, AT-2, MIFF | | | DM-31, AT-2, MIFF |
| Romania | | | | | | MC-71 | |

**Table G-5. Foreign side-attack AT mines**

| Origin | Command | Break Wire | Influence | AHD |
|---|---|---|---|---|
| Russia | MON-100/200/500, TM-83 | LMG | TM-73/83 | |
| Yugo | | TMRP-6, COBRA | | |
| France | MIACAH F1 | MIACAH F1 | MIACAH F1 | |
| UK | L14 | L14 | AJAX | AJAX |
| Germany | | | DM-12 | |
| Sweden | | FFV-016, AT-4 | | |

**Table G-6. Foreign pressure-fused AP mines**

| Origin | Scatterable | Low Metal | Simple Pressure | Small Diameter | Low-Operating Threshold | Pronged Fuse | Blast Hardened | AHD |
|---|---|---|---|---|---|---|---|---|
| Russia | PFM-1 | | PMN, PMD-6 | | PMN, PMD-6, PFM-1 | OZM series | PMN-2 | |
| China | GLD-111 | Type 72A, GLD-111 | Type 72A, Type 58, GLD-111 | GLD-111 | Type-72, Type 58, GLD-111 | | | Type 72A/B |
| Yugo | | PMA-1/2/3 | PMA-1, PMD-1 | | PMA-1/3, PMD-1 | PROM-1/2, PMR-3 | PMA-2 | |
| Czech | | | PP-MI-D, PP-MI-BA | | PP-MI-D | PP-MI-SR, PP-MI-SR II | | |
| Italy | TS-50, VS-50, SB-33, VS-MK2 | TS-50, SB-33, VS-MK2 | | | | Valmara 59/69 | TS-50, VS-50, SB-33, VS-MK2 | VS-50, SB-33, VS-MK2 |
| France | | M-59 | | M-59 | M-59 | | M-59 | |
| UK | Ranger | | Ranger | Ranger | | | | |
| Germany | | DM-11 | PPM-2 | | DM-11 | | DM-11 | |

Table G-7. Foreign trip-wire/break-wire-fused AP mines

| Origin | Scatterable | Trip Wire | Break Wire |
|---|---|---|---|
| Russia | POM-2 | POMZ-2, OZM series, MON series | OZM series, MON series |
| China | | Type 58, Type 59, Type 66, Type 69 | |
| Yugo | | PROM-1/2/KD, PMR-2/3/4, MRUD | PROM-1/2/KD, PMR-2/3/4, MRUD |
| Czech | | PP-MI-SR, PP-MI-SR II, PP-MI-SK | |
| Italy | VS-SAPFM3 | Valmara 59/69, BM-85, P-25, P-40, VS-DAFM1 | |
| France | | M1951/1955, MAPED F1 | M1951/1955 |
| UK | | PAD | |
| Germany | | DM-31, SM-70 | |

**Table G-8. Foreign emplaced minefields**

| | Russia | Iraq | North Korea | Bosnia |
|---|---|---|---|---|
| **Manually** | | | | |
| Antitrack AT | 1 | 1 | 1 | 1 |
| Antihull AT | 2 | 3 | 3 | 2 |
| Side-attack AT | 3 | 4 | 3 | 4 |
| Blast AP | 1 | 1 | 1 | 1 |
| Directional AP | 1 | 4 | | 1 |
| Bounding AP | 1 | 1 | 2 | 1 |
| Stake AP | 1 | 4 | 1 | 1 |
| **Mechanically** | | | | |
| Antitrack AT | 1 | 1 | 1 | 4 |
| Antihull AT | 2 | 3 | | 1 |
| **Vehicle Scattered** | | | | |
| Antitrack AT | 1 | | | |
| Antihull AT | 1 | | | |
| Blast AP | | | | |
| Directional AP | | | | |
| Bounding AP | | | | |
| Simple Frag | 1 | | | |
| **Artillery/MLRS** | | | | |
| Antitrack AT | 1 | 4 | | |
| Antihull AT | | 1 | | 1 |
| Blast AP | | | | |
| Directional AP | | | | |
| Bounding AP | | | | |
| Stake AP | 1 | | | |
| **Helicopter/Fixed-Wing Aircraft** | | | | |
| Antitrack AT | 1 | 1 | | |
| Antihull AT | 1 | | | |
| Blast AP | 1 | 1 | | |
| Directional AP | | | | |
| Bounding AP | | | | |
| Stake AP | 1 | | | |
| **NOTE: Scored from 1 (most common) to 4 (very rare).** | | | | |

**Table G-9. Foreign mine delivery systems**

| Origin | System | Type[1] | Range | Mine Payload | Density[2] | Remarks |
|---|---|---|---|---|---|---|
| Russia | BM-21 | R | 20 km | PTM-1/3, POM-2 | 1.5 | 1,000 x 500 m per battery |
| | BM -22 | R | 35 km | PTM 1/3, POM-2 | 1.5 | |
| | PKPI | F, H | NA | PTM 1/3, PFM-1, POM-2 | 1.5 | Lays relatively narrow strips |
| | KSMU | F | NA | PTM 1/3, PFM-1, POM-2 | 1.5 | |
| | UMZ | V | NA | PTM 1/3, POM-2 | 1.5 | |
| | GMZ/PMZ-4 | MP | NA | Varies | 0.4 to 1.0 | 3 systems can emplace 1,200 m with 624 mines in 26 minutes |
| | VMR-1/2 | H | NA | Varies | 0.4 to 1.0 | 3 Mi-8s can lay 1,200 to 3,000 m with 1,200 mines |
| China | Type 83 | R | 15 km | Type 84 | 0.5 | 900 x 800 m per battery |
| Yugo | M87 | R | 50 km | KB-2 | 0.5 | 900 x 500 m per battery |
| Italy | FIROS 25 | R | 22 km | VS-HCL, VS-Mk, VS-SAPFM 3 | 0.5 | 1,000 x 500 m per battery |
| | FIROS 30 | R | 35 km | VS-HCL, VS-Mk, VS-SAPFM 3 | 0.5 | |
| | DAT | H | NA | Varies | 1.2 | Lays relatively narrow strips of up to 320 AT mines |
| | Istrice | V | NA | SATM1, VS-Mk 2 EL, VS-SAPFM 3 | Max 1.6 | Typically 360 x 140 m |
| | ST-AT/V | MP | NA | Varies | 0.3 to 1.5 | 240 mines per hour |
| | Grillo 90 | MD | NA | SATM1, VS-Mk 2 EL | 0.5 | 5 AT or 15 AP mines per dispenser |
| Sweden | FFV-5821 | MP | NA | FFV-028 | 0.4 to 1.4 | 300 buried mines per hour or 500 surface-laid mines per hour |
| France | Minotaur | V | NA | Lance | 0.7 | 1,200 x 600 m with 800 mines |
| | EBG | V | NA | Lance | 1.5 | 60 x 600 m with 40 mines |
| | 155-mm How | A | 18 km | Lance | | 8 mines per round |
| | Matenin | MP | NA | HPD series | 0.4 to 1.4 | 500 mines per hour |
| UK | JP-233 | F | NA | HB 876 | | 430 mines per Tornado aircraft |
| | Barmine layer | MP | NA | L9 Barmine | 0.4 to 0.9 | Up to 700 mines per hour |
| | Ranger | V | NA | L10 | | 1,296 mines per dispenser |

**Table G-9. Foreign mine delivery systems (continued)**

| Origin | System | Type[1] | Range | Mine Payload | Density[2] | Remarks |
|---|---|---|---|---|---|---|
| Germany | Skorpion | V, H | NA | AT-2 | 0.4 | 1,500 x 50 m with 600 mines |
| | MW-1 | F | NA | MIFF, MUSPA | 0.5 | 55 to 500 m wide and 200 to 2,500 m long with 892 AT mines |
| | LARS | R | 14 km | AT-2 | 0.5 | 1,836 AT mines per launcher, 1,000 x 500 m per battery |
| | MARS | R | 30 km | AT-2 | 0.3 | 1,000 x 500 m with 336 AT mines per launcher |
| | FFV-5821 | MP | NA | DM-31 | 0.4 to 1.4 | 300 buried mines per hour or 500 surface-laid mines per hour |

[1] A = artillery, H = helicopter, F = fixed-wing aircraft, R = rocket, V = vehicle dispensed, MD = man-portable dispenser, MP = mechanical planter

[2] AT mines per meter of front

Appendix H

# Metric Conversion Chart

This appendix complies with current Army directives which state that the metric system will be incorporated into all new publications. Table H-1 is a conversion chart.

**Table H-1. Metric conversion chart**

| US Units | Multiplied By | Metric Units |
|---|---|---|
| Feet | 0.3048 | Meters |
| Inches | 2.54 | Centimeters |
| Inches | 0.0254 | Meters |
| Inches | 25.4001 | Millimeters |
| Miles | 1.6093 | Kilometers |
| Pounds | 453.59 | Grams |
| Pounds | 0.4536 | Kilograms |
| **Metric Units** | **Multiplied By** | **US Units** |
| Centimeters | 0.3937 | Inches |
| Grams | 0.03527 | Ounces |
| Kilograms | 2.2046 | Pounds |
| Kilometers | 0.62137 | Miles |
| Meters | 3.2808 | Feet |
| Meters | 39.37 | Inches |
| Meters | 1.0936 | Yards |
| Millimeters | 0.03937 | Inches |

# Glossary

| | |
|---|---|
| **1LT** | first lieutenant |
| **1SG** | first sergeant |
| **2LT** | second lieutenant |
| **AA** | avenue of approach |
| **ACE** | armored combat earthmover |
| **ACR** | armored cavalry regiment |
| **AD** | armor division |
| **ADA** | air-defense artillery |
| **ADAM** | area-denial artillery munition |
| **ADE** | assistant division engineer |
| **AHD** | antihandling device |
| **AI** | area of interest |
| **ALO** | air liaison officer |
| **ammo** | ammunition |
| **AO** | area of operations |
| **AP** | antipersonnel |
| **APB** | antipersonnel blast |
| **APC** | armored personnel carrier |
| **APDS** | armor-piercing discarding sabot |
| **APF** | antipersonnel fragmentation |
| **APL** | antipersonnel land mine |
| **APOBS** | Antipersonnel Obstacle Breaching System |
| **AR** | Army regulation |
| **ASP** | ammunition supply point |

| | |
|---|---|
| **ASTAMIDS** | Airborne Standoff Minefield Detection System |
| **AT** | antitank |
| **atk** | attack |
| **ATP** | ammunition transfer point |
| **attn** | attention |
| **AVLB** | armored vehicle-launched bridge |
| **AVLM** | armored vehicle-launched MICLIC |
| **BAI** | battlefield air interdiction |
| **bde** | brigade |
| **BICC** | battlefield information control center |
| **BLPS** | ballistic and laser protective spectacles |
| **bn** | battalion |
| **BOS** | Battlefield Operating System |
| **BP** | battle position |
| **BRDEC** | US Army Belvoir Research, Development, and Engineering Center |
| **C** | Celsius |
| **$C^2$** | command and control |
| **CA** | civil affairs |
| **cal** | caliber |
| **CAS** | close air support |
| **cdr** | commander |
| **CEV** | combat engineer vehicle |
| **CG** | commanding general |
| **CHS** | combat health support |
| **CL** | centerline |
| **cm** | centimeter(s) |
| **cmd** | command |
| **CMMC** | corps material-management center |

| | |
|---|---|
| **co** | company |
| **COA** | course of action |
| **COB** | civilian on the battlefield |
| **comp** | composition |
| **CS** | combat support |
| **CSR** | controlled supply rate |
| **CSS** | combat service support |
| **CTCP** | combat trains command post |
| **cu** | cubic |
| **Czech** | Czechoslovakia |
| **d** | day |
| **DA** | Department of the Army |
| **DAO** | division ammunition officer |
| **DCU** | dispenser control unit |
| **det** | detonated |
| **det** | detonating |
| **dia** | diameter |
| **div** | division |
| **DMMC** | division material-management center |
| **DOD** | Department of Defense |
| **DODIC** | Department of Defense identification code |
| **DST** | decision support template |
| **DTG** | date-time group |
| **DZ** | drop zone |
| **EA** | engagement area |
| **ea** | each |
| **EBA** | engineer battlefield assessment |
| **EDD** | explosive detector dog |

| | |
|---|---|
| **EFP** | explosive-formed penetrating |
| **EID** | end item code |
| **EM** | enlisted member |
| **engr** | engineer |
| **ENS** | Explosive Neutralization System |
| **EOD** | explosive ordnance disposal |
| **ESMB** | explosive standoff minefield breacher |
| **F** | Fahrenheit |
| **FCP** | forward command post |
| **FCS** | fire control system |
| **FD** | firing device |
| **FECS** | Field-Expedient Countermine System |
| **FIST** | fire-support team |
| **FLOT** | forward line of own troops |
| **FM** | field manual |
| **FM** | frequency-modulated |
| **FMFM** | Fleet Marine Force manual |
| **FPF** | final protective fires |
| **frag** | fragmentation |
| **FRAGO** | fragmentary order |
| **FSB** | forward support battalion |
| **FSC** | federal stock classification |
| **FSCL** | fire-support coordination line |
| **FSCOORD** | fire-support coordinator |
| **FSE** | fire-support element |
| **ft** | foot, feet |
| **ft** | fort |
| **FWF** | former warring faction |

| | |
|---|---|
| **g** | gram(s) |
| **G2** | Assistant Chief of Staff, G2 (Intelligence) |
| **G3** | Assistant Chief of Staff, G3 (Operations and Plans) |
| **G4** | Assistant Chief of Staff, G4 (Logistics) |
| **GDP** | general defense plan |
| **Ger** | Germany |
| **GPS** | global-positioning system |
| **H** | hour of execution |
| **HE** | high-explosive |
| **HEI-T** | high-explosive incendiary tracer |
| **HEMMS** | hand-emplaced minefield marking set |
| **HEMTT** | heavy expanded mobility tactical truck |
| **HEXJAM** | hand-emplaced, expandable jammer |
| **HMMWV** | high-mobility, multipurpose, wheeled vehicle |
| **HQ** | headquarters |
| **hr** | hour(s) |
| **HTF** | how to fight |
| **hvy** | heavy |
| **IBASIC** | Improved Body Armor System, Individual Countermine |
| **ID** | infantry division |
| **IDA** | improved dog-bone assembly |
| **IFV** | infantry fighting vehicle |
| **ILLUM** | illumination |
| **IMU** | inertial measurement unit |
| **in** | inch(es) |
| **inf** | infantry |
| **IOE** | irregular outer edge |
| **IP** | improved plow |

| | |
|---|---|
| **IPB** | intelligence preparation of the battlefield |
| **IR** | infrared |
| **ITV** | improved TOW vehicle |
| **IVMMD** | interim vehicle-mounted mine detector |
| **JAAT** | joint air-attack team |
| **Jan** | January |
| **Jul** | July |
| **K-Kill** | catastrophic kill |
| **kg** | kilogram(s) |
| **km** | kilometer(s) |
| **kph** | kilometers per hour |
| **KW** | Kiowa Warrior |
| **lb** | pound(s) |
| **LBE** | load-bearing equipment |
| **LE** | low explosive |
| **LED** | light-emitting diode |
| **LIN** | line item number |
| **LM** | landmark |
| **LOC** | lines of communication |
| **LOGPAC** | logistics package |
| **LRA** | local reproduction authorized |
| **LRP** | logistics release point |
| **LRSD** | long-range surveillance device |
| **LZ** | landing zone |
| **m** | month |
| **m** | meter(s) |
| **M-Kill** | mobility kill |
| **M-S** | Miznay-Schardin |

| | |
|---|---|
| **mag** | magnetic |
| **Mar** | March |
| **MBA** | main battle area |
| **MC** | mobility corridor |
| **MCAP** | mine-clearing/armor-protection kit |
| **MCB** | mine-clearing blade |
| **MCD** | magnetic coupling device |
| **MCO** | Marine Corps order |
| **MCOO** | modified combined obstacle overlay |
| **MCR** | mine-clearing roller |
| **MCRP** | Marine Corps reference publication |
| **MDI** | modernized demolition initiation |
| **MDV** | mine-detection vehicle |
| **Met+VE** | meteorological data/velocity error |
| **METL** | mission-essential task list |
| **METT-TC** | mission, enemy, terrain, troops, time available, and civilian considerations |
| **MHE** | material handling equipment |
| **MICLIC** | mine-clearing line charge |
| **MIDAP** | minefield-detection algorithm and processor |
| **MILSTD** | military standard |
| **min** | minute(s) |
| **min** | minimum |
| **MLRS** | Multiple-Launched Rocket System |
| **mm** | millimeter(s) |
| **MOBA** | military operations in built-up areas |
| **MOD** | mobile obstacle detachment |
| **MOPMS** | Modular Pack Mine System |
| **MOPP** | mission-oriented protective posture |

| | |
|---|---|
| **MOOTW** | military operations other than war |
| **MOUT** | military operations on urbanized terrain |
| **MP** | military police |
| **MRB** | motorized rifle battalion |
| **MRC** | motorized rifle company |
| **MRR** | motorized rifle regiment |
| **MSD** | movement support detachment |
| **MSR** | main supply route |
| **NA** | not applicable |
| **NAI** | named area of interest |
| **NATO** | North Atlantic Treaty Organization |
| **NAVSEA** | Naval Sea Systems Command |
| **NBC** | nuclear, biological, chemical |
| **NCO** | noncommissioned officer |
| **NCOIC** | noncommissioned officer in charge |
| **NE** | northeast |
| **NGO** | nongovernment organization |
| **NLT** | no later than |
| **nm** | nautical mile |
| **no** | number |
| **NSN** | national stock number |
| **OBSTINTEL** | obstacle intelligence |
| **Oct** | October |
| **OIC** | officer in charge |
| **OP** | observation post |
| **OPCON** | operational control |
| **OPLAN** | operation plan |
| **OPORD** | operation order |

| | |
|---|---|
| **P** | pace(s) |
| **pam** | pamphlet |
| **PIR** | priority intelligence requirement |
| **PIRS** | passive infrared sensor |
| **PL** | platoon leader |
| **PL** | phase line |
| **PLS** | palletized load system |
| **plt** | platoon |
| **PSG** | platoon sergeant |
| **psi** | pounds per square inch |
| **PSYOP** | psychological operations |
| **pt** | point |
| **PVO** | private volunteer organization |
| **QASAS** | quality assurance ammunition specialist |
| **R&S** | reconnaissance and surveillance |
| **RAAM** | remote antiarmor mine |
| **RAC** | risk-assessment code |
| **RCU** | remote-control unit |
| **RDX** | cyclonite |
| **recon** | reconnaissance |
| **rep** | representative |
| **RF** | radio frequency |
| **ROKUS** | Republic of Korea, United States |
| **RP** | reference point |
| **RSO** | range safety officer |
| **RTO** | radio-telephone operator |
| **S&A** | safing and arming |

| | |
|---|---|
| **S&H** | safety and handling |
| **S&T** | supply and transport |
| **S2** | Intelligence Officer (US Army) |
| **S3** | Operations and Training Officer (US Army) |
| **S4** | Supply Officer (US Army) |
| **SAW** | squad automatic weapon |
| **SCATMINE** | scatterable mine |
| **SCATMINWARN** | scatterable minefield warning |
| **sct** | scout |
| **SD** | self-destruct |
| **SE** | southeast |
| **SEAD** | suppression of enemy air defense |
| **sec** | second(s) |
| **SEE** | small emplacement excavator |
| **SFC** | sergeant first class |
| **SFC** | sergeant first class |
| **SFF** | self-forging fragmentation |
| **SIR** | serious incident report |
| **SITEMP** | situation template |
| **SLAM** | selectable lightweight attack munition |
| **SOEO** | scheme of engineer operations |
| **SOF** | special-operations forces |
| **SOP** | standard operating procedure |
| **SOSR** | suppress, obscure, secure, and reduce |
| **SPECS** | Special Protective Eyewear Cylindrical System |
| **SPOTREP** | spot report |
| **SSG** | staff sergeant |
| **SSN** | social security number |

| | |
|---|---|
| **STANAG** | International Standardization Agreement |
| **STP** | soldier training publication |
| **T/MDV** | towing/mine-detection vehicle |
| **TACC** | tactical air control center |
| **TAHQ** | theater army headquarters |
| **TAI** | target area of interest |
| **TC** | track commander |
| **TC** | training circular |
| **TCP** | traffic-control post |
| **TCP** | tactical command post |
| **TF** | task force |
| **tm** | team |
| **TM** | technical manual |
| **TMD** | tactical-munition dispenser |
| **TNT** | trinitrotoluene |
| **TO** | theater of operations |
| **TO** | Air Force technical order |
| **TOC** | tactical operations center |
| **TOW** | tube-launched, optically tracked, wire-guided missile |
| **TRADOC** | United States Army Training and Doctrine Command |
| **TRP** | target reference point |
| **TTP** | tactics, techniques, and procedures |
| **UAV** | unmanned aerial vehicle |
| **UK** | United Kingdom |
| **UL** | Underwriters Laboratories, Incorporated |
| **US** | United States |
| **USAF** | United States Air Force |
| **USN** | United States Navy |

| | |
|---|---|
| **UTM** | Universal Transverse Mercator |
| **UXO** | unexploded ordnance |
| **vs** | versus |
| **W** | width |
| **w/** | with |
| **WLGH** | weapon-launched grapnel hook |
| **WO** | warning order |
| **WP** | white phosphorus |
| **XO** | executive officer |
| **y** | year |
| **Yugo** | Yugoslavia |

# References

## SOURCES USED

These are the sources quoted or paraphrased in this publication.


**Joint and Multiservice Publications**

AR 385-63. *Policies and Procedures for Firing Ammunition for Training, Target Practice and Combat {MCO P3570.1A}*. 15 October 1983.

FM 20-11. *Military Diving {NAVSEA 0910-LP-708-8000}*. 20 January 1999.

FM 90-3. *Desert Operations {FMFM 7-27}*. 24 August 1993.

FM 90-13. *River-Crossing Operations {MCWP 3-17.1}*. 26 January 1998.

FM 101-5-1. *Operational Terms and Graphics {MCRP 5-2A}*. 30 September 1997.

STANAG 2036. *Land Mine Laying, Marking, Recording and Reporting Procedures*. Edition 5. 12 February 1987.

STANAG 2889. *Marking of Hazardous Areas and Routes Through Them*. Edition 3. 26 March 1984.

STANAG 2990. *Principles and Procedures for the Employment in Land Warfare of Scatterable Mines With a Limited Laid Life - ATP-50*. Edition 1. 15 September 1988.

TM 5-6665-298-10. *Operator's Manual AN/PSS-12 Mine Detecting Set {TM 5-6665-298-10}*. 28 February 1995.

TM 9-1345-203-12&P. *Operator's and Organizational Maintenance Manual (Including Repair Parts and Special Tools List) for Land Mines {TO 11A8-3-1}*. 14 January 1977.

TM 9-2590-509-10. *Operator's Manual Mine Clearing Blade for M1, IPM1, or M1A1 Abrams Tank (NSN 2590-01-230-8862) {TM 2590-10/1}*. 12 April 1990.

**Army Publications**

AR 40-10. *Health Hazard Assessment Program in Support of the Army Materiel Acquisition Decision Process*. 1 October 1991.

AR 385-10. *The Army Safety Program*. 23 May 1988.

AR 385-16. *System Safety Engineering and Management*. 3 May 1990.

DA Pam 350-38. *Standards in Weapons Training*. 3 July 1997.

FM 1-113. *Utility and Cargo Helicopter Operations*. 12 September 1997.

FM 3-34.2 (formerly FM 90-13-1). *Combined-Arms Breaching Operations*. 31 August 2000.

FM 5-7-30. *Brigade Engineer and Engineer Company Combat Operations (Airborne, Air Assault, Light)*. 28 December 1994.

FM 5-10. *Combat Engineer Platoon*. 3 October 1995.

FM 5-34. *Engineer Field Data*. 30 August 1999.

FM 5-71-2. *Armored Task-Force Engineer Combat Operations*. 28 June 1996.

FM 5-71-3. *Brigade Engineer Combat Operations (Armored)*. 3 October 1995.

FM 5-100. *Engineer Operations*. 27 February 1996.

FM 5-170. *Engineer Reconnaissance*. 5 May 1998.

FM 5-250. *Explosives and Demolitions*. 30 July 1998.

FM 90-5 (HTF). *Jungle Operations (How to Fight)*. 16 August 1982.

FM 90-7. *Combined Arms Obstacle Integration*. 29 September 1994.

FM 90-10 (HTF). *Military Operations on Urbanized Terrain (MOUT) (How to Fight)*. 15 August 1979.

FM 90-10-1. *An Infantryman's Guide to Combat in Built-up Areas*. 12 May 1993.

FM 101-5. *Staff Organization and Operations*. 31 May 1997.

MILSTD 882C (Revision). *System Safety Program Requirements.* 19 January 1993.

STP 5-12B1-SM. *Soldier's Manual: MOS 12B, Combat Engineer, Skill Level 1*. 14 February 2001.

TC 25-8. *Training Ranges*. 25 February 1992.

TC 31-34. *Demining Operations*. 24 September 1997.

TM 1-1520-237-10. *Operator's Manual for UH-60A Helicopters, UH-60L Helicopters, and EH-60A Helicopters*. 31 October 1996.

TM 9-1095-208-10-1. *Operator's Manual for Dispenser, Mine: M139 (NSN 1095-01-235-3139) (EIC: 3V8) with Mounting Kits for 5-Ton Vehicle (1095-01-252-2818) (EIC: 3V9) and M548A1 Vehicle (1095-01-331-6755) (Ground Volcano)*. 10 July 1992.

TM 9-1300-206. *Ammunition and Explosives Standards*. 30 August 1973.

TM 9-1345-203-12. *Operator's and Unit Maintenance Manual for Land Mines*. 30 October 1995.

TM 9-1345-209-10. *Operator's Manual for Modular Pack Mine System (MOPMS) Consisting of Dispenser and Mine, Ground: M131 (NSN 1345-01-160-8909) Control, Remote, Land Mine System: M71 (1290-01-161-3662) and Dispenser and Mine, Ground Training: M136 (6920-01-162-9380)*. 31 March 1992.

TM 9-1375-213-12. *Operator's and Unit Maintenance Manual (Including Repair Parts and Special Tools List): Demolition Materials*. 30 March 1973.

TM 9-1375-215-14&P. *Operator's, Unit, Direct Support and General Support Maintenance Manual (Including Repair Parts and Special Tools List) for Demolition Kit, Mine Clearing Line Charge (MICLIC)*. 31 January 1992.

TM 43-0001-36. *Army Ammunition Data Sheets for Land Mines (FSC 1345)*. 1 September 1994.

## DOCUMENTS NEEDED

These documents must be available to the intended users of this publication.

DA Form 1355. *Minefield Record.* March 1987.

DA Form 1355-1-R. *Hasty Protective Row Minefield Record (LRA).* September 2001.

DA Form 2028. *Recommended Changes to Publications and Blank Forms.* February 1974.

## READINGS RECOMMENDED

These readings contain relevant supplemental information.

FM 5-490. *Engineer Diving Operations.* 31 March 1992.

FM 6-20-40. *Tactics, Techniques, and Procedures for Fire Support for Brigade Operations (Heavy).* 5 January 1990.

FM 6-20-50. *Tactics, Techniques, and Procedures for Fire Support for Brigade Operations (Light).* 5 January 1990.

TM 5-2590-214-10. *Operator's Manual for Roller Kit, Mine (NSN 2590-01-134-3724) (EIC: AKZ) and Mounting Kit, Mine Clearing Roller (2590-01-235-5458) (EIC: AKZ).* 30 March 1993.

TM 9-1095-209-10. *Operator's Manual for Mine Dispenser: M138 (Flipper).* 29 March 1990.

# Index



I

## T



## U



## V



## W